Methods in Enzymology

Volume 129

PLASMA LIPOPROTEINS

Part B

Characterization, Cell Biology, and Metabolism

METHODS IN ENZYMOLOGY

EDITORS-IN-CHIEF

Sidney P. Colowick   Nathan O. Kaplan

Methods in Enzymology

Volume 129

Plasma Lipoproteins

Part B
Characterization, Cell Biology, and Metabolism

EDITED BY

John J. Albers

DEPARTMENTS OF MEDICINE AND PATHOLOGY
UNIVERSITY OF WASHINGTON SCHOOL OF MEDICINE
SEATTLE, WASHINGTON

Jere P. Segrest

DEPARTMENTS OF PATHOLOGY AND BIOCHEMISTRY
UNIVERSITY OF ALABAMA AT BIRMINGHAM
BIRMINGHAM, ALABAMA

1986

ACADEMIC PRESS, INC.
Harcourt Brace Jovanovich, Publishers
Orlando   San Diego   New York   Austin
London   Montreal   Sydney   Tokyo   Toronto

ACADEMIC PRESS, INC.
Orlando, Florida 32887

United Kingdom Edition published by
ACADEMIC PRESS INC. (LONDON) LTD.
24–28 Oval Road, London NW1 7DX

LIBRARY OF CONGRESS CATALOG CARD NUMBER: 54-9110

ISBN 0–12–182029–7

PRINTED IN THE UNITED STATES OF AMERICA

86 87 88 89   9 8 7 6 5 4 3 2 1

Table of Contents

Section I. Characterization of Plasma Lipoproteins

v

Section II. Cell Biology of Plasma Lipoproteins

Section III. Metabolism of Plasma Lipoproteins

A. *In Vivo* Metabolism of Lipoproteins

B. Cellular and Tissue Metabolism of Plasma Lipoproteins

C. Enzymes of Lipoprotein Metabolism

D. Methods

Contributors to Volume 129

Article numbers are in parentheses following the names of contributors.
Affiliations listed are current.

JOHN J. ALBERS (5, 45, 48), *Departments of Medicine and Pathology, University of Washington School of Medicine, and Northwest Lipid Research Clinic, Harborview Medical Center, Seattle, Washington 98104*

RICHARD G. W. ANDERSON (12), *Department of Cell Biology and Anatomy, The University of Texas Health Science Center, Dallas, Texas 75235*

PAUL S. BACHORIK (5), *Departments of Pediatrics and Laboratory Medicine, The Johns Hopkins University School of Medicine, Baltimore, Maryland 21205*

MARK BAMBERGER (37), *Department of Biological Chemistry, The Johns Hopkins University School of Medicine, Baltimore, Maryland 21205*

DEBENDRANATH BANERJEE (18), *Lindsley Kimball Research Institute of the New York Blood Center, New York, New York 10021*

DOUGLAS M. BENSON (51), *Department of Medicine, Baylor College of Medicine, Houston, Texas 77030*

ROY A. BORCHARDT (32), *Cell Biology Unit, Louisiana State University Medical Center, New Orleans, Louisiana 70112*

RAYMOND C. BOSTON (22), *Department of Computer Science, Murdoch University, Murdoch, Western Australia 6150*

WILLIAM A. BRADLEY (20), *Department of Medicine, Baylor College of Medicine, Houston, Texas 77030*

H. BRYAN BREWER, JR. (29), *Molecular Disease Branch, National Heart, Lung, and Blood Institute, National Institutes of Health, Bethesda, Maryland 20892*

W. VIRGIL BROWN (11, 23, 27), *Department of Medicine, Mount Sinai School of Medicine, New York, New York 10029*

JOSEPH BRYAN (51), *Department of Cell Biology, Baylor College of Medicine, Houston, Texas 77030*

PHILLIP R. BUKBERG (27), *Department of Medicine, St. Vincent's Hospital, New York, New York 10011*

LAWRENCE CHAN (19), *Departments of Cell Biology and Medicine, Baylor College of Medicine, Houston, Texas 77030*

CHING-HONG CHEN (45), *Department of Medicine, University of Washington School of Medicine, and Northwest Lipid Research Clinic, Harborview Medical Center, Seattle, Washington 98104*

MARIAN C. CHEUNG (8), *Department of Medicine, University of Washington School of Medicine, and Northwest Lipid Research Clinic, Harborview Medical Center, Seattle, Washington 98104*

BYUNG HONG CHUNG (42), *Department of Pathology, University of Alabama at Birmingham, Birmingham, Alabama 35294*

GARY L. CUSHING (10), *Department of Medicine, Baylor College of Medicine, Houston, Texas 77030*

ROGER A. DAVIS (17, 32), *Cell Biology Unit, Louisiana State University Medical Center, New Orleans, Louisiana 70112*

PAUL A. DAWSON (16), *Department of Pharmacological Sciences, Health Sciences Center, State University of New York at Stony Brook, Stony Brook, New York 11794*

RALPH B. DELL (23), *Department of Pediatrics, College of Physicians and Surgeons, Columbia University, New York, New York 10032*

RUDY A. DEMEL (44), *The Biochemical Laboratory, The State University of Utrecht, Utrecht, The Netherlands*

LADISLAV DORY (39), *Department of Pharmacology, University of Tennessee Cen-*

ix

ter for the Health Sciences, Memphis, Tennessee 38163

HANS A. DRESEL (13), Department of Medicine, University of Heidelberg, Heidelberg, West Germany

CHRISTIAN EHNHOLM (43), National Public Health Institute, SF-00280 Helsinki, Finland

SHLOMO EISENBERG (21), Lipid Research Laboratory, Department of Medicine B, Hadassah University Hospital, Jerusalem 91120, Israel

JEFFREY D. ESKO (15), Department of Biochemistry, University of Alabama at Birmingham, Birmingham, Alabama 35294

NOEL H. FIDGE (26), Baker Medical Research Institute, Melbourne, Victoria 3181, Australia

CHRISTOPHER J. FIELDING (46), Cardiovascular Research Institute, and Department of Physiology, University of California Medical Center, San Francisco, California 94143

DAVID M. FOSTER (22, 24), Center for Bioengineering, University of Washington, Seattle, Washington 98195

JOHN W. GAUBATZ (10), Department of Medicine, Baylor College of Medicine, Houston, Texas 77030

SANDRA H. GIANTURCO (20), Department of Medicine, Baylor College of Medicine, Houston, Texas 77030

JOYCE COREY GIBSON (11, 27), Department of Medicine, University of Miami School of Medicine, Miami, Florida 33101

HENRY N. GINSBERG (11, 23, 27), Department of Medicine, College of Physicians and Surgeons, Columbia University, New York, New York 10032

ROBERT M. GLICKMAN (31), Department of Medicine, College of Physicians and Surgeons, Columbia University, New York, New York 10032

LAURA A. GLINES (2), Lawrence Berkeley Laboratory, Donner Laboratory, University of California, Berkeley, California 94720

ANTONIO M. GOTTO, JR. (13, 51), Department of Medicine, Baylor College of Medicine, Houston, Texas 77030

RICHARD E. GREGG (29), Molecular Disease Branch, National Heart, Lung, and Blood Institute, National Institutes of Health, Bethesda, Maryland 20892

JOHN T. GWYNNE (40), Department of Medicine, Division of Endocrinology, University of North Carolina, Chapel Hill, North Carolina 27514

GIDEON HALPERIN (49), Lipid Research Laboratory, Department of Medicine B, Hadassah University Hospital, Jerusalem 91120, Israel

ICHIRO HARA (4), Scientific Instrument Division, Toyo Soda Manufacturing Company, Hayakawa, Ayase-shi, Kanagawa 252, Japan

RICHARD J. HAVEL (35), Cardiovascular Research Institute, and Department of Medicine, University of California, San Francisco, California 94143

THOMAS L. INNERARITY (33), Gladstone Foundation Laboratories for Cardiovascular Disease, Cardiovascular Research Institute, Department of Pathology, University of California, San Francisco, California 94140

PER-HENRIK IVERIUS (41), Department of Internal Medicine, University of Utah School of Medicine, Endocrinology Section, and Veterans Administration Medical Center, Salt Lake City, Utah 84148

RICHARD L. JACKSON (44), Merrell Dow Research Institute, Cincinnati, Ohio 45215, and University of Cincinnati, Cincinnati, Ohio 45206

FRED L. JOHNSON (3), Department of Comparative Medicine, Bowman Gray School of Medicine, Winston-Salem, North Carolina 27103

TALWINDER S. KAHLON (2), Western Regional Research Center, United States Department of Agriculture, Agricultural Research Service, Albany, California 94710

JOHN P. KANE (7), *Department of Medicine, Cardiovascular Research Institute, and Department of Biochemistry and Biophysics, University of California, San Francisco, California 94143*

MONTY KRIEGER (14), *Department of Biology and the Whitaker College of Health Sciences, Technology, and Management, Massachusetts Institute of Technology, Cambridge, Massachusetts 02139*

TIMO KUUSI (43), *Third Department of Medicine, University of Helsinki, SF-00290 Helsinki, Finland*

ANDRAS G. LACKO (45), *Department of Biochemistry, Texas College of Osteopathic Medicine, and Division of Biochemistry, North Texas State University, Fort Worth, Texas 76107*

NGOC-ANH LE (23, 27), *Departments of Biomathematical Sciences and Medicine, Mount Sinai School of Medicine, New York, New York 10029*

CHIN-TARNG LIN (19), *Department of Physiology, The Milton S. Hershey Medical Center, The Pennsylvania State University, Hershey, Pennsylvania 17033*

FRANK T. LINDGREN (2), *Lawrence Berkeley Laboratory, Donner Laboratory, University of California, Berkeley, California 94720*

ARTHUR M. MAGUN (31), *Department of Medicine, College of Physicians and Surgeons, Columbia University, New York, New York 10032*

DARIEN D. MAHAFFEE (40), *Department of Medicine, Division of Endocrinology, University of North Carolina, Chapel Hill, North Carolina 27514*

ROBERT W. MAHLEY (9, 33), *Gladstone Foundation Laboratories for Cardiovascular Disease, Cardiovascular Research Institute, Departments of Medicine and Pathology, University of California, San Francisco, California 94140*

JULIAN B. MARSH (30), *Physiology and Biochemistry Department, The Medical College of Pennsylvania, Philadelphia, Pennsylvania 19129*

CAROL A. MARZETTA (3), *Department of Medicine, Division of Metabolism, University of Washington, Seattle, Washington 98195*

LARRY R. MCLEAN (44), *Merrell Dow Research Institute, 2110 East Galbraith Road, Cincinnati, Ohio 45215*

JOEL D. MORRISETT (10), *Departments of Medicine and Biochemistry, Baylor College of Medicine, Houston, Texas 77030*

PAUL J. NESTEL (26), *CSIRO, Division of Human Nutrition, Adelaide, South Australia 5000*

MITSUYO OKAZAKI (4), *Department of General Education, Tokyo Medical and Dental University, Kohnodai, Ichikawashi, Chiba 272, Japan*

JOHN F. ORAM (38), *Division of Metabolism, Endocrinology, and Nutrition, Department of Medicine, University of Washington, Seattle, Washington 98195*

JOSE M. ORDOVAS (25), *Lipid Metabolism Laboratory, United States Department of Agriculture, Human Nutrition Research Center on Aging at Tufts University, Endocrinology Division, New England Medical Center, Boston, Massachusetts 02111*

ANN-MARGRET ÖSTLUND-LINDQVIST (41), *AB Hässle, S-43183, Mölndal, Sweden*

C. J. PACKARD (34), *Department of Biochemistry, Royal Infirmary, Glasgow G4 0SF, Scotland*

JOSEF R. PATSCH (1), *Department of Medicine, Baylor College of Medicine, Houston, Texas 77030*

WOLFGANG PATSCH (1), *Department of Medicine, Baylor College of Medicine, Houston, Texas 77030*

MICHAEL C. PHILLIPS (37), *Physiology and Biochemistry Department, The Medical College of Pennsylvania, Philadelphia, Pennsylvania 19129*

ROBERT E. PITAS (33), *Gladstone Foundation Laboratories for Cardiovascular Dis-*

ease, Cardiovascular Research Institute, Department of Pathology, University of California, San Francisco, California 94140

RAY C. PITTMAN (36), *Department of Medicine, University of California, San Diego, La Jolla, California 92093*

MARY E. POAPST (24), *Departments of Medicine and Physiology, University of Toronto, and Division of Endocrinology and Metabolism, Toronto General Hospital, Toronto, Ontario, Canada M5G 2C4*

RAJASEKHAR RAMAKRISHNAN (23), *Department of Pediatrics, College of Physicians and Surgeons, Columbia University, New York, New York 10032*

PAUL S. ROHEIM (39), *Department of Physiology, Louisiana State University Medical Center, New Orleans, Louisiana 70112*

GEORGE H. ROTHBLAT (37), *Physiology and Biochemistry Department, The Medical College of Pennsylvania, Philadelphia, Pennsylvania 19129*

ARDON RUBINSTEIN (11), *Municipal Government Medical Center, Ichilove Hospital, Tel Aviv, Israel*

LAWRENCE L. RUDEL (3), *Department of Comparative Medicine, Bowman Gray School of Medicine, Winston-Salem, North Carolina 27103*

ERNST J. SCHAEFER (25), *Lipid Metabolism Laboratory, United States Department of Agriculture, Human Nutrition Research Center on Aging at Tufts University, Endocrinology Division, New England Medical Center, Boston, Massachusetts 02111*

JERE P. SEGREST (42), *Department of Pathology, University of Alabama at Birmingham, Birmingham, Alabama 35294*

J. SHEPHERD (34), *Department of Biochemistry, Royal Infirmary, Glasgow G4 0SF, Scotland*

MASAKI SHINOMIYA (44), *The Second Department of Internal Medicine, Chiba University School of Medicine, Chiba 280, Japan*

STEVEN L. SHUMAK (24), *Division of Endocrinology and Metabolism, Toronto General Hospital, Toronto, Ontario, Canada M5G 2C4*

CHARLES H. SLOOP (39), *Department of Physiology, Louisiana State University Medical Center, New Orleans, Louisiana 70112*

LOUIS C. SMITH (50, 51), *Department of Medicine, Baylor College of Medicine, Houston, Texas 77030*

LILIAN SOCORRO (44), *Department of Medicine, Washington University School of Medicine, St. Louis, Missouri 63110*

OLGA STEIN (49), *Hubert Humphrey Center for Experimental Medicine and Cancer Research, Hebrew University–Hadassah Medical School, Jerusalem 91000, Israel*

YECHEZKIEL STEIN (49), *Lipid Research Laboratory, Department of Medicine B, Hadassah University Hospital, Jerusalem 91120, Israel*

GEORGE STEINER (24), *Departments of Medicine and Physiology, University of Toronto, and Division of Endocrinology and Metabolism, Toronto General Hospital, Toronto, Ontario, Canada M5G 2C4*

P. V. SUBBAIAH (47), *Division of Endocrinology and Metabolism, Presbyterian-St. Luke's Medical Center, Chicago, Illinois 60612*

ALAN R. TALL (28), *Department of Medicine, College of Physicians and Surgeons, Columbia University, New York, New York 10032*

CLINTON A. TAYLOR, JR. (36), *Department of Medicine, University of California, San Diego, La Jolla, California 92093*

JOHN H. TOLLEFSON (48), *Department of Medicine, University of Washington, and Northwest Lipid Research Clinic, Harborview Medical Center, Seattle, Washington 98104*

DAVID P. VIA (13, 50), *Department of Medicine, Baylor College of Medicine, Houston, Texas 77030*

G. RUSSELL WARNICK (6), *Northwest Lipid*

Research Clinic, Lipoprotein Laboratory, University of Washington, Seattle, Washington 98104

KARL H. WEISGRABER (9), *Gladstone Foundation Laboratories for Cardiovascular Disease, Cardiovascular Research Institute, Department of Pathology, University of California, San Francisco, California 94140*

DAVID L. WILLIAMS (16), *Department of Pharmacological Sciences, Health Sciences Center, State University of New York at Stony Brook, Stony Brook, New York 11794*

LOREN A. ZECH (22), *Laboratory of Mathematical Biology, National Cancer Institute, National Institutes of Health, Bethesda, Maryland 20892*

Preface

Methodology development has played a central role in understanding the structure, biosynthesis, and physiological functions of the plasma lipoproteins. The ultracentrifuge played a major role in the discovery and characterization of the plasma lipoproteins, and the instrument continues to be a methodologic mainstay. One result has been a progressively better appreciation of the metabolic interrelationships of the different plasma lipoprotein species.

The discovery of the LDL receptor less than 15 years ago ushered in the molecular era in plasma lipoprotein physiology. Simultaneously, the identification, isolation, and amino acid sequence analyses of the different plasma apolipoproteins laid the foundation for a molecular revolution in our understanding of plasma lipoprotein structure and metabolism. This revolution has been accelerated in the past five years by explosive advances in recombinant DNA technology. The structures of the genes encoding for apolipoproteins, lipoprotein receptors, and enzymes involved in the regulation of lipoprotein metabolism are emerging. With the added availability of new techniques for the isolation and analysis of plasma lipoprotein subspecies, an exciting new era in lipoprotein research beckons. This era promises a detailed understanding of the molecular basis for genetic and metabolic regulation of the plasma lipoproteins.

The exciting state of growth in lipoprotein research, the complete lack of a recent comprehensive treatise on the central methodology presently being used in the field, and the relatively limited treatment of lipoproteins in earlier volumes of *Methods in Enzymology* warranted the assembly of this two-volume work. Volume 128, Part A, deals with the preparation, structure, and molecular biology of the plasma lipoproteins; Volume 129, Part B, deals with the characterization, cell biology, and metabolism of the plasma lipoproteins. These volumes should serve as convenient handbooks for all investigators involved in lipoprotein research.

We wish to acknowledge our indebtedness to the contributors. We also want to thank Dr. Leon W. Cunningham, Dr. John M. Taylor, and Dr. Antonio M. Gotto, Jr., for their help in the conception and organization of these volumes. We express our appreciation to the staff of Academic Press for their pleasant and efficient assistance.

JERE P. SEGREST
JOHN J. ALBERS

METHODS IN ENZYMOLOGY

EDITED BY

Sidney P. Colowick and Nathan O. Kaplan

VANDERBILT UNIVERSITY
SCHOOL OF MEDICINE
NASHVILLE, TENNESSEE

DEPARTMENT OF CHEMISTRY
UNIVERSITY OF CALIFORNIA
AT SAN DIEGO
LA JOLLA, CALIFORNIA

METHODS IN ENZYMOLOGY

EDITORS-IN-CHIEF

Sidney P. Colowick and Nathan O. Kaplan

Section I

Characterization of Plasma Lipoproteins

[1] Zonal Ultracentrifugation

By JOSEF R. PATSCH and WOLFGANG PATSCH

Overview

Zonal ultracentrifugation is a form of density gradient centrifugation. With this technique, macromolecules or particles are separated as discrete bands in a preformed supporting density gradient. The basis for separation is either difference in sedimentation rate or in density. Since Brakke[1] first used the zonal centrifuge principle, it has become a widely used technique for separating and purifying macromolecules, subcellular organelles, and viruses. For years after its introduction, zonal centrifugation was employed using small, swinging bucket-type rotors. This technology was applied occasionally for separating lipoproteins.[2-4] However, wider and continuous use of zonal ultracentrifugation to isolate and analyze plasma lipoproteins did not occur until the advent of large-capacity, high-performance zonal rotors. These rotors, designated B-XIV and B-XV, were developed by Anderson and co-workers at the Oakridge National Laboratories. As opposed to the earlier elongated cylinder-shaped zonal rotors, they were more flat and allowed operation without upper bearings.[5-7] Using the B-XIV and B-XV titanium zonal rotors, Wilcox and Heimberg[8] experimented with gradient material such as sodium chloride, potassium bromide, sucrose, and sodium bromide, alone or in combination, and succeeded in complete separation of very low-density lipoproteins (VLDL) and low-density lipoproteins (LDL) from a peripheral rotor fraction containing the high-density lipoproteins (HDL) and bulk plasma proteins. After their development by Anderson and co-workers, the B-XIV and B-XV rotors were made available from commercial manufacturers, such as International Equipment Company, Nytime Heights, MA; Measuring and Scientific Equipment (MSE), Ltd., London, England; and the Spinco Division of Beckman Instruments, Palo Alto, CA. Utilizing the

[1] M. K. Brakke, *J. Am. Chem. Soc.* **73,** 1847 (1951).
[2] J. L. Oncley, K. W. Walton, and D. G. Cornwell, *J. Am. Chem. Soc.* **79,** 4666 (1957).
[3] S. Light and F. R. N. Gurd, *Vox Sang.* **5,** 92 (1960).
[4] G. H. Adams and V. N. Schumaker, *Nature (London)* **202,** 490 (1964).
[5] N. G. Anderson, *J. Phys. Chem.* **66,** 1984 (1962).
[6] N. G. Anderson, *Science* **154,** 103 (1966).
[7] N. G. Anderson, D. A. Walters, W. D. Fisher, G. B. Cline, C. E. Nunely, L. H. Elrod, and C. T. Rankin, Jr., *Anal. Biochem.* **21,** 235 (1967).
[8] H. G. Wilcox and M. Heimberg, *J. Lipid Res.* **11,** 7 (1970).

Beckman versions Ti-14 and Ti-15, in connection with higher g forces and denser gradient material, Wilcox et al.[9] separated HDL as a distinct lipoprotein zone in addition to VLDL and LDL. Similar results were reported by Viikari et al. using the MSE version of the B-XV rotor.[10] Patsch et al.[11] experimented with discontinuous gradients in the Beckman Ti-14 rotor and succeeded in separating the two major HDL subfractions, HDL_2 and HDL_3, by a single step procedure. Subsequently, the same gradient[11] was employed in the newer Beckman Z-60 zonal rotor.[12] Patsch et al.[13] have also described a method for subfractionating VLDL. Standardization of the method using the analytical ultracentrifuge allowed correlation of zonal rotor volume with flotation rates.[13] The same was achieved for flotation rates of HDL particles.[14] Zonal ultracentrifugation was also employed to isolate and characterize pathologic lipoproteins of low and high density occurring in obstructive liver disease.[15–18]

The primary advantage of zonal ultracentrifugation over other ultracentrifugation methods used in isolating lipoproteins is its superior power of resolution, its large capacity capability, but most importantly, its simultaneous analytical and preparative capabilities. This feature gives immediate information on abundance of lipoproteins and the quality of separation. It is most useful for detecting anomalous lipoproteins such as those occurring in cholestatic liver disease. In the hands of a skilled worker, zonal ultracentrifugation also allows one to detect subtle but most informative alterations in the biophysical properties of normal lipoproteins. This chapter reviews and describes the application and execution of zonal ultracentrifugation for analysis, isolation, and characterization of human plasma lipoproteins and describes the major characteristics of lipoprotein particles isolated by this method.

[9] H. G. Wilcox, D. C. Davis, and M. Heimberg, J. Lipid Res. **12,** 160 (1971).

[10] J. Viikari, E. Haati, T.-T. Helela, K. Juva, and T. Nikkari, Scand. J. Clin. Lab. Invest. **23,** 85 (1969).

[11] J. R. Patsch, S. Sailer, G. Kostner, F. Sandhofer, A. Holasek, and H. Braunsteiner, J. Lipid Res. **15,** 356 (1974).

[12] P. Laggner, H. Stabinger, and G. Kostner, Prep. Biochem. **7,** 33 (1977).

[13] W. Patsch, J. Patsch, G. M. Kostner, S. Sailer, and H. Braunsteiner, J. Biol. Chem. **253,** 4911 (1978).

[14] W. Patsch, G. Schonfeld, A. M. Gotto, and J. R. Patsch, J. Biol. Chem. **255,** 3178 (1980).

[15] B. Danielsson, B. G. Johansson, and B. C. Petersson, Clin. Chim. Acta **47,** 365 (1973).

[16] J. R. Patsch, W. Patsch, S. Sailer, and H. Braunsteiner, Biochim. Biophys. Acta **434,** 419 (1976).

[17] J. R. Patsch, K. C. Aune, A. M. Gotto, and J. D. Morrisett, J. Biol. Chem. **252,** 2113 (1977).

[18] B. Danielsson, R. Ekman, and B. G. Petersson, FEBS Lett. **50,** 180 (1975).

Technique of Zonal Ultracentrifugation

Rate or Isopycnic Zonal Ultracentrifugation

Zonal ultracentrifugation allows separation of lipoproteins into discrete zones on the basis of their difference in flotation rate and/or buoyant density. Rate zonal centrifugation depends on the lipoprotein flotation rate. Lipoproteins with similar flotation rates due to similarities in hydrodynamic properties such as hydrated density, particle weight, and shape form distinct zones in the density gradient and migrate at different rates against the ultracentrifugal field. The lipoprotein zones or bands are usually not allowed to reach equilibrium in the density gradient, but are isolated at gradient densities higher than those of the respective lipoprotein bands because ultracentrifugation is usually discontinued when various lipoprotein bands are separated best from one another. In Fig. 1, for example, VLDL has migrated through the entire path length of the zonal rotor, but LDL, due to its lower flotation rate, has moved only about half the distance available. If ultracentrifugation were not interrupted at this point, LDL would continue to migrate toward the center of the rotor and eventually merge with VLDL. Although rate flotation is the principle usually employed for lipoprotein separation in zonal rotors, isopycnic separation of lipoproteins has also been used in the past. With this technique, migration of lipoprotein particles is continued until they reach a portion of the gradient where the gradient density and the buoyant density of the lipoprotein particles are equal, and no further flotation will occur. At this point, lipoprotein particles may be considered to be at or near equilibrium ("quasi-equilibrium"). The density at which the lipoprotein particles will band isopycnically is a function of their density in the medium employed. Apparent density can be altered when the particles are permeable to the solutes used, or when the solutes are adsorbed or bound to the particles. Therefore, one must consider that the buoyant densities of lipoprotein particles are, in part, a function of the gradient material used. However, all normally occurring plasma lipoproteins exhibit very similar structure; they are pseudomicelles containing an apolar core. Also, the gradient material used is usually potassium bromide or sodium bromide. Therefore, the above cautionary statements are not usually applicable when different plasma lipoproteins are compared. However, vesicular lipoproteins such as the pathological lipoprotein of cholestatic liver disease, LP-X, are likely to be permeated by the gradient material and therefore their buoyant density may be altered in a somewhat different way than that of the other lipoproteins. Isopycnic banding of lipopro-

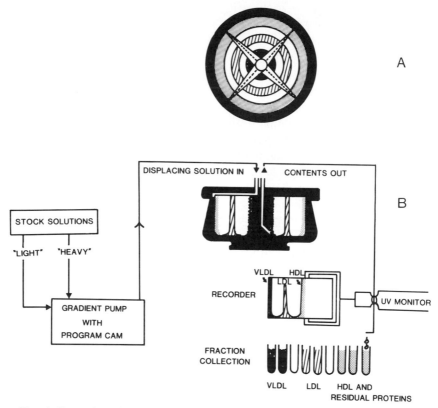

FIG. 1. Separation of VLDL and LDL in the zonal rotor. (A) Overhead view. (B) Side view of zonal rotor interior after ultracentrifugal separation of VLDL (banded at center of rotor) from LDL (banded at half the path length of rotor) and from HDL and residual plasma proteins (at periphery of zonal rotor).

teins is a convenient and very reliable technique to determine buoyant density of particles; however, the buoyant density of lipoproteins, because of the reasons discussed above, is not necessarily exactly the same as the reciprocal of the partial specific volume. Isopycnic separation of lipoproteins has been employed occasionally in the past. The three subspecies of the cholestatic lipoprotein LP-X can be separated best isopycnically under quasi-equilibrium conditions.[17]

Choice of Rotor

Zonal rotors are hollow centrifuge heads made of aluminum or titanium. Isolation of lipoproteins, particularly HDL, requires g forces that can be generated only by titanium rotors. Unlike aluminum rotors, titanium rotors are also resistant to corrosive gradient materials such as potassium bromide or sodium bromide, which are required for isolation of lipoproteins. Commercially available versions (for instance, the Beckman Ti-14 and Ti-15 rotor) have evolved from the B-XIV and B-XV rotors developed at the Oakridge Laboratories.[7] In contrast to their predecessors, both the Ti-14 and Ti-15 rotors are characterized by a height-to-diameter ratio of less than 1 and are distinguished only by their different capacities.

The volume of the Ti-14 rotor is 665 ml with a path length of 5.3 cm, and that of the Ti-15 rotor is 1675 ml with a path length of 7.5 cm. The smaller Ti-14 rotor accommodates a sample volume of up to 50 ml, can be spun at higher rpm (48,000 rpm), and can generate higher g forces (171,800 g maximum). Hence this rotor will require, for similar results, shorter centrifugation times and lesser gradient material. The higher g forces of the Ti-14 rotor will counteract particle diffusion more efficiently and, thus, produce narrower band widths and better resolution of particles where diffusion is a major factor, as is the case for the small HDL particles and their subfractions. The larger Ti-15 rotor has the advantage of greater capacity (1675 ml) and therefore will accept a greater sample volume (up to 200 ml). Because of its lower maximum speed (32,000 rpm), it generates lesser g forces (102,000 g maximum), but offers the advantage of a longer path length (7.5 cm). Therefore, better resolution can be expected with larger lipoprotein particles where diffusion is no major problem because of their size. Indeed, for isopycnic separation of LP-X_1, LP-X_2, and LP-X_3, which range in diameter from 58 to 68 nm, the large Ti-15 rotor gives the best results.[17] The newer Beckman Z-60 zonal rotor has a capacity of 330 ml with a path length of 4.8 cm. Typical sample volume is 15 ml or less. It can generate g forces of 265,000 g at 60,000 rpm and therefore appears most useful for work with HDL. Indeed, when using the discontinuous density gradient developed by Patsch et al.[11] it affords excellent resolution of HDL subclasses.[12,19] We have had extensive experience with all three types of rotors. If one had to choose from the three rotor types for work in the lipoprotein laboratory, we would recommend the Ti-14 rotor. In our hands, it combines best performance, general applicability, convenience, and economy; acceleration and deceleration times are shortest; and it can be filled and emptied rapidly. It is also the least expensive of the three rotor types discussed.

[19] G. Schmitz and G. Assmann, *J. Lipid Res.* **23,** 903 (1982).

Formation of Density Gradient

A characteristic of all zonal ultracentrifugal separations is the requirement for a preformed supporting density gradient. To prevent conductive currents and undesirable mixing, the fluids filling the zonal rotor cavity must always increase in density with increasing radius. For lipoprotein work, the required density gradients are prepared by a gradient maker from small-molecule solutes such as potassium bromide and sodium bromide, which are introduced into the rotor cavity at the start of the experiment. Cesium chloride would also be applicable but is not used much in the larger zonal rotors because of its high cost. Sucrose solutions, a widely used gradient material in ultracentrifugal separation of macromolecules and cellular components, are not appropriate for lipoprotein work because of their high viscosity.

The gradient shape most commonly used in zonal rotors is a simple linear gradient, in which the density changes linearly with the volume of solution delivered to the rotor (in a sector-shaped centrifuge compartment and in hollow centrifuge heads such as in zonal rotors, the plot of density vs radius is not the same as a plot of density vs volume). In the cylindrical zonal rotors, a gradient linear with volume has a contour which is slightly concave with radius; when compared with the area near the center, the density at the edge of the rotor increases more rapidly. The exact shape of the gradient delivered to the rotor will change during ultracentrifugation and the gradient material concentration that is usually indicated on zonal effluent patterns is that measured on fractions of rotor effluent collected at the end of an experiment. For replicate zonal ultracentrifugal experiments with plasma lipoproteins, it is important to generate gradients which are highly reproducible. Gradients linear with zonal rotor volume are used to separate VLDL subclasses; VLDL, IDL, and LDL; LP-X subfractions; and total HDL. For optimal resolution of HDL subfractions, however, a discontinuous gradient shape is necessary. The shape of this sodium bromide density gradient consists of three major steps and three very shallow linear components. This gradient configuration keeps the plasma proteins at the periphery and VLDL and LDL at the center of the rotor, thus providing a maximum path length for separation of HDL_2 and HDL_3.[11]

Besides linear and step density gradients, there have been described a number of density gradients used for special purposes. These include "isokinetic" gradients, in which the sedimentation velocity of sample components is independent of radius,[20] linear-log gradients which simplify

[20] J. Steensgard, *Eur. J. Biochem.* **16,** 166 (1970).

calculations of sedimentation coefficients,[21] and accelerating gradients.[22] Equivolumetric gradients[23] are the sectorial analogs of isokinetic gradients allowing particles to pass through a constant volume of gradient per unit of time. To our knowledge, none of these latter gradients has been used for work with plasma lipoproteins.

To fill the zonal rotor with a desired density gradient, one must first form the appropriately shaped density gradient from two stock solutions, and then transport the gradient solution into the rotor. This requires a gradient-forming device and a pump. These two functions are combined in the Beckman model 141 gradient pump. This high-capacity piston gradient pump controls flow rate, shape of the gradient, and volume of liquid delivered to the zonal rotor. It varies the density of the solution in accordance with a preselected program and pumps the mixed liquid at a constant flow rate into the rotor, providing thereby the proper relation of volume to density. The use of this pump allows the formation of gradients of virtually any shape, including ones which are linear, concave, convex, or discontinuous. For this purpose, it is necessary to cut a cam with an appropriately shaped contour which limits the stroke of the cam follower and provides in this way the desired gradient. Gradients of a wide variety of shapes and total volumes can be generated by the gradient pump when the cam is cut appropriately. The total volume of the mixed gradient liquid can be varied from 250 to 2000 ml. The metal cam blanks can be obtained by the manufacturer of the pump (Beckman Instruments, Palo Alto, CA) or similar program cam blanks can be made of sheet metal. They are cut with ordinary scissors. The desired shape of the program cam can be computed and plotted first on a plastic template provided by the manufacturer of the pump (Fig. 2). The horizontal scale of these templates represents fractional parts of the total programmed volume as determined by the gear selection for the cam drive mechanism. Examples for total programs of 1 liter are illustrated in Fig. 2, where each numbered division of the horizontal scale represents 100 ml. The vertical divisions represent the proportions of the two stock solutions in the pump effluent. Thus, for a contour height one-third of the vertical distance, the effluent will be a mixture of one-third (by volume) of the denser stock solution ("heavy" solution) and two-thirds of the less dense ("light") stock solution. The two stock solutions have a density equal to the respective extremes of the density gradient. The "light" stock solution usually has a density of 1.00

[21] M. K. Brakke and N. Van Pelt, *Anal. Biochem.* **38,** 56 (1970).
[22] R. Kaempfer and M. Meselson, this series, Vol. 20, p. 521.
[23] M. S. Pollack and C. A. Price, *Anal. Biochem.* **42,** 38 (1971).

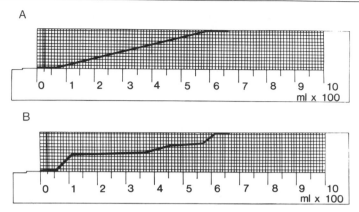

FIG. 2. Contour of program cams to provide a density gradient linear with respect to the rotor's volume (A), and the three-step density gradient to isolate HDL_2 and HDL_3 (B). The cam contours limit the stroke of the program cam follower of the Beckman Model 141 gradient pump. The contours have been drawn on original plastic templates provided by the producer of the pump (Beckman Instruments, Palo Alto, California).

g/ml and the "heavy" solution for lipoprotein work usually varies from 1.15 to 1.40 g/ml. Both solutions contain 0.35 mM disodium EDTA, pH 7.6. The "heavy" stock solution, usually containing NaBr, should be filtered prior to use to remove any insoluble matter. Figure 2 illustrates two cam contours for the zonal rotor Ti-14 (665 ml volume); one used for a linear density gradient and one for the three-step gradient to separate HDL_2 and HDL_3.[11] Particular gradients are described in the sections where individual lipoprotein classes are discussed.

If a Beckman model 141 gradient pump is not available, one can use alternatively a gradient former in connection with a pump. For instance, the gradient former Ultrograd (LKB Instruments, Bromma, Sweden) and Multiperpex pump (LKB Instruments, Bromma, Sweden) have been used[12] to create the three-step density gradient designed for the isolation of HDL subfractions.[11]

Preparation of Sample

The sample used for zonal ultracentrifugation must be at the dense end of the gradient when the experiment is started. Therefore, the sample is introduced at the periphery of the rotor immediately after it has been filled with the density gradient. The sample should have a density equal to or slightly higher than the densest part of the gradient. Adjustments of the sample to the desired density can be achieved by adding solid salt, potassium bromide, or sodium bromide to the sample. The amount of salt to be added can be calculated using the following formula:

$$X = \frac{V(d_f - d_i)}{1 - \bar{v}[(d_f + d_i)/2]}$$

where X is grams sodium bromide or potassium bromide added, V is the initial volume of the sample in milliliters, d_f is the desired final density, d_i is the initial density of the sample, and $\bar{v}$ is the partial specific volume of the salt added ($\bar{v}$ of sodium bromide at concentrations in the region of 250 g/liter and at 20° is 0.228; $\bar{v}$ of potassium bromide under these conditions is about 0.267). If plasma is used as the sample, an initial density of 1.025 g/ml can be assumed. If a lipoprotein sample is contained in a saline solution, the initial sample density is 1.006 g/ml. Otherwise, the density of a given sample can be determined hydrometrically. If this is not possible for some reason, a sample of unknown density can be adjusted by equilibrium dialysis against the "heavy" stock solution.

Loading the Rotor

At the beginning of an experiment, the empty assembled zonal rotor equilibrated at room temperature is put into the chamber of the ultracentrifuge. The rotor is then accelerated to a speed of 2500 to 3000 rpm to ensure stability of the density gradient throughout the filling procedure. Inside the rotor is a core with four vanes, each of which has an inlet and outlet port for loading and unloading the rotor with gradient material and sample through the periphery and the center. The two stock solutions, i.e., the "light" and "heavy" solutions which feed the gradient maker, should be at 15–20°. The density gradient is pumped into the spinning rotor through the peripheral line beginning at the low-density end of the gradient. When the lightest part of the gradient enters the rotor cavity, it forms a thin cylindrical layer on the wall, supported there by the centrifugal force generated by spinning the rotor at about 3000 rpm. As increasingly dense solution enters the cavity, new layers are formed which displace the preceding layer so that the lightest part of the gradient is gradually pushed to the center of the rotor cavity. During the filling procedure, the air will be displaced from the zonal rotor through the central line. This line should be submerged in distilled water so that the appropriate filling of the zonal rotor can be monitored by observing the emerging air bubbles. Completion of the filling process is signaled by the appearance of lowest density fluid leaving the rotor through the central line. At this point, the pump is stopped. The rotor will contain at its center an overlay consisting of 50 ml of the light end of the density gradient and the entire gradient. The sample, whose density is adjusted to that of the high-density end of the gradient, is slowly injected into the rotor from the periphery using a plastic hypodermic syringe. The sample, usually less than 10% of the total rotor volume, is then chased by a cushion of high-density solution. The cushion, about 5–7% of total rotor volume, should be no less than 25 ml when using rotors Ti-14 or Ti-15. This cushion will ensure that all of the sample has left the core and is actually in the high-density end of the gradient. The cushion is required for optimum recovery of lipoproteins from the zonal rotor and to avoid obstruction of lines in the core of the rotor due to proteins sedimenting during ultracentrifugation. As the sample and the cushion solutions are introduced into the rotor, an equal volume of the overlay solution is displaced from the rotor through the center line connection. Thus, after the filling procedure, the rotor contains only the density gradient, followed by the sample and, at the very periphery, the cushion. After filling is completed, the seal assembly is removed from the spinning zonal rotor and the rotor cap is placed on the lid. The chamber of the ultracentrifuge is closed and vacuum is generated. Ultracentrifugal separation is initiated by increasing the rotor speed to obtain the desired g forces.

Unloading the Rotor

After appropriate ultracentrifugation time, i.e., at the end of the high-speed part of the experiment, the rotor is decelerated to 2500–3000 rpm using the brakes, the vacuum is released, and the chamber is opened. The rotor cap is removed and replaced by the seal assembly. High-density solution is introduced into the rotor from the periphery at a constant rate, displacing the contents with its separated zones of lipoproteins through the center of the core. From here, the rotor effluent is passed through a flow cell where its absorbance is monitored and recorded continuously and then fractionated manually. In Fig. 1 an experiment is illustrated where VLDL has migrated through the entire rotor path from the periphery to the center, separated from LDL, which is located at about half the path length, and HDL, which in the given example is still located at the periphery of the rotor.

Analyses of Contents of Rotor

The first analysis of the rotor contents is usually the continuous monitoring of the absorbance at 280 nm during the collection of fractions. After this routine analysis, selected fractions of the rotor can be subjected to different analytical procedures. The density of the fractions can be determined by refractive index using an Abbe refractometer, by precision hydrometry, or by precision densimetry using the mechanical oscillator technique.[24]

For chemical analysis, fractions constituting a lipoprotein peak are pooled and, if necessary, concentrated by pressure filtration using membranes of appropriate size exclusion. After this procedure the gradient salt can be removed by dialysis, after which chemical analyses for protein, lipid phosphorus, cholesterol and cholesteryl esters, and triglycerides can be performed using standard assay procedures. Lipoprotein fractions can also be extracted using an appropriate organic solvent and analyzed by thin-layer chromatography. Apolipoprotein composition can be determined by various electrophoretic methods and radioimmunoassay procedures. Radioimmunoassay has been used for evaluating the distribution of various apolipoproteins across the zonal rotor path. The contents of the zonal rotor can also be analyzed for radioactivity. Lipoproteins labeled with ^{131}I or ^{125}I on its protein moiety can be easily monitored for their distribution by gamma counting.

Lipoproteins eluted from the zonal rotor can also be analyzed for their flotation behavior. As will be discussed below, the effluent volume for a particular lipoprotein class in the zonal rotor can be related directly to its flotation rate in the analytical ultracentrifuge (see sections on VLDL and HDL below). The contents of the zonal rotor has also been analyzed for LCAT activity.[25]

Advantages and Disadvantages

Since its introduction to the field of plasma lipoprotein research, zonal ultracentrifugation has contributed significantly to our understanding of lipoprotein structure and metabolism. A major advantage of the procedure is that lipoproteins can be isolated in much shorter running times than conventional ultracentrifugal techniques. Running time to isolate VLDL in the zonal rotor is 45 min. For LDL, 140 min is required; and for HDL_2 and HDL_3, 24 hr is sufficient. Another major advantage is that zonally isolated lipoprotein density classes are entirely free from contamination by undesired lipoproteins and plasma proteins. The technique also offers the advantage that changes in quantitative distribution of lipoprotein density classes in a given plasma sample can be directly observed from the effluent pattern monitored at 280 nm. In addition, abnormal lipoprotein patterns can be readily detected and the quality of separation is immediately apparent. Since the effluent volumes of subfractions of

[24] O. Kratky, H. Leopold, and H. Stabinger, this series, Vol. 27a, p. 98.
[25] B. Danielsson, R. Ekman, and B. G. Johansson, *Prep. Biochem.* **8,** 295 (1978).

VLDL[13] and HDL[14] subclasses have been calibrated with the analytical ultracentrifugation, flotation rates for these lipoproteins are immediately available as they are isolated. The conventional methods available for measuring flotation rates rely on optical properties of their solutions; however, the zonal procedure allows the use of other properties also. The potential of this approach has been demonstrated where radioimmunoassays were employed to monitor the rotor effluent.[14] Similar approaches can be employed for determining the diffusion coefficient as well as buoyant densities of lipoprotein particles under study.[14,17,26] Also, radioactive markers can be included as internal standards for detecting minute alterations in flotation behavior as a result of dietary or metabolic perturbations.[26] An additional advantage is that flotation behavior of lipoproteins in the zonal rotor is independent of sample concentration over a wide range.[14,27] The technique, at its present state of sophistication, involves considerable capital, equipment, and operator training. A significant disadvantage is that only one sample can be processed at a time and that in the case of equipment or power failure, samples loaded onto the zonal rotor are difficult to recover. While limitations and disadvantages will probably prevent zonal ultracentrifugation from becoming widely used in routine clinical lipoprotein laboratories, the method has been and should continue to be a powerful and, in many instances, an indispensable analytical and preparative technique in the lipoprotein research laboratory.

Application for Studying Lipoprotein Structure and Metabolism

VLDL

Triglyceride-rich (TG-rich) lipoproteins are secreted by the liver and the intestine as spherical particles that consist of lipids and apolipoproteins.[28,29] Chylomicrons, the largest TG-rich lipoproteins, originate in the intestine. They have a density (*d*) of <0.96 g/ml and can be separated from other TG-rich particles by a short ultracentrifugal step in zonal, angle head, and swinging bucket rotors.[9,30,31] The chylomicron-free fraction is still heterogeneous and contains particles which may vary consid-

[26] J. R. Patsch, A. M. Gotto, Jr., T. L. Olivecrona, and S. Eisenberg, *Proc. Natl. Acad. Sci. U.S.A.* **75,** 4519 (1978).

[27] J. R. Patsch and A. M. Gotto, Jr., *in* "Report of the High Density Lipoprotein Methodology Workshop" (K. Lippel, ed.), Publ. No. 79-1661, Dept. of Health, Education and Welfare, Washington, D.C., 1979.

[28] A. Gangel and R. K. Ockner, *Gastroenterology* **68,** 167 (1975).

[29] H. Glaumann, A. Bergstrand, and J. L. E. Ericsson, *J. Cell Biol.* **64,** 356 (1975).

[30] V. P. Dole and H. T. Hamlin, III, *Physiol. Rev.* **42,** 674 (1962).

[31] W. J. Lossow, F. T. Lindgren, J. C. Murchio, G. R. Stevens, and L. C. Jensen, *J. Lipid Res.* **10,** 68 (1969).

erably in size, flotation properties, chemical composition, and apolipoprotein content. Depending on the metabolic condition, both liver and intestine secrete various TG-rich particles that float between 0.96 and 1.006 g/ml, but vary considerably in overall properties. In addition, heterogeneity among the $d < 1.006$ g/ml lipoproteins is caused by the fact that TG-rich lipoproteins, after secretion in nascent form, are continuously modified after entering the circulation. Thus, the $d < 1.006$ g/ml fraction referred to as VLDL contains a wide spectrum of lipoproteins.

VLDL can be isolated and subfractionated by zonal ultracentrifugation on the basis of their flotation rates. The procedure involves a single ultracentrifugation step and a NaBr gradient linear with rotor volume over the density range of 1.00–1.15 g/ml.[13] Ultracentrifugation is performed at 42,000 rpm at 14° for 45 min using a Beckman Ti-14 zonal rotor. In postabsorptive plasma of normal subjects, VLDL elute in the first 360 ml of the rotor. Flotation rates of VLDL subspecies correlate directly with their Stokes radii and TG content, and inversely with their proportion of cholesterol, cholesteryl ester, phospholipids, protein, and apoB. Analysis of small-volume fractions (10–20 ml) by analytical ultracentrifugation provided a calibration of the zonal rotor volume in terms of S_f rates (Fig. 3). Thus, the procedure allows determination of the S_f rate of VLDL as they are isolated. Since the selection of volume fractions is not bound to fixed limits of flotation rates, one can prepare any fraction of interest. Although this procedure is solely based on differential flotation of VLDL and does therefore not distinguish among secretory products and remnants, it nevertheless facilitates the selection of functional entities, based on the ap-

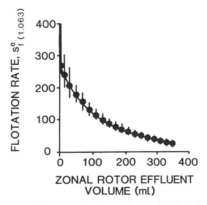

FIG. 3. Calibration curve of the zonal rotor effluent for flotation rates of VLDL.[13] The curve has been constructed using 19 VLDL subfractions from the zonal rotor for determining their flotation rates in the analytical ultracentrifuge. Individual points represent the peak flotation rate for each subfraction; vertical bars indicate the range of flotation for the material contained in each individual sample.

pearance of a VLDL pattern. Alterations of the VLDL pattern as it occurs in the cholesterol-fed rabbit are easily detectable by zonal ultracentrifugation.[32] In the plasma of rats fed cholesterol and propylthiouracil, the smaller VLDL and IDL increase dramatically. Such lipoprotein fractions not only differ compositionally from control VLDL in terms of lipid and apolipoprotein content, but functionally also in that they interact more avidly with macrophages and fibroblasts.[33] Thus, studies on *in vitro* interaction of these lipoproteins with cells offer a plausible explanation for the atherogenic potential of such diets.

VLDL patterns are also altered in human dyslipoproteinemias. In type IV phenotype the most abundant VLDL species are $VLDL_3$ and $VLDL_4$, whereas in normal plasma one finds a more even distribution. In type V plasma, the predominant VLDL fractions are $VLDL_1$ and $VLDL_2$, and $VLDL_4$ to $VLDL_7$ are present only in minimal amounts (unpublished).

The zonal procedure for VLDL isolation is particularly suited to follow changes in *in vitro* experiments, since analytical and preparative techniques are combined. Incubation of VLDL with bovine lipoprotein lipase and HDL_3 causes accumulation of VLDL remnants; the excess surface material liberated during lipolysis of parent VLDL is transferred to HDL_3 to form HDL_2-like particles.[26] When $VLDL_1$ are incubated with bovine LPL and albumin in the absence of HDL, the parent $VLDL_1$ disappears and remnant fractions with reduced S_f rates are generated. These VLDL remnants, when compared to the parent VLDL, express different epitopes of apoB on their surface and are more effective in competing with LDL for the receptor on fibroblasts.[34]

LDL

LDL is defined as the lipoprotein fraction isolated ultracentrifugally in the density interval of 1.006 to 1.063 g/ml. Major subfractions of these cuts were designated LDL_1 (1.006 to 1.019 g/ml) and LDL_2 (1.019–1.063 g/ml). The LDL_1 fraction is now more commonly known as intermediate-density lipoprotein (IDL). In the postabsorptive state, rather small amounts of IDL are present in normolipidemic subjects (0–40 mg/dl).[35] In contrast, LDL_2 is one of the two major lipoprotein fractions in normal plasma and transports the majority of plasma cholesterol.

[32] R. I. Roth, J. W. Gaubatz, A. M. Gotto, Jr., and J. R. Patsch, *J. Lipid Res.* **24,** 1 (1983).
[33] T. G. Cole, I. Kuisk, W. Patsch, and G. Schonfeld, *J. Lipid Res.* **25,** 593 (1984).
[34] G. Schonfeld, W. Patsch, B. Pfleger, J. L. Witztum, and S. W. Weidman, *J. Clin. Invest.* **64,** 1288 (1979).
[35] P. N. Herbert, A. M. Gotto, Jr., and D. S. Fredrickson, *in* "The Metabolic Basis of Inherited Disease" (J. B. Stanbury, J. B. Wyngaarden, and D. S. Fredrickson, eds.), 4th Ed., p. 588. McGraw-Hill, New York, 1978.

For isolation of LDL subclasses, a zonal procedure has been described.[36] A linear sodium bromide gradient spanning the density of 1.00 to 1.30 g/ml is used. Up to 50 ml of serum or plasma is adjusted to 1.30 g/ml and subjected to ultracentrifugation in a Beckman Ti-14 rotor at 42,000 rpm and 15° for 140 min. VLDL elute in the first 100 ml of rotor effluent volume, while the peak elution volume of LDL is between 200 and 290 ml. IDL elutes between VLDL and LDL (120–180 ml). Flotation rates and molecular weights of LDL are inversely related to the rotor effluent volume.[37] In a kindred with familial hypercholesterolemia, the amounts of free cholesterol, cholesteryl esters (CE), and phospholipids per LDL particle were directly related to LDL size. No such correlation was observed for the amount of TG. LDL protein was nearly constant throughout. More than 95% of the protein moiety of zonally isolated LDL is apoB as judged by solubility of protein in tetramethylurea. No B-48 was present; the majority was B-100 (80–100%) as judged by SDS electrophoresis in 3.5% acrylamide gels. Thus, similar amounts of apoB are associated with different amounts of lipids in various subjects. Within subjects, LDL size and the proportion of protein to lipid is relatively constant.[38] Changes in the lipid content of LDL, however, do occur in patients treated with probucol. Flotation rates of LDL slightly decrease and the amount of CE per particle is reduced.[39] The lipid-to-protein ratio in LDL appears to affect the expression of apoB epitopes.[40] Alternatively, genetic heterogeneity of apoB sequence could also explain the differing immunoreactivity of apoB in various LDL preparations when tested with monoclonal antisera.[41] Subjects with manifest hypertriglyceridemia have a preponderance of the smaller sized LDL. Latent hypertriglyceridemia, which becomes manifest upon eating a carbohydrate-rich diet, is also associated with smaller LDL (unpublished). These LDL are relatively rich in TG, but poor in CE. Cholesterol-rich diet does not change LDL particle size or stoichiometry in normal subjects, but increases the number of LDL particles in the circulation.[42] In contrast, LDL of patients with familial hypercholesterolemia contain more cholesteryl esters at the expense of triglycerides.[37]

[36] J. Patsch, S. Sailer, and H. Braunsteiner, *Eur. J. Clin. Invest.* **5**, 45 (1975).

[37] W. Patsch, R. Ostlund, I. Kuisk, R. Levy, and G. Schonfeld, *J. Lipid Res.* **23**, 1196 (1982).

[38] W. R. Fisher, M. G. Hammond, M. C. Mengel, and G. L. Warmke, *Proc. Natl. Acad. Sci. U.S.A.* **72**, 2347 (1975).

[39] D. R. Lock, I. Kuisk, B. Gonen, W. Patsch, and G. Schonfeld, *Atherosclerosis* **47**, 271 (1983).

[40] M. J. Tikkanen, T. G. Cole, and G. Schonfeld, *J. Lipid Res.* **24**, 1494 (1983).

[41] V. N. Schumaker, M. T. Robinson, L. K. Curtiss, R. Butler, and R. S. Sparks, *J. Biol. Chem.* **259**, 6423 (1984).

[42] G. Schonfeld, W. Patsch, L. L. Rudel, C. Nelson, M. Epstein, and R. Olson, *J. Clin. Invest.* **69**, 1072 (1982).

The method of isolating and characterizing LDL subfractions was also useful in defining the lipoprotein abnormality in type III hyperlipoproteinemia. Patients suffering from this familial disorder have in their VLDL fraction not only significant amounts of chylomicron remnants, called β-VLDL, but also large amounts of IDL, termed Lp-III. This lipoprotein fraction bands between VLDL and LDL, and its overall composition falls between that of VLDL and LDL as does the size of this lipoprotein.[36]

The zonal procedure has been successfully employed in identifying anomalous LDL fractions in the cord blood of patients suffering from familial hypercholesterolemia (FH)[43] as well as in clinical studies where the efficacy of hypolipidemic drugs was tested. It was also useful in demonstrating that most LDL are derived by VLDL processing,[44,45] but some LDL can be synthesized *de novo* by the liver.[46] The production of LDL-like particles by liver has also been demonstrated in primary rat hepatocyte cultures.[47]

High-Density Lipoproteins

High-density lipoproteins (HDL) have been isolated from human plasma by ultracentrifugation in the density interval of 1.063 to 1.210 g/ml. This lipoprotein material, when analyzed in the analytical ultracentrifuge by the moving boundary technique, exhibits a pattern consisting of a major peak with a shoulder on its leading edge indicating the existence of at least two HDL particle populations with distinct flotation rates. The two HDL species that cause the bimodal distribution of HDL under sedimentation velocity conditions have been designated HDL$_2$ and HDL$_3$.[48] According to this definition, HDL$_2$ has a flotation rate at density 1.20 g/ml ($F_{1.20}$) of 3.5 to 9, and HDL$_3$ has flotation rates of 0 to 3.5. Using sequential ultracentrifugation, HDL$_2$ has been isolated traditionally in the density interval of 1.063 to 1.125 g/ml, and HDL$_3$ in the interval of 1.125 to 1.210 g/ml. Using zonal ultracentrifugation and a linear density gradient, Wilcox *et al.*[9] isolated HDL as a major peak. They observed that the leading edge of this peak had a chemical composition similar to that of HDL$_2$ isolated by sequential ultracentrifugation and differed significantly

[43] W. Patsch, J. L. Witztum, R. Ostlund, and G. Schonfeld, *J. Clin. Invest.* **66,** 123 (1980).
[44] J. R. Patsch, D. Yeshrun, R. L. Jackson, and A. M. Gotto, Jr., *Am. J. Med.* **63,** 1001 (1977).
[45] S. R. Behr, J. Patsch, T. Forte, and A. Bensadoun, *J. Lipid Res.* **22,** 443 (1981).
[46] N. Nakaya, B. H. Chung, J. R. Patsch, and O. D. Taunton, *J. Biol. Chem.* **252,** 7530 (1977).
[47] W. Patsch, S. Franz, and G. Schonfeld, *J. Clin. Invest.* **71,** 1161 (1983).
[48] O. F. De Lalla and T. W. Gofman, *Methods Biochem. Anal.* **1,** 459 (1954).

from the overall composition of the major HDL peak. In an effort to obtain optimum resolution between HDL_2 and HDL_3, Patsch et al.[11] evaluated a series of gradients and found that a particular three-step density gradient brought about the desired separation; the nonlinear gradient spans the density range of 1.00 to 1.40 g/ml sodium bromide. Ultracentrifugation is performed at 41,000 rpm for 24 hr using a Beckman Ti-14 zonal rotor. The procedure allows separation of HDL_2 and HDL_3 from each other, from other lipoproteins, and from the residual plasma proteins in a single ultracentrifugal spin. HDL_2 is distinguished from HDL_3 by a higher relative abundance of lipid and lower abundance of protein which confers on the particle a density lower than that of HDL_3. HDL_2 is also distinguished from HDL_3 by a lower relative content of apoA-II, resulting in an apoA-I/apoA-II molar ratio greater than that of HDL_3. However, there are great individual variations in the relative apoA-II content in HDL_2: we have observed apoA-I/apoA-II molar ratios as low as 2.1 and as high as 18.5 (unpublished). In contrast to HDL_2, the apoA-I/apoA-II molar ratio in HDL_3 shows little variation among individuals averaging at 1.8 (range 1.3–2.4). HDL_2 is larger (average diameter 9.8 nm) than HDL_3 whose particle diameter ranges from 6.5 to 9.5 nm.[49] The molecular weight of HDL_2 is about twice that of HDL_3.[14]

HDL_2 elutes from the zonal rotor in a relatively small volume of about 70 ml, while the rather wide HDL_3 peak elutes in a much larger volume of 150–200 ml, which has been interpreted to indicate that HDL_3 particles are more heterogeneous in one or more of their hydrodynamic properties. As determined by electron microscopy, zonally isolated HDL_2 represents a rather homogeneous particle population with respect to size, while HDL_3 shows signs of polydispersity.[49] HDL_2 and HDL_3 were evaluated for homogeneity also by compositional, physical, and immunochemical criteria.[14] This and other studies suggested that HDL in man contain density subfractions whose relative plasma concentrations can vary; an ultracentrifugally homogeneous HDL_2 with an average density of 1.088–1.098 g/ml and a heterogeneous HDL_3 fraction containing primarily particles with an average density of 1.13–1.14 g/ml, and in lesser abundance, particles with an average density of 1.15–1.16 g/ml. Both HDL_3 populations are flanked by closely related particles in lesser amounts suggesting the existence of at least two distinct subclasses of HDL_3 termed HDL_{3L} and HDL_{3D}. HDL_{3D} represents only a minor fraction in normolipidemic plasma and thus is not readily detectable in a typical HDL zonal analysis due to its proximity to the less dense HDL_{3L} species. However, HDL_{3D} is of considerable metabolic significance since it can be the predominant

[49] J. R. Patsch, S. Sailer, H. Braunsteiner, and T. Forte, Dtsch. Med. Wochenschr. **101**, 1607 (1976).

HDL fraction in the plasma of patients with very low HDL levels and hypertriglyceridemia.[27]

One has to keep in mind, however, that HDL_2, when compared with HDL_3, appears to be more homogeneous only with respect to its hydrodynamic properties. Clearly, there has to exist some heterogeneity within HDL_2 because of the great individual variations in the molar ratio of apoA-I and apoA-II in this fraction. Also, estimation of the molecular weight does not allow for the presence of only one type of particle with the apolipoprotein composition that is obtained for the entire HDL_2 fraction in those subjects whose apoA-I/apoA-II molar ratio in HDL_2 exceeds 5. Additional proof for the metabolic heterogeneity is provided by recent experiments where HDL_2 from the postprandial phase was treated with hepatic lipase *in vitro*.[50] Only the HDL_2 particles with appreciable content of apoA-II, but not those lacking apoA-II, are converted into HDL_3.

Zonally isolated HDL_2 and HDL_3 are metabolically distinct species and have fundamentally different effects on cholesterol metabolism. It has been shown that HDL_2 suppresses while HDL_3 stimulates the activity of HMG-CoA reductase, the rate-limiting enzyme for cholesterol biosynthesis in fibroblasts.[51] The zonal ultracentrifugal procedure for isolation of HDL_2 and HDL_3 proved to be a most powerful analytical and preparative tool in studies on HDL metabolism and structure. The method was used to investigate the effects of sex, diet, and nicotinic acid therapy on the levels of HDL_2 and HDL_3 and to correlate these findings with turnover kinetics of the two major apolipoproteins of HDL, apoA-I and apoA-II.[44,52–54] The method was also used to study the metabolism of HDL species in animals such as the rabbit,[32] the rat,[55] and the chicken.[45] The method was very helpful in characterizing HDL subfractions in hyperalphalipoproteinemia, a condition which affords resistance to the development of atherosclerosis. Hyperalphalipoproteinimics are distinguished only by their very high levels of HDL_2.[56] The method of isolating HDL subfractions by zonal

[50] J. R. Patsch, S. Prasad, A. M. Gotto, Jr., and G. Bengtsson-Olivecrona, *J. Clin. Invest.* **74,** 2017 (1984).

[51] W. H. Därr, S. H. Gianturco, J. R. Patsch, L. C. Smith, and A. M. Gotto, Jr., *Biochim. Biophys. Acta* **619,** 287 (1980).

[52] J. Shepherd, C. J. Packard, J. R. Patsch, A. M. Gotto, and O. D. Taunton, *J. Clin. Invest.* **61,** 1582 (1978).

[53] J. Shepherd, C. J. Packard, J. R. Patsch, A. M. Gotto, and O. D. Taunton, *J. Clin. Invest.* **63,** 858 (1979).

[54] J. Shepherd, J. R. Patsch, C. J. Packard, O. D. Taunton, and A. M. Gotto, Jr., *J. Lipid Res.* **19,** 383 (1978).

[55] W. Patsch, K. Kim, W. Wiest, and G. Schonfeld, *Endocrinology* **107,** 1085 (1980).

[56] W. Patsch, I. Kuisk, C. Glueck, and G. Schonfeld, *Arteriosclerosis* **1,** 156 (1981).

ultracentrifugation was particularly valuable in studying the effects of lipolytic degradation of VLDL on metabolism and structure of HDL subfractions both *in vitro*[26] and *in vivo*.[45] The large-scale preparative capability of the method was realized when highly purified HDL subfractions in large amounts were needed for detailed physical studies using electron paramagnetic resonance and nuclear magnetic resonance.[57]

Postprandial Lipoproteins

The most notable lipoproteins occurring in plasma postprandially are the chylomicrons. They transport dietary triglycerides which are rapidly hydrolyzed by lipoproteins lipase attached to endothelial cells of many tissues. After most of the triglyceride is lost, smaller particles called remnants remain. These remnants contain a smaller core which is greatly enriched in cholesteryl esters. Most of the chylomicron remnants are quickly removed by the liver.[58] Chylomicrons can be separated by a variety of ultracentrifugal techniques, including zonal ultracentrifugation.[9,30,31] When subjected to the zonal ultracentrifugal procedure for isolating VLDL,[13] remnants float mostly in the $VLDL_1$ class. If remnants are present in this fraction, electrophoretic analysis reveals both VLDL (migrating as pre-β lipoproteins) and remnants migrating as β lipoproteins.[59]

The second lipoprotein class which is substantially affected by ingestion of a fat-rich meal is the HDL class. As chylomicrons are degraded to remnants by the action of lipolytic enzymes, excess surface material, mostly phospholipids, are transferred to both HDL_2 and HDL_3.[60,61] As a result of this incorporation of phospholipids into the surface of HDL particles, both HDL_2 and HDL_3 float at higher rates in the ultracentrifuge.[50,60,61] An example of this subtle but important postprandial alteration in ultracentrifugal behavior is illustrated in Fig. 4. As can be seen from this figure, both HDL_2 and HDL_3 display, in the postprandial phase, a lower density because of enrichment with phospholipids. Clearly, this figure illustrates a very valuable application of the zonal ultracentrifugal technique since both HDL_2 and HDL_3 have different flotation behavior depending on the state of absorption. The analytical capability of the method allows one to detect these lower densities and, thus, fractionate

[57] J. R. Brainard, R. D. Knapp, J. R. Patsch, A. M. Gotto, Jr., and J. D. Morrisett, *Ann. N.Y. Acad. Sci.* **348,** 299 (1980).

[58] T. G. Redgrave, *J. Clin. Invest.* **49,** 465 (1970).

[59] J. R. Patsch, G. M. Kostner, and W. Patsch, *in* "Handbook of Electrophoresis III" (L. A. Lewis, ed.), p. 67. CRC Press, Boca Raton, Florida, 1983.

[60] J. R. Patsch and A. M. Gotto, Jr., *Clin. Res.* **32,** 197A (1984).

[61] A. R. Tall, C. B. Blum, G. P. Forester, and C. A. Nelson, *J. Biol. Chem.* **257,** 198 (1982).

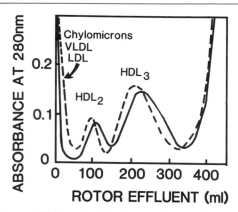

FIG. 4. Distribution of HDL_2 and HDL_3 in the plasma of a healthy volunteer in the postabsorptive (——) and in the postprandial (------) state. The two plasma specimens were obtained after a 14-hr overnight fast and 6 hr after ingestion of a fatty standardized test meal,[50] respectively. Directional flotation in the zonal rotor is from right to left. The effluent patterns suggest that HDL_2 cannot be separated from HDL_3 at varying states of absorption by using preselected density cuts.

uncontaminated postprandial HDL_2 and HDL_3; the preparative capability allows isolation of amounts sufficient for their characterization. Indeed, it has been shown recently that the HDL_2 in individuals with pronounced postprandial lipemia and low HDL_2 levels[62] become enriched in the postprandial phase not only with phospholipids but also with triglycerides.[50] In contrast, the HDL_2 from individuals with low-level postprandial lipemia and high HDL_2 levels do not show postprandial enrichment with triglycerides. When treated with human hepatic lipase *in vitro,* only the triglyceride-enriched postprandial HDL_2 is converted to HDL_3. Thus, the differences in postprandial modification of HDL_2 may help to explain the negative correlation between the magnitude of postprandial lipemia and HDL_2 levels in normolipidemic individuals.[62]

Pathologic Lipoproteins

Severe cholestatic liver disease is often accompanied by elevated plasma levels of unesterified cholesterol and phosphatidylcholine. The accumulation of these lipids is reflected in a gross abnormality of all the major lipoprotein density classes. The VLDL in cholestatic liver disease show an abnormal electrophoretic behavior in that they migrate to a β position rather than into the usual pre-β position. The LDL density class

[62] J. R. Patsch, J. B. Karlin, L. W. Scott, L. C. Smith, and A. M. Gotto, Jr., *Proc. Natl. Acad. Sci. U.S.A.* **80,** 1449 (1983).

usually contains two fundamentally different types of lipoprotein particles. One contains the major LDL apolipoprotein, i.e., apoB just as normal LDL do. However, the core of these LDL particles is uncommonly enriched with triglyceride at the expense of cholesteryl esters.[17,63,64] The second type of lipoprotein particles with the density of LDL occurring in severe cholestatic liver disease is known as LP-X.[65] This particle contains no apoB; rather, its major protein component is albumin, but apoC, apoA-I, and apoE are also present.[17] The concentrations of HDL are usually extremely low in severe cholestatic liver disease. Occasionally, an abnormal HDL particle can be found which has been shown by zonal ultracentrifugation to have higher flotation rates and lesser densities than normal HDL_2.[18]

LP-X was initially regarded as a rather homogeneous lipoprotein fraction. However, zonal ultracentrifugation indicated considerable heterogeneity.[15] This heterogeneity was also observed by Patsch et al.,[16] who attempted to isolate purified subfractions of LP-X for detailed structural and metabolic studies. They subjected cold plasma to an ethanolic fractionation procedure. Seven volumes of cold ethanolic acetate buffer at $-6°$ (19% ethanol, 38 mM sodium acetate, 3 mM acetic acid, pH 5.8) was added to 1 vol of plasma at $1°$. The mixture was stirred for 30 min at $-6°$ and centrifuged at that same temperature at 20,000 g for 30 min. The supernatant, Cohn fractions IV–VI, was dialyzed against a standard buffer containing 100 mM NaCl, 1 mM NaN$_3$, pH 7.4. Cohn fractions IV–VI were then subjected to rate zonal ultracentrifugation in a Beckman Ti-14 zonal rotor. A sodium bromide density gradient linear with volume was used ranging from 1.00 to 1.18 g/ml. Using this procedure, two density fractions of LP-X, designated LP-X$_1$ and LP-X$_2$, could be isolated.[16] When the Cohn fractions IV–VI were subjected to isopycnic zonal ultracentrifugation under quasi-equilibrium conditions, the existence of an additional subfraction became apparent. Using the Beckman Ti-15 zonal rotor and a linear density gradient in the range of 1.00 to 1.15 g/ml at 32,000 rpm, three subfractions of LP-X can be isolated after 24 hr of ultracentrifugation. These fractions were designated LP-X$_1$, LP-X$_2$, and LP-X$_3$.[17] All three populations are rich in phospholipids and unesterified cholesterol but poor in cholesteryl esters, triglycerides, and protein. There are slight differences in chemical composition resulting in buoyant

[63] P. Müller, R. Felin, J. Lambrecht, B. Agostini, H. Wieland, W. Rost, and D. Seidel, *Eur. J. Clin. Invest.* **4**, 419 (1974).
[64] G. M. Kostner, P. Laggner, J. H. Prexl, A. Holasek, E. Ingolic, and W. Geymayer, *Biochem. J.* **157**, 401 (1976).
[65] D. Seidel, P. Alaupovic, R. H. Furman, and W. J. McConathy, *J. Clin. Invest.* **49**, 2396 (1970).

densities of 1.038, 1.049, and 1.058 g/ml, respectively, allowing their separation under quasi-equilibrium conditions. LP-X$_1$, LP-X$_2$, and LP-X$_3$ have Stokes radii of 34, 35, and 29 nm, respectively, and apparent flotation rates [$S_{f(1.063)}$] of 17.3, 9.7, and 3.2 S. All of the particles contain serum albumin and C protein as major protein constituents, but only LP-X$_2$ and LP-X$_3$ contain appreciable amounts of apoA-I and apoE. As determined from circular dichroic measurements, the protein constituents of all three populations possess a high degree of α-helical structure (41–65%). Each of the LP-X subfractions exhibits abnormally low fluidity as evaluated by electron paramagnetic resonance.[17] This low degree of fluidity was also apparent from ^{13}C nuclear magnetic resonance studies, which indicated that the motion of the cholesterol rings and phospholipid fatty acid chains in the particles are far more restricted than in normal LDL and HDL.[66] Proton and ^{31}P nuclear magnetic resonance studies indicate that all three LP-X particles are single bilayer vesicle structures containing the above proteins.[67] The rate zonal ultracentrifugal procedure described above[16] has been used to study the effect of purified LCAT on LP-X.[68] The study showed that LCAT can indeed form cholesteryl esters from the unesterified cholesterol contained in LP-X particles, albeit in relatively small amounts. The procedure was also used to demonstrate that LP-X, when added to whole plasma, contributes to the cholesteryl ester production through LCAT. During this process, all the LP-X is consumed and gives rise to the formation of HDL$_2$-like particles.[69]

Animal Lipoproteins

Zonal ultracentrifugation is particularly useful in isolating and analyzing lipoproteins from various animal species whose density distributions differ from those of human lipoproteins. The above-described procedures for isolating VLDL subfractions, for resolving VLDL, IDL, and LDL, and for separating HDL subfractions, proved quite useful for work with plasma lipoproteins of rabbit, pig, chicken, and rat. In the rabbit, LDL are somewhat less dense than in humans, and therefore float in the zonal

[66] J. R. Brainard, E. H. Cordes, A. M. Gotto, Jr., J. R. Patsch, and J. D. Morrisett, *Biochemistry* **19**, 4273 (1980).

[67] J. R. Brainard, J. A. Hamilton, E. H. Cordes, J. R. Patsch, A. M. Gotto, Jr., and J. D. Morrisett, *Biochemistry* **19**, 4266 (1980).

[68] J. R. Patsch, A. K. Soutar, J. D. Morrisett, A. M. Gotto, Jr., and L. C. Smith, *Eur. J. Clin. Invest.* **7**, 213 (1977).

[69] W. Patsch, J. R. Patsch, F. Kunz, S. Sailer, and H. Braunsteiner, *Eur. J. Clin. Invest.* **7**, 523 (1977).

ultracentrifuge[32] as well as in the analytical ultracentrifuge[70] at somewhat higher rates. Upon cholesterol feeding, LDL and HDL can disappear and all the lipoproteins present in this condition float in the chylomicron–VLDL–IDL density range.[32] In the pig, LDL consists of two density fractions as determined by zonal ultracentrifugation.[71,72] Supplementation of the diet with cholesterol in these animals causes an increase of the plasma concentration of LDL. This rise in LDL is accompanied by a decrease in density and an increase in flotation rates of VLDL and an increase in cholesteryl esters in these LDL at the expense of triglycerides. When examined by differential scanning calorimetry, LDL of the cholesterol-fed pigs have phase transitions above body temperature whereas the LDL from control animals have phase transitions below 37°. The different thermal behavior of the LDL after cholesterol feeding is due to differences in triglyceride content which are a secondary effect of cholesterol feeding. In the chicken, zonal ultracentrifugation demonstrated that VLDL and LDL distribution is similar to that observed in man.[45] However, chicken HDL differ from human HDL in that they consist of only one major density fraction which resembles human HDL_2. When lipoprotein lipase is immunologically blocked in these animals, VLDL with high flotation rates accumulate, LDL disappear from the circulation, and the plasma levels of HDL decrease within 8 hr by about 50%. The remaining HDL have a smaller size and higher density, thus resembling human HDL_3.[45] Zonal ultracentrifugation is also very useful for the work with rat plasma lipoproteins. In this animal species, the major lipoprotein fraction is HDL, whose flotation properties resemble human HDL_2.[55] Rat LDL is about one-tenth as abundant as its human counterpart.[73,74] Between HDL_2 and LDL, a distinct lipoprotein population with varying flotation properties is present which has been termed HDL_1.[55,74,75] HDL_1 differs from HDL_2 and LDL in that it contains apoE as the most abundant apolipoprotein. Due to similarities in hydrodynamic properties, HDL_1 tends to contaminate LDL when LDL is prepared at predetermined density limits by conventional ultracentrifugation. The lipoprotein pattern of the rat can be easily altered by dietary and hormonal perturbations.[33,55,73]

[70] M. J. Chapman, *J. Lipid Res.* **21,** 789 (1980).
[71] G. D. Calvert and P. S. Scott, *Atherosclerosis* **22,** 583 (1975).
[72] H. J. Pownall, R. L. Jackson, R. I. Roth, A. M. Gotto, Jr., J. R. Patsch, and F. A. Kummerow, *J. Lipid Res.* **21,** 1108 (1980).
[73] W. Patsch and G. Schonfeld, *Diabetes* **30,** 530 (1981).
[74] T. Tamai, W. Patsch, D. Lock, and G. Schonfeld, *J. Lipid Res.* **25,** 1568 (1983).
[75] B. Danielsson, R. Ekman, B. G. Johannson, P. Nilsson-Ehle, and B. G. Peterson, *FEBS Lett.* **86,** 299 (1978).

Acknowledgments

The authors wish to thank Irene Perez for typing the manuscript and Susan Kelly for providing the line drawings. Some of the authors' work discussed in this chapter was supported by Grants HL-24759 and HL-27341, and by a grant-in-aid from the American Heart Association (83-1098), with funds contributed in part by the American Heart Association, Texas Affiliate, Inc. Josef R. Patsch is an Established Investigator of the American Heart Association.

[2] Analytic Ultracentrifugation of Plasma Lipoproteins

By TALWINDER S. KAHLON, LAURA A. GLINES, and FRANK T. LINDGREN

Introduction

The use of the analytic ultracentrifuge to study plasma lipoproteins began with the earliest studies of Mutzenbecher, McFarlane, and Pedersen, described in the Gofman *et al.* paper.[1] There, analytic flotation was first described and introduced as a relatively crude quantitative procedure. The historical developments, including the technical achievements since these early studies, are given in some detail elsewhere.[2] A full discussion of the routine lipoprotein fraction preparation in our laboratory, salt solution preparation, and fourth-place density monitoring is also given in detail elsewhere.[3] The important preparative procedures for fractionation and subfractionation of the major lipoprotein classes prior to analytical ultracentrifugal (AnUC) analysis are presented in articles [6, 7 and 8], Volume 128 of this series. Recently, subfractions of HDL and LDL have been described by Anderson *et al.*[4,5] and Shen *et al.*,[6] respectively. The latter subfractions of LDL have been used for partial specific volume ($\bar{v}$) studies, lipoprotein compressibility, molecular weights by sedimentation equilibrium, and evaluation of preferential hydration.[6a]

[1] J. W. Gofman, F. T. Lindgren, and H. Elliott, *J. Biol. Chem.* **179**, 973 (1949).

[2] F. T. Lindgren, *Ann. N.Y. Acad. Sci.* **348**, 1 (1980).

[3] F. T. Lindgren, *in* "Analysis of Lipids and Lipoproteins" (E. G. Perkins, ed.), p. 204. American Oil Chemists' Society, Champaign, Ill., 1975.

[4] D. W. Anderson, A. V. Nichols, T. M. Forte, and F. T. Lindgren, *Biochim. Biophys. Acta* **493**, 55 (1977).

[5] D. W. Anderson, A. V. Nichols, S. S. Pan, and F. T. Lindgren, *Atherosclerosis* **29**, 161 (1978).

[6] M. M. S. Shen, R. M. Krauss, F. T. Lindgren, and T. Forte, *J. Lipid Res.* **22**, 236 (1981).

[6a] T. S. Kahlon *et al.*, *Lipids*, in press (1986).

Special Analytic Ultracentrifuge Cell Construction and Alignment

Although commercial analytic cells are available, they are deficient in three respects. First, lipoprotein flotation from the bottom of the cell is very sensitive to cell misalignment, resulting in striations and distortion of the Schlieren areas throughout the entire course of the AnUC runs. As shown in Fig. 1, we pin our rotors and appropriately slot the cell housing for precise radial alignment. All assembled and torqued cells are checked for this radial alignment with a special fixture. Second, lipoprotein flotation is routinely done in high salt concentrations of NaCl, or the appropriate monovalent salt. During the run at 52,640 rpm, there is significant and progressive sedimentation of the salt, resulting in elevation and curvature of the salt solution baseline in the reference sector and in the sector containing the lipoprotein solution. In spite of the use of calibrated syringes, there is usually a mismatch of the two menisci. This will introduce an insidious baseline error, especially in the 1.20 g/ml HDL run. The solution to this problem is to drill a 0.050-in. hole and scribe a 0.001-in. scratch across the top of the cell partition, as shown in Fig. 2. By slightly overfilling each sector, the solution in each sector then drains into the

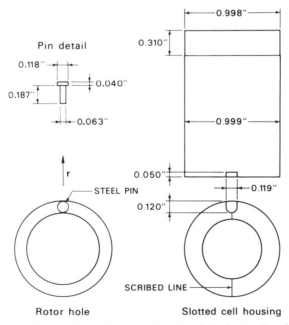

FIG. 1. Details of slotted cell housing and pinned analytical rotor. Reprinted, with permission of the publishers, from Lindgren et al.[9]

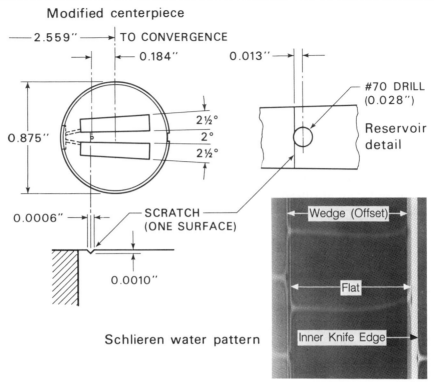

FIG. 2. Modified centerpiece showing special reservoir. In this case, the Schlieren water pattern illustrates precise meniscus equalization for both the flat (offset) and wedge cells. Reprinted, with permission of the publishers, from Lindgren *et al.*[9]

"reservoir" leaving a near-perfect meniscus, matching at the scribe line level. Finally, since AnUC runs are expensive, two runs are normally done, one in a "flat" cell and the other in a "wedge" cell. Since all the flotation measurements are made from the base-of-cell position, it is important not to have a fuzzy double cell bottom which will introduce variable errors. To resolve this, we utilize a 0.020-in. offset centerpiece in the wedge cell, clearly allowing identification in each cell of the true base of cell (see Fig. 2).

Another feature of the above-described cell alignment is that no silicone grease need be used in the rotor holes (to prevent rotation on acceleration), thus allowing cells to be weighed to ±0.1 mg before and after the run. This clearly allows evaluation of leakage or microleakage, if present, during each run.

Analytic Ultracentrifugal Cells for Lipoprotein Analysis

Care in Analytical Cell Management

Analytical ultracentrifugal cells are constructed to rigorous dimensional standards and subjected to great stress during torqueing and during the analytical run itself. The following are step-by-step instructions and guidelines in assembly and cell manipulations to optimize cell performance and to minimize distortion and wear. Although these instructions relate directly to our specially designed and laboratory-fabricated cells, the principles apply to the Al-filled Epon double-sectored Beckman cell system (Beckman Instruments, Palo Alto, CA). We recommend the latter-type centerpiece (12 mm), for mechanical strength, dimensional stability, and noncorrosive qualities where high salt concentrations are routinely used, as in low- and high-density lipoprotein flotation analysis.

Care in the Assembly of Analytical Ultracentrifuge Cells

1. A standard AnUC run utilizes one wedge (X) and one flat (F) cell. The components are alike in design, except for the wedge window and its sector cup in the X cell. The F sector cup and window are flat, with both flat surfaces parallel. The X sector cup and window surfaces are not parallel but have a 1ᵘ wedge to displace the Schlieren pattern vertically on the film or plate.

2. Before assembling a pair of cells, blow the housings and retaining rings clean with compressed air. Apply a thin coat of fluorocarbon spray to the retaining rings and Bakelite gaskets. Make sure the sector cups are dust free and insert one new polyethylene gasket in each cup. Put a comparable Bakelite strip in each sector cup with the center of the strip closest to the outer cell keyway. Now, insert the dried, dust-free, fingermark-free windows into the sector cups. It is important to note that some parts can be placed in any position when assembled, but consistent positioning will minimize the chance of variation and error. It is desirable to have positive identification on all cell parts since the F windows may face in either direction or be rotated in any way, and still fit properly. Thus, it is a good idea to mark the windows with a diamond stylus for conventional positioning of each assembly. The X window must be positioned such that it sits flat in the sector cup, and alignment must be exact. When the centerpieces and quartz disks are cool and dry, any dust should be blown off each cell part with an ear syringe. Assemble completely, but do not fully torque cells until ready to use.

Preparation of Assembled Cells for Analytic Ultracentrifuge

1. Before loading the AnUC cells, turn the Beckman E-AnUC light source on and check the screen for (1) horizontal streaks, usually due to a light source problem, and (2) vertical streaks, usually due to oil or dirt on the collimating lenses. Eliminate these streaks since they may interfere with the quality and resolution of pictures taken throughout the run.

2. Next, weigh each of the empty pair of assembled cells (approximately 35 g). They must be within 100 mg of each other for a low- (D) and high (G)-density run. The empty F cell which holds the less dense sample (LDL and VLDL) should be the heavier.

3. Then use a 1-in.-diameter collet fixture and torque wrench to tighten the retaining ring to desired foot-pounds (usually 6–10 ft-lb). This is the most crucial step in a successful AnUC run. If the cell is torqued too much, the quartz windows will crack. Yet, if there is insufficient torque, the cell will leak. Turn the torque wrench clockwise until the desired torque is attained. The desired torque may differ slightly for each cell, but the sensation in the hands to attain the desired torque should be the same from cell to cell. Remove the cell and check the seal between the centerpiece and the two windows. The contact surfaces should be clean and uniform, free of scratches and dust. Again, check torque to ensure a stabilized torqued cell assembly.

4. Fill the cells using 1.0-cc glass syringes and stainless steel #23 needles. The needles are sawed off and rounded at the ends. We use two lengths: (1) $\frac{3}{8}$-in. (short), used to fill the cells, and (2) $\frac{3}{4}$-in. (long), to remove sample from cell. The long-length needle is used to remove and/or save sample from the cell, and is just long enough to not quite reach the bottom of the centerpiece (and introduce scratches) when fully inserted into a fill hole. The desired amount of sample and baseline is 0.420 cc. Since each numbered syringe-plunger varies somewhat, calibration at 0.420 cc is essential. Use a different syringe for each sample, but the same syringe may be used for baseline of the same type, i.e., the D and G run. Overfill the syringe sufficiently to clear all bubbles. Adjust the amount to the desired reading for 0.420 cc. With the cell screw ring facing you, inject sample slowly into the left side of the cell. Inject baseline slowly into the right side of the cell, following the same procedure. For a standard run, the X cell is filled with the G sample and G baseline ($\rho = 1.200$ g/ml). The F cell is filled with the D sample and D baseline ($\rho = 1.061$ g/ml).

5. Seal the cell fill holes by inserting, in order: two polyethylene gaskets and one Teflon gasket. Then insert a delrin screw plug. Use a screw-holding screwdriver to start the screw plug, then tighten with a screwdriver as tightly as possible without stripping threads. Shave off the

excess screw plug protruding from the housing with a sharp, single-edge razor blade so as to leave it flush with the outer surface. Weigh each filled cell. For each pair, the weight difference should be no more than 140 mg.

6. To control temperature, E-machine rotors should be stored such that, when removed, the measured temperature is 26.3° ± 0.1° and the running temperature approximately 26.0°. While the rotor is cooling down to 26.0°, place the cells in rotor holes with the screw-plug fill holes facing toward the center of the rotor. Use a special wrench to be sure of fit, turning the cell slowly until the scribe mark on the bottom of the housing matches the scribe mark on the rotor. Turn in the direction of tightening, so as not to accidentally loosen the cell retaining ring. When rotor temperature reaches 26.0°, place the rotor in the vacuum chamber immediately.

7. When the E-machine vacuum has dropped sufficiently (0.5–1.0 μm), start advancing drive variac. Record zero time when the current reaches approximately 15 A, then accelerate at 15 A until rotor reaches nearly full speed (52,640 rpm after 5.20 min). For a standard D and G run, the essential pictures are at 0, 2, 6, 30, and 64 min up to speed (UTS).

8. At the end of the run, apply full brake and shut off the diffusion pump. The temperature is checked immediately after the rotor stops. Then weigh the cells. In our experience, double-sectored Al-filled Epon cells routinely leak 2–4 mg during each run. The loss should not be greater than 4 mg to ensure a successful run. Should the loss exceed 4 mg, further examination of the developed film will determine if a rerun is necessary.

9. Before the cells are disassembled, the film (or plate) must be developed. Check the film for proper run photos. The patterns of the baseline and sample should be completely visible and have a minimal number of striations (due to minor misalignment of the cell in the rotor or misalignment of the centerpiece in the cell). If there is a problem with the film, the cells may be rerun after thoroughly shaking up the cell to resolubilize material. The shaking motion should be done holding the cell with the screw plug facing upward; shake up and down vertically so as to minimize transfer of any sample or baseline across centerpiece sectors.

Dismantling, Cleaning, and Reassembly of Analytic Ultracentrifuge Cells

1. Carefully remove the delrin screw plug without damaging the screw threads of the housing. Use a probe to remove the Teflon and polyethylene gaskets, being careful not to scratch the centerpieces. Evacuate the cells with vacuum into a waste flask using a long needle. If sample is to be saved, shake up the cell thoroughly (to resolubilize lipoproteins); then

withdraw sample with the same syringe previously used for filling the cell, but change the needle to a long one ($\frac{3}{4}$-in.). Use a collet and torque wrench to loosen cell parts. Remove the retaining ring and Bakelite gasket and, if necessary, drift out the cell parts with a special plastic rod. Remove the windows from the sector cups and discard the polyethylene gaskets, but save the Bakelite strips. In an organized fashion (for reassembly), set the parts that do not usually have to be washed on a developer tray, i.e., retaining rings with comparable Bakelite gaskets, sector cups, Bakelite strips, and cell housings. With minimal cell leakage these parts are not usually exposed to sample material. In the case of a cell leak, these parts should be rinsed thoroughly with distilled water and then dried. Excessive heat may cause permanent distortion of parts.

2. Wash the syringes and needles with distilled water. Under running distilled water, rub the window between your clean thumb and forefinger until it is squeaky clean. Set on wet Kimwipes and cover until ready to dry them. (It is harder to remove foreign matter from a window that has dried.) Wipe the wet windows dry to prevent water droplet marks. To clean the centerpieces, use a pipe cleaner (size B, 5 mm diameter), bent double, whose wire is completely covered with fibers. Under running distilled water, wash faces of centerpiece and sectors until the centerpiece is thoroughly clean. Be careful not to abrade its polished surfaces. Also, with a fine plastic wash bottle, squirt distilled water through the meniscus overflow hole and the fill holes to rinse away all salt or lipoprotein material. Exposure of the epoxy centerpieces to organic solvents should be minimized, since prolonged exposure may cause softening. With an ear syringe, blow dry the centerpiece to remove water droplets, then set the centerpiece, covered, under heated blow dryer for 20–40 min to ensure complete drying.

3. After all parts are dry, assemble each cell, making sure that all parts are from the right cell and are properly oriented. After each part is stacked, blow away any dust with an ear syringe. Then push the cell housing down to assemble the upper and lower sector cups and centerpiece. Finally, invert the cell assembly—pushing parts to the base of the cell housing, inserting the Bakelite gasket, and loosely tightening the screw ring. Do not fully torque until next usage. As a final check for X cells, hold the cell up to make sure deflection is upward toward the fill hole(s).

Calibration of the Analytic Ultracentrifuge

One of the features of the analytic ultracentrifuge using the Schlieren system is that it measures the invariant property of refractive index increment. Thus, if the optical system is properly calibrated, data can be compared quantitatively over an indefinite time period. We use a Beckman calibration cell (Beckman part #306386), which is run at 5000–6000 rpm. The formula relating area to Δn [Eq. (1)] is given in Fig. 3. Alternatively, and also as a means of validating the calibration cell, a specially scribed double-sector boundary-forming cell is shown in Fig. 3. Thus, the calibration cell Δn can be accurately compared with a Schlieren sucrose boundary. The latter true Δn of the sucrose solution above water is determined by precision or differential refractometry. Thus, from the examples in Fig. 4, absolute Δn calibration of the AnUC can be validated to less than 1%. However, whenever a drive change occurs or when there are any manipulations of the optics, a new calibration run must be made.

Calibration of the AnUC not only involves the optical evaluation of absolute Δn but conversion to the lipoprotein fraction concentrations being measured. Specifically, the major lipoprotein fractions of most importance needing quantification are the "atherogenic" LDL and the "anti-atherogenic" HDL fractions. Although absolute calibration of the ultracentrifuge has been validated to ±1%, the important conversion of Δn to lipoprotein concentration requires precise evaluation of the specific refractive increment (SRI) for each lipoprotein class. In the past, we have used 0.00154 Δn/g/100 ml for all very-low-density lipoproteins (VLDL)

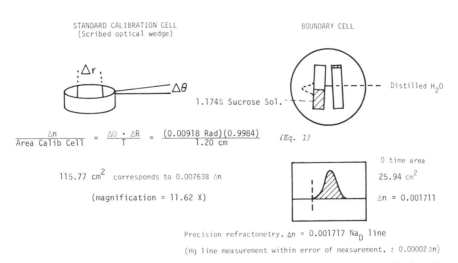

FIG. 3. Comparison of refractive increment calibration by the Beckman scribed wedge cell and the special sucrose boundary-forming cell. Equation (1) gives the relationship between Δn and the calibration cell parameters. Reprinted, with permission of the publishers, from Kahlon *et al.*[7]

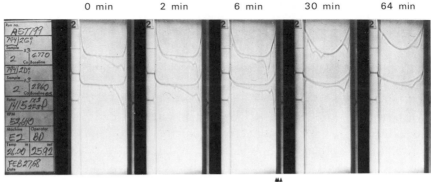

HIGH-DENSITY RUN (1.200 g/ml), WEDGE ⫯
LOW-DENSITY RUN (1.061 g/ml), FLAT ⫯
KNIFE EDGE ⫟

FIG. 4. Essential Schlieren photographs for a total low-density and high-density AnUC run. Reprinted, with permission of the publishers, from F. T. Lindgren, *in* "Fundamentals of Lipid Chemistry" (R. M. Burton and F. C. Guerra, eds.), p. 475. BI-Science Publication Div., Webster Groves, Missouri, 1972.

and LDL as measured in 1.061 g/ml NaCl and 0.00149 $\Delta n/g/100$ ml for HDL measured in 1.200 g/ml NaBr. A recent evaluation[7] of the refractive increment for total LDL and HDL in the above solutions has given values of 0.00142 and 0.00135 $\Delta n/g/100$ ml, respectively. Thus, conversion of older data for proper comparison with these newer data can be made using factors of 1.087 and 1.106, respectively.

All Schlieren photographs (see Fig. 4) are routinely traced at a magnification of 5× using a calibrated precision enlarger. This is best done in a darkroom after appropriate dark eye adaptation. Magnification from the analytical cell is approximately 2.32× to the film, and the total magnification is approximately 11.62×. An enlarger test film consisting of a scribed 2-cm-diameter circle is checked before and after any Schlieren film tracing. Such tracing on calibrated flotation rate templates for the low- and high-density runs are done for the 0-, 2- (or 6-), and 30-min up-to-speed (UTS) frames, and on either the 48- or 64-min UTS frames for the high-density spectra. A typical Schlieren film tracing is shown in Fig. 5. These dimensions are fed into a computer program, as originally described[8] and later modified[9] to include moving boundary flotation rate determinations,

[7] T. S. Kahlon, G. L. Adamson, L. A. Glines, F. T. Lindgren, M. A. Laskaris, and V. G. Shore, *Lipids* **19,** 558 (1984).

[8] A. M. Ewing, N. K. Freeman, and F. T. Lindgren, *Adv. Lipid Res.* **3,** 25 (1965).

[9] F. T. Lindgren, L. C. Jensen, and F. T. Hatch, *in* "Blood Lipids and Lipoproteins" (G. J. Nelson, ed.), p. 181. Wiley (Interscience), New York, 1972.

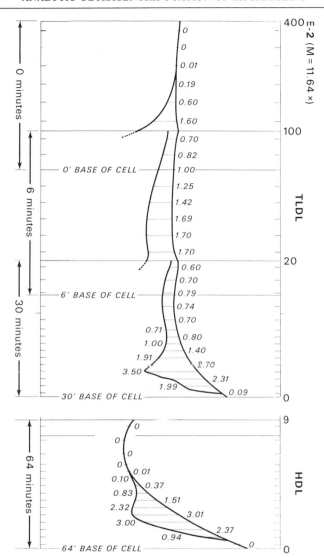

Fig. 5. Typical Schlieren film tracing of a total low- and high-density AnUC run. Reprinted, with permission of the publishers, from F. T. Lindgren, *in* "Fundamentals of Lipid Chemistry" (R. M. Burton and F. C. Guerra, eds.), p. 475. BI-Science Publication Div., Webster Groves, Missouri, 1972.

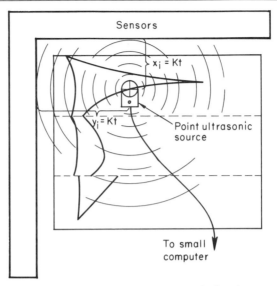

FIG. 6. Schematic diagram of improved Schlieren analysis using a sonic digitizer computer system. Reprinted, with permission of the publishers, from Lindgren *et al.*[10]

$\eta F°$ versus ρ analysis, minimum hydrated molecular weights of the major LDL component, and graphical presentation of the data. The latter includes plots of mean populations and difference plots magnified to visualize more clearly subtle differences within the full lipoprotein spectra in both individual samples or between experimental and population means.

Instead of the older manual techniques for Schlieren pattern analysis, we now use a sonic digitizer computer system to generate an indefinite number of (x_i, y_i) points (accurate to ± 0.1 mm) on each traced Schlieren curve.[10] These stored and edited results of each sample analysis are sent via telephone modem to a larger CDC 7600 computer for data processing using a 50K field length program. In the near future, one of the VAX 790 or 8600 series computers will be substituted for this latter data processing. Schematically, the digitizer system is shown in Fig. 6. Using all the points generated in the primary Schlieren tracing analysis program will shortly allow a more accurate curve resolution, particularly within the human HDL spectra. At present, this is resolved into three components, HDL_{2a}, HDL_{2b}, and HDL_{3a}, essentially by the procedure described by Anderson *et al.*[5] It should be noted that the older HDL_2 ($F_{1.20}$ 3.5–9) and HDL_3 ($F_{1.20}$

[10] F. T. Lindgren, S. B. Hulley, G. L. Adamson, and T. S. Kahlon, *in* "Dietary Fats and Health" (E. G. Perkins and W. J. Visek, eds.), p. 564. American Oil Chemists' Society, Champaign, Ill., 1983.

0–3.5) subdivisions are not equivalent to the newer HDL_{2a+2b} and HDL_{3a}, respectively; rather, the subdivisions for HDL_{2b}, HDL_{2a}, and HDL_{3a} are for $F^{\circ}_{1.20}$ 4.5–9, $F^{\circ}_{1.20}$ 2.4–4.5, and $F^{\circ}_{1.20}$ 0–2.4, respectively. The peak $F^{\circ}_{1.20}$ rates for these components are approximately 5.4, 3.2, and 1.6 $F^{\circ}_{1.20}$ units, respectively.[5]

Sedimentation Equilibrium of Lipoprotein Subfractions

Theory of the Rayleigh Optical System

The Rayleigh optical system involves the use of a point source of light in the analytical ultracentrifuge in which the light source is rotated 90° as compared to the Schlieren configuration. The analytic cell used has double-slit interference masks at both ends. A typical interference fringe pattern obtained using a double-sectored intereference cell and counterbalance reference cell, as used in meniscus depletion sedimentation equilibrium, is shown in Fig. 7. Each fringe is actually a proportional plot of the lipoprotein concentration c vs r (the distance from the axis of rotation). The theory and practice of this procedure has been discussed earlier.[11–13]

Preparation of Human LDL Subfractions

Fasting blood is drawn into evacuated tubes containing EDTA (100 mg/100 ml) and plasma is prepared. Then garamycin (7.5 mg/100 ml) and ε-amino caproic acid (130 mg/100 ml) are added. The plasma density is adjusted to 1.0190 g/ml[14] by mixing 4 ml plasma with 2 ml of an NaBr solution of density 1.0426 g/ml containing 0.196 m NaCl and 10 mg/100 ml EDTA. The resulting 6-ml solution(s) are centrifuged at 17° in a 40.3 rotor at 40,000 rpm for 18 hr. The top 1-ml VLDL–IDL fraction is removed, and the second 1 ml is taken as a background. The subnatant 4 ml is adjusted to 1.0670 g/ml by adding 2 ml of a NaBr solution, $\rho = 1.1566$ g/ml, containing 0.196 m NaCl and 10 mg/100 ml EDTA. Again these tubes are similarly centrifuged for 24 hr, and the top 1-ml fraction containing the total LDL is recovered. These top fractions are pooled and dialyzed 3× against a NaBr solution, $\rho = 1.0400$ g/ml, overnight at 4°. The remaining

[11] T. Svedberg and K. O. Pedersen, in "The Ultracentrifuge" (R. H. Fowler and P. Kapitza, eds.), p. 5. Clarendon, Oxford, 1940.

[12] D. A. Yphantis, Biochemistry **3**, 297 (1964).

[13] C. H. Chervenka, "A Manual of Methods for the Analytical Ultracentrifuge." Beckman Instruments Co., Palo Alto, Calif., 1973.

[14] R. J. Havel, H. A. Eder, and J. H. Bragdon, J. Clin. Invest. **34**, 1345 (1955).

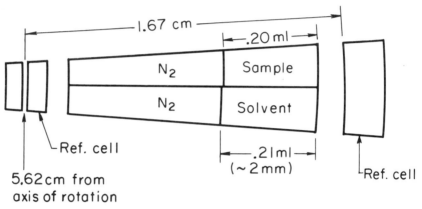

FIG. 7. Typical low-density lipoprotein subfraction interference fringe pattern.

density gradient procedure is essentially that given by Shen *et al.*[6] in which 2 ml of the LDL (ρ = 1.0400 g/ml) fraction is layered dropwise at a 45° angle over 2.5 ml of a NaBr–NaCl solution (ρ = 1.0540 g/ml); then 2.5 ml of a NaBr–NaCl solution (ρ = 1.0275 g/ml) is dropwise overlayered above the LDL fraction. Five such 7-ml tubes and one salt reference are then centrifuged at 40,000 rpm (22°) in a Beckman SW 45 rotor. After 40 hr the run is stopped without braking, and a densitometric scan of each tube is obtained[10] to validate the fractionation. One top 0.5-, six 1-, and one bottom 0.5-ml fractions are taken. The individual fractions are pooled, the 1-ml fractions #1–#5 quantified by precision refractometry, and the densities of the corresponding salt reference gradient fractions determined by precision refractometry[3] or by use of a Paar-Mettler DM 30 density meter (Mettler Instrument Co., Princeton, NJ).

The effect of diffusion in this salt reference gradient during rotor deceleration was determined by again accelerating to 40,000 rpm and decelerating the rotor. This effect is nonsignificant.

Partial Specific Volume of LDL Subfractions

Each lipoprotein subfraction along with its corresponding equilibrium gradient background salt solution are dialyzed together at 4° against a 1.0063 g/ml NaCl solution (1 mg/dl EDTA, pH 7.2–7.4) in the same 250-ml cylinder. The CO_2 present in this dialysate solution is displaced by bubbling N_2 before use. The dialyzed density gradient background gives an approach-to-equilibrium of the final (3×) 24-hr dialysate. At this time, the background salt gradients and the lipoprotein subfractions are only at equilibrium to the final dialysate to the fourth decimal place. Such an error would result in a lipoprotein $\bar{v}$ error of two parts in the second place. True equilibrium dialysis to the sixth place would require approximately 1 week, but by that time significant lipoprotein degradation would occur. Thus, approach-to-equilibrium of each lipoprotein solution must be quantified by the difference in density between the dialyzed background and the final dialysate. The dialyzed sample and corresponding background are removed in sequence, and the solution and solvent density are determined using a Paar-Mettler DMA-60 sixth-place density meter (Mettler Instrument Co., Princeton, NJ). The mass concentration (g/ml) of each LDL solution is determined from lipoprotein composition analysis[7] to ±0.5%[9] by a modified carbon, hydrogen, and nitrogen (CHN) analyzer (Model 185, Hewlett-Packard, Palo Alto, CA).

Apparent partial specific volume ($\bar{v}$, ml/g) of each LDL subfraction is calculated using the equation

$$\bar{v} = \frac{1}{\rho_0} - \frac{1}{x}\left(\frac{\rho - \rho_0}{\rho}\right) \tag{2}$$

where $\bar{v}$ = partial specific volume (ml/g)
     ρ_0 = solvent density (g/ml)
     ρ = lipoprotein solution density (g/ml)
     x = mass concentration of LDL solution (g/ml)

Each solvent background and LDL solution densities are measured to ±0.000001 g/ml, while the temperature of the 1.8-ml density measuring cell is stabilized to 20 ± 0.001°. Density readings are taken only during small and equal upward and downward 0.0003° deviations from 20°, as monitored by a microammeter. Reference standards used for density calibrations are dry air and the purest deionized water, boiled for 10 min to remove dissolved CO_2, cooled to 20°, and used only once. Dry air density readings are measured before and after each sample, and barometric pressure is monitored to correct for any fluctuations in dry air density.

Using the Analytic Ultracentrifuge Interference Cell to Obtain Equilibrium Fringes

1. The interference cell has a 12-mm-thick double-sectored centerpiece ($2\frac{1}{2}°$) with interference masks on both window holders. The cell is assembled, manipulated, disassembled, and cleaned as described in the section, Care in the Assembly of Analytical Ultracentrifuge Cells. Just before filling, it is torqued to 11 ft-lb. Each sector is first thoroughly flushed with N_2.

2. With the screw ring facing you, inject in the left sector (using a $\frac{3}{4}$-in. blunted #23 needle) 0.20 ml of the dialyzed lipoprotein solution, diluted with the final dialysate to a concentration of 3 mg/ml. Similarly, inject 0.21 ml of the final dialysate into the right sector. Place two polyethylene gaskets in each fill hole and tighten the screw plugs snugly. Adjust the weight of the counterbalance reference cell to 0.2–0.3 g less than the filled sample cell.

3. Equilibrate the rotor, sample, and baseline at running temperature (20°). Use a calibrated RTIC temperature control system for stability at 20° during the AnUC run.

4. Place the sample cell and counterbalance in either an An D or an An J rotor, carefully aligning rotor and cell scribe lines (coincident or parallel).

5. Equilibrium speed for each LDL subfraction depends upon the density difference between the lipoprotein and the solvent as well as the molecular weight. Subfractions of density 1.0492–1.0267 g/ml (1 g) corresponding to S_f values of S_f 3.7–9.5, respectively, were run at rotor speeds of from 5000 to 10,000 rpm.

6. As soon as the rotor attains equilibrium speed, take the first photograph using 2 × 10-in. Kodak spectroscopic plates. After development, check for baseline optical distortion using an appropriate microcomparator (Gaertner Instrument Co., Chicago, IL).

7. Increase rotor speed 20% for 3 hr, then return to the equilibrium speed so as to enhance earlier attainment of equilibrium.

8. After 72 hr, visually check the screen and take the first picture of the equilibrium fringes. Confirm by taking another picture 24 hr later. In each case, a picture of zero-order fringes is obtained with a clear glass, 10-mm-thick filter. Align the microcomparator, using the middle of the three dark zero-order fringes, and take three final pictures, using the green 10-mm filter with variable exposures to achieve best photographic contrast. Read approximately 20 or more (x_i, y_i) points along the concentration

gradient (which should occupy the outer half of the fluid column) and then plot $\ln(y_i - y_0)$ vs r^2. If the correlation coefficient of this $\ln kc$ vs r^2 is greater than 0.999, monitor the final speed during this time for 1 hr, then shut the run off.

Developing Spectroscopic Plates

The Kodak (Type II-GCIKO) spectroscopic plates are processed in a dark room without use of any "safe" lights. Further, if there are any fluorescent lights in the dark room, turn them off at least 5 min before opening the Beckman plate holder. We use glass cylinders for each solution in which the plates are inserted vertically. The cylinders sit in a 68°F water bath. The sequence is as follows:

1. Using fresh Kodak (D-19), develop 4 min.
2. Rinse in H_2O, dipping 10 times in 3% glacial acetic acid.
3. Place in Kodak fixer 13 min, then rinse in H_2O under cold tap, then 2 min in Hypoclear, and finally 5 min under cold H_2O.
4. Dry plates, using warm air from a hair dryer.

Reading Spectroscopic Plates

Using a clear glass filter, align the zero-order middle dark fringe from the two reference holes in the counter balance. The center wire of the left reference fringe is 5.62 cm, and the start of the outer fringe in the reference cell is 7.29 cm from the axis of rotation. Magnification used is therefore equal to (measured distance between wire and the start of the outer fringes)/1.67. Figure 7 shows appropriate equilibrium fringes from LDL subfraction #3.

With the plate aligned on the x axis ($y = 0$), bring the crosshair to the beginning of the straight part of the equilibrium fringe. Then determine vertical positions (y_i) at horizontal (radial) intervals of 0.05 cm where fringes are straight. Record each pair of (x_i, y_i) coordinates. When $y > 70$ μm, reduce x interval to 0.01 cm and later to 0.005 cm. Continue to measure y_i of the center of the same fringe until it becomes unresolved. Normally, 20–40 (x_i, y_i) points are measured for one fringe.

The radial measurements (x_i) are converted to the actual distances (in centimeters) from the axis of rotation by:

$$r_i \text{ (cm)} = 5.62 \text{ cm} + \left(\frac{\text{measured distance from center of inner wire}}{\text{magnification}} \right) \quad (3)$$

Then vertical (y_i) distances from the initial (y_0) measurement are converted to $\ln(y_i - y_0)$ and plotted against r_i^2. Correlation coefficients of $r = 0.999+$ were normally obtained, as shown in Fig. 8, suggesting physical

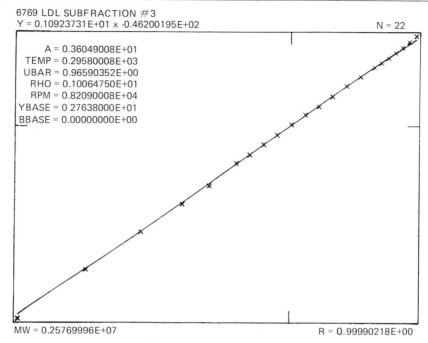

6769 LDL SUBFRACTION #3
Y = 0.10923731E+01 x -0.46200195E+02   N = 22

A = 0.36049008E+01
TEMP = 0.29580008E+03
UBAR = 0.96590352E+00
RHO = 0.10064750E+01
RPM = 0.82090008E+04
YBASE = 0.27638000E+01
BBASE = 0.00000000E+00

MW = 0.25769996E+07   R = 0.99990218E+00

FIG. 8. Plot of $\ln kc_i$ vs r_i^2 for LDL subfraction of subject 6769 (see the table).

monodispersity. The table shows this typical set of data and results for LDL subfraction 3 (case 6769).

If fringes near the meniscus are not straight, the meniscus has not been depleted, thus higher rotor speed is required to achieve true meniscus depletion equilibrium. If the solute is polydisperse, the $\ln(y_i - y_0)$ vs r_i^2 plot curves upward toward the base of the cell; if it curves downward, it indicates nonideality of the solution system.

Finally, molecular weight is obtained by the following equation:

$$MW_{SE} = \frac{2RT(\ln c)/r^2}{(1 - \bar{v}\rho)\omega^2} \qquad (4)$$

where R = gas constant, 8.314×10^7 erg/mol/degree
T = absolute temperature, Kelvin (°C + 273)
$(\ln c)/r^2$ = slope of the fringe plot at equilibrium
$\bar{v}$ = partial specific volume of lipoprotein, obtained from CHN lipoprotein mass and sixth-place density meter
ρ = final dialysate background solution (1.0063 g/ml)

$$\omega^2 = (\text{radial velocity})^2 = \frac{\text{rpm}}{60} \times 2\pi \ (\text{radians/sec})^2$$

INTERFERENCE EQUILIBRIUM FRINGE DATA FROM LDL
SUBFRACTION 3, CASE 6769

Number	Equilibrium fringes			$(y_i-y_0) \times 10^4$ (ln c)
	x_i	y_i	r^2 (cm)^a	
1	2.1938	2.7638 (y_0)		
2	2.6800	2.7825	47.0806	5.2311
3	2.7300	2.7909	47.3990	5.6021
4	2.7700	2.7997	47.6545	5.8833
5	2.8000	2.8076	47.8466	6.0822
6	2.8200	2.8138	47.9749	6.2146
7	2.8400	2.8225	48.1033	6.3750
8	2.8500	2.8263	48.1676	6.4378
9	2.8600	2.8309	48.2319	6.5088
10	2.8700	2.8356	48.2963	6.5765
11	2.8800	2.8411	48.3607	6.6503
12	2.8900	2.8467	48.4251	6.7202
13	2.9000	2.8517	48.4896	6.7788
14	2.9100	2.8581	48.5542	6.8491
15	2.9200	2.8655	48.6187	6.9246
16	2.9300	2.8727	48.6834	6.9930
17	2.9400	2.8801	48.7480	7.0588
18	2.9450	2.8843	48.7804	7.0942
19	2.9500	2.8889	48.8127	7.1317
20	2.9550	2.8938	48.8451	7.1701
21	2.9600	2.8989	48.8775	7.2086
22	2.9650	2.9049	48.9099	7.2521
23	2.9700	2.9109	48.9423	7.2937

^{a} $r = 0.999902189$, slope $= 1.092373107$; $(y_i-y_0) \times 10^4$ gives the ln c with no change in slope.

Results of Sedimentation Equilibrium of LDL Subfractions

Figure 9 shows molecular weight results obtained by flotation velocity and by sedimentation equilibrium on the same five LDL subfractions isolated by the Shen et al.[6] procedure. In addition, the same fractions were compared using $\bar{v}$ obtained by two methods, 1-g densities of the final density gradient and by CHN mass analysis and sixth-place density measurements. Where appropriate, analogous values obtained by others also are given but are limited to those results where the flotation rates of the major LDL component were available. These results were by flotation velocity, with f/f_0 evaluated by diffusion measurements.[15–17] Note the

[15] G. H. Adams, "A New Approach to the Study of Serum Lipoproteins," p. 60. Ph.D. thesis, University of Pennsylvania, Philadelphia, 1966.
[16] M. G. Hammond and W. R. Fisher, J. Biol. Chem. 246, 5454 (1971).
[17] D. Lee and P. Alaupovic, Biochem. J. 137, 155 (1974).

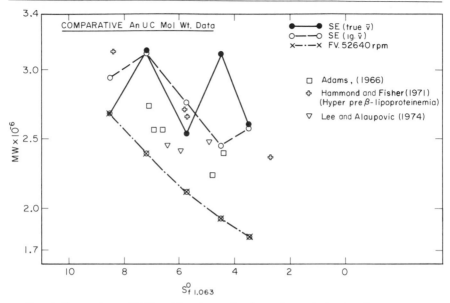

FIG. 9. Comparison of LDL subfraction molecular weights obtained by flotation velocity and sedimentation equilibrium with results obtained by others.

general trend of lower molecular weight with lower S_f rate, except for the sedimentation equilibrium measurements. The latter results suggest that various species of LDL may not result from the continuous conversion of larger, less dense particles to smaller, denser LDL. Rather, specific species of LDL subfractions may result from independent VLDL–IDL conversion and/or independent synthesis and release from other metabolic sites, e.g., the liver.

Preferential Hydration of Lipoproteins

Lipoprotein molecules in H_2O-salt solutions may preferentially bind some pure water, yielding a lower density as compared with the same molecules in a D_2O system. The assumption is that the same number of pure H_2O and pure D_2O molecules are bound. In the latter case, the ratio of H_2O/D_2O in the solution would be the same as that preferentially bound, resulting in minimal change of $\bar{v}$. A theoretical treatment of a three-component sedimentation equilibrium system by Hearst and Vinograd[18] can be considered where H_2O and/or D_2O is the third component. If

[18] J. E. Hearst and J. Vinograd, *Proc. Natl. Acad. Sci. U.S.A.* **47**, 999 (1961).

this is developed, the final expression for wt% preferential hydration is as follows:

$$\Gamma' = \frac{\bar{v}_{LDL} - \bar{v}_{H\text{-}LDL}}{\bar{v}_{H\text{-}LDL} - \bar{v}_{H_2O}} \tag{5}$$

where Γ' = preferential hydration expressed at wt% of total macromolecule

$\bar{v}_{LDL}$ = ρ intercept^{-1} of $\eta F°$ vs ρ using D_2O (unhydrated)

$\bar{v}_{H\text{-}LDL}$ = ρ intercept^{-1} of $\eta F°$ vs ρ using NaCl, NaBr, etc. (hydrated)

$\bar{v}_{H_2O}$ = partial specific volume of H_2O at 20°, i.e., 1.00177 ml/g

In two samples of LDL subfraction #3, values of preferential hydration in NaCl–H_2O solutions were 4.1 and 5.0 wt%, and in NaBr, 0.194 m NaCl–H_2O solutions were 0.8 and 0.4 wt%, respectively. Previous data of Fisher et al.[19] for LDL have given essentially no preferential hydration in a H_2O–KBr system. Recently a value as high as 10% has been reported.[20]

Acknowledgments

This research was supported by NIH NHLBI National Research Service Award Grant HL 07279 and Program Project Grant HL 18574. Special thanks to Gerald L. Adamson and Dr. Verne Schumaker for extensive discussion.

[19] W. R. Fisher, M. E. Granade, and J. L. Mauldin, *Biochemistry* **10**, 1622 (1971).
[20] V. Schumaker, personal communication, 1984.

[3] Separation and Analysis of Lipoproteins by Gel Filtration

By LAWRENCE L. RUDEL, CAROL A. MARZETTA, and
FRED L. JOHNSON

Introduction and Principles

Plasma lipoproteins consist of families of particles that differ in size and density. The differences in density have been used as preparative tools for separation of families, thus the names very-low-density (VLDL), intermediate-density (IDL), low-density (LDL), and high-density lipoproteins (HDL). In general, size and density are inversely proportional (see the table) so that the largest lipoproteins (VLDL) are the least dense while the smallest lipoproteins (HDL) are the most dense. Lipoproteins appear

SIZE AND DENSITY CHARACTERISTICS OF PLASMA LIPOPROTEINS

Family of lipoprotein	Isolation density (g/ml)	Average diameter (Å)[a]	Apparent molecular weight (Da)[b]
Chylomicrons[c]	$d < 1.006$	>1000[d]	>50 × 10[6d]
VLDL	$d < 1.006$	300–800	10–80 × 10[6]
IDL	1.006–1.019	~300	5–15 × 10[6]
LDL	1.019–1.063	220	3 × 10[6]
Lp(a)	1.05–1.09	280	5 × 10[6]
HDL₂	1.063–1.125	120	3 × 10[5]
HDL₃	1.125–1.21	85	1.5 × 10[5]

[a] As determined by electron microscopy after negative staining.
[b] As seen by agarose column chromatography.
[c] As isolated from lymph.
[d] All values were determined in the author's laboratory using human or monkey lipoproteins and are rounded to the nearest significant figure.

to be spherical in most cases, therefore their apparent size and molecular weight are closely related. It should be noted that lipoproteins are particles consisting of hundreds to thousands of molecules, so that the term molecular weight is used only to convey the concept; particle weight would be a more correct term. The size differences among lipoproteins can be used to separate families. The principal method used for separation of biological molecules by size is gel filtration or gel chromatography. This method is also sometimes termed molecular sieve chromatography. In theory molecules of different size may also undergo some ion exchange or adsorption effects in addition to size separation during gel chromatography and this is probably true for lipoproteins. However, for the purposes of this presentation, chromatographic separation of lipoprotein particles appears to occur primarily due to differences in size and the term gel filtration will be used.

Agarose gel is the support medium most readily available for the separation of particles with the sizes of lipoproteins. Chromatography columns of agarose gel can be constructed that provide the ability to preparatively separate the major classes of lipoproteins, VLDL, LDL, and HDL[1] or to subfractionate VLDL.[2] Separation of lipoproteins by agarose gel filtration offers several advantages over those of other techniques, including (1) the procedure is gentle and nondestructive, (2) the size distribution of particles can easily be visualized during separation for

[1] L. L. Rudel, J. A. Lee, M. D. Morris, and J. M. Felts, *Biochem. J.* **139,** 89 (1974).
[2] T. Sata, R. J. Havel, and A. L. Jones, *J. Lipid Res.* **13,** 757 (1972).

comparison among samples, (3) the procedure is simultaneously preparative and analytical, and (4) the completeness of recovery after separation
is high. The main disadvantages are (1) the lipoprotein samples are diluted
during separation instead of concentrated as during preparative ultracentrifugation and (2) the size–density relationships are not completely understood within and among many lipoprotein families and preparation by
gel filtration often requires further characterization. Overall, separation of
lipoproteins by size is a useful and instructive technique that, when used
under well-defined conditions, provides concepts and information about
lipoprotein particles that are not readily apparent from other commonly
used procedures.

The Gel Filtration Medium

The selection of support media is determined by the particle separation required. Agarose gel beads consisting of 4 and 6% agarose are the
best preparations to use to separate lipoprotein families. In our experience the sized preparations of gel beads (BioGel A) sold by Bio-Rad Laboratories, Richmond, CA, offer the best opportunity to preparatively
separate lipoprotein families. We checked the available products from
Pharmacia and LKB but were unable to obtain resolution comparable to
that achieved with BioGel A5m (6%) and A15m (4%). We use the smallest
sized BioGel A beads (200–400 mesh) available for the maximum degree
of separation. Both the 4 and 6% agarose gels will separate the major
lipoprotein families, i.e., VLDL, LDL, and HDL. The resolution between LDL and HDL is better on 6% gels while on 4% gels resolution
from LDL of lipoproteins larger than LDL is much improved. With 4%
gels, populations of particles in the size range of 300–500 Å in diameter
have consistently been detected that are separated from the LDL and
from the larger VLDL that elute at the excluded or void volume, V_0, of
the column.

For subfractionation of VLDL, 2% agarose gel beads (BioGel A50m)
are most useful.[2] For most VLDL preparations, particles are resolved
into two peaks, one at the void volume and one broad peak within the
included volume. LDL and HDL elute near the total column volume, V_t,
and their separation is incomplete on 2% preparations. We have attempted to subfractionate HDL on 8% agarose gel beads (BioGel A1.5m);
although some resolution is achieved, distinct subfractions of HDL are
apparently incompletely resolved. One difficulty we have encountered
with the 8% gels is that sorption effects are appreciably higher. We have
attempted to use the 1% agarose gel (A150m) to fractionate lymph chylomicron particles. Many of these particles elute at the exclusion volume of

the column and fractionation is incomplete at best; however, separation of smaller chylomicrons from larger chylomicrons does occur. Some investigators have used agarose gel filtration to remove albumin from lymph chylomicrons.[3] This procedure should be used with caution. One group has described the selective loss of a particular apolipoprotein, termed apoA-V, in addition to albumin, when chylomicrons were chromatographed on agarose gels.[4] We have noted that several apolipoproteins in addition to albumin are isolated near the V_t when isolated lymph chylomicrons were chromatographed on BioGel A50m. These data suggest that although chromatography on agarose gels is a relatively gentle technique, it does not appear that one can automatically assume that the apolipoproteins present on particles isolated by this procedure are necessarily those that were present *in vivo*.

Procedure for Separation of Lipoprotein Families

The gel filtration procedure to be selected depends on the goals of the research. There is some evidence that ultracentrifugation modifies lipoprotein particles[5] but the actual extent to which this occurs is unknown. Lipoproteins can be separated by chromatographing whole plasma on agarose columns[6] but this is primarily an analytical technique as the isolated fractions usually contain plasma protein contaminants. We have developed a procedure that employs a single ultracentrifugation of plasma to isolate all of the lipoproteins after which lipoproteins are separated into families by agarose gel filtration.[1] This procedure is simultaneously preparative and analytical since essentially pure lipoprotein fractions are available for further analyses while mass distributions among lipoprotein families can be measured as can the relative size properties of the individual lipoprotein classes. Examples of the usefulness of this procedure in deriving these data are published[7-9] and the details of the procedure are as follows.

Blood donors are fasted overnight to minimize the changes in lipoproteins that occur postprandially. Blood is collected by venipuncture into

[3] R. M. Glickman and K. Kirsch, *J. Clin. Invest.* **52,** 2910 (1973).

[4] N. H. Fidge and P. J. McCullagh, *J. Lipid Res.* **22,** 138 (1981).

[5] S. T. Kunitake and J. P. Kane, *J. Lipid Res.* **23,** 936 (1982).

[6] J. C. Gibson, A. Rubinstein, P. R. Bukberg, and W. V. Brown, *J. Lipid Res.* **24,** (1983).

[7] L. L. Rudel, R. Shah, and D. Greene, *J. Lipid Res.* **18,** 55 (1979).

[8] L. L. Rudel and L. L. Pitts, II, *J. Lipid Res.* **19,** 992 (1978).

[9] L. L. Rudel, C. W. Leathers, M. G. Bond, and B. C. Bullock, *Arteriosclerosis* **1,** 144 (1981).

tubes containing EDTA as an anticoagulant (1 mg/ml final concentration), DTNB as an inhibitor of lecithin : cholesterol acyl transferase (0.4 mg/ml final concentration), and NaN$_3$ as an antibacterial agent (0.1 mg/ml final concentration). The tubes are kept on ice until plasma is isolated at 4° by centrifugation at 900 g for 10 min. Solid KBr that has been predried in a desiccator is then added to plasma in the proportion of 0.3517 g/ml to raise the solvent density to 1.225 g/ml. The KBr is dissolved over a 10- to 15-min period by gentle inversion of the plasma container. The density-adjusted plasma (up to 10 ml/tube) is pipetted into the bottom of an ultracentrifuge tube. The tube is then filled by carefully overlayering (without mixing) the plasma with a solution of density 1.225 g/ml that is prepared by adding solid KBr to a 0.9% NaCl solution, pH 7.4, containing the same concentrations of EDTA and NaN$_3$ as given above for blood. The overlayering procedure is important for preventing serum protein contamination of the isolated lipoprotein mixture. Either the SW-40 rotor or the Ti-70.1 rotor (Beckman Instruments, Spinco Div., Irvine, CA) is used; heat-sealed tubes are used with the latter rotor. The SW-40 rotor is centrifuged for 40 hr at 40,000 rpm at 15°; the Ti-70.1 rotor is centrifuged 24 hr at 50,000 rpm at 15°. The lipoprotein fraction is removed after the tube is sliced in a tube slicer. Greater than 99% of plasma lipoprotein cholesterol is floated to the top 1–2 ml of the ultracentrifuge tube with this procedure. This lipoprotein mixture is stored at 4° and/or it is directly applied to an agarose column.

The agarose gel columns are prepared as follows. Agarose gel slurry as purchased from the supplier, is mixed 1 : 1 with saline and poured into a 1.6 × 100 cm glass column to fill the column. Additional slurry is added as needed until a settled bed height of 90 cm is attained. The column is kept running during preparation with a hydrostatic head of approximately 40 cm. The column is prepared and maintained at 4° in the presence of NaN$_3$ (0.01%) to prevent bacterial growth. With these precautions some preparations of column packing have been in use in our laboratory for 10 years. It is not necessary to run the column at 4°, and for some samples this temperature is contraindicated[10] in which case room temperature is used. The total column volume (V_t) is about 180 ml; V_0 is usually about 0.35 × V_t. The column is fitted with a flow adapter so that the sample can be automatically loaded directly on the top of the gel bed. The lipoprotein sample is introduced into an in-line loop (maximum volume of 4 ml). The mass of sample (measured as cholesterol) loaded is usually between 2 and

[10] D. L. Puppione, S. T. Kunitake, R. L. Hamilton, M. L. Phillips, V. N. Schumaker, and L. D. Davis, *J. Lipid Res.* **23,** 283 (1982).

70 mg. Smaller or larger amounts require smaller or larger columns, respectively. To start a separation, the three-way valves to the sample loop are turned and the sample is delivered onto the column.

The column is eluted by gravity at a flow rate of 5–8 ml/hr. The eluant is 0.9% NaCl containing 0.01% EDTA and 0.01% NaN_3, pH 7.4. A Mariotte bottle is used to maintain the hydrostatic head constant (usually about 50 cm in height). We have attempted to pump columns with a peristaltic pump but have encountered persistent problems with packing of the bed during forced flow. We have tried reversing the direction of flow but no improvement in column performance was found. In particularly hypercholesterolemic samples, such as those of cholesterol-fed pigeons and rabbits, it is frequently necessary to use an eluant with a higher salt content such as that described by Margolis[11] to prevent particle aggregation. Column life varies but in all cases after approximately 50 procedures, columns are disassembled and the agarose gel matrix is cleaned in running water on a 36-mesh nylon screen. The agarose gel is then reused. The eluant from the column is continuously monitored at 280 nm and optical density or %T is recorded. It is then collected in plastic tubes changed at 15- or 20-min intervals in a fraction collector. Lipoprotein fraction volumes are measured after pooling several tubes. For more concentrated samples the volume per tube usually decreases during elution of the LDL peak, then returns to the starting volume after this peak.

The molecular weight of the LDL can be measured during elution. A method for this using the column elution parameters, V_0, V_t, and V_e, the elution volume at the center of the sample peak, has been described by Margolis.[11] We have developed an alternative method for LDL molecular weight determination that is carried out during preparative separation of the plasma lipoprotein mixture.[12] For this procedure ^{125}I-labeled LDL of a known molecular weight is used as a tracer for each sample. It is added directly to the isolated lipoprotein mixture before column elution. The ratio of the V_e of the ^{125}I-labeled LDL to the V_e of the sample LDL is termed the relative size index, r_I. The molecular weights of 25 different LDL samples ranging in size from 2.5 to 8×10^6 were measured in the analytical ultracentrifuge by flotation equilibrium analysis. A graph of the r_I versus the log of the molecular weight defined a linear regression with a correlation coefficient of $r = 0.987$. The regression equation is log mol. wt. $= 2.6215 \, r_I - 2.127$. This equation has been used to calculate LDL molecular weights for a wide variety of lipoprotein samples. A high corre-

[11] S. Margolis, *J. Lipid Res.* **8,** 501 (1967).
[12] L. L. Rudel, L. L. Pitts, II, and C. A. Nelson, *J. Lipid Res.* **18,** 211 (1977).

lation has been found between the relationship of LDL molecular weight measured in this way and the extent of coronary artery atherosclerosis in nonhuman primates,[9,13] thus this method has been used extensively in our laboratory.

The LDL preparation that is iodinated and used as the reference marker has been repeatedly prepared from the same group of six male rhesus monkeys. These animals have been set aside for this purpose. Several animals are used so that natural variations that may occur will tend to be averaged and their effect thereby minimized. Blood from these animals is pooled and the LDL is prepared as above. It is then iodinated using the iodine–monochloride procedure of MacFarlane[14] as modified by Bilheimer,[15] and finally repurified by agarose gel filtration. The molecular weight range of this LDL as measured by flotation equilibrium analysis has only been from $3.0–3.2 \times 10^6$ over the 8 years we have used the procedure. Each preparation of ^{125}I-labeled LDL is directly compared with the preceding preparation by measuring molecular weights of the LDL in several samples with both ^{125}I-labeled LDL preparations. Variation has never been outside the 10% confidence interval when comparing data for LDL molecular weights of samples calculated using "old" and "new" ^{125}I-labeled LDL preparations.

After iodination, each ^{125}I-labeled LDL preparation is stored as a dilute solution (<0.25 mg/ml) in the saline, EDTA, NaN₃ solution and is kept at 4° under nitrogen. The ^{125}I-labeled LDL standard is remade at 6-month intervals. The specific activity is approximately 10^5 DPM/μg of protein, and approximately 10^4 DPM is added to each sample as an internal standard. This means that it is truly a tracer and contributes essentially no real mass to the sample. For a representative column, across a 1-year period, $n = 68$, the mean $\pm$SD for the V_e of the ^{125}I-labeled LDL preparation was 115.6 ± 2.8 ml, yielding a coefficient of variation of 2.5%. This included two preparations of ^{125}I-labeled LDL and the column was repoured once during this time. For 10 repeat determinations of 4 different LDL samples the coefficient of variation was 4.2%. As these numbers demonstrate, this procedure for LDL molecular weight measurement is as accurate as the procedure used for standardization and is precise. We believe this is due to the fact that the reference ^{125}I-labeled LDL as well as the sample is subjected to all of the idiosyncracies of the chromatographic procedure, thereby maximizing the reproducibility of the technique.

[13] L. L. Rudel, in "Use of Nonhuman Primates in Cardiovascular Research" (S. S. Kalter, ed.), p. 37. Univ. of Texas Press, 1980.
[14] L. L. Rudel, D. G. Greene, and R. Shah, J. Lipid Res. 18, 734 (1977).
[15] R. M. Krauss and D. J. Burke, J. Lipid Res. 23, 97 (1982).

Results and Applications

The utility of the procedure is illustrated by the typical agarose gel filtration profiles for lipoproteins from several animals of one experiment as shown in Fig. 1.[16] The rather remarkable shift in distribution of lipoproteins that occurs with the hypercholesterolemic response to an atherogenic diet in these nonhuman primates is clearly apparent even though these data are not strictly quantitative since transmittance at 280 nm is decreased by turbidity and the larger lipoprotein particles are represented out of proportion to their mass. The shift to a smaller elution volume at the center of the LDL peak that occurs in hypercholesterolemic animals can be seen in Fig. 1, and the real decrease in the amount of HDL is also apparent.

For quantitation, the column profile is inspected and fractions containing material within the four main peaks are pooled as illustrated in Fig. 2. The LDL peak is often remarkably shifted among animals as shown in this figure, so the selection of fractions to be pooled is done by inspection. For quantitation purposes, the mass of cholesterol in each of the pooled fractions is then determined and the cholesterol distribution among fractions is calculated. The recovery of cholesterol from the column is generally good; for example, in 1 study including 100 samples, the mean $\pm$SD for recovery was $89.5 \pm 8.8\%$.[8] We have monitored cholesterol as a means to determine distribution and recovery of lipoproteins since it is a constituent of all lipoproteins; it is effectively insoluble in the absence of lipoproteins, and it is a molecule that is conveniently assayed. Further characterization of the plasma lipoproteins of each column region has been carried out in a number of studies.[1,7-9] All of the material in the column region labeled VLDL can be isolated at $d < 1.006$ g/ml. In the elution profiles from 4% agarose gel columns, a size population termed the intermediate-sized low-density lipoproteins (ILDL) is usually defined. This material occupies the region between the VLDL and LDL and consists of a mixture of lipoprotein classes, including VLDL, IDL, LDL, and Lp(a), as described previously.[7,14] The material in the peak labeled LDL has the main characteristics of LDL, i.e., enriched in cholesteryl ester, apoB$_{100}$ as the primary apolipoprotein, and density between $1.006-1.063$ g/ml.[7-9] Likewise, the material in the peak labeled HDL has the same characteristics as material isolated by ultracentrifugation between densities $1.063-1.2$ g/ml, i.e., apoA-I and apoA-II were the primary apolipoproteins, 40–50% protein, and cholesteryl ester as the primary neutral lipid. The HDL peak infrequently appears to be heterogeneous in shape, as shown for the

---

[16] L. L. Pitts, II, Ph. D. thesis, Wake Forest University, Winston-Salem, N.C., 1976.

MALES M. FASCICULARIS

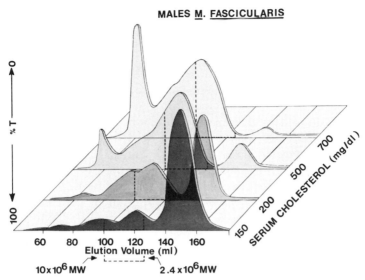

FIG. 1. Column elution profiles for lipoproteins from four male cynomolgus monkeys (*Macaca fascicularis*) that were fed a cholesterol-enriched diet as reported.[8] The lipoproteins from 8 ml of plasma were applied to a 4% agarose column in each case. The cholesterol concentration in each plasma sample is indicated. This work is taken from Pitts.[16]

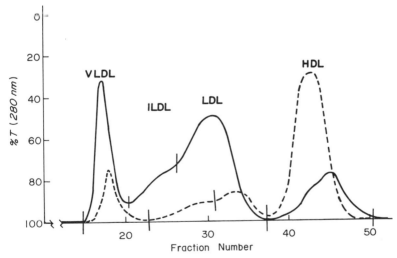

FIG. 2. Column elution profiles for lipoproteins from two rhesus monkeys (*Macaca mulatta*), one fed a control diet (dotted line) and one fed a cholesterol-enriched diet (solid line). The lipoproteins from 8 ml of plasma were applied to the column in each case. The material between the vertical lines was pooled, concentrated, and analyzed as described.[7,14]

solid line of Fig. 2, but by electron microscopy, gradient gel electrophoresis, and density gradient ultracentrifugation, it can be seen to be made up of particles of various sizes from 70–130 Å in diameter.

As lipoproteins elute from gel filtration columns, the material is collected in separate tubes in a fraction collector, thus many potentially different subfractions of lipoproteins are prepared separately. We have analyzed the material in individual tubes to help define the degree of heterogeneity in some of the separated fractions. Gradient gel electrophoresis in 2–12% polyacrylamide gels has been used to demonstrate the degree of heterogeneity across the ILDL and LDL density regions as shown in Fig. 3. In this example, plasma lipoproteins from a cholesterol-fed vervet monkey were separated by agarose gel filtration, and the material from individual tubes was analyzed on gradient gel electrophoresis performed on 2–12% gels according to procedures similar to those described by Krauss and Burke.[15] It is clear from the densitometric scans of the gel electrophoresis separations that individual tubes from the column contain material that is often heterogeneous in size. The data demonstrate

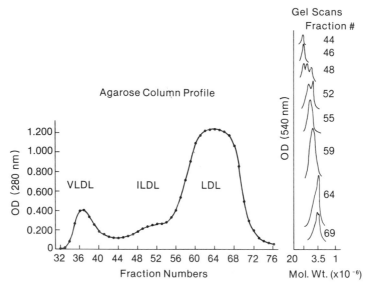

FIG. 3. Separation of lipoprotein subfractions from gel filtration columns by gradient gel electrophoresis. Plasma lipoproteins from a cholesterol-fed vervet monkey were isolated by ultracentrifugation at $d < 1.063$ g/ml and then were separated on a BioGel A15m column, and collected in 1.3-ml fractions in tubes in a fraction collector. The material from single tubes, as indicated by number, was then subjected to electrophoresis in 2–12% polyacrylamide gradient gels. The gels were scanned densitometrically at 540 nm after staining with Coomassie Blue. The molecular weights were determined from LDL and protein standards applied to the gel; the tubes selected for evaluation included those of the ILDL and LDL regions of the elution profile.

the decreasing average size for the lipoproteins eluted in successive tubes. In this particular example, it is clear that even the lipoproteins within a single tube of both the ILDL and LDL regions are made up of multiple components that are heterogeneous in size. These data confirm the fact that lipoprotein classes are very heterogeneous in size and that the gel filtration procedure is separating lipoproteins by size. The size separation occurs even within a peak as defined by OD at 280 nm. The agarose gel filtration procedure offers excellent size discrimination that is not often recognized but can be useful in certain types of experiments.

The ^{125}I-labeled LDL can be used as a molecular weight reference marker because the apoB is an apolipoprotein that is nonexchangeable and stays on the original particle. However, when ^{125}I-labeled LDL are injected into the circulation, lipid transfer and metabolism reactions are known to occur that could modify the size of the LDL particle. Nevertheless, it is possible to use gel filtration to monitor and detect differences in the size of the ^{125}I-labeled LDL after injection. In Fig. 4, for example, an

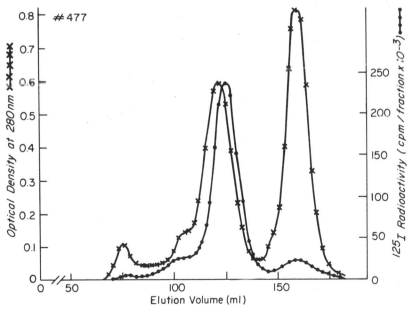

FIG. 4. Column elution profiles for plasma lipoproteins isolated from a rhesus monkey 24 hr after injection of ^{125}I-labeled LDL. LDL was isolated from pooled plasma of several normocholesterolemic rhesus monkeys by the procedure described herein, after which it was iodinated and injected into a hypercholesterolemic rhesus monkey fed an atherogenic diet. Twenty-four hours after injection, a blood sample was taken and the lipoproteins were isolated by ultracentrifugation and separated on a BioGel A15m column. Radioactivity and optical density were monitored as indicated.

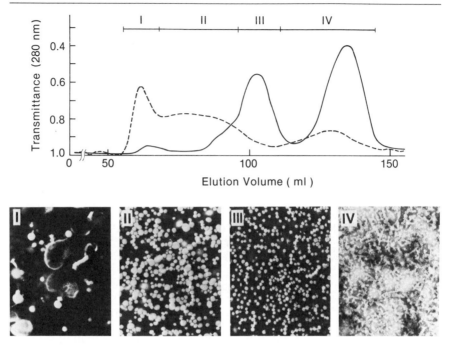

FIG. 5. Elution profiles of lipoproteins isolated and separated from the plasma (solid line) of a vervet monkey compared to those from the perfusate (dotted line) of the same animal's liver. Plasma lipoproteins were isolated from the blood sample taken from the animal just prior to removing the liver for perfusion. Recirculating perfusion was performed using a synthetic medium containing 22% human red blood cells.[17] After a 4-hr perfusion period, lipoproteins were floated in the ultracentrifuge as described herein for plasma, and were then applied and separated on a BioGel A15m column, 1.5 × 90 cm as described above. The fractions containing perfusate lipoproteins in the regions indicated by the Roman numerals were pooled, and aliquots were then subjected to negative staining and electron microscopy.

agarose column elution profile is shown for the isolated lipoproteins from a plasma sample taken from a rhesus monkey 24 hr after injection of ^{125}I-labeled LDL. The labeled LDL were prepared from normocholesterolemic monkeys and were smaller than the LDL of the recipient who was hypercholesterolemic. The data show that the average size of the ^{125}I-labeled LDL has remained smaller than the average size of the recipient's LDL, even though the labeled LDL has been present in the circulation of the animal for 24 hr. Some of the radioactivity that was present on the LDL at the time of injection has become associated with other lipoprotein classes, but the majority of this material was non-apoB apolipoproteins. This example illustrates the potential of agarose gel filtration in analyzing data on changes occurring in LDL size during LDL turnover studies.

One final application for gel filtration of lipoproteins that will be mentioned is its use in the analysis of newly formed or nascent lipoproteins secreted by the isolated, perfused monkey liver, as has been reported by Johnson et al.[17] The plasma lipoproteins from the liver donor monkey were separated by gel filtration and their elution profile was compared to that of the lipoproteins isolated from the liver perfusate after 4 hr of recirculating perfusion (Fig. 5). The size distribution of perfusate lipoproteins was remarkably different from that in plasma lipoproteins and furthermore, as shown in the electron micrographs for the negatively stained material from different regions of the perfusate lipoproteins, the structures seen were also different. For example, the material in region I (at V_0) contained numerous vesicular structures, and in addition to many smaller 100-Å-diameter round particles the material in the HDL region (region IV) contained many discoidal particles that formed rouleaux upon negative staining. The material in regions II and III was primarily spherical, but with an apparent excess of surface material appearing as electron-lucent "flaps" or "tabs." These data illustrate that even though the monkey liver perfusate lipoprotein spectrum represents a different group of particles than is the case for the plasma lipoproteins, the agarose gel filtration column is effective at separating the size populations that are present.

Acknowledgment

This work has been supported by NIH Grants HL-14164 and HL-24736.

[17] F. L. Johnson, R. W. St. Clair, and L. L. Rudel, *J. Clin. Invest* **72**, 221 (1983).

[4] High-Performance Liquid Chromatography of Serum Lipoproteins

By Ichiro Hara and Mitsuyo Okazaki

High-performance liquid chromatography (HPLC) has been widely used for the separation and analysis of many substances due to the short experimental time, small sample for analysis, high resolution, and excellent reproducibility. There are three types of HPLC systems currently being used for the analytical and preparative separation of proteins and polypeptides: reverse-phase chromatography (RP/HPLC) using alkylated.

silica or synthetic polymer, ion exchange chromatography (HP/IEC), and size-exclusion chromatography (HPSEC) or gel permeation chromatography (HPGPC). The separation or quantitation of serum lipoproteins is usually carried out by HPSEC or HPGPC because of the large molecular size of lipoproteins and great differences in the molecular sizes of lipoprotein classes. However, in the case of the analysis of apolipoproteins of lipoproteins, RP/HPLC and HPGPC systems are used.

Recently, a simple and rapid method for lipoprotein analysis has been developed using HPLC with the gel permeation column (TSK GEL, Toyo Soda Mfg. Co., Japan) by Hara et al.[1-3] and has been found reliable by other investigators.[4-6] Moreover, a direct quantitation method for cholesterol in each lipoprotein class using a very small amount of intact serum (10–20 μl) has been developed by Hara et al.[7-9] through a combination of the methods of (1) the separation by HPLC with gel permeation column and (2) the selective detection of cholesterol in a postcolumn effluent using an enzymatic reaction reagent. This method has been applied to the analysis of other lipid components in serum such as triglycerides[10] and phospholipids.[11,12] A prestaining procedure of cholesterol has been proposed by Busbee et al.[4] Since the absorption of a dye used for prestaining may cause a decrease in the column lifespan, a poststaining method by the enzymatic reaction is recommended for the detection and quantitation of lipid components in serum lipoproteins. The HPLC technique developed by Hara et al. should prove useful for lipoprotein analysis, as demonstrated in this chapter.

Methodology

Column. A silica gel or synthetic polymer-based gel permeation column for HPLC is commercially available as the TSK GEL, SW type or PW type, from Toyo Soda Mfg. Co. (Tokyo). Various columns with varying pore size are available, and the specification of the TSK GEL columns

[1] I. Hara, M. Okazaki, Y. Ohno, and K. Sakane, *J. Am. Oil Chem. Soc.* **57**, 88A (1980).
[2] M. Okazaki, Y. Ohno, and I. Hara, *J. Chromatogr.* **221**, 257 (1980).
[3] Y. Ohno, M. Okazaki, and I. Hara, *J. Biochem.* **89**, 1675 (1981).
[4] D. L. Busbee, D. M. Payne, D. W. Jasheway, S. Carlisle, and G. Lacko, *Clin. Chem.* **27**, 2052 (1981).
[5] R. Vercaemst, M. Rosseneu, and J. P. Van Biervliet, *J. Chromatogr.* **276**, 174 (1983).
[6] R. M. Carrol and L. L. Rudel, *J. Lipid Res.* **24**, 200 (1983).
[7] I. Hara, M. Okazaki, and Y. Ohno, *J. Biochem.* **87**, 1863 (1980).
[8] M. Okazaki, Y. Ohno, and I. Hara, *J. Biochem.* **89**, 879 (1981).
[9] M. Okazaki, K. Shiraishi, Y. Ohno, and I. Hara, *J. Chromatogr.* **223**, 285 (1981).
[10] I. Hara, K. Shiraishi, and M. Okazaki, *J. Chromatogr.* **239**, 549 (1982).
[11] M. Okazaki, N. Hagiwara, and I. Hara, *J. Biochem.* **91**, 1381 (1982).
[12] M. Okazaki, N. Hagiwara, and I. Hara, *J. Chromatogr.* **231**, 13 (1982).

used for lipoprotein analysis is listed in Table I. The particle size at the exclusion limit of the column and separation range of human serum lipoprotein classes for each column of Table I are determined from the column calibration curve using standard lipoproteins and proteins of known particle diameter. The TSK GEL columns with various commercial names are supplied from other companies, as presented in Table I. Column systems used for lipoprotein analysis in this method are as follows: G5000PW + G3000SW + G3000SW, G5000PW + G5000PW, G4000SW + G3000SW, G3000SW + G3000SW + G3000SW, G5000PW, and G4000SW. The dimension of each column is 7.5 mm i.d. × 600 mm, and each system has a guard column (GSWP, 7.5 × 75 mm) unless otherwise noted in the text.

Apparatus. HPLC is carried out with a high-speed chemical derivatization chromatograph (HLC 803, Toyo Soda Mfg. Co.) equipped with two detectors of ultraviolet and visible wave lengths, but an enzymatic reaction is performed using a Teflon tube (0.5 or 0.4 mm i.d. × 10,000 or 20,000 mm) in the thermostatted water bath shown in Fig. 1.

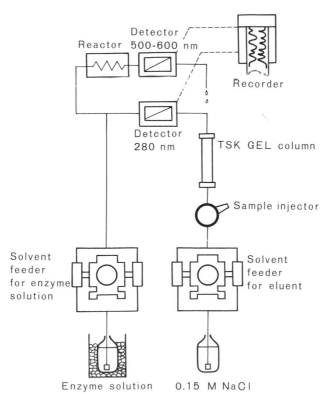

FIG. 1. Flow diagram of lipoprotein analysis by HPLC developed by Hara *et al.*

TABLE I

SPECIFICATION OF TSK GEL COLUMNS USED FOR LIPOPROTEIN ANALYSIS[a]

Grade	Particle size (μm)	Exclusion limit			Applicable range of lipoprotein fraction[b]	Commercial name from various companies
		Protein (mol. wt.)	Dextran (mol. wt.)	Lipoprotein[b] [particle diameter (nm)]		
SW type						
G3000SW	10 ± 2	3×10^5	1×10^5	18.0	-LDL, HDL$_{sub}$, VHDL	MicroPak TSK-3000SW,[c] Spherogel TSK-3000SW,[d] Bio-Sil TSK-250,[e] ProteinPak 300SW,[f] UltroPac TSK SW 3000[g]
G4000SW	10 ± 2	1×10^6	4×10^5	44.0	-VLDL, LDL, HDL	MicroPak TSK-4000SW,[c] Spherogel TSK-4000SW,[d] Bio-Sil TSK-400,[e] UltroPac TSK SW 4000[g]

PW type	PEG[h] (mol. wt.)	Dextran (mol. wt.)	Lipoprotein[b] [particle diameter (nm)]		
G3000PW	13 ± 2	5×10^4	1×10^5	—	-LDL, HDL	MicroPak TSK-3000PW,[c] Spherogel TSK-3000PW,[d] BioGel TSK-30[e]
G4000PW	13 ± 2	3×10^5	7×10^5	40.0	-VLDL, LDL, HDL	MicroPak TSK-4000PW,[c] Spherogel TSK-4000PW,[d] BioGel TSK-40[e]
G5000PW	17 ± 2	1×10^6	2×10^6	100.0	C, VLDL, LDL, HDL	MicroPak TSK-5000PW,[c] Spherogel TSK-5000PW,[d] BioGel TSK-50[e]
G6000PW	17 ± 2	$8 \times 10^{6\,i}$	$2 \times 10^{7\,i}$	300.0	C, VLDL, LDL, HDL	MicroPak TSK-6000PW,[c] BioGel TSK-60[e]

[a] From the brochure offered by Toyo Soda Mfg. Co. (Tokyo).
[b] Determined from the calibration curve obtained for the standard samples of known particle diameter by Okazaki and Hara.[18] C, Chylomicrons; VLDL, very-low-density lipoproteins; LDL, low-density lipoproteins; HDL, high-density lipoproteins; VHDL, very-high-density lipoproteins; HDL_sub, HDL subfractions.
[c] Varian Instrument Group (USA).
[d] Beckman Instruments (USA).
[e] Bio-Rad Laboratories (USA).
[f] Waters Associates (USA).
[g] LKB Produkter AB (Sweden).
[h] PEG, Polyethylene glycol.
[i] Estimated value offered by Toyo Soda.

Samples. Human sera used in this method are obtained from normal male and female subjects, hyperlipidemic and other patients with various diseases after 12–16 hr of fasting. Sera are stored at 4°, and analyzed within 1 week. Sera stored at −20° are analyzed within 1 day after thawing. The individual lipoproteins, chylomicrons + VLDL (d < 1.006), LDL (d 1.006–1.063), HDL$_2$ (d 1.063–1.125), and HDL$_3$ (d 1.125–1.210) are isolated from human serum by the sequential ultracentrifugal flotation method of Havel *et al.*[13] and dialyzed against a 0.15 M NaCl solution. The total lipoprotein fraction is obtained from human serum as a supernatant following ultracentrifugation at 105,000 g for 24 hr at d 1.210, and dialyzed against 0.15 M NaCl solution. The standard globular proteins used for column calibration are HMW and LMW calibration kits (Pharmacia Fine Chemicals, Sweden), and particle diameter of standard lipoproteins used for column calibration is determined by electron microscopy using the negative staining method.

Loaded Volume. The individual lipoproteins or the total lipoprotein fraction containing 20–100 μg of protein are applied onto the HPLC apparatus and monitored by A_{280}. The loaded volume of normolipidemic human serum for analysis of total cholesterol (TC), choline-containing phospholipids (PL), and triglycerides (TG) in a postcolumn effluent are 10, 20, and 50 μl, respectively, due to the detection limit of each lipid in this HPLC method, as will be discussed later. In analyzing the hyperlipidemia, the loaded volume of serum decreases with an increase in serum lipid concentration.

Eluent and Flow Rate. A 0.15 M NaCl, pH 7.0, solution is used as the eluent for the separation of serum lipoproteins. Flow rate of the eluent is 1.0 ml/min for three connected columns, 0.60 ml/min for two connected columns, and 0.50 ml/min for single columns, unless otherwise noted in the text. The separation of lipoproteins by HPLC is carried out at ambient temperature.

Detection of Serum Lipoproteins by HPLC. Lipoproteins are detected by A_{280} in the case of the total lipoprotein fraction or individual lipoproteins isolated from serum by ultracentrifugation. Lipoprotein analysis using intact serum is performed by poststaining the lipid components with the enzymatic reaction. The selective detection of TC, TG, and PL in a postcolumn effluent is carried out by A_{550} or A_{500} following the enzymatic reaction in an on-line system using the high-speed chemical derivatization chromatograph presented in Fig. 1 (see Chapter [6]). The enzymatic reaction is performed by passing a mixed solution of the column effluent and enzyme solution through a reaction tube at an appropriate temperature. The enzyme solution is prepared from a commercially available reagent

[13] R. J. Havel, H. A. Eder, and J. H. Bragden, *J. Clin. Invest.* **34**, 1345 (1955).

kit for the clinical laboratory test: TC, Determiner TC"555" (Kyowa Medex Co., Tokyo); TG, Determiner TG (Kyowa Medex); PL, PL kit K"f" (Nippon Shoji Co., Osaka, Japan). Each reagent kit is obtained as a premixed lyophilized vial (reagent A) and buffer solution (reagent B). The reaction scheme for determining cholesterol and choline-containing phospholipids is identical to that illustrated in Chapter [6], Figs. 1 and 4, except that color production is monitored at 555 nm for cholesterol. The content of reagents A and B for each lipid and the reaction scheme for triglyceride are as follows.

Total Cholesterol (Determiner TC"555")[14]

Reagent A: Cholesterol esterase, 80 U/vial; cholesterol oxidase, 150 U/vial; peroxidase, 540 U/vial; 4-AA, 0.4 mmol/vial; stabilizing agents
Reagent B: Potassium hydrogen phthalate buffer (pH 6.0), 0.025 M; EMAE, 0.3 mM; potassium phosphate, 0.010 M; detergents
4-AA: 4-Aminoantipyrine
EMAE: N-ethyl-N-(3-methylphenyl)-N'-acetylethylenediamine

Triglycerides (Determiner TG)[15]

$$\text{Glycerides} + H_2O \xrightarrow{\text{lipoprotein lipase}} \text{glycerol} + \text{fatty acid}$$

$$\text{Glycerol} + O_2 \xrightarrow{\text{glycerol oxidase}} \text{glyceraldehyde} + H_2O_2$$

$$\text{4-AA} + \text{EMAE} + 2H_2O_2 \xrightarrow{\text{peroxidase}} \text{quinone diimine dye} + 4H_2O$$
$$(\lambda_{max}, 555 \text{ nm})$$

Reagent A: Lipoprotein lipase, 85 U/vial; glycerol oxidase, 2840 U/vial; peroxidase, 850 U/vial; 4-AA, 0.042 mmol; stabilizing agents
Reagent B: Good buffer (pH 6.75), 0.1 M; EMAE, 0.9 mM; detergents; aldehyde-trapping agent

Choline-Containing Phospholipids (PL Kit K"f")[16]

Reagent A: Phospholipase D, 40 U/vial; choline oxidase, 130 U/vial; peroxidase, 110 U/vial; 4-AA, 0.025 mmol/vial; stabilizing agents
Reagent B: Tris-HCl buffer (pH 7.8 ± 0.1), 0.05 M; Triton X-100, 200 mg/100 ml; phenol, 20 mg/100 ml; NaN$_3$

[14] T. Uwajima, H. Akita, K. Ito, A. Mihara, K. Aisaka, and O. Terada, *Agric. Biol. Chem.* **37,** 2345 (1973); **38,** 1149 (1974).
[15] T. Uwajima, H. Akita, K. Ito, A. Mihara, K. Aisaka, and O. Terada, *Agric. Biol. Chem.* **43,** 2633 (1979); **44,** 399 (1980).
[16] M. Takayama, S. Itoh, T. Nagasaki, and I. Tanimizu, *Clin. Chim. Acta* **79,** 93 (1977).

Preparation of Enzyme Solutions and Reaction Conditions of HPLC

Total Cholesterol Determination. Reagent A is dissolved in 80 ml of reagent B. Optimum reaction conditions for this determination by the on-line system are as follows: reaction temperature, 40°; dimensions of a reaction tube, 0.4 mm i.d. × 10,000 mm; mixing volume ratio of the enzyme solution to column effluent, i.e., ratio of the flow rate of the enzyme solution to that of the eluent (0.15 M NaCl), above 0.30. Since the enzymatic reaction proceeds to completion within 1 min at 40°, sufficient reaction time can be obtained under above-optimum conditions in the flow diagram.

Triglyceride Determination. Reagent A is dissolved in 85 ml of reagent B. Optimum reaction conditions for this determination by the on-line system are as follows: reaction temperature, 45°; dimensions of reaction tube, 0.5 mm i.d. × 20,000–30,000 mm; mixing volume ratio of the enzyme solution to column effluent, above 0.80. Under these conditions, the end point of the reaction cannot be obtained, since more than 10 min is required for the reaction to go to completion.

Choline-Containing Phospholipid Determination. Reagent A is dissolved in 25 ml of 0.05 M Tris-HCl buffer solution containing 93.8 mg/100 ml of Triton X-100, 44.4 mg/100 ml of phenol, and 3 mg/100 ml of NaN_3 for the HPLC.[17] Optimum reaction conditions for this determination by the on-line system are as follows: reaction temperature, 39°; dimensions of reaction tube, 0.5 mm i.d. × 20,000 mm; mixing volume ratio of the enzyme solution to column effluent, 0.40–0.50. Since the enzymatic reaction proceeds to completion within 4 min at 39°, sufficient reaction time can be obtained under above-optimum conditions in the flow diagram. When the enzyme solution is prepared using a commercial buffer solution (reagent B), reagent A is dissolved in 25 ml of the buffer solution diluted with an equal volume of distilled water, so as to approximate as much as possible the above optimum reaction conditions.

Elution Pattern of Serum Lipoproteins

To examine the elution volume of each lipoprotein class, reference standards were prepared by the sequential ultracentrifugation according to the method of Havel *et al.*[13] After examining their homogeneities by immunoelectrophoresis, the reference standards were applied onto the column system of G5000PW + G3000SW + G3000SW, and eluted with 0.15 M NaCl at a flow rate of 1.0 ml/min. The elution pattern of each

[17] N. Hagiwara, M. Okazaki, and I. Hara, *Yukagaku* **31**, 262 (1982).

lipoprotein monitored at A_{280} showed good separation for chylomicrons + VLDL, LDL, and HDL as evident from Fig. 2. Separation of HDL_2 and HDL_3 could be expected to some extent.

The total lipoprotein fraction ($d < 1.210$) prepared from serum by ultracentrifugation was applied onto the same column system in Fig. 2, and eluted with 0.15 M NaCl, pH 7.0, at a flow rate of 1.0 ml/min. Well-separated peaks corresponding to chylomicrons + VLDL, LDL, HDL_2, and HDL_3 were obtained by monitoring A_{280}. The elution volume of each lipoprotein peak in Fig. 3 is consistent with that of each lipoprotein in Fig. 2. Although a separation pattern was obtained by monitoring A_{280}, the concentration of each class could not be calculated from the peak area monitored by A_{280}. Thus, the total cholesterol, triglycerides, and phospholipids (phosphorus) in each milliliter of eluate aliquot was determined and the concentration of each lipid in the eluted fraction is plotted in Fig. 3. For all lipoprotein classes, the peak positions of protein coincide with those of the total cholesterol, triglycerides, and phospholipids. Recovery of the total lipoprotein fraction and individual lipoproteins from the TSK GEL columns was examined on the basis of total cholesterol determination, and was found satisfactory as follows: about 90% for total lipopro-

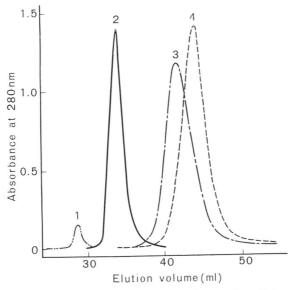

FIG. 2. Elution patterns of individual serum lipoproteins. Column: G5000PW + G3000SW + G3000SW; eluent: 0.15 M NaCl; flow rate: 1.0 ml/min, detector: A_{280}. Sample: 1, chylomicrons + VLDL ($d < 1.006$); 2, LDL (d 1.006–1.063); 3, HDL_2 (d 1.063–1.125); 4, HDL_3 (d 1.125–1.210).

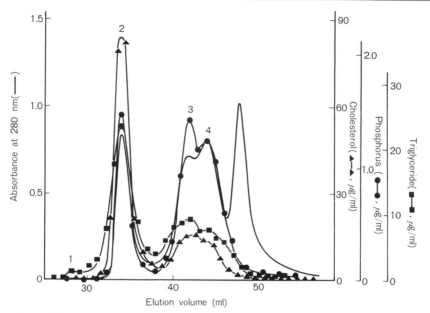

FIG. 3. Elution pattern of total lipoprotein fraction ($d < 1.210$) from human serum. Peaks: 1, chylomicrons + VLDL; 2, LDL; 3, HDL$_2$; 4, HDL$_3$. Other HPLC conditions as in Fig. 2.

tein fraction, 85.3 ± 6.0% for chylomicrons + VLDL, 94 ± 5% for LDL, 91 ± 5% for HDL$_2$, and 90 ± 7% for HDL$_3$. The chemical composition of protein, total cholesterol, triglycerides, and phospholipids recovered from the TSK GEL columns for individual lipoproteins was consistent with that of the original lipoproteins prepared by ultracentrifugation.[3] These results indicate that no degradation of lipoproteins occurs during the HPLC procedure and that the quantitation of lipid component for each lipoprotein class can be performed through determination of lipid content in the postcolumn effluent.

Elution Pattern of Serum Cholesterol

The quantitation of lipoproteins in intact serum by monitoring A_{280} is not practical, since the amounts of lipoproteins are small compared to those of albumin or immunoglobulins, which elute overlapping with the LDL and HDL fractions.

We have developed an on-line continuous lipid-monitoring system using an enzyme reaction. Elution patterns monitored by total cholesterol and proteins, using the G5000PW + G3000SW + G3000SW column system, are shown in Fig. 4. In our HPLC flow diagram equipped with two

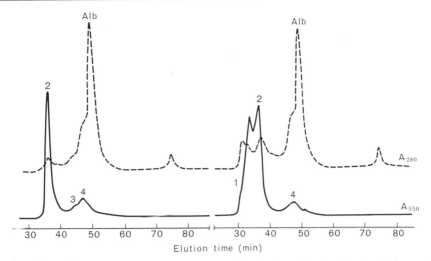

FIG. 4. Elution patterns of human serum monitored by proteins (A_{280}) and cholesterol (A_{550}) using flow diagram of Fig. 1. Sample: serum from normal female (left) and hyperlipidemia (right). Loaded volume: 10 μl. Peaks and other HPLC conditions as in Fig. 3.

detectors, as shown in Fig. 1, the elution patterns of the protein and one of the lipid components can be obtained by one analysis. The recovery of lipids from the TSK GEL columns is very high (above 90%) and is the case for all lipoprotein classes. Each peak position monitored by the total cholesterol corresponds to that of individual lipoproteins in Fig. 2. Thus, the elution pattern monitored by total cholesterol for intact serum gives precise quantitative as well as qualitative information about lipoprotein distributions.

Various Column Systems and Elution Patterns of Serum Lipids

The combination of TSK GEL columns consisting of G5000PW + G3000SW + G3000SW was best for the separation of all lipoprotein classes in our early studies.[1-3,7-9] Subsequently, G4000SW with high resolution is now commercially available. Evaluation of TSK GEL columns in lipoprotein analysis was carried out using standard proteins and lipoproteins of known particle diameter,[18] and the applicable range of lipoproteins separated by each column is presented in Table I. G4000SW is suitable for the separation of VLDL, LDL, and HDL, G3000SW for that

[18] M. Okazaki, and I. Hara, *in* "Handbook of HPLC for the Separation of Amino Acids, Peptides and Proteins" (W. S. Hancock, ed.), Vol. 11, p. 393. CRC Press, Boca Raton, Florida, 1984.

of HDL subfractions, G5000PW or G6000PW for that of chylomicrons, VLDL, and LDL. At the present, the best column system for the separation of all lipoprotein classes may be the combined system of G4000SW + G3000SW,[11,12] by which the separation of serum lipoproteins into the following classes is possible: chylomicrons + VLDL, LDL, HDL$_2$, and HDL$_3$. Carrol and Rudel[6] reported the best resolution of individual lipoproteins to be possible with the G5000PW + G4000SW + G3000SW system.

Serum lipoproteins contain cholesterol, cholesterol ester, triglycerides, and phospholipids as the major lipid components. Nearly 95% of the total phospholipids consists of choline-containing phospholipids: phosphatidylcholine, lysophosphatidylcholine, and sphingomyelin. Thus, elution patterns of serum lipoproteins using intact serum can be obtained by the selective detection of triglycerides or choline-containing phospholipids other than total cholesterol. According to differences in chemical composition in each lipoprotein class, the selective detection of triglycerides is suitable for analysis of chylomicrons and VLDL, total cholesterol for LDL and HDL, and choline-containing phospholipids for all lipoprotein classes, respectively. Therefore, we can select an appropriate analytical system with two factors: column system and lipid components for selective detection, according to the objective of the analysis.

The elution patterns of human serum (hyperlipidemia of type IV) obtained for various analytical systems are shown in Fig. 5. The elution positions of the individual lipoproteins are presented for various column systems using numbered arrows: 1, chylomicrons; 2, VLDL; 3, LDL; 4, HDL$_2$; 5, HDL$_3$; 6, VHDL. Free choline or free glycerol in the serum is detected at the total permeation of the column (number 7 for G5000PW) using the reagent kits described in Methodology.

The elution patterns monitored by triglycerides for serum of hyperlipidemia were compared for the single G5000PW and G5000PW + G5000PW columns.[10] As shown in Fig. 6, elongation of the column improves the resolution of all lipoprotein classes. A detailed examination of subclasses of VLDL can be carried out by using the G5000PW + G5000PW system.

The elution patterns of serum cholesterol and choline-containing phospholipids obtained for various liver diseases using the G4000SW + G3000SW column system are shown in Fig. 7.[12] The elution patterns of liver diseases differ greatly from those of normal subjects: disappearance of major lipoprotein peaks and/or appearance of peaks other than major classes (see upward arrows in Fig. 7). Since abnormal lipoproteins in certain liver diseases contain large amounts of phospholipids, their detection can be performed by monitoring choline-containing phospholipids.

The elution pattern of serum cholesterol or choline-containing phos-

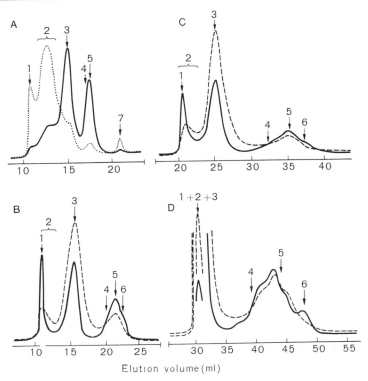

Elution volume (ml)

FIG. 5. Elution patterns of cholesterol (------), triglycerides (······), and choline-containing phospholipids (——) for human serum in various column systems. Column: (A) G5000PW, (B) G4000SW, (C) G4000SW + G3000SW, and (D) G3000SW + G3000SW. Sample: serum from a hyperlipidemia of type IV. Loaded volume: 5 or 10 μl for TC monitor, 15 μl for TG monitor, 10 or 20 μl for PL monitor. Flow rate of eluent (0.15 M NaCl): 0.50 ml/ min for (A) and (B), 0.60 ml/min for (C) and (D). Detector: A_{550} for TC and TG monitor, A_{500} for PL monitor. Peak positions: 1, chylomicrons; 2, VLDL; 3, LDL; 4, HDL_2; 5, HDL_3; 6, VHDL; 7, total permeation. Other HPLC conditions as described in the text.

pholipids using the G3000SW + G3000SW + G3000SW system gives many peaks in HDL fraction, as presented in Fig. 5. The frequency distribution of peaks in the elution patterns of serum choline-containing phospholipids for a group ($n = 71$) of normal males and females showed five maxima, as indicated in Fig. 8.[19] Their particle diameters determined from the calibration curve of the column are also shown in Fig. 8. Moreover, these five HDL subfractions were confirmed by rechromatography using this HPLC technique. HDL subfractions have been examined by various

[19] M. Okazaki, N. Hagiwara, and I. Hara, *J. Biochem.* **92,** 517 (1982).

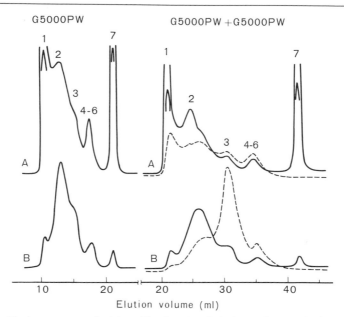

FIG. 6. Elution patterns of triglycerides (——) and cholesterol (------) for human serum. Column: G5000PW, G5000PW + G5000PW. Sample: serum from hyperlipidemia of type V (A) and type IV (B). Loaded volume: 20 μl. Elution positions as in Fig. 5. Other HPLC conditions as described in text.

analytical methods,[19-21] as presented in Table II. Experimental procedures for examining HDL subfractions other than our HPLC technique involves ultracentrifugation. Since the physical and chemical properties of lipoproteins may change during ultracentrifugation, our method is very useful for the examination of HDL subfractions.

Lipoproteins Measured by HPLC and Sequential Ultracentrifugal
    Flotation Compared[22]

A correlation of serum lipoprotein analysis by HPLC to that by the sequential ultracentrifugal flotation method is of fundamental importance.

The sera of 15 normolipidemia and 12 hyperlipidemia were examined by the HPLC method and the results were compared with those by the

[20] D. W. Anderson, A. V. Nichols, T. M. Forte, and F. T. Lindgren, *Biochim. Biophys. Acta* **493,** 55 (1977).

[21] P. J. Blanche, E. L. Gong, T. M. Forte, and A. V. Nichols, *Biochim. Biophys. Acta* **665,** 408 (1981).

[22] M. Okazaki, H. Itakura, K. Shiraishi, and I. Hara, *Clin. Chem.* **29,** 768 (1983).

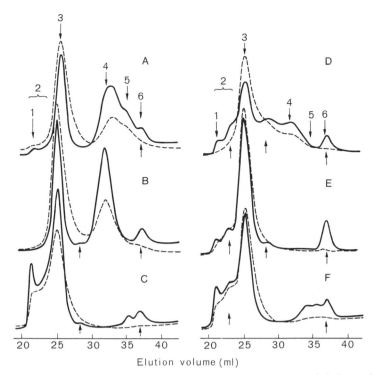

FIG. 7. Elution patterns of choline-containing phospholipids (——) and cholesterol (-----) for serum from patients with various liver diseases. Column: G4000SW + G3000SW. Sample: serum of normal female (A), patients with liver cirrhosis (B), acute hepatitis (C), primary biliary cirrhosis (D), intrahepatic cholestasis (E), and drug-induced liver injury (F). Loaded volume: 20 μl for PL monitor, 10 μl for TC monitor. Elution positions and other HPLC conditions as in Fig. 5.

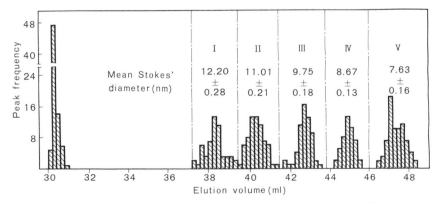

FIG. 8. Frequency distribution of peaks in the elution patterns of choline-containing phospholipids for a group of normal human subjects ($n = 71$). A definition of HDL subfractions of groups I–V is described in text and Ref. 19.

TABLE II

PARTICLE DIAMETERS (nm) OF HDL SUBFRACTIONS BY
VARIOUS METHODS

Okazaki et al.[a] (1982)	Anderson et al.[b] (1977)	Blanche et al.[c]
Group I: 12.20 ± 0.28	HDL$_{2b}$: 10.8–12.0	(HDL$_{2b}$)$_{gge}$: 10.57
Group II: 11.01 ± 0.21	HDL$_{2a}$: 9.7–10.7	(HDL$_{2a}$)$_{gge}$: 9.16
Group III: 9.75 ± 0.18	HDL$_3$: 8.5– 9.6	(HDL$_{3a}$)$_{gge}$: 8.44
Group IV: 8.67 ± 0.13		(HDL$_{3b}$)$_{gge}$: 7.97
Group V: 7.63 ± 0.16		(HDL$_{3c}$)$_{gge}$: 7.62

[a] From Okazaki et al.[19] Subfractions were obtained by the peak
frequency analysis of HPLC patterns of human serum monitored by choline-containing phospholipids (see Fig. 8 in text).
Particle diameter of each fraction was determined from the
elution volume by using the calibration curve of the column
obtained for standard samples of known particle diameter.

[b] From Anderson et al.[20] Subfractions were obtained by the gradient gel electrophoresis of the total HDL fraction isolated
from human serum by ultracentrifugation. Particle diameter
was determined by electrophoresis on pore-gradient polyacrylamide gels calibrated for Stokes' diameter as a function of
migration distance.

[c] From Blanche et al.[21] Subfractions were obtained by the frequency distribution of subpopulation peaks obtained by gradient gel electrophoresis for the ultracentrifugal fraction ($d \leq$
1.200) from human serum. Particle diameter was derived from a
calibration of peak relative migration distance versus Stokes'
diameter for standard proteins.

sequential ultracentrifugal flotation method of Havel et al.[13] Serum (10 or
20 μl) was applied onto the G4000SW + G3000SW column system, and
the elution pattern was obtained by monitoring total cholesterol and choline-containing phospholipids. The concentration of total cholesterol and
choline-containing phospholipids in each lipoprotein class was calculated
from the peak area of their elution patterns. Each lipoprotein fraction was
defined on the basis of the following particle diameter range: chylomicrons + VLDL, larger than 30.0 nm; LDL, 30.0–16.0 nm; HDL$_2$, 16.0–
10.0 nm; HDL$_3$, 10.0–8.0 nm; VHDL, less than 8.0 nm. Peak division of
each fraction was made drawing a line perpendicular to the base line at the
elution volume corresponding to the particle diameter of each fraction, as
shown in Fig. 9. The concentration of total cholesterol and choline-containing phospholipids was calculated from the percentage of the relative
peak area and their concentration in the serum. The lipid concentration of
each fraction isolated by the ultracentrifugation of Havel et al.[13] was

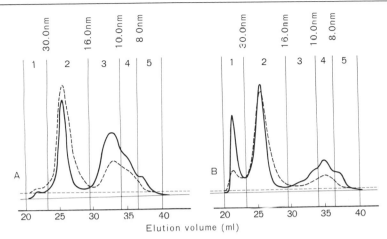

FIG. 9. Quantitation of each lipoprotein fraction from elution patterns of cholesterol (------) and choline-containing phospholipids (———) for individual human serum. Column: G4000SW + G3000SW. Sample: serum from normal female (A) and hyperlipidemia of type IV (B). Loaded volume: 10 μl for TC monitor, 20 μl for PL monitor. Fraction: 1, chylomicrons + VLDL; 2, LDL; 3, HDL$_2$; 4, HDL$_3$; 5, VHDL. Other HPLC conditions as in Fig. 5.

measured with the same enzymatic reagent kits for the HPLC analysis. After volume correction, the concentration of total cholesterol and choline-containing phospholipids was calculated from the percentage distribution of lipid among these fractions and their concentration in the serum.

The results from the two methods were compared and a correlation for the corresponding fractions in these methods was very good for all lipoprotein fractions in both groups except for the VHDL fraction. The correlation coefficients of least-squares analysis are as follows: 0.91–0.94 for chylomicrons + VLDL, 0.97–0.98 for LDL, 0.96 for HDL$_2$, and 0.83–0.87 for HDL$_3$ in the normolipidemic group (n = 15); 0.98 for chylomicrons + VLDL, 0.99 for LDL, 0.98 for HDL$_2$, and 0.88–0.97 for HDL$_3$ in the hyperlipidemic group (n = 12). The individual data for the concentration of choline-containing phospholipids for each lipoprotein fraction as measured by the two methods are shown in Fig. 10.

However, this good correlation between the two methods ceases to exist when the size of the lipoproteins no longer corresponds to the density, as was evidenced by a comparison of the results for a liver disease group by Okazaki et al.[22] and by a study of lecithin : cholesterol acyltransferase deficiency by Kodama et al.[23]

[23] T. Kodama, Y. Akanuma, M. Okazaki, H. Aburatani, H. Itakura, K. Takahashi, M. Sakuma, F. Takaku, and I. Hara, *Biochim. Biophys. Acta* **752**, 407 (1983).

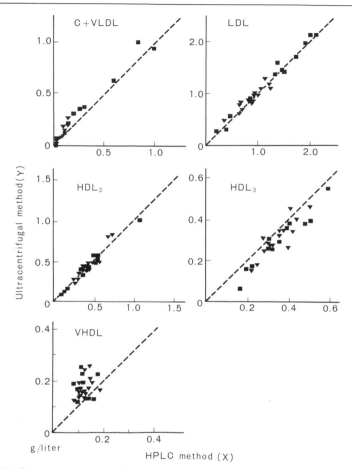

Fig. 10. Concentration of choline-containing phospholipids in each lipoprotein fraction as measured by the HPLC method (X) and sequential ultracentrifugal flotation method (Y). ■, A group of 15 normolipidemia; ▼, a group of 12 hyperlipidemia. (------), Ideal correlation ($Y = X$).

Detection Limits of Serum Lipids and Reproducibility for HPLC Method

The detection limit for each lipid by the enzymatic reaction under our experimental conditions is very high: 0.5 μg for total cholesterol,[9] 1.0 μg for choline-containing phospholipids[12] and triglycerides.[10] But that for protein by A_{280} is about 3 μg.

The relationship of peak area of the elution pattern to lipid concentration in applied sample was examined using the LDL fraction prepared by ultracentrifugation for these three lipids, and the linearity was obtained in the following concentration range: 5–54 μg for total cholesterol, 17–51 μg for triglycerides, and 3–85 μg for choline-containing phospholipids.

Reproducibility of the HPLC method was found to be very good. The elution volume was reproducible within ±0.01 ml for each lipoprotein peak using the individual lipoproteins or serum lipid monitoring system. Reproducibility of the concentration of each lipoprotein as shown in Fig. 9 was very good. The standard deviation from the mean value for each fraction was 0.1 to 0.4 g/liter, and the coefficient of variation (CV) was less than 2% for the concentration of choline-containing phospholipids.

Comments on the HPLC Method

1. *Samples:* Serum is usually used for lipoprotein analysis by this method. Plasma can be also used in this method. But it should be noted that plasma might form fibrin during the HPLC procedure, so that serum is preferable to plasma.

2. *Sample storing:* Serum stored at $-20°$ can be used for analysis. But in the case of hyperlipidemic serum containing a large amount of triglycerides, the elution pattern of freezing–thawing serum might give a larger peak of chylomicrons and VLDL than that of fresh sample.[10] Hyperlipidemic subjects should be analyzed as soon as possible after obtaining the blood samples.

3. *Eluent:* A 0.15 M NaCl, pH 7.0, solution is usually used as the eluent for lipoprotein separation in this method. However, various aqueous solutions containing appropriate amounts of inorganic salts can be used as eluents, provided the pH is less than 7.5. Buffering of the eluent, salt concentration, and pH may alter the separation extent of the lipoproteins, and in some cases may improve the resolution of lipoproteins. Effects of the eluent properties (buffering agents, salt concentration, and pH) on lipoprotein resolution have been examined[2,6] and found negligibly small on the separation of lipoproteins larger than LDL. An increase in the buffering agent or salt concentration results in peak broadening, and in the case of HDL possibly causes resolution of HDL into multiple subfractions.[6] The optimum concentration of the Tris-PO$_4$ buffer was reported to be 0.25 M by Carrol and Rudel.[6] Elevation of pH lessens the elution volume of HDL.[2,6]

4. *Flow rate of eluent:* A slower flow rate of the eluent gives better lipoprotein resolution.[9] But this is a matter of balance between the increase in resolution and time required for analysis.

5. *Flow rate of enzyme solution:* An increase in the flow rate of the enzyme solution reduces the peak intensity and lessens the reaction time,

since the serum lipid in a postcolumn effluent is diluted with the enzyme solution, and the reaction time is inversely proportional to the sum of the flow rate of the eluent and that of the enzyme solution. Therefore a slower flow rate of the enzyme solution is recommended so long as the concentration of enzyme reagent is sufficient for the reaction to proceed to completion in the flow diagram.

6. *Reactor for enzymatic reaction:* A Teflon or stainless tube in a thermostated water bath is used for this reaction. Peak broadening in the reaction tube was found negligible as long as the tube had an appropriate diameter, i.e., less than 0.5 mm i.d.[12] Thus the reaction time can be made sufficient by elongating the reaction tube without a decrease in peak resolution.

7. *Wave length for lipid detection:* By this method, serum lipid in a postcolumn effluent is detected by A_{550} or A_{500} using an enzyme reagent kit. In the elution pattern monitored by A_{500}, about 20–30% of the descending part of the HDL peak is due to the absorption of pigments in human serum, such as bilirubin. In the case of the elution pattern monitored by A_{550}, this may be negligibly small. Therefore, a wavelength longer than 500 nm is preferable for detection of serum lipid. Enzymatic reagent other than those used in this method may be used for lipid detection. But a proper adjustment must be made of the optimum reaction condition in the flow diagram.

8. *Quality of TSK GEL columns for lipoprotein analysis:* Recovery of lipoproteins from the column is satisfactory, and the lipid concentration in each fraction can be directly obtained from the peak area of the elution pattern of the serum lipid. However, the elution volume of the lipoproteins may slightly change according to the particular column used. Thus, a precise estimation of the molecular weight or particle diameter from the elution volume should be made using a calibrated column with standard samples. At the present time, resolution in the low particle size range of the SW type column is better than that of the PW type column. Therefore we use a column system containing G4000SW and/or G3000SW to study HDL distribution. Some of the SW type column may be unsuitable for lipoprotein analysis due to low recovery of lipids from the column, and it is better to use G4000SW or G3000SW selected for lipoprotein analysis by the supplier.

9. *Maintenance of column:* Denatured samples of serum or lipoproteins cause deterioration of the column. Since entrance of the column deteriorates first, use of a guard column may increase the lifespan of the column: TSK guard column SW (7.5 mm i.d. × 75 mm) for the SW type column system and TSK guard column PWH (7.5 mm i.d. × 75 mm) for the PW type column system.

10. *Quantitation of lipid in each fraction:* With this method, the end point of the enzyme reaction can be obtained in the case of total cholesterol and choline-containing phospholipids. Therefore the peak area of an elution pattern monitored by these lipids directly represents the concentration of each lipid. Lipid concentration can be calculated from the peak area using the apparatus constant obtained with a standard sample. However, the end point of the enzyme reaction cannot be obtained when analyzing for triglycerides. The concentration of triglycerides in each fraction must be calculated from the percentage peak area and concentration of triglycerides in the serum.

Conclusion

The HPLC technique of monitoring lipids in a postcolumn effluent holds promise as a means of serum lipoprotein analysis from the following standpoints: simplicity of procedure, an analysis time of less than 1 hr, no need for sample pretreatment, small sample amount required (less than 20 μl of intact serum), high resolution, and good reproducibility. Moreover,

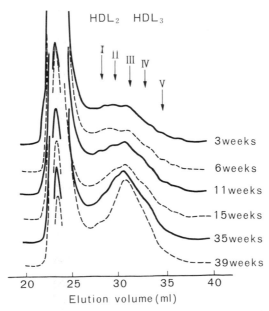

FIG. 11. Elution patterns of cholesterol for human serum: change of HDL subfractions by drug therapy. Column: G3000SW + G3000SW. Sample: serum from a patient with coronary heart disease. Loaded volume: 10 μl. Flow rate of eluent (0.15 M NaCl): 0.40 ml/min. Other HPLC conditions as described in the text.

the elution pattern by this method not only resolves the lipoprotein peaks of the major classes quantitatively but indicates whether lipoprotein variants are present or not; that is not possible by the ultracentrifugal method. Although our HPLC method cannot be used for the preparation of serum lipoproteins, this can be made by applying the total lipoprotein fraction, isolated by ultracentrifugation, onto the column. Our HPLC method has many clinical applications, some of which have been reported.[23-25] One example of the clinical application is shown in Fig. 11. Changes in the HDL subfractions can be observed by drug therapy. Correlation between our HPLC and sequential ultracentrifugal flotation methods for lipoprotein analysis is very good for all lipoprotein classes except for the VHDL fraction. Therefore, the HPLC method may come to be recognized as the best means for lipoprotein analysis.

Acknowledgments

The authors are grateful to Dr. Shinji Yokoyama, National Cardiovascular Center Research Institute, for valuable comments.

[24] M. Okazaki, I. Hara, A. Tanaka, T. Kodama, and S. Yokoyama, *New Engl. J. Med.* **304,** 1608 (1981).
[25] Y. Takashima, T. Kodama, H. Iida *et al., Pediatrics* **75,** 541 (1985).

[5] Precipitation Methods for Quantification of Lipoproteins

By PAUL S. BACHORIK and JOHN J. ALBERS

Introduction

The purpose of this chapter is to discuss the quantitation of plasma lipoproteins by precipitation methods. The procedures generally involve the preliminary separation of lipoproteins from each other with various precipitating agents, followed by the measurement of a lipoprotein component, generally cholesterol, as an index of lipoprotein concentration.

The plasma lipoproteins are composed primarily of cholesterol, cholesteryl esters, triglycerides, phospholipids (principally phosphatidylcholine and sphingomyelin), and several specific proteins that in the lipid-free state are called apolipoproteins. Structurally, the plasma lipoproteins are normally spherical macromolecular complexes in which the core consists primarily of cholesteryl esters and triglycerides. Unesterified cholesterol and phospholipids are concentrated near the surface with the polar groups oriented toward the aqueous medium. The apolipoproteins are at the surface and are associated with the lipids through noncovalent interactions.

Four major classes of lipoproteins are found in plasma: chylomicrons, very-low-density lipoproteins (VLDL), low-density lipoproteins (LDL), and high-density lipoproteins (HDL). They differ in their metabolic origins and functions as well as in their chemical composition, apolipoprotein content, hydrated density, and other physicochemical properties (see Chapter [1], Volume 128). They are readily separated from other serum proteins and from each other by ultracentrifugation at appropriate densities. Chylomicrons are large, triglyceride-rich particles that are synthesized in the intestine in response to fat-containing meals and serve to transport dietary fat through the lymphatics and into the circulation. There they are converted to smaller, cholesterol-enriched remnant particles that are rapidly taken up by the liver. The overall process of chylomicron synthesis, secretion, and clearance is essentially complete within 5 or 6 hr after a meal, and chylomicrons are not usually detected in fasting plasma.

VLDL, LDL, and HDL are present in plasma at all times. VLDL are triglyceride-rich lipoproteins that are synthesized in the liver, undergo a series of intravascular changes, and are ultimately converted to LDL. LDL transports most of the plasma cholesterol in man. Small amounts of intermediate-density lipoproteins (IDL, d 1.006–1.019 g/ml) can generally be found in plasma. They are formed during conversion of VLDL to LDL. HDL participates in the catabolism of triglyceride-rich lipoproteins, and may also facilitate the removal of cholesterol from peripheral tissues and delivery to the liver for disposal or reuse.

The precipitation of plasma lipoproteins by sulfated polysaccharides was first reported in the mid-1950s[1–3] and precipitation methods were developed initially for the preparative isolation of serum lipoproteins.[4,5] During the past 30 years these methods have been developed further and have been adapted for lipoprotein quantitation. They are rapid and relatively inexpensive to perform. Lipoproteins can generally be separated from each other within 10 min to 2 hr, depending on the precipitation conditions and lipoprotein class of interest, and these methods can there-

[1] M. Burstein and J. Samaille, C. R. Hebd. Seances Acad. Sci. 241, 664 (1955).
[2] P. Bernfeld, Fed. Proc., Fed. Am. Soc. Exp. Biol. 14, 182 (1955).
[3] J. L. Oncley, K. W. Waltan, and D. G. Cornwell, Am. Chem. Soc. 128th Meeting Abstr. 41c (1955).
[4] M. Burstein and J. Samaille, J. Physiol. (Paris) 49, 83 (1957).
[5] M. Burstein, H. R. Scholnick, and R. Morfin, J. Lipid Res. 11, 583 (1970).

fore be readily applied for screening, clinical analysis, and other purposes for which analyses are required rapidly or in a large number of samples. It should be noted that certain non-lipid-containing plasma proteins can also be precipitated under conditions used for lipoprotein separations.[6] This is not a problem when lipoprotein–cholesterol or other lipid component is to be measured, but must be considered if lipoprotein–protein is to be analyzed.

Despite the speed and simplicity of the precipitation methods, they have not been widely used except for the measurement of HDL. This apparently has resulted from a combination of factors that are not particularly related to the adequacy of the precipitation methods for the analysis of other lipoprotein classes. First, the lipoproteins have been classically defined in terms of their behavior in the ultracentrifuge, and for the most part have been measured in research laboratories that have access to this equipment. One widely used procedure for the measurement of the major lipoprotein classes is that described by Fredrickson et al.,[7] which uses a combination of ultracentrifugation and polyanion precipitation to prepare lipoprotein-containing fractions. The cholesterol content of the fractions is measured and used to calculate lipoprotein–cholesterol concentrations. This method was used extensively in several large-scale studies, including those of the Lipid Research Clinics Program.[8] In this method polyanion precipitation was used only to remove apoB-containing lipoproteins from plasma. HDL remains soluble and the cholesterol content of the supernatant was used as a measure of total HDL concentration. A separate aliquot of plasma was ultracentrifuged without density adjustment, and the floating layer of VLDL (and chylomicrons, if present) was removed. The infranatant contained IDL, LDL, HDL, and small but measurable amounts of Lp(a) lipoprotein, a minor class of lipoprotein that contains apoB and whose density overlaps LDL and HDL.[9] The VLDL–cholesterol concentration was measured as the difference between total plasma cholesterol and that in the ultracentrifugal infranatant. Similarly, LDL–cholesterol was measured as the difference between cholesterol in the ultracentrifugal infranatant and HDL–cholesterol. Since IDL and Lp(a) are precipitated with apoB-containing lipoproteins, IDL- and Lp(a)-cholesterol were included in the measurement of LDL–cholesterol.

[6] M. Burstein and P. Legmann, in "Monographs on Atherosclerosis" (T. B. Clarkson, D. Kritchevsky, O. J. Pollak, eds.). Karger, New York, 1982.
[7] D. S. Fredrickson, R. I. Levy, and F. P. Lindgren, J. Clin. Invest. 47, 2446 (1968).
[8] Lipid Research Clinics Program, "Manual of Laboratory Operations." DHEW Publ. No. (NIH) 75-628, May 1974, revised 1982.
[9] J. J. Albers, J. L. Adolphson, and W. R. Hazzard, J. Lipid Res. 18, 331 (1977).

Second, for many purposes adequate estimates of VLDL- and LDL–cholesterol concentrations can be obtained from measurements of total cholesterol, triglycerides, and HDL–cholesterol using the empirical relationship of Friedewald et al.[10]:

$$[\text{LDL–chol}] = [\text{total chol}] - [\text{HDL–chol}] - \frac{[\text{TG}]}{5} \qquad (1)$$

where [TG]/5 corresponds to an estimation of [VLDL–chol] and all measurements are expressed in milligrams/deciliter.

Third, it has long been appreciated that an increased plasma LDL level is an important risk factor for the development of coronary atherosclerosis, and in the 1960s and early 1970s HDL–cholesterol was assayed primarily because the measurement was needed to calculate the LDL–cholesterol concentration. Beginning in the mid-1970s, however, it became widely appreciated that elevated HDL levels were independently associated with reduced cardiovascular risk.[11] This created the need for the measurement of HDL–cholesterol for its own sake and a demand for more accurate and reproducible HDL methods. As a result, considerable effort has been spent during the past decade developing and evaluating precipitation methods for HDL–cholesterol, and more recently for HDL_2 and HDL_3, the major subfractions of HDL, and the application of these methods to the analysis of other lipoprotein classes has been largely neglected.

Sample Collection

A number of factors related to blood drawing and storage of the samples before analysis can influence lipoprotein analyses.[12-14] Such factors include changes in the subject's posture prior to venipuncture, prolonged venous occlusion during blood drawing, large osmotic shifts of water between red cells and plasma produced by certain anticoagulants, and various enzymatic and structural and compositional changes that occur during storage. Lipoprotein levels and composition can also be influenced by medications such as sex steroids; recent ingestion of a fat-containing

[10] W. T. Friedewald, R. I. Levy, and D. S. Fredrickson, Clin. Chem. 18, 499 (1972).
[11] G. Heiss, N. J. Johnson, S. Reiland, C. E. Davis, and H. A. Tyroler, Circulation 62 (Suppl. IV), 116 (1980).
[12] C. Alper, in "Clinical Chemistry, Principals and Techniques" (R. J. Henry, D. C. Cannon, and J. Winkelman, eds.), 2nd Ed., p. 373. Harper, New York, 1974.
[13] P. S. Bachorik, in "Report of the High Density Lipoprotein Methodology Workshop" (K. Lippel, ed.), p. 164. NIH Publ. No. 79-1661, Bethesda, Maryland, 1979.
[14] P. S. Bachorik, Clin. Chem. 28, 1375 (1982).

meal, which transiently raises triglyceride levels due to the presence of chylomicrons; recent vigorous exercise, which can cause a rapid transient decrease in triglycerides; and anticoagulant therapy using heparin, which promotes the release of several lipases into the circulation, produces dramatic decreases in triglycerides and alterations in the composition of the lipoproteins. For these reasons, the condition of the patient should be noted, posture should be standardized as much as possible for blood drawing, and lipoprotein fractions should be prepared as soon after venipuncture as possible. Cholesterol itself is stable during long periods of storage, and in general, if the lipoprotein–cholesterol analyses must be delayed, it is preferable to store the separated fractions rather than unfractionated plasma or serum.

Patient Preparation

Instruct the subject to fast for 12 hr prior to venipuncture. Water can be taken and prescribed medications should not be withheld during this period. At the time of sampling, make a note of whether the subject is pregnant or being treated with sex steroids or heparin, and the actual period of fasting.

The sitting position is most commonly used for drawing blood and should be used if possible. Let the subject sit quietly for 10 min before drawing blood. Blood can be obtained from an antecubital vein or some other convenient arm vein. Use a tourniquet, but release it prior to drawing the blood. Blood can be drawn with a syringe and then transferred to appropriate tubes, or can be collected directly into evacuated blood collection tubes, which then also serve as centrifuge tubes when the cells are removed. If a syringe is used, remove the needle before transferring the blood to prevent hemolysis.

Sample Preparation

Lipoproteins can be analyzed either in plasma or serum, but most workers prefer to use EDTA plasma (see below).

Plasma. Use disodium EDTA to give a final concentration of 1–1.5 mg/ml. Fill the collection tube completely. Mix the blood gently by inversion five or six times and immediately cool to 2–4° in an ice bath. As soon as possible, preferably within 3 hr, centrifuge the sample at 4° for 30 min at 1500 g. Transfer the plasma to a clean storage vial that can be closed to prevent evaporation, and store at 4°. If several tubes of blood are drawn, the storage vial should be large enough to permit the plasma to be pooled and mixed.

Serum. Collect blood as described above but without the anticoagulant. Allow the sample to stand at room temperature for 45 min in a glass tube. Transfer the serum to a storage vial as above, cool to 2–4°. If the analyses must be delayed, add EDTA (final concentration 1×10^{-3} M) to the serum to inhibit oxidative changes in the lipoproteins. This is not necessary if the analyses are to be performed promptly.

Notes

Anticoagulants such as oxalate or citrate should not be used because they produce large osmotic shifts of water from cells to plasma. Osmotic effects with EDTA are observable but small, and produce apparent decreases of about 3% in the total cholesterol and triglyceride concentrations of EDTA plasma compared to serum,[15] although little if any difference was observed for HDL–cholesterol.[16] Nonetheless, EDTA plasma is generally preferred, first because the samples can be cooled immediately to slow storage-related changes, and second because EDTA complexes metals such as Cu^{2+} that promote autooxidation of lipids and consequent degradation of the lipoproteins.

If EDTA-containing evacuated blood drawing tubes are used, they should be filled completely to ensure the proper final concentration of EDTA. Undue concentration of EDTA can promote hemolysis, increase osmotic changes in lipoprotein concentration, and interfere with lipoprotein precipitation methods by chelating the divalent cation reagents.

In general, the samples should be stored at 4° and the analyses should be performed as soon as possible. If necessary, HDL–cholesterol analyses can be performed in samples that have been frozen for a few days. For longer periods of frozen storage it is preferable to separate the lipoprotein fractions before freezing. In addition, samples are more stable at −70° than at −20°.

Lipoprotein Precipitation

Lipoproteins can be selectively precipitated with a variety of agents, including combinations of particular polyanions and divalent cations (e.g., heparin sulfate, dextran sulfate, or sodium phosphotungstate in combination with Mn^{2+}, Mg^{2+}, Ca^{2+}), anionic detergents such as sodium

[15] Laboratory Methods Committee, Lipid Research Clinics Program, *Clin. Chem.* **23,** 60 (1977).
[16] A. R. Folsom, K. Kuba, R. V. Leupker, D. R. Jacobs, and I. D. Frantz, *Clin. Chem.* **29,** 505 (1983).

decyl sulfate, polyethylene glycol, protamine, tetracyclines, and others. Of these, perhaps the most widely used to precipitate apoB-containing lipoproteins have been sulfated polysaccharide-divalent cation combinations. The specific precipitants used, conditions of reagent concentration, temperature, and time vary depending on the lipoprotein class of interest. The lipoprotein specificity and completeness of precipitation are affected by a number of factors including pH, ionic strength, the presence of other serum proteins, the presence of polyols such as sucrose, and the lipid-to-protein ratio of the lipoprotein. In general, precipitation with sulfated polysaccharides occurs more readily (i.e., at lower precipitant concentrations) at low ionic strength and in the absence of other serum proteins. The specificity of precipitation with polyanions–divalent cations is apparently conferred by the lipid-to-protein ratio of the lipoprotein rather than by the particular lipid or apoprotein present. For example, chylomicrons are more readily precipitated than VLDL, which in turn precipitate more easily than LDL. HDL generally remain soluble under conditions in which the lower desity lipoproteins precipitate. The reader is referred to an excellent recent review ([6]) for a comprehensive discussion of factors that govern the precipitation of major lipoprotein classes.

Analysis of HDL

A wide variety of precipitating agents have been used to remove apoB-containing lipoproteins from serum or plasma. We make no attempt to discuss them all here. Rather, we have described several methods that have been studied in some detail, used frequently, and which we consider to be among the most accurate currently available.

Total HDL

Heparin–MnCl₂. The heparin–MnCl$_2$ method of Burstein and Samaille[17] has been perhaps the widest used and most intensively evaluated of the precipitation methods for HDL, and much of the epidemiologic data that relates HDL–cholesterol concentration with cardiovascular risk has been accumulated using this method.[11]

Samples. The conditions for this method were originally developed for serum, but the method has been widely used in EDTA plasma also.

Reagents. MnCl$_2 \cdot 4H_2O$, 1.0 *M*: The crystalline reagent is slightly deliquescent and should therefore be stored in a tightly closed container. Dissolve 19.791 g MnCl$_2 \cdot 4H_2O$ in distilled H$_2$O, transfer quantitatively to a 100-ml volumetric flask, and dilute to volume with distilled H$_2$O. The

[17] M. Burstein and J. Samaille, *Clin. Chim. Acta* **5**, 609 (1960).

solution should be stored in the refrigerator and should be prepared fresh monthly.

Heparin (Porcine Intestinal Mucosa). Heparin is a polydisperse sulfated polysaccharide that is obtained from several sources, generally bovine lung or porcine intestinal mucosa. The preparations from lung and intestine differ somewhat in size, chemical structure,[18-20] and lipoprotein-precipitating properties,[21] although porcine intestinal heparin from various manufacturers has been reported to give equivalent HDL–cholesterol measurements.[21] Heparin sodium can be obtained either as a powder or as an injectible solution in 0.15 M NaCl. The dry powder is dissolved in 0.15 M NaCl to a concentration of 35 mg/ml. Injectable heparin sodium is generally provided with concentrations expressed in USP units of anticoagulability/milliliter. Dilute the solution to a final concentration of 5000 USP units/ml with 0.15 M NaCl. This provides a concentration of about 35 mg/ml. The solution is stored in the refrigerator, and in the author's experience, can be used for at least a month.

Procedure. Allow the samples to reach room temperature, then measure 2.0 ml of sample into a transparent 15-ml conical centrifuge tube. Use a class A volumetric pipet, or other device of equal or greater accuracy, for the volume measurements. Add 80 μl heparin (35 mg/ml) to each sample and mix thoroughly but smoothly to avoid foaming. If a vortex mixer is used, mix the sample intermittantly several times to avoid layering of the reagent. Add 100 μl 1.0 M MnCl$_2$ to each sample and mix as above. A heavy white precipitate will form immediately. Place the samples in an ice bath and allow them to stand for 30 min, then sediment the precipitate by centrifuging at 1500 g for 30 min at 4°. The precipitate should be packed tightly at the bottom of the tube and the supernatant should be clear. Using a Pasteur pipet equipped with a rubber bulb, carefully remove an aliquot of the clear supernatant for cholesterol analysis. If the analysis will not be performed immediately, the supernatant should be stored in a screw-capped culture tube, scintillation vial, or other container that can be sealed to prevent evaporation. The supernatants can be stored at 4° for a few days before analysis, but should be frozen, preferably at −70° or lower, if longer periods of storage are required.

Cholesterol in the heparin–Mn^{2+} supernatant can be analyzed either by chemical or enzymatic methods (see below). The measured cholesterol concentration is multiplied by the factor 1.09 to correct for the sample dilution that results from the addition of the reagents.

[18] S. E. Lasker, *Fed. Proc., Fed. Am. Soc. Exp. Biol.* **36,** 92 (1977).
[19] N. Sugishahi and F. J. Petracek, *Fed. Proc., Fed. Am. Soc. Exp. Biol.* **36,** 89 (1977).
[20] L. B. Jaques, *Adv. Exp. Med. Biol.* **52,** 139 (1975).
[21] C. Mayfield, G. R. Warnick, and J. J. Albers, *Clin. Chem.* **25,** 1309 (1979).

Notes

Accuracy: as mentioned above, the heparin–$MnCl_2$ method has been used extensively and it has been investigated in some detail. The accuracy of the method has been estimated by comparison with a procedure in which LDL and lower density lipoproteins are removed by ultracentrifugation at d 1.063 g/ml and HDL–cholesterol then measured in the $d >$ 1.063 g/ml fraction. On the average, the two methods agree within about 2–4 mg/dl.[22,23] The difference is related in part to the presence of some Lp(a) in the $d >$ 1.063 g/ml fraction. Lp(a) is found in a density range that overlaps those of LDL and HDL, but this lipoprotein is precipitated with heparin and $MnCl_2$. The two methods agreed within about 1 mg/dl or less in samples with low Lp(a) concentrations.[22,23]

The heparin–$MnCl_2$ method is generally reliable, but there are conditions in which inaccurate results are obtained. Although the following comments refer specifically to the heparin–$MnCl_2$ method, a number of the issues that are raised would apply equally regardless of the precipitant used.

Reagent Concentration. The reagent concentrations specified above provide final concentrations of 0.046 M $MnCl_2$ and about 1.3 mg/ml heparin. This is sufficient to precipitate apoB-containing lipoproteins without precipitating HDL_2 or HDL_3. It should also be mentioned that HDL has a minor subfraction of apoE-containing particles that accounts for at most a few milligrams per deciliter HDL–cholesterol. While heparin–$MnCl_2$ has been used to remove apoE-containing particles from isolated HDL,[24] various studies have revealed the presence of apoE in heparin–$MnCl_2$ supernatants of unfractionated plasma or serum,[6] suggesting that apoE-containing HDL was not precipitated. Recent findings in plasma indicate that heparin–$MnCl_2$ preferentially precipitates apoE that is associated with apoB-containing lipoproteins and leaves the apoE-containing fraction of HDL in solution.[25] In contrast, two other precipitants, dextran sulfate–$MgCl_2$ and phosphotungstate–$MgCl_2$, readily precipitate a large proportion of apoE-containing HDL from plasma along with the apoB-containing lipoproteins.[25]

Maximal precipitation of apoB-containing lipoproteins is actually complete at heparin concentrations of about 1 mg/ml or higher, and as much as 10 mg/ml does not precipitate HDL. An $MnCl_2$ concentration of 0.046 M is sufficient for the complete precipitation of apoB-containing

[22] G. R. Warnick and J. J. Albers, *J. Lipid Res.* **19**, 65 (1978).
[23] P. S. Bachorik, P. D. Wood, J. J. Albers, P. Steiner, M. Dempsey, K. Kuba, R. Warnick, and L. Karlsson. *Clin. Chem.* **22**, 1828 (1976).
[24] K. H. Weisgraber and R. W. Mahley, *J. Lipid Res.* **21**, 316 (1980).
[25] J. C. Gibson, A. Rubinstein, and W. V. Brown, *Clin. Chem.* **30**, 1784 (1984).

lipoproteins from serum, but is just a little higher than the minimum required in EDTA plasma, probably due to chelation of some of the Mn^{2+} by EDTA. As a consequence, traces of apoB can be detected in most clear heparin–Mn^{2+} supernatants of EDTA plasma with sufficiently sensitive immunochemical methods. In most cases the unprecipitated apoB would account for about 1.5–2.5 mg/dl apoB-associated cholesterol.[22,26] It should be mentioned that these estimates are based on studies in which plasma samples were stored in the refrigerator for about 4 days before being treated with heparin–$MnCl_2$. ApoB-containing lipoproteins, however, are precipitated more readily when the samples are fresh,[27] and unprecipitated apoB-associated cholesterol was reported to average 0.5 mg/dl in a study in which the precipitation was performed the same day to within 1 day after sample collection.[28] In the author's experience, precipitation should be performed as soon as possible, preferably on the day the sample is drawn.

More complete precipitation of apoB-containing lipoproteins can be achieved using a higher concentration of $MnCl_2$. Based on extensive evaluations of the heparin–$MnCl_2$ method, the use of 0.092 M $MnCl_2$ was recommended for use in EDTA plasma.[22,28] Under these conditions, unprecipitated apoB-associated cholesterol was reduced to 0.5 mg/dl or less,[22,28] even in samples that were stored at 4° for several days. Some HDL is also precipitated, but the loss is less than 1 mg/dl HDL cholesterol.[22] If this modification is used, the procedure is performed as described above, except that 2.0 M $MnCl_2$ is substituted for 1.0 M $MnCl_2$. Note that 0.046 M $MnCl_2$ should be used with serum, since larger amounts of HDL are precipitated at the higher $MnCl_2$ concentration and HDL–cholesterol is therefore underestimated.

Samples with Elevated Triglycerides. Two kinds of difficulties can arise in samples with high triglyceride levels. One fairly common problem is that in which the heparin–Mn^{2+} lipoprotein complex does not sediment completely, and the supernatant remains turbid. This occurs most frequently in samples with triglyceride levels exceeding 300–400 mg/dl, but can occur in the presence of lower triglyceride levels also.[22] Since most plasma cholesterol is associated with apoB-containing lipoproteins, even small amounts of turbidity can lead to grossly inaccurate HDL–choles-

[26] J. J. Albers, G. R. Warnick, D. Wiebe, P. King, P. Steiner, L. Smith, C. Breckenridge, A. Chow, K. Kuba, S. Weidman, H. Arnet, P. Wood, and A. Shlagenhaft, *Clin. Chem.* **24,** 853 (1978).
[27] P. S. Bachorik, R. Walker, K. D. Brownell, A. J. Stunkard, and P. O. Kwiterovich, *J. Lipid Res.* **21,** 608 (1980).
[28] G. R. Warnick and J. J. Albers, *in* "Report of the High Density Lipoprotein Methodology Workshop," p. 53. NIH Publ. No. 79-1661, Bethesda, Maryland, 1979.

terol measurements. Several approaches have been used with hypertriglyceridemic samples. First, triglyceride-rich lipoproteins are first removed by ultracentrifugation and the precipitation then performed on the $d > 1.006$ g/ml fraction. HDL–cholesterol values measured in this fraction are slightly lower than those in unfractionated plasma, due in part to losses incurred in preparing the ultracentrifugal fraction. In another approach, the turbid heparin–$MnCl_2$ supernatant is filtered through 0.22-μm filters to remove the unsedimented lipoprotein complexes, and HDL–cholesterol is analyzed in the filtrate.[29] In a third approach, turbid plasma is diluted with 0.15 M saline before being treated with heparin–$MnCl_2$.[22] This reduces the triglyceride and plasma protein concentrations of the sample, and generally produces clear heparin–$MnCl_2$ supernatants. Dilutions greater than 2-fold are not recommended, however, because the HDL–cholesterol concentrations become too low to measure accurately, and dilution errors become excessive. The fourth approach that has been used has been simply to centrifuge the turbid supernatant at higher speed,[22] which usually clears the supernatant. In hypertriglyceridemic samples, however, some of the precipitate may float to the top of the tube rather than sediment, which can make it difficult to recover the clear HDL-containing fraction. Finally, the incidence of turbid supernatants is about 3-fold lower when the precipitation is performed with 0.092 M $MnCl_2$ rather than 0.046 M reagent.[22,26] All of the methods that are used to clear turbid heparin–$MnCl_2$ supernatants require additional steps to prepare the sample for analysis. The precision of the HDL–cholesterol analyses will therefore necessarily be somewhat less than when the additional manipulations are not required.

Incomplete sedimentation of the heparin–$MnCl_2$–lipoprotein complex is usually evident when it occurs. The more serious difficulties are encountered when the precipitated lipoproteins are sedimented satisfactorily but apoB-containing lipoproteins are incompletely precipitated. This problem has been discussed above. While it may only be a few milligrams per deciliter in normotriglyceridemic samples, the error can be much larger in hypertriglyceridemic samples. This problem is particularly troublesome because it is not readily detected without assessing the heparin–$MnCl_2$ supernatant for complete precipitation of apoB-containing lipoproteins. Such an assessment can be made with electrophoretic or immunochemical methods, but when large numbers of samples must be analyzed, this might not be feasible in every sample. Incomplete precipitation should be suspected in samples with HDL–cholesterol concentrations that exceed about 80 mg/dl, and the analyses should be confirmed in

[29] G. R. Warnick and J. J. Albers, *Clin. Chem.* **24**, 900 (1978).

these samples. In samples with apparent HDL–cholesterol concentrations in the normal range, however, there might be no reason to suspect that a particular analysis was in error. For example, incomplete precipitation of apoB-containing lipoproteins from a sample with an HDL–cholesterol concentration of 45 mg/dl might lead to an apparent concentration of 55 mg/dl and the error would go undetected. It is therefore advisable to assess routinely the completeness of precipitation of apoB-containing lipoproteins in a subset, perhaps 5 or 10%, of samples with HDL–cholesterol concentrations in the normal range.

Analysis of Stored Samples. The lipoproteins are subject to a variety of alterations during storage, including autooxidation, changes in composition mediated by lipase or lecithin cholesterol acyl transferase, exchange of cholesteryl esters and triglycerides between HDL and other plasma lipoproteins, bacterial contamination, proteolytic cleavage, and others.[13,14] These changes apparently influence the precipitability of the lipoproteins. For example, the apoB-containing lipoproteins become increasingly more difficult to precipitate completely, and HDL becomes progressively easier to precipitate as the length of storage increases.[27,30] The net effect of these opposing trends depends on the HDL–cholesterol concentration. In samples with low HDL–cholesterol levels the incomplete precipitation of apoB-containing lipoproteins apparently predominates and there is a tendency for the apparent HDL–cholesterol concentration to increase with storage. In samples with high HDL–cholesterol concentrations, the inappropriate precipitation of HDL apparently predominates and the measured HDL–cholesterol concentrations tend to decrease somewhat in stored samples.[27,30] These effects are most marked when samples are stored at refrigerator temperatures, but are also observable in samples that are stored at $-20°$ and to some extent at $-70°$ as well.[27,31] At refrigerator temperatures the effects were observable but small in samples stored up to 4 days but became more pronounced after 1 and 2 weeks of storage. In our experience it is best to precipitate samples as soon as possible after they are drawn, preferably on the day of venipuncture. If necessary, the heparin–$MnCl_2$ supernatants can be stored at $-70°$ for at least 1 year prior to analysis of HDL–cholesterol.[31]

Enzymatic Cholesterol Analysis. The sources of inaccuracy and imprecision discussed above relate to the preparation of the HDL-containing fractions, and the information provided was based on the analysis of lipid extracts of samples using chemical cholesterol methods such as those

[30] L. I. Gidez, G. J. Miller, M. Burstein, S. Slagle, and H. A. Eder, *J. Lipid Res.* **23,** 1206 (1982).
[31] P. S. Bachorik, R. E. Walker, and P. O. Kwiterovich, *J. Lipid Res.* **23,** 1236 (1982).

based on the Liebermann–Burchard reaction. Under these conditions, the precipitating reagents do not interfere with the cholesterol analyses. In the past several years chemical methods have been largely replaced by enzymatic cholesterol methods that are more rapid, more specific for cholesterol, and which can be performed directly in serum or plasma. $MnCl_2$, however, interferes with the enzymatic analysis of cholesterol and gives falsely high HDL–cholesterol values in heparin–$MnCl_2$ supernatants.[32–35] The mechanism of the interference is uncertain, but may be due in part to the precipitation of Mn^{2+} by phosphate-containing buffers that are commonly used in the enzyme reagents and in part to the direct interference of Mn^{2+} with one or more of the enzymatic reactions themselves. Several approaches have been used to avoid this problem. In one, interference was eliminated by including EDTA ($4 \times 10^{-3}\ M$) in the enzymatic reaction system to chelate Mn^{2+}.[32] Workers using a 2.5-fold higher concentration of the chelating agent, however, reported that EDTA itself interferes with enzymatic cholesterol analysis.[34] In another recently described modification, Mn^{2+} is first removed from the heparin–$MnCl_2$ supernatant by precipitation with HCO_3^-.[35] This procedure is as follows.

Prepare the heparin–$MnCl_2$ supernatant as described above. Accurately measure a convenient aliquot (0.50 or 1.0 ml) of the heparin Mn^{2+} supernatant into a 1.4-ml polypropylene microcentrifuge tube. The volume measurements should be made with a micropipet or positive displacement-type pipet of similar accuracy. Add 0.1 vol of 1 M $NaHCO_3$, mix well, and allow the sample to stand at room temperature for at least 30 min. A white precipitate of $Mn(HCO_3)_2$ forms immediately and precipitation is complete within this period. If the cholesterol analyses are not to be performed immediately the samples can be sealed to prevent evaporation and stored for 2 or 3 days in the refrigerator. The $Mn(HCO_3)_2$ precipitate is then sedimented by centrifuging for 10 min at 10,000 g in a benchtop microcentrifuge, and cholesterol is measured in the clear supernatant.

With this modification the measured HDL–cholesterol values must be multiplied by 1.199 to correct for dilution by the heparin, $MnCl_2$, and $NaHCO_3$ reagents.

Both of the foregoing modifications represent promising attempts to adapt the heparin–$MnCl_2$ method for use with enzymatic cholesterol

[32] B. W. Steele, D. F. Koehler, M. M. Azar, T. P. Blaszkowski, K. Kuba, and M. E. Dempsey, *Clin. Chem.* **22**, 98 (1976).

[33] P. N. M. Demacker, H. E. Vos-Janssen, A. P. Jansen, and A. van't Laar, *Clin. Chem.* **23**, 1238 (1977).

[34] P. N. M. Demacker, H. E. Vos-Janssen, A. G. M. Hijmans, A. van't Laar, and A. P. Jansen, *Clin. Chem.* **26**, 1780 (1980).

[35] P. S. Bachorik, R. E. Walker, and D. G. Virgil, *Clin. Chem.* **30**, 839 (1984).

methods. To date, however, neither has been as widely used or as intensively studied as the heparin–$MnCl_2$ method itself. The addition of EDTA to the cholesterol assay system introduces no additional steps into the preparation of heparin–$MnCl_2$ supernatants and would therefore not be expected to influence the precision of the assays. The direct interference of EDTA with different enzymatic assay systems, however, deserves further investigation. The preliminary removal of Mn^{2+} introduces another step in the preparation of the sample and would be expected to increase the variability of the assays somewhat. The practical importance of this effect must be evaluated further.

The third approach has been to use precipitating agents that do not interfere with the enzymatic cholesterol assays, as discussed below.

Dextran Sulfate–MgCl_2 Method

As mentioned earlier, a variety of precipitants have been used to remove apoB-containing lipoproteins, the more common of which include dextran sulfate–Mg^{2+}, phosphotungstate–Mg^{2+}, polyethylene glycol, and heparin–Ca^{2+}.[6] The completeness with which these agents remove apoB-containing lipoproteins and leave HDL in solution differ somewhat, and can lead to results for HDL–cholesterol that are as much as 5–10% higher or lower than those obtained with heparin and $MnCl_2$.[36] Nonetheless, with the advent of completely enzymatic cholesterol methods, there has been a growing interest in developing these procedures further, and recent modifications of methods that use the three nonheparin precipitants mentioned above give HDL–cholesterol values that are quite similar to those obtained with the heparin–$MnCl_2$ method.[37] The dextran–sulfate–$MgCl_2$ method is described here because of its usefulness both with enzymatic cholesterol methods and for the separation of HDL subclasses (see below).

The following method is that described by Warnick et al.[38]

Samples. The method can be used with either EDTA plasma or serum.

Reagents

Preservative solution: Dissolve 5.0 g NaN_3, 50 mg gentamicin sulfate, and 100 mg chloramphenicol in distilled, deionized H_2O and adjust to 100 ml with H_2O.

[36] G. R. Warnick, M. C. Cheung, and J. J. Albers, *Clin. Chem.* **25**, 596 (1979).

[37] G. R. Warnick, T. Nguyen, and A. A. Albers, *Clin. Chem.* **31**, 217 (1985).

[38] G. R. Warnick, J. Benderson, and J. J. Albers, *Clin. Chem.* **28**, 1379 (1982).

Dextran sulfate (M_r 50,000 ± 5000), 20 g/liter: The dry reagent is stored at 4° in a desiccator. Dissolve 2.0 g dextran sulfate in 80 ml distilled, deionized H_2O and adjust to pH 7.0 with HCl. Transfer the solution quantitatively to a 100-ml volumetric flask. Add 1.0 ml of the preservative solution and adjust to volume with H_2O. Store the solution at 4°.

$MgCl_2 \cdot 6H_2O$, 1.0 M: This reagent is hygroscopic and should be stored in a tightly closed container or a desiccator. Dissolve 20.3 g $MgCl_2 \cdot 6H_2O$ in 80 ml H_2O and adjust to pH 7.0 with a solution of NaOH. Dilute to 100 ml with H_2O as described above and store the solution at 4°.

Accurately measure equal volumes of the dextran sulfate and $MgCl_2$ solutions into a suitable container and mix. Store at 4°. The working solution is stable for at least 4 months.

Procedure. Allow the samples to reach room temperature. Using a volumetric pipet or device of similar accuracy, place 1.0 ml sample in a transparent test tube or conical centrifuge tube. Add 100 μl working reagent and mix thoroughly. The final reagent concentrations are 0.91 mg/ml dextran sulfate and 0.045 M $MgCl_2$. Allow the mixture to stand at room temperature for 10 min, then sediment the precipitate by centrifuging at 1500 g for 30 min at 4°. Using a Pasteur pipet, carefully remove an aliquot of the clear supernatant for cholesterol analysis. If necessary, the supernatant can be stored at 4° in a sealed container for a few days before analysis.

The precipitants do not interfere either with chemical cholesterol methods such as the Liebermann–Burchard reaction, or any of the available enzymatic cholesterol methods, and any of these methods can be used to measure cholesterol in the dextran sulfate–$MgCl_2$ supernatants. The measured cholesterol values are multiplied by 1.10 to correct for dilution by the reagent.

Notes

Accuracy. This method described here was modified from earlier procedures that used dextran sulfate preparations of higher (M_r 500,000)[39,40] or lower (M_r 15,000)[41] molecular weight. The higher molecular weight material apparently precipitated a significant amount of HDL[36] and the lower molecular weight preparations incompletely removed apoB-con-

[39] G. M. Kostner, *Clin. Chem.* **22,** 695 (1976).
[40] P. R. Finley, R. B. Schifman, R. J. Williams, and D. A. Lichti, *Clin. Chem.* **24,** 931 (1978).
[41] M. Burstein and H. R. Scholnick, *Adv. Lipid Res.* **11,** 67 (1973).

taining lipoproteins.[38] The use of dextran sulfate of M_r 50,000 gave HDL–cholesterol values that averaged 1–2 mg/dl lower than those obtained with the heparin–$MnCl_2$ method, in part because of the presence of less residual apoB-containing lipoproteins in the dextran sulfate–$MgCl_2$ supernatants. The method precipitated slightly more apoA-I than heparin–$MnCl_2$, but this accounted for only about 0.5 mg/dl HDL–cholesterol.[38]

Based on paired comparisons of the dextran sulfate–$MgCl_2$ and heparin–$MnCl_2$ methods in EDTA plasma, the two methods were highly correlated ($r = 0.98$) and gave similar results over the concentration range 25–75 mg/dl.[38]

Samples with Elevated Triglycerides. As with the heparin–$MnCl_2$ method, turbid dextran sulfate–$MgCl_2$ supernatants are occasionally encountered and can give grossly inaccurate HDL–cholesterol values. This occurs most frequently in samples with high triglyceride levels, but occurs less than half as often with this method as with heparin–$MnCl_2$. When it does occur, clear supernatants can be prepared by ultrafiltration, sample dilution or preliminary removal of triglyceride-rich lipoproteins by ultracentrifugation at d 1.006 g/ml, as described above for the heparin–$MnCl_2$ method.

HDL Subclasses

It is known that the variation in total HDL concentration is generally due primarily to variations in HDL_2 concentration. Thus, HDL_2 levels in women are higher than in men, and the changes in total HDL levels induced by diet, exercise, various medications, and other factors are due primarily to changes in HDL_2 levels. Until recently, HDL_2 and HDL_3 were measured following separation by analytical or preparative ultracentrifugal methods. Since these methods are not readily applied to large numbers of samples, there has been recent interest in developing precipitation methods to separate HDL_2 and HDL_3. Both of the methods presented below are rapid, relatively easy to perform, and provide reasonable estimates of HDL_2– and HDL_3–cholesterol. Both are dual precipitation methods in which apoB-containing lipoproteins are first precipitated. HDL_2 is then precipitated from the clear HDL-containing supernatant. Cholesterol is then analyzed in the first (total HDL) and second (HDL_3) supernatants and HDL_2–cholesterol is calculated as the difference between the two.

Heparin–$MnCl_2$/Dextran Sulfate Method

This method was devised by Gidez *et al.*[30]
Samples. This method has been used in EDTA plasma.

Reagents

$MnCl_2 \cdot 4H_2O$, 1.12 *M*: Dissolve 22.166 g $MnCl_2 \cdot 4H_2O$ in distilled H_2O, transfer quantitatively to a 100-ml volumetric flask, and dilute to volume with H_2O. Store at 4°.

Heparin, 135 mg/ml: Dissolve 2.025 g heparin sodium powder in H_2O and dilute to 15 ml. If injectable heparin is used, dilute to 20,000 USP units/ml with H_2O. This should provide a concentration of approximately 135 mg/ml. Store at 4°.

Combined heparin–$MnCl_2$ reagent: Mix 6.0 ml heparin solution with 50.0 ml $MnCl_2$ solution. The precipitant concentrations of the combined reagent are 1.0 *M* $MnCl_2$ and 14.5 mg/ml heparin. Store at 4°.

Dextran sulfate (M_r 15,000), 14.3 mg/ml: Dissolve 1.430 g dextran sulfate in 0.15 *M* NaCl. Transfer quantitatively to a volumetric flask and dilute to 100 ml. Store at 4°.

Procedure. Precipitation of apoB-containing lipoproteins: Using a volumetric pipet, transfer 3.0 ml plasma to a transparent 15-ml conical centrifuge tube. Add 300 μl combined heparin–$MnCl_2$ reagent and mix thoroughly but avoid foaming. The final concentrations of precipitant are 0.091 *M* $MnCl_2$ and 1.3 mg/ml heparin. Allow the sample to stand at room temperature for 20 min, then centrifuge at 1500 *g* for 1 hr at 4°. Immediately remove an aliquot of the clear heparin–$MnCl_2$ supernatant for the analysis of total HDL–cholesterol and another for HDL_2 precipitation.

HDL_2 precipitation: Using a volumetric pipet, transfer 2.0 ml clear heparin–$MnCl_2$ supernatant to a centrifuge tube. Add 200 μl dextran sulfate solution to the sample and mix well. The final concentration of dextran sulfate is 1.3 mg/ml. Let the sample stand at room temperature for 20 min, then centrifuge at 1500 *g* for 30 min at 4°. The clear soluble fraction represents HDL_3. An aliquot is immediately removed for cholesterol analysis.

The measured value for total HDL–cholesterol is multiplied by 1.10, and that for HDL_3 is multiplied by 1.21 to correct for dilution by the reagents. HDL_2 is then calculated as the difference between total HDL–cholesterol and HDL_3–cholesterol.

Notes

Identity of the subfractions and accuracy of the measurements: HDL_2 and HDL_3 are defined in terms of their flotation rates in the analytical ultracentrifuge. HDL_2 is found in the density range 1.063–1.12 g/ml and HDL_3 in the range 1.12–1.21 g/ml, and the two subfractions can be separated from each other by preparative ultracentrifugation at *d* 1.12 g/ml. Under these conditions HDL_2 accumulates as a floating layer and HDL_3

sediments. The separation of analogous fractions by polyanion precipitation depends on properties other than the density of the molecules, presumably on charge–charge interactions between the lipoproteins and the precipitants, and is affected by factors such as pH, ionic strength, protein–lipid ratio, and others discussed above. For this reason, some investigators have been reluctant to equate ultracentrifugally isolated HDL_2 and HDL_3 with the corresponding polyanion precipitation fractions. Gidez et al.[30] referred to their precipitation fractions as "HDL_2" and "HDL_3," and others[42] distinguish the fractions with a subscript that indicates the precipitated (HDL_p) and soluble (HDL_s) fractions. In the original method, Gidez et al.[30] found that the chemical compositions of "HDL_2" and "HDL_3" agreed well with those of ultracentrifugally isolated HDL_2 and HDL_3. Several other laboratories using rate zonal ultracentrifugation, however, found some unprecipitated HDL_2 in the soluble fraction and some HDL_3 in the precipitated fraction.[42,43] These findings are not surprising since the precipitation of HDL_2 by dextran sulfate might be expected to be influenced by the amount of heparin and $MnCl_2$ in the heparin–$MnCl_2$ supernatant. This would depend on the amount of reagents that was removed with apoB-containing lipoproteins, which in turn would depend on the plasma lipoprotein concentration. Thus, the separation of HDL_2 from HDL_3 may not be equally complete in all samples.

These difficulties notwithstanding, the values for HDL_2– and HDL_3–cholesterol that are obtained with the dual precipitation method apparently agreed within about 2 mg/dl of those obtained by preparative ultracentrifugation[30] and agreed well with values that can be calculated from HDL subfraction distributions determined with the analytical ultracentrifuge.[44] Furthermore, HDL_2 levels measured with this method show the expected variations with respect to various factors known to influence HDL_2 levels. For example, higher HDL_2 levels were found in women[30,45] and in physically active subjects (runners),[30] and lower levels were found in patients with coronary heart disease and hypertriglyceridemic patients.[30,45] Also, as expected, HDL_3 levels measured with the precipitation method vary less than HDL_2 levels.[30,45]

[42] H. S. Simpson, F. C. Ballantyne, C. J. Packard, H. G. Morgan, and J. Shepherd, Clin. Chem. 28, 2040 (1982).

[43] W. H. Daerr, E. E. T. Windler, H. B. Rohwer, and H. Greten, Atherosclerosis 49, 211 (1983).

[44] D. W. Anderson, A. V. Nichols, S. S. Pan, and F. T. Lindgren, Atherosclerosis 29, 161 (1978).

[45] S. Martini, G. Baggio, L. Boroni, G. B. Enzi, R. Fellin, M. R. Baiocchi, and G. Crepaldi, Clin. Chim. Acta 137, 291 (1984).

Analysis of Cholesterol. In the original method, the cholesterol analyses were performed using the reference method of Abell *et al.*[46] with which Mn^{2+} does not interfere. The method has been used with enzymatic cholesterol analysis following the addition of 8 mM EDTA.[42]

Dextran Sulfate–MgCl₂ Method

The following method, developed by Warnick *et al.*,[47] is a two-step precipitation procedure using dextran sulfate and $MgCl_2$.

Samples. The method was applied to EDTA plasma. The precipitations were performed in fresh samples on the day of venipuncture.

Reagents

Preservative solution: Dissolve 5.0 g NaN_3, 50 mg gentamicin sulfate, and 100 mg chloramphenicol in distilled, deionized H_2O and adjust to 100 ml with H_2O.

Dextran sulfate (M_r 50,000 ± 5,000), 20 g/liter: Dissolve 2.0 g dextran sulfate in 80 ml distilled, deionized H_2O and adjust to pH 7.0 with HCl. Transfer the solution quantitatively to a 100-ml volumetric flask. Add 1.0 ml of the preservative solution and dilute to 100 ml with H_2O. Store at 4°.

Dextran sulfate (M_r 50,000 ± 5,000), 40 g/liter: Dissolve 4.0 g dextran sulfate in H_2O as above, add 1.0 ml of the preservative solution and adjust to 100 ml with H_2O. Store at 4°.

$MgCl_2 \cdot 6H_2O$, 1.0 M: Dissolve 20.3 g $MgCl_2 \cdot 6H_2O$ in 80 ml H_2O and adjust to pH 7.0 with a solution of NaOH. Dilute to 100 ml with H_2O. Store at 4°.

$MgCl_2 \cdot 6H_2O$, 2.0 M: Dissolve 40.3 g $MgCl_2 \cdot 6H_2O$ in H_2O as described above and dilute to 100 ml. Store at 4°.

Combined Reagent A. Mix equal volumes of dextran sulfate (20 g/liter) and $MgCl_2$, 1.0 M. The reagent concentrations are as follows: $MgCl_2$, 0.5 M and dextran sulfate, 10 g/liter. Store at 4°.

Combined Reagent B. Mix 1 vol of dextran sulfate (40 g/liter) with 3 vol $MgCl_2$ (2.0 M). The reagent concentrations are $MgCl_2$, 1.5 M, and dextran sulfate, 10 g/liter. Store at 4°.

Procedure. Precipitation of apoB-containing lipoproteins: Allow the samples to reach room temperature. Accurately measure 2.0 ml plasma into a transparent 15-ml conical centrifuge tube. Add 200 μl combined reagent A and mix immediately and thoroughly, using a vortex mixer. Avoid foaming. The final reagent concentrations are 0.91 mg/ml dextran

---

[46] L. L. Abell, B. B. Levy, B. B. Brodie, and F. E. Kendall, *J. Biol. Chem.* **195**, 357 (1952).
[47] G. R. Warnick, J. M. Benderson, and J. J. Albers, *Clin. Chem.* **28**, 1574 (1982).

sulfate and 0.045 M MgCl$_2$. Let the mixture stand at room temperature for 10 min, then centrifuge at 1500 g for 30 min at 4°. Remove aliquots of the clear supernatant for the analysis of total HDL–cholesterol and for the precipitation of HDL$_2$.

Precipitation of HDL$_2$. Measure 1.0 ml of the above supernatant into a transparent conical centrifuge tube. Add 100 μl of combined reagent B and mix thoroughly, using a vortex mixer. Avoid foaming. Combined reagent B increases the residual reagent concentrations of the supernatant by 0.91 mg/ml for dextran sulfate and 0.136 M for MgCl$_2$. Allow the mixture to stand at room temperature for 15 min, then centrifuge at 1500 g for 30 min at 4°. Remove an aliquot of the clear supernatant for the analysis of HDL$_3$.

The measured values for total HDL are multiplied by 1.10, and those for HDL$_3$ are multiplied by 1.21 to correct for dilution of the samples by the reagents. HDL$_2$–cholesterol is then calculated as the difference between total HDL–cholesterol and HDL$_3$–cholesterol.

Notes

HDL$_2$ values measured with this method correlate highly with those measured by analytical ultracentrifugation or by the heparin–MnCl$_2$/dextran sulfate method described above ($r > 0.95$ for both methods). The correlations for HDL$_3$ are apparently a little lower; about 0.7 with respect to analytical ultracentrifugation and 0.91 for the heparin–MnCl$_2$/dextran sulfate method. The method gave HDL$_2$– and HDL$_3$–cholesterol values that were similar to those obtained with the heparin–MnCl$_2$/dextran sulfate method and provided reasonable estimates of levels obtained by ultracentrifugation.[47]

Cholesterol Analysis. The dextran sulfate method can be used with either chemical or enzymatic cholesterol analyses.

Measurement of Other Lipoproteins

As mentioned earlier, precipitation methods have not been too widely used to measure lipoproteins other than HDL. Consequently, while potentially useful procedures have been suggested for the quantitation of VLDL and LDL, their performance characteristics have not been as extensively evaluated as those for HDL. Two general approaches have been used. In the first, the precipitation conditions have been varied to allow the separation of VLDL-, LDL-, and HDL-containing fractions and the lipoproteins have been measured quantitatively, generally in terms of their cholesterol content. This approach deserves further development and rigorous evaluation compared to accepted reference methods because of its potential usefulness in making complete quantitative lipoprotein analysis available in the routine clinical laboratory. The second approach has been used to screen for hyperlipidemia by measuring the turbidity developed when VLDL plus LDL are precipitated. This approach does not distinguish between elevations in VLDL or LDL, but rather identifies subjects who must be evaluated further. One example of each of these methods is presented here. The reader is referred to a recent review for a more complete discussion of the various precipitants and precipitating conditions that have been used.[6]

Quantitation of VLDL, LDL, and HDL by Selective Precipitation

This procedure is based on that proposed by Wilson and Spiger.[48]
Samples. This method was applied to EDTA plasma.

Reagents

$MnCl_2 \cdot 4H_2O$, 1.0 M: Dissolve 19.791 g $MnCl_2 \cdot 4H_2O$ in distilled H_2O, and dilute quantitatively to 100 ml. Store at 4°.

Heparin, 35 mg/ml: Dissolve 1.75 g heparin sodium in 0.15 M NaCl and dilute quantitatively to 50 ml. Store at 4°.

Sodium dodecyl sulfate, 10%: Dissolve 10 g sodium dodecyl sulfate in 0.15 M NaCl. Adjust to pH 9.0 with NaOH, and dilute to 100 ml with 0.15 M NaCl.

Procedure. The method requires the measurement of total plasma cholesterol as well as that in two plasma fractions.

VLDL Precipitation. Allow samples to reach room temperature. Measure 2.0 ml sample into a centrifuge tube and add 150 μl 10% sodium dodecyl sulfate solution. Mix well, then cover the samples to minimize evaporation and incubate in a water bath for 2 hr at 37°. The sample is then centrifuged at room temperature for 10 min at 10,000 g. Aggregated VLDL forms a pellicle that floats at the top of the tube. The clear subnatant contains LDL and HDL, and an aliquot is removed for the analysis of cholesterol. The measured cholesterol value is multiplied by 1.075 to correct for dilution by the reagent solution.

VLDL + LDL Precipitation

A separate aliquot of sample is treated with heparin and $MnCl_2$ as described earlier, and an aliquot of the clear heparin–$MnCl_2$ supernatant

[48] D. E. Wilson and J. J. Spiger, *J. Lab. Clin. Med.* **82**, 473 (1973).

is removed for the analysis of HDL–cholesterol. The value is multiplied by 1.09 to correct for dilution by the reagents.

Calculations. The lipoprotein concentrations are calculated from the following relationships:

$$[\text{VLDL–chol}] = [\text{TC}] - [(\text{LDL} + \text{HDL})\text{–chol}] \tag{2}$$
$$[\text{LDL–chol}] = [((\text{LDL} + \text{HDL})\text{–chol}) - (\text{HDL–chol})] \tag{3}$$

where TC = total plasma cholesterol and (LDL + HDL)–chol = cholesterol in the sodium dodecyl sulfate-soluble fraction.

Notes

The method was reported to give VLDL– and LDL–cholesterol values that correlated well ($r > 0.96$) with those obtained by preparative ultracentrifugation.[48] In samples with high triglyceride levels, however, VLDL may be incompletely precipitated by sodium dodecyl sulfate, or the precipitate may not be completely removed, resulting in the underestimation of VLDL–cholesterol and the overestimation of LDL–cholesterol.[49] Turbidity and incomplete precipitation in heparin–$MnCl_2$ supernatants was discussed earlier.

Turbidimetric Screening for Hyperlipidemia

This approach is perhaps most useful when large numbers of samples are to be screened to identify hyperlipidemic subjects who will then be further evaluated with other methods. In this method, VLDL plus LDL is precipitated and the resulting turbidity is measured. The procedure described here is based on that of Berenson *et al.*[50]

Samples. This procedure was performed in serum.

Reagents

Heparin, 2.5 mg/ml: Dissolve 250 mg heparin in distilled H_2O and dilute to 100 ml.
$CaCl_2$, 0.5 *M*: Dissolve 5.050 g anhydrous $CaCl_2$ or 7.351 g $CaCl_2 \cdot 2H_2O$ in distilled H_2O and dilute to 100 ml.

Procedure. Measure 3.2 ml distilled H_2O into a test tube. Add 200 μl sample and mix. Then add 0.1 ml heparin solution and 0.5 ml $CaCl_2$ solution, mix, and let stand at room temperature for 15 min. A blank is prepared in which distilled H_2O is substituted for the heparin solution. The turbidity of the sample and blank is measured at 600 nm, and the

[49] D. E. Wilson and G. A. Done, *Clin. Chem.* **20**, 394 (1974).
[50] G. S. Berenson, S. R. Srinivasan, P. S. Pargaonhar, B. Radhakrishnamurthy, and E. R. Dalferes, *Clin. Chem.* **18**, 1463 (1972).

difference between the two reflects the concentration of VLDL plus LDL.

Notes

The turbidity measurements were reported to correlate reasonably well with VLDL plus LDL concentrations measured with the analytical ultracentrifuge ($r = 0.9$). The level of turbidity above which subjects would be recalled can be set as desired by establishing the relation between turbidity and lipoprotein level using samples of known VLDL and LDL concentration.

As a screening method this procedure has the advantage that lipid measurements are not required. It is therefore quite rapid and inexpensive to perform. It does not distinguish between elevations in VLDL and LDL, however, and further evaluation by more specific methods is required.

Summary

In the foregoing discussion we have included precipitation methods for HDL analysis that we consider to be among the most accurate and extensively characterized compared to analyses performed with the ultracentrifuge. The emphasis has been on the precipitation steps rather than on the analysis of cholesterol in the lipoprotein fractions. It is evident, however, that the accuracy of the cholesterol method itself will influence the overall reliability of the measurements even when the lipoprotein fraction has been separated adequately. Ideally, the cholesterol measurements would be made with reference methods. Since this is usually not practical with large numbers of samples, the accuracy and precision of the method employed should be established with respect to a reference method, such as that of Abell *et al.*,[46] before being used for routine cholesterol measurements.

Finally, it should be mentioned that the conditions used in the methods discussed here were developed for human plasma or serum, and in the authors' experience may have to be modified for the adequate separation of plasma lipoproteins in other species.

Acknowledgments

The authors wish to thank Mrs. Carol McGeeney for her assistance in preparing the manuscript. This work was supported in party by NIH Contract (1-HV-1-2158L) and NIH Grants (1 R01 HL31450-01A1) and (HL 30086).

[6] Enzymatic Methods For Quantification of Lipoprotein Lipids

By G. RUSSELL WARNICK

The term "lipid," by definition, refers to the fatlike molecules that are generally insoluble in water but soluble in organic solvents. The major circulating lipids, insoluble or marginally soluble in the blood, are transported in lipoprotein particles. The more polar lipids, phospholipids, and unesterified cholesterol reside with certain proteins on the surface of the particles, oriented to provide solubility in the aqueous medium, while the nonpolar lipids, the triglycerides, and esterified cholesterol are in the core. Cholesterol and phospholipids are essential constituents of cell membranes, whereas the triglycerides are an important fuel source.

Association of Cholesterol with Coronary Artery Disease

Much attention has been focused on the lipoproteins and their lipid constitutents because of their association with the process of atherosclerosis and coronary artery disease (CAD). The direct association between blood cholesterol levels and risk of developing CAD has long been recognized. The low-density lipoprotein (LDL) particles, which carry approximately 75% of the cholesterol in plasma, are primarily responsible for this association. The high-density lipoprotein (HDL) particles contain most of the remaining cholesterol with small amounts in the very low-density (VLDL) class. Measurement of total cholesterol has served as a reasonable approximation of the atherogenic LDL cholesterol, because the LDL comprising most of the cholesterol are highly correlated with total plasma cholesterol.

Recently an inverse association between the levels of HDL cholesterol and risk of CAD has been recognized. The differential association of these lipoprotein classes with CAD risk has made their separation and quantification important. The lipoproteins generally have been quantified in terms of their cholesterol content.

The plasma triglyceride level also has been implicated as a risk factor for CAD, although the relationship is not as clear as that for cholesterol. In some epidemiological studies, triglyceride levels were significantly associated with risk, whereas in other studies this was not the case. Certain genetic disorders may be expressed with triglyceride elevation and may confer increased risk. Therefore, measurement of plasma triglycerides is important in the routine clinical laboratory as well as in the research laboratory. Triglycerides of lipoprotein fractions may also be of interest. Plasma levels of the phospholipids, major constituents of lipoproteins, correspond closely to cholesterol levels, and their measurement usually has not been considered necessary in assessing risk of CAD. Measurement of total phospholipids is useful in the research laboratory, and there are applications in the routine laboratory.

Population distributions for the major lipid classes have been reported. The multicenter, collaborative Lipid Research Clinics program established age- and sex-specific distributions for total plasma cholesterol and triglycerides on a total of nearly 50,000 subjects in 10 populations throughout North America.[1] Distributions were reported for cholesterol in the major lipoprotein classes on a subset of approximately 5000 subjects.[2] Lipoprotein fraction triglyceride distributions were reported by one center of the Lipid Research Clinics.[3] Cholesterol measurements were made by a Liebermann–Burchard reagent method for the AutoAnalyzer II.[4] EDTA plasma specimens were extracted into isopropyl alcohol and treated with zeolite mixture before analysis. The method was in good agreement with the Abell–Kendall reference method.[5] Triglycerides were measured in the same extracts by a fluorescent method that quantitated the glycerol moeity.[4] The extraction procedure removed approximately half the free glycerol and there was no correction in patient specimens for the remaining free glycerol; thus, higher triglyceride levels would be expected for a method that quantified total triglyceride (triglyceride plus glycerol) and lower results for a method corrected for the free glycerol contribution (explained later in more detail).

A summary of the recommendations and considerations for patient preparation, specimen collection, and processing are given by Bachorik and Albers (Article [5], this volume).

Advantages of Enzymatic Methods

Compared to the older chemical methods, enzymatic methods have significant advantages. Among these are improved specificity, which per-

[1] The Lipid Research Clinics Program Epidemiology Committee, *Circulation* **60**, 427 (1979).
[2] G. Heiss, I. Tamir, C. E. Davis, H. A. Tyroler, B. M. Rifkind, G. Schonfeld, D. Jacobs, and I. D. Frantz, *Circulation* **61**, 302 (1980).
[3] P. W. Wahl, G. R. Warnick, J. J. Albers, J. J. Hoover, C. E. Walden, R. O. Bergelin, J. T. Ogilvie, W. R. Hazzard, and R. H. Knopp, *Atherosclerosis* **39**, 111 (1981).
[4] A. Hainline, J. Karon, and K. Lippell, eds., "Manual of Laboratory Operations, Lipid Research Clinics Program, Lipid and Lipoprotein Analysis," 2nd Ed. NIH, Department of Health and Human Services, Washington, D.C., 1982.
[5] I. W. Duncan, A. Mather, and G. R. Cooper, "The Procedure for the Proposed Cholesterol Reference Method." Centers for Disease Control, Atlanta, 1982.

mits in many cases direct measurement of the analyte without pretreatment, i.e., hydrolysis or extraction. Although the direct enzymatic methods are not totally free from interference effects, with reasonable precautions they can give acceptable results for most specimens. The enzymatic methods are usually more sensitive, permit measurement in small specimen volumes, and are versatile with a variety of different color reactions. The reagents are generally mild and well suited to modern chemical analyzers. In discussing the enzymatic systems, it is difficult to avoid mentioning commercial reagents and equipment. The recent rapid expansion of the clinical laboratory and the use of laboratory methods for diagnostic purposes has generated remarkable progress in reagents and equipment. Much of the development has been in the realm of technology and not recorded in the scientific literature. One area of great progress has been in the instruments for chemical analysis. The modern automated or semiautomated chemical analyzers have important advantages over manual techniques. Besides achieving the expected savings in personnel time, the analyzers often use less reagent, an important consideration with the relatively expensive enzyme reagents. Many of the analyzers provide highly accurate pipetting even for specimen volumes of less than 5μl. In addition, the analytical parameters, such as incubation time and temperature, are tightly controlled. These features result in good reproducibility.

There are unique characteristics of certain types of analyzers that must be considered in using enzymatic procedures for the lipids. For example, a bichromatic or dual wavelength instrument has advantages for glycerol blanking in the triglyceride assay because the secondary wavelength compensates at least partially for turbidity present in the reaction mixture from large lipoprotein particles. However, the secondary wavelength can affect some of the interference characteristics. For example, interference from bilirubin in the cholesterol assay might be enhanced because the secondary wavelength subtracts more of the color from the cholesterol chromogen than from the bilirubin. Another common type of instrument, the centrifugal analyzer, will sediment any precipitate in the specimen. Therefore, a heparin–manganese HDL supernatant, which on standing develops a manganese carbonate precipitate, might give less interference with a centrifugal analyzer than with a noncentrifugal type. Factors such as these must be considered in selecting a suitable chemical analyzer and appropriate reagents.

Because the chemical analyzers and reagents are generally developed for the clinical laboratories, which constitute a much larger market than the research laboratories, not all are suitable for or affordable by the latter. This laboratory has used with good success the ABA 200 bichromatic analyzer (Abbott Laboratories, Irving, TX 75015). This instrument is versatile, a necessary feature for the research laboratory. It also has

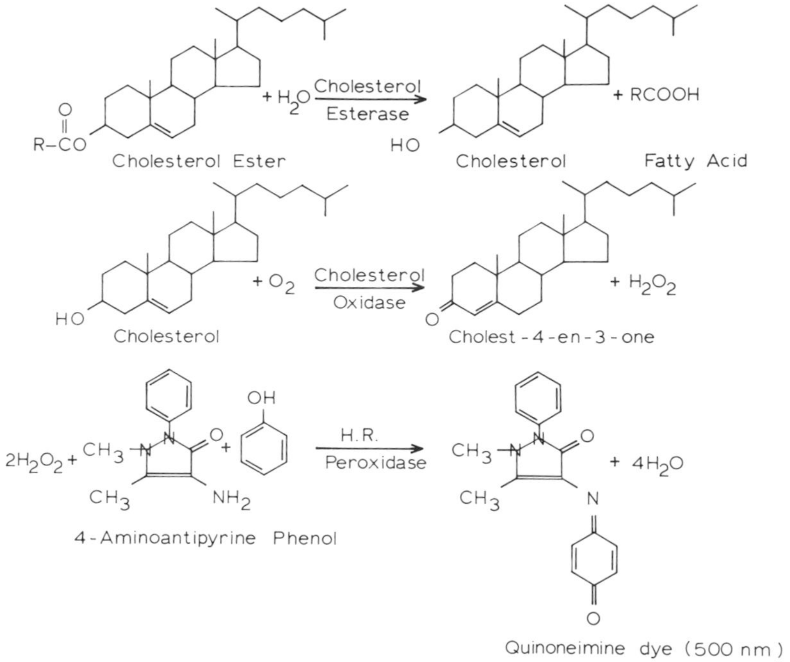

FIG. 1. Enzymatic cholesterol reaction sequence.

temperature-controlled incubation and microprocessor control. Reproducibility is acceptable. (Note: Mention of commercial systems or reagents does not constitute endorsement; information is simply provided for the convenience of others.)

Cholesterol Quantification

Cholesterol, a constituent of cell membranes, is the most common sterol in human tissue and fluids. About 70% of cholesterol in plasma is esterified with a fatty acid group at the 3-hydroxy position. In enzymatic measurement hydrolysis is necessary to free this hydroxy group for reaction with cholesterol oxidase. Hydrolysis can be accomplished with a cholesterol ester hydrolase. These two enzymatic steps are usually coupled to the Trinder reaction with the peroxidase enzyme, as illustrated in Fig. 1.[6,7] Color production is monitored at approximately 500 nm.

[6] W. Richmond, *Clin. Chem.* **19,** 1350 (1973).
[7] C. C. Allain, L. S. Poon, C. S. G. Chan, W. Richmond, and P. C. Fu, *Clin. Chem.* **20,** 470 (1974).

In the past, colorimetric methods with strong acid reagents were used for cholesterol quantification. Because these reagents were relatively nonspecific and subject to various interferences, cholesterol was usually separated from other plasma constituents by an extraction step. A common strong acid reagent, that described by Liebermann–Burchard, produced more color with esterified cholesterol than free cholesterol, and for best accuracy hydrolysis of cholesteryl esters was required. A method[8] employing these principles (Abell–Kendall) has been validated as a reference procedure for cholesterol.[5] Another common Liebermann–Burchard reagent method,[4] that used in the Lipid Research Clinics Program, eliminated the hydrolysis step by using a serum-based calibrator to correct for the increased color production.

The enzymatic cholesterol methods are relatively free of interference effects and usually can be applied to whole plasma without extraction or other pretreatment. Analysis of HDL supernatants is more of a problem because the lipoprotein precipitation methods leave in solution only about 75% of the cholesterol with all of the original interfering substances. Therefore, hemoglobin, bilirubin, ascorbic acid, and other reducing substances at levels insignificant for total cholesterol may produce interference in the HDL cholesterol assay. Bilirubin has an especially complex interference[9]; besides exhibiting absorbance at 500 nm, bilirubin competes with the color reaction for H_2O_2. These two opposing interference components may compensate or give net negative or positive interference, depending on the reaction conditions. In addition, some precipitation reagents, e.g., manganese, have been reported to interfere with certain enzymatic methods.[10]

Other potential problems have been described for enzymatic cholesterol reagents. Some of the cholesteryl ester hydrolase enzymes used do not completely cleave cholesteryl esters.[11] Therefore, such methods, which are calibrated with a primary standard of free cholesterol, may underestimate cholesterol in specimens because the remaining cholesteryl esters are not reactive in the subsequent cholesterol oxidase step. The detergents or alcohol used to solubilize free cholesterol in a primary standard may inhibit the enzymes in the reaction mixture, resulting in errors.[12] Some cholesterol reagents contain Carbowax (polyethyleneglycol), which has been reported to produce turbidity from aggregation of lipoproteins

[8] L. L. Abell, B. B. Levy, B. B. Brodie, and F. E. Kendall, *J. Biol. Chem.* **195,** 357 (1951).
[9] M. W. McGowan, J. D. Artiss, and B. Zak, *Microchem. J.* **27,** 564 (1982).
[10] B. W. Steele, D. F. Koehler, M. M. Azar, T. P. Blaszkowski, K. Kupa, and M. E. Dempsey, *Clin. Chem.* **22,** 98 (1976).
[11] D. A. Wiebe and J. T. Bernert, Jr., *Clin. Chem.* **30,** 352 (1984).
[12] W. Miner-Williams, *Clin. Chim. Acta* **101,** 77 (1980).

and interference.[13] In spite of these and other potential problems, there are enzymatic methods that have been demonstrated to give excellent accuracy and precision.

The methods described below were primarily developed for cholesterol measurement in human plasma and lipoprotein fractions and are well suited for high-volume screening applications. Two reagents are described, one a commercial kit method[14] (Boehringer–Mannheim High Performance) and one prepared in-house from the constituents (modified CDC).[15] For both reagents an application for the ABA 200 chemical analyzer and a manual procedure are given.

Modified CDC Method

We adapted this enzymatic cholesterol method from a procedure[16] developed by the Clinical Chemistry Standardization Section of the Centers for Disease Control (CDC) in an unsuccessful effort to define an enzymatic reference method for cholesterol. Our primary objective was to define an optimized reagent suitable for the ABA 200 bichromatic analyzer. The major modifications from the CDC version were the following:

1. A *Streptomyces* cholesterol oxidase enzyme (Calbiochem-Behring, San Diego, CA) was substituted for the *Nocardia* oxidase (Boehringer–Mannheim, Indianapolis, IN) used by CDC. The *Streptomyces* enzyme is similar to the latter in stability but considerably less expensive. Half the original CDC concentration was adequate for reaching a stable endpoint in 10 min.

2. A 2-fold increase in horseradish peroxidase concentration seemed to minimize the overall bilirubin interference. This change may diminish the negative component of the bilirubin interference by diverting relatively less H_2O_2 to oxidation of bilirubin.

3. A decrease in phenol concentration eliminated a problem of sporadic turbidity in the reaction mixture.

4. A 2-fold increase in 4-aminoantipyrine concentration improved color yield and sensitivity.

[13] P. N. M. Demacker, G. J. M. Boerma, H. Baadenhuijsen, R. van Strik, B. Leijnse, and A. P. Jansen, *Clin. Chem.* **29**, 1916 (1983).

[14] J. Siedel, E. O. Hagele, J. Ziegenhorn, and A. W. Wahlefeld, *Clin. Chem.* **29**, 1075 (1983).

[15] G. R. Warnick, J. Benderson, and J. J. Albers, *in* "Selected Methods of Clinical Chemistry" (G. R. Cooper, ed.), Vol. 10, p. 91. Am. Assoc. Clin. Chem., Washington, D.C., 1983.

[16] G. R. Cooper, P. H. Duncan, J. S. Hazelhurst, D. T. Miller, and D. D. Bayse, *in* "Selected Methods for the Small Clinical Laboratory" (W. R. Faulkner and S. Meites, eds.), p. 165. Am. Assoc. Clin. Chem., Washington, D.C., 1982.

5. A PIPES buffer preparation from Research Organics (Cleveland, OII) was pH adjusted and more easily dissolved.

Although the cholesteryl ester hydrolase used in this method does not achieve complete hydrolysis of cholesteryl esters, results are accurate, provided an appropriate secondary serum-based standard is used as described subsequently.

Reagents

Microbial cholesterol ester hydrolase (EC 3.1.13) was obtained from Boehringer–Mannheim Biochemicals, Indianapolis, IN (cat. no. 393916), and microbial cholesterol oxidase (EC 1.1.3.6) from Calbiochem-Behring, San Diego, CA (cat. no. 228250). PIPES buffer (1,4-piperazinediethane-sulfonate) was obtained from Research Organics, Cleveland, OH. Horseradish peroxidase, type VI (EC 1.11.1.7), 4-aminoantipyrine, and sodium cholate were purchased from Sigma Chemical Co., St. Louis, MO; phenol was obtained from J. T. Baker Chemical Co., Phillipsburg, NJ; and Triton X-100 from Rohm and Haas, Philadelphia, PA.

Prepare the following reagents:

PIPES Buffer, 50 mM, pH 6.9. Into a 1000-ml glass beaker containing approximately 800 ml of distilled water, add 17.7 g of PIPES buffer and stir for approximately 10 min until the solution clears. Equilibrate the solution to 37° and adjust, if necessary, to pH 6.9. Quantitatively transfer the solution to a 1000-ml volumetric flask; after cooling to ambient temperature, adjust the volume to 1000 ml.

Stock Mixed Reagent. The following reagents are dissolved in 400 ml of PIPES buffer in a 500-ml volumetric flask: 4-aminoantipyrine, 0.508 g; sodium cholate, 1.292 g; potassium chloride, 7.46 g; and Triton X-100, 1.0 ml. Adjust the final volume to 500 ml with PIPES buffer. (*Note:* This solution is stable for at least 1 month at 4°.)

Stock Phenol Reagent. Prepare from phenol stored in a desiccator. Carefully but quickly weigh 2.9 g of phenol crystals and dissolve in PIPES buffer in a 500-ml volumetric flask; bring to volume with the buffer. (*Note:* This reagent has been stored successfully for as long as a month at 4° in a tightly closed glass container.)

Working Reagent. Mix 50 ml of the stock mixed reagent and 50 ml of the stock phenol reagent. Add 12.5 U of cholesterol oxidase, 25 U of cholesterol esterase, and 2500 U of peroxidase (based on the respective specific enzyme activities), either from concentrated solutions of the enzymes or by weighing the dry enzyme preparations.

Alternatively, the enzymes can be added to the stock mixed reagent at the time of initial preparation. In this instance, however, the stock mixed reagent should be prepared fresh daily. Volumes can be adjusted, depending on the number of specimens to be analyzed.

Calibrators

Note. In our experience calibration of the enzymatic cholesterol methods with primary standards of unesterified cholesterol in water or alcohol solution results in low cholesterol values for plasma specimens, as stated previously. Therefore, we recommend the use of secondary serum-based calibration standards with analyte and matrix properties similar to those of the specimens. The secondary standard must have an accurate target value which has been determined by an accepted reference technique. We recommend at the least, high-normal and mid-level standards, each analyzed in duplicate.

A suitable secondary calibrator is Standard Reference Material 909, a human serum pool available from the National Bureau of Standards, Washington DC. After reconstitution, aliquots can be frozen in sealed vials. This material has a target value for cholesterol assigned by an assay involving isotope dilution-mass spectrometry, which is a candidate Definitive Method for cholesterol.[17] Laboratories performing analyses for support of research programs of the U.S. National Heart, Lung, and Blood Institute (NHLBI) may be eligible to receive calibration and control materials from the Clinical Chemistry Standardization Section, Centers for Disease Control, Atlanta, GA. Alternatively, an in-house pool of human donor or pooled serum or plasma can be prepared as described previously[18]; if necessary, dilute with albumin solution (60 g/liter) to obtain appropriate cholesterol levels. Establish a target value by a suitable reference method. A secondary standard for HDL analysis can be prepared from the commercial pool, Lipid Fraction Control Serum (Cooper Biomedical, Inc., Malvera, PA). After reconstituting, dilute with an equal volume of 60 g/liter albumin, 0.15 M NaCl solution, and freeze aliquots in sealed vials. The target value for this pool is approximately 160 mg/dl as determined by the CDC modified Abell–Kendall method[5] and the dilution yields a value of 80 mg/dl, appropriate for HDL specimens.

Manual Analysis Procedure

1. Turn on the spectrophotometer and adjust the wavelength to 500 nm.

---

[17] A. Cohen, H. S. Hertz, J. Mandel, R. C. Paule, R. Schaffer, L. T. Sniegoski, T. Sun, M. J. Welck, and E. White, *Clin. Chem.* **26,** 854 (1980).

[18] G. R. Warnick, C. Mayfield, and J. J. Albers, *Clin. Chem.* **27,** 116 (1981).

2. Label tubes for water blank, calibrators, control materials, and specimens.
3. Dispense 2.0 ml of enzymatic cholesterol reagent into each of the tubes and place them in an ice bath.
4. Equilibrate specimens to room temperature. Add 20 μl of water, calibrator, control material, or supernatant to the appropriate tubes and mix thoroughly. (Add proportionately larger volumes for low cholesterol specimens; 100 μl for HDL supernatants.)

Note: Mix specimens thoroughly before pipetting to assure homogeneity.

5. Transfer tubes to a water bath at 37° for 10 min.
6. Cool tubes to room temperature and within 15 min read absorbance after adjusting spectrophotometer to zero with the water blank.
7. Calculate cholesterol in specimens and control materials in relation to the calibrator.

$$\text{Cholesterol unknown} = \frac{\text{absorbance of unknown}}{\text{absorbance of calibrator}} \times \text{cholesterol calibrator}$$

ABA 200 Procedure

Instrument Parameters. Sample volume, 5 μl for total and 10 μl for HDL cholesterol; reagent volume, 500 μl for total and 250 μl for HDL cholesterol; filter, 500/600 nm; temperature, 37°; analysis time, 10 min. Use a water reagent blank.

Analysis Procedure. Equilibrate EDTA plasma specimens and pools to room temperature. Pipet 50 μl of each specimen into specimen cups and overlay with 10 μl Sera Seal (silicone oil). Continue the analysis according to the usual operating procedures for this instrument.

Performance

Performance on quality control materials for this method applied on the ABA 200 analyzer is shown in Table I. The method demonstrates good precision with coefficients of variation (CVs) for total cholesterol analysis of 1–2%. For HDL cholesterol analysis, standard deviations are in the 1–2 mg/dl range. The method meets the standardization requirements of the CDC NHLBI Lipoprotein Standardization Program[4] (Table I). In terms of accuracy, the method demonstrates good agreement with other established procedures. On 309 plasma specimens, the relationship by linear

TABLE I
PERFORMANCE ON QUALITY CONTROL MATERIALS AND STANDARDIZATION PROGRAM

A. Performance on bench-quality control pools

Quality control pool	Reference value (RV)	Number	Mean	Deviation from RV	SD	Coefficient of variation (CV)
			Total cholesterol			
Q9	188.0 mg/dl	83	187.2 mg/dl	−0.8 mg/dl	2.1 mg/dl	1.1%
Q12	286.0	79	283.1	−2.9	4.4	1.6
			Low total cholesterol			
MQ4	52.0	122	51.0	−1.0	1.4	2.6
			HDL cholesterol			
AQ8	54.0	220	52.4	−1.6	1.6	3.1
DS-C		120	49.0		1.5	3.1

B. Performance on CDC/NHLBI lipoprotein standardization program (in mg/dl)

	Part I					Part II			
Pool ID	Accepted mean	Our mean	Accepted SD	Our SD	Pool ID	Accepted mean	Our mean	Accepted SD	Our SD
				Total cholesterol					
30	162 ± 5	160	4.86	1.90	30	162 ± 5	162	5.0	1.70
34	229 ± 7	226	6.87	0.49	34	229 ± 7	229	7.0	2.02
46	330 ± 10	326	9.90	3.37	40	305 ± 9	303	9.0	3.07
					48	349 ± 10	350	10.0	2.87
				Low total cholesterol					
41	28.0 ± 2.0	27.9	2.0	1.02	41	28.0 ± 2.0	28.1	2.0	0.57
61	40.4 ± 2.0	39.6	2.0	1.37	37	42.8 ± 2.0	42.9	2.0	0.81
63	70.2 ± 3.0	68.8	3.0	1.73	43	50.8 ± 3.0	50.9	3.0	1.09
					39	72.7 ± 3.0	71.9	3.0	1.33
				HDL cholesterol					
35	26.5 ± 2.5	27.6	2.50	1.01	35	26.5 ± 2.6	26.6	2.5	0.72
47	35.1 ± 2.5	34.0	2.50	1.20	47	35.1 ± 3.5	33.1	2.5	0.74
49	50.7 ± 3.0	48.1	3.00	0.84	49	50.2 ± 5.0	46.5	3.0	1.01
					51	63.0 ± 6.3	58.2	3.5	0.73

regression with the Lipid Research Clinics Liebermann–Burchard reagent method for the Auto Analyzer II[4] was as follows:

$$\text{Modified CDC} = 1.03 \text{ LRC} + 1.4 \text{ mg/dl}, r = 0.997$$

Compared to the CDC Abell–Kendall reference procedure[5] on 54 specimens, the relationship was as follows:

$$\text{Modified CDC} = 1.02 \text{ Abell–Kendall} - 1.5 \text{ mg/dl}, r = 0.999$$

On 129 HDL cholesterol supernatants (dextran sulfate–Mg^{2+} precipitation)[15] the relationship by linear regression with the LRC cholesterol method was as follows:

$$\text{Modified CDC} = 1.04 \text{ LRC} + 0.7 \text{ mg/dl}, r = 0.990$$

In addition, HDL cholesterol in dextran sulfate–Mg^{2+} supernatants[15] analyzed by the modified CDC enzymatic method was compared to HDL cholesterol values obtained in heparin–Mn^{2+} supernatants analyzed by the CDC Abell–Kendall reference procedure:[5]

$$\text{DS} - \text{modified CDC} = 1.01 \text{ HM–Abell–Kendall} + 0.9 \text{ mg/dl}, r = 0.992$$

In summary, this reagent gives excellent performance when used as described with a bichromatic analyzer. The reagent could be used successfully with other analyzers. With manual application, precision may be somewhat less. A secondary serum-based standard must be used for calibration to obtain accurate results. Calibration with a primary standard of free cholesterol in water–detergent solution produced results about 5% below reference values.

Commercial Kit Method for Cholesterol Analysis

A commercial kit method, High-Performance Monotest Cholesterol (Boehringer–Mannheim Diagnostics, Indianapolis, IN), is available with a *Pseudomonas* cholesterol ester hydrolase and reaction conditions that achieve nearly complete hydrolysis of cholesterol esters.[11,14] This method appears to give accurate results calibrated with a primary standard of free cholesterol.[19] (*Note:* We have evaluated some others, but not all of the many commercial kit methods for cholesterol. There may be other commercial reagents that achieve full hydrolysis and demonstrate similar good performance. This particular kit was selected for evaluation because the significant aspects of the development and validation were reported in

[19] G. R. Warnick and C. Lum, *in* "Recent Advances in Lipid and Lipoprotein Analysis" (G. R. Warnick, ed.), p. 17. National Meeting, Am. Assoc. Clin. Chem., Indianapolis, 1983.

the scientific literature and supported the manufacturer's claims of full hydrolysis and improved performance.)

Reagent

According to the package insert the constituents of the reagent are as follows:

Constituent	Concentrations
Tris buffer	100 mM; pH 7.7
Magnesium aspartate	50 mM
4-Aminophenazone	1 mM
Phenol	6 mM
3,4-Dichlorophenol	4 mM
Hydroxypolyethoxy-N-alkanes	0.3%
Cholesterol esterase	≥0.4 U/ml
Cholesterol oxidase	≥0.25 U/ml
Peroxidase	≥0.2 U/ml

Manual Procedure

This reagent can be used with a manual procedure similar to that described previously for the modified CDC method. According to the package insert and our experience, incubation for 5 min at 37° or 10 min at 25° is sufficient to achieve full color development.

ABA 200 Procedure

Instrument parameters are identical to those described above for the modified CDC method. Calibration was accomplished with a commercial primary standard of free cholesterol in water–detergent solution (Preciset Series, Boehringer–Mannheim Diagnostics). Standards of 300 and 150 mg/dl were used for total cholesterol and 100 and 50 mg/dl for HDL cholesterol.

Performance

The precision of the high-performance reagent used with the ABA-200 instrument is shown in Table II. For total cholesterol on bench quality control pools, Q9 and Q12, the accuracy was within approximately 1% of reference values and the precision was excellent (with CVs near 1%), similar to that observed with the modified CDC enzymatic method. On the low total cholesterol pool, MQ4, which has a cholesterol level in the HDL range but is not subjected to the precipitation step, the accuracy was

TABLE II
BMD High-Performance Enzymatic Performance on Bench Quality
Control Pools

Quality control pool	Reference value (RV) (mg/dl)	Number	Mean (mg/dl)	Deviation from RV (mg/dl)	SD (mg/dl)	CV (%)
Total cholesterol						
Q9	188	25	190	2	2.5	1.3
Q12	286	25	286	0	3.3	1.2
Low total cholesterol						
MQ4	52	17	52	0	2.0	3.8
HDL cholesterol						
AQ8	54	12	53	1	1.5	2.8
DS-C		8	49		1.5	3.0

good and the CV slightly higher than for total cholesterol. On the HDL cholesterol pools, AQ8 and DS-C, which involve both precipitation and cholesterol quantification, the CVs ranged from 2 to 3%, similar to that observed with the previous enzymatic method.

In terms of accuracy, the high-performance method was in good agreement with the CDC reference method,[5] with the LRC method,[4] and with the modified CDC enzymatic method[15] (Table III).

Triglycerides

Triglyceride methods generally quantify the glycerol moiety released by either chemical or enzymatic hydrolysis of the triglycerides. Any free glycerol in specimens, unless removed by extraction or corrected for by a blanking procedure, is included in the estimation of the triglycerides level. Mean levels for free glycerol ranging to approximately 0.15 M (approximately 10–14 mg/dl as triolein)[20,21] have been reported. In certain instances, free glycerol levels may be considerably higher, for example, in subjects receiving glycerol-containing medications or infusions, in diabetics, and in certain animal species.[22]

[20] K. Stinshoff, D. Weishaar, F. Staehler, D. Hesse, W. Gruber, and E. Steier, *Clin. Chem.* **23,** 1029 (1977).
[21] H. F. ter Welle, T. Baartscheer, and J. W. T. Fiolet, *Clin. Chem.* **30,** 1102 (1984).
[22] H. K. Naito and J. A. David, *in* "Lipid Research Methodology" (J. A. Story, ed.) p. 1. Liss, New York, 1984.

TABLE III
COMPARISON OF BOEHRINGER–MANNHEIM METHOD (y) WITH
REFERENCE PROCEDURE

Comparison method (x)	(n)	Slope	y-Intercept (mg/dl)	Correlation coefficient
Total cholesterol				
Modified CDC	127	1.01	−1.66	0.996
LRC	188	1.07	−7.65	0.996
CDC Abell–Kendall	43	1.03	−2.85	0.999
HDL cholesterol				
Modified CDC[a]	48	.95	1.56	0.991
LRC[a]	67	1.02	1.07	0.992
CDC Abell–Kendall[b]	42	1.02	−1.04	0.993

[a] Cholesterol analyzed by both methods in dextran sulfate–Mg^{2+} supernatants.

[b] Cholesterol in dextran sulfate–Mg^{2+} supernatants by Boehringer–Mannheim method and in heparin–Mn^{2+} supernatants by Abell–Kendall reference method.

Some investigators[20] have suggested that glycerol correction is unnecessary or can be accomplished by a constant correction factor. While this may be valid for the routine clinical laboratory, a specialized lipid research laboratory would be advised to obtain true or net triglyceride values. Free glycerol blank correction is essential to meet the accuracy requirements of the CDC Lipoprotein Standardization Program.[4]

Since the specificity of the enzymatic triglyceride methods is such that extraction is necessary, a practical approach to blanking is to measure and subtract the free glycerol value. This can be accomplished by a sec-

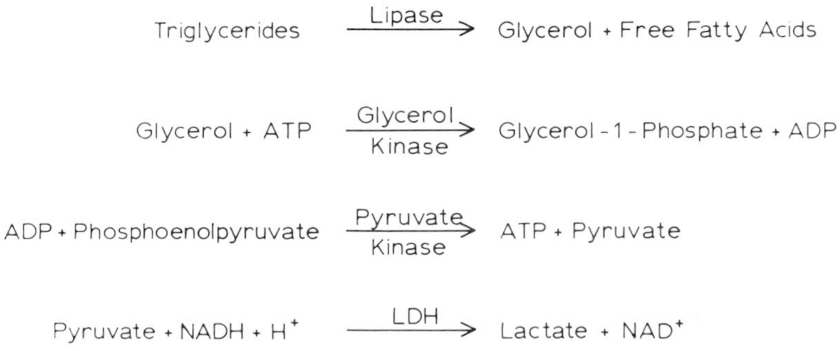

Triglycerides $\xrightarrow{\text{Lipase}}$ Glycerol + Free Fatty Acids

Glycerol + ATP $\xrightarrow{\text{Glycerol Kinase}}$ Glycerol-1-Phosphate + ADP

ADP + Phosphoenolpyruvate $\xrightarrow{\text{Pyruvate Kinase}}$ ATP + Pyruvate

Pyruvate + NADH + H$^+$ $\xrightarrow{\text{LDH}}$ Lactate + NAD$^+$

FIG. 2. Abbott Enzymatic triglyceride method (UV 340 method).

Triglycerides $\xrightarrow{\text{Lipase}}$ Glycerol + Free Fatty Acids

Glycerol + ATP $\xrightarrow[\text{Kinase}]{\text{Glycerol}}$ Glycerol-1-PO$_4$ + ADP

Glycerol-1-PO$_4$ + NAD$^+$ $\xrightarrow[\text{Dehydrogenase}]{\text{Glycerol PO}_4}$ NADH + H$^+$ + Dihydroxyacetone PO$_4$

NADH + H$^+$ + INT $\xrightarrow{\text{Diaphorase}}$ Formazan (pink) + NAD$^+$

FIG. 3. INT-colorimetric method.

ond analysis deleting the lipase enzyme from the usual triglyceride reagent. Total triglyceride plus free glycerol is measured with the complete reagent, the glycerol blank measured with a lipase-free reagent, and net triglyceride calculated by difference. One problem is that triglyceride-related turbidity is not cleared in the specimen–reagent mixture in the absence of the lipase enzyme. The turbidity may produce photometric interference and an error in the blank measurement. A bichromatic or dual wavelength analyzer, which corrects to some extent for specimen turbidity, appears to be less subject to this type of interference and well suited for the enzymatic triglyceride methods.

The enzymatic triglyceride methods, in addition to using a lipase for triglyceride hydrolysis, all include a second enzyme, glycerol kinase, which produces glycerol 3-phosphate. The first commonly available enzymatic reagent[23] included pyruvate kinase and lactate dehydrogenase enzymes (Fig. 2). The reaction was monitored in terms of NADH disappearance at 340 nm; hence the usual designation for this reaction sequence is UV.

The next common enzymatic reaction sequence for triglycerides included the lipase and glycerol kinase enzymes, together with glycerol phosphate dehydrogenase and diaphorase enzymes,[24] which produce a pink formazan product detected at 500 nm (Fig. 3). This reaction is commonly designated INT, the acronym for the color reaction substrate. In our experience, this chromophore is unstable and sensitive to matrix effects, such as the concentrations of protein and detergents in the specimen. In an evaluation of various enzymatic reagents, results with this type of reagent were more variable than those by the UV method and in poor agreement with a reference method.

[23] G. Bucolo and H. David, *Clin. Chem.* **19,** 476 (1973).
[24] R. E. Megraw, D. E. Dunn, and H. G. Biggs, *Clin. Chem.* **25,** 273 (1979).

Another recently reported reaction sequence uses lipase and glycerol kinase enzymes coupled with glycerol phosphate oxidase to produce hydrogen peroxide.[25] This sequence (designated GPO) is coupled to the common Trinder reaction with the peroxidase enzyme (Fig. 4).

We have adapted for glycerol blanking a commercial reagent of the UV type, A-gent triglyceride reagent (Abbott Laboratories, South Pasadena, CA). A lipase-free version of the standard reagent, now available from Abbott, is used for free glycerol measurement. This method has demonstrated excellent performance in our laboratory, and when applied on the ABA 200 has met the requirements of the CDC-NHLBI Lipoprotein Standardization Program. Another commercial reagent, the Enzymatic Triglycerides-Glycerol 3 Vial Stat Pack (Calbiochem-Behring, La Jolla, CA) is of the same reaction type and can be reconstituted with and without the lipase enzyme for glycerol blanking. We have not evaluated the latter method; an application on another bichromatic analyzer reportedly met the CDC Standardization requirements.[26]

A mixture of constituents for a GPO method, with and without the lipase enzyme, can be obtained from Fermco Diagnostics, Elk Grove Village, IL. An application for a UV method on the ABA 200 bichromatic analyzer and a manual procedure for a GPO reagent will be presented here.

ABA 200, UV Method

Reagents. According to the package insert, the A-gent Triglycerides Kit has the following constituents:

Constituent	Concentrations	
Lipase	250,000	U/l
Lactic dehydrogenase	1,000	U/l
Pyruvate kinase	1,667	U/l
Glycerol kinase	667	U/l
NADH, Na$_2$	0.37	U/l
Phosphenolypyruvate (tricyclohexylammonium salt)	0.72	mM
ATP, Na$_2$	0.06	mM
MgSO$_4$	5.5	mM
Tris (hydroxymethyl aminomethane)	101	mM
Succinic acid	26	mM

[25] M. W. McGowan, J. D. Artiss, D. R. Strandbergh, and B. Zak, *Clin. Chem.* **29,** 538 (1983).

[26] D. F. Koehler, B. W. Steele, M. M. Azar, K. Kuba, and M. E. Dempsey, *Clin. Chem.* **24,** 326 (1978).

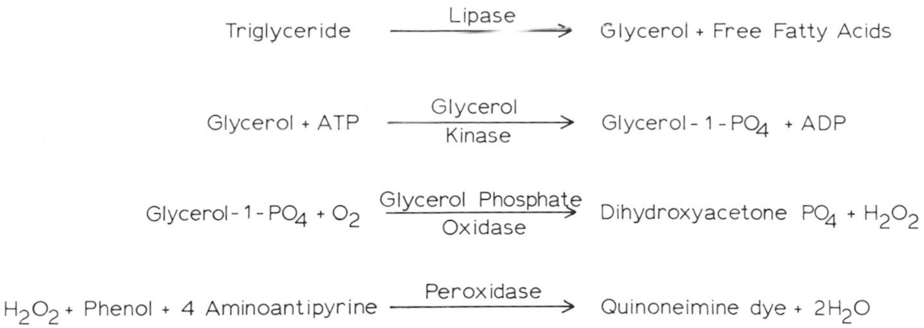

Triglyceride $\xrightarrow{\text{Lipase}}$ Glycerol + Free Fatty Acids

Glycerol + ATP $\xrightarrow[\text{Kinase}]{\text{Glycerol}}$ Glycerol-1-PO_4 + ADP

Glycerol-1-PO_4 + O_2 $\xrightarrow[\text{Oxidase}]{\text{Glycerol Phosphate}}$ Dihydroxyacetone PO_4 + H_2O_2

H_2O_2 + Phenol + 4 Aminoantipyrine $\xrightarrow{\text{Peroxidase}}$ Quinoneimine dye + $2H_2O$

FIG. 4. GPO-colorimetric method.

The glycerol blanking reagent (A-gent free glycerol) is the same as above, except the lipase enzyme has been deleted.

According to the package insert, these reagents are stable at 2–8° for 10 days; in our experience use within 2–3 days is recommended.

Calibration Standards. Glycerol standards of 100, 300, and 500 mg/dl (triolein equivalent) are prepared. Dissolve glycerol (Fisher Scientific, Pittsburgh, PA, is suitable) at the rate of 5.20 g/liter in distilled water, adjusting appropriately for purity, to give a 50,000 mg/dl as triolein stock solution. Dilute with distilled water to give the above concentrations. We have found it convenient to freeze aliquots of approximately 150 μl in securely sealed vials, which are stored at −15° to avoid deterioration. Thaw and mix aliquots thoroughly before use.

Instrument Parameters. Sample volume, 5 μl; reagent volume, 250 μl; filter, 340/380 nm; temperature, 37°; analysis time, 10 min. Use a water reagent blank and a sample blank.

Analysis Procedure. Equilibrate specimen and standards to room temperature and pipet 50 μl into ABA 200 cups. Cover with 10 μl Sera Seal (silicone oil). Continue the analysis according to the usual operating procedure for this instrument. Total triglyceride values (triglyceride plus free glycerol) are determined with the lipase-containing reagent. After completing the analysis, this reagent is flushed from the system and replaced with the lipase-free reagent. Net triglyceride is calculated as the difference between total triglyceride (with lipase) and free glycerol (without lipase).

According to the manufacturer, this reagent can be used in a manual procedure, which is described in the package insert. Since we have not evaluated such a procedure, the details are not given here.

Performance. Performance of the UV reagent on the ABA 200 instrument was acceptable. Accuracy and precision on known quality control

TABLE IV
ANALYTICAL PERFORMANCE OF THE ABBOTT ENZYMATIC
TRIGLYCERIDE METHOD WITH THE ABA 200
BICHROMATIC ANALYZER

A. Bench (known)-quality control pools (in mg/dl)

Quality control pool	Reference value (RV)[a]	Number	Mean	Deviation from RV	SD	CV %
Q9	164	107	161.8	−2.2	3.8	1.3
Q12	251	114	258.7	7.1	5.4	1.9
Q14	222	26	224.5	2.5	5.8	2.0

B. CDC/NHLBI Lipoprotein Standardization Program (in mg/dl)

Pool ID	Accepted mean[a]	Our mean	Accepted SD	Our SD
Part I				
30	73 ± 9	70	7	1
34	140 ± 10	141	8	1.8
46	270 ± 13	296	13	8
Part II				
44	55 ± 9	49	7	2
70	96 ± 10	89	8	4
88	121 ± 10	123	8	3
42	245 ± 12	256	12	8

[a] By Loffland method at Lipid Standardization Laboratory, Centers for Disease Control.

pools is shown in Table IV. Precision was excellent with CVs in the range of 1 to 2%. This level of precision was obtained in the net triglyceride result, which is the difference between two measurements, total and free glycerol. In terms of accuracy, the method was in good agreement with the Loffland procedure,[27] a reference for triglycerides, performed by the Clinical Chemistry Standardization Section of the Centers for Disease Control. Accuracy on patient specimens was excellent compared to the Lipid Research Clinics method[4]; the relationship by linear regression was as follows: Abbott = 1.00 LRC − 4.0 mg/dl with a correlation coefficient of 0.997 on 87 specimens.

[27] H. B. Lofland, Jr., *Anal. Biochem.* **9,** 303 (1964).

Manual Method (from McGowan et al.)[25]

Reagents. Lipase from *Chromobacterium viscosum* and L-α-glycerophosphate oxidase from *Aerococcus viridans* were obtained from Fermco Biochemics, Inc., Elk Grove Village, IL. Glycerokinase from *Escherichia coli*, peroxidase from horseradish, 4-aminoantipyrine, and glycerol were from Sigma Chemical Co., St. Louis, MO.

Sodium 2-hydroxy-3,5-dichlorobenzenesulfonate were obtained from either Research Organics, Inc., Cleveland, OH, or Biosynth International, Skokie, IL.

Working Reagents. The triglyceride reagent was prepared in Tris-HCl buffer (50 mM, pH 7.6) to contain per liter, 0.1 g of Triton X-100, 1 mmol of 4-aminoantipyrine, 1.5 mmol of sodium 2-hydroxy-3,5-dichloroben-zenesulfonate, 5 mmol of $MgCl_2$, 0.5 mmol of ATP, 10 kU of peroxidase, 4 kU of glycerophosphate oxidase, 0.25 kU of glycerokinase, and 100 kU of lipase. This reagent is stable for 24 hr at 4°.

Standards were prepared from glycerol in deionized water.

Procedure

1. Pipet 1.0 ml of the enzymatic triglyceride reagent into appropriately labeled test tubes.
2. Add 10 μl of specimen or glycerol standard.
3. Mix well and incubate at 37° for at least 13 min.
4 Measure the absorbances of samples and standard at 510 nm vs a reagent blank in which deionized water is substituted for sample.

The method reportedly is linear to 1000 mg/dl triglycerides. Dilute overrange samples with a 9.0 g/liter saline solution.

A prepared mixture of the reagents for this method can be obtained from Fermco Diagnostics, Elk Grove Village, IL. A lipase-free reagent for glycerol blanking is also available.

Performance. McGowan et al.[25] reported CVs of 2.3 to 3.3% for the manual method in total triglyceride analysis. The precision of the manual method with a glycerol blanking step would be somewhat poorer. In terms of accuracy this method was compared to the Abbott UV reagent method for total triglyceride with the following relationship:

$$\text{Abbott} = 1.06 \text{ GPO} + 1.96 \text{ mg/dl}, \, r = 0.995$$

The authors[25] attributed the negative bias to overestimation by the Abbott method, resulting from triglyceride turbidity-related interference. We have not evaluated the manual version, but on the ABA 200, the GPO method gives consistently low results for net triglyceride compared to the

$$\text{Phospholipids} \begin{pmatrix} \text{Lecithin} \\ \text{Sphingomyelin} \\ \text{Lysolecithin} \end{pmatrix} + H_2O$$

$$\xrightarrow{\text{Phospholipase D}} \text{Choline} + \begin{cases} \text{Phosphatidic acid} \\ \text{N-Acylsphingocyl phosphate} \\ \text{Lysophosphatidic acid} \end{cases}$$

$$\text{Choline} \xrightarrow{\text{Choline Oxidase}} 2H_2O_2 + \text{Betaine}$$

$$\text{4 - Aminoantipyrine} + \text{Phenol} + 2H_2O_2 \xrightarrow{\text{Peroxidase}} \text{Quinoneimine} + 4H_2O$$

FIG. 5. Enzymatic phospholipid reactions.

Abbott method, the latter having demonstrated excellent accuracy as shown above. The observed negative bias of the GPO method may be due to interference from reducing substances. Interference effects would be expected to be more severe compared to the cholesterol assay, because the usual molar concentration of triglycerides is lower.

Measurement of Phospholipids

The major types of phospholipids in human serum are the glycerophospholipids, including lecithin and lysolecithin, and sphingomyelin, a sphingophospholipid. Comprising approximately 95% of total serum phospholipids, these three types all contain choline, which is the basis for an enzymatic assay[28,29] approximating total plasma phospholipids (Fig. 5).

Phospholipase D from *Streptomyces chromofuscus,* which cleaves choline from the three aforementioned phospholipids, as well as ethanolamine from phosphotidyl ethanolamine, is stimulated by the presence of Ca^{2+} and Triton X-100. The second enzyme in the sequence, choline oxidase, from *Arthobacter globiformis,* produces H_2O_2, which is coupled to the well-known Trinder reaction. Different phospholipases can be coupled with other enzyme systems to measure the individual phospholipids. However, only the above mentioned for total choline containing phospholipids will be described in detail here.

Because this phospholipid reaction sequence is coupled to the Trinder reaction, the system is subject to interferences similar to those described previously in relation to the enzymatic cholesterol method. Bilirubin has both positive and negative components of interference and other reducing substances; e.g., ascorbic acid may give negative interference.

[28] M. Takayama, S. Itoh, T. Nagasaki, and I. Tanimizu, *Clin. Chim. Acta* **79,** 93 (1977).
[29] M. W. McGowan, J. D. Artiss, and B. Zak, *J. Clin. Chem. Clin. Biochem.* **20,** 807 (1982).

Enzymatic reagent kits for phospholipid analysis are available from commercial sources (Wako Pure Chemical Industries, Osaka, 541 Japan, and Dallas, TX; Boehringer–Mannheim GmbH, Mannheim, D-6800 West Germany, and Indianapolis, IN).

An in-house reagent adapted from one described by McGowan et al.[29] will be explained here. Procedures are given for manual assay and for the ABA 200 analyzer with dilutions appropriate for assay of plasma specimens. The method can be adjusted for other types of specimens and for a variety of automated chemical analyzers.

Reagents

Phospholipase D and choline oxidase can be obtained from Fermco Biochemics, Inc. (Elk Grove Village, IL). Horseradish peroxidase, 4-aminoantipyrine, phenol, L-α-phosphatidylcholine, Tris-hydrochloride, and Triton X-100 are available from Sigma Chemical Co. (St. Louis, MO). Biosynth International (Skokie, IL) is the source for sodium 2-hydroxy-3,5-benzenesulfurate. Calcium chloride dihydrate is available from Mallinckrodt, Inc. (Paris, KY).

Stock Mixed Reagent. Prepare a Tris-HCl buffer, 50 mM, pH 7.8, containing 3.6 g/liter Triton X-100, 5.3 mM 4-aminoantipyrine (1.076 g/liter), and 909 μM CaCl$_2$ · 2H$_2$O (0.1336 g/liter). This reagent is stable at 4° for at least 1 month.

Stock Phenol Reagent. Prepare in Tris-HCl buffer (pH 7.8, 50 mM) a solution of phenol (60 mM, stored in a desiccator). Quickly weigh phenol crystals (4.9 g/liter); dissolve and adjust to volume with Tris buffer. The sensitivity of the color reaction can be increased approximately 4-fold by substituting 2-hydroxy-3,5-dichlorobenzene sulfonate (HDCBS) for phenol. The latter color reagent can be recommended for the analysis of phospholipid in HDL supernatants or other preparations with relatively low phospholipid concentrations. Prepare a 28 mM solution (7.42 g/liter) of the reagent, which can be stored for at least 1 month in a tightly closed glass container.

Working Reagent. Mix equal volumes of the stock-mixed reagent and stock phenol (or HDCBS) reagent. Add the enzymes at the following concentrations: phospholipase D, 1 kU/liter, choline oxidase 1 kU/liter, peroxidase 10 kU/liter. The working reagent is stable at 4° for 1 week.

Calibration Standards. Dissolve in distilled water containing 5 g/liter Triton X-100, L-α-phosphatidylcholine (dipalmitoyl, synthetic) to a concentration of 6 g/liter. Some lecithin preparations are difficult to dissolve; alternatively, a choline standard may be used. Prepare dilutions with the same Triton solution of 4, 2, and 1 g/liter. For analysis of low phospholipid specimens, e.g., HDL supernatants, prepare solutions of 1.0, 0.5, 0.2, and 0.1 g/liter. This standard of lecithin must be kept at room temperature to avoid precipitation and prepared fresh weekly.

Control Materials. There is no generally accepted reference procedure that has been thoroughly validated for quantification of phospholipids. Thus, the accuracy base for phospholipid measurement is not established to the same extent as that for cholesterol. Some commercial control materials have assay values for total phospholipids and can be used as a guideline for assessment of accuracy.

Manual Analysis Procedure

1. Turn on spectrophotometer and set to 500 nm.
2. Label tubes for water (reagent) blank, calibrators, control materials, and specimens.
3. Dispense 1 ml of working phospholipids reagent into tube and place tube in an ice bath.
4. Add 10 μl each of water (reagent blank) calibrator, control material, or specimen to the appropriate tubes and mix thoroughly. Increase specimen volume or use HDCBS reagent for low-level specimens.
5. Transfer tubes to water bath at 37° for 15 min.
6. Cool tubes to room temperature, and within 15 min measure absorbance at 500 nm after adjusting spectrophotometer to zero with the reagent blank.
7. Calculate cholesterol in unknowns in relation to the calibrator.

Instrument Parameters for ABA 200 Assay. Sample volume, 5 μl for total and 15 μl for HDL supernatants; reagent volume, 500 μl for total and 250 μl for HDL supernatants; filter, 500/600 nm; temperature, 37°; analysis time, 15 min. Use a water reagent blank.

Analysis was conducted as described previously for the cholesterol and triglyceride methods.

Performance. Performance in phospholipid analysis by the reagent described here with the ABA 200 instrument was compared to a modified Bartlett procedure.[30] For among-run precision ($n = 6$), coefficients of variation (CVs) of 4.9 and 1.7% were obtained for quality control pools, with mean levels of 151 and 179 mg/dl, respectively.

The enzymatic method was in good agreement with the modified Bartlett method on specimens. The relationship by linear regression was as follows: Enzymatic = 1.01 Bartlett + 11 mg/dl. The correlation coefficient was 0.964, as determined from the comparison of 28 specimens in 3 runs.

[30] G. R. Bartlett, *J. Biol. Chem.* **234,** 466 (1959).

This enzymatic method appears to be acceptably accurate and precise for routine use. The enzymatic method, which does not require extraction or other pretreatment of specimens, is considerably more convenient than the older methods requiring extraction and digestion.

Acknowledgment

Author's research referred to in this chapter was supported by LRC Contract NO1-HV-12157-L and Grant HL-30086.

[7] Characterization of Apolipoprotein B-Containing Lipoproteins

By JOHN P. KANE

Introduction

In recent years it has become apparent that the lipoproteins that contain the B-apolipoproteins, i.e., chylomicrons, very-low-density lipoproteins (VLDL), remnant particles derived from triglyceride rich lipoproteins, low-density lipoproteins (LDL), and the Lp(a) particles, are all structurally heterogeneous. Further, considerable evidence is at hand in support of the existence of multiple discrete lipoprotein subspecies which may differ in either lipid or apolipoprotein content, within given density intervals. Thus, studies of lipoprotein metabolism must be designed to appreciate changes either in composition of individual lipoproteins or in the relative proportions of subspecies within classes of lipoproteins.

The lipoproteins in circulating blood are all in thermodynamic disequilibrium and hence are subject to certain continuing processes, such as esterification and transfer of sterol, and transfer of triglycerides, phospholipids, or apolipoproteins, after blood is removed from the body. It is important to arrest these processes as soon as blood samples are obtained if we are to perceive the true state of lipoprotein systems in vivo.

The association of the protein and lipid constituents of plasma lipoproteins is dependent upon a large number of noncovalent interactions of low energy. Hence, the methods employed in isolation of lipoproteins may introduce important changes in their composition or even cause the obliteration of natural speciation. Though the shear to which lipoproteins are subjected during ultracentrifugation is no greater than that due to diffusion, the high ionic strength and pressure to which the particles are subjected may indeed alter them. There is abundant evidence that some apoE and apoA-I, at least, is dissociated from lipoproteins during ultracentrifugation. The effects of high ionic strength can be eliminated by use of D_2O in place of KBr in adjusting the solvent density for ultracentrifugation; however, there is no practical way to reduce pressure during preparative ultracentrifugation. Polyanion precipitation, on the other hand, appears to coprecipitate some high-molecular-weight proteins which may not be normal constituents of lipoproteins in blood. A new strategy, selected affinity immunosorption,[1] may prove to be the method best suited for isolation of naturally occurring lipoprotein complexes because it allows the use of relatively nonperturbing elution systems.

Characterization of intact lipoproteins involves the application of a number of physical and chemical methods, many of which are described in other chapters in this volume. In this chapter, the focus is upon means of preventing artifactual modification of apoB-containing lipoproteins during preparation, and upon selected aspects of their characterization.

Mechanisms by Which Chemical or Physical Modification of Lipoproteins Can Occur

Shear Denaturation. Denaturation of lipoproteins by shear forces is demonstrable in capillary viscometers where the apparent intrinsic viscosity increases as a function of the number of passes through the capillary. Comparable forces exist when plasma or lipoprotein solutions are drawn rapidly into Pasteur pipets. Rapid aggregation of lipoproteins subjected to shear is observed at very low ionic strengths, presumably reflecting electroviscous effects.

Denaturation at Interface. Lipoproteins appear to be particularly susceptible to this kind of disruption, presumably due to the important role of noncovalent forces in the organization of the three-dimensional structures of the particles. The chief thermodynamic drives to disruption appear to come from the orientation of hydrophobic domains away from the aqueous phase, and the enormous electrical field discontinuity at interface which causes the reorientation of charged regions of proteins. Experiments thus should be designed to minimize the surface exposure of lipoproteins. The formation of bubbles while separating serum or manipulating lipoprotein solutions is a major cause of surface denaturation.

Disruption of Structure Attributable to Phase Transitions of Lipids. This occurs primarily in special circumstances where increased content of saturated fatty acids raises transition temperatures. Appropriate adjust-

---

[1] J. P. McVicar, S. T. Kunitake, R. L. Hamilton, and J. P. Kane, *Proc. Natl. Acad. Sci. U.S.A.* **81,** 1356 (1984).

ment of the temperature of centrifugation will generally prevent this artifact.

Hydroperoxidative Reactions Leading to Modification of Proteins and Lipids. These reactions involve polyolefinic sites, producing free radicals which can lead to cross-linking or scission of proteins. Azide catalyzes scission of proteins.[2] Because sequestration of heavy metals, particularly copper, suppresses these reactions, lipoproteins should be kept in the presence of 1 mM EDTA at all times.

Esterification of Cholesterol by LCAT. In the presence of plasma this reaction will proceed to completion at room temperature. The reaction is greatly inhibited at refrigerator temperatures, or by the addition of 1.4 mM DTNB [5,5′-dithiobis(2-nitrobenzoic acid)].

Scission of Apolipoproteins by Microbial Endoproteases. Several endoproteases of bacterial origin rapidly cleave apolipoproteins in native lipoprotein complexes. ApoB-100 is especially sensitive to such cleavage. Also, some bacteria produce lipases, phospholipases, and neuroaminidases which rapidly attack lipoproteins. At least one of these organisms (*Pseudomonas fluorescens*) is able to degrade lipoproteins rapidly at refrigerator temperatures. The author has found that virtually all bacterial degradation of lipoproteins can be prevented by inclusion of azide (0.05%) and gentamycin sulfate (0.5 mg/ml) to all fluid media with which lipoproteins come in contact. It is also very important to eliminate sources of bacterial contamination of stock solutions, centrifuge caps, etc., because preformed bacterial enzymes contaminating these solutions can degrade lipoproteins, even though bacterial growth is inhibited by the antimicrobials.

Cleavage of Apolipoproteins by Endoproteases in Plasma. ApoB-100 is known to undergo cleavage to the complementary fragments B-74 and B-26.[3] This scission, which occurs largely if not wholly *in vitro,* is most likely due to a kallikrein-like enzyme.[2] The addition of benzamidine (2 mM) to blood and to solutions in contact with the lipoproteins provides a practical means of suppressing this reaction. After purification of the lipoproteins, benzamidine can be removed by dialysis.

Dissociation of Constituents during Ultracentrifugation. Because many of the apolipoproteins exist in an equilibrium between lipoprotein-bound and -dispersed phases, dissociation tends to be favored as the dispersed phase is subjected to sedimentation. This phenomenon has been demonstrated for apoA-I and for apoE, but may apply to other constitu-

---

[2] A. D. Cardin, K. R. Witt, J. Chao, H. S. Margolius, V. H. Donaldson, and R. L. Jackson, *J. Biol. Chem.* **259,** 8522 (1984).

[3] J. A. Glomset, K. Applegate, T. M. Forte, W. C. King, C. D. Mitchell, K. R. Norum, and E. Gjone, *J. Lipid Res.* **21,** 1116 (1980).

ents of chylomicrons and VLDL. Similarly, dissociation could be favored by exposure of lipoprotein particles to an increased volume of aqueous medium as in column chromatography.

Characterization of Lipoproteins

Determination of Particle Size Distribution. Electron microscopy of negatively stained preparations, assisted by computerized digitization,[3,4] has proved a facile method for generating histograms of apparent particle diameter. In this technique, error due to flattening of particles is difficult to correct precisely, but for all but the largest triglyceride-rich particles, it is probably of a low order. (This technique is described in Chapter [26], Volume 128.)

Determination of particle weights in the fluid phase can be achieved using a combination of density gradient centrifugation and laser light scattering techniques.[5] Flotation coefficients are obtained from behavior in the gradient, whereas the diffusion coefficients are derived from intensity fluctuations of scattered laser light. Observations made in pairs of gradients with different average densities permit the calculation of particle weights and buoyant densities for each subspecies of lipoprotein as well as the concentration at every point on the gradients. This technique has the advantage that it can identify species which have the same apparent diameter but different densities, or vice versa.

Electrophoresis of Intact Lipoproteins. Study of electrophoretic heterogeneity of lipoproteins within a given ultracentrifugal density fraction has led to the discovery of important classes of lipoproteins, such as Lp(a), HDL_c, and the pre-β HDL. Except for chylomicrons, separation of intact lipoproteins in agarose gel is entirely a reflection of net particle charge.[6] It should be noted that the acquisition of charged molecular species by lipoproteins, such as the free fatty acids resulting from heparin-induced lipolysis, greatly influences the electrophoretic mobility of these complexes. Likewise, lipolysis of triglycerides in VLDL by contaminating bacteria can easily be appreciated by the appearance of sharp bands of increased mobility. Modification of the agarose gel technique has permitted the selective recovery for analysis of pre-β lipoproteins of different mobilities.[7] In general, the electrophoretic behavior of lipoproteins

[4] G. C. Chen, L. S. S. Guo, R. L. Hamilton, V. Gordon, E. G. Richards, and J. P. Kane, *Biochemistry* **23,** 6530 (1984).
[5] S. T. Kunitake, E. Loh, V. N. Schumaker, S. K. Ma, C. N. Knobler, J. P. Kane, and R. L. Hamilton, *Biochemistry* **17,** 1936 (1978).
[6] R. P. Noble, *J. Lipid Res.* **9,** 693 (1968).
[7] A. Pagnan, R. J. Havel, J. P. Kane, and L. Kotite, *J. Lipid Res.* **18,** 613 (1977).

in agarose gels can be reproduced precisely on a preparative scale in electrophoresis in starch blocks[8] or in Geon pevikon, with the exception that chylomicrons have pre-β mobility in both these media. Because the hydrated particles of these block media present an aqueous surface layer in which the lipoproteins migrate and because buffer systems of physiologic ionic strength and pH can be employed, they probably present the least perturbing environment of the available modes of separation.

Identification of Subspecies of Apo-B-Containing Lipoproteins. The identification of subspecies within fractions of lipoproteins prepared from plasma by ultracentrifugation, gel permeation, or immunosorption yields information of potential metabolic importance. Analytical ultracentrifugation has shown the existence of hydrodynamically distinct subspecies of LDL which can show distinctive patterns of distribution in the plasma of different subjects.[9] Several bands of LDL are observed on ultracentrifugation of LDL in an NaBr gradient in swinging bucket rotors.[10] Significant differences in lipid composition can be demonstrated in sequential gradient fractions. Circular dichroism spectra also show differences in secondary structure among such fractions (G. C. Chen, R. M. Krauss, and J. P. Kane, unpublished data). Particle fractions of VLDL differing significantly in diameter can be produced either by density gradient ultracentrifugation[11] or by gel permeation chromatography on agarose gel beads. Chromatography on heparin agarose columns is capable of separating fractions on the basis of their content of apoE. These fractions differ with respect to their ability to accept LCAT-derived cholesteryl esters.[12] Though chylomicron-rich fractions can be obtained from lipemic serum by short ultracentrifugal runs, neither this method nor the aforementioned techniques can effect a categorical separation of chylomicrons and large VLDL.

Methods employing immunosorption may become the most effective means of isolating subfractions of apoB-containing lipoproteins. Monoclonal antibodies to the nonhomologous region of apoB-100 allow complete separation, by column immunosorption, of chylomicrons and their remnants from VLDL.[13] Separation of these particles was achieved with

[8] A. G. Olsson, *Scand. J. Lab. Invest.* **39**, 229 (1979).

[9] W. R. Fisher, M. G. Hammond, and G. L. Warmke, *Biochemistry* **11**, 519 (1972).

[10] M. M. S. Shen, R. M. Krauss, F. T. Lindgren, and T. M. Forte, *J. Lipid Res.* **22**, 236 (1981).

[11] F. T. Lindgren, A. V. Nichols, F. T. Upham, and R. D. Wills, *J. Phys. Chem.* **66**, 2007 (1962).

[12] C. J. Fielding, G. M. Reaven, G. Lui, and P. E. Fielding, *Proc. Natl. Acad. Sci. U.S.A.* **81**, 2512 (1984).

[13] R. W. Milne, P. K. Weech, L. Blanchette, J. Davignon, P. Alaupovic, and Y. L. Marcel, *J. Clin. Invest.* **73**, 816 (1984).

the triglyceride-rich lipoproteins from serum of patients with familial dysbetalipoproteinemia, where considerable overlap of particle diameter exists. In this case, a significantly greater content of apoE and cholesteryl
esters was thus demonstrable in the apoB-48-containing particles than in
the particles containing apoB-100. Monoclonal antibodies also appear to
be able to differentiate between apoB of VLDL and IDL, presumably
reflecting changes in exposure of epitopes as VLDL are modified during
intravascular lipolysis.[14]

Determination of Apolipoprotein Content. Measurement of the total
protein content of triglyceride-rich lipoproteins by the Lowry method
presents two problems: light scattering lipids must be removed by extraction with chloroform or dispersed by a detergent such as Triton X-100 or
SDS,[15] and the absorbance of apoVLDL must be corrected for differences
in chromogenicity of apoB-100 and non-B proteins.[16,17]

ApoB. The relative apoB-100 content of VLDL varies with particle
diameter because of variations in content of non-B protein. About 30–
40% of apoVLDL from normal subjects is apoB-100. The apoB content of
lipoproteins can be determined by techniques employing radial immunodiffusion, radioimmunoassay, or enzyme-linked immunoassay or immunonephelometry. In all of these techniques, the presence of apoB-100
on particles of widely different diameter and lipid content produces non-
uniformity in the interaction between antigen and antibody. This can be
overcome only if apoB is separated from the lipid moiety by the action of
an amphiphile which forms a uniform product with the protein.

Two chemical methods separate apoB from the other apolipoproteins
of VLDL. Tetramethylurea (4.2 M) solubilizes the non-B proteins such
that the B protein can be readily estimated by the difference of soluble
protein in treated and nontreated samples.[16] Isopropanol solubilizes
non-B proteins and lipids, leaving apoB-100 insoluble.[18] Both methods are
suitable for quantitative separation of radiolabeled apoB-100 from other
VLDL proteins in reinfusion experiments.[19] When the content of apoE is
very high, as in VLDL of dysbetalipoproteinemia, estimates of apoB may
be erroneously high using the isopropanol technique. With precipitation
of apoB by tetramethylurea at 30°, however, no coprecipitation occurs.
Differential determination of radiolabel in apoB-48 and apoB-100 in rein-

[14] B. P. Tsao, L. K. Curtiss, and T. Edgington, *J. Biol. Chem.* **257**, 15222 (1984).
[15] K. L. Kayshap, B. A. Hynd, and K. Robinson, *J. Lipid Res.* **21**, 491 (1980).
[16] J. P. Kane, T. Sata, R. L. Hamilton, and R. J. Havel, *J. Clin. Invest.* **56**, 1622 (1975).
[17] G. L. Vega and S. M. Grundy, *J. Lipid Res.* **25**, 580 (1984).
[18] L. Holmquist and K. Carlson, *Biochim. Biophys. Acta* **493**, 400 (1977).
[19] G. Egusa, D. W. Brady, S. M. Grundy, and B. S. Howard, *J. Lipid Res.* **24**, 1261 (1983).

fusion turnover studies can be accomplished in gradient gels using modifications of technique which give suitable recoveries.[20]

Because of the small content of B-48 protein in chylomicrons and because of the coprecipitation of non-B high-molecular-weight proteins, neither the tetramethylurea or isopropanol technique is capable of measuring the apoB content of chylomicrons or of yielding apoB-48 suitable for measurement of specific activity.

Non-B proteins. The principal non-B proteins are, in quantitative order, apoC-III, apoE, apoC-II, and apoC-I. A number of techniques employing radioimmunoassay, radial immunodiffusion, and immunonephelometry have been developed for analysis of the content of individual non-B proteins. Again, uniformity of response is often dependent upon the manner in which the antigen is solubilized. The nonsialated, and mono- and disialated forms of apoC-III and the isoforms of apoE and apoC-II are resolved in isoelectric focusing gels.[21] The major non-B apolipoproteins of VLDL can be measured quantitatively in a single polyacrylamide electrophoresis gel by a quantitative densitometric technique.[16]

Determination of Secondary Structure of the Protein Moieties of Intact Lipoproteins by Circular Dichroism. Circular dichroic (CD) analysis allows the appreciation of both secondary structure of the protein moieties and order in the lipid domains of intact lipoproteins. The circular dichroic contribution of lipid chromophores is discussed in Chapter [30], Volume 128.

Helical, β-pleated sheet, and random coil conformations have been demonstrated in the protein moiety of LDL.[22-24] The limits of CD measurement in particles of larger diameter which are imposed by light scattering have been systematically defined.[24] For instance, accurate measurements can be made down to 194 nm on particles with diameters of 450 Å or less, and to 210 nm on particles of 520 Å. Because the CD contribution of triglyceride is so much smaller than that of cholesteryl esters, the total CD contribution of lipids in these particles is no greater than for LDL. Both human VLDL and IDL have substantially more helical content ($\sim$50%) than do LDL ($\sim$30%), whereas LDL contain more β structure ($\sim$26%) than do IDL ($\sim$19%) or VLDL ($\sim$14%).

[20] A. F. H. Stalenhoef, M. J. Malloy, J. P. Kane, and R. J. Havel, *Proc. Natl. Acad. Sci. U.S.A.* **81,** 1839 (1984).
[21] G. Uterman, M. Jaeschke, and J. Menzel, *FEBS Lett.* **56,** 352 (1975).
[22] A. M. Gotto, R. I. Levy, and D. S. Fredrickson, *Proc. Natl. Acad. Sci. U.S.A.* **60,** 1436 (1968).
[23] A. Scanu and R. Hirz, *Nature (London)* **218,** 200 (1968).
[24] G. C. Chen and J. P. Kane, *J. Lipid Res.* **20,** 481 (1979).

[8] Characterization of Apolipoprotein A-Containing Lipoproteins

By MARIAN C. CHEUNG

Introduction

Lipoproteins are conventionally defined based on their hydrated density and electrophoretic mobility. Chylomicrons, the least dense of all the lipoproteins, have densities less than 0.94 g/ml and remain at the origin when subjected under an electrical field on paper, cellulose acetate, or agarose. Very-low-density lipoproteins (VLDL) have densities between 0.94 and 1.006 g/ml and migrate with pre-beta mobility in zone electrophoresis. Low-density lipoproteins (LDL) with densities between 1.006 and 1.063 g/ml and high-density lipoproteins (HDL) with densities between 1.063 and 1.21 g/ml migrate with β and α electric mobility, respectively. Each class of lipoproteins defined by these two physical properties represent a heterogeneous population of particles differing not only in size, but also in lipid and apolipoprotein composition. The major apolipoprotein (apo) components of chylomicrons are A-I, B, and the C's, that of VLDL are B, E, and the C's. Low-density lipoproteins contain mostly apoB, while HDL contains apoA-I, A-II, A-IV, B, C's, D, and E.

Apolipoproteins are essential in maintaining the structural integrity of a lipoprotein particle. Their importance in lipoprotein formation is best illustrated in Tangier disease and abetalipoproteinemia where genetic defects in apoA metabolism and apoB production, respectively, result in a very low level of HDL in the former, and absence of chylomicrons, VLDL, and LDL in the latter.[1] Apolipoproteins also play a major role in lipoprotein metabolism. In addition to their transport functions, some of them are activators or inhibitors of lipolytic enzymes,[2–4] while others are determinants in receptor-mediated lipoprotein catabolism.[5–7] The recogni-

[1] D. S. Fredrickson, J. L. Goldstein, and M. Brown, *in* "The Metabolic Basis of Inherited Disease," (J. B. Stanbury, J. B. Wyngaarden, and D. S. Fredrickson, eds.), 4th Ed., p. 604. McGraw-Hill, New York, 1978.

[2] C. J. Fielding, V. G. Shore, and P. E. Fielding, *Biochem. Biophys. Res. Commun.* **46,** 1493 (1972).

[3] J. C. LaRosa, R. I. Levy, P. N. Herbert, S. E. Lux, and D. S. Fredrickson, *Biochem. Biophys. Res. Commun.* **41,** 57 (1970).

[4] W. V. Brown and M. L. Baginsky, *Biochem. Biophys. Res. Commun.* **46,** 375 (1972).

[5] T. L. Innerarity, R. W. Mahley, K. H. Weisgraber, and T. P. Bersot, *J. Biol. Chem.* **253,** 6289 (1978).

tion of the importance of apolipoproteins in the structure, function, and metabolism of lipoproteins led to the concept of classification of lipoproteins by their apolipoprotein component, a concept originally proposed by Alaupovic and co-workers over a decade ago.[8] Based on this classification, lipoprotein particles are identified by the presence of certain specific groups of apolipoproteins, such as A, B, or C. This chapter discusses the method employed in isolating apoA-containing lipoproteins and the characteristics of these lipoproteins.

ApoA-I, A-II, and A-IV are the three A apolipoproteins in humans. The intestine is a common site of synthesis for these proteins. In addition, apoA-I and A-II are synthesized in the liver. The physical, chemical, and metabolic properties of these proteins are described by Brewer *et al.* (chapter [10], Volume 128) and Schaefer *et al.* (chapter [25], this volume). Immunochemical studies using antisera specific for apoA-I and A-II have showed that in fasting normolipidemic human plasma, less than 1% of A-I and A-II is associated with lipoproteins with densities less than 1.063 g/ml. About 90% of plasma A-I and greater than 96% of plasma A-II are found within the density 1.063–1.21 g/ml ultracentrifuge fractions.[9] Unlike apoA-I and A-II, 98% of apoA-IV in fasting, normolipidemic human plasma and 90% of apoA-IV in lipemic plasma are not associated with lipoproteins prepared by ultracentrifugation.[10] However, recently, using a combination of gel filtration chromatography and immunolocalization, Bisgaier *et al.* showed that approximately 20% of plasma A-IV normally exist in HDL particles in association with A-I.[11]

Although the majority of apoA-I and A-II in normolipidemic fasting plasma are found within the d 1.063–1.21 g/ml range, apoA-containing lipoproteins should not be considered synonymous with HDL because within the HDL density range there are lipoprotein particles such as the Lp(a) lipoproteins, which do not contain the A apolipoproteins. Furthermore, it is likely that in plasma of subjects with lipoprotein disorders a significant portion of apoA-containing lipoproteins can be found in the $d < 1.063$ g/ml and the $d > 1.21$ g/ml ultracentrifuge fractions. Indeed it has

[6] J. L. Goldstein and M. S. Brown, *Science* **212,** 628 (1981).
[7] J. F. Oram, E. E. Brinton, and E. L. Bierman, *J. Clin. Invest.* **72,** 1611 (1983).
[8] P. Alaupovic, *Protides Biol. Fluids Proc. Colloq.* **19,** 9 (1972).
[9] M. C. Cheung and J. J. Albers, *J. Clin. Invest.* **60,** 43 (1977).
[10] P. H. R. Green, R. M. Glickman, J. W. Riley, and E. Quinet, *J. Clin. Invest.* **65,** 911 (1980).
[11] C. L. Bisgaier, O. P. Sachdev, L. Megna, and R. M. Glickman. *J. Lipid Res.* **26,** 11 (1985).

been shown that in Tangier disease most of the apoA-containing lipoproteins are found in the $d > 1.21$ g/ml fraction.[12]

Types of ApoA-Containing Lipoproteins

ApoA-containing lipoproteins can be divided into two broad categories—those associated with apoB and those not associated with apoB. The former lipoproteins generally would have a density less than 1.063 g/ml and normally represent less than 1% of plasma apoA-containing lipoproteins. The latter are lipoproteins with a density greater than 1.063 g/ml. They represent the bulk of plasma apoA-containing lipoproteins.

The B-unassociated apoA-containing lipoproteins can be subdivided into two types with respect to their apoA constituents: lipoprotein particles that contain both A-I and A-II [Lp(A-I with A-II)] and particles that contain A-I but no A-II [Lp(A-I without A-II)]. Several laboratories have isolated the two types of apoA-containing lipoprotein particles. The characteristics of these particles, however, appear to vary with the isolation procedure. To date, there is no evidence of existence in normal plasma of lipoprotein particles which contain apoA-II but not A-I. Such lipoprotein particles have only been identified in the plasma of subjects with Tangier disease.[12]

Methods

ApoA-containing lipoproteins have been isolated from ultracentrifugally prepared HDL by a single or combination of chromatographic techniques. These include DEAE, QAE, hydroxylapatite, gel permeation, affinity column chromatography, and chromatofocusing.[13-17] Since ultracentrifugation can cause redistribution of apolipoproteins and lipid in lipoproteins and dissociation of substantial amounts (35%) of plasma A-1 from lipoprotein particles,[18] an isolation procedure which does not require ultracentrifugation would be preferable. Immunoaffinity chromatography employing antibodies specific for apoA-I and A-II is the only currently

[12] G. Assmann, P. N. Herbert, D. S. Fredrickson, and T. Forte, *J. Clin. Invest.* **60**, 242 (1977).

[13] R. Rubenstein, *Atherosclerosis* **33**, 415 (1979).

[14] G. M. Kostner and A. Holasek, *Biochim. Biophys. Acta* **34**, 257 (1977).

[15] S.-O. Olofsson and A. Gustafson, *Scand. J. Clin. Invest.* **34**, 257 (1974).

[16] P.-I. Norfeldt, S.-O. Olofsson, G. Fager, and G. Bondjers, *Eur. J. Biochem.* **118**, 1 (1981).

[17] A. C. Nestruck, P. D. Niedmann, H. Wieland, and D. Seidel, *Biochim. Biophys. Acta* **753**, 65 (1983).

[18] S. T. Kunitake and J. P. Kane, *J. Lipid Res.* **23**, 936 (1982).

available method which permits the isolation of apoA-containing lipoproteins[19,20] and their two subclasses, Lp(A-I with A-II) and Lp(A-I without A-II)[19] directly from plasma without prior ultracentrifugation. The success of this method relies greatly on the nature of antibodies used. The antibodies must be monospecific to either apoA-I or A-II and should preferably dissociate from A-I and A-II under a mild condition. The following is a detailed description of this method as used in our laboratory.

Preparation of Antigens

High-density lipoproteins of d 1.090–1.21 g/ml is isolated from fasting plasma of individual normolipidemic donors by sequential ultracentrifugation. The nonprotein solvent density of plasma is adjusted to d 1.090 with solid KBr. Ultracentrifugation is performed in a 60 Ti rotor (Beckman Instruments, Inc., Palo Alto, CA) at 50,000 rpm at 10° for 24 hr. After removing the top one-third by tube slicing, the bottom two-thirds is brought to d 1.21 g/ml by solid KBr and centrifuged under the same conditions for 48 hr. The top one-third of each tube containing the d 1.090–1.21 g/ml HDL is pooled, diluted with an equal volume of KBr solution of d 1.21 g/ml containing 1 mM disodium EDTA, pH 7.4, and recentrifuged as before for 24 hr. The top one-third of each tube is pooled, dialyzed against 1 mM EDTA, pH 7.4, brought to 6 M guanidine hydrochloride, and incubated at 37° for 3 hr. This 6 M guanidine hydrochloride treatment selectively dissociates most A-I from HDL particles. At the end of the 2 hr, the guanidine hydrochloride is removed by dialyzing against 1 mM disodium EDTA, pH 7.4. The HDL is again brought to d 1.21 g/ml with solid KBr and centrifuged in a 60 Ti rotor at 50,000 rpm at 10° for 26 hr. The resulting top third fraction containing HDL particles with primarily apoA-II and the apoC's is dialyzed against 1 mM disodium EDTA, pH 7.4, lyophilized, and delipidated with ether–ethanol (3 : 1). The delipidated preparation is dissolved in 0.03 M Tris-HCl, pH 8.0, containing 6 M urea and chromatographed on DEAE–Sepharose CL-6B (Pharmacia, Uppsala, Sweden) equilibrated with the same buffer. ApoA-II is eluted by a linear gradient of NaCl from 0 to 0.125 M. The bottom third fraction of the 1.21 g/ml centrifugation containing mainly apoA-I is dialyzed against 0.03 M Tris-HCl, pH 8.0, containing 6 M urea and chromatographed similarly on DEAE–Sepharose CL-6B to obtain purified A-I. In both cases, a column size of 1.6 cm internal diameter by 40 cm height is used for a sample size of 200-mg proteins. The total volume of the elution

[19] M. C. Cheung and J. J. Albers, *J. Biol. Chem.* **259**, 12201 (1984).
[20] J. P. McVicar, S. T. Kunitake, R. L. Hamilton, and J. P. Kane, *Proc. Natl. Acad. Sci. U.S.A.* **81**, 1356 (1984).

gradient is 1000 ml. Samples are loaded at 40 ml/hr and eluted at a flow rate of 60 ml/hr.

Production of Antisera

Polyacrylamide gel electrophoresis (PAGE) is performed on purified apoA-II according to the method of Davis[21] using an acrylamide monomer concentration of 7.5% in 8 M urea. Sodium dodecyl sulfate (SDS)–PAGE is performed on purified apoA-I according to the method of Laemmli[22] using a 12% running gel. The protein bands corresponding to apoA-I and A-II in each case are sliced from each gel, ground with a mortar and pestle, and emulsified with Freund's complete adjuvant for injection into rabbits or goats. Each rabbit receives 30–50 μg while each goat receives 300–500 μg of apoA-I or A-II. Subsequent boosters are performed with antigen emulsified with Freund's incomplete adjuvant. In our experience apoA-I is a much better antigen than apoA-II in both rabbits and goats. All of the animals immunized with apoA-I succeed in producing useable antibodies. However, about 30% of the rabbits immunized with apoA-II fail to produce useable antibodies even after the fourth or fifth booster.

Isolation of Specific Anti-A-I and Anti-A-II Immunoglobulins for Immunosorbent Preparation

Antibodies specific to either apoA-I or A-II are isolated from the respective antisera by HDL affinity chromatography. HDL (d 1.090–1.21 g/ml) are prepared as described previously and dialyzed into 0.1 M NaHCO$_3$, 0.5 M NaCl, pH 8.0 (NaHCO$_3$–NaCl), before conjugation. About 8–10 mg of HDL proteins is conjugated per milliliter of CNBr-activated Sepharose 4B (Pharmacia Fine Chemicals, Uppsala, Sweden) according to the procedure provided by the manufacturer. The efficiency of conjugation of HDL to CNBr-activated Sepharose 4B is usually over 90% in our hands.

One hundred milliliters of antiserum is mixed with 10–15 ml of HDL–Sepharose 4B and rotated end-over-end for a minimum of 1 hr at 4°. The HDL–Sepharose is then washed on a sintered glass funnel (coarse grade) with 0.01 M Tris, 0.15 M NaCl, 1 mM EDTA, pH 7.4 (Tris–NaCl–EDTA), to remove unbound protein until the A_{280} of the wash is below 0.02 units. Nonspecifically bound proteins are removed by washing the gel with NaHCO$_3$–NaCl with 1 mM EDTA (NaHCO$_3$–NaCl–EDTA). The gel is then packed into a column and reequilibrated in Tris–NaCl–EDTA.

---

[21] B. J. Davis, *Ann. N.Y. Acad. Sci.* **121**, 404 (1964).
[22] U. K. Laemmli, *Nature (London)* **227**, 680 (1970).

Antibodies of relatively low affinity to be used for making immunosorbent are eluted with 0.1 M HAc, 0.5 M NaCl, pH 3.5, and immediately buffered with solid Tris or 1 M Tris-HCl, pH 8.6, and dialyzed against 0.02 M sodium phosphate buffer, pH 6.8, for conjugation to CNBr–Sepharose 4B. High-affinity antibodies are eluted with 1 M HAc, pH 2.6, and then with 3 M NaSCN in 0.02 M sodium phosphate, pH 7.0 (NaSCN). Although these antibodies are not suitable for the present purpose, they must be dissociated from the HDL–Sepharose 4B to regain the affinity gel capacity. In this and all subsequent affinity columns, we use a flow rate of 60 ml for all elutions. When all antibodies are removed, the HDL–Sepharose 4B is stored in Tris–NaCl–EDTA containing 0.02% Merthiolate. We have used HDL affinity gel repeatedly without loss of capacity for over a year. To avoid the possibility of contamination, different batches of HDL–Sepharose should be used for isolating anti-A-I and anti-A-II antibodies.

Comment: In this procedure, the amount of HDL–Sepharose 4B required to remove all antibodies from an antiserum, as well as the quantity of low-affinity antibodies recovered from an isolation, vary from bleeding to bleeding. The volume of HDL affinity gel required for isolation increases with the titer of the antiserum. Yet, the percentage of low-affinity antibodies actually recovered from the isolation usually decreases with increasing antibody titer. Hence, it is preferable on an economical basis to use antiserum of low titer for this purpose. We find that antibodies which are too weak for use in immunoassay usually can be used for immunosorbent preparation.

The use of HDL affinity gel to isolate specific anti-A-I and anti-A-II immunoglobulins from antiserum is recommended only if the antiserum is monospecific for these proteins. If one has an antiserum which cross-reacts with other HDL proteins, one will have to rely on purified apoA-I and apoA-II affinity gel for the isolation of specific anti-A-I and anti-A-II immunoglobulins, respectively.

Antibody Specificity

Before a batch of antibodies is ready for use in immunosorption, its specificity should be carefully checked. If various purified radiolabeled apolipoproteins are available, specificity of the antibodies can be checked by immunoprecipitation, using excess first antibody to ensure detection of trace contaminants. If various purified apolipoproteins are not available, the immunoblotting technique of Towbin et al.[23] is an alternative method

[23] H. Towbin, T. Stachelin, and J. Gordon, *Proc. Natl. Acad. Sci. U.S.A.* **79**, 4350 (1979).

which provides comparable sensitivity. ApoHDL and apoVLDL are electrophoretically separated by SDS–PAGE, transferred to nitrocellulose paper, and allowed to react with testing antibodies. Binding of antibodies to the proteins on nitrocellulose can be visualized by commercially available horseradish peroxidase conjugated antibody specific for the species of the testing immunoglobulins and the substrate 4-chloro-1-naphthol.

Preparation of Anti-A-I and Anti-A-II Immunosorbent

Antibodies monospecific for A-I and A-II are conjugated to CNBr–Sepharose 4B at a ratio of 7–8 mg immunoglobulin per milliliter of gel in 0.2 M sodium phosphate buffer, pH 6.8, with 0.5 M NaCl. Antibodies precipitated during acid elution should be centrifuged and removed before conjugation. Conjugation efficiency is usually greater than 95%. The gel is stored in Tris-NaCl with 0.02% Merthiolate at 4°. Before using for the first time, the gel is washed with 1.0 M HAc, pH 2.6, and NaSCN to ensure that the conjugated antibodies can withstand these harsh conditions. The binding capacity of the anti-A-I and anti-A-II immunosorbents is estimated by mixing 1 ml of plasma with 2 ml of immunosorbent for an hour at 4°. Nonbinding proteins are washed from the immunosorbent by Tris–NaCl–EDTA and concentrated to 1 ml. The apoA-I and A-II contents of plasma before and after immunosorption are quantitated by immunoassay. One milliliter of our immunosorbent usually binds about 0.3 mg A-I and 0.05 mg A-II. Some of our immunosorbents have been used more than 30 times over a year without loss of binding capacity. Upon repeated use, leaching of antibodies could eventually occur. The immunosorbent should then be discarded.

Isolation of ApoA-Containing Lipoproteins

Total ApoA-Containing Lipoproteins [Lp(A-I)]. As stated earlier, all apoA-containing lipoproteins contain A-I. Therefore, to isolate these lipoproteins, fresh plasma or serum is incubated with anti-A-I immunosorbent for 1 hour at 4°. [To minimize apoB-associated A-I-containing lipoproteins, fasting (12–14 hr) plasma should be used.] The immunosorbent is packed into a borosilicate glass column of 1.5 or 1.6 cm internal diameter. The amount of immunosorbent used depends on the capacity of the immunosorbent and the volume of plasma. Proteins that do not bind to the immunosorbent are removed by washing with Tris–NaCl–EDTA and concentrated to a volume equal to or less than the starting plasma volume. This and all subsequent concentrations are performed by Micro-Conflit concentrator using protein dialysis membrane of 10,000–15,000 molecular weight cut-off (Biomolecular Dynamics, Beaverton, OR). A concentration method utilizing high pressure is not acceptable because it can result in aggregation and denaturation of lipoprotein particles. Nonspecifically bound proteins are removed with $NaHCO_3$–NaCl–EDTA. ApoA-containing lipoproteins are eluted with either 0.5 M HAc, 0.5 M NaCl, 1 mM EDTA, pH 3.0, and immediately buffered with solid Tris, or with NaSCN, dialyzed against Tris–NaCl–EDTA, and concentrated, if necessary, for physical–chemical characterization. Usually, two-column to three-column volumes of desorbing agent are necessary to elute the lipoprotein particles. Lipoprotein recovery from the immunosorbent based on apoA-I, A-II, and D is quantitative (over 90%).

Comment: While the recovery of apoA-I from the immunosorbent is comparable whether thiocyanate or acetic acid is used to dissociate lipoproteins from the immunosorbent, data from our laboratory indicate that these two desorbing agents differ in their effect on lecithin–cholesterol acyltransferase (LCAT) and cholesteryl ester transfer (CET) activities. Recovery of LCAT activity is higher when acetic acid is used as the desorbing agent. However, CET activity can only be detected in A-I-containing lipoprotein eluted by thiocyanate.

A-I-Containing Lipoproteins with A-II [Lp(A-I with A-II)]. To isolate the subfraction of apoA-containing lipoproteins which contain both apoA-I and A-II, fresh plasma (or serum) is incubated with anti-A-II immunosorbent for 1 hr at 4°. The immunosorbent is packed into a borosilicate glass column of 1.5 or 1.6 cm internal diameter. As before, the amount of immunosorbent needed depends on its capacity and the volume of plasma. Nonbinding proteins, nonspecifically bound proteins, and A-II-containing lipoproteins are sequentially washed from the immunosorbent and processed as described above.

A-I-Containing Lipoproteins without A-II [Lp(A-I without A-II)]. Plasma devoid of A-II (the nonbinding plasma proteins from B) is incubated with anti-A-I immunosorbent for 1 hr at 4°. Again, nonbinding proteins, nonspecifically bound proteins, and A-I-containing lipoproteins without A-II are sequentially washed from the immunosorbent and processed as previously described.

Comment: In the isolation procedure reported by McVicar et al.,[20] apoA-containing lipoproteins were eluted from an anti-A-I affinity column with 1 M acetic acid, pH 3.0. (Unlike our antibodies, the anti-A-I antibodies used were isolated by A-I–Sepharose affinity gel.) This single elution step recovered over 90% apoA-I and all the apoA-II applied to the column. The column was used over 300 times without apparent diminution in capacity.

The use of immunoaffinity chromatography for preparative isolation of lipoproteins is a relatively new approach. Furthermore, unlike a chemical reagent, characteristics of antibodies vary considerably from bleeding to bleeding. In the course of developing the methodology, I have found it necessary to periodically adjust the antibody selection and lipoprotein elution conditions. Availability of a permanent supply of appropriate low-affinity monoclonal antibodies should alleviate some of the technical difficulties encountered with polyclonal antibodies. The methods described here are intended to be used mainly as basic guidelines upon which other investigators can adjust and improve to suit the characteristics of their antibodies, and the physical, chemical, and/or biological studies the isolated lipoproteins are to be used for. Finally, it should be pointed out that the procedure as described isolates apoB-associated as well as non-apoB-associated apoA-containing lipoproteins. The former can be removed from the latter simply by passing the lipoproteins through an anti-B immunosorbent.

Characterization of ApoA-Containing Lipoproteins Isolated from Plasma by Immunoaffinity Chromatography

We have used immunoaffinity chromatography to isolate apoA-containing lipoproteins and their subfractions from over 20 individual and pooled plasmas. The general characteristics of these lipoproteins are described here.

Electrophoretic Mobility and Density

When fresh plasma is incubated with anti-A-I immunosorbent to remove all apoA-I, no apoA-II can be detected. This further confirms observations to date that all plasma apoA-II is associated with A-I. Agarose gel electrophoresis of the A-I- and A-II-free plasma followed by lipid staining with Sudan Red 7B reveals no α-migrating lipoproteins. Furthermore, when the A-I- and A-II-free plasma is centrifuged in a Beckman VTi 80 vertical rotor at 80,000 rpm for 90 min at 10°, and cholesterol concentration along the centrifuge tube is measured by a continuous flow cholesterol monitor,[24] there is essentially no cholesterol in the HDL region when compared to the cholesterol profile of the original plasma (Fig. 1A) (Cheung, Segrest, and Cone, unpublished data). Hence, anti-A-I immunosorbent removes from plasma the bulk of lipoproteins with α electrophoretic mobility and with density corresponding to plasma HDL.

When plasma is incubated with anti-A-II immunosorbent to remove all apoA-II, substantial amounts of apoA-I can still be detected in the A-II-

[24] B. H. Chung, J. P. Segrest, J. T. Cone, J. Pfau, J. C. Geer, and L. A. Duncan, *J. Lipid Res.* **22,** 1003 (1981).

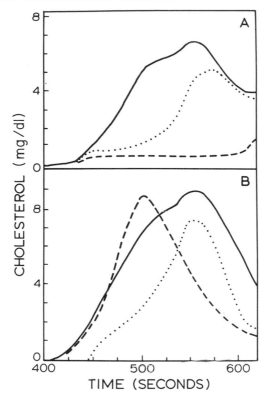

FIG. 1. Cholesterol profile of plasma (A) and lipoproteins isolated by immunoaffinity chromatography (B) from a female subject with an HDL cholesterol level of 84 mg/dl after a single vertical spin ultracentrifugation. Plasma or lipoprotein samples (1.4 ml) were adjusted to density 1.21 g/ml and overlayered with 3.5 ml of a d 1.06 g/ml KBr solution and spun at 80,000 rpm for approximately 90 min in a Beckman VTi80 rotor at 10°. After centrifugation, cholesterol concentration along the centrifuge tube was measured by a continuous flow cholesterol monitor. The profiles in (A) represent the HDL region of plasma (——); plasma after all apoA-II was removed (······); and plasma after all apoA-II and A-I were removed (------). The profiles in (B) represent plasma lipoproteins adsorbed and acid eluted from anti-A-I immunosorbent, Lp(A-I) (——), from anti-A-II immunosorbent, Lp(A-I with A-II) (------), and from anti-A-I immunosorbent when A-II-negative plasma was used, Lp(A-I without A-II) (······).

free plasma. This indicates that not all plasma apoA-I is associated with A-II. The percentage of plasma A-I which does not associate with A-II varies (between 20–50%) from subject to subject. When agarose gel electrophoresis is performed on the A-II-free plasma, α-migrating lipoproteins can still be detected. Vertical rotor centrifugation shows that while a considerable amount of lipoproteins corresponding to the dense HDL region is absent, most of the lipoproteins corresponding to the lighter density HDL region are still present in the A-II-free plasma (Fig. 1A). Hence, anti-A-II immunosorbent removes mainly lipoproteins within the denser HDL region. When the A-II-free plasma is further subjected through an anti-A-I immunosorbent prior to agarose electrophoresis, all α-migrating lipoproteins are removed.

In most subjects, apoA-containing lipoprotein particles eluted from anti-A-I immunosorbent show only α electrophoretic mobility in the Paragon Lipo Gel system (Beckman). However, in some subjects, small quantities of lipoproteins with pre-β mobility are also seen. These pre-β apoA-containing lipoproteins are removed by anti-B immunosorbent, indicating they are the B-associated apoA-containing lipoproteins. Vertical rotor centrifugation and continuous cholesterol profiling along the centrifuge tube shows that the apoA-containing lipoproteins [Lp(A-I)] isolated from plasma had density profiles comparable to the HDL region of plasma. The lipoproteins eluted from the anti-A-II immunosorbent [Lp(A-I with A-II)] float mostly in the region of the denser HDL with a small quantity found in the lighter density HDL region. On the other hand, the cholesterol profile of the lipoproteins containing apoA-I but no A-II [Lp(A-I without A-II)] shows two distinct density regions. Some of the particles are lighter, while others are denser than the mean particle density of Lp(A-I with A-II) (Fig. 1B). The proportion of Lp(A-I without A-II) particles in these two density regions vary considerably among subjects.

The lack of any obvious change in density of the cholesterol profile of the immunoaffinity isolated apoA-containing lipoproteins compared to the HDL region of the original plasma in Fig. 1A and B indicates that desorption of lipoproteins from the immunosorbent has not significantly altered the protein–lipid composition of the lipoprotein particles. The morphology of these particles viewed under negative-staining electron microscopy also indicates no particle fusion or aggregation (Cheung, Segrest, and Brouilette, unpublished data).

Particle Size

The particle sizes of the various apoA-containing lipoproteins have been studied by nondenaturing gradient polyacrylamide gel electrophoresis using Pharmacia precasted PAA 4/30 gels coupled with Coomassie Brilliant Blue G 250 protein staining.[25,26] In gradient polyacrylamide gel electrophoresis, nine differently sized species of apoA-containing lipopro-

[25] M. C. Cheung and J. J. Albers, *J. Lipid Res.* **20**, 200 (1979).
[26] P. J. Blanche, E. L. Gong, T. M. Forte, and A. V. Nichols, *Biochim. Biophys. Acta* **665**, 408 (1981).

teins can be distinguished (Fig. 2). The mean Stokes diameters of these particles based on observations from 20 subjects are (1) 7.6 nm; (2) 7.8 nm; (3) 8.0 nm; (4) 8.5 nm; (5) 8.9 nm; (6) 9.6 nm; (7) 10.8 nm; (8) 12.2 nm; and (9) 13.2 nm. These diameters are calculated by using the hydrated Stokes diameters of thyroglobulin (17 nm), apoferritin (12.2 nm), catalase (10.4 nm), lactate dehydrogenase (8.2 nm), and bovine albumin (7.5 nm) as standards. Particles of sizes 3–7 constitute the bulk (over 95%) of the plasma apoA-containing lipoproteins in normolipidemic plasma. The ma-

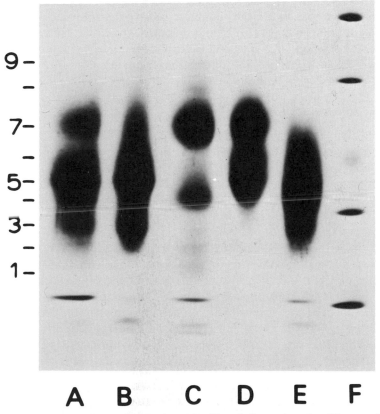

FIG. 2. Nondenaturing gradient polyacrylamide gel electrophoresis of lipoproteins isolated by immunoaffinity chromatography and ultracentrifugation from a male subject with an HDL cholesterol level of 69 mg/dl. Lane A, apoA-I containing lipoproteins Lp(A-I); lane B, lipoproteins containing both apoA-I and A-II, Lp(A-I with A-II); lane C, lipoproteins containing A-I but no A-II, Lp(A-I without A-II), lane D, HDL_2 (d 1.063–1.125 g/ml); lane E, HDL_3 (d 1.125–1.21 g/ml) and lane F, calibration proteins (from top to bottom: thyroglobulin, ferritin, catalase, lactate dehydrogenase, and bovine albumin). The positions of size subspecies 1–9 are indicated on the left.

jority of Lp(A-I without A-II) particles belong to size subspecies 4 (8.5 nm) and 7 (10.8 nm). The proportion of these two sizes, however, varies widely among subjects. Some have predominantly the larger 10.8 nm size, while others have about equal proportions of the two particle sizes. Lipoprotein particles containing both apoA-I and A-II consist mostly of size subspecies 3 (8.0 nm), 5 (8.9 nm), and 6 (9.6 nm), with particles of 8.9 nm (subspecies 5) being the dominant size subspecies in all normolipidemic subjects studied. When the size distribution of Lp(A-I with A-II) and Lp(A-I without A-II) are compared to the standard flotation subfractions HDL_2 (d 1.063–1.125 g/ml) and HDL_3 (d 1.125–1.21 g/ml) of the same subjects, it appears that HDL_2 contains mostly particle size subspecies 5–7 with a small quantity of size subspecies 4, while HDL_3 contains particle size subspecies 2–5 with a trace of size 6. In other words, each of these two major HDL density subfractions contains a mixture of Lp(A-I with A-II) and Lp(A-I without A-II).

Lipid and Protein Composition

Lipid analysis of Lp(A-I without A-II) and Lp(A-I with A-II) particles shows that Lp(A-I without A-II) has a slightly higher lipid/protein ratio (1.18) than Lp(A-I with A-II) (0.92). However, their lipid compositions are identical with percentages of cholesterol, phospholipid, and triglycerides being 33, 61, and 6, respectively. (The weight of total unesterified cholesterol after enzymatic hydrolysis of cholesteryl ester is used in these calculations.)

The Lp(A-I with A-II) particles isolated from 20 subjects by our immunoaffinity chromatography procedure have a fairly constant molar A-I/A-II ratio of 1.8 ± 0.2 ($\bar{x}$ ± SD) irrespective of the original plasma A-I/A-II ratio. This confirms our earlier suggestion[27] that the variation in A-I/A-II ratio observed at various hydrated densities does not reflect particles with differing A-I/A-II ratios, but reflects the differing proportions of two compositionally discrete particle types: Lp(A-I without A-II) and Lp(A-I with A-II) containing A-I and A-II at a nearly constant ratio. Other investigators[28] have also reported that the A-I/A-II molar ratio in lipoprotein particles containing both A-I and A-II is fairly constant, but the ratio is 1.2. The difference in the ratios observed is probably due primarily to the differences in the immunoassays used to quantitate these apolipoproteins.

With respect to non-apoA protein components, Lp(A-I with A-II) particles contain also most (about 60%) of the apoD and E in plasma, and 20% of plasma LCAT enzymes based on both mass and activity. Small

---

[27] M. C. Cheung and J. J. Albers, *J. Lipid Res* **23,** 747 (1982).
[28] R. F. Atmeh, J. Shepherd, and C. J. Packard, *Biochim. Biophys. Acta* **751,** 175 (1983).

quantities of the apoC proteins are also seen using analytical isoelectric focusing gel electrophoresis and polyacrylamide gel electrophoresis which contains 8 M urea in the gel. Lipoproteins containing A-I but no A-II contain also about 30% of plasma apoD and 70% LCAT enzyme. Only a small percentage (less than 10%) of plasma E is found in these particles. However, based on gel electrophoresis, proportionally more apoCs are found in these particles than in the Lp(A-I with A-II) particles. Furthermore, the majority (80%) of plasma CET activity is associated with these particles.[29]

Characterization of ApoA-Containing Lipoproteins Isolated from Ultracentrifuged HDL by Various Chromatographic Techniques

ApoA-containing lipoprotein subfractions have also been isolated from ultracentrifugally isolated HDL (d 1.063–1.21 g/ml) by various chromatographic techniques. Using hydroxylapatite column chromatography and stepwise elution with phosphate buffers of molarity between 0.001 to 0.15 M at 4°, Olofsson and Gustafson[15] consistently obtained three subfractions (Ia, II, and III) which contained apoA-I, A-II, C-I, C-II, and D. Subfractions Ia and III also contained apoC-III. Subfraction II, which did not contain C-III, had a relatively high lysolecithin content and a low ratio of unesterified-to-esterified cholesterol. All lipoprotein particles isolated at 4° by this procedure contained both A-I and A-II

By subjecting HDL through QAE-Sephadex A-50 column chromatography to obtain an unretained and a retained fraction, Norfeldt et al.[16] isolated two types of apoA-containing lipoproteins. When the QAE-unretained fraction was further subfractionated by hydroxylapatite and another QAE-Sephadex A-50 column, these investigators obtained a lipoprotein fraction which contained apoA-I as its sole apolipoprotein component. This fraction was homogeneous in size, had a diameter of 8.7 nm based on BioGel A5m column chromatography, and had the density of an HDL$_2$ (d 1.063–1.105 g/ml) fraction. From the unretained QAE fraction, again by using hydroxylapatite column chromatography followed by anti-D-immunosorption and ultracentrifugation at d 1.21 g/ml, they isolated a lipoprotein fraction which contained apoA-I and A-II but no other apolipoproteins. This lipoprotein fraction was also homogeneous in size with a diameter of 7.9 nm estimated from BioGel A5m column chromatography. It also had the hydrated density of an HDL$_2$ fraction. Based on electroimmunoassay and amino acid compositions, the molar ratio between A-I and A-II in this particle was 1.3. The recovery of lipoproteins

[29] M. C. Cheung, A. C. Wolf, K. D. Lum, J. H. Tollefson, and J. J. Albers, *Arteriosclerosis* **5**, 546a (1985).

containing solely A-I and containing only A-I and A-II from 100 mg HDL was 2 and 12 mg, respectively. It is possible that these two types of apoA-containing lipoproteins represent one of the subspecies of Lp(A-I without A-II) and Lp(A-I with A-II). Because both types of particles floated within the d 1.063–1.105 g/ml density range, it is tempting to speculate that they may be corresponding to the size subspecies 7 of Lp(A-I without A-II) and size subspecies 6 of Lp(A-I with A-II), respectively.

Chromatofocusing has also been used to isolate various subfractions of apoA-containing lipoproteins.[17] This method used Polybuffer-Exchanger 94 (Pharmacia, Uppsala, Sweden) as the chromatography media. The exchanger was equilibrated with 25 ml piperazine hydrochloride, pH 5.8, and packed into 1.6 × 30 cm column. HDL (45–55 mg protein in 9 ml) was applied to the column at a flow rate of 20 cm hr^{-1} and eluted with approximately 400 ml Polybuffer 74, diluted 1 : 15 with water, pH 4.0. Application of HDL was preceded by 5 ml of elution buffer. Based on cholesterol analysis, recovery from the column was 85%.

This chromatofocusing procedure isolated six fractions (1–6) from HDL, pH between 5.1 and 4.2. Fractions 1 and 2 floated at d 1.105 g/ml, while fractions 4–6 sedimented at that density. Fraction 3 distributed equally between the top and bottom of the density. The protein/lipid ratio as well as cholesteryl ester content increased from fraction 1–6. However, the unesterified cholesterol content and mean particle diameter decreased from fractions 1–6. The particle diameters of fraction 1–6 based on electron microscopy and negative staining with 1% phosphotungstate were 8.0, 7.6, 6.6, 6.1, 6.3, and 6.0 nm, respectively. With respect to apolipoprotein composition, fractions 1 and 2 contained exclusively apoA-I. Fraction 3 contained only apoA-I and A-II. Fraction 4 contained apoA-I, A-II, and D, while fractions 5 and 6 contained apoA-I, A-II, C, D, and E particles.

Comment: Isolation of apoA-containing lipoproteins using chromatofocusing, hydroxylapatite, and ion-exchange chromatography all require the preisolation of HDL by ultracentrifugation. Because of the losses and redistribution of lipid and proteins during ultracentrifugation, one cannot be sure whether certain subfractions of apoA-containing lipoproteins are representative of native lipoproteins or methodologically derived lipoprotein.

Conclusion

Evidence for the existence of the two major types of apoA-containing lipoproteins with respect to their apoA constituents Lp(A-I with A-II) and Lp(A-I without A-II) have been presented by several investigators over the past 15 years. Their actual isolation from plasma, however, was not achieved until recently. The characteristics of the apoA-containing lipoproteins covered in this chapter represent primarily lipoproteins isolated from normolipidemic plasma (or serum). Little is known about the characteristics of these lipoprotein particles in subjects with lipid disorders or with other diseases. Atmeh *et al.*[28] suggested that Lp(A-I with A-II) and Lp(A-I without A-II) particles are metabolically distinct, and that the A-I in these two lipoprotein species do not completely equilibrate. Therefore, it is likely that Lp(A-I with A-II) particles and Lp(A-I without A-II) particles have different origins, metabolic pathways, and functions. The various methodological schemes described in this chapter for the isolation of these two types of particles should permit one to further understand their physical, chemical, metabolic, and functional characteristics. Of particular importance is to elucidate their relationship to human lipoprotein pathophysiology.

Acknowledgment

Author's research referred to in this chapter was supported by the NIH grant HL-30086.

[9] Characterization of Apolipoprotein E-Containing Lipoproteins

By KARL H. WEISGRABER and ROBERT W. MAHLEY

Apolipoprotein E (apoE) occurs in several animal species and is an apolipoprotein component of a variety of plasma lipoproteins, including very-low-density lipoproteins (VLDL), intermediate-density lipoproteins (IDL), chylomicrons and their remnants, and a subfraction of high-density lipoproteins (HDL).[1-3] The apoE-containing HDL subclasses include HDL_1 and HDL_c, which are more generally referred to as HDL-with apoE. Apolipoprotein E also occurs in a lipoprotein class referred to as β-migrating VLDL (or β-VLDL). These lipoproteins are present in the

[1] R. W. Mahley, *in* "Disturbances in Lipid and Lipoprotein Metabolism" (J. M. Dietschy, A. M. Gotto, Jr., and J. A. Ontko, eds.), p. 181. American Physiological Society, Bethesda, MD, 1978.

[2] R. W. Mahley, T. L. Innerarity, S. C. Rall, Jr., and K. H. Weisgraber, *J. Lipid Res.* **25**, 1277 (1984).

[3] R. W. Mahley and T. L. Innerarity, *Biochim. Biophys. Acta* **737**, 197 (1983).

$d < 1.006$ g/ml fraction of plasma from cholesterol-fed animal models or from humans affected with the genetic lipid disorder, type III hyperlipoproteinemia (familial dysbetalipoproteinemia).[4,5]

An important property of apoE is its ability to interact with cell surface lipoprotein receptors.[3] Lipoproteins that contain apoE can bind to hepatic and extrahepatic apoB,E(LDL) receptors and to hepatic apoE receptors. Several structural variants of apoE are known, some of which have differing receptor binding activities (for review see Ref. 3). Apolipoprotein E is known to be responsible for the hepatic clearance of chylomicron remnants. In addition, in some species the HDL-with apoE subclasses have been postulated to play a role in reverse cholesterol transport, an important process for maintaining cholesterol homeostasis.[3] It has been demonstrated that mutant forms of apoE, which are defective in their ability to interact with receptors, are an underlying factor in type III hyperlipoproteinemia, with defective receptor interaction most likely contributing to the accumulation of hepatic and intestinally derived remnants in the plasma of these subjects.[6–9] Because of the central role played by apoE in lipoprotein metabolism, it is important to be able to separate and isolate the various apoE-containing lipoproteins so that their physical, chemical, and metabolic properties may be characterized. In this chapter, various techniques used to separate apoE-containing lipoproteins will be presented, as well as methods used to characterize their properties.

Isolation Methods

Ultracentrifugation

In the majority of cases, the separation of apoE-containing lipoproteins, as well as most other lipoprotein classes, involves ultracentrifuga-

[4] R. W. Mahley and B. Angelin, Adv. Intern. Med. 29, 385 (1984).
[5] M. S. Brown, J. L. Goldstein, and D. S. Fredrickson, in "The Metabolic Basis of Inherited Disease" (J. B. Stanbury, J. B. Wyngaarden, D. S. Fredrickson, J. L. Goldstein, and M. S. Brown, eds.), 5th ed., p. 655. McGraw-Hill, New York, 1983.
[6] K. H. Weisgraber, T. L. Innerarity, and R. W. Mahley, J. Biol. Chem. 257, 2518 (1982).
[7] S. C. Rall, Jr., K. H. Weisgraber, T. L. Innerarity, and R. W. Mahley, Proc. Natl. Acad. Sci. U.S.A. 79, 4696 (1982).
[8] S. C. Rall, Jr., K. H. Weisgraber, T. L. Innerarity, T. P. Bersot, R. W. Mahley, and C. B. Blum, J. Clin. Invest. 72, 1288 (1983).
[9] W. J. Schneider, P. T. Kovanen, M. S. Brown, J. L. Goldstein, G. Utermann, W. Weber, R. J. Havel, L. Kotite, J. P. Kane, T. L. Innerarity, and R. W. Mahley, J. Clin. Invest. 68, 1075 (1981).

tion, using solid KBr to adjust the density of the solutions.[10] To ensure that chylomicrons are not present, it is necessary to begin with fasted plasma (1 mg EDTA/ml). Serum can also be used; however, plasma is preferable because of the recent concern over the effect of the various coagulation proteases on apolipoproteins. Ultracentrifugation is most often carried out with high-capacity fixed-angle rotors such as the type 60 Ti (Beckman Instruments, Fullerton, CA). It is preferable to begin centrifugation on the day that the plasma is obtained. Several laboratories add various preservatives to plasma before beginning lipoprotein isolation.[11–13] Whether or not this is done, it is important to keep the plasma and lipoprotein fractions at 4° throughout the isolation procedure to prevent deterioration. Isolation of triglyceride-rich lipoproteins that contain high levels of long-chain saturated fatty acids incorporated into the triglyceride represents an exception to this general rule. In these special cases, the high saturated fatty acid content of the triglyceride core imparts liquid crystalline properties to the core, which results in a phase transition from a supercooled liquid state to a crystalline state in the range of 4 to 20°. Isolation of these lipoproteins below their phase transition affects the physical properties.[14] Therefore, the centrifugal isolation and handling of lipoproteins is done above the transition temperature (usually >15°).[14] However, for most applications, isolation, handling, and storage of plasma lipoproteins at 4° is recommended. The source of plasma and the apoE phenotype of the donor represent another concern, which will be discussed in a later section (Lipoprotein Receptor Binding Properties).

It is known that the high shearing forces to which lipoprotein particles are subjected during ultracentrifugation can result in the loss of loosely associated apolipoproteins from the particles. This is a particular concern for apoE,[15] and it appears to be more of a problem with apoE contained in VLDL or with lipoproteins contained in the $d < 1.006$ g/ml fraction than with apoE-containing lipoproteins in the higher-density fractions. For this reason, VLDL are isolated at a lower ultracentrifugal speed. This has been demonstrated to improve the recovery of apoE in VLDL; however, some apoE is still lost to the $d > 1.21$ g/ml fraction. Typical lengths of

[10] R. J. Havel, H. A. Eder, and J. H. Bragdon, *J. Clin. Invest.* **34,** 1345 (1955).
[11] J. P. Kane, D. A. Hardman, and H. E. Paulus, *Proc. Natl. Acad. Sci. U.S.A.* **77,** 2465 (1980).
[12] D. M. Lee, A. J. Valente, W. H. Kuo, and H. Maeda, *Biochim. Biophys. Acta* **666,** 133 (1981).
[13] W. A. Bradley, E. B. Gilliam, A. M. Gotto, Jr., and S. H. Gianturco, *Biochem. Biophys. Res. Commun.* **109,** 1360 (1982).
[14] J. S. Parks, D. Atkinson, D. M. Small, and L. L. Rudel, *J. Biol. Chem.* **256,** 12992 (1981).
[15] M. Fainaru, R. J. Havel, and K. Imaizumi, *Biochim. Med.* **17,** 347 (1977).

time required to isolate lipoprotein classes with a type 60 Ti rotor are as follows: VLDL ($d < 1.006$ g/ml) or $d < 1.02$ g/ml lipoproteins, 16 hr at 50,000 rpm followed by a recentrifugation (wash) at the same density, time, and speed to remove plasma contaminants; $d = 1.006–1.019$ or $1.006–1.020$ g/ml fraction (IDL), 16 hr at 59,000 rpm, followed by a wash; $d = 1.020–1.050$ or $1.020–1.063$ g/ml fraction [low-density lipoproteins (LDL)], 18 hr at 59,000 rpm, followed by a wash; $d = 1.063–1.125$ g/ml fraction (HDL$_2$), 24 hr at 59,000 rpm, followed by a wash; and $d = 1.125–1.21$ g/ml fraction (HDL$_3$), 40 hr at 59,000 rpm, followed by a 24-hr wash. All centrifugation steps are carried out at 4°. When processing small volumes of plasma, similar density fractions as those described above can be obtained in a single spin using density gradient ultracentrifugation.[16]

An alternative to sequential density ultracentrifugation or density gradient centrifugation is zonal centrifugation.[17] This technique is particularly useful for HDL subfractionation and has been used to subfractionate $d < 1.006$ g/ml lipoproteins from cholesterol–fat-fed rabbits.[18] The disadvantages of this method are that only a limited amount of plasma can be processed, and the subfractions are obtained in a relatively dilute form. However, in special cases the disadvantages are offset by the resolution that can be obtained.

After centrifugation, the lipoprotein fractions are typically dialyzed against 0.15 M NaCl, pH 7.4, containing 0.01% EDTA. This is a good general solution that maintains the integrity of the lipoprotein particles; it may be buffered with 10 mM phosphate. If lipoproteins are to be used for apolipoprotein preparation (see Chapter [13], Volume 128), they are dialyzed against distilled water containing 0.01% EDTA, pH 7.4, and then lyophilized and delipidated. In many instances, it is necessary to concentrate a dilute lipoprotein fraction before it can be used. This is conveniently done with an Amicon Diaflo apparatus (Amicon, Lexington, MA) using PM30 membranes.

Pevikon Block

The Pevikon block method is a preparative electrophoretic procedure that is extremely useful for a wide variety of lipoprotein separations. The procedure is based on the method of Barth,[19] and electrophoresis is car-

[16] T. G. Redgrave, D. C. Roberts, and C. E. West, *Anal. Biochem.* **65,** 42 (1975).
[17] J. R. Patsch, S. Sailer, G. Kostner, F. Sandhofer, A. Holasek, and H. Braunsteiner, *J. Lipid Res.* **15,** 356 (1974).
[18] R. I. Roth, J. W. Gaubatz, A. M. Gotto, Jr., and J. R. Patsch, *J. Lipid Res.* **24,** 1 (1983).
[19] W. F. Barth, R. D. Wochner, T. A. Waldmann, and J. L. Fahey, *J. Clin. Invest.* **43,** 1036 (1964).

ried out in a flat bed containing inert Pevikon as support medium. In the original description of the method for lipoprotein isolation, a 50 : 50 mixture of Pevikon and Geon was used.[20] Variations among the batches of Pevikon and Geon have been found over the years, the most noticeable being with Geon. Presently, we recommend that Pevikon C870 be used alone. This modification gives satisfactory results in all applications of the method.

Critical to the success of the method is construction of the electrophoretic apparatus according to specific plans (plans available from the authors).[20] Proper construction ensures that the correct height of the bed relative to the buffer tanks is maintained. This height affects the moisture content of the Pevikon bed. If the bed is too wet or too dry, the lipoprotein separations will be adversely affected.

The apparatus is made from Plexiglas and can be built for a modest cost. It consists of two 4-liter buffer chambers ($6 \times 12 \times 6\frac{1}{2}$ in.) and an electrophoretic tray ($13\frac{3}{8} \times 15 \times \frac{5}{8}$ in.), which contains the Pevikon bed (Fig. 1). The tray and buffer chambers are constructed of $\frac{1}{4}$-in. Plexiglas and both have $\frac{1}{8}$-in. Plexiglas covers.

Equipment/Reagents

1. Electrophoretic apparatus (plans available from the authors)
2. Pevikon C870, 600 g (Mercer Consolidated Corp., Yonkers, NY)
3. Power supply to deliver 50 mA constant current
4. Electrode wicks, 3×8-in. Telfa surgical dressing pads (Kendall Hospital Products, Boston, MA)
5. Barbital buffer, pH 8.6, ionic strength 0.06, 8 liters (16.56 g barbituric acid plus 92.4 g sodium barbiturate in 8 liters of distilled water)
6. Parafilm, large roll
7. Coarse-porosity sintered glass funnel (125-mm i.d.) and vacuum flask
8. Coarse-porosity glass funnels (45-mm i.d.) and side-arm test tubes, 200×25 mm
9. Sudan Black B (lipid stain)
   a. Stock solution: 1 g Sudan Black B in 100 ml diethylene glycol (prepared by heating to 110° and filtering)
   b. Prestaining solution: 0.5 ml stock solution diluted to 10 ml with H_2O. Take 1 ml and add 0.1 ml Tween 20 (1% stock solution)
10. Towels for draining

[20] R. W. Mahley and K. H. Weisgraber, *Biochemistry* **13**, 1964 (1974).

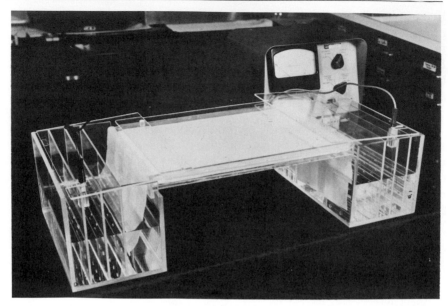

FIG. 1. Photograph of the Pevikon block apparatus. Plans for construction may be obtained from the authors. From R. W. Mahley and K. H. Weisgraber, *in* "CRC Handbook of Electrophoresis" (L. A. Lewis and H. K. Naito, eds.), Vol. IV: Lipoproteins of Nonhuman Species, p. 151. CRC Press, Boca Raton; reproduced with permission.

Procedure. The Pevikon is washed prior to use by adding 900 ml of distilled H_2O to 600 g Pevikon in a 2-liter beaker and stirring to form a slurry. After standing for 10 min, the fine particles are aspirated by vacuum from the slurry. The slurry is then transferred to a large coarse-porosity sintered glass funnel and washed twice more with H_2O (900 ml each time) with the aid of vacuum. After the last wash, most of the water is removed by suction, and the Pevikon is resuspended in the funnel with 800 ml of barbital buffer (pH 8.6, ionic strength 0.06). Most of the buffer is also removed by suction, and the Pevikon is resuspended a second time in 500 ml of barbital buffer. Parafilm is placed over the stem of the funnel to prevent further draining, and the slurry is transferred (avoiding generation of bubbles) to a 2-liter beaker.

The electrophoretic tray is lined with Parafilm, and four Telfa pad wicks, moistened with buffer, are placed across each end of the tray. The wicks are pressed firmly onto the bottom of the tray (~1 in. from the end) and extend over the ends of the tray. The Pevikon slurry (~1100 ml is needed) is poured carefully into the tray (avoid generating bubbles and dislodging the wicks from the bottom of the tray). The excess buffer is allowed to drain from the bed through the wicks onto towels for approximately 15 min (Fig. 2). To check for proper consistency, a slit is made in the medium. The walls of the slit should not collapse. Overdrying the block can be avoided by rolling up the Telfa wicks and placing them on top of the block.

A slit for loading the lipoproteins is made in one end of the block, approximately 7 cm from the end. The slit can extend across most of the block and should penetrate to within approximately 2 mm of the bottom of the support medium. A slit-forming aid and modified spatula that extend to within 2 mm of the bottom (pictured in Fig. 2) are helpful. The lipoproteins, usually in saline–EDTA solution, are applied in a volume of 10–12 ml to the sample loading slit. Up to 100 mg of a lipoprotein mixture can be resolved. The sample slit is flanked by two smaller slits (~1 cm); an aliquot of the lipoprotein mixture that has been prestained with Sudan Black B is added to each of these smaller slits (0.2 ml lipoprotein plus 0.2 ml stain). The stained aliquot aids in the location of lipoprotein bands, particularly in cases where the lipoprotein load is low or the lipoproteins are not colored.

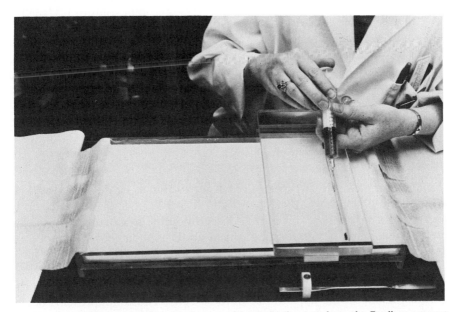

FIG. 2. Photograph of the technique used to apply the sample to the Pevikon support medium contained in the electrophoretic tray. From R. W. Mahley and K. H. Weisgraber, *in* "CRC Handbook of Electrophoresis" (L. A. Lewis and H. K. Naito, eds.), Vol. IV: Lipoproteins of Nonhuman Species, p. 151. CRC Press, Boca Raton; reproduced with permission.

The cover is placed on the tray, and the tray is placed on the buffer chambers, each containing 3 liters of chilled (4°) barbital buffer. Contact between the block and the buffer chambers is made through the Telfa wicks (Fig. 1). It is important that buffer chambers are placed on a level surface and that each chamber contains exactly the same amount of buffer. This is to avoid a unidirectional siphoning of buffer through the Pevikon bed, which will affect the migration distances and resolution of the lipoproteins. Use of a rubber tube siphon between the buffer chambers prior to placing the electrophoretic tray on them is a convenient way to ensure that the buffer levels are equal. The lipoproteins migrate toward the positive electrode. Therefore, the block should be placed on the buffer chambers with the end containing the site of lipoprotein application situated farthest from the positive electrode. After electrophoresis overnight (18 hr) at 4° and 50 mA constant current, the excess moisture is drained from the block through the buffer wicks onto towels. Placing the block tray on a light box aids in the visualization of the lipoprotein bands. The bands within the Pevikon are removed from the block with a spatula and transferred to the small coarse-porosity sintered glass funnel. The lipoproteins are eluted from the Pevikon with several saline rinses with the aid of gentle suction (avoid foaming of the lipoprotein solution). Usually, a total volume of 50 ml is sufficient. Recoveries are usually greater than 75%. Any Pevikon particles that pass through the sintered glass filter are removed by low-speed centrifugation. The eluted lipoproteins can be concentrated with an Amicon Diaflo apparatus. Typically, normal human LDL migrates as a sharp band ~5–6 cm from the point of application. The major portion of HDL (HDL-without apoE) migrates ~12–15 cm and the apoE-containing HDL ~7–11 cm.

β-VLDL Isolation

The first example of Pevikon block separation to be discussed will be the separation of β-VLDL from VLDL. The β-VLDL from cholesterol-fed animals or humans with type III hyperlipoproteinemia are cholesterol-rich lipoproteins that contain apoB and apoE as their major apolipoprotein moieties.[1-3] An important metabolic characteristic of β-VLDL is their ability to cause massive cholesteryl ester accumulation in macrophages.[3] This ester accumulation is in the form of lipid droplets with the macrophages taking on the appearance of foam cells. These lipoproteins occur in the $d < 1.006$ g/ml density fraction along with VLDL and are distinguished from the VLDL in that they exhibit β (LDL) migration, while the VLDL migrate in a pre-β position (Fig. 3). Thus, the Pevikon method is

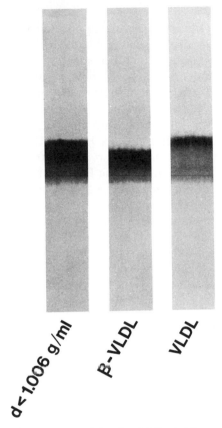

FIG. 3. Paper electrophoretograms of the $d < 1.006$ g/ml lipoprotein fraction from the plasma of a type III hyperlipoproteinemic individual and the β-VLDL and pre-β-VLDL fractions that were separated by Pevikon block electrophoresis.

ideal for their separation. The β-VLDL are composed of two subclasses that can be resolved by agarose column chromatography (to be discussed in a later section).

The β-VLDL migrate on the Pevikon block as a 3-cm-wide band with the center of the band ~6–8 cm from the origin. In type III subjects, the β-VLDL are orange. The VLDL appear yellow-orange and migrate as an ~2-cm-wide band just ahead of the β-VLDL band. The separation of the β-VLDL and VLDL is demonstrated in Fig. 3. Identical separations can be achieved with the $d < 1.006$ g/ml density fraction of cholesterol-fed animal models.

Rat and Canine HDL₁ Isolation

In isolating and studying the plasma lipoproteins in animal models, it is often assumed that the density ranges established for human lipoproteins will apply to the lipoproteins of other species. It is now clear that the application of human density ranges to define animal plasma lipoproteins is inadequate. This is particularly true for LDL (d = 1.02–1.063 g/ml). As is illustrated in Fig. 4, the d = 1.02–1.063 g/ml fraction of rat plasma contains two distinctly different lipoproteins, a β-migrating LDL and an α_2-migrating lipoprotein referred to as HDL₁ (HDL-with apoE). Canine plasma also contains HDL₁ in the d = 1.02–1.063 g/ml fraction. These lipoproteins are easily separated on the Pevikon block. The LDL migrate ~6 cm from the origin and the HDL₁ ~10–12 cm. The presence of these two lipoprotein classes in the rat has important metabolic implications. The LDL contain apoB as their exclusive apolipoprotein component, whereas the HDL₁ contain primarily apoE.[21] Thus, simply using the d = 1.02–1.063 ultracentrifugal density fraction to define the LDL of the rat is inadequate, because it is in fact a mixture of two lipoproteins, each with distinct chemical and metabolic characteristics. A lipoprotein with properties similar to the rat HDL₁ described above has also been isolated using zonal ultracentrifugation.[22,23] A summary of their chemical and physical properties is presented in Table I.

Canine Apolipoprotein E HDLc Isolation

The separation of HDLc is another instance in which the Pevikon method can be used to separate apoE-rich lipoproteins. These lipoproteins appear to be equivalent to the HDL₁ of the dog and rat and are induced by cholesterol feeding.[1–3] For this reason, they are distinguished from the HDL₁ by the use of the subscript "c." The HDL₁ and HDLc are HDL subclasses generally designated as HDL-with apoE. The HDLc occur in the density range 1.006 to 1.090 g/ml and represent a spectrum of lipoprotein particles that are characterized on the low-density end as being enriched in apoE and on the high-density end as being enriched in apoA-I.[2] The d = 1.006–1.02 g/ml fraction of cholesterol-fed dogs is particularly enriched with HDL-with apoE particles, referred to as apoE HDLc (see Table I for chemical composition). In many instances, apoE is

[21] K. H. Weisgraber, R. W. Mahley, and G. Assmann, *Atherosclerosis* **28,** 121 (1977).
[22] B. Danielsson, R. Ekman, B. G. Johansson, P. Nilsson-Ehle, and B. G. Petersson, *FEBS Lett.* **86,** 299 (1978).
[23] L. T. Lusk, L. F. Walker, L. H. DuBien, and G. S. Getz, *Biochem. J.* **183,** 83 (1979).

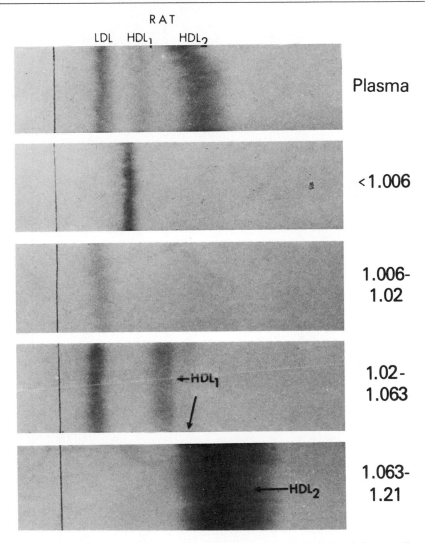

FIG. 4. Paper electrophoretograms of normal rat plasma and the isolated ultracentrifugal density fractions. The d = 1.02–1.063 g/ml fraction contains both the β-migrating LDL and the α_2-migrating HDL$_1$. From K. H. Weisgraber, R. W. Mahley, and G. Assmann, *Atherosclerosis* **28,** 121 (1977); reproduced with permission.

TABLE I

SUMMARY OF CHEMICAL AND PHYSICAL PROPERTIES OF ApoE-CONTAINING LIPOPROTEINS

Lipoprotein	Density (g/ml)	Chemical composition[a]				Size (Å)	Electrophoretic mobility	Additional apolipoproteins	Method of isolation
		TG	TC	PL	Protein				
Human VLDL	$d < 1.006$	56.4	14.6	22.8	6.2	>750	Pre-β	B-100, C	Ultracentrifugation Pevikon block
Human β-VLDL	$d < 1.006$	34.6	34.4	23.1	7.9	300–1000	β	B-100, B-48, C	Ultracentrifugation Pevikon block
Human β-VLDL-I	$d < 1.006$	46.2	26.0	24.4	3.4	~800	Origin	B-48, C	Agarose A5M
Human β-VLDL-II	$d < 1.006$	25.7	35.6	26.6	12.1	~400	β	B-100, C	Agarose A5M
Canine apoE HDL$_c$	$d = 1.006$–1.02	2.3	50.8	31.4	15.3	~230	α_2	A-I, C	Ultracentrifugation Pevikon block
Canine apoE HDL$_c$	$d = 1.006$–1.02	4.4	41.2	33.4	21.2	220–250	α_2	A-I, C	Agarose A15M Pevikon block
Canine HDL$_c$	$d = 1.02$–1.063	3.0	38.5	37.4	20.8	150–300	α_2	A-I, C	Ultracentrifugation Pevikon block
Canine HDL$_1$	$d = 1.02$–1.063	2.2	34.6	40.6	22.6	120–350	α_2	A-I, C	Ultracentrifugation Pevikon block
Rat HDL$_1$	$d = 1.02$–1.063	4.0	39.9	26.0	30.1	140–290	α_2	A-I, C	Ultracentrifugation Pevikon block
Human HDL-with apoE	$d = 1.063$–1.125	3.2	22.3	33.4	41.0	100–150	α_2	A-I, A-II, C	Ultracentrifugation Pevikon block
Human HDL-with apoE	$d = 1.063$–1.21	1.3	25.8	35.1	36.8	100–150	α_2	A-I, A-II, C	Ultracentrifugation Heparin–Sepharose

[a] Expressed as a percentage of the total. TG, triglyceride; TC, total cholesterol; PL, phospholipid.

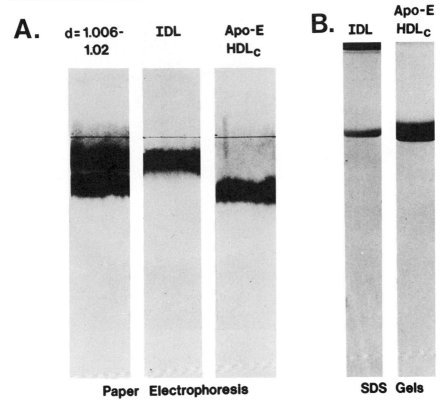

FIG. 5. Paper electrophoretograms of the $d = 1.006-1.02$ g/ml density fraction and the Pevikon block isolated fractions (A). The sodium docecyl sulfate–polyacrylamide gels (B) of the isolated IDL and apoE HDL$_c$. From R. W. Mahley and K. H. Weisgraber, *in* "CRC Handbook of Electrophoresis" (L. A. Lewis and H. K. Naito, eds.), Vol. IV: Lipoproteins of Nonhuman Species, p. 151. CRC Press, Boca Raton; reproduced with permission.

the exclusive apolipoprotein component of apoE HDL$_c$. As shown in Fig. 5, the apoE HDL$_c$ can be isolated from the β-migrating lipoproteins by Pevikon electrophoresis.

Human HDL-with Apolipoprotein E Isolation

The Pevikon block can also be used to isolate the HDL-with apoE subclass of human HDL.[24] A Pevikon block of the 1.063–1.125 g/ml fraction from normal human plasma is shown in Fig. 6. Zone I lipoproteins represent HDL that do not contain apoE. These lipoproteins are charac-

[24] K. H. Weisgraber and R. W. Mahley, *J. Biol. Chem.* **253**, 6281 (1978).

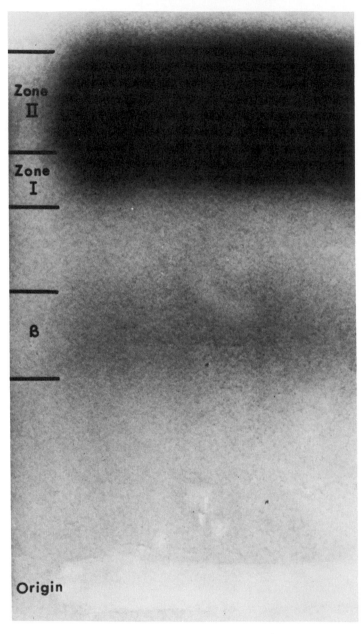

FIG. 6. Photograph of a Pevikon block electrophoretic separation of the d = 1.063–1.125 g/ml fraction from normal human plasma. From K. H. Weisgraber and R. W. Mahley, *J. Biol. Chem.* **253,** 6281 (1978); reproduced with permission.

terized by the presence of apoA-I and apoA-II (Table I). Zone II represents the HDL-with apoE subclass. In addition to apoA-I and apoA-II, this subclass contains apoE and may contain the disulfide-linked heterodimer apo(E–A-II). The β zone contains LDL-like particles with apoB as the main apolipoprotein constituent. The human HDL-with apoE can also be obtained by heparin–Sepharose affinity chromatography from this density fraction (to be discussed in a later section).

Agarose Chromatography

The β-VLDL fraction isolated from the Pevikon block consists of two subclasses representing remnants derived from both the intestine and liver.[25] Because the intestinally derived particles are significantly larger (>700 Å) than the hepatic remnants (~400 Å), these subclasses can be separated by agarose column chromatography. A procedure for the use of this separation technique with either β-VLDL from cholesterol–fat-fed animal models or type III subjects has been described by Fainaru et al.[25] Briefly, the β-VLDL in a saline–EDTA buffer are passed through an agarose A5M (Bio-Rad, Richmond, CA) column (2.5 × 95 cm). This results in the separation of two peaks. The first peak, designated fraction I, elutes at the void volume of the column and is characterized by the presence of apoE and apoB-48, the intestinal form of apoB. The second peak, fraction II, elutes later and contains apoE and apoB-100, the hepatic form of apoB. The chemical and physical properties are summarized in Table I. It is noteworthy that fraction I is more effective in the delivery of cholesterol to macrophages than fraction II,[25] suggesting a potential role for these lipoproteins in the deposition of lipid in the arterial wall during the formation of an atherosclerotic plaque. Fraction I is also cleared in vivo at a faster rate than fraction II.[25]

A combination of agarose chromatography and Pevikon block electrophoresis can be used to isolate canine apoE HDL$_c$ directly from plasma.[26] With this approach the purified apoE HDL$_c$ can be obtained without the use of ultracentrifugation. Chromatography of 20 ml of plasma from a high fat/high cholesterol-fed dog on an agarose A15M column (4.0 × 130 cm, equilibrated with saline–EDTA) results in the separation of a fraction enriched in β-migrating lipoproteins and apoE HDL$_c$. In this fraction there is only a moderate amount of contamination with other plasma proteins. The apoE HDL$_c$ can be purified from this mixture directly by

[25] M. Fainaru, R. W. Mahley, R. L. Hamilton, and T. L. Innerarity, J. Lipid Res. **23,** 702 (1982).
[26] H. Funke, J. Boyles, K. H. Weisgraber, E. H. Ludwig, D. Y. Hui, and R. W. Mahley, Arteriosclerosis **4,** 452 (1984).

Pevikon block electrophoresis. Apolipoprotein E is the major apolipoprotein component of the isolated lipoprotein, and the chemical, physical, and metabolic properties of this preparation are identical to the properties of apoE HDL$_c$ isolated with an ultracentrifugation step.[26]

Heparin–Sepharose Affinity Chromatography

Of the various apolipoproteins, only apoB and apoE are known to interact with heparin and other proteoglycans. The subfractionation of HDL by heparin–Sepharose affinity chromatography takes advantage of this property. As discussed in a previous section, human HDL contains a subclass that is characterized by the presence of apoE, the HDL-with apoE subclass. The heparin–Sepharose subfractionation method was developed to provide a reproducible and quantitative method to isolate this subclass. It was also the first method used to subfractionate human HDL based on a metabolic property, i.e., lipoprotein receptor binding activity, rather than physical properties.[27]

Preparation of Heparin–Sepharose

The heparin–Sepharose is prepared by a modification of the method of Iverius and Östund-Lindqvist.[28] All operations should be carried out in a well-ventilated hood. To 100 ml of settled Sepharose 6B CL (Pharmacia Fine Chemicals, Uppsala, Sweden), 50 ml of deionized water is added. The pH is adjusted to 11.0 with 2 M NaOH, and CNBr (9 g dissolved in 300 ml distilled water) is added and stirred with a magnetic stirrer. The pH is maintained at ~11.0 for 6 min with 2 M NaOH, then the gel is transferred to a large sintered glass funnel (medium porosity) and washed with 1 liter chilled (4°) distilled water, followed by 1 liter chilled 0.1 M NaHCO$_3$. Excess CNBr is destroyed with NaOH pellets. For 75 ml of CNBr-activated gel, 75 ml chilled 0.1 M NaHCO$_3$ and 160 mg sodium heparin (Hynson, Westcott and Dunning Inc., Baltimore, MD) are added and stirred gently with a magnetic stirrer for 18 hr at 4°. The excess unreacted activated sites on the Sepharose are blocked by the addition of 7.5 ml ethanolamine with continued stirring for 4 hr at 4°. The gel is transferred to a sintered glass funnel and washed with 1 liter distilled water, 0.5 liter 0.15 M NaCl, and then 3 liters distilled water. The gel is stored at 4° in distilled water containing 0.01% Merthiolate until ready for use. Heparin–Sepharose prepared in this manner has consistently given

[27] K. H. Weisgraber and R. W. Mahley, *J. Lipid Res.* **21**, 316 (1980).
[28] P. H. Iverius and A. M. Östund-Lindqvist, *J. Biol. Chem.* **251**, 7791 (1976).

reliable separations. Commercial preparations of heparin–Sepharose gels and those prepared by other procedures have not produced uniform results in our hands.

Separation of HDL-with Apolipoprotein E

The affinity separation of human HDL-with apoE is carried out on the $d = 1.063–1.125$ or $d = 1.063–1.21$ g/ml density fractions prepared by ultracentrifugation. The HDL density fraction is dialyzed against 5 mM Tris–HCl, pH 7.4, containing 25 mM NaCl. For a 1.0×30 cm column, up to 30 mg of HDL protein in 2.5–3.5 ml can be fractionated. Just prior to applying the sample to the column, solid MnCl$_2$ is added to the sample to give a final Mn^{2+} concentration of 25 mM. The column is equilibrated with 2.5 mM Tris, pH 7.4, containing 25 mM NaCl and 25 mM MnCl$_2$. In preparing the column buffer, the Tris–HCl solution is adjusted to pH <7.8 before adding the MnCl$_2$. This prevents the formation and precipitation of Mn(OH)$_2$. Chromatography is carried out at 4° with a peristaltic pump operating at a rate of 24 ml/hr, and 3.6-ml fractions are collected. The sample is pumped on the column, and the tubing is rinsed with column buffer to ensure that all of the sample is in contact with the column. The sample is then allowed to equilibrate with the column overnight. Two methods can be used to elute the columns (method A or B described below). Both give an identical HDL-with apoE fraction; however, method B partially resolves the apoB-containing lipoproteins that are present in the HDL density fraction into two subclasses.

Method A. Following the overnight equilibration, the column is eluted with the Mn^{2+}-containing buffer until fraction 11 has been collected (3.6 ml/fraction). This results in the elution of the unbound lipoproteins. This fraction represents the majority (>85%) of the HDL and does not contain either apoE or apoB. At fraction 11, the elution buffer applied to the top of the column is changed to 2.5 mM Tris–HCl, pH 7.4, containing 95 mM NaCl (Mn^{2+} is not present). This buffer has the same ionic strength as the Mn^{2+}-containing buffer. Elution with this buffer is continued until tube 27 and results in the elution of the HDL-with apoE (approximately tubes 24 to 29). The third eluting buffer is started at tube 27 and contains 0.29 M NaCl and 5 mM Tris–HCl (pH 7.4). This buffer elutes the apoB-containing lipoproteins, which may contain variable amounts of apoE (approximately tubes 37 to 43). A 0.6 M Tris–NaCl buffer containing 0.01% Merthiolate is used to strip the column (start at tube 37). The column is stored in this buffer in between runs and is reequilibrated with the initial column buffer just prior to applying sample. The elution is monitored at 280 nm. The peak fractions are pooled and immediately dialyzed against NaCl–

EDTA (0.15 M NaCl, 0.01% EDTA, pH 7.0). Recoveries of lipoprotein protein applied to the column are greater than 90%.

Method B. This method is identical to method A up to the end of the second elution buffer (tube 27). At this point a 0.115 M Tris–NaCl HCl buffer is applied and continued for 10 tubes. At tube 37 the NaCl concentration is increased to 0.29 M, and the column is stripped (0.6 M NaCl) at tube 47. Method B results in partial resolution of the apoB-containing lipoproteins into beta- (β_1) and pre-β-migrating (β_2) fractions.

It is important to note that with each new batch of heparin–Sepharose, it is necessary to determine that the ionic strength of the second buffer (the one that elutes the HDL-with apoE) is giving the maximal resolution of the apoE- and apoB-containing lipoproteins. If apoB is eluting with the HDL-with apoE, it is necessary to lower the ionic strength of the second elution buffer. If the second buffer is not eluting all of the HDL-with apoE, then the ionic strength should be increased. These adjustments are usually minor (0.075–0.100 M NaCl) but are necessary to ensure maximum resolution.

Characterization of E-Containing Lipoproteins

Sodium Dodecyl Sulfate–Polyacrylamide Gel Electrophoresis

The apolipoprotein composition of a lipoprotein class can be assessed by sodium dodecyl sulfate–polyacrylamide gel electrophoresis. Although the molecular weights of homologous apolipoproteins differ slightly in the various species, the apparent molecular weights on sodium dodecyl sulfate gels can be used as a first step in the identification of a particular apolipoprotein in most cases. The molecular weights for apoE, A-I, A-II monomer, E–A-II, A-IV, C's, B-100, and B-48 are 34,000, 28,000, 8000, 46,000, 46,000, 8000–12,000, 550,000, and 225,000, respectively. In all species examined to date, only human and chimpanzee apoA-II contain the amino acid cysteine and are found as a dimer. Human apoE-2 and E-3 also contain cysteine (see Chapter [13], Volume 128) and can exist as dimers and heterodimers linked with apoA-II. The disulfide-linked multimers are easily identified by comparing the molecular weights of a nonreduced sample with a sample that has been reduced with a disulfide reducing agent, such as β-mercaptoethanol or dithiothreitol. For example, the $M_r = 46,000$ apo(E–A-II) complex can be distinguished from apoA-IV by reduction to its monomeric components: apoE ($M_r = 34,000$) and apoA-II ($M_r = 8500$).

The lipoproteins should be delipidated prior to application to the gels.

This is particularly important for the lipid-rich lipoproteins such as VLDL and β-VLDL. Delipidation can be done with CHCl₃ : MeOH (2 : 1) prior to addition of the sodium dodecyl sulfate sample buffer. As discussed above, it is informative to run both a reduced and nonreduced sample to identify disulfide-linked structures. The major apolipoproteins contained in several of the apoE-containing lipoproteins are presented in Table I.

Chemical and Physical Characteristics

The lipid composition of an isolated lipoprotein class is an important chemical characteristic that can be used to compare the similarity of lipoprotein classes isolated in other laboratories or by other methods. Most commonly, the three major lipid classes are analyzed by standard procedures: triglycerides and total cholesterol by enzymatic assay (Bio-Dynamics, Boehringer–Mannheim, Corp., Indianapolis, IN) and phospholipids by phosphorus analysis.[29] The lipid values and the protein content[30] of a particular lipoprotein class are usually normalized and expressed as a percentage of the total composition of the lipoprotein. The percentage of cholesteryl ester can also be determined.[31] In special cases, it may be useful to determine the phospholipid classes[32] or the fatty acid composition of the triglycerides, phospholipids, or cholesteryl esters.[33] The lipid compositions of many of the apoE-containing lipoproteins discussed in previous sections are presented in Table I.

A useful physical characteristic that helps to define a lipoprotein class is particle diameter, which can be determined by negative staining electron microscopy.[34] The size distribution allows determination of particle homogeneity in an isolated fraction. For example, with the Pevikon block-isolated β-VLDL, the size distribution reveals the presence of two subpopulations of particles;[25] this information was useful in devising the agarose column method to separate the two populations of lipoprotein particles (see Agarose Column Chromatography, above). Typical sizes of the various apoE-containing lipoproteins are also presented in Table I.

[29] G. R. Bartlett, *J. Biol. Chem.* **234,** 466 (1959).

[30] O. H. Lowry, N. J. Rosebrough, A. L. Farr, and R. J. Randall, *J. Biol. Chem.* **193,** 265 (1951).

[31] R. E. Pitas, G. J. Nelson, R. W. Jaffe, and R. W. Mahley, *Lipids* **14,** 469 (1979).

[32] W. W. Christie, "Lipid Analysis-Isolation, Separation, Identification, and Structural Analysis of Lipids," 2nd Ed., p. 63. Pergamon, New York, 1973.

[33] W. W. Christie, "Lipid Analysis-Isolation, Separation, Identification, and Structural Analysis of Lipids," 2nd Ed., p. 107. Pergamon, New York, 1973.

[34] R. L. Hamilton, D. M. Regen, M. E. Gray, and V. S. LeQuire, *Lab. Invest.* **16,** 305 (1967).

Electrophoretic Properties

Paper electrophoresis is also an aid in characterizing lipoprotein classes and assessing their purity. We most commonly use paper electrophoresis, employing the method of Hatch and Lees.[35] Electrophoresis on agarose gels or cellulose acetate is widely used. Examples of paper electrophoretograms are presented in Figs. 3–5. With the exception of chylomicrons and their remnants, the relative migration of lipoproteins on paper electrophoresis corresponds with their relative migration on Pevikon block electrophoresis. Thus, paper electrophoresis is a convenient method to determine if the Pevikon block will resolve a particular lipoprotein mixture; it may subsequently be used as a method to monitor the effectiveness of the purification (see Figs. 3 and 5).

Gradient gel electrophoresis on polyacrylamide slab gels (Pharmacia electrophoresis apparatus) is another method used to characterize lipoprotein classes. This method is sensitive to the size of lipoprotein particles and can be used to estimate the molecular weight of lipoprotein particles.[36] Electrophoresis is performed on 4 to 30% or 2 to 16% gels (Pharmacia Fine Chemicals), depending on the particle sizes of interest. The lipoproteins are electrophoresed at 12° in a Tris–borate buffer as described.[27] The gels can be stained either with oil red O to stain for lipid or with Coomassie Blue to stain for protein.

Lipoprotein Receptor Binding Properties

An important characteristic of apoE-containing lipoproteins is their ability to bind to lipoprotein receptors. The most common assay system used to measure receptor binding is the apoB,E(LDL) receptor binding assay using cultured human fibroblasts. This assay as well as other receptor binding assays are discussed in detail in Chapter [33], this volume. Two important points need to be considered when isolating human apoE-containing lipoproteins for characterization of their receptor binding properties. First, it is highly desirable to use plasma from a single individual as a source of the lipoproteins. This eliminates the possibility of dealing with a complex mixture of apoE structural variants, each potentially having differing receptor binding characteristics. (See Table II for a summary of the receptor binding activities of known apoE variants compared to apoE-3, the most common form in the population.) Second, the apoE phenotype of the donor should be known, including the cysteine content

[35] F. T. Hatch and R. S. Lees, *Adv. Lipid Res.* **6,** 1 (1968).
[36] D. W. Anderson, A. V. Nichols, T. M. Forte, and F. T. Lindgren, *Biochim. Biophys. Acta* **493,** 55 (1977).

TABLE II
RELATIVE RECEPTOR BINDING ACTIVITIES OF KNOWN APOLIPOPROTEIN
E VARIANTS COMPARED TO APOLIPOPROTEIN E-3

Isoelectric focusing position	Substitution(s) relative to E-3 (most common form)	Receptor binding activity relative to apoE-3
E-3	—	100%
E-4	$Cys_{112} \rightarrow Arg$	100%
E-2	$Arg_{158} \rightarrow Cys$	<2%
E-2	$Arg_{145} \rightarrow Cys$	45%
E-2	$Lys_{146} \rightarrow Gln$	40%
E-3	$Cys_{112} \rightarrow Arg, Arg_{142} \rightarrow Cys$	<20%
E-3	$Ala_{99} \rightarrow Thr, Ala_{152} \rightarrow Pro$	Unknown
E-1	$Gly_{127} \rightarrow Asp, Arg_{158} \rightarrow Cys$	4%

of the individual's apoE (see Chapter [13], Volume 128) for a more complete discussion of apoE phenotypes and apoE phenotyping). This consideration relates, in part, to the first. If the subjects are heterozygotes, the two forms of apoE may have different receptor binding activities. This is particularly true when apoE-2 is present, as all known apoE-2 variants have been demonstrated to have a receptor binding defect (Table II). Thus, the apoE phenotype should be considered as an essential part of the characterization of apoE-containing lipoproteins. The importance of establishing the cysteine content of the subject's apoE relates to the presence or absence of the apo(E–A-II) complex and dimeric apoE. It has been determined that when apoE is complexed with apoA-II, nearly all of the receptor binding activity of the apoE is masked.[37] Thus, the relative amounts of uncomplexed and complexed apoE become an important consideration. If necessary, methods are available to reduce and alkylate the apoE-containing lipoproteins to eliminate this complication.[37]

One very important difference in receptor binding properties exists between apoE- and apoB-containing lipoproteins. This difference relates to the fact that apoE-containing lipoproteins are capable of binding to multiple receptor binding sites on apoB,E(LDL) receptors, whereas apoB-containing lipoproteins cannot. The model in Fig. 7 illustrates this point and was developed from the results of Scatchard analysis of the binding data of apoE HDL$_c$ and LDL and the use of radiation inactivation to determine the molecular weight of the apoB,E(LDL) receptor.[3] The model proposes that each apoB,E(LDL) receptor unit has a molecular

---

[37] T. L. Innerarity, R. W. Mahley, K. H. Weisgraber, and T. P. Bersot, *J. Biol. Chem.* **253**, 6289 (1978).

The Apoprotein B,E Cell Surface Receptor Unit

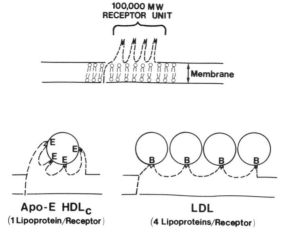

FIG. 7. Model of the hypothetical structure of the apoB,E(LDL) receptor unit depicting the multiple binding sites, which can accommodate one apoE HDL$_c$ particle or four LDL particles. From R. W. Mahley and T. L. Innerarity, *Biochim. Biophys. Acta* **737**, 197 (1983); reproduced with permission.

weight of ~100,000 and contains multiple binding sites. Each receptor molecule is capable of binding one apoE HDL$_c$ particle by interacting with four apoE molecules on the apoE HDL$_c$ particle. In contrast, each receptor can bind up to four LDL particles, presumably because there is only one binding site on apoB per LDL particle or because the binding sites are spaced such that multiple interactions are not possible. As a consequence, apoE-containing lipoproteins are more effective competitors than LDL for apoB,E(LDL) receptors. Because of the multiple interactions, apoE-containing lipoproteins saturate at lower concentrations and bind with a higher affinity than LDL.

Acknowledgments

The authors would like to thank Kerry Humphrey for manuscript preparation, Barbara Allen and Sally Gullatt Seehafer for editorial assistance, and James Warger for graphic arts.

[10] Quantitation, Isolation, and Characterization of Human Lipoprotein (a)

By JOHN W. GAUBATZ, GARY L. CUSHING, and JOEL D. MORRISETT

Introduction

Discovery of the Apo(a) Antigen

The apo(a) antigen was discovered in 1963 by Berg while searching for genetic polymorphisms related to purified β-lipoprotein or LDL.[1] He demonstrated that the Lp(a) antigen was unrelated to the Ag antigen, previously shown to represent the β-lipoprotein. Using double immuno-diffusion, Berg further demonstrated that these two antigens in fact represented two distinct populations of lipoprotein particles.[2] Early screening of human sera for Lp(a) by double diffusion utilized antisera which had been raised in rabbits, then carefully absorbed with Lp(a)-negative serum to ensure relative purity. With these absorbed antisera, it was possible to distinguish between two classes of human sera: those exhibiting a precipitin line in double diffusion and those not; these were designated Lp(a$^+$) and Lp(a$^-$), respectively. Using electrophoresis, Seegers *et al.*[3] found a lipoprotein which they reported as a genetic variant of the β-migrating lipoprotein, LDL. This particle was later shown to be identical to the lipoprotein exhibiting pre-β mobility upon agarose electrophoresis as described by Dahlen and co-workers.[4] Rider *et al.*,[5] in 1970, observed a "sinking pre-beta lipoprotein" which was demonstrable by a combination of ultracentrifugation and electrophoresis. It is now known that these lipoproteins, first observed independently by different techniques, are identical and contain the Lp(a) antigen described originally by Berg.

Factors Affecting Lp(a) Blood Levels

The original hypothesis of a simple autosomal dominant inheritance[1] has been suggested in several family studies. This hypothesis has been

[1] K. Berg, *Acta Pathol. Microbiol. Scand.* **59**, 369 (1963).
[2] K. Berg, *Ser. Haematol.* **1**, 111 (1968).
[3] W. Seegers, K. Hirschhorn, L. Burnett, E. Robson, and H. Harris, *Science* **149**, 303 (1965).
[4] G. Dahlen, C. Erickson, C. Furberg, L. Lundkvist, and K. Svardsudd, *Acta Med. Scand. Suppl.* **531**, 11 (1972).
[5] A. K. Rider, R. I. Levy, and D. S. Fredrickson, *Circulation* **42** (Suppl. 3), 10 (1970).

extended to include several modifying factors which affect the expression of the alleles at a single autosomal locus.[6] Population studies have shown a wide variation of the gene frequencies, with the highest in United States whites and the lowest in Labrador Indians.[2] Linkage analysis of Lp(a) to red cell esterase D, a known genetic marker, is suggestive but not conclusive. This association tentatively assigns the major gene locus to chromosome 13. A single locus could account for 87% of the variation, with persons in the upper mode of the bimodal distribution being either homozygous or heterozygous for the dominant allele.[7,8] Studies involving a large family pedigree[9] and a large volunteer population of nonrelatives[10] yielded evidence to support a major locus with dominant inheritance accounting for 95 to 98%, and polygenic effects accounting for the remaining 2 to 5%. Negligible evidence has been obtained for environmental effects on plasma Lp(a) levels, including studies with monozygotic and dizygotic twin pairs. Hence, it appears that heritability is near 100%. This contrasts with other lipoprotein parameters which do not exhibit such a high degree of heritability and whose serum levels are significantly altered by environmental influences.[10,11] One study has indicated a consistent increase in Lp(a$^+$) frequency with increasing age.[6] Dietary manipulation commonly known to affect other lipoproteins apparently has no effect on serum Lp(a) levels.[11,12] Hewitt et al.[10] have found some evidence for a lower mean serum Lp(a) in subjects who reported drinking alcohol; the amount of intake was unspecified, however. Marth et al.[13] have studied heavy alcohol consumers ingesting greater than 200 g of ethanol per day and found markedly reduced levels of Lp(a) in persons both with and without documented alcoholic liver disease. Cholestyramine has been demonstrated to be ineffective in lowering serum Lp(a) levels, although it does lower the serum level of LDL.[14] Studies in rabbits indicate that cholestyramine causes an increase in LDL receptor-mediated uptake by the

[6] J. S. Schultz, D. C. Shreffler, C. F. Sing, and N. R. Harvie, Ann. Hum. Genet. 38, 39 (1974).

[7] J. S. Schultz and D. C. Shreffler, Proc. Int. Congr. Hum. Genet. 4th, Paris, 1971. 345 (1972).

[8] C. F. Sing. J. S. Schultz, and D. C. Shreffler, Ann. Hum. Genet. 38, 47 (1974).

[9] S. J. Hasstedt, D. E. Wilson, C. Q. Edwards, W. N. Cannon, D. Carmelli, and R. R. Williams, Am. J. Med. Genet. 16, 179 (1983).

[10] D. Hewitt, J. Milner, A. R. Owens, W. C. Breckenridge, G. F. Maguire, G. J. Jones, and J. A. Little, Clin. Genet. 21, 301 (1982).

[11] J. R. Patsch, J. B. Karlin, L. W. Scott, L. C. Smith, and A. M. Gotto, Jr., Proc. Natl. Acad. Sci. U.S.A. 80, 1449 (1983).

[12] F. Krempler, G. M. Kostner, A. Roscher, F. Haslauer, K. Bolzano, and F. Sandhofer, J. Clin. Invest. 71, 1431 (1983).

[13] E. Marth, G. Cazzalato, B. G. Bittolobon, P. Avogaro, and G. M. Kostner, Ann. Nutr. Metab. 26, 56 (1982).

[14] B. Vessby, G. Kostner, H. Lithell, and J. Thomis, Atherosclerosis 44, 61 (1982).

liver,[15] suggesting this may not be a major route of metabolism for Lp(a). Lp(a) levels do not correlate with total cholesterol, LDL, HDL, or other known lipoprotein levels.[11,14] The metabolic half-life of Lp(a) in man is comparable to that of LDL; furthermore, both lipoproteins appear to be metabolized by the same LDL receptor in cultured fibroblasts.[12] From limited study, it has been concluded that elevated levels of Lp(a) are the result of increased rates of lipoprotein synthesis, and not from decreased rates of catabolism.[16] Lp(a) is not converted to other known serum lipoproteins,[16] nor is it a metabolic product of other lipoproteins containing apoB.[17]

Evidence from recent studies[18] has shown that for Lp(a$^+$) individuals, the risk of developing coronary heart disease is approximately three times greater than for those persons typed Lp(a$^-$). Relatively intact Lp(a) has been observed in the intima of atherosclerotic plaques,[19] which is thought to be entrapped by a mechanism similar to that demonstrated for LDL. Because Lp(a) usually contributes less than 15% of the total plasma cholesterol, many persons with previously considered "normal" levels of total cholesterol may have atherogenic levels of Lp(a). Most recently, Dahlen et al.[20] tested the strength of the association of plasma Lp(a) levels with coronary artery disease in direct comparison to other, better known, atherogenic risk factors. Lp(a) correlated positively to the coronary score, with a strength of association about one-half that of age, sex, or total plasma cholesterol. However, among patients aged 55 and younger, Lp(a) level was the strongest risk factor, surpassing any other known lipoprotein. The strength of Lp(a) association with coronary artery disease, occurring at relatively low plasma cholesterol concentrations, suggests that Lp(a) may be the most potent atherogenic lipoprotein in man.

Methods for Detection/Identification of the Apo(a) Antigen Directly in Plasma

Agarose Electrophoresis. Lp(a) is demonstrable by electrophoresis of serum in 0.5% agarose gel. The electrophoretic mobility of Lp(a) after

[15] H. R. Slater, C. J. Packard, S. Bicker, and J. Shepherd, *J. Biol. Chem.* **255**, 10210 (1980).

[16] F. Krempler, G. M. Kostner, K. Bolzano, and F. Sandhofer, *J. Clin. Invest.* **65**, 1483 (1980).

[17] F. Krempler, G. Kostner, K. Bolzano, and F. Sandhofer, *Biochim. Biophys. Acta* **575**, 63 (1979).

[18] K. Berg, *in* "Inherited Lipoprotein Variation and Atherosclerotic Disease" (A. M. Scanu, R. W. Wissler, and G. S. Getz, eds.), p. 419. Dekker, New York, 1979.

[19] K. W. Walton, J. Hitchins, H. N. Magnani, and M. Khan, *Atherosclerosis* **20**, 323 (1974).

[20] G. H. Dahlen, M. Attar, J. R. Guyton, J. A. Kautz, and A. M. Gotto, Jr., *Arteriosclerosis* **3**, 478a (1983).

staining with Oil Red O or Sudan Black is shown by the pre-β_1 band (Fig. 1).[4] The use of fresh sera is essential for accurate scoring of samples as Lp(a$^+$) or Lp(a$^-$); sera which have been stored at 4° for periods even as short as 48 hr become difficult to score.[21] The problem is thought to be due to atypically migrating VLDL and presumably could be circumvented by removing the "sinking pre-β" ($d > 1.006$) lipoproteins,[22,23] although this does add an ultracentrifugal step to the analysis. Numerous studies have used agarose electrophoresis either alone[24,25] or in combination with other methods[26,27] for assessing the presence of Lp(a). A highly positive correlation has been found between samples exhibiting precipitin lines for Lp(a), as determined by the double diffusion immunological test, and samples exhibiting a pre-β_1-lipoprotein in agarose electrophoresis. However, the number of individuals assessed as Lp(a$^+$) by agarose electrophoresis was always less than that determined by other methods. Whereas agarose electrophoresis demonstrates pre-β_1-lipoprotein in 25% of the population,[22,25] double diffusion detects about 40% positive, electroimmunoassay about 60%, and radioimmunoassay 90%. All individuals with Lp(a) levels greater than 40 mg/dl have detectable pre-β_1-lipoprotein by agarose electrophoresis.[22] In many clinical laboratories, this technique is a routine procedure for the determination of serum lipoproteins, and provides a simple means of recognizing individuals with elevated Lp(a) levels. Additionally, the technique has been used to study alterations in electrophoretic mobility produced by various chemical treatments of Lp(a).[23]

Immunochemical Techniques

Double Diffusion. The Lp(a)-specific antigen was first detected by Berg using the double diffusion test procedure in agarose gel as developed by Ouchterlony.[28] This has been the most widely used method for classifying plasma samples as Lp(a$^+$) or Lp(a$^-$). One may use commercially prepared immunodiffusion plates (Hyland Diagnostics, Deerfield, IL),

[21] G. Dahlen, K. Berg, and M. H. Frick, *Clin. Genet.* **9,** 558 (1976).
[22] J. J. Albers, V. G. Cabana, G. R. Warnick, and W. R. Hazzard, *Metabolism* **24,** 1047 (1975).
[23] G. Dahlen, K. Berg, T. Gillnäs, and C. Ericson, *Clin. Genet.* **7,** 334 (1975).
[24] M. H. Frick, G. Dahlen, C. Furberg, C. Ericson, and M. Wiljasalo, *Acta Med. Scand.* **195,** 337 (1974).
[25] A. D. Postle, J. M. Darmady, and D. C. Siggers, *Clin. Genet.* **13,** 233 (1978).
[26] G. Dahlen, M. H. Frick, K. Berg, M. Vallé, and M. Wiljasalo, *Clin. Genet.* **83,** 183 (1975).
[27] F. Krempler, G. M. Kostner, A. Roscher, K. Bolzano, and F. Sandhofer, *J. Lipid Res.* **25,** 283 (1984).
[28] O. Ouchterlony, *Prog. Allergy* **5,** 1 (1958).

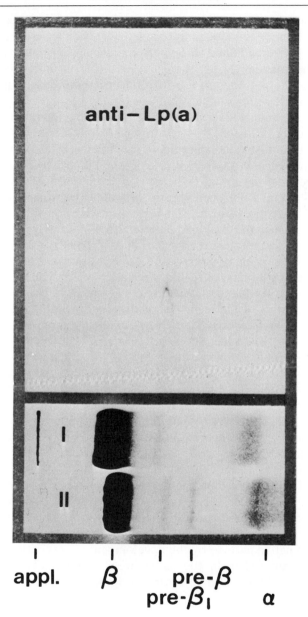

FIG. 1. Two-dimensional immunoelectrophoresis of Lp(a). This lipoprotein migrates with pre-β_1 mobility, and is readily detected after electrophoresis in the second dimension into agarose containing anti-Lp(a).

consisting of a central well and five peripheral wells, all spaced 5 mm apart. The antibody is added to the central well and the serum samples for testing are applied undiluted to the peripheral walls. The plate is placed in a humid chamber for 24 hr, then viewed using an immunoilluminator (Hyland) for the presence of a precipitin line representing the antigen–antibody complex. For samples that are weakly positive, factors such as the volume of serum added to the wells, the incubation temperature, the time allowed for diffusion, the well-to-well distance, and unique characteristics of the antibody are all critical to reproducibility in scoring. Serum samples which have been divided into 250-μl aliquots and frozen at $-20°$ in small screw-cap glass vials (Wheaton) give the same results (for at least 12 months) as fresh unfrozen serum, provided the frozen serum is not repeatedly thawed and frozen. In addition to simple positive or negative scoring of samples, one may use double diffusion to assess the purity of an antibody preparation where formation of a single precipitin line is the objective. This procedure can also be used to assay chromatographic or ultracentrifugal fractions, or to study the effect of enzymatic or chemical treatments upon the immunological reactivity of the Lp(a) preparation.[29] The low sensitivity of the technique is its primary limitation.

Immunoelectrophoresis. Standard immunoelectrophoresis in one dimension[30] or as crossed immunoelectrophoresis[31] in two dimensions combines detection of an antigenic component by a specific antibody with its separation based on charge. Crossed immunoelectrophoresis has been used to demonstrate that the pre-β_1 band obtained in agarose electrophoresis represents the material immunoreactive for the specific Lp(a) antibody (Fig. 1). In addition, crossed immunoelectrophoresis has been used to test the purity of Lp(a) preparations and possible charge heterogeneity of Lp(a)-immunoreactive species. Alterations in charge after experimental modifications of Lp(a) can also be examined.

Standard one-dimensional immunoelectrophoresis has been used to test whether serum is Lp(a$^+$), to test for the absence of LDL from purified Lp(a) preparations,[32] and to test whether other proteins sometimes detectable in Lp(a) preparations (e.g., albumin, apoCs, apoA-III) are associated with the native particle by the development of precipitin lines against the different antibodies in the same pre-β position characteristic for Lp(a). A principal limitation of this method is its relative insensitivity; the

[29] G. Utermann and H. Wiegandt, *Hoppe-Seyler's Z. Physiol. Chem.* **352,** 938 (1971).
[30] P. Grabar and C. A. Williams, *Biochim. Biophys. Acta* **10,** 193 (1953).
[31] C. B. Laurell and U. Persson, *Biochim. Biophys. Acta* **310,** 500 (1974).
[32] K. Simons, C. Ehnholm, O. Renkonen, and B. Bloth, *Acta Pathol. Microbiol. Scand., Sect. B* **78,** 459 (1970).

requirement for Lp(a) lipoprotein concentration is greater than 20 mg/dl. A second limitation is the relatively large amount of antibody required.

Enzyme-Linked Immunosorbent Assay (ELISA). A microplate method for detecting apo(a) has been developed in our laboratory to meet the need to determine whether hybridomas were producing antibodies to apo(a) in the preparation of monoclonal antibodies. By this technique, Lp(a) can be detected in concentrations down to 0.10 mg/dl. There is now a voluminous literature describing various types and applications of the ELISA assay.[33,34] The following protocol provides instructions for establishing a functioning ELISA for Lp(a) with minimum effort even for someone unfamiliar with this technique.

1. 100 μl of Lp(a) at 1.0 μg/ml of lipoprotein is added to the wells of an Immulon I U-bottom plate (Dynatech Labs, Inc., Alexandria, VA). Two wells are left uncoated to serve as controls. The Lp(a) is added to 0.075 M phosphate, 0.075 M NaCl, pH 7.2 (PBS). This plate is incubated at room temperature (RT) for 2 hr, overnight at 4°, and then thoroughly washed, initially with PBS made 0.5% in bovine serum albumin (PBS–BSA), then with PBS alone.

2. 250 μl of PBS–BSA is added to each well and the plate incubated for 2 hr at RT. Plates can be used immediately or stored overnight at 4°. PBS–BSA is removed just before use of the plate.

3. Appropriate dilutions of unknown samples are prepared in PBS–BSA. Antibody to Lp(a) is diluted in PBS–BSA. This dilution will depend upon the particular antibody; typical dilutions range from 1 : 5000 to 1 : 25,000. Equal volumes of unknown and antibody are transferred to a culture tube and mixed gently. Of this mixture, 110 μl is then added to each well of the microtiter plate. The mixture can be added in duplicate or triplicate as desired. It is essential that certain specified control wells contain only PBS–BSA, or only PBS–BSA + antibody. To the two wells not coated originally with lipoprotein, PBS–BSA + antibody is added. Plates are incubated for 2 hr at RT, then overnight at 4°. They are then washed thoroughly just before the next step.

4. A peroxidase-conjugated second antibody (Cappel Labs, Cochranville, PA), immunoreactive against the IgG species used in the first incubation, is then added. The dilution, typically 1 : 2500, is prepared in PBS–BSA and 110 μl of this is added to each well. The plate is incubated 2 hr at RT, then washed thoroughly.

[33] D. Seidel, H. P. Geisen, and D. Roelcke, *FEBS Lett.* **18,** 43 (1971).
[34] A. Voller, D. E. Bidwell, and A. Bartlett, "The Enzyme Linked Immunosorbent Assay (ELISA): A Guide with Abstracts of Microplate Applications." Dynatech Laboratories, Alexandria, Va., 1979.

5. Substrate solution is freshly prepared by dissolving 10 mg of o-phenylenediamine in 25 ml of phosphate–citrate buffer, pH 5.0, to which 10 μl of 30% H_2O_2 is finally added.

The phosphate–citrate buffer is made up from 6.075 ml of 0.1 M citrate, 6.425 ml of 0.2 M phosphate, and 12.50 ml of deionized water. The substrate solution is light sensitive and should be protected from light and used soon after preparation. To each well is added 160 μl of this substrate solution. The reaction is allowed to proceed 30–60 min at RT, then quenched by addition of 40 μl of 2.5 N sulfuric acid. The plates are read automatically by a Flow Lab plate reader equipped with a 492-nm filter and interfaced to an Apple II+ computer. The lower the optical density, the more apo(a) present in the sample. A quantitative value can be determined by comparing the optical densities of test samples to those of standards of known Lp(a) concentration. The sensitivity range of the assay is 1–20 μg/ml Lp(a) lipoprotein. Essentially the same procedure can be used to quantitate apoB by substituting LDL and its antibodies for Lp(a).

In addition to the simple detection and quantitation of Lp(a) in samples, this technique can be used to study the effect of enzymatic, chemical, and/or physical treatments upon the immunoreactivity of apo(a).[35]

Western Blot Transfer. Transfer of electrophoretically resolved protein from SDS–PAGE gels to nitrocellulose as exact replicas, followed by localization of the separated protein components with specific antibodies, has become a powerful technique.[36] Electrophoretic patterns unique for the apoproteins of Lp(a) have been demonstrated by this technique.[37] The total apoprotein load per lane required by this technique is 1.0 μg. Individual differences in apoprotein pattern revealed by SDS–PAGE can be studied. Enzymatic and chemical fragmentation as well as protein aggregation can be detected. Monoclonal antibodies with reactivities for specific epitopes can be utilized since the formation of an immunoprecipitate is not required. Peptide mapping after limited proteolysis with trypsin is highly suitable for this technique.[38]

Immunochemical Quantitation of the Apo(a) Antigen

1. *Production of Antibodies.* Rabbits, goats, and mice have all been used successfully for producing antibodies specific to Lp(a). The follow-

[35] D. Altschuh and M. H. V. Van Regenmortel, *J. Immunol. Methods* **50,** 99 (1982).
[36] H. Towbin, T. Stachelin, and J. Gordon, *Proc. Natl. Acad. Sci. U.S.A.* **76,** 4350 (1979).
[37] J. W. Gaubatz, C. Heideman, A. M. Gotto, Jr., J. D. Morrisett, and G. H. Dahlen, *J. Biol. Chem.* **258,** 4582 (1983).
[38] D. W. Cleveland, S. G. Fischer, M. W. Kirschner, and U. K. Laemmli, *J. Biol. Chem.* **252,** 1102 (1977).

ing procedure has been used in the authors' laboratory to obtain polyclonal antisera specific to Lp(a).

An emulsion composed of 0.6 ml of Freund's complete adjuvant and 0.4 ml of purified Lp(a) solution containing 1 mg/ml protein is injected subcutaneously at multiple sites (~10) into male New Zealand white rabbits weighing 6 kg. Ten days after the initial injection, each animal is boosted with the same antigen emulsion. Subsequently, the rabbits are boosted at 10-day intervals with an emulsion composed of 0.6 ml of Freund's incomplete adjuvant and 0.4 ml of Lp(a) at a protein concentration of 0.5 mg/ml. Ten days following the second boost, the titer is usually 1 : 2, as determined by double diffusion. The rabbits are then bled from the central artery of the ear at 10-day intervals. From 35 to 70 ml of blood can be obtained at each bleeding, and the titer generally increases to 1 : 4 or 1 : 8 and is maintained at that level for several months with continued boosting. When a satisfactory amount of antibody has been obtained or when the titer begins to drop, the animals are exsanguinated. The sera are stored frozen at −20° until needed. Sera obtained in this manner contain antibodies against both protein components of Lp(a), apo(a), and apoB. For this reason, the sera must be made monospecific for apo(a) before they can be used for qualitative or quantitative determination of Lp(a). This is generally done by removal of the antibodies to apoB from the antisera. In our experience, adding excess purified LDL to anti-Lp(a) and removing the resulting LDL/anti-apoB immunoprecipitate has certain disadvantages: (1) the antisera are somewhat diluted, (2) residual LDL becomes an antiserum contaminant, and (3) soluble LDL/anti-apoB complexes may be formed which are not removed by centrifugation. These soluble complexes may later dissociate to yield antisera not truly monospecific for Lp(a). A Sepharose 4B column to which LDL has been coupled (LDL–Seph) is an excellent immunoabsorber for removing the anti-apoB. The serum is not diluted, no contaminating LDL is added, and the possibility of forming a soluble antigen–antibody complex is avoided. A further advantage is that the anti-apoB bound to the column can be eluted subsequently as pure IgG and the immunoabsorber reused. The purified anti-apoB IgG can be used for a number of purposes including the preparation of an immunosorbent affinity column. The antisera, now monospecific for apo(a), are aliquoted into small ampoules and frozen at −20°.

Monoclonal antibodies secreted by hybridomas recognize single antigenic determinants or epitopes and are homogeneous with respect to specificity and affinity. Even when Lp(a) is used as antigen for injecting mice, the individual cell lines produce antibodies to specific immunogenic sites on apoB or apo(a). Therefore, the steps needed to remove anti-apoB from polyclonal antibodies are not necessary with monoclonal antibodies, and no question arises as to monospecificity, if the dilution subcloning has been carefully performed. Also, monoclonal antibodies can be obtained in large amounts. However, monoclonal antibodies do have certain disadvantages: (1) they often exhibit low affinity, and (2) they do not form antibody–antigen precipitates, hence they cannot be used in techniques such as double diffusion where precipitation is a requirement. Our laboratory has isolated seven monoclonal antibodies specific for the apo(a) apoprotein of Lp(a). Another laboratory has reported isolation of a mouse monoclonal antibody specific for the apo(a) component, and devised an ELISA technique with it.[39]

2. *Preparation of Immunosorbent Columns.* Affinity chromatography is a method whereby the macromolecule or multimolecular complex to be purified or removed is specifically and reversibly absorbed by a complementary binding substance immobilized on (e.g., covalently coupled to, mechanically trapped in) a solid support. Separations can often be obtained in a single chromatographic step with significant savings of time, material, and expense. An LDL immunosorbent column provides an ideal way to remove antibodies against apoB from immune serum produced by injecting animals with purified Lp(a). Anti-apoB obtained in this manner can be coupled to Sepharose to prepare an immunosorbent column. This column can then be used for specific removal of apoB-containing material from heterogeneous preparations. Such a procedure has been used to remove immunoreactive apoB-containing material from homogenates of human aortic intimal plaques.[40] The preparation of a column with either monoclonal or polyclonal antibody reactive to apo(a) allows the separation of Lp(a) directly from plasma samples. The combination of this column with an anti-apoB column allows the separation of apo(a) and apoB from delipidated, reduced preparations of Lp(a).

The procedure as currently used in our laboratory is as follows:

1. Cyanogen bromide-activated Sepharose 4B (Pharmacia P-L Biochemicals, Piscataway, NJ) is used for coupling. For a 0.9×20 cm column 6 g is required, and for a 1.5×20 cm column 10 g are needed.

2. The gel is swollen and washed for 15 min with 200 ml/g gel of 2 mM HCl using a medium-porosity fritted glass filter. A vacuum pump can be attached to a sidearm flask to speed washing but great care must be used to avoid drying the gel. The gel slurry should be stirred gently with a glass rod to avoid crushing the beads. The acid rinse is necessary to remove dextran and lactose, which were added to stabilize the gel.

3. The material to be coupled (either affinity purified IgG or purified

[39] C. R. Duvic, G. M. Smith, W. E. Sledge, L. T. Lee, M. D. Murray, P. S. Roheim, W. R. Gallaher, and J. J. Thompson, *J. Lipid Res.* **26,** 540 (1985).
[40] H. F. Hoff and J. W. Gaubatz, *Atherosclerosis* **42,** 273 (1982).

lipoprotein or apoprotein) is dissolved in or dialyzed against 0.1 M NaHCO$_3$, 0.5 M NaCl (coupling buffer). Approximately 5–10 mg of protein to be coupled per gram gel is used.

4. To remove HCl from the gel, it should be rinsed briefly with coupling buffer just before adding the material to be coupled. The coupling is performed at 4° overnight in a tightly capped container that is rotated slowly end over end using an infiltration wheel.

5. To determine the efficiency of coupling, the gel mixture is filtered again on a fritted glass filter, and the protein concentration of the filtrate is determined. The efficiency of coupling as described here is usually greater than 90%. One molar ethanolamine, pH 8.0, is mixed with the gel and the mixture rotated end over end at room temperature for 1–2 hr. Ethanolamine covalently couples to any remaining active groups on the gel.

6. Three washing cycles are used to remove noncovalently adsorbed protein, each cycle consisting of a wash with two different buffers:

   a. 0.1 M sodium acetate, 1 M NaCl (pH 4.0)
   b. 0.1 M sodium borate, 1 M NaCl (pH 8.0)

7. The gel is washed with 0.05 M Tris, 0.15 M NaCl, 0.01% azide, pH 7.4 (column buffer); the column is packed, stored, and run at 4°.

Running the Column

1. Apply sample to column in column buffer
2. Wash with column buffer until A_{280} of eluate is zero.
3. Wash with saline adjusted to pH 10.5 with NH$_4$OH. This releases most bound material. A_{280} should peak, then return to zero. Immediately dialyze eluted sample to neutral pH.
4. Wash column with starting buffer or go directly to step 5.
5. Wash column with 6 M urea, 10–20 ml/g gel.
6. Equilibrate column with column buffer for next run.

The following materials have been successfully coupled and used as affinity columns in our laboratory: LDL, rabbit anti-human LDL, goat anti-human LDL, human apoA-I, goat anti-human apoA-I, and heparin. These columns have all performed satisfactorily for up to 5 years from the date of preparation even with frequent usage. High flow rates (up to 60 ml/hr) and large sample volumes (e.g., 100 ml) can be used since the separations depend upon the specific adsorbing capacity of the column rather than upon physical properties (e.g., pore size, charge) of the gel material itself.

3. *Quantitative Measurements.* A number of techniques have been utilized to quantify Lp(a). All of these are immunochemical techniques in which the antibody utilized is apparently directed against the specific

apo(a) component. However, the assay results are expressed either as total Lp(a) lipoprotein or total Lp(a) protein per unit volume since the preparation of purified apo(a) as a primary standard is not yet a straightforward operation. The need for a reliable standard is common to all of these methods. Freshly isolated Lp(a) for which compositional data have been obtained is used as the primary standard. Such samples are generally stable at 4° for up to 1 month.[41,42] It is desirable to use Lp(a) from individual rather than pooled plasmas because of possible heterogeneity.[43] Several fresh serum samples are then assayed using the primary Lp(a) standard. These serum samples (secondary standards) are then aliquoted into small glass screw-cap vials and frozen at −20°. When stored in this manner, the standards give reproducible results (as determined in our laboratory by electroimmunoassay) for up to 1 year. One study has made use of lyophilized reference standard serum [containing human Lp(a)] obtained from Immuno Diagnostika (Vienna, Austria).[44] Dilutions of primary or secondary standard are generally made in buffer containing 0.5–3% BSA or lipoprotein-deficient serum.

The simplest immunoassay method available is radial immunodiffusion.[45] However, it suffers from a lack of sensitivity (range of the assay is 3–25 mg/dl) and a rather high coefficient of variation (both intra- and interassay) which is very concentration dependent. It is probably best used for initial screening before a more precise assay is performed, or when precise measurements are not necessary. Several investigators have utilized radioimmunoassay for Lp(a) quantitation, although only one group has rigorously examined statistical criteria.[46] Radioimmunoassay is technically more difficult and time consuming than most other immunoassays but meets exacting requirements of specificity and precision with the added advantage of extreme sensitivity allowing measurement of Lp(a) values down to 0.5 mg/dl. The coefficient of variation is 6% (intraassay) and 7.3% (interassay).

Several recent procedures which have been developed for quantitation are nephelometry[42] and zone immunoelectrophoresis.[44] The range of these assays is approximately 1–20 mg/dl, which is about the same as electroimmunoassay. The coefficients of variation for these methods are

[41] F. Krempler, G. M. Kostner, K. Bolzano, and F. Sandhofer, *J. Clin. Invest.* **65,** 1483 (1980).
[42] G. Cazzolato, G. Prakasch, S. Green, and G. M. Kostner, *Clin. Chim. Acta* **135,** 203 (1983).
[43] G. M. Fless, C. A. Rolih, and A. M. Scanu, *J. Biol. Chem.* **259,** 11470 (1984).
[44] W. März and W. Gross, *Clin. Chim. Acta* **134,** 265 (1983).
[45] J. J. Albers and W. R. Hazzard, *Lipids* **9,** 15 (1974).
[46] J. J. Albers, J. L. Adolphson, and W. R. Hazzard, *J. Lipid Res.* **18,** 331 (1977).

reasonably low. These assays seem to offer little advantage over electroimmunoassay and would be preferable only if such methods were already routinely used for other applications.

A microplate ELISA procedure has been used in our laboratory not only for detection but also for quantitation of Lp(a). This procedure is described above in the section, Enzyme-Linked Immunosorbent Assay (ELISA). It is a highly sensitive assay and is easily set up in laboratories where related techniques are already in use.

The most widely used method for Lp(a) quantitation, either directly or as a means of evaluating new assay procedures, is electroimmunoassay.[19,37,41,42,44] The assay can be adjusted so that the sensitivity range is 1–10 mg/dl, and a satisfactory coefficient of variation ($< 10\%$) is easily attainable. The requirements for equipment and supplies are not extensive. One percent agarose (type C, Calbiochem-Behring, La Jolla, CA) is prepared fresh weekly in the same sodium barbiturate buffer (54 mM, pH 8.6) that is used in the electrophoresis chambers. The agarose mixture is rapidly brought to boiling and after being suspended is tightly capped and placed in a 65° oven until used. A mold is prepared from a 1-mm-thick Plexiglass spacer sandwiched between two 110 × 205 mm glass plates. Clamps are used to secure the mold. Gel-Bond plastic film (Marine Colloids, Rockland, ME) can be used during the gel casting step. The antibody is warmed to RT while the mold is warmed in a 65° oven. The agarose (20 ml) is removed from the oven. When its temperature reaches 50–55°, antibody is added, the volume depending on antibody titer. After mixing by inversion, the agarose mixture is then quickly poured into the upright-positioned glass mold, just removed from the oven. The agarose sandwich is allowed to gel at RT for 15 min, then placed in an air-tight humid chamber in the refrigerator until wells are punched. A 3.0-mm punch (Bio-Rad) is used in conjunction with an electrophoresis punching template (Bio-Rad) to produce 36 evenly spaced sample wells. The gel is then placed on a water-cooled (10°) horizontal electrophoresis cell (Behring Diagnostics, Sommerville, NJ). The chambers are filled with electrophoresis buffer and fitted with ultrawicks (LKB, Gaithersburg, MD), trimmed to fit the gel exactly, so as to complete electrical contact between the anodic and cathodic chambers, and the agarose gel. The field strength is adjusted to 3.0 V/cm at constant current, and tested for homogeneity at several places along the width of the gel. Dilutions of standards [1–10 mg/dl Lp(a)] are made using barbiturate electrophoresis buffer containing 0.5% bovine serum albumin. The field strength is reduced to 0.5 V/cm for sample loading. A 5-μl sample is added to each well, and the field strength is then increased to 3.0 V/cm. Electrophoresis is stopped after 15 hr, and the gels are dried, stained, and destained. Generally, plots of peak height versus concentration are linear. Any unknown samples with peak heights not in the range of the standard curve are rerun at an appropriate dilution. Materials known to perturb the formation of the precipitin rockets are detergents (e.g., sodium dodecyl sulfate, deoxycholate), denaturants (e.g., urea, guanidine hydrochloride), high salt concentration, and certain enzymes (e.g., collagenase, trypsin). To increase sensitivity, the antibody added to the agarose can be decreased as long as the resulting precipitin reaction is strong enough to allow unambiguous measurement of rocket peak heights.

Isolation Procedures for Human Lp(a)

Suitable donors for obtaining whole plasma for isolation of Lp(a) are individuals with high Lp(a) concentrations (>75 mg/dl). Preliminary screening is done using double diffusion tests, followed by a more quantitative measurement using electroimmunoassay. Typically, one may obtain up to 800 ml of plasma by plasmapheresis. Trasylol (Mobay Chemical Corporation, 10,000 kallikrein inactivator U/ml) is introduced into the collecting bag at the start of the procedure (1.0 ml of Trasylol/dl of plasma). At the end of the collection, the plasma is made 1 mM in EDTA, 0.01% in sodium azide, and 0.0001% in PMSF. Some published procedures involve precipitation of all apoB-containing lipoproteins either with sodium phosphotungstate/MgCl$_2$[47–49] or with dextran sulfate/CaCl$_2$[19] as a first step in the isolation. The primary purpose of this step is to concentrate the Lp(a)-containing fraction and simultaneously eliminate the HDL. However, it has been observed in studies with other lipoproteins that phosphotungstate precipitation can be attended by low recoveries and degradation of one or more constituent apolipoproteins.[50] Most of Lp(a) is isolated in the $d = 1.05–1.12$ g/ml fraction.[50,51] It has been shown that during the ultracentrifugal steps considerable losses of Lp(a) occur as determined by rocket immunoelectrophoresis.[37] It is unclear whether this is a result of apo(a) dissociation from the lipoprotein, proteolysis of the apolipoprotein, real mass losses, or other factors. However, the losses can be significantly reduced by ultracentrifuging first at $d = 1.12$ rather than $d = 1.05$. This floats most of the Lp(a)-containing material along with some HDL, LDL, and VLDL. This $d = 1.12$ fraction is then dialyzed

[47] G. Jurgens and G. M. Kostner, Immunogenetics 1, 560 (1975).
[48] G. Jurgens, E. Marth, G. M. Kostner, and A. Holasek, Artery 3, 13 (1977).
[49] C. Ericson, G. Dahlén, and K. Berg. Clin. Genet. 11, 433 (1977).
[50] J. D. Morrisett, D. Wenkert, A. M. Gotto, Jr., G. H. Dahlén, S. Gianturco, C. Heideman, and J. W. Gaubatz, "Proceedings of the Workshop on Apolipoprotein Quantitation." NIH Publication No. 83-1286, p. 158, 1983.
[51] G. Utermann and H. Wiegandt, Humangenetik 8, 39 (1969).

extensively, adjusted to $d = 1.05$, and again ultracentrifuged. The infranatant from this step is the $d = 1.05–1.12$ density fraction most commonly used for subsequently separating Lp(a) from HDL and LDL by gel filtration chromatography on BioGel A5m or A15m (Bio-Rad).[37,45,47] Better separation of Lp(a) from LDL occurs if 1.0 M NaCl is added to the usual column running buffer. Even so, the Lp(a$^+$) and Lp(a$^-$) peaks are not completely separated on a 90-cm column so that only part of the Lp(a) can be recovered free of LDL.

A second method used by the authors and others[19] offers a number of advantages over gel filtration chromatography, but requires an anti-apo(a) affinity column. The second ultracentrifugal step at $d = 1.05$ can be omitted and the $d = 1.12$ supernatant chromatographed directly, without the need for concentration. Lp(a) binds specifically to the column and can be eluted after washing with pH 10.5 saline. The eluted pure Lp(a) is immediately dialyzed to neutral pH. The passage of the isolated material (0.5 mg/ml) through a short agarose or glass wool column prior to storage removes any visible particulates and usually prolongs its stability during storage. Under these isolation and storage conditions, we have found little tendency for Lp(a) to aggregate or precipitate as reported by others.[32,19,48,52,53] When Lp(a) is concentrated to 5–20 mg protein/ml, there is a definite tendency for the lipoprotein to precipitate at room temperature. Such a concentration-dependent tendency to precipitate has also been reported for purified Lp(a) when cooled to 4° as is normally done for storage; solutions of Lp(a) at 0.5–2 mg protein/ml were found to be stable at 4°.[54]

A third method for isolating Lp(a) involves the use of rate zonal density gradient centrifugation.[43] Either a slower, small-scale procedure or a faster, large-scale procedure may be used. The essential steps of these procedures are shown in the flow diagram of Fig. 2. This methodology enjoys the capability of separating different density populations of Lp(a) as illustrated in Fig. 3. However, it may fail to completely resolve LDL from Lp(a), in which case an additional chromatofocusing or heparin–Sepharose chromatography step is required.

Criteria for Determination of Lp(a) Purity

A number of early studies reported that human Lp(a) preparations contained not only apo(a) and apoB, but also albumin[53,54] apoC,[33] and apoA-III.[47] Most recent publications, however, agree that apo(a) and apoB are the only constituent apoproteins in purified human

[52] C. Ehnholm, H. Garoff, K. Simons, and H. Aro, *Biochim. Biophys. Acta* **236**, 431 (1971).
[53] C. Ehnholm, H. Garoff, O. Renkonen, and K. Simons, *Biochemistry* **11**, 3229 (1972).
[54] G. Utermann, K. Lipp, and H. Wiegandt, *Humangenetik* **14**, 142 (1972).

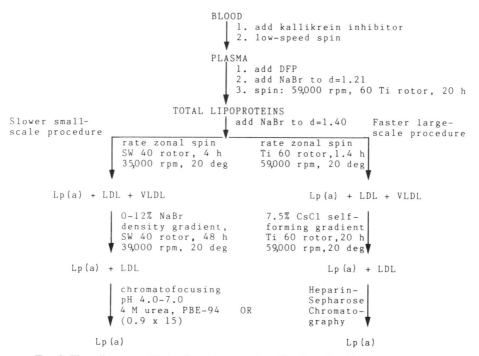

Fig. 2. Flow diagram outlining the isolation and purification of Lp(a) from human plasma as described by Fless *et al.*[43]

Lp(a).[37,43,48,55] Therefore, apoA-I, apoA-II, apoC, or albumin in an Lp(a) preparation should be considered as contaminating impurities. Immunological methods can be used for detection of these impurities. Also, SDS–PAGE can detect apolipoproteins other than apo(a) or apoB if they differ in molecular weight. The most likely contaminant of Lp(a) and the one most difficult to detect is LDL. It overlaps Lp(a) in density and size, and its only apolipoprotein, B-100, is essentially indistinguishable from the apoB of Lp(a). However, contamination of Lp(a) preparations with other native lipoproteins is most easily detected by standard polyacrylamide gel electrophoresis.[45,49,51] Although Lp(a) and LDL have similar sizes, they are readily separated on 2–16% gradient gels (4.9 × 82 × 82 mm, Pharmacia). To detect small amounts of contaminating lipoproteins, the Lp(a) preparation is overloaded (usual load = 10 μg protein) so that 20–50 μg

[55] G. Utermann and W. Weber, *FEBS Lett.* **154,** 357 (1983).

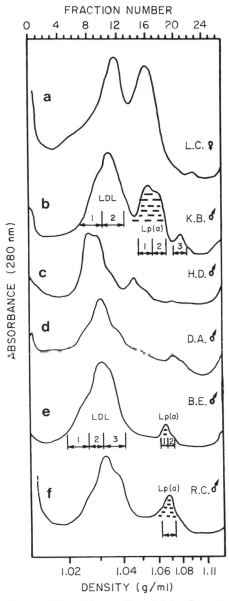

FIG. 3. Density gradient profiles of the apoB-containing lipoproteins of six individuals. Lp(a)-containing fractions are indicated by the shaded areas of the gradient profiles. Ultracentrifugation conditions: Beckman SW-40 rotor, 0–12% NaBr gradient, 39,000 rpm, 48 hr, 20°.

protein is added per lane. As little at 1–2 μg protein band is detectable after staining.

A method specifically used for the detection of LDL as a contaminant in Lp(a) samples is double-decker rocket immunoelectrophoresis.[37] In principle, its sensitivity equals or exceeds that of standard PAGE for detecting LDL. It requires antibodies to both apo(a) and apoB. Samples containing only Lp(a) and no LDL produce only one rocket; while those containing LDL as a contaminant develop two rockets (Fig. 4).

Properties of Lp(a)

During the earlier era of Lp(a) investigation, there were conflicting reports about its physical and chemical properties. However, many of the earlier studies were done on Lp(a) prepared from pooled plasmas. Recently, it has been shown that Lp(a) can exist as several discrete populations of particles that differ in mass, composition, and hydrated density.[43] This heterogeneity can occur both within and between individuals, and probably is the reason for the wide variations in some of the published Lp(a) data. Some of these discrepancies may also be due to proteolytic degradation of the lipoprotein which can easily occur when appropriate inhibitors are not included during isolation.[50] Only in the past few years have some of these discrepancies been resolved and a consensus reached.[12,37,43,55] Generally, Lp(a) exhibits a greater molecular weight, hydrated density, and size than LDL, which it closely resembles. Whereas LDL contains the single apoprotein apoB, Lp(a) contains apoB and apo(a), which usually are linked by one or more SS bonds to form a mixed disulfide.

The apoB in Lp(a) released by reductive disulfide cleavage is indistinguishable from B-100 of LDL as determined by immunochemical characterization with polyclonal antibodies to LDL and by electrophoretic mobility.[37] No B-48 has been reported to occur in any Lp(a) preparation. Utermann and Weber[55] have reported a single apo(a) species, having an apparent molecular weight about 1.4 times that of the B-100 component after disulfide cleavage. Others[37,43] have reported some heterogeneity of apo(a), with the molecular weight of the three or four detected species ranging from slightly less than that of B-100 to 1.5 times it. The lowest molecular weight apo(a) species has been detected only in the Lp(a) fraction isolated from the low-density range ($d = 1.019–1.063$). The number and proportion of these species vary among individuals. Apo(a) is a glycoprotein containing about 19–22% carbohydrate of which a significant fraction is N-acetylgalactosamine[50]; in contrast apoB contains none of this sugar. Within the intact lipoprotein, apoLp(a) exhibits 29% α-helical

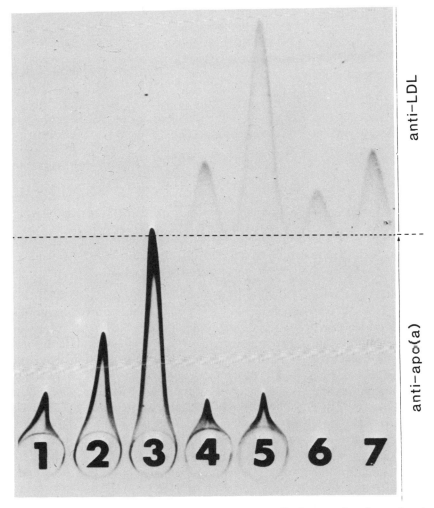

anti-LDL

anti–apo(a)

FIG. 4. Rocket immunoelectrophoresis with a two-antibody procedure for testing the exclusion of LDL from Lp(a) samples. The sample wells are punched in the agarose containing anti-apo(a). Over this portion toward the anode was layered agarose containing anti-LDL. To the respective wells were added the following: (1–3) Purified Lp(a); (4–5) mixture of Lp(a) and differing amounts of LDL; (6–7) pure LDL.

structure, 20% β-pleated sheet, and 51% random coil.[50] Although one of the first cell studies on Lp(a) claimed that the lipoprotein was not metabolized through the apoB/E receptor,[56] several other laboratories have sub-

[56] K. Maartmann-Moe and K. Berg, *Clin. Genet.* **20**, 352 (1981).

sequently reported data indicating that Lp(a) is indeed metabolized through this receptor in human fibroblasts.[12,50,57,58] In mouse peritoneal macrophages, neither Lp(a) nor LDL stimulates cholesteryl ester formation. However, when these lipoproteins are complexed with dextran sulfate, both cause significant increase in this process.[27]

[57] C.-H. Floren, J. J. Albers, and E. L. Bierman, *Biochem. Biophys. Res. Commun.* **102,** 636 (1981).
[58] L. Havekes, B. J. Vermeer, T. Brugman, and J. Emeis, *FEBS Lett.* **132,** 169 (1981).

[11] Isolation of Apolipoprotein E-Containing Lipoproteins by Immunoaffinity Chromatography

By Joyce Corey Gibson, Ardon Rubinstein, Henry N. Ginsberg, and W. Virgil Brown

ApoE is one of a group of specific lipid-binding apolipoproteins associated with the plasma lipoproteins. Although apoE makes a quantitatively minor contribution to the protein mass of the lipoproteins, data *in vitro* and *in situ*[1,2] have suggested that a potentially major physiological role exists for this protein. It is known that apoE interacts with two cell surface receptors[3] and its presence on a lipoprotein may direct the transfer of cholesterol to cells. Recent studies have shown that apoE has a discrete association with four lipoprotein subclasses.[4] Two of these are subclasses of the triglyceride-rich lipoproteins, chylomicrons and very low-density lipoproteins (VLDL). Two others are similar in size but clearly distinct from the major cholesterol-carrying lipoproteins, low-density (LDL) and high-density lipoproteins (HDL). Column chromatographic separation and analysis of apolipoprotein mass by radioimmunoassay (RIA) has allowed description of these subclasses in terms of size. The changes in the lipoprotein distribution of apoE which occur in response to metabolic perturbations[5] have been studied with this tech-

[1] E. Windler, Y. Chao, and R. J. Havel, *J. Biol. Chem.* **255,** 8303 (1980).
[2] F. Shelburne, J. Hanks, W. Meyers, and S. Quarfordt, *J. Clin. Invest.* **65,** 652 (1980).
[3] D. Y. Hui, T. L. Innerarity, and R. W. Mahley, *J. Biol. Chem.* **256,** 5646 (1981).
[4] J. C. Gibson, A. Rubinstein, P. R. Bukberg, and W. V. Brown, *J. Lipid Res.* **24,** 886 (1983).
[5] A. Rubinstein, J. C. Gibson, J. R. Paterniti, Jr., and W. V. Brown, *Arteriosclerosis* **2,** 443a (1982).

nique. However, this approach alone is not capable of isolating apoE-containing lipoproteins free of the milieu of similarly sized, non-apoE-containing lipoproteins. Such a preparative technique is a necessary preliminary to further study of the composition or metabolism of these lipoproteins.

Immunoaffinity chromatography, exploiting the unique attraction of antigen and specific antibody has extended the field of preparative bio-chemistry.[6,7] In lipoprotein research, this method is no less appealing since it provides the opportunity of isolating, with a minimum of manipulation, intact lipoprotein subclasses based solely upon the presence or absence of one apolipoprotein. This chapter describes the utilization of immunoaffinity chromatography for the specific isolation of apoE-containing lipoprotein subclasses. While the data relate only to apoE-containing lipoproteins, the potential extension of this technique to other apolipoprotein-specific subclasses is clear and has already been exploited by other investigators.[8–10]

Preparation of the Immunoadsorbent Matrix: Preparation of Antigen Matrix for Isolation of Specific IgG

The anti-apoE immunoadsorbent is prepared by the coupling of a supporting matrix to specific immunoglobulins directed against human apoE. The utilization of the IgG directed specifically against apoE increases the matrix capacity for apoE, reducing the volume of immunoadsorbent needed and thus minimizing the nonspecific protein adsorption. These specific IgG are prepared from the whole antiserum by immunoadsorption with another matrix to which apoE has been coupled. For this purpose we have used Sepharose CL-4B activated with cyanogen bromide (60 mg/ml gel).[6] The cross-linked matrix gives a more stable gel than non-cross-linked Sepharose. The gel matrix of 4% agarose is recommended since it provides good porosity and capacity. Some triglyceride-rich lipoproteins (chylomicrons and large VLDL) are largely excluded from the interior of this gel bead, however. The reduced access of these lipoproteins to IgG molecules may reduce the capacity below that desired. Using 2% agarose should address this problem. The procedure for coupling both apoE and the specific anti-apoE IgG is outlined in Table I.

[6] P. Cuatrecasas and C. B. Anfinsen, this series, Vol. 22, p. 345.
[7] I. Silman and F. Katchalsi, *Annu. Rev. Biochem.* **35,** 873 (1966).
[8] J. P. Vicar, S. T. Kunitake, R. L. Hamilton, and J. P. Kane, *Proc. Natl. Acad. Sci. U.S.A.* **81,** 1356 (1984).
[9] M. C. Cheung and J. J. Albers, *J. Lipid Res.* **21,** 747 (1982).
[10] G. R. Castro and C. J. Fielding, *J. Lipid Res.* **25,** 58 (1984).

TABLE I

PREPARATION OF AFFINITY MATRIX

1. Activate Sepharose CL-4B with CNBr (60 mg/ml gel)
2. Equilibrate activated gel in coupling buffer[a]
3. Add apoE or IgG in coupling buffer at a concentration of 2–5 mg protein/ml to a volume of gel sufficient to give a final concentration of ligand = 2–5 mg/ml gel
4. Couple with gentle shaking overnight at 4°
5. Remove supernatant and save. Wash gel once with coupling buffer
6. Add 1 M glycine, pH 8.6. Shake 1 hr at 4°
7. Transfer gel to sintered glass funnel and wash with coupling buffer
8. Wash gel with three cycles of 200 ml each of:
   - a. 0.1 M sodium acetate, 0.5 M NaCl, pH 4.0
   - b. 0.1 M NaHCO$_3$, 0.5 M NaCl, pH 9.0
9. Wash gel with phosphate-buffered saline (PBS) and store after addition of sodium azide (0.02%)

[a] For apoE matrix: 0.5 M NaCl, 0.1 M NaHCO$_3$, 5 mM decylsulfate, pH 8.3. For IgG matrix: 0.5 M NaCl, 0.05 M phosphate, pH 6.5.

For coupling of apoE, an alkaline buffer of 0.5 M NaCl, 0.1 M NaHCO$_3$, pH 8.3, is used. Sodium decyl sulfate (5 mM) is added to reduce aggregation and adsorption of the purified apoE. Routinely, apoE is prepared from hypertriglyceridemic human serum by previously published techniques.[4] Absolute purity is not a requirement when a monospecific antiserum is to be used. Sufficient apoE to provide 2–5 mg for each milliliter of gel matrix is dialyzed in the alkaline coupling buffer with decyl sulfate. When using lyophilized apoE, dissolving in buffer containing 20 to 50 mM decyl sulfate prior to dialysis is preferable in view of the potential insolubility of apoE after lyophilization. Prior to coupling, the cyanogen bromide-activated Sepharose CL-4B is washed with decyl sulfate coupling buffer (above). If preactivated gel is used, then the lyophilized gel is reconstituted in the same coupling buffer. The protein solution (2–5 mg/ml) is then added and coupling is allowed to occur overnight with gentle shaking at 4°. After coupling and washing (Table I), antigen density may be estimated indirectly by protein analysis of the supernatant. The gel should then be washed on a filter in phosphate-buffered saline (PBS) and stored at 4° in 0.02% sodium azide.

To minimize nonspecific adsorption of protein to the affinity matrix, the gel is treated with an equal volume of either an albumin solution (3 g/100 ml) or with nonimmune goat serum, diluted (1 : 1) with PBS before application to the column. Subsequent treatment is exactly as that outlined in Table I. Two or three cycles with irrelevant protein are advised before using the column for isolating IgG.

Preparation of Specific IgG

The specific IgG is prepared from 15–20 ml of antiserum by dilution in an equal volume of PBS and incubation with 5 ml of the antigen matrix (Table II). After incubation for 2 hr (4°) with gentle shaking, the mixture is poured into a column and the gel is washed with 20 bed volumes of PBS followed by 20 vol of 25 mM phosphate, 0.5 M NaCl, pH 7.4. The specifically bound IgG is eluted from the gel matrix by treatment with 3 M sodium thiocyanate. These steps are summarized in Table II. In practice, this elution is performed by allowing slightly less than one column volume of the chaotropic agent to enter the gel, and after 1 hr, collecting the thiocyanate fraction in several small aliquots (2–3 ml). All fractions are immediately dialyzed against PBS until thiocyanate is no longer detectable in the dialysate by addition to a crystal of ferric chloride (red color indicates presence of thiocyanate).

Other methods of dissociation of bound immunoglobulin which are commonly used in affinity chromatography are (1) exposure to acid pH by elution with 0.2 M glycine or 1 M acetic acid; (2) exposure to basic pH (e.g., 0.15 M NaCl adjusted to pH 11); or (3) exposure to denaturing agents such as urea or guanidine. The choice of dissociating agent should represent a balance between maximum yield and minimum disruption of the bound material. Selective elution of lower affinity antibodies may be achieved by treatment with the milder agents cited above. In our hands, thiocyanate elutes more total IgG than does pH change and we routinely use this procedure. This thiocyanate is immediately removed by dialysis. A yield of 2–3 mg specific anti-apoE IgG is obtained per 1 ml gel when prepared as above. The yields may differ with antisera which differ in titer and affinity.

TABLE II
Utilization of Affinity Gels

1. Binding
   a. For preparation of specific IgG, incubate 15–20 ml antiserum and an equal volume of PBS with 5 ml apoE–Sepharose matrix for 2 hr at 4°
   b. For isolation of apoE-containing lipoproteins, apply serum or lipoprotein fraction to column containing the apoE IgG–Sepharose matrix
2. Collect unbound fraction
3. Wash with minimum of 20 bed volumes of PBS
4. Wash with 20 bed volumes of 25 mM sodium phosphate in 0.5 M NaCl, pH 7.4
5. Elute bound fraction with 3 M NaSCN and wash gel with a minimum of 10 bed volumes of sodium thiocyanate
6. Dialyze eluted fraction(s) exhaustively vs PBS
7. Reequilibrate gel in PBS with 0.02% sodium azide

An immunoadsorbent which is capable of binding 80–90 μg lipoprotein bound apoE/ml gel can be obtained by binding the specific anti-apoE IgG purified as above at a density of 4–5 mg/ml gel matrix. The coupling procedure is that outlined in Table I. A modification in coupling buffer pH has been made for IgG coupling in order to reduce the multipoint attachment which can lead to poor immune binding.[6] The pH at which coupling takes place is a major factor determining coupling efficiency, since the reactive groups of the protein are the unprotonated forms of the amino groups. When large proteins such as IgG are coupled at a pH of 6–6.5, however, only a small number of the potentially activatable groups will be capable of reaction with the activated matrix. As with the apoE gel, it is recommended that the gel containing IgG be pretreated with the diluted (1 : 1) serum of the type from which apoE is to be isolated. Two or three cycles should be adequate treatment to reduce nonspecific adsorption. Store in 0.02% azide and 50 KIU of Trasylol (aprotinin).

ApoE-containing lipoproteins are prepared by the interaction of a crude lipoprotein preparation with the anti-apoE IgG immunoadsorbent utilizing a column procedure rather than a batch technique. Buffers routinely contain Trasylol (50 KIU/ml) to inhibit proteolytic activity. An amount of antibody gel representing a 50% excess of that required on theoretical grounds to bind all of the apoE in the sample is packed into a small (1–20 ml) column as needed. Prior to chromatography, the immune gel is always prepared by a preliminary washing with approximately 20 bed volumes of 3 M sodium thiocyanate followed by reequilibration in PBS or in the buffer of the sample to be applied. If EDTA is present in the sample, it is removed by dialysis, since EDTA reduces the binding of the antigens to the immunoadsorbent gel. The sample is then applied to the antibody gel at a slow rate (approximately 0.5 column volumes per hour or less). The subsequent washing is identical to that outlined for preparation of the specific IgG.

Problem Areas

There are several problems which should be recognized. Some of these are inherent in the technique of immunoaffinity chromatography and others result from the structural qualities of lipoproteins. Nonspecific binding to the immunoabsorbent has been alluded to previously. It can be a particularly serious problem when dealing with apolipoproteins, since they are present in quite low concentrations relative to those of the major plasma proteins, particularly albumin. Pretreatment of the gel may reduce this by tying up reactive sites. The problem is minimized by using the smallest volume of gel compatible with the goals of the experiment. Other important considerations include (1) selection of a nonreactive gel matrix such as agarose, (2) selection of a buffer of moderately high ionic strength such as the 25 mM phosphate, 0.5 M NaCl suggested above, and (3) preliminary partial purification of the material to be isolated. If the problem remains, one might consider introducing a precolumn containing a matrix with specific affinities for the contaminants.

With very high-affinity antisera, removal of the specifically bound protein from its immune complex may be difficult. Increased time of exposure to the eluting buffer may help. The use of strong denaturing agents (urea, guanidine–HCl) may be tried. Dilution of the gel by removing it to a tube or beaker and treatment with high molarity salt or a pH change may help.[6] However, such procedures are incompatible with maintenance of lipoprotein structure. An alternative approach is to select antibodies of low affinity for the initial preparation of the immunoadsorbent. An opportunity to do this conveniently is provided by the use of the antigen–Sepharose matrix, since elution of specific IgG from the antigen column (after interaction with whole antiserum) may be attempted with a variety of buffer conditions such as a change in ionic strength, pH, or specific chaotropic agents. The particular population of antibodies so removed and bound to the support matrix should yield the antigen under the same conditions.

A third problem which must be considered in applying this technique to lipoproteins has to do with the availability of epitopes in the lipid-bound state. Triglyceride-rich lipoproteins have a three-dimensional structure which appears to mask antigenic sites on certain of the apolipoproteins.[11] Also, apoA-I, the major protein of high-density lipoproteins, is more completely expressed immunochemically after exposure to heating[12] or detergents.[13] However, isolation of intact lipoproteins precludes heating or exposure to detergents prior to immunoabsorption. Incomplete exposure of apolipoprotein epitopes creates artifacts since selection of particular populations of lipoproteins may result. By testing the capacity of the immunoadsorbent for apoE with each class of lipoproteins, this problem should be uncovered. Also, repeated application of the fraction which is poorly retained can distinguish between a general reduction in availability of the antibody from a selective rejection of a specific subset of molecules which do not correctly present the epitope to the antibody. It is also possible that the use of strong chaotropic agents such as 3 M

[11] G. Schonfeld, W. Patsch, B. Pfleger, J. L. Witztum, and S. W. Weidman, *J. Clin. Invest.* **64,** 288 (1979).

[12] J. B. Karlin, D. J. Juhn, J. I. Starr, A. M. Scanu, and A. H. Rubinstein, *J. Lipid Res.* **17,** 30 (1976).

[13] S. J. T. Mao and B. A. Kottke, *Biochim. Biophys. Acta* **620,** 447 (1980).

thiocyanate may separate the protein components of lipoproteins if the exposure is prolonged. This possibility can be examined by monitoring for gel filtration during elution. However, rapid dialysis appears to prevent this, as demonstrated by studies of lipoprotein integrity (below). The possibility that dissociation or reassociation occurs during dialysis has not been ruled out by the experiments outlined, however, and subtle changes in protein conformation might result from these manipulations.

Validation of the Technique for Isolation of Apo E-Containing Lipoproteins

Column chromatography utilizing 2.5 × 100 cm columns of 4% agarose beads (BioGel A15M, 200–400 mesh) effectively separates the lipoproteins into size classes which correspond to the major lipoprotein groups defined by the ultracentrifuge: chylomicrons, VLDL, IDL, LDL, and HDL.[4] This is the preferred method for separation of apoE-containing lipoproteins prior to immunoaffinity isolation because techniques such as ultracentrifugation or precipitation dissociate apoE. When the column eluate is analyzed for apolipoproteins B, A-I, E, and C-III, profiles are obtained which strongly suggest that subpopulations of lipoproteins exist which are relatively enriched in apoE and are distinct in size from the major cholesterol-carrying lipoproteins (Fig. 1).

The following studies were undertaken to validate the efficacy of the apoE immunoadsorbent for isolation of these apoE-enriched fractions from the milieu of similarly sized VLDL, IDL, LDL, or HDL. These studies examine (1) the capacity of the immunoadsorbent to bind delipidated and lipoprotein-associated apoE and (2) the extent of nonspecific binding of other apolipoproteins.

The capacity for lipoprotein-bound apoE was determined by application of the whole lipoprotein fraction ($d < 1.21$ g/ml) of serum to 1 ml of immunoadsorbent. The capacity of the gel for lipoprotein-bound apoE proved to be lower (24 μg/ml) than that determined by application of delipidated apoE (33 μg/ml). It is possible that lipid partially inhibits immune binding by either (1) masking immunoreactive sites, or (2) preventing access of large, intact lipoprotein particles to IgG molecules bound to the interior of the agarose beads. In practice, the capacity of the immunoadsorbent for apoE associated with triglyceride-rich VLDL and chylomicrons is less than for that associated with HDL-sized lipoproteins. It is also possible that delipidated apoE is self-associating, increasing the capacity of the column through binding of dimers or larger units.

There is a tendency for low levels of other apolipoproteins to be nonspecifically retained by the immunoadsorbent. ApoA-I, apoC-III, and

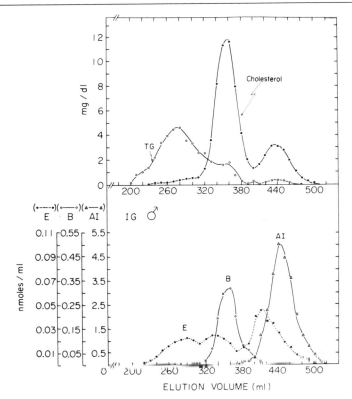

FIG. 1. Lipid and apolipoprotein profiles of normal human plasma after 4% agarose column chromatography. ApoB and apoA-I cochromatograph with LDL and HDL cholesterol, respectively. ApoE elutes in three peaks. Peak I is coincident with VLDL (220–300 ml), peak II is intermediate in size between VLDL and LDL (300–370 ml) and peak III is slightly larger than the major HDL fraction (380–460 ml).

apoB (in apoE-free LDL) all bound to a limited extent (0.112, 0.07, and 0.750 μg/ml gel, respectively) despite the fact that none cross-reacted in an RIA utilizing the same source of anti-apoE. Nevertheless, the extent of this nonspecific adherence was less than 3% of the mass of apoE binding and was less than 5% of the amount of apoA-I, apoC-III, or apoB when compared to that bound as a part of native lipoproteins also containing apoE. When the lipoprotein loads are near the capacity for apoE, the nonspecific binding should be negligible for all apolipoproteins. The gel would then provide the basis for evaluating adsorption of other apolipoproteins present in the lipoprotein complex with apoE. Caution should be taken, however, in interpreting binding of apolipoproteins at levels near those which bind nonspecifically. If a very heterogeneous mixture of proteins (e.g., whole serum) is applied, it would also be important to evaluate the potential contamination by other proteins, many of which are present in extremely high concentrations relative to the apolipoproteins. The usual application of polyacrylamide gel electrophoresis or immuno-chemical techniques might be used in this evaluation.

We have tested the stability of lipoprotein complexes containing apoE and other apolipoproteins by examining whether the same group of apoli-poproteins would remain associated during a second immunoadsorbent exposure. The lipoproteins of intermediate size containing apoE (peak II, Fig. 1) were separated by 4% agarose chromatography and were pooled and applied to a volume of immunoadsorbent gel (10 ml) sufficient to bind all of the apoE present. After elution in 15 ml and dialysis against a total of 6 liters of buffer (3 × 2 liters), a portion of the bound fraction was then reapplied to 1 ml of the anti-apoE gel and the apoE, apoB, and apoC-III originally bound by the anti-apoE column were assayed in the void vol-ume after reapplication as a measure of that dissociated by the affinity procedure. These masses represented 2.9, 5.6, and 1.1% of the apoE, apoB, and apoC-III applied, respectively. Thus, the lipoprotein appears to remain as a unit by these criteria.

Two additional types of studies were used to examine the integrity of the apoE-containing lipoprotein subclasses isolated by immunoaffinity chromatography. In these studies fractions bound and eluted from affinity columns were chromatographed on 4% agarose. In the first group of ex-periments, whole serum from a nonfasted subject was subjected to immu-noaffinity chromatography and the total bound fraction was eluted and applied to a column of 4% agarose in order to compare the apolipoprotein profile with those seen in whole plasma. The chromatographic profiles of apoE, apoB, and apoA-I in whole serum (Fig. 2a), in the unbound (Fig. 2b) and in the bound fraction (Fig. 2c) after immune binding, are shown for one of two subjects studied. In whole serum, apoB and apoA-I chro-matographed with the two major cholesterol peaks which were thus iden-tified as LDL and HDL, respectively. In contrast, the apoE distribution was described by a complex profile including (1) a fraction in the void volume representing chylomicrons and large VLDL (fraction I), (2) a large peak in the region of intermediate-sized lipoproteins (fraction II), and (3) a peak in the region of large HDL (fraction III). When the whole serum was passed over the immunoadsorbent, 90.3% (3.64 mg) of the apoB, 98.8% (5.59 mg) of apoA-I, and 5.6% (0.0088 mg) of the apoE applied did not bind and were collected in the unretained fraction. The less than complete binding of apoE reflects the fact that this nonfasting sample had a relatively high proportion of apoE in very large triglyceride-rich lipoproteins (as discussed above).

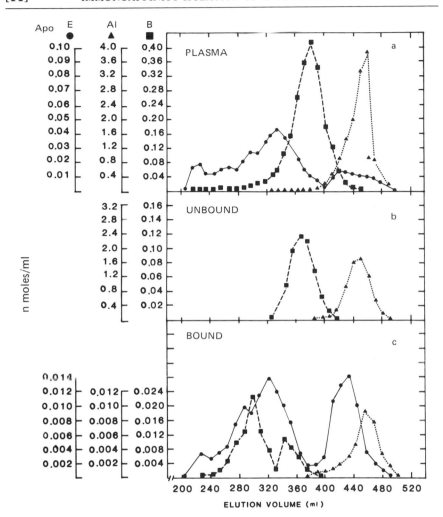

FIG. 2. Apolipoprotein E, B, and A-I profiles after 4% agarose chromatography of normal human plasma before (a) and after immunoaffinity chromatography on anti-apoE IgG–Sepharose CL-4B. (b) Profiles of apoB and apoA-I which did not bind to the anti-apoE immunoadsorbent. (c) Profiles of apoE, apoB, and apoA-I which were retained by and eluted from the immunoadsorbent.

Agarose chromatography of this unretained fraction verified that apoB and apoA-I were eluted within the same elution volumes as were apoB and apoA-I in the original serum. In contrast, 9.4% (0.418 mg) of the apoB, 1.2% (0.072 mg) of the apoA-I, and 94.9% (0.149 mg) of the apoE applied were bound by the apoE immunoadsorbent. As in the original serum, apoE described a complex profile across the lipoprotein spectrum with the peaks of the major fractions eluting at 287, 320, and 431 ml. These were approximately the same elution volume as these same fractions (298, 334, and 420 ml, respectively) in the original serum (Fig. 2a). Thus, a clear size distinction between the major apoE-containing lipoproteins was maintained. However, there was an apparent reduction in the relative proportion of apoE in chylomicrons, VLDL, and the intermediate-size fraction II after immunoaffinity isolation. This probably reflects reduced binding of the more triglyceride-rich lipoproteins. It is noteworthy that the apoB which was retained with the apoE rechromatographed in two peaks within the same elution profile as the apoE in fraction II. This is in contrast to the bulk of the apoB in the original serum (Fig. 2a) which chromatographed with the LDL cholesterol in a peak clearly separate from the apoE-containing lipoproteins. The bound apoA-I rechromatographed in the region of HDL. This profile is complex, suggesting populations of particles with different ratios of E/AI. This observation has been confirmed in four studies with the particles of largest size having the highest E/AI ratio. It should be reemphasized that the fractions of both apoB and apoA-I which were retained were only a small proportion of that applied (note different scales in Fig. 2).

The physical integrity of the lipoproteins in fraction III (large HDL) was also studied after preliminary isolation from plasma by 4% agarose chromatography. Using fraction III from the plasma of two normolipoproteinemic subjects, the specific apoE-containing lipoproteins were isolated by immunoaffinity chromatography and the bound fractions were reapplied to the 4% agarose column and analyzed in a protocol analogous to that described for whole serum (see above). The results from one subjects are detailed in Fig. 3. For this study, the peak apo E-containing tubes from the agarose column (Fig. 3a) (404–432 ml) were pooled and applied to a volume of anti-apoE immunoadsorbent sufficient to bind all apoE. After binding and washing, the bound apoE-containing fraction was eluted in 10 ml of 3 M thiocyanate, made 0.1% with respect to ovalbumin, dialyzed against column buffer, and 5 ml was reapplied to the gel filtration column. The profiles of the original and rechromatographed samples of one subject are shown in Fig. 3a and b. This subject had 90% of the plasma apoE in this HDL fraction. After immunoaffinity binding, all apoE (87.1 μg) and only 0.23% (3.6 μg) apoA-I applied were retained. After rechromatography of this bound fraction, the peak of apoE eluted in a similar volume to that in the original pool (390–440 ml), providing strong evidence for the maintenance of lipoprotein particles of similar size. In addition, the apoA-I which was bound to the anti-apoE column chromatographed near apoE, but again with a profile displaced from that of apoE,

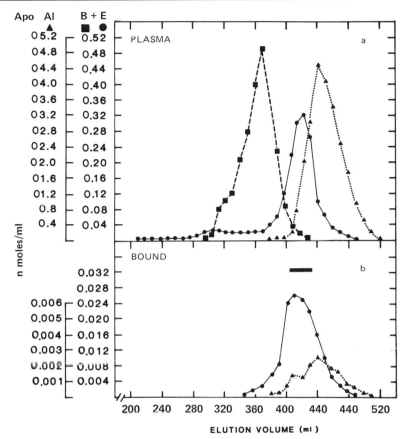

FIG. 3. Apolipoprotein profiles after 4% agarose chromatography of normal human plasma (a) and of apoE-containing fraction III (404–432 ml) after binding to and elution from the anti-apoE IgG–Sepharose CL-4B (b).

suggesting changing ratios of apoE to apoA-I within the lipoprotein pool originally adsorbed to the affinity column.

Summary

These data suggest that subfractionation of lipoproteins by immunoaffinity chromatography according to apolipoprotein content provides a valuable method for physical isolation of intact lipoprotein subclasses. While this study has focused specifically on apoE-containing lipoproteins, the technique of immunoaffinity isolation clearly has wider application. It must be emphasized that antibodies differ with regard to affinity and avidity, and that both binding and elution conditions may need to be adjusted for each group of antibodies. Maintenance of physical integrity of lipoproteins thus isolated must also be evaluated independently for each antibody. Nevertheless, immunoaffinity isolation of apolipoprotein-specific subclasses offers a powerful tool for preparing lipoproteins which should reflect physical and/or metabolic properties conferred by the apolipoprotein of interest.

Section II

Cell Biology of Plasma Lipoproteins

[12] Methods for Visualization of the LDL Pathway in Cultured Human Fibroblasts

By RICHARD G. W. ANDERSON

Introduction

Receptor-mediated endocytosis is a complex process utilized by cells to take up metabolically important macromolecules. The specificity and efficiency of internalization depend upon cell surface receptors that rec-ognize a specific macromolecule and upon clathrin-coated pits. Once a molecule binds to a receptor and the receptor–ligand complex is clustered over coated pits, efficient internalization takes place by the transformation of the coated pit into a coated endocytic vesicle. The internalized macromolecule then proceeds through various endocytic compartments and is eventually delivered to a target site within the cell. Usually the target site is the lysosome, where the molecules are degraded, or a transcellular site, where the molecule is exocytosed.[1]

The low-density lipoprotein (LDL) receptor is one of the best studied receptors involved in receptor-mediated endocytosis. Normal LDL receptors contain a high-affinity binding site for LDL and a recognition site for coated pits on the cell surface.[2] Once LDL binds to receptors that are in coated pits, it is rapidly internalized and delivered to the lysosome. Within the lysosome, acid hydrolases digest the cholesteryl ester core of LDL, liberating free cholesterol for use in various biosynthetic processes within the cell.[3]

The LDL receptor is extremely important for normal cholesterol metabolism in the body. In its absence, the regulation of LDL synthesis and clearance from the plasma are severely impaired. As a result, individuals who lack functional LDL receptors develop atherosclerosis at a very early age.[3]

The LDL receptor has been studied using both biochemical[4] and microscopic techniques.[5] Each of these techniques has made a unique con-

[1] R. G. W. Anderson and J. Kaplan, *Mod. Cell Biol.* **1**, 1 (1983).
[2] J. L. Goldstein, R. G. W. Anderson, and M. S. Brown, *Nature (London)* **279**, 679 (1979).
[3] J. L. Goldstein and M. S. Brown, *Annu. Rev. Biochem.* **46**, 897 (1979).
[4] J. L. Goldstein, S. K. Basu, and M. S. Brown, this series, Vol. 98, p. 241.
[5] R. G. W. Anderson, J. L. Goldstein, and M. S. Brown, *in* "Receptor-Mediated Binding and Internalization of Toxins and Hormone" (J. L. Middlebrook and L. D. Kohn, eds.), p. 1. Academic Press, 1981.

METHODS IN ENZYMOLOGY, VOL. 129

tribution to the understanding of LDL receptor structure and function. The initial discovery of the LDL receptor and its absence in individuals with the genetic disorder familial hypercholesterolemia was based on biochemical techniques. On the other hand, our understanding of the dynamics of the receptor on the cell surface and the relationship of receptor to coated pits has depended on microscopic techniques.

Microscopic methods for studying receptor activity employ either the light microscope or the electron microscope. In general, light microscopy, using fluorescent probes, offers the opportunity to rapidly acquire information about the surface and intracellular distribution of receptors and their behavior under various experimental conditions. Fluorescence microscopy does not give quantitative information, which often is important for understanding relationships within the cell. Moreover, because of low resolution, fluorescence images are sometimes difficult to interpret. Electron microscopy, by contrast, yields both quantitative information about the distribution of receptors and high-resolution information about the relationship of receptors to important cellular organelles.

In the application of these techniques to the study of receptors, usually both light microscopy and electron microscopy are used in tandem: fluorescence microscopy is used to determine the best experimental conditions, whereas electron microscopy is employed to obtain the high-resolution, quantitative information required for a complete analysis.

To apply these techniques, specific probes must be available that will label the receptor and make it visible. The most direct approach is to conjugate the specific ligand to a marker that can be visualized by either fluorescence microscopy or electron microscopy. However, in the case of LDL receptors, often antibodies against the receptor are used in situations where LDL conjugates will not bind. The antibodies are then visualized by the use of indirect immunocytochemistry.

The majority of the microscopic studies on LDL receptor function have employed cultured cells; therefore, the methods developed in this laboratory apply almost exclusively to their use for studying receptor biology of cultured fibroblasts.

Preparation of Fluorescent LDL [r(PMCA Oleate) LDL]

LDL is a large spherical particle with an average diameter of 21 nm and a molecular weight of 3×10^6.[6] Each LDL particle is composed of an apolipoprotein–phospholipid shell and a cholesteryl ester-rich core that

---

[6] R. L. Jackson, J. D. Morrisett, and A. M. Gotto, Jr., *Physiol. Rev.* **56**, 259 (1976).

contains 1300 molecules of cholesteryl ester.[7] The preparation of fluorescent LDL takes advantage of the procedure developed by Krieger *et al.*[8] for removing the endogenous cholesteryl ester and replacing it with a fluorescent derivative of cholesteryl ester.[9] This permits the incorporation of greater than a thousand molecules of the fluorescent cholesteryl ester into the core of each LDL particle. This highly fluorescent LDL particle can be visualized both on the cell surface and after internalization by fluorescence microscopy.[10]

Reagents

Human LDL (d = 1.019–1.063 g/ml)
PMCA oleate (3-pyrenemethyl-23,24-dinor-5-cholen-22-oate-3β-yl oleate)

Procedure. LDL is dialyzed against 0.3 mM sodium EDTA (pH 7); 1.9 mg of dialyzed LDL is lypholized in the presence of potato starch (starch-to-protein ratio, 12:1, w/w) in Siliclade-treated glass tubes and then extracted two times with 5 ml of heptane at −10°. The heptane-extracted LDL is then reconstituted by the addition of 200 μl of benzene containing 6 mg of PMCA oleate. The tubes are incubated for 1 hr at −10°. Benzene is then evaporated under nitrogen at 0°. The reconstituted LDL is then solubilized by incubating in 1 ml of 10 mM Tricine (pH 8.4) for 12 hr at 4°, and then the starch and unincorporated PMCA oleate are removed by centrifugation. The mass ratio of PMCA oleate to protein (mg/mg) in the final solubilized preparation should be about 1.24. This fluorescent LDL is referred to as r(PMCA oleate) LDL.

Preparation of LDL-Ferritin

A portion of the amino groups in the apolipoprotein (apoB) component of LDL can be modified without destroying biological activity.[11] Therefore, macromolecules such as ferritin can be covalently attached to these amino groups. The LDL–ferritin conjugate can then be used to visualize LDL receptor distribution by electron microscopy.[12]

[7] V. P. Skipski, *in* "Blood Lipids and Lipoproteins: Quantitation, Composition, and Metabolism" (G. J. Nelson, ed.), p. 471. Wiley (Interscience), New York, 1972.
[8] M. Krieger, M. J. McPhaul, J. L. Goldstein, and M. S. Brown, *J. Biol. Chem.* **254,** 3845 (1979).
[9] M. Krieger, L. C. Smith, R. G. W. Anderson, J. L. Goldstein, Y. J. Kao, H. J. Pownall, A. M. Gotto, Jr., and M. S. Brown, *J. Supramol. Struct.* **10,** 467 (1979).
[10] R. G. W. Anderson, J. L. Goldstein, and M. S. Brown, *J. Receptor Res.* **1,** 17 (1980).
[11] K. H. Weisgraber, T. L. Innerarity, and E. W. Mahley, *J. Biol. Chem.* **253,** 9053 (1978).
[12] R. G. W. Anderson, J. L. Goldstein, and M. S. Brown, *Proc. Natl. Acad. Sci. U.S.A.* **73,** 2434 (1976).

Reagents

Human LDL (d = 1.019–1.063 g/ml)
EM-grade ferritin (6× crystallized and cadmium free)
EM-grade glutaraldehyde
Buffer A (0.1 M sodium phosphate, pH 7.3)

Procedure. The procedure is based on the method of Kishida *et al.*,[13] where ferritin is activated with a 1200-fold excess of glutaraldehyde per amino group in the ferritin protein. The excess glutaraldehyde activates the amino group of the ferritin without causing ferritin to aggregate. Once the activated ferritin is removed from the glutaraldehyde, it will covalently bind to the free amino groups of LDL.

Ten milligrams of ferritin in 2.0 ml of buffer A is mixed with 1 ml of 50% glutaraldehyde. The mixture is incubated, with stirring, at room temperature for 30 min and centrifuged at 20,000 g for 10 min to remove small amounts of precipitated ferritin. The activated ferritin is separated from glutaraldehyde by gel filtration on a Sephadex G-25 column (coarse, using a 2.5 × 90 cm column). The column is equilibrated and rapidly eluted at 4° with buffer A. The activated ferritin is mixed with purified LDL at a ferritin-to-LDL molar ratio of approximately 0.6 (based on a molecular weight of 9×10^5 for ferritin and 5×10^5 for the protein component of LDL). This mixture is incubated at 4° for 60 to 72 hr on a slowly rotating table. The density of the mixture is then adjusted to 1.085 g/ml with solid potassium bromide and centrifuged at 214,000 g for 18 hr at 4° in a 1.6 × 7.6 cm tube using a Beckman type 65 ultracentrifuge rotor. After the centrifugation, the free ferritin forms a pellet at the bottom of the tube and the free LDL floats to the top of the tube. Immediately above the ferritin pellet is a yellow-orange zone approximately 1 cm in height that contains LDL conjugated to ferritin. This fraction, which typically contains 20% of the initial LDL cholesterol, is aspirated from the tube and dialyzed extensively against 0.15 M NaCl. By negative-stain electron microscopy, > 90% of the LDL is complexed to ferritin. On the average, each LDL is labeled with two ferritin molecules. The concentrations of LDL–ferritin used in each experiment is based on the concentration of LDL protein.

Preparation of LDL–Gold

Gold conjugated to either ligands or immunoglobulins has become a popular method for localizing receptors and antigens in tissues. Such conjugates are easily prepared with a minimum of time and effort. Forma-

[13] Y. Kishida, B. R. Olsen, R. A. Berg, and D. J. Prockop, *J. Cell Biol.* **64,** 331 (1975).

tion of the conjugate takes advantage of the fact that gold is negatively charged due to surface hydroxide ions and readily adsorbs a variety of different proteins. The method for conjugating LDL to gold was developed by Handley et al.[14]

Reagents

Gold chloride ($HAuCl_4$)
Human LDL ($d = 1.019–1.063$ g/ml)
Sodium citrate
Buffer A (0.05 M EDTA, pH 5.5)
Buffer B (0.15 M NaCl, 0.1 M sodium phosphate, pH 7.3)
BioGel

Procedure. Colloidal gold is prepared by heating 50 ml of 0.01% $HAuCl_4$ to boiling and then adding 1.6 ml of 1% sodium citrate. The mixture is further boiled for 5 min, giving a clear red solution (pH 5.5) that contains 1×10^{12} particles/ml with an average diameter of 19 nm. To 0.5 ml of LDL (3.0 mg/ml) in buffer A is added rapidly with mixing 5 ml of colloidal gold. The mixture is incubated for approximately 30 min and excess LDL is removed by centrifuging the mixture at 9000 g for 20 min onto a 35% sucrose cushion (SW 25.1 rotor). The pellet containing the conjugate is dialyzed extensively against buffer B to remove sucrose and EDTA. The conjugate can be further concentrated using BioGel dehydration. Typically, conjugates contain 7 to 9 LDL molecules per gold particle. If large amounts of conjugate are needed, small preparations of 5 to 8 ml are prepared and then pooled. These conjugates are stable for 24 hr after preparation.

Antibodies against the LDL Receptor

Antibodies against the LDL receptor are useful for visualizing the LDL pathway in situations where LDL cannot be used to identify the location of the receptor. Three main types of antibodies have been utilized: (1) polyclonal antibodies that are monospecific against the LDL receptor,[15] (2) monoclonal antibodies directed against the LDL receptor,[16] and (3) synthetic polypeptide antibodies made against synthetic peptides

---

[14] D. A. Handley, C. M. Arbeeny, L. D. Witte, and S. Chien, *Proc. Natl. Acad. Sci. U.S.A.* **78,** 368 (1981).
[15] U. Beisiegel, T. Kita, R. G. W. Anderson, W. J. Schneider, M. S. Brown, and J. L. Goldstein, *J. Biol. Chem.* **256,** 4071 (1981).
[16] U. Beisiegel, W. J. Schneider, J. L. Goldstein, R. G. W. Anderson, and M. S. Brown, *J. Biol. Chem.* **256,** 11923 (1981).

that correspond to a specific amino acid sequence within the LDL receptor protein.[17]

The polyclonal antibodies have been used to study receptor distribution on the surface of cells fixed with glutaraldehyde, a fixative that destroys LDL binding.[18] Furthermore, because LDL cannot be used in detergent solutions, these antibodies will detect intracellular receptors following permeabilization of the cells with detergent. Monoclonal antibodies work well for electron microscopic localization of receptor because they can be readily conjugated to ferritin.[18] Synthetic polypeptide antibodies have been used to determine the orientation of the LDL receptor in the plane of the membrane. All three types of antibodies have been used to visualize the endocytic pathway of the LDL receptor.[15–17]

Preparation of Antigens. For the production of both polyclonal and monoclonal antibodies, LDL receptors from bovine adrenal cortex are solubilized with octylglucoside, partially purified by DEAE–cellulose and agarose gel chromatography, and precipitated with acetone in the presence of phosphatidylcholine according to previously described methods.[19] This procedure gives about 350-fold purification of the LDL receptor. This antigen is used to immunize animals by standard procedures.[15,16]

Preparation of Anti-Peptide Antibodies. Anti-peptide antibodies are prepared according to the methods developed by Lerner *et al.*[20] and Walter *et al.*[21] As an example, the 16-residue peptide corresponding to the NH_2 terminal sequence to the LDL receptor was synthesized by Peninsula Laboratories, Inc., Belmont, CA. The peptide was coupled to keyhole limpet hemocyanin at room temperatures using the following procedure: a solution of hemocyanin and 50% glycerol (80 μl containing 10 mg protein) is mixed with 420 μl of a buffer A (0.14 M NaCl, 1.6 mM KCl, 1.1 mM KH_2PO_4 and 8 mM Na_2HPO_4, pH 7.4). A freshly prepared solution (20 μl) of 0.1 M disuccinimidyl suberate in dimethylsulfoxide is added and the mixture stirred for 10 min. Then 6.3 mg of solid receptor peptide is added and stirred continuously for 90 min. Then 500 μl of buffer A is added and stirred for 30 min.

To obtain antibodies, 160 μl of receptor peptide–hemocyanin conju-

[17] W. J. Schneider, C. J. Slaughter, J. L. Goldstein, R. G. W. Anderson, J. D. Capra, and M. S. Brown, *J. Cell. Biol.* **97,** 1635 (1983).
[18] R. G. W. Anderson, M. S. Brown, U. Beisiegel, and J. L. Goldstein, *J. Cell Biol.* **93,** 523 (1982).
[19] W. J. Schneider, J. L. Goldstein, and M. S. Brown, *J. Biol. Chem.* **255,** 11442 (1980).
[20] R. A. Lerner, J. G. Sutcliff, and T. M. Shinnick, *Cell* **23,** 309 (1981).
[21] G. Walter, K.-H. Scheidtmann, A. Carbone, A. P. Lavdano, and R. F. Doolittle, *Proc. Natl. Acad. Sci. U.S.A.* **77,** 5197 (1980).

gate is used for injection. Rabbits are injected subcutaneously according to standard procedures.[17]

Distribution of Surface Membrane Receptors

A first consideration in localizing receptors by microscopic techniques is to establish conditions for detecting the native distributions of the receptor in the plane of the membrane. Endocytosis is inhibited at 4° and receptors are relatively quiescent since ligand entry and receptor recycling are not in progress. Therefore, regardless of the receptor specific probe, binding at 4° gives a measure of the native distribution of this class of receptors.

Since receptors are detected by the presence of labeled ligand, the ligand itself may induce a change in receptor distribution. For example, the ligand may induce receptor clustering, even at 4°. Therefore, in some cases prefixing cells with either formaldehyde or glutaraldehyde fixatives before ligand binding may give a clearer picture of the native distribution of receptors. The fixatives may affect the ability of the natural ligand to bind; LDL will not bind to cells following glutaraldehyde fixation. However, receptor-specific antibody probes often can be used in these situations.[18]

Receptor localization is easier when cells are expressing the maximum number of receptors. Therefore, to study the LDL receptor, cells are grown in the absence of lipoprotein cholesterol for 2 days prior to the experiment. This treatment causes a 10-fold increase in the number of receptors, which is required to have sufficient signal to detect receptors.

The above consideration should be applied in any type of LDL receptor localization study. Regardless of the receptor-labeling method, whether modified ligand or antibody, the same protocol is employed. Subsequent processing of the cells depends on whether receptor localization is to be visualized by fluorescence microscopy or electron microscopy.

Reagents

r(PMCA oleate) LDL
LDL–ferritin
Anti-LDL receptor antibody (polyclonal or monoclonal)
Monoclonal anti-LDL receptor antibody conjugated to ferritin (IgG C7–ferritin)
Anti-human LDL antibody
LDL–gold

Human fibroblast grown in Eagles minimum essential medium that contains 10% fetal calf serum. Cells are routinely grown for 7 days and 2 days before each experiment medium is replaced with medium that contains 10% (v/v) human lipoprotein-deficient serum (LPPS) ($d > 1.215$ g/ml)

Buffer A (0.15 M NaCl, 2 mM MgCl$_2$, 10 mM sodium phosphate, pH 7.3)

Buffer B (buffer A, 5 mM NH$_4$Cl)

Buffer C (Dulbecco's PBS, 1% bovine serum albumin)

Fixative A (3% formaldehyde in buffer A)

Fixative B (2% glutaraldehyde, 0.1 M sodium cacodylate, pH 7.4)

Fixative C (2% OsO$_4$, 0.1 M sodium cacodylate, pH 7.4, 4.5% sucrose)

Goat anti-rabbit IgG coupled to rhodamine tetramethylisothiocyanate

Rabbit anti-goat IgG coupled to fluorescein isothiocyanate

Goat anti-mouse IgG coupled to fluorescein isothiocyanate

Procedure. To localize LDL receptors by fluorescence microscopy, either a direct method, using fluorescent LDL, or an indirect method, using unlabeled LDL or anti-LDL receptor antibodies, is employed. Regardless of whether direct or indirect labeling, two basic protocols are used to delineate these receptors.

Procedure A: Human fibroblasts cultured on coverslips for 7 days (final 2 days in 10% LPPS) after an initial seeding density of 7.5×10^4 cells are chilled to 4° by placing the dishes on ice at 4° for 30 min. One of the following ligands is added directly to the dish and incubated at 4° for 1–2 hr: (1) LDL (20 μg/ml), (2) r(PMCA oleate) LDL (10 μg/ml), (3) polyclonal anti-LDL receptor antibody (0.2 mg/ml), or (4) monoclonal anti-receptor antibody (50 μg/ml).

Procedure B: Cultured cells on coverslips are initially fixed for 15 min at room temperature with either fixative A or fixative B, washed with buffer B, and incubated with the ligand, in buffer A, for 1–2 hr at 37°. Cells fixed with fixative A can be labeled with r(PMCA oleate) LDL (50 μg/ml), LDL (50 μg/ml), or polyclonal anti-receptor antibody (0.2 mg/ml). Cells fixed with fixative B can be labeled with polyclonal anti-receptor antibody (0.2 mg/ml). All incubations are carried out by overlaying the coverslip with 60–100 μl of buffer A containing the ligand.

In both procedures A and B, the cells are washed three times (10 min each) with buffer C (at the appropriate temperature) and fixed 15 min at room temperature with fixative A.

Whereas fluorescent LDL can be observed directly in the fluorescence microscope, using a microscope equipped with the appropriate filter package,[9] the other receptor-specific ligands must be localized by indirect immunofluorescence. To localize LDL the coverslips are overlayed with 60–100 μl of goat anti-human LDL (0.5 mg/ml) in buffer A for 1 hr at 37°. The coverslips are washed four times (5 min each) with 3 ml of buffer C and incubated at 37° for 1 hr with 60–100 μl of buffer A containing 0.5 mg/ml of rabbit anti-goat IgG coupled to fluorescein isothiocyanate. To localize anti-receptor antibodies, the same procedure is used except 0.5 mg/ml goat anti-rabbit IgG coupled to rhodamine tetramethylisothiocyanate detects polyclonal anti-receptor antibodies and 0.5 mg/ml goat anti-mouse IgG coupled to fluorescein isothiocyanate detects monoclonal anti-receptor antibodies. Detection of monoclonal anti-receptor antibody on the cell surface requires a third incubation with rabbit anti-goat IgG coupled to fluorescein isothiocyanate (0.5 mg/ml). The coverslips are washed after each antibody treatment four times (5 min each) with buffer C and mounted on glass slides with 90% glycerol in 0.1 M Tris/chloride, pH 9.4, and viewed with a fluorescence microscope using the appropriate filter package for rhodamine or fluorescein.

To localize receptors on the cell surface by electron microscopy, either LDL–ferritin, LDL–gold, or Ig-C7–ferritin is used. Cells are grown in the absence of coverslips as above and chilled to 4° for 30 min before addition to the media. The additions include either LDL–ferritin (50 μg/ml), IgG-C7–ferritin (50 μg/ml), or LDL–gold (1 ml of conjugate containing 1 × 10^{13} conjugates/ml). After incubation at 4° for 2 hr, each monolayer is washed five times with ice-cold buffer C and fixed with fixative B for 1 hr. The cells are then fixed with fixative C and processed for electron microscopy by standard procedures.[14,22]

Expected Results. These techniques should yield a characteristic distribution pattern for the LDL receptor. By fluorescence microscopy one should observe numerous focal spots of fluorescence on the surface of the cell (Fig. 1). In some cells, those that have well-delineated stress fibers, these dots will be lined up along the stress fibers. By electron microscopy, most of the label should be associated with coated pits (Fig. 2).

Distribution of Intracellular Receptors

In contrast to the visualization of surface receptors, intracellular receptors must be visualized with anti-receptor antibodies. This is because intracellular receptors can only be detected after permeabilization of cells with detergent, a treatment that interferes with LDL–receptor interaction.

[22] R. G. W. Anderson, M. S. Brown, and J. L. Goldstein, *Cell* **10**, 351 (1977).

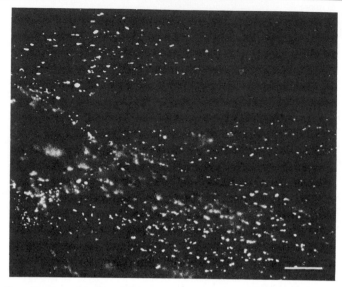

FIG. 1. Localization of surface LDL receptors in human fibroblasts by indirect immuno-fluorescence. Human fibroblasts grown on coverslips were chilled to 4° and received a direct addition of 20 μg/ml LDL. After 60 min at 4°, the cells were washed, fixed, and processed for indirect immunofluorescence localization of LDL. 900×. Bar = 11 μm. (Reprinted from Larkin et al.[26] with permission.)

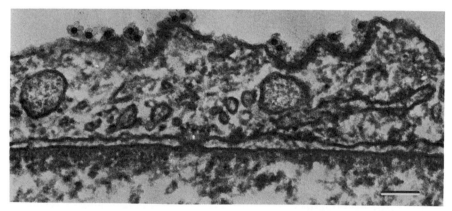

FIG. 2. Localization of surface LDL receptors in human fibroblasts by electron micros-copy. Human fibroblasts were chilled to 4° and incubated in the presence of LDL–gold (3 × 10^{11} conjugates/ml) at 4° for 1 hr, washed with culture medium that contain human lipopro-tein-deficient serum and warmed to 37° for 3 min before fixation [3% (v/v) glutaraldehyde in 0.15 M sodium cacodylate, pH 7.38, for 30 min followed by 2 hr in 1.0% OsO_4 containing 0.1% Ruthenium Red]. 100,000×. Bar = 0.1 μm. (Courtesy of Hadley et al.[14])

All of the studies on the distribution of intracellular LDL receptors have utilized indirect immunofluorescence techniques.[23] The procedure, however, can be adapted to use a peroxidase-labeled second antibody,[24] thus allowing the detection by electron microscopy of intracellular receptors.

Reagents

Polyclonal anti-LDL receptor antibody
Goat anti-rabbit IgG coupled to fluorescein isothiocyanate
Buffer A (0.15 M NaCl, 2 mM MgCl$_2$, 10 mM sodium phosphate, pH 7.4)
Buffer B (buffer A plus 5 mM NH$_4$Cl)
Buffer D (Dulbecco's PBS, 1% bovine serum albumin)
Buffer C (Buffer A plus 0.05% Triton X-100)
Fixative A (Buffer A plus 3% formaldehyde)
Human fibroblasts grown for 7 days on coverslips (see Surface Distribution of LDL Receptors)

Procedure. Human fibroblasts, grown on coverslips, are fixed with 2 ml of fixative A for 15 min at room temperature and washed twice with 2 ml of buffer B. Each coverslip is then treated with 3 ml of buffer C for 10 min at −10°. The coverslip is then placed in a fresh Petri dish (cell side up), covered with 60 μl of rabbit anti-LDL receptor IgG (0.2 mg/ml), and incubated for 1 hr at 37°. Following this incubation, the cells are washed four times with buffer D (15 min each) and incubated with 60 μl of goat anti-rabbit IgG coupled to fluorescein isothiocyanate (0.5 mg/ml) for 1 hr at 37°. The coverslips are then washed at room temperature four times (15 min each) with buffer D and mounted on a glass slide with 90% glycerol in 0.1 M Tris-HCl (pH 9.4).

Expected Results. With gentle permeabilization, anti-receptor antibody staining is seen in three compartments of the cell: punctate fluorescent dots on the surface of the cell, various medium-sized vesicles in the body of the cell, and a reticular pattern near the nucleus of the cell (Fig. 3A). The reticular pattern is associated with the Golgi apparatus, because in the same cell (Fig. 3B), this region of receptor also binds wheat germ agglutinin coupled to rhodamine tetramethylisothiocyanate, a marker for the Golgi apparatus.[25]

[23] S. K. Basu, J. L. Goldstein, R. G. W. Anderson, and M. S. Brown, *Cell* **24**, 493 (1981).
[24] R. G. W. Anderson, E. Vasile, R. J. Mello, M. S. Brown, and J. L. Goldstein, *Cell* **15**, 919 (1978).
[25] A. M. Tartakoff and P. Vassalli, *J. Cell Biol.* **97**, 1243 (1983).

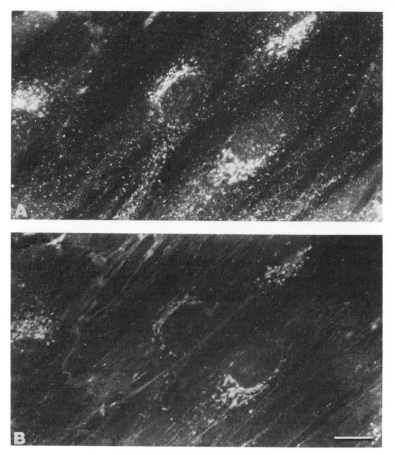

FIG. 3. Localization of LDL receptors (A) and wheat germ agglutinin-binding sites (B) in the same cell by fluorescence microscopy. Human fibroblasts were grown on coverslips and fixed with 3% formaldehyde. Following permeabilization with 0.05% Triton X-100, cells were sequentially incubated with polyclonal anti-LDL-receptor antibody (0.2 mg/ml), fluorescein-labeled anti-rabbit IgG (0.5 mg/ml), and rhodamine-labeled wheat germ agglutinin (0.5 mg/ml). The coverslips were mounted on glass slides and the same cell was viewed with either fluorescein (A) or rhodamine (B) fluorescence filter packages. 700×. Bar = 14 μm.

Visualization of Receptor-Mediated Endocytosis

As in the case of LDL receptor distribution on the cell surface, receptor-mediated endocytosis can be visualized with both the fluorescence microscope and the electron microscope.[10,22] For the light microscope, fluorescent LDL is the most useful probe; for the electron microscope, either LDL–ferritin, LDL–gold, or IgG-C7–ferritin is used.

The best way to visualize the endocytosis sequence, by both light microscopy and electron microscopy, is to first label the receptors at 4° (see Distribution of Surface Receptors). The cells are then shifted to 37° and fixed after various time intervals (fixative A or B, see Distribution of Intracellular Receptors). This permits identification and characterization of the organelles involved in the uptake process. For some purposes, cells can be incubated with the appropriate ligand at 37° prior to processing for either light or electron microscopy.

Expected Results. After human fibroblasts are exposed to r(PMCA oleate) LDL at 4°, numerous fluorescent foci are seen linearly arranged on the cell surface (Fig. 4A). Following warming for 3 min at 37°, the linear foci transform into numerous circular fluorescent vacuoles (Fig. 4B). These vacuoles appear larger in diameter than the surface binding sites. After 10 min at 37° many of the fluorescent vacuoles fuse to form larger vacuoles (Fig. 4C). Finally, after 20 min (Fig. 4D) the fluorescent vacuoles accumulate in the perinuclear region of the cell.

Using a probe such as LDL–ferritin or IgG-C7–ferritin, the initial distribution of the label is in coated pits. However, with time at 37°, LDL–ferritin is found progressively in coated vesicles (Fig. 5A), partially coated vesicles (Fig. 5B), uncoated vesicles (Fig. 5C), and multivesicular bodies (Fig. 5D). Of particular interest, the shape of the uncoated vesicle (endosome) is quite variable. Sometimes they have a tubular shape (Fig. 5C); other times they have a circular shape. The functional significance of the different shaped vesicles is not known.

Summary and Conclusions

A number of different light and electron microscopic probes have been developed for visualizing the LDL pathway. These probes give a consistent picture of the endocytic process and allow the identification of various cellular organelles through which ligand traffics on the way to the lysosome. Moreover, the pathway, with some variation, appears to be the same for other ligands that are taken up by receptor-mediated endocytosis.[1,2] These methodologies have been the foundation for studying such issues as the role of coated pits in endocytosis,[24,26] the route of receptor recycling during endocytosis,[23] and the role of coated pits in clustering of

[26] J. M. Larkin, M. S. Brown, J. L. Goldstein and R. G. W. Anderson, *Cell* **33**, 273 (1983).

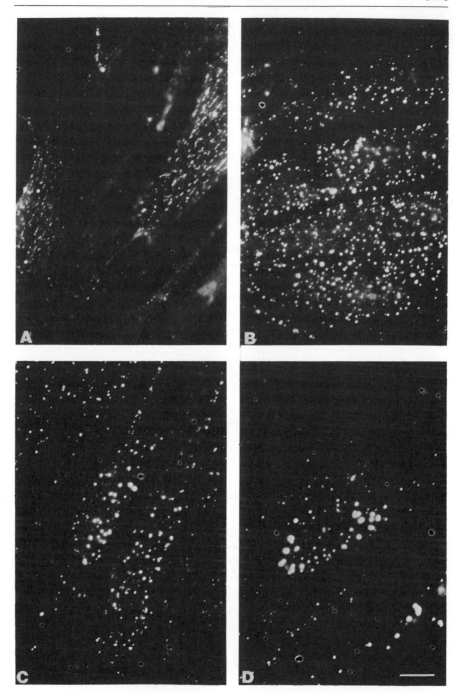

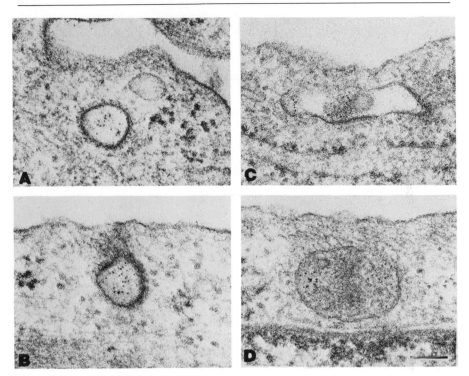

FIG. 5. Visualization of receptor-mediated endocytosis by electron microscopy. Mono-layers of human fibroblasts were chilled to 4° and incubated with LDL–ferritin (50 μg/ml) for 2 hr at 4°. The cells were washed and incubated with fresh culture medium at 37° for 5 (A–C) and 15 min (D). At each time point the cells were fixed and processed for electron micros-copy.[26] 81,000×. Bar = 0.12 μm.

FIG. 4. Visualization of the internalization at 37° of r(PMCA oleate)LDL previously bound to the cell surface of normal fibroblasts at 4°. On day 7 of cell growth, each cell monolayer was prechilled to 4° for 30 min, after which the medium was replaced with 1.5 ml of ice-cold buffer containing r(PMCA oleate)LDL (10 μg/ml). After incubation for 2 hr at 4°, the monolayers were washed at 4°, then either fixed directly with 3% formaldehyde without warming (A), or warmed to 37° before fixation with 3% paraformaldehyde. For those dishes that were warmed to 37° (B–D), the cold medium was replaced with warm medium before the cells were placed in the incubator and incubated for 3 (B), 10 (C), and 20 min (D). 750×. Bar = 13 μm. (Reprinted from Anderson et al.[10] with permission.)

receptors on the cell surface.[22,27] These reagents should prove useful in future studies to understand better the cell biology of receptor-mediated endocytosis.

Acknowledgments

The methods outlined in this chapter were developed in collaboration with Drs. J. L. Goldstein, M. S. Brown, and co-workers. I would like to thank Ms. Mary Surovik for help in preparing this manuscript.

[27] R. G. W. Anderson, J. L. Goldstein, and M. S. Brown, *Nature (London)* **270,** 695 (1977).

[13] Isolation and Assay of the Ac-LDL Receptor

By DAVID P. VIA, HANS A. DRESEL, and ANTONIO M. GOTTO, JR.

Introduction

An alternative pathway for lipoprotein metabolism has been described in macrophages and endothelial cells from a variety of species.[1-4] This pathway in the macrophage has been termed the scavenger cell pathway.[5,6] When macrophages are incubated *in vitro* with chemically modified lipoproteins such as acetylated low-density lipoprotein, massive accumulation of cholesteryl esters results.[7] Cells treated in this manner deposit the accumulated lipid as free droplets in the cytoplasm, taking on the appearance of the foam cells of atherosclerotic lesions.[7] Isolated foam cells from atherosclerotic lesions of rabbits readily metabolize acetoacetylated LDL through this receptor.[8] The receptor also appears to be oper-

[1] J. L. Goldstein, Y. K. Ho, S. K. Basu, and M. S. Brown, *Proc. Natl. Acad. Sci. U.S.A.* **76,** 333 (1979).
[2] R. W. Mahley, T. L. Innerarity, K. H. Weisgraber, and S. Y. Oh, *J. Clin. Invest.* **64,** 743 (1979).
[3] A. M. Fogelman, I. Schechtor, J. Seager, M. Hokom, J. S. Child, and P. A. Edwards, *Proc. Natl. Acad. Sci. U.S.A.* **77,** 2214 (1980).
[4] O. Stein and Y. Stein, *Biochim. Biophys. Acta* **620,** 631 (1980).
[5] M. S. Brown, S. K. Basu, J. R. Falck, Y. K. Ho, and J. L. Goldstein, *J. Supramol. Struct.* **13,** 67 (1980).
[6] M. S. Brown and J. L. Goldstein, *Annu. Rev. Biochem.* **52,** 223 (1983).
[7] M. S. Brown, J. L. Goldstein, M. Kreiger, Y. K. Ho, and R. G. W. Anderson, *J. Cell Biol.* **82,** 597 (1979).
[8] R. E. Pitas, T. L. Innerarity, and R. W. Mahley, *Arteriosclerosis* **3,** 2 (1983).

TABLE I
LIGANDS BINDING TO THE SCAVENGER RECEPTOR

Ligand	Modifying agent	Reference
Ac-LDL	Acetic anhydride	Goldstein et al.[a]
Ac-Ac-LDL	Diketene	Mahley et al.[b]
MDA-LDL	Malondialdehyde	Fogelman et al.[c]
EC-LDL	Endothelial cells	Henriksen et al.[d]
MAL-BSA	Maleic anhydride	Goldstein et al.[a]

[a] J. L. Goldstein, Y. K. Ho, S. K. Basu, and M. S. Brown, *Proc. Natl. Acad. Sci. U.S.A.* **76,** 333 (1979).
[b] R. W. Mahley, T. L. Innerarity, K. H. Weisgraber, and S. Y. Oh, *J. Clin. Invest.* **64,** 743 (1979).
[c] A. F. Fogelman, I. Schector, J. Seager, M. Hokom, J. S. Child, and P. A. Edwards, *Proc. Natl. Acad. Sci. U.S.A.* **77,** 2214 (1980).
[d] T. Henriksen, E. M. Mahoney, and D. Steinberg, *Proc. Natl. Acad. Sci. U.S.A.* **78,** 6499 (1981).

ative *in vivo*. The endothelial cells of the liver rapidly clear any modified lipoproteins injected into the circulation.[9] The uptake of the modified lipoproteins is mediated by a specific membrane protein. A variety of lipoprotein ligands, and at least one nonlipoprotein ligand, compete for binding to the receptor (Table I). We have recently shown that Ac-LDL, MDA-LDL, and MAL-BSA all bind to the same specific peptide having an approximate M_r of 260,000 by ligand blotting of partially purified receptor protein.[10]

To effectively study the scavenger cell pathway receptor, one needs a convenient source of receptor, a means for purification of the receptor, and methods for assay of receptor function. Each of these points is dealt with in detail in the remainder of this chapter.

Sources of Receptor

Murine Macrophage Cell Lines

A convenient source of Ac-LDL receptor is murine macrophage cell lines. Such cell lines have been used extensively to study other macro-

[9] J. F. Nagelkerke, K. P. Barto, and T. J. C. van Berkel, *J. Biol. Chem.* **258,** 12221 (1983).
[10] D. P. Via, H. A. Dresel, S. L. Cheng, and A. M. Gotto, Jr., *J. Biol. Chem.* **260,** 7379 (1985).

phage functions.[11] We and others have noted that the cell lines P388D₁, J774.1, and 1C-21 express considerable scavenger pathway activity.[12-14] The origin and Ac-LDL receptor status of a number of the macrophage cell lines is shown in Table II. None of the human cell lines tested thus far (U-937, HL-60) has scavenger receptor activity.[12] All of the murine macrophage cell lines in Table II, except 1C-21, are currently available from the American Type Tissue Culture Collection. All except the PU51.8 line will grow loosely attached to the growth surface. These lines are passaged by gently scraping with a sterile rubber policeman. The cells grow rapidly with generation times of less than 24 hr. We routinely cultivate the cells in 100-mm tissue culture dishes in a 5% CO_2 atmosphere. Lines are subcultivated every 3–4 days at split ratios of 1/20 to 1/40. For receptor isolations, 30–50 plates (100 mm) of the P388D₁ line are utilized to provide sufficient material for receptor blots.

Advantage can also be taken of the tumorigenic nature of the murine macrophage cell lines to provide large quantities of material for receptor isolations. P388D₁ cells are scraped from the dishes, washed once in sterile saline and resuspended in serum-free RPMI-1640. Injection of 1×10^6–1×10^7 P388D₁ cells subcutaneously into multiple sites of DBA/2 mice (Jackson Laboratory), the syngeneic mouse of origin, produces solid tumors within 3 to 4 weeks. Tumor burdens of 5 g/mouse can be readily achieved. Tumors contain Ac-LDL receptor activity derived predominantly from the injected cell line, with a minor contribution from endogenous macrophages migrating to the tumor site.[10] Tumors from the J774.1 line can be induced in BALB/c mice in a similar fashion.

Murine Peritoneal Macrophages

Murine and rat peritoneal macrophages cultured *in vitro* were used to initially describe the existence of the scavenger cell pathway receptor. A major disadvantage to this source of receptor is the limited number of cells that can be obtained. Unstimulated mice yield approximately 1×10^6 macrophages/mouse. Enhancement of this number is readily achieved by prior injection of the mice with thioglycolate broth to act as a peritoneal irritant. After such treatment, approximately 1×10^7 cells are obtained per mouse with 80–90% of the cells being macrophages. We have found

[11] P. S. Morohan and W. S. Walker, *J. Reticuloendothel. Soc.* **27**, 223 (1980).

[12] D. P. Via, A. L. Plant, I. F. Craig, A. M. Gotto, Jr., and L. C. Smith, *Biochim. Biophys. Acta* **833**, 417 (1985).

[13] T. Henriksen, E. M. Mahoney, and D. Steinberg, *Proc. Natl. Acad. Sci. U.S.A.* **78**, 6499 (1981).

[14] M. G. Traber, V. Defendi, and H. J. Kayden, *J. Exp. Med.* **154**, 1852 (1981).

TABLE II
MURINE MACROPHAGE CELL LINES

Cell line	Origin	Culture medium[a]	Ac-LDL receptor
P388D$_1$	3-Methylcholanthrene treatment of Balb/c mice	RPMI-1640, 10% FCS	Yes
J774.1	Mineral oil injection of Balb/c mice	DMEM, 10% FCS	Yes
RAW 264	Albelson leukemia virus infection of Balb/c mice	DMEM, 10% FCS	Yes
1C-21	SV40 transformation of Balb/c peritoneal macrophage	DMEM, 10% FCS	Yes
PU51.8	Spontaneous lymphoma of Balb/c mice	DMEM, 10% HS	No

[a] DMEM, Dulbecco's modified Eagles medium; FCS, fetal calf serum; HS, horse serum. All media supplemented with 1 mM glutamine and 50 μg/ml garamycin.

no major difference in the levels of scavenger receptor activity in resident versus thioglycolate-elicited macrophages.

Balb/c mice, 6–8 weeks old, are injected ip with 3 ml of sterile Brewer thioglycolate broth (Difco #0236-1). The thioglycolate broth is stored in the dark at room temperature and is most effective when aged for several weeks after preparation. Three days postinjection, the mice are dispatched by cervical dislocation and immersed in 95% ethanol. The abdominal skin is retracted to expose the abdominal wall, which is rinsed with ethanol. Using a 22-gauge needle, 5 ml of sterile 0.34 M sucrose is injected rapidly into the peritoneal cavity. The chances of puncturing an intestine during the procedure are minimized if food is withheld from the animals for 24 hr prior to harvest. With practice, 80% of the injected volume can be recovered using a syringe and 22-gauge needle. The collected cells are placed in ice-cold centrifuge tubes and centrifuged at 4° for 10 min at 200 g. The cells are washed once in cold serum-free Dulbecco's modified Eagles medium and used for receptor isolation. Further washes are unnecessary and can result in a loss of as much as 30% of the cells per wash. Cells from approximately 40 mice provide sufficient material for liposome binding and ligand blot assays.

Receptor Isolation

A general scheme for partial purification of the Ac-LDL receptor is seen in Fig. 1. Purification is dependent upon the appropriate combination of detergent solubilization and ion-exchange chromatography in the pres-

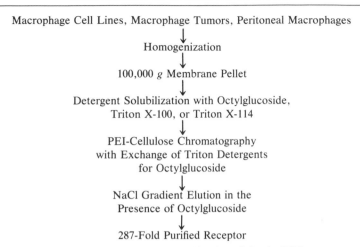

Macrophage Cell Lines, Macrophage Tumors, Peritoneal Macrophages
↓
Homogenization
↓
100,000 *g* Membrane Pellet
↓
Detergent Solubilization with Octylglucoside,
Triton X-100, or Triton X-114
↓
PEI-Cellulose Chromatography
with Exchange of Triton Detergents
for Octylglucoside
↓
NaCl Gradient Elution in the
Presence of Octylglucoside
↓
287-Fold Purified Receptor

FIG. 1. General scheme for partial purification of the Ac-LDL receptor.

ence of detergent. A wide variety of detergents can be used to successfully solubilize the receptor. Octylglucoside (40 mM) is a nonionic detergent that has been used for solubilization of the LDL receptor[15,16] and works equally well for solubilizing the Ac-LDL receptor.[11] The detergent has the advantage of not interfering with the receptor–liposome binding assay (see Assay of Receptor Activity) and can be used throughout the purification scheme. However, the detergent is quite expensive and less expensive detergents can be utilized to achieve the initial solubilization. Triton X-100, Triton X-114, and CHAPS at 1% (v/v) all solubilize the Ac-LDL receptor. However, these detergents do interfere with the receptor–liposome assay. These detergents are therefore exchanged for octylglucoside during the ion exchange step using the procedure for Schneider *et al.*[17] Triton X-114 has the additional advantage in that it forms a two-phase system when warmed above 25°.[18] This allows for separation of the LDL receptor (lower phase) from the Ac-LDL receptor (upper phase) and eliminates one centrifugation step.[9] A detailed protocol for partial purification of the receptor using the Triton X-114 method follows.

Materials

Mouse tumors, 20–80 g
Sorvall Omnimixer

[15] W. J. Schneider, S. K. Basu, M. J. McPhaul, J. L. Goldstein, and M. S. Brown, *Proc. Natl. Acad. Sci. U.S.A.* **76,** 5577 (1980).
[16] W. J. Schneider, J. L. Goldstein, and M. S. Brown, *J. Biol. Chem.* **255,** 11442 (1980).
[17] W. J. Schneider, J. L. Goldstein, and M. S. Brown, *J. Biol. Chem.* **257,** 2664 (1980).
[18] C. J. Bordier, *J. Biol. Chem.* **256,** 1604 (1981).

Beckman 60 Ti rotor and tubes

Conical polypropylene centrifugation tubes, 50 ml

Homogenization buffer [50 mM Tris-HCl, 150 mM NaCl, 1 mM EDTA, 0.5 mM phenylmethylsulfonyl fluoride (PMSF), 10 U/ml aprotinin, pH 8.0]

Solubilization buffer [50 mM Tris-HCl, 150 mM NaCl, 1 mM EDTA, 1% (v/v) Triton X-114, 0.5 mM PMSF, 10 U/ml aprotinin, pH 8.0]

Sucrose cushion (50 mM Tris-HCl, 1 mM EDTA, 0.6% Triton X-114, 6% sucrose, 0.5 mM PMSF, 10 U/ml aprotinin)

PEI-cellulose equilibrated in column buffer (50 mM Tri-HCl, 10 mM NaCl, 1 mM EDTA, 1% Triton X-114, 0.5 mM PMSF, 10 U/ml aprotinin, pH 8.0)

Detergent exchange buffer (50 mM Tris-HCl, 10 mM NaCl, 40 mM octylglucoside, 1 mM EDTA, 0.5 mM PMSF, 10 U/ml aprotinin, pH 8.0)

Elution buffer, high salt (50 mM Tris-HCl, 1.0 M NaCl, 1 mM EDTA, 40 mM octylglucoside, 0.5 mM PMSF, 10 U/ml aprotinin, pH 8.0)

Elution buffer low salt (same as detergent exchange buffer)

Procedure

1. P388D$_1$ tumors are placed in homogenization buffer (4 ml/g) and disrupted with 6–10 pulses of 15 sec each in a precooled Sorval Omnimixer. The homogenate is spun at 4° for 10 min at 1500 g to remove nuclei and debris. The supernatant is centrifuged at 40,000 rpm (100,000 g) in a Beckman 60 Ti rotor to yield a crude membrane pellet. The pellet may be stored in liquid nitrogen for up to 1 week if not used immediately.

2. Membrane pellets are suspended in ice-cold solubilization buffer (3 ml/g tissue). It is imperative to maintain the low temperature during the solubilization as the cloud point of the detergent is 25°. The solution is stirred on ice for 10 min and then placed over an equal volume of cold sucrose cushion in a chilled, 50-ml conical centrifuge tube. The tube is placed in a 37° water bath for 10 min to allow the micelles of detergent to form. The tube is then centrifuged for 10 min at room temperature at 800 g. The upper aqueous phase containing the Ac-LDL receptor is removed and diluted 3-fold with column buffer to reduce to NaCl concentration to 10 mM. The diluted material is then applied to the column of PEI-cellulose.

3. PEI-cellulose is prepared by cycling once with 0.5 M NaOH followed by 0.5 M HCl. The resin is then equilibrated in column buffer. We generally prepare the resin the same day it is used as the pH will often drop with storage. Columns are used only once. A 20-ml column is suffi-

cient for chromatographing the material from up to 40 g of tumor. Solubilized receptor is applied to the column at a rate of 40–60 ml/hr by either gravity flow or pump. The column is washed with 10 column volumes of column buffer, followed by 3 column volumes of detergent exchange buffer to replace the Triton X-114 with octylglucoside. The receptor is eluted with a linear gradient (40 ml for a 20-ml column) of 0.10–1.0 M NaCl in column buffer containing octylglucoside. Receptor activity elutes between 600–800 mM NaCl. Samples are taken for conductivity, protein determination, liposome binding assay, and ligand blots.

Assay of Receptor Activity

To evaluate receptor activity in ion exchange fractions, or in octylglucoside-solubilized tissue and cell culture samples, a modification of the method of Schneider *et al.*[17] is utilized as detailed below.

Liposome Receptor Asssay

Materials

Assay buffer [50 mM Tris-HCl, 50 mM NaCl, 2 mM EDTA, 1% (w/v) BSA, pH 8.0]
Wash buffer (20 mM Tris-HCl, 50 mM NaCl, 2 mM EDTA, 1% BSA, pH 8.0)
Liposomes [2 mg/ml egg phosphatidylcholine (Avanti #840051) in liposome buffer]
Liposome buffer (50 mM Tris-HCl, 2 mM EDTA, pH 8.0)
Acetone, reagent grade
1.5-ml conical polypropylene centrifuge tubes
Sorvall SM-24 rotor with #374 inserts and RC2B centrifuge or Beckman Microfuge
1-ml tuberculin syringes with 26-gauge needle
Nalgene 0.2-μm filter unit
"Nuflow" cellulose acetate filters, 0.45 μm, 25 mm diameter (Oxoid, USA #N25/45UP)
[125]I-labeled Ac-LDL (specific activity 100–300 cpm/ng)
Hoeffer 10-place filtration manifold (#FH 224)

Procedure

1. Liposomes are prepared by placing 40 mg of egg phosphatidylcholine in chloroform in a 50-ml flask and evaporating the solvent under

nitrogen. The dried lipid is desiccated under vacuum for 1–2 hr and 20 ml of liposome buffer is added. The flask is shaken for 5–10 min vigorously by hand or vortexed vigorously until no trace of dried lipid is visible. The cloudy suspension of liposomes is active in the assay for at least 3 weeks.

2. Assay buffer is prepared and filtered through the Nalgene filter unit. A shallow dish is filled with buffer and the cellulose acetate filters to be used in the assay are hydrated and allowed to soak for 1 hr in the buffer. Filters should be handled only with blunt filter forceps to prevent tearing or piercing of the filters.

3. Samples are added to prechilled 1.5-ml microfuge tubes, followed by liposome suspension diluted one-third in liposome buffer. Acetone cooled to 4° is then added to 33% (v/v). In a typical assay of PEI fractions, 100 μl of sample is mixed with 300 μl of liposomes, followed by the addition of 300 μl of acetone. Samples are vortexed and allowed to sit for 5 minutes at 4°. Samples are then centrifuged at 15,000 rpm for 20 min at 4° in a Sorval SM-24 rotor (24,000 g) with rubber adaptors. Alternatively, samples are centrifuged for 10 min in a Beckman Microfuge for 10 min at 4°. After centrifugation, the supernatant is carefully removed from the tubes without disturbing the small white liposome pellet. This is best accomplished by vacuum aspiration using the tip of a Pasteur pipet that has been flame drawn to a very fine point. To each tube is added 100 μl of assay buffer and each pellet is resuspended by repeated aspiration through a 1-ml tuberculin syringe with 26-gauge needle.

4. ^{125}I-Labeled Ac-LDL prepared by the iodine monochloride method[19,20] is diluted to 10–30 μg/ml in assay buffer. This material is then immediately filtered through a layer of five cellulose acetate filters on the filtration device, with the filtrate being collected into a glass or plastic scintillation vial. This step is crucial for removing a small fraction of the iodinated Ac-LDL which binds nonspecifically to the filters and can lead to extremely high backgrounds. The filtered ^{125}I-labeled Ac-LDL is aliquoted into 1.5-ml microfuge tubes (100 μl/tube) and 50 μl of the resuspended liposome pellets added. The tubes are gently mixed and allowed to sit at room temperature for 1 hr.

5. Presoaked cellulose acetate filters are placed in the filtration device and covered with 3 ml of wash buffer. To the 3 ml of buffer is added 120 μl of the incubation mixture. The vacuum is applied to each well immediately after the sample aliquot is added. Filters are washed five times with 3-ml aliquots of wash buffer and then placed in tubes for gamma counting.

[19] H. A. Dresel, I. Otto, H. Weigel, G. Schettler, and D. P. Via, *Biochim. Biophys. Acta* **795**, 452 (1984).
[20] J. Shepherd, O. K. Bedford, and H. G. Morgan, *Clin. Chim. Acta* **66**, 97 (1976).

Specific Ligand Blotting

To detect the specific receptor peptide responsible for binding Ac-LDL or other charge-modified ligands, sensitive ligand blotting techniques are employed.[21] Ligand blotting may be performed in several ways after SDS–gel electrophoresis of samples and electrophoretic transfer of nitrocellulose paper. Direct ligand blots which involve treating the nitrocellulose with ^{125}I-labeled Ac-LDL followed by autoradiography are possible, but in our experience the background is often unacceptably high. Alternatively, indirect ligand blots can be performed by treating the nitrocellulose strips with unlabeled Ac-LDL, followed by anti-LDL and either ^{125}I-labeled anti-IgG or horseradish peroxidase-conjugated (HRP) anti-IgG. For the Ac-LDL receptor, the best results in our hands are achieved using the indirect ligand blots and visualization of HRP-conjugated second antibody. A protocol for this approach is given below.

Materials

5% or 7% polyacrylamide slab gels (1.5–3.0 mm thick) prepared for the Laemmli buffer system[22]

SDS sample buffer (0.063 M Tris-HCl, 1.25% SDS, 12% glycerol, 0.006% Bromophenol Blue, pH 6.8)

Sephadex G-25 swollen in SDS sample buffer

1-ml tuberculin syringes with glass wool plugs

Nitrocellulose (Schlicher and Schuell, 0.22 μm, #BA83)

Transfer buffer [25 mM Tris-HCl, 192 mM glycine, 20% (v/v) methanol, pH 8.3]

Bio-Rad Transblot apparatus

BSA blocking buffer (0.01 M Tris-HCl, 0.15 M NaCl, 1 or 5% BSA, pH 8.0)

Ac-LDL, 30 μg/ml in 5% BSA blocking buffer

Anti-LDL IgG, 25 μg/ml in 1% BSA blocking buffer

HRP-conjugated rabbit anti-goat IgG (Cappel), 1/1000 dilution in 1% BSA blocking buffer

0.0025% o-diansidine in 0.1 M Tris-HCl, 0.01% H_2O_2, pH 7.4

0.018% 4-chloro-1-naphthol in 20 mM Tris, 500 mM NaCl, 0.015% H_2O_2, 20% methanol, pH 7.4

Prestained molecular weight standards (Bethesda Research Laboratory #6041SA)

[21] T. O. Daniel, W. J. Schneider, J. L. Goldstein, R. G. W. Anderson, and M. S. Brown, *J. Biol. Chem.* **258**, 4606 (1983).
[22] U. K. Laemmli, *Nature (London)* **227**, 680 (1970).

Procedure

1. Ion-exchange fractions should be desalted prior to electrophoresis due to the high salt content and large sample loads usually necessary to achieve a good blot. Desalting is accomplished just prior to electrophoresis by a modification of the methods of Fry *et al.*[23] and Tuszynski *et al.*[24] A 1-ml syringe is filled with Sephadex G-25 equilibrated in SDS sample buffer. The syringe is centrifuged at 200 *g* for 3 min and the sample in electrophoresis buffer is added to the top of the dry resin. The sample is centrifuged again at 200 *g* and the eluate used for electrophoresis on 5 or 7% polyacrylamide slab gels prepared according to Laemmli[22] at 30 mA per 1.5-mm-thick gel. Samples are not treated with 2-mercaptoethanol since this results in a loss of receptor activity.[10] Prestained molecular weight standards are included in at least one lane. These standards have the advantage that one can visually monitor the migration of proteins during the electrophoresis. They are also immediately visible on the nitrocellulose after transfer and circumvent the problem of trying to match blots with protein standards on nitrocellulose strips that have shrunk due to conventional protein staining procedures.

2. After electrophoresis, separated proteins are transferred to nitrocellulose paper at 90 V for 4–5 hr at 10° or overnight at 50 V at 10°. Nonspecific binding sites on the nitrocellulose are then blocked by incubation in 5% BSA blocking buffer for 30 min at 37°.

3. Nitrocellulose strips are next incubated for 8–10 hr at 4° with 30 μg/ml Ac-LDL. The strips are then washed 4 × 15 min in 1% BSA blocking buffer, followed by incubation for 1 hr at room temperature with 25 μg/ml anti-LDL IgG in 1% BSA blocking buffer. The antibody is best prepared in the user laboratory but is also commercially available as whole antiserum from CalBiochem. Nitrocellulose is next washed 4 × 15 min as above and incubated with diluted HRP-rabbit anti-goat IgG for 60 min at room temperature. After washing the strips 4 × 15 min as before, the blots are ready to be developed.

4. Two different substrates may be used for the color development. *o*-Diansidine gives a brown reaction product while the 4-chloro-1-naphthol gives a purple reaction product. In many cases a stronger signal is obtained with the 4-chloro-1-naphthol. Stock solutions of 1% *o*-diansidine in water are stable at 4° in the dark for 1 week. The 4-chloro-1-naphthol is dissolved in cold methanol just prior to use and added to room-tempera-

[23] D. W. Fry, J. C. White, and I. D. Goldman, *Anal. Biochem.* **90**, 809 (1978).

[24] G. P. Tsuzynski, L. Knight, J. R. Piperino, and P. N. Walsh, *Anal. Biochem.* **106**, 118 (1980).

FIG. 2. Ligand blot detection of Ac-LDL receptor. P388D₁ tumor membranes were solubilized with Triton X-100 and the receptor partially purified by PEI-cellulose chromatography. An aliquot of the fraction of peak ^{125}I-labeled Ac-LDL binding activity, containing 50 μg of protein, was electrophoresed on a 5% polyacrylamide gel. Separated proteins were electrophoretically transferred to nitrocellulose and sequentially treated with Ac-LDL, anti-LDL, and HRP-anti-IgG. The blot was developed using o-diansidine as the enzyme substrate. The arrow indicates the receptor and the position of a myosin molecular weight marker is indicated. The small square denotes the top of the separating gel.

ture aqueous buffer such that the final methanol concentration is 20%. The reaction product from both reagents is light sensitive, so blots should be protected by aluminum foil while developing. The development is stopped in either case by washing the nitrocellulose strip with distilled water. Developed blots, if protected from light, will maintain their color for many months.

A typical blot from a PEI-cellulose fraction of P388D₁ tumor material is seen in Fig. 2. The Ac-LDL receptor is identified a single protein of M_r 260,000. Similar results are obtained with blots of partially purified peritoneal macrophage or cultured P388D₁ cell protein.

[14] Isolation of Somatic Cell Mutants with Defects in the Endocytosis of Low-Density Lipoprotein

By MONTY KRIEGER

Receptor-mediated endocytosis is a process by which animal cells bind and internalize physiologically active extracellular macromolecules, including serum transport proteins such as low-density lipoprotein (LDL) and peptide hormones.[1] The LDL pathway is one of the most thoroughly studied endocytic pathways.[2] LDL receptors reside in thickened and indented regions of the plasma membrane, called coated pits. LDL bound to its receptor is internalized when the coated pits invaginate and pinch off to form coated endocytic vesicles (see Fig. 1). Inside the cell, the vesicles lose their coats, LDL dissociates from the receptors, and the receptors recycle back to the surface to mediate another round of endocytosis. The LDL in endocytic vesicles is subsequently transferred to and degraded in lysosomes. The hydrolysis of the cholesteryl esters of LDL produces unesterified cholesterol, which can be used to satisfy the cholesterol needs of the cell, including the biosynthesis of membranes and the regulation of cellular cholesterol metabolism. Cells also can derive cholesterol from *de novo* synthesis (Fig. 1). The cholesterol content and requirements of a cell are detected by an as yet unknown mechanism, and in response, cholesterol biosynthesis [specifically, the rate controlling enzyme 3-hydroxy-3-methylglutaryl-coenzyme A (HMG-CoA) reductase], cholesterol storage (acyl cholesteryl : acyl transferase), and cholesterol uptake (LDL receptor) are coordinately regulated.

Much of our current understanding of the LDL pathway is due to detailed comparisons of LDL processing by human fibroblasts derived from normal individuals and from patients with disorders in cholesterol metabolism, particularly familial hypercholesterolemia (FH).[2] FH, a relatively common autosomal codominant genetic disease, is a consequence of defects in the LDL receptor's structural gene. Because of the inherent limitations of relying on naturally occurring human mutations to extend the study of the LDL pathway, we have developed a somatic cell culture system for the genetic and biochemical analysis of endocytosis.[3–6]

---

[1] J. L. Goldstein, S. K. Basu, and M. S. Brown, this series, Vol. 98, p. 241.

[2] J. L. Goldstein and M. S. Brown, *in* "The Metabolic Basis of Inherited Disease" (J. B. Stanbury, J. B. Wyngaarden, J. L. Goldstein, and M. S. Brown, eds.), p. 672. Mc-Graw-Hill, New York, 1983.

[3] M. Krieger, M. S. Brown, and J. L. Goldstein, *J. Mol. Biol.* **150**, 167 (1981).

METHODS IN ENZYMOLOGY, VOL. 129

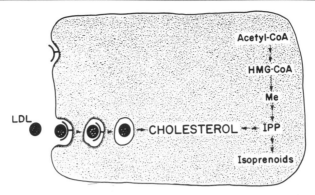

FIG. 1. The LDL pathway of receptor-mediated endocytosis and the pathway of cellular cholesterol biosynthesis. Most mammalian cells can obtain exogenous cholesterol via the LDL pathway or endogenous cholesterol by biosynthesis. HMG-CoA reductase catalyzes the conversion of HMG-CoA to mevalonate, the rate-controlling step in cholesterol biosynthesis. Mevalonate serves as a precursor to both sterols and nonsteroidal isoprenoids. Abbreviations: LDL, low-density lipoprotein; Me, mevalonic acid; IPP, isopentenyl pyrophosphate.

Using Chinese hamster ovary (CHO) cells and the two selection techniques described below[3,4] we have assembled a large collection of mutants which are defective in the receptor-mediated endocytosis of LDL. We also have developed a selection method which has permitted the isolation of spontaneous, mutagen-induced and DNA-transfection-induced revertants of some of these mutants[3,5,6] (and unpublished data). These studies have defined four genes (*ldlA-ldlD*) required for the expression of LDL receptor activity.[6,7] One of these, *ldlA,* is the structural gene for the LDL receptor. The *ldlA* mutants are phenotypically and genotypically analogous to cells from human homozygous and heterozygous FH patients.[6] Another gene, *ldlD,* is defined by a mutant that exhibits a very unusual property.[7] The LDL receptor-deficient phenotype of an *ldlD* mutant can be corrected when *ldlD* mutant cells are cocultivated with other mammalian cells (CHO, human, mouse). This complementation of the *ldlD* mutation by cocultivation occurs only when the complementing cells are in close proximity and cannot be mimicked by conditioned medium. Isolation of additional classes of endocytosis mutants and detailed analyses of

[4] M. Krieger, J. Martin, M. Segal, and D. Kingsley, *Proc. Natl. Acad. Sci. U.S.A.* **80,** 5607 (1983).
[5] R. D. Sege, K. Kozarsky, D. L. Nelson, and M. Krieger, *Nature (London)* **307,** 742 (1984).
[6] D. Kingsley and M. Krieger, *Proc. Natl. Acad. Sci. U.S.A* **81,** 5454 (1984).
[7] M. Krieger, *Cell* **33,** 413 (1983).

the structures and functions of the *ldl* genes and their products should provide new insights into the mechanism of receptor-mediated endocytosis.

Mutant Selection Using Reconstituted LDL (Method 1)

Principle. This method takes advantage of the unusual structure of LDL. LDL is a large particle (approximately 2.5×10^6 Da) which contains a core of neutral lipid surrounded by an amphipathic shell.[8] LDL's amphipathic shell contains phospholipid, unesterified cholesterol, and a large protein called apolipoprotein B. LDL's core is composed mainly of fatty acyl esters of cholesterol (primarily cholesteryl linoleate) and some triglyceride. The lipids in the core of LDL cannot enter cells unless the LDL is taken up by receptor-mediated endocytosis. We have developed a method, called the reconstitution of LDL, to replace the endogenous lipids in the core of LDL with exogenous hydrophobic compounds.[9] The resultant reconstituted LDL, r-(compound)LDL, resembles native LDL in many of its chemical and physical properties and is processed normally by the LDL pathway of cultured cells.

Reconstituted LDL and a two-step selection protocol can be used to isolate mutant CHO cells with defects in endocytosis.[3] First, mutagen-treated CHO cells are incubated with LDL reconstituted with a toxic lipid. This "Trojan horse" LDL enters and kills most of the cells. Survivors will include cells with defective LDL pathways (toxin not released within the cells) and cells which are intrinsically resistant to the toxic molecule carried in the core of the reconstituted LDL. These two types of mutants can be differentiated by a second screening step using LDL reconstituted with a fluorescent dye. The survivors of the toxic reconstituted LDL selection are incubated with fluorescent reconstituted LDL and examined *in situ* using an inverted fluorescence microscope. Endocytosis mutants which cannot accumulate the fluorescent dye are easily recognized, isolated, and cloned. Activity of the LDL receptor pathway in wild-type and mutant cells can be assayed as previously described.[1,3,7]

Mutants bearing defects in each of the four *ldl* genes (*ldlA–ldlD*)[6] have been isolated using the reconstituted LDL selection. The efficiency of this selection will depend on the mechanism of action and potency of the toxic lipid incorporated into the core. When the toxic lipid used is 25-hydroxycholesteryl oleate, an inhibitor of cellular cholesterol biosynthesis, the selection is most effective when cells are plated at relatively low densities (200,000 cells/100-mm dish).

[8] R. J. Deckelbaum, G. G. Shipley, and D. M. Small, *J. Biol. Chem.* **252**, 744 (1977).
[9] M. Krieger, this series, Vol. 128 [34].

Reagents

Chinese hamster ovary (CHO-K1) cells (American Type Culture Collection, #CCL61)

Medium A: Ham's F-12 medium containing penicillin (100 U/ml), streptomycin (100 μg/ml), and 2 mM glutamine (standard cell culture reagents can be obtained from Gibco)

Phosphate-buffered saline (PBS)

Newborn calf (bovine) serum (Gibco)

3% NCLPDS: Medium A supplemented with 3% (v/v) newborn calf lipoprotein-deficient serum

25-Hydroxycholesteryl oleate (Steraloids, Inc.)

1-Pyrenemethyl 3β(*cis*-9-octadecenoyloxy)-22,23-bisnor-5-cholenate (PMCA oleate, Molecular Probes)

1,1'-Dioctadecyl-3,3,3',3'-tetramethylindocarbocyanine iodide (DiI, Molecular Probes)

Ethyl methanesulfonate (Sigma)

LDL, prepared as previously described[1,9]

Preparation of Newborn Calf (Bovine) Lipoprotein-Deficient Serum (NCLPDS). To ensure maximal expression of LDL receptor activity, selections are conducted in media which do not contain significant amounts of serum-derived sterol [medium A supplemented with newborn calf lipoprotein-deficient serum (NCLPDS)]. Lipoproteins are removed from newborn calf serum (NCS) by KBr density centrifugation.[1] Whole serum (initial density, 1.006 g/ml) is adjusted to a final density of 1.215 g/ml with solid potassium bromide (0.3366 g of KBr/ml of serum), subjected to ultracentrifugation and isolated as previously described for the preparation of lipoprotein-deficient serum from whole plasma.[1] In contrast to lipoprotein-deficient plasma, the newborn calf lipoprotein-deficient serum need not be treated with thrombin and the final protein concentration should be adjusted to 60 mg/ml. Ultracentrifugation may be performed in polyallomer tubes using standard rotors (e.g., 60 Ti, Beckman) or on a larger scale using a KBr gradient in a zonal rotor with a reorienting core.[10]

Preparation of Reconstituted LDL. LDL is reconstituted with 25-hydroxycholesteryl oleate (dissolved in heptane) or PMCA oleate (dissolved in benzene) as previously described.[9]

Cell Culture. Standard cell culture techniques are used. Unless otherwise noted, all incubations are at 37° in a cell culture incubator with a 5% CO_2 atmosphere. CHO cells are maintained as stock cultures in medium A supplemented with 10% (v/v) NCS. All media should be sterile filtered

[10] Byung H. Chung, Jere P. Segrest, Marjorie J. Ray, Ken Beaudrie, and John T. Cone, this series, Vol. 128 [8].

after addition of lipoproteins. It is important to seed cells into experimental dishes as uniform single-cell suspensions; clumps of cells are notoriously resistant to selection and can survive even the most stringent selections described below. Colonies of cells can be isolated using cloning cylinders (e.g., penicylinders, Fisher Scientific, can be used without grease) and trypsin/EDTA. Alternatively, the colonies can be washed twice with PBS without calcium or magnesium to loosen their attachment to the dish and then be removed by aspiration into a pipet[10a] (e.g., Gilson Pipetman with sterile tip). Clones can be isolated by dilution plating into the wells of microtiter dishes (0.5 cells/0.2 ml of 10% NCS/well; examine wells after 7 days of incubation).

Mutagenesis. On day 0, wild-type CHO cells are seeded into 250-ml flasks (3×10^5 cells/flask) in 10 ml of 3% NCLPDS. On day 1, the medium is changed to 3% NCLPDS containing 400 μg/ml of ethyl methanesulfonate. After 18–19 hr of incubation, the monolayers are washed three times with PBS and refed with 3% NCLPDS. After an expression period of 5–7 days, the cells are harvested with trypsin and set into dishes for selection (see below). Alternative mutagens include the frameshift mutagen ICR 191 (1–3 μg/ml, overnight incubation) and gamma irradiation (600–800 rad dose). Mock-treated control cells should be carried through all procedures.

Procedure

Step 1. Selection with toxic r-(25-hydroxycholesteryl oleate)LDL. Mutagen-treated cells are seeded into 100-mm dishes at 2×10^5 cells/dish in 6.5 ml of 3% NCLPDS containing 10–20 μg protein/ml of r-(25-hydroxycholesteryl oleate)LDL. The medium should be replaced every other day for the first 6 days and every third day thereafter. After a total of 16–20 days under selection, macroscopic colonies should be visible by direct inspection of the dishes. At this stage, colonies can be isolated, grown into mass culture, cloned, and assayed for LDL receptor activity.[1] If an inverted fluorescence microscope is available, it is prudent to examine the phenotypes of the colonies *in situ* as described below.

Step 2. Screening with fluorescent r-(PMCA oleate)LDL. After selection with toxic reconstituted LDL, the surviving colonies are washed once with PBS and refed with 3% NCLPDS containing 5–10 μg protein/ml of r-(PMCA oleate)LDL. After incubation for 1–2 days, the colonies are refed with fresh 3% NCLPDS and examined *in situ* with a Leitz inverted fluorescence microscope (or its equivalent) equipped with the following filters: exciter filter, 340–380 nm; chromatic beam splitter, 400

[10a] L. H. Thompson, this series, Vol. 58, p. 308.

nm; barrier filter, 430 nm. Colonies which do not accumulate fluorescent dye may be isolated, grown into mass culture, cloned, and assayed for LDL receptor activity. Alternatively, LDL labeled with DiI[11] may be used in place of the r-(PMCA oleate)LDL. The accumulation of DiI-LDL (1–5 μg protein/ml) after a 5-hr incubation may be examined using the rhodamine filter package of an inverted fluorescence microscope (e.g., exciter filter, 560/30 nm; chromatic beam splitter, 580 nm; barrier filter, 580 nm).

Mutant and Revertant Selection using MeLoCo-Amphotericin B (Method 2)

Principle. This two-step mutant selection method takes advantage of two fungal metabolites, amphotericin B and compactin.[4,11a,13] Compactin is a potent inhibitor of mevalonate production by HMG-CoA reductase and thus inhibits cholesterol biosynthesis (Fig. 2). Mevinolin, a derivative of compactin, is somewhat more potent than compactin[12] and may be used in its place. Amphotericin B is a polyene antibiotic that forms complexes with sterols in membrane bilayers. These complexes form pores which kill the cells in a manner analogous to killing by serum complement. The MeLoCo-amphotericin B selection is based in part on the previous use of polyene antibiotics to isolate mutants with defects in sterol biosynthesis.[14,15]

When cells are grown overnight in medium containing trace levels of mevalonate ("Me"), LDL ("Lo"), and compactin ("Co"), or "Me-LoCo" medium, the cells become cholesterol auxotrophs and only cells that express the LDL pathway contain normal amounts of cholesterol[13] (see Fig. 2). The trace level of mevalonate in MeLoCo medium permits the synthesis of adequate amounts of essential nonsteroidal isoprenoid metabolites of mevalonate (e.g., dolichol) but does not permit the synthesis of significant amounts of sterol. A 6- to 8-hr incubation with amphotericin B will then kill the cholesterol-replete wild-type cells, but not mutant cells that are cholesterol depleted because of defective endocytosis of LDL's cholesterol (see Fig. 3). The MeLoCo-amphotericin B selec-

---

[11] R. E. Pitas, T. L. Innerarity, J. N. Weinstein, and R. W. Mahley, *Arteriosclerosis* **1,** 177 (1981).

[11a] A. Endo, M. Kuroda, and K. Tanzawa, *FEBS Lett.* **72,** 323 (1976).

[12] A. W. Alberts, J. Chen, G. Kuron, V. Hunt, J. Huff, C. Hoffman, J. Rothrock, M. Lopez, H. Joshua, E. Harris, A. Patchett, R. Monaghan, S. Currie, E. Stapley, G. Albers-Schonberg, O. Hensens, J. Hirshfield, K. Hoogsteen, J. Liesch, and J. Springer, *Proc. Natl. Acad. Sci. U.S.A.* **77,** 3957 (1980).

[13] M. Krieger, *Anal. Biochem.* **135,** 383 (1983).

[14] Y. Saito, S. M. Chou, and D. F. Silbert, *Proc. Natl. Acad. Sci. U.S.A.* **74,** 3730 (1977).

[15] T.-Y. Chang and C. C. Y. Chang, *Biochemistry* **21,** 5316 (1982).

CONDITIONS

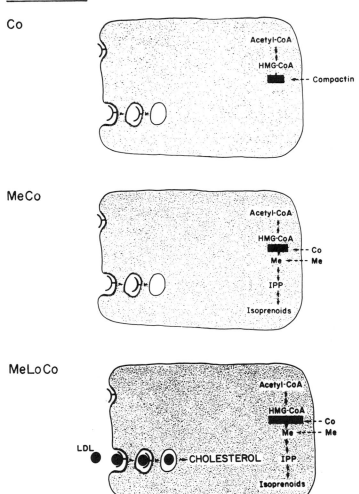

Fig. 2. MeLoCo selection medium. Cells are grown in medium (Hams F-12) containing newborn calf lipoprotein-deficient serum and the indicated additions. Top panel: When cells are incubated with compactin (Co) alone, there is no source of sterols or nonsteroidal isoprenoids. Middle panel: When cells are incubated with compactin and trace levels of mevalonate (250 μM), called MeCo, adequate amounts of nonsteroidal isoprenoids are synthesized by relatively high-affinity pathways; however, sterol biosynthesis is inadequate. Under MeCo conditions, the cells are essentially sterol auxotrophs. Bottom panel: When cells are incubated with compactin, mevalonate, and LDL (MeLoCo), they are provided with adequate amounts of both sterols and nonsteroidal isoprenoids.

RECEPTOR PATHWAY EXPRESSED

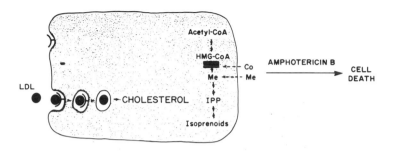

RECEPTOR PATHWAY NOT EXPRESSED

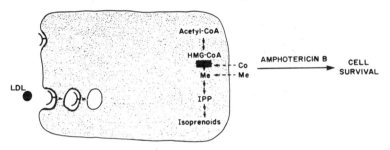

FIG. 3. MeLoCo-amphotericin B selection of mutants with defects in the endocytosis of LDL. Cells are preincubated for 24 hr in MeLoCo and then subjected to amphotericin B treatment for 4–8 hr. Cholesterol-replete cells (cells with intact LDL pathways) are killed by amphotericin B. Cholesterol-depleted cells survive because there is not enough sterol in the membranes to form toxic complexes with amphotericin B.

tion is efficient even when cells are initially plated at high density (1–3 × 10^6/100 mm dish).

The timing of preincubation in MeLoCo is critical. When CHO cells are incubated in the presence of MeLoCo for extended periods (> 2 days), the LDL pathway-defective cells stop growing and die slowly because of cholesterol deprivation while the wild-type cells grow.[3,6,13,16] Thus, MeLoCo can be used to select mutant cells with defects in the LDL pathway by overnight incubation followed by treatment with amphotericin B or to select revertants of these mutants (by long-term incubation without amphotericin B). The MeLoCo revertant selection is most effec-

[16] J. L. Goldstein, A. S. Helgeson, and M. S. Brown, *J. Biol. Chem.* **254,** 5403 (1979).

tive when the cells are seeded at relatively low densities (best results when density is <500,000 cells/100-mm dish).

The MeLoCo-amphotericin B selection appears to be genotype specific in that all mutants isolated using this technique have fallen into the *ldlA* complementation group.[6] This specificity was unanticipated and may be related to either the stringency of the selections used to date (see below) or an unrecognized feature of the mechanism of amphotericin B toxicity.

The MeLoCo-amphotericin B selection also has led to the development of simple methods for the rapid screening of potential cholesterol biosynthesis inhibitors and for detecting the presence of LDL receptor activity in cultured cells.[4,13]

Reagents

LDL

Amphotericin B (Sigma, do not use commercial preparations which contain detergents): Amphotericin B should be prepared freshly as a 25 mg/ml stock solution in dimethyl sulfoxide (mild warming with hot tap water will speed the dissolution of the amphotericin B). Amphotericin B is somewhat light sensitive; therefore, all manipulations with amphotericin B-containing solutions should be conducted under subdued light

Mevalonic acid stock solution (0.2 M): Dissolve 1 g of DL-mevalonic acid lactone (Sigma) in 25 ml of 0.5 N NaOH, heat at 37° for 30 min, adjust pH to 8.0 with 1 N HCl, add water to 38 ml, filter sterilize and store frozen in 1- to 2-ml aliquots

Compactin stock solution (10 mM): Dissolve 181 mg of compactin (gift of A. Endo) in 3.6 ml ethanol in a 50-ml conical glass tube. Dissolution is facilitated by warming with hot tap water and vigorous vortexing. Slowly add 1.8 ml of 0.6 N NaOH while vortexing. Slowly (dropwise) add 7.2 ml of water while gently shaking, then add water to 36 ml. Adjust to pH 8 by dropwise addition of 0.6 N HCl to the continuously stirring solution, add water to 45 ml, filter sterilize, and store frozen in 2-ml aliquots. Mevinolin (Merck) can be used in place of compactin.

Mutant Selection Procedure

Step 1. Preincubation with MeLoCo to deplete mutant cells of cholesterol. Mutagen-treated or control CHO cells are seeded into 100-mm Petri dishes at a concentration of 10^6 cells/dish in 3% NCLPDS and allowed to grow for 2 days to induce high levels of LDL receptor activity. Each dish is then refed with 8 ml of 3% NCLPDS containing MeLoCo (0.25 mM mevalonate, 30 μg protein/ml of LDL, and 5 μM compactin) and incubated for 24 hr. The stringency of the selection may be increased by either increasing the amount of LDL in the medium (the K_m of LDL receptors in CHO cells is approximately 30 μg of LDL protein/ml), increasing the incubation period by as much as 24 hr, or both. Conversely, the stringency may be reduced by lowering the LDL concentration and/or reducing the preincubation period.

Step 2. Incubation with amphotericin B to kill cholesterol-replete cells. After preincubation with MeLoCo, the cells are washed once with 5 ml of PBS and refed with 5 ml of 1% NCLPDS containing 300 μg/ml of amphotericin B. Amphotericin B is only sparingly soluble in aqueous solution; thus, a fine, dense, orange precipitate will form in the cell culture medium. The amphotericin B should be uniformly suspended prior to addition to each dish. After incubation with amphotericin B for 6–8 hr, the cells are washed three times with PBS and refed with 8 ml of 3% NCLPDS. The last traces of insoluble amphotericin B are removed and the background of surviving wild-type cells is lowered when the liquid in the dishes is vigorously swirled immediately before and during each PBS wash. After an additional 1 or 2 days of incubation in 3% NCLPDS, a second round of the two-step selection should be conducted to kill the low background of wild-type colonies that survive the first round of selection. Virtually no wild-type cells survive two rounds of the MeLoCo-amphotericin B selection. Colonies are grown for an additional 3–6 days in 3% NCLPDS prior to isolation, growth to mass culture, cloning, and assaying for LDL receptor activity. The frequency at which mutant clones arise will depend on the stringency of the selection and the type of mutagen employed; we have typically observed frequencies of 1–26 × 10^{-6}.

Minor modifications of this procedure (e.g., mutagenesis and cell growth in whole rather than lipoprotein-deficient sera) permit the isolation of mutants with defects in sterol biosynthesis.[4,14]

Revertant Selection Procedure

The revertant selection procedure can be applied to stock populations of mutants to isolate spontaneous revertants or to mutagen-treated (see above) or DNA-transfected[5] mutants. The cells are seeded into 100-mm dishes at a concentration of 500,000–750,000 cells/dish in 10 ml of 3% NCLPDS containing MeLoCo (0.25 mM mevalonate, 2–3 μg of LDL protein/ml, and 40 μM compactin). To ensure adequately stringent conditions, higher concentrations of compactin and lower concentrations of LDL are used for revertant selection than for mutant selection. Dishes are refed every 3 days for 2 to 3 weeks until macroscopic colonies are visible. It is highly unlikely that revertants will be found in a selection dish if large colonies are not detected after 3 weeks of incubation. The background of mutant cells that survive this revertant selection is substantially higher than the background of wild-type cells that survives the MeLoCo-amphotericin B mutant selection. Most background mutant cells are found in small colonies which are easily distinguished from the larger revertant colonies; however, if clumps of mutant cells are seeded into the dishes, they will form relatively large colonies. These unwanted colonies can be identified by incubating the cultures with r-(PMCA oleate)LDL or DiI-LDL and examining the colonies *in situ* using an inverted fluorescence microscope as described above. Colonies are isolated, grown to mass culture, cloned, and assayed for LDL receptor activity. The frequency at which revertant colonies arise will depend on the nature of the mutation and the treatment the cells receive prior to selection. For example, DNA-transfection-induced revertants of *ldlA* mutant clone 7 were observed at a frequency of 2×10^{-7} (Ref. 5). Hybrids of complementing mutants also can be isolated using this procedure.[6]

Acknowledgments

This work was supported by an NIH Research Career Development Award and grants from the NIH, March of Dimes, and the Whitaker Health Sciences Fund. I am grateful to Robert Sege, David Kingsley, Karen Kozarsky, and Lawrence Hobbie for their constructive comments and their collaboration in this work. A. Endo kindly provided compactin.[11a]

[15] Detection of Animal Cell LDL Mutants by Replica Plating

By JEFFREY D. ESKO

Introduction

Low-density lipoprotein (LDL) is the major carrier of cholesterol in plasma and is internalized by endocytosis after binding to high-affinity receptors on the surfaces of cells. Genetic studies of human subjects afflicted with familial hypercholesterolemia (FH) indicates that at least three mutant alleles of the receptor gene exist.[1] These define a binding site

[1] J. L. Goldstein, B. A. Kottke, and M. S. Brown, *in* "Human Genetics, Part B: Medical Aspects" (B. Bonne-Tamir and T. Cohen, eds.), p. 161. Liss, New York, 1982.

for LDL, a domain involved in posttranslational processing and secretion, and a region required for receptor clustering in coated pits on the cell surface. Studies of induced mutants of Chinese hamster ovary (CHO) cells indicate that the products of other genetic loci help determine the proper orientation and function of LDL receptors.[2] Because essential genes undoubtedly participate in the uptake of LDL, suitable cell culture systems are needed in which conditionally lethal mutations can be rescued by growth at low temperature or by nutrient supplementation. Biochemically defined cell mutants defective in LDL metabolism provide cellular models for further study of the FH phenotype and form a starting point for reconstructing whole animal models for the disease.[3]

A major obstacle to the genetic analysis of somatic cells has been the lack of rapid screening procedures, particularly clonal replica plating, for detecting relatively rare variants. In order to assure a high probability of obtaining mutants, most investigators employ enrichment or selective techniques which quite often presuppose a knowledge of the biochemistry and cell physiology of potential mutants.[4] To circumvent this problem, replica plating procedures were developed using limiting dilution of cells in microtiter dishes or colonies established on agar.[5-9] Another approach, developed by Stamato and Hohman, involved the transfer of established colonies onto nylon cloth,[10] but unfortunately, all of these methods met with limited success due to poor fidelity and low capacity.

To circumvent these problems new methods were devised which permit animal cell colonies to be replicated onto filter paper and cloth of woven polyester with higher fidelity and superior resolution than obtained by previous methods.[11,12] These procedures adapt well to a variety of indirect screening strategies, including classical Lederberg-style replica

[2] D. M. Kingsley and M. Krieger, *Proc. Natl. Acad. Sci. U.S.A* **81**, 5454 (1984).

[3] B. Mintz and K. Illmensee, *Proc. Natl. Acad. Sci. U.S.A.* **72**, 3585 (1975).

[4] L. H. Thompson, this series, Vol. 58, p. 308.

[5] T. Kuroki, *in* "Methods in Cell Biology" (D. M. Prescott, ed.), Vol. 9, p. 157. Academic Press, New York, 1975.

[6] J. A. Robb, *Science* **170**, 857 (1970).

[7] F. Suzuki and M. Korikawa, *in* "Methods in Cell Biology" (D. M. Prescott, ed.), Vol. 11, p. 127. Academic Press, New York, 1975.

[8] M. Brenner, R. L. Dimond, and W. F. Loomis, *in* "Methods in Cell Biology" (D. M. Prescott, ed.), Vol. 14, p. 187. Academic Press, New York, 1976.

[9] R. A. Goldsby and E. Zisper, *Exp. Cell Res.* **54**, 271 (1969).

[10] L. K. Hohmann, *in* "Methods in Cell Biology" (D. M. Prescott, ed.), Vol. 20, p. 247. Academic Press, New York, 1978.

[11] J. D. Esko and C. R. H. Raetz, *Proc. Natl. Acad. Sci. U.S.A.* **75**, 1190 (1978).

[12] C. R. H. Raetz, M. M. Wermuth, T. M. McIntyre, J. D. Esko, and D. C. Wing, *Proc. Natl. Acad. Sci. U.S.A.* **79**, 3223 (1982).

plating, autoradiography of colonies labeled with radioactive precursors, and direct enzymatic assay of colonies performed *in situ*. Although the method was originally developed for isolating Chinese hamster ovary (CHO) cell mutants altered in lipid biosynthesis,[11-16] minor modification of the technique has permitted the identification of mutants defective in the metabolism of LDL and other molecules.

Choice of Cell Lines

In the 1950s and 1960s, Puck and collaborators developed a cell line from Chinese hamster ovary that could be cloned and propagated from single cells with high efficiency.[17,18] Most animal cells, including CHO, double *in vitro* every 12 to 24 hr when bathed in a medium consisting of various nutrients and serum. As the cells become confluent on their growth surface, proliferation gradually stops and the pH of the medium falls. Proteolytic treatment of such monolayer cultures yields single-cell suspensions which can resume exponential growth after appropriate dilution and addition of fresh medium.[19] Many permanent cell lines can be propagated indefinitely by repeated trypsin treatment and dilution, while normal cells, such as skin fibroblasts, have a limited life span *in vitro*.[20]

Several features of CHO cells make them ideal for genetic studies. (1) CHO cells grow both in monolayer and in liquid culture, permitting the generation of up to 10 g of cell protein. When diluted to low density, single CHO cells will give rise to colonies. These characteristics permit the identification of mutants in a clonal fashion and the subsequent characterization of interesting strains in monolayer and liquid culture. (2) Compared to other immortal cell lines, CHO cells have a very stable karotype,[21] and mutants tend to retain their phenotypes during subsequent cell division. (3) CHO cells can be grown in defined medium[22] containing intact sera and sera that have been pretreated to remove lipoprotein or

[13] J. D. Esko and C. R. H. Raetz, *J. Biol. Chem.* **255**, 4474 (1980).

[14] J. D. Esko and C. R. H. Raetz, *Proc. Natl. Acad. Sci. U.S.A.* **77**, 5192 (1980).

[15] J. D. Esko, M. M. Wermuth, and C. R. H. Raetz, *J. Biol. Chem.* **256**, 7388 (1981).

[16] M. A. Polokoff, D. C. Wing, and C. R. H. Raetz, *J. Biol. Chem.* **256**, 7687 (1981).

[17] T. T. Puck, "The Mammalian Cell as a Microorganism—Genetic and Biochemical Studies *In Vitro*." Holden-Day, San Francisco, CA, 1972.

[18] T. T. Puck, S. J. Cieciura, and A. Robinson, *J. Exp. Med.* **108**, 945 (1958).

[19] J. Litwin, *in* "Tissue Culture—Methods and Applications" (P. F. Kruse and M. K. Patterson, eds.), p. 188. Academic Press, New York, 1973.

[20] L. Hayflick, *Exp. Cell Res.* **37**, 614 (1965).

[21] R. G. Worton, C. C. Ho, and C. Duff, *Somat. Cell Genet.* **3**, 27 (1977).

[22] R. G. Ham, *Proc. Natl. Acad. Sci. U.S.A.* **53**, 288 (1965).

total lipid.[23] Growth medium lacking serum has also been described.[24] (4) CHO cells will proliferate between 32 and 40°, making possible the isolation of conditionally lethal mutants altered in essential genes.

Other cell types may lend themselves to the screening procedures described in this chapter. For example, teratocarcinoma stem cells, capable of differentiation and implantation into mice, carry fully functional LDL receptors similar to those described on human fibroblasts.[25] Thus, mutants derived from teratocarcinoma cells may permit the engineering of inbred mice-bearing lesions in the LDL pathway. Some cells, such as macrophages, do not express the typical LDL receptor found on fibroblasts, but instead express a receptor that recognizes modified forms of LDL.[26] The availability of macrophage-like cell lines[27] should render this receptor system amenable to genetic characterization.

Conditions of Mutagenesis

A variety of mutagens, including alkylating agents and nucleoside analogs, can be used to increase the likelihood of finding desirable mutants of CHO cells.[4] Treatment with ethyl methanesulfonate (EMS) has proven adequate for the identification of many mutants and successfully yielded CHO strains deficient in LDL metabolism. In this procedure, a tissue culture flask (75 cm^2 surface area) containing $1-2 \times 10^6$ cells in the exponential phase of growth is treated with 200–400 μg/ml of EMS (density = 1.17 g/ml) in normal growth medium at 37°. After 16 hr, the medium is removed and the flask is rinsed with fresh medium. Subsequent growth of the cells at 33° is desirable for rescuing potential temperature-sensitive strains bearing mutations in essential genes. After 3–4 days, the cells are harvested with trypsin[19] and dispersed into two 150 cm^2 flasks at 33°. Another 5 days of growth yields approximately $1-2 \times 10^7$ cells. At this time the cells are placed in ampoules containing growth medium supplemented with 8% glycerol or 10% DMSO (vol/vol), and the cells are stored in liquid nitrogen according to procedures recommended by the American Type Culture Collection (Bethesda, MD).

The effectiveness of the mutagen-treatment is assessed by measuring

[23] J. D. Esko and K. Y. Matsuoka, *J. Biol. Chem.* **258**, 3051 (1983).

[24] P. Pohjanpelto, I. Virtanen, and E. Holtta, *Nature (London)* **293**, 475 (1981).

[25] J. L. Goldstein, M. S. Brown, M. Krieger, R. G. W. Anderson, and B. Mintz, *Proc. Natl. Acad. Sci. U.S.A.* **76**, 2843 (1979).

[26] M. S. Brown and J. L. Goldstein, *Annu. Rev. Biochem.* **52**, 223 (1983).

[27] Available through Cell Distribution Center, The Salk Institute for Biological Studies, San Diego, CA and the American Type Culture Collection, Bethesda, Md.

the incidence of mutants resistant to a cytotoxic drug, such as ouabain.[28] After revival of the cells from storage in liquid nitrogen, they are permitted to grow at 33° for 4 to 5 days to allow for phenotypic expression of any mutations. At this time, 10^6 cells are added to multiple 100-mm-diameter culture dishes containing 15 ml of complete growth medium supplemented with 1 mM ouabain at 37°. Another set of cultures are seeded with 1000, 3000, and 10,000 cells per plate and incubated without ouabain. After 10 to 12 days of growth, the medium from each dish is removed, and the colonies present on the plates are rinsed with phosphate-buffered saline (PBS). Fixation with 10% trichloroacetic acid and staining with 0.05% Coomassie Brilliant Blue in methanol/water/acetic acid (45 : 45 : 10, v/v) renders the colonies visible. The colonies present on plates incubated in the absence of ouabain are counted, and the fraction of viable cells plated is calculated. Those plates containing ouabain are stained in a similar fashion, and the number of ouabain-resistant colonies is enumerated. Dividing the number of ouabain-resistant colonies by the number of viable cells plated in the dish yields the incidence of ouabain-resistant strains in the population. The spontaneous frequency of ouabain-resistant CHO cells is less than 1×10^{-7}, while after mutagenesis the incidence rises to approximately $1-5 \times 10^{-4}$.[28] Those mutagen-treated cell stocks in which the incidence of ouabain resistance is greater than 10^{-4} are retained for subsequent mutant screenings.

Immobilization of CHO Cells on Polyester Cloth

Materials

1. Polyester cloth—Tetko, Inc. (Elmsford, NY) provides polyester bolting cloth by the yard; 17-μm (PeCap HD7-17) and 1-μm (HD7-1) pore diameter cloth should be cut into disks that fit snugly inside a 100-mm-diameter tissue culture dish
2. Glass beads—4 mm diameter
3. Plastic tissue culture vessels, growth medium, and supplies

Methods. When single CHO cells are added to a standard tissue culture dish, they attach to the plastic within 4–24 hr and grow into macroscopic colonies within 8–16 days, depending on the temperature. However, CHO cells do not adhere very tightly to their growth supports, and late in each division cycle cells detach and migrate, forming secondary colonies around the original innoculum. If 100 to 200 CHO cells are placed

[28] R. M. Baker, D. M. Brunette, R. Mankovitz, L. H. Thompson, G. F. Whitmore, L. Siminovitch, and J. E. Till, *Cell* **1,** 9 (1974).

in a culture dish containing growth medium, small colonies are increasingly obscured by secondary colonies if growth is allowed to continue for several days longer.

The normal detachment of cells during cell division can be used to produce replica copies of colonies on overlying supports.[11,12] In this procedure, a sterile disk of polyester cloth is floated in the medium bathing the cells, and a single layer of 4-mm-diameter glass beads is poured on top of the disk. The beads act as a ballast, facilitating even contact between the disk and the bottom of the dish. The overlay does not interfere with the development of colonies and permits the growth medium to be changed periodically without disturbing the colony pattern on the plate. If overlayed at the single-cell stage, the formation of secondary colonies is completely eliminated, yielding much larger colonies than ordinarily feasible. Because of the improved resolution, the colonies can be quantitated by automatic and manual methods, and the subsequent purification of individual colonies is greatly simplified.

After 8 to 16 days, staining both the overlay and the plastic dish with Coomassie Brilliant Blue reveals that over 98% of the colonies present on the plate have transferred to the polyester cloth. Other stains such Fuchsins, Giemsa, and Methylene Blue can be used to locate the colonies, but Coomassie Blue is the most sensitive. The small number of variants that do not transfer to the overlay arises from cells that do transfer well with a high spontaneous frequency, and this trait appears to be heritable. Thus, the technique is limited to the predominant fraction of cells that are capable of growing from one surface to the other.

Transferred colonies are about 0.1 to 0.5 cm in diameter on the cloth, and examination of the colony pattern on the plate reveals that most of the cells in each colony have transferred.[12] The cells remaining on the plate can be recovered by allowing the colonies to "fill in" during subsequent incubation under standard growth conditions. Treatment of the plate with 1 μg/ml of poly-L-lysine in phosphate-buffered saline for 5 min prior to inoculation facilitates the more equal division of the cells between the plastic and the overlying support.[12]

Microscopic examination of the disk reveals that the cells are not only attached to the surface adjacent to the plastic, but actually grow along the filaments of the cloth and appear on the side of the disk adjacent to the glass beads. Subsequent detachment of the cells can cause formation of satellite colonies on the disk, especially when large-pore-diameter cloth is employed. If a second disk of smaller pore diameter (usually 1 μm) is applied over the primary (large-pore) overlay, the formation of satellite colonies is virtually eliminated.

CHO cells will also form colonies through stacks of polyester cloth containing as many as four layers.[12] Staining each disk reveals that the majority of colonies have grown through the stack. This phenomenon greatly simplifies multiple screenings, because each colony can be assayed for as many as four biochemical defects, and may be applicable to cells that do not grow well on plastic surfaces, such as myelomas. Raetz *et al.*[12] hypothesized that the vertical growth of CHO cells in polyester stacks may reflect chemotropic movement.

The immobilization technique is applicable to several common cell lines, such as HeLa, mouse L, and 3T3 cells. The mouse muscle cell line, C2,[29] and nonadherent cell lines, such as the myeloma SP2/0 and derivative hybridomas,[12] transfer to polyester cloth. Hybrids formed by fusing CHO cells with polyethylene glycol[30] also transfer, which simplifies complementation analysis between strains. In general, transfer occurs most successfully with cells like CHO which are not tightly adherent *in vitro*. It may be possible to improve the transfer of difficult cells lines to supports by varying the composition of the growth medium or by treating established colonies with mild proteases or glycosidases before the overlay is applied. Alternatively, coating the disks with collagen, poly-L-lysine or other cell-adherent substances may prove advantageous.[31]

Transferred colonies are viable by several criteria. First, they exclude stains used to assess the viability of cells, such as Trypan Blue. Secondly, they retain the capacity to incorporate radioactive precursors into macro molecules. Finally, the immobilized colonies continue to grow and can be utilized to generate replica copies on fresh polyester disks. Since 300–500 colonies can be grown on to a single disk, as many as 100,000 colonies (200–300 plates) can be readily sorted for interesting mutants.

Supports. Polyester disks of the proper diameter (8.2-cm-diameter disks for 100-mm-diameter tissue culture plates) can be generated with a suitable punch or with a metal template and scissors. New cloth disks are notched, numbered with a soft-lead pencil, interleaved with Whatman No. 1 paper in a glass culture dish, and sterilized in an autoclave for 20 min.[12] Residual moisture is removed from the sterile disks by opening the dish in a laminar flow hood for 1–2 hr. Used polyester disks can be recycled by soaking them in concentrated HCl (10 ml/disk, 3–4 changes over as many days). The disks are then rinsed with phosphate-buffered saline, water, and finally ethanol.[12] If necessary, the disks can be pressed

---

[29] R. A. Black and Z. W. Hall, *Proc. Natl. Acad. Sci. U.S.A.* **82**, 124 (1985).

[30] C. M. Croce, *in* "Growth, Nutrition, and Metabolism of Cells in Culture" (G. H. Rothblat and V. J. Cristafalo, eds.), p. 81. Academic Press, New York, 1977.

[31] A. Coelho-Macieira and S. Avrameas, *in* "Tissue Culture—Methods and Applications" (P. F. Kouse and M. K. Patterson, eds.), p. 379. Academic Press, New York, 1973.

with a household iron to remove wrinkles. The notch is used to mark the disk's orientation in the plate.

Faithful replicas of CHO cells can be generated on cloth of 1- to 25-μm pore diameter. Other cell types that differ from CHO cells in morphology and growth characteristics may require specific varieties of cloth. For example, the fusion and transfer of muscle cells occurs effectively on 10-μm pore diameter cloth,[29] while the cloning of CHO cells in polyester stacks occurs optimally when large-pore cloth (~17 μm) is employed.

CHO cells will also attach to smooth varieties of Whatman filter paper (Nos. 50, 52, 540, etc.).[11] Other grades (e.g., No. 42) do not accept cells as well, but completely reduce satellite colony formation. Nitrocellulose membranes and cellulose acetate filters also accept cells. Although these alternative supports are not as reliable as polyester cloth, they each have their advantages, depending upon the characteristics of the cell line used and the type of screening to be employed (see Other Uses of Colony Replica Methods).

Beads. Four-millimeter-diameter glass beads poured in a single even layer forces the disk to remain opposed to the cells attached to the plastic bottom of the dish. Beads varying from 2 to 6 mm in diameter are also effective, and after use they can be washed in acid and sterilized. For simplicity, test tubes containing enough sterile beads for one plate are prepared in advance of the experiment. Beads are better than other ballasts because they allow the medium to percolate through the disk and around the cells, and they conform to the contours of the plastic dishes. Interestingly, when glass rings cut from different diameter test tubes are used to keep disks submerged, transfer only occurs where the ring is opposed to the disk, but the formation of secondary colonies both inside and outside the ring is completely prevented. Therefore, transfer requires a certain amount of contact pressure and it is not simply associated with the reduced cell migration. The basis of the contact pressure phenomenon is not clear, but it deserves further study because of its possible relationship to cell adhesion.

Media. Although cloning conditions were originally developed using Ham's F12 medium[22] supplemented with 10% (v/v) fetal bovine serum,[11,12] other growth media permit cell transfer to polyester cloth. For example, bovine, newborn calf, and horse sera will replace fetal bovine serum. Lipid-deficient serum prepared by solvent extraction[23,32] or by density centrifugation[23] supports the growth and transfer of CHO cells. The addition of 20 μg/ml bovine pancreatic insulin[11] sometimes improves the adherence of CHO cells to overlying supports, especially filter paper.

---

[32] G. H. Rothblat, L. Y. Arbogast, L. Ovellette, and B. V. Howard, *In Vitro* **12**, 554 (1976).

Completely defined growth medium composed of a 0.1% albumin in a 1 : 1 mixture of α-MEM and F12 media has been described,[24] but the transfer of colonies to polyester cloth under these conditions has not been investigated.

Autoradiographic Sorting of LDL Mutants

Materials

1. Polyester disks containing approximately 300 immobilized colonies derived from mutagen-treated cells.
2. ^{125}I-labeled LDL, specific activity 100–300 cpm/ng: Human low-density lipoproteins can be iodinated with Iodogen or Iodobeads (Pierce Chemical Company, Rockford, IL) to reproducibly yield ^{125}I-labeled LDL of desired radioactive and biological activities. In this procedure, a scintillation vial is mechanically etched and washed in acid prior to use. Approximately 0.5 mg of Iodogen dissolved in methylene chloride is added to the vial and the solvent is evaporated under a stream of nitrogen. The vial is rinsed vigorously with Tris-buffered saline (pH 9.9) to remove particulate Iodogen, and 5–10 mg of purified LDL adjusted to pH 9.9 is added along with 3 mCi of ^{125}I-labeled sodium iodide. After incubation for 10 min, the mixture is dialyzed against phosphate-buffered saline containing 0.3 mM EDTA and 1% NaI at 4°. The buffer should be changed to phosphate-buffered saline containing 0.3 mM EDTA after 2 hr. When less than 500 cpm is present in the dialysate (usually 10 buffer changes), the iodinated LDL should be filter sterilized (0.45-μm pore diameter). This preparation is stable at 4° for 6 to 8 weeks.
3. Lipoprotein-deficient fetal calf serum: Removal of lipoproteins from fetal calf serum is accomplished by ultracentrifugation of serum after adjustment to an electrolyte density of 1.21 g/ml with solid NaBr.[23] Alternatively, large volumes of serum can be delipidated by extraction with acetone and ethanol according to the method of Rothblatt *et al.*[32] as modified by Esko and Matsuoka.[23] Lipoprotein-deficient human serum obtained during the generation of LDL will substitute for delipidated fetal calf serum.

Methods. Mutagen-treated cells should be grown for approximately 1 week in growth medium supplemented with lipid-deficient serum in order to eliminate strains that might require factors normally present in intact fetal bovine serum which are missing in the deficient medium. This step is not essential and will reduce the chances of finding strains that might require LDL for growth, such as cholesterol auxotrophs (see below). If auxotrophic mutants are desired the cells should be grown for several days in medium supplemented with dialyzed serum to eliminate strains that require small, dialyzable molecules. After trypsin treatment enough single cells are added to multiple 100-mm-diameter dishes to give rise to approximately 300 colonies per plate at 33° in complete growth medium. After 1 day the polyester and glass bead overlay is employed and the plates are incubated for another 16 days. At days 7–10, the medium should be renewed to keep the cells dividing rapidly. If temperature-sensitive strains are not desired, the entire screening can be performed at 37°. Under these conditions, the formation of colonies of sufficient size is attained within 10 days.

Removal of the polyester disks from the master plates must be performed under sterile conditions. After aspiration of the growth medium, the beads are decanted by inversion of the plate over a suitable vessel. The plate is immediately returned upright, and sterile tweezers are used to transfer the individual polyester disks to fresh culture dishes containing 10–15 ml of growth medium supplemented with lipoprotein-deficient medium. Bacterial culture dishes can be used in this step and for all subsequent manipulations in order to minimize cost. Incubation of the cells in lipoprotein-deficient medium for 2–3 additional days results in adequate induction of LDL receptors.

The original master plates should be rinsed with growth medium or phosphate-buffered saline to remove loosened cells and refilled with fresh growth medium supplemented with 20 U/ml nystatin and 2.5 μg/ml of Fungizone (GIBCO, NY) in order to prevent subsequent contamination with yeast and fungi. Overlaying the master cultures with Whatman No. 42 filter paper and glass beads[11] eliminates the formation of satellite colonies in the master plates while the screening of the polyester disks is in progress. The masters are incubated for 2–3 days at 33° to permit the colonies to fill in, and then the plates are held at 28° under otherwise normal growth conditions until the screening is complete. Colonies have been held for as long as 30 days with complete retention of viability.

After 2–3 days incubation in lipoprotein-deficient medium, the induction of LDL receptors in the immobilized colonies is adequate for the autoradiographic detection of ^{125}I-labeled LDL binding. The disks are shifted to 40° overnight to allow for phenotypic expression of any temperature-sensitive mutations. The next day, the disks are transferred to fresh bacterial dishes filled with 5 ml of growth medium supplemented with lipoprotein-deficient serum and 10 μg/ml of ^{125}I-labeled LDL. Incubation at 40° for 4 hr with continuous stirring (60 rpm) results in homogeneous labeling of the colonies. If incubators without interior electrical connections are used, the dishes should be manually stirred every hour to obtain even labeling of the colonies.

After incubation with ^{125}I-labeled LDL, the disks are immediately dipped in phosphate-buffered saline at 4° in order to remove LDL that is nonspecifically absorbed to the disk. Treatment of the colonies with a few milliliters of 10% trichloroacetic acid results in the precipitation of ^{125}I-labeled LDL associated with the cells. Acid-soluble ^{125}I-labeled catabolites are removed by washing the disks on a filter paper support in a Buchner funnel under reduced pressure with 100 ml of 2% trichloroacetic acid.[11] The disks should be dried overnight at room temperature or for a few hours at 50° and then mounted on a piece of paper. Exposure of the disks to Kodak XAR-5 X-ray film for 4 days reveals the position of colonies that contain ^{125}I-labeled LDL. The image on the film represents the sum of the LDL bound to the cell surface (less than 10%) and undegraded LDL in the interior of the cell (about 90%).

In order to detect mutants that fail to bind and take up LDL, the disks are stained with Coomassie Brilliant Blue after autoradiography.[11,12] Superimposition of the autoradiogram and the stained disks reveals that the intensity of the image on the autoradiogram is approximately proportional to the staining intensity of the colony on the disk. The 2-fold deviation from linearity that is observed is presumably due to incomplete induction of LDL receptors in isolated colonies, and in practice this variation is ignored. Instead, one examines the film for occasional colonies that entirely lack an autoradiographic halo. These strains are recovered from the original master dish with glass cloning cylinders and localized trypsin treatment.[33] After dilution, putative mutants are passed through the screening procedure again in order to achieve their purification and to confirm their phenotype. Purified strains are expanded in monolayer culture and stored in liquid nitrogen for further biochemical characterization.

The incidence of mutants that fail to bind and take up ^{125}I-labeled LDL is 10^{-3} to 10^{-4} in mutagen-treated populations that were not challenged in lipoprotein-deficient medium prior to screening. The majority of these strains are auxotrophs for cholesterol and therefore require LDL for growth. These interesting mutants are found because incubation of colonies in lipoprotein-deficient medium, required for the induction of LDL receptors, results in the loss of viability of strains that require LDL–cholesterol for growth. Nonviable CHO colonies remain attached to the polyester, but they fail to incorporate ^{125}I-labeled LDL. Thus, the screening procedure permits the isolation of mutants in receptor-mediated

[33] L. Jacobs and R. DeMars, in "Handbook of Mutagenicity Test Procedures" (B. J. Kilbey, M. Legator, W. Nichols, and C. Ramel, eds.), p. 193. Elsevier, Amsterdam, 1977.

uptake of LDL in addition to mutants defective in *de novo* cholesterol genesis.

Occasionally, colonies are observed which yield an autoradiographic signal in excess of that expected from the Coomassie Blue staining intensity of the colony. Several strains identified in this way bind LDL normally and accumulate 3- to 5-fold more undegraded LDL in the interior of the cell compared to the wild-type parent (see Ref. 36). Thus, the screening technique also permits the identification of mutants that potentially bear lesions in steps lying distal to the receptor, such as vesicle translocation, fusion of vesicles with lysosomes, and subsequent catabolism of LDL.

Other Uses of Colony Replica Methods

The general usefulness of the colony immobilization technique is documented by the large number of mutants that have been isolated in this fashion (see the table). Strains carrying lysosomal enzyme deficiencies, lesions in cell surface receptors for LDL, mannose 6-phosphate, and acetylcholine, enzyme defects in the biosynthesis of lipids, glycoproteins, and glycosaminoglycans, and alterations in the transport of amino acids and the repair of DNA have been reported. Conceivably, any ligand or macromolecular precursor that can be labeled and discriminated from its metabolic product can be used to identify interesting mutants. Because the screening procedure involves the analysis of replicate colonies, the identification of mutants being lesions in essential genes has proved feasible without prior knowledge of the phenotypes of the mutants.

Some of the variants listed in the table were isolated using modifications of the cloning technique described in the previous sections. Because these alternate screenings can be adjusted to permit the isolation of mutants altered in LDL and cholesterol metabolism, a short discussion of these techniques is appropriate.

Generation of Replica Plates. The efficient transfer of animal cells to filter paper and polyester cloth suggested that it might be possible to use the copies to generate authentic replica plates.[11,12] In this procedure, the medium is aspirated from the master plate and the beads are decanted. Removing the overlay, however, divides the colonies, spreading loosened cells across the disks. Since these cells would subsequently form secondary colonies and obscure the original colony pattern, they are removed by rinsing the disk. If filter paper copies are employed the disks are placed cell side up in a metal or glass pan leaning at about a 60° angle in a laminar flow hood. The pan is equipped with an aspiration outlet, so that liquid draining from the filter is removed rapidly. The papers are then forcibly

Mutation	Cell type	Method used	Reference
myo-Inositol auxotroph	CHO cells	Filter paper Replica plating	a
Cholesterol auxotroph	CHO cells	Polyester cloth Replica plating ^{125}I-labeled LDL binding	b
UV-induced DNA repair	CHO cells	Filter paper Replica plating	c, d
Ethanolamine phosphotransferase	CHO cells	*In situ* enzymatic assay (radiochemical)	e
Alkaline phosphatase	CHO cells	*In situ* enzymatic assay (fluorometric)	f
α-Mannosidase	CHO cells	*In situ* enzymatic assay	g
Choline exchange enzyme	CHO cells	*In situ* enzymatic assay (radiochemical)	h
CDP-choline synthetase	CHO cells	Incorporation of [^{14}C]choline	i
N-Acetylglucosaminyl transferase	CHO cells	Incorporation of [^{14}C]fucose	j
Xylosyltransferase	CHO cells	Incorporation of ^{35}SO$_4$	k
Acyl-CoA synthetase	Mouse fibrosarcoma	^{3}H suicide followed by incorporation of [^{3}H]arachidonate	l
Uptake via LDL receptors	CHO cells	Uptake of ^{125}I-labeled LDL	b
Uptake via mannose 6-phosphate receptors	CHO cells	Uptake of [^{35}S]methionine-labeled lysosomal hydrolases	m
Acetylcholine receptor mutants	Muscle cells	Binding of α-[^{125}I]bungarotoxin or anti-receptor anti-bodies	n
Amino acid transport	CHO cells	Uptake of [^{14}C] amino-isobutyric acid	o

a J. D. Esko and C. R. H. Raetz, *Proc. Natl. Acad. Sci. U.S.A.* **75**, 1190 (1978); *b* J. D. Esko, in preparation (1986); *c* D. B. Busch, J. E. Cleaver, and D. A. Glaser, *Somat. Cell Genet.* **6**, 407 (1980); *d* M. Stefanini, A. Reuser, and D. Bootsma, *Somat. Cell Genet.* **8**, 635 (1982); *e* M. A. Polokoff, D. C. Wing, and C. R. H. Raetz, *J. Biol. Chem.* **256**, 7687 (1981); *f* J. R. Gum, Jr. and C. R. H. Raetz, *Proc. Natl. Acad. Sci. U.S.A.* **80**, 3918 (1983); *g* A. R. Robbins, *Proc. Natl. Acad. Sci. U.S.A.* **76**, 1911 (1979); *h* M. Nishijima, O. Kuge, M. Maeda, A. Nakano, and Y. Akamatsu, *J. Biol. Chem.* **259**, 7101 (1984); *i* J. D. Esko and C. R. H. Raetz, *Proc. Natl. Acad. Sci. U.S.A.* **77**, 5192 (1980); *j* C. B. Hirschberg, M. Perez, M. Snider, W. L. Hanneman, J. D. Esko, and C. R. H. Raetz, *J. Cell. Physiol.* **111**, 255 (1982); *k* J. D. Esko, T. E. Stewart, and W. H. Taylor, *Proc. Natl. Acad. Sci. U.S.A.* **82**, 3197 (1985); *l* E. J. Neufeld, T. E. Bross, and P. W. Majerus, *J. Biol. Chem.* **259**, 1986 (1984); *m* A. R. Robbins, S. S. Peng, and J. L. Marshall, *J. Cell Biol.* **96**, 1064 (1983); *n* R. A. Black and Z. W. Hall, *Proc. Natl. Acad. Sci. U.S.A.* **82**, 124 (1985); *o* A. H. Dantzig, C. W. Slayman, and E. A. Adelberg, *Somat. Cell Genet.* **8**, 509 (1982).

rinsed with saline or growth medium lacking serum using a 10-ml pipet and an air-driven pipet aid (Drummond Scientific Co., Broomal, PA) to produce a very powerful stream of fluid. Washing the disks evenly with about 30 ml of liquid adequately removes loosened cells without disturbing the colonies,[11] which are firmly attached to the disk. If polyester disks are employed, loosened cells can be removed by gently shaking the replicas in a culture dish containing growth medium or phosphate-buffered saline.[12] Care must be taken to prevent both filter paper and polyester disks from drying out during the washing procedure.

After washing filter paper replicas, the disks are placed cell side down in fresh dishes filled with growth medium and weighed down with glass beads as described previously. Depending on the temperature, the cells require 3 to 6 days to grow back down to the plastic surface.[11] When a polyester disk is placed in contact with a fresh plastic surface, marginal transfer of the colonies occurs,[12] presumably because the cells have higher affinity for the polyester than for the culture dish. However, colonies established on polyester cloth will transfer with high fidelity to a fresh polyester disk. Optimal results are obtained when the fresh disk is placed over the original disk and glass beads are used to maintain even contact. Complete transfer of colonies occurs within 2–3 days at 37°.

Unlike the original transfer of cells from plastic to paper or polyester, the transfer of cells from the disk to the replica copy is *entirely quantitative*.[11,12] In practice, therefore, it is necessary to make two replica plates from each disk, one under permissive conditions (e.g., complete growth medium) and one under restrictive conditions (e.g., growth medium lacking a specific nutrient for which an auxotroph is being sought). Before replica plating, the immobilized colonies may be placed in a dish containing growth medium and incubated under selective conditions (e.g., high temperature, or in the absence of a specific nutrient). In this way, the phenotype of a potential mutant can be accentuated by giving time for nutrient depletion or enzyme inactivation.

Using this methodology, the proline requirement[34] of parental CHO-K1 cells has been demonstrated. The replica plating technique has yielded auxotrophic mutants requiring myo-inositol,[11,13] lysophosphatidylcholine,[35] and cholesterol.[36] Busch et al. used the paper replica plating technique to isolate UV-sensitive variants of CHO cells.[37] These investigators irradiated the master plates and used the replica plates to recover poten-

[34] F.-T. Kao and T. T. Puck, *Genetics,* **55,** 513 (1967).
[35] J. D. Esko, M. Nishijima, and C. R. H. Raetz, *Proc. Natl. Acad. Sci. U.S.A.* **79,** 1698 (1982).
[36] J. D. Esko, in preparation (1986).
[37] D. B. Busch, J. E. Cleaver, and D. A. Glaser, *Somat. Cell Genet.* **6,** 407 (1980).

tial mutants. Although not tested yet, it should be possible to isolate strains temperature sensitive for growth by preparing paper or polyester copies at 33°, the permissive temperature, and using these to generate replicate plates or disks at 40°. By analogy, cold-sensitive variants should also be obtainable by reversing the incubation and screening temperatures.

Enzyme-Specific Sorting of Immobilized Colonies. The discovery of cell transfer to paper and polyester disks not only solves the problem of animal cell replica plating, but also creates the possibility of isolating mutants by *in situ* biochemical analyses,[11] as originally described for microbial colonies.[38,39] This strategy represents the most exciting aspect of paper and polyester sorting, and further unifies biochemistry with somatic cell genetics. Mutants altered in predetermined enzymes or subcellular components can be identified among large-colony populations without any prior assumptions concerning phenotypes or growth properties.

In this procedure, the replicate colonies are rendered permeable by simply freezing and thawing the disks in a sucrose-containing buffer.[11] The colony homogenates generated in this way remain well localized, permitting the measurement of specific enzymes *in situ*. For example, the formation of phosphatidylinositol can be measured by incubating disks in an assay mixture containing CDP-diglyceride and myo-[U-^{14}C]inositol.[11] After reaction, treating the paper with trichloroacetic acid precipitates any radioactive lipid formed and unreacted radioactive precursor is removed by washing the disk on a Buchner funnel with additional acid. Autoradiography of the disk reveals a halo of radioactive phosphatidylinositol produced around each colony homogenate. Superimposing the autoradiogram on the disk stained with Coomassie Brilliant Blue permits the identification of mutants as blue spots on the disk lacking an autoradiographic halo on the film, while normal colonies are present on both the disk and the film. Mutants in ethanolamine phosphotransferase were identified in this way by screening approximately 10,000 colonies.[16] Fluorometric colony assays (rather than autoradiography) were successfully employed to obtain several CHO variants defective in α-mannosidase,[40] a lysosomal enzyme, and alkaline phosphatase, a secretory enzyme.[41] Presumably, many of the colorimetric assays collated by Brenner

[38] B. M. Olivera and F. Bonhoeffer, *Nature (London)* **250,** 513 (1974).
[39] C. R. H. Raetz, *Proc. Natl. Acad. Sci. U.S.A.* **72,** 2274 (1975).
[40] A. R. Robbins, *Proc. Natl. Acad. Sci. U.S.A.* **76,** 1911 (1979).
[41] J. R. Gum, Jr. and C. R. H. Raetz, *Proc. Natl. Acad. Sci. U.S.A.* **80,** 3918 (1983).
[42] B. M. Brenner, R. L. Dimond, and W. F. Loomis, *in* "Methods in Cell Biology" (D. M. Prescott, ed.), Vol. 16, p. 187. Academic Press, New York, 1976.

et al.[42] can be adapted to immobilized colonies, so that a very broad range of mutants should now be obtainable by direct enzymatic assays.

Autoradiographic Sorting of Viable Colonies. Since the animal cells attached to filter paper and polyester are viable, omitting the freezing and thawing step described above allows the *in situ* detection of macromolecular synthesis.[11] Immobilized colonies incubated in a few milliliters of growth medium can be labeled effectively with radioactive thymidine, uridine, or leucine to measure DNA, RNA, or protein synthesis, respectively.[11] Following radiochemical incorporation (which may take hours), the paper is treated with trichloroacetic acid to precipitate newly made radioactive macromolecules as described above for measuring the uptake of ^{125}I-labeled LDL.

An advantage in using an early precursor to label the final end product of a multistep metabolic pathway is the fact that several enzymes are screened simultaneously, increasing the likelihood that mutants of interest will be found. The main disadvantage of using intact cells in this way is that mutations affecting the transport or activation of the precursor may also be obtained along with the desired specific mutations. However, mutants altered in transport or energy generation may also be useful, since the biochemical basis of the defect can probably be unraveled. Dantzig *et al.*[43] have shown that *in situ* autoradiography with intact cells immobilized on paper can be employed in conjunction with radiation suicide to isolate variants unable to transport specific small molecules across the plasma membrane.

In Situ Hybridization and Immunochemical Colony Sorting. The availability of colonies immobilized on solid supports makes possible the biochemical detection of many other cellular components besides enzymes. The method of *in situ* colony hybridization for detecting specific DNA sequences in *E. coli* colonies[44,45] should now be applicable to animal cells as well. This will permit the detection of nucleic acid sequences, arising from the presence of tumor viruses or other foreign DNA molecules introduced by transformation or cloning vectors. Like colony hybridization, the *in situ* detection of specific antigens is also feasible,[12] permitting the assay of proteins lacking enzymatic activity. Black and Hall recently demonstrated that muscle cell mutants lacking acetylcholine receptors can be obtained using either radioactive antireceptor antibodies or by *in situ* binding of α-[^{125}I]bungarotoxin.[29] Antibody secretion by hy-

[43] A. H. Dantzig, C. W. Slayman, and E. A. Adelberg, *Somat. Cell Genet.* **8**, 509 (1982).
[44] M. Grunstein and D. S. Hogness, *Proc. Natl. Acad. Sci. U.S.A.* **72**, 3961 (1975).
[45] J. S. Beckman, P. F. Johnson, and J. Abelson, *Science* **196**, 205 (1977).

bridomas cloned between layers of polyester cloth has been reported.[12] The availability of anti-LDL receptor antibodies[46] should make it possible to identify antigenic variants of the LDL receptor in cloned cells.

Detection of Rare Mutants. It must be emphasized that biochemical screenings of animal cells on polyester and paper supports are compatible with any preselection technique, such as enrichments based on radiation suicide, BUdR incorporation, or thymidine starvation.[4] Automated cell analyzers have recently become available which can detect fluorescence signals or scattered light generated by individual animal cells flowing past a high-intensity laser source.[47] Large populations (up to 10^8 cells) can be examined and sorted, and in some cases viable cells can be separated after such treatments, provided that the cell components of interest can be labeled selectively without killing the cells. Fluorescent derivatives of LDL are available and have been used to sort cells automatically.[25,48] The subpopulation resulting from such an enrichment step would need to be examined further by more specific biochemical assays, such as those made possible by immobilizing cells on polyester cloth and filter paper. The combined use of an enrichment step and replica cloning may provide the order of magnitude and biochemical specificity necessary to identify exceedingly rare mutations.

In conclusion, cell sorting on polyester cloth and filter paper, unlike costly fluorescence cell analyzers or other devices designed to mechanize the isolation of mutants, is readily accessible to any investigator with a biochemical assay applicable to crude cell extracts or a labeling scheme for whole cells. The glass beads (approximately $1.50/dish) and the polyester cloth (approximately $1.00/disk) are the most expensive components of the system, but they can be recycled, while the dish (ca. 30¢) and growth medium (ca. 50¢) are discarded after use. We calculate that to screen approximately 50,000 colonies costs less than $1000 in materials, a very reasonable price for the fundamental information obtained from defined mutants.

[46] U. Beisiegel, W. J. Schneider, J. L. Goldstein, R. G. W. Anderson, and M. S. Brown, *J. Biol. Chem.* **256,** 11923 (1981).
[47] M. R. Melamed, P. F. Mullaney, and M. Mendelsohn, "Flow Cytometry and Sorting." Wiley, New York, 1979.
[48] M. Krieger, this volume [14].

[16] Immunochemical Measurement of Apolipoprotein Synthesis in Cell and Organ Culture

By DAVID L. WILLIAMS and PAUL A. DAWSON

Measurement of Apolipoprotein Synthesis

Immunoprecipitation of radiolabeled proteins can be used to detect and quantitate the synthesis of specific proteins in tissues and cells. The basic strategy is to carry out metabolic labeling with radiolabeled amino acid or carbohydrate precursors in cell culture or short-term organ culture. A tissue extract is prepared, and the protein of interest is precipitated by reaction with specific antibody. The immunoprecipitate can be analyzed in a qualitative or quantitative fashion for the incorporation of radioactivity into the specific protein. This general approach can be used in three broad areas. In the first case the tissue sites of synthesis of a protein can be established with short-term organ cultures. This approach was used to show that apolipoprotein (apo) A-I is synthesized in many peripheral tissues in chickens[1] and apoE is synthesized in peripheral human tissues.[2] In a second case radiolabeling and immunoprecipitation methods can be used to quantitate the synthesis of specific proteins in response to drugs, hormones, nutritional status, or during specific physiological transitions such as embryonic development or aging. For example, the regulation of avian hepatic apoB synthesis by estrogenic hormones has been studied in this fashion.[3,4] Similarly, the ontogeny of hepatic apoB synthesis has been elucidated by immunoprecipitation of pulse-labeled apoB derived from short-term liver cultures prepared at various stages of embryonic development.[5] A third general application of these methods is in the study of protein biosynthesis and posttranslational modifications. For example, the complex biosynthetic pathway of the vitellogenins in the avian liver has been elucidated in this fashion. The use of pulse-labeling and pulse-chase protocols with different precursors combined with appropriate immunoprecipitation and gel analysis methods

---

[1] M.-L. Blue, P. Ostapchuk, J. S. Gordon, and D. L. Williams, *J. Biol. Chem.* **257**, 11151 (1982).

[2] M.-L. Blue, D. L. Williams, S. Zucker, S. A. Khan, and C. B. Blum, *Proc. Natl. Acad. Sci. U.S.A.* **80**, 283 (1983).

[3] F. Capony and D. L. Williams, *Biochemistry* **19**, 2219 (1980).

[4] F. Capony and D. L. Williams, *Endocrinology* **108**, 1862 (1981).

[5] S. A. Nadin-Davis, C. B. Lazier, F. Capony, and D. L. Williams, *Biochem. J.* **192**, 733 (1980).

permitted temporal ordering of the glycosylation and phosphorylation steps in vitellogenin biosynthesis.[6] In general, immunoprecipitation methods provide a very powerful approach to the study of the metabolism of specific proteins.

Metabolic Labeling of Apolipoproteins

Short-Term Organ Culture

In this procedure tissue slices are incubated *in vitro* with radiolabeled amino acid to label newly synthesized proteins. The tissue incubation is brief (0.5–1 hr) in order that the spectrum of newly synthesized proteins is representative of the proteins being made at the time the tissue is removed from the animal. The following procedure has been used to label newly synthesized apoA-I, apoB, and apoE in a variety of mammalian and avian tissues.[1-4]

Freshly isolated tissue is dissected free of fat and connective tissue, chopped into 5-mg slices with a razor blade, and rinsed twice with bicarbonate-buffered Krebs–Ringer solution[7] (KRB), pH 7.4, containing 50 U/ml penicillin G and 50 μg/ml streptomycin. The washes remove blood and minimize the content of plasma lipoproteins. The sample containing 20–40 mg of tissue slices is incubated for 1 hr at 37° (40° for chicken tissues) in 5 vol (vol/tissue wt) KRB containing antibiotics, radiolabeled amino acid, and 0.1% glucose under an atmosphere of 95% O_2/5% CO_2. The tissue slices are washed twice with 2 ml iced KRB to stop incorporation and kept on ice until homogenized. The sample can be quick frozen at this point for later processing. The sample is transferred from the incubation flask to a tissue homogenizer with a small spatula or with a wide-bore Pasteur pipet using 15 vol (vol/tissue wt) of homogenization buffer (HB) [0.02 M sodium phosphate, pH 7.4, 0.15 M NaCl, 0.005 M EDTA, 2% Triton X-100, 200 μl/ml phenylmethylsulfonyl fluoride (PMSF)] as the transfer vehicle. The sample is homogenized in the 15 vol HB used for transfer with a motor-driven tissue grinder while the grinder is held in an ice or salt–ice bath. Homogenization is carried out in 30-sec bursts separated by 30-sec cooling until the homogenate is smooth and homogeneous. This generally requires 2–4 grinding periods depending upon the consistency of the tissue. Particular care is taken to keep the sample at 0–2° during homogenization and thereafter. The homogenate is transferred

[6] S.-Y. Wang and D. L. Williams, *J. Biol. Chem.* **257**, 3837 (1982).
[7] W. W. Umbreit, R. H. Burris, and J. F. Stauffer, "Manometric Techniques and Tissue Metabolism." Burgess Publ., Minneapolis, 1945.

to an Oak Ridge-type polycarbonate ultracentrifuge tube and centrifuged for 1 hr at 226,000 g at 0–2° to prepare a high-speed supernatant (HSS). The HSS is removed and transferred to a small (0.5–1.5 ml) cryotube in an ice-water bath. After removal of duplicate aliquots (10 μl) for the measurement of total protein radioactivity, the remainder of the sample is frozen in liquid N_2 and stored at −70° until analysis.

Comments

1. Tissue incubations are usually carried out with 0.5–1 mCi/ml of high specific activity radiolabeled amino acid. The reason for using a high concentration of high specific activity precursor with a small amount of tissue (20–40 mg) is to minimize the actual mass of tissue protein that must be analyzed to detect the newly synthesized protein of interest. In many cases the final stage of analysis is polyacrylamide gel electrophoresis (PAGE) of the protein which has been immunoprecipitated from the HSS. Since the capacity of analytical PAGE is limited to approximately 100 μg of total protein (including the antibody), it is essential that the newly synthesized protein be labeled to a high specific activity. This is particularly important for apolipoprotein measurements because it is the preexisting intracellular apolipoprotein and the blood apolipoproteins within the tissue slice that actually determine the parameters for the immunoprecipitation reaction. The mass of newly synthesized apolipoprotein made during the labeling period is very small in comparison. For example, if the minimum detectable radioactive band on a polyacrylamide gel contained 100 dpm and the protein in question represented 0.1% of total protein synthesis, it would be necessary to immunoprecipitate and analyze a minimum of 100,000 dpm of protein radioactivity in the HSS irrespective of the specific activity of total HSS protein. Thus, if the specific activity is high, the total amount of HSS protein that must be analyzed is minimized. If the specific activity is low, the sensitivity of the analysis will be limited by the protein mass that can be loaded on the gel.

High specific activity [^{3}H]leucine (60–120 Ci/mmol) or [^{35}S]methionine (600–1200 Ci/mmol) have been used for apolipoprotein measurements. [^{35}S]Methionine is useful because of its very high specific activity and high energy level, which add to the speed and sensitivity of analysis. Commercial preparations are often stored in mercaptoethanol to prolong stability. The mercaptoethanol should be removed by lyophilization before dissolving in the incubation medium. Similarly, leucine and other radiolabeled amino acids are frequently stored in dilute HCl–ethanol, which should be removed by lyophilization before use. On rare occasions [^{35}S]methionine will artifactually label tissue proteins, presumably through an oxidation adduct of the sulfur. Should this occur, all tissue proteins including apolipoproteins from contaminating blood will be labeled. This nonspecific labeling is not seen with [^{3}H]leucine, which serves as a control for this artifact. An additional control to ensure that the incorporation of radioactivity into protein is actually due to protein synthesis is to block protein synthesis with a specific inhibitor such as cycloheximide.[1,2]

2. A variety of incubation media can be used for tissue slice incubations ranging from the simple KRB noted above to complex tissue culture media with or without serum additives. For the analysis of apoB synthesis in avian liver slices, little or no difference was seen among a variety of media when the incorporation of precursor amino acid into apoB was expressed relative to incorporation into total protein.[3] It is clear that this would not be the case with long periods of organ culture, but for short-term organ culture KRB is adequate to support the linear incorporation of precursor into protein for at least 2 hr. With liver slice incubations greater than 95% of the protein synthesized in 1 hr remains intracellularly or within the tissue slice.[3] With a 2-hr incubation 15–25% of newly synthesized protein is found in the incubation medium.

3. Due to the short duration of the tissue slice incubations, conditions of sterility employed with cell and tissue culture are not as important. Nevertheless, incubation medium and incubation flasks are sterilized and antibiotics are included as a precaution. Tissue slice incubations have been carried out in small (10-ml) Ehrlenmeyer flasks stoppered with rubber or silicone stoppers. With an incubation volume of up to 0.2 ml, such flasks provide a large surface area for gas exchange. After the addition of the tissue slices and the medium containing the radiolabeled precursor, the flask is gassed with a stream of 95% O_2/5% CO_2 to equilibrate the medium, stoppered, and placed in a shaking water bath or metabolic incubator at 80–100 cycles/min. Equilibration of the medium with the 95% O_2/5% CO_2 is easily monitored by including Phenol Red (8 mg/liter) as a pH indicator. All KRB solutions used for washes are similarly equilibrated with 95% O_2/5% CO_2. Tissue incubations of up to 0.1 ml also have been carried out in 15-ml plastic disposable round bottom culture tubes with snap caps.

4. The homogenization buffer includes nonionic detergent to solubilize the endoplasmic reticulum membranes containing newly synthesized secretory proteins. In addition, the detergent disrupts lipoprotein particles and solubilizes even relatively insoluble apolipoproteins such as apoB. In studies with avian liver apoB was recovered in the HSS with an efficiency of approximately 95% while total tissue protein was recovered with an efficiency of 90%.[3] Homogenization with a glass–Teflon tissue grinder is effective for the preparation of HSS from soft tissues such as

liver and kidney. However, for fibrous tissues such as lung or muscle, ground-glass tissue grinders are required for efficient extraction.

5. Protease inhibitors and careful maintenance of low temperature are important to minimize proteolytic breakdown of apolipoproteins. Apolipoprotein B, in particular, is very susceptible to proteolytic breakdown.[3,8] Proteolysis of apoB is effectively inhibited in liver extracts with PMSF and EDTA. A freshly made stock of PMSF in dimethylsulfoxide (100 mg/ml) is made immediately before use and added to HB to 200 μg/ml. This saturating concentration of PMSF is used because the serine-sulfonyl ester formed at the protease active site is subject to rapid hydrolysis in aqueous solution.[9]

Cell Culture

The following labeling procedure has been used to study apolipoprotein synthesis in cultured avian skeletal muscle cells[1] as well as avian and monkey fibroblasts and aortic smooth muscle cells.

Approximately 5×10^5 cells are transferred to a 60-mm culture dish and grown to near confluency with appropriate culture conditions. For labeling, the monolayers are washed twice with prewarmed 0.02 M sodium phosphate, pH 7.4, 0.15 M NaCl (PBS), and incubated in 1.5 ml minimal essential medium lacking methionine but containing 200 μCi/ml [^{35}S]methionine (600–1200 Ci/mmol) and 10% dialyzed calf serum for 4 hr at 37°. The plates are placed on ice, the medium is removed and adjusted to 100 μg/ml PMSF, and centrifuged at 2000 g for 3 min to remove any cells. The medium is dialyzed against three changes of 100 vol of PBS containing 100 μg/ml PMSF, and 0.001 M methionine for 18 hr at 0–4°, frozen in liquid N_2, and stored at −70° until analysis. Cell monolayers are washed twice with cold PBS and removed from the dish in HB using a rubber policeman. Cells are homogenized and processed to prepare an HSS as described above for tissue slices.

Immunoprecipitation

Direct Immunoprecipitation

Both direct and indirect immunoprecipitation have been used to study apolipoprotein synthesis.[1–3] With direct immunoprecipitation the HSS is reacted with excess antibody to the protein of interest and the resultant immunoprecipitate is isolated by centrifugation. The advantages of direct

[8] D. L. Williams, *Biochemistry* **18**, 1056 (1979).
[9] A. M. Gold and D. Fahrney, *Biochemistry* **3**, 783 (1964).

immunoprecipitation are that it is simple, requires only one antibody, and yields very low background values because the mass of the immunoprecipitate is small. The disadvantages are that it usually requires carrier antigen for efficient precipitation and cannot be used with a nonprecipitating polyclonal or a monoclonal antibody. The following procedure has been used to immunoprecipitate newly synthesized apoB from avian and mammalian liver.

HSS is thawed on ice and centrifuged at 10,000 g for 3 min at 0–2° to remove debris. After removal of the necessary HSS, the remainder of the sample is aliquoted and refrozen in liquid N_2. Subsequently, the aliquots are used only once to avoid repeated freezing and thawing. Antiserum is adjusted to 1% Triton X-100, 100 μg/ml PMSF, incubated for 10 min at room temperature, cleared by centrifugation for 5 min at 10,000 g, and placed on ice. All subsequent operations are carried out on ice. HSS (30 μl) is mixed with 20 μl antiserum and 0.5 μg of carrier very-low-density lipoprotein (VLDL) (as protein) in a 1.5-ml polypropylene centrifuge tube. After incubation for 1 hr, 1 ml of PBS containing 1% Triton X-100 and 100 μg/ml PMSF (PBS-T) is added, the sample is mixed, and centrifuged for 4 min at 10,000 g. While the tube is held in front of a bright lamp in order to visualize the small white precipitate, the supernatant is carefully aspirated. The pellet is vigorously vortexed to smear the immunoprecipitate over a large area of the tube bottom. This facilitates the washing of the immunoprecipitate. PBS-T (1 ml) is added, the sample is vortexed, and centrifuged for 4 min at 10,000 g. The immunoprecipitate is washed three more times in the same fashion. The final pellet may be dissolved for gel analysis or counted directly for radioactivity as described below.

Comments

1. Immunoprecipitation of apoB occurs rapidly and efficiently even at reduced temperature as used here. This is likely due to the large size of apoB and its numerous determinants that facilitate precipitate formation. With other apolipoproteins it may be necessary to use much longer incubation periods or elevated temperatures although the latter is likely to promote proteolysis in tissue extracts. We have used incubation periods of 15–24 hr on ice with apoA-I and apoE with no evidence of proteolysis.

2. Whole antiserum is suitable for most immunoprecipitations without further fractionation. Fresh serum is centrifuged for 1 hr at 226,000 g, adjusted to 0.02% NaN_3, and stored at −70°. Repeated thawing and freezing causes aggregation and leads to high nonspecific precipitation in the immunoprecipitation assay. To avoid this, sufficient antiserum for about

1 month is thawed and stored at 4°. Antiserum sufficient for each day is treated with Triton X-100 and PMSF as described above. Control sera and secondary antisera are treated in the same fashion.

3. In the example given above the apoB in the HSS represents (1) radiolabeled apoB synthesized during the tissue slice incubation, (2) unlabeled intracellular apoB, and (3) extracellular apoB that was present in the blood lipoproteins within the tissue slice. The antibody must be present in excess to the sum of the apoB to ensure complete immunoprecipitation. In most cases one will not know the mass amount of total apoB in the HSS and thus will not know how much antibody is required. As a result, it is necessary to titrate the HSS directly to establish conditions of antibody excess. If the assay will be used to measure radiolabeled apoB in a series of samples, the titration should be done with an HSS sample that is likely to contain the most antigen. The assay conditions will then be valid for any sample containing the same amount of antigen or less. First, determine the volume of HSS to be routinely assayed on the basis of the desired sensitivity and the precursor incorporation into total protein radioactivity. For example, if the assay is constructed to measure apoB that is greater than or equal to 0.1% of protein synthesis and the minimum detectable signal is 100 dpm, it is necessary to assay at least 100,000 dpm of protein radioactivity. Assume that 100,000 dpm of protein radioactivity is present in 20 μl of HSS. The first step is to titrate 20-μl HSS samples with increasing volumes of antiserum in the standard assay given above. The immunoprecipitates are dissolved and analyzed by sodium dodecyl sulfate (SDS)–PAGE and autoradiography (or fluorography) as described below. The radioactivity in the immunoprecipitated apoB band can be assessed by visual inspection of the autoradiogram, densitometry, or direct counting of the gel. Visual inspection is usually adequate at this stage. The expected result is that little or none of the radiolabeled apoB should be detected with low amounts of antiserum (region of antigen excess), apoB radioactivity should increase markedly near equivalence, and the radioactivity should reach a maximum with high amounts of antiserum (region of antibody excess). Select an amount of antiserum that is approximately 50% in excess to the amount that yielded maximum precipitation of apoB from the HSS. In the present example assume that this is 10 μl of antiserum. The second step is to determine the amount of carrier antigen to be used in the assay. It is prudent to use carrier antigen because HSS samples containing very low amounts of total antigen do not form immunoprecipitates efficiently. In this case 20 μl HSS is incubated with 10 μl antiserum in the presence of increasing quantities of authentic antigen. The radiolabeled apoB recovered in the immunoprecipitate is assayed as above. Figure 1 is a typical titration showing complete precipi-

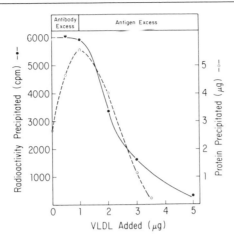

Fig. 1. Determination of carrier antigen for avian apoB immunoprecipitation. Anti-apoB serum (10 μl) was reacted with [³H]leucine-labeled liver HSS (20 μl) and the indicated amount of VLDL protein in the standard immunoprecipitation assay. The washed immunoprecipitate was dissolved in 0.2 ml 0.1 N NaOH and divided for liquid scintillation spectrometry and protein determination.

tation of newly synthesized apoB with up to 1 μg of added VLDL antigen (region of antibody excess) followed by decreasing apoB radioactivity as additional VLDL is added (region of antigen excess). Note that measurement of total protein in the immunoprecipitate should yield a classical precipitin curve. For the routine assay select an amount of carrier antigen that is 50% of equivalence in this titration. From these two titrations the routine assay would contain 0–20 μl HSS, 10 μl antiserum, and 0.5 μg carrier VLDL. Each sample is assayed in duplicate at three concentrations of HSS. The maximum volume of HSS assayed should be equal to or less than the HSS volume used to construct the assay. The radioactivity in the immunoprecipitated protein should behave as a linear function of the HSS volume assayed.

Indirect Immunoprecipitation

Indirect immunoprecipitation employs a second reagent to precipitate or immobilize the primary antibody–antigen complex. Advantages of indirect immunoprecipitation are that it can be used with nonprecipitating polyclonal or monoclonal antibodies and does not require the addition of carrier antigen. The disadvantage of indirect immunoprecipitation is that the mass of the immunoprecipitate is usually much larger than with direct immunoprecipitation. Consequently, the background of nonspecific precipitation is usually greater with indirect immunoprecipitation. Indirect immunoprecipitation has been used to detect and quantitate the synthesis of apoA-I,[1] apoE,[2] and apoB[10] in a variety of tissues. The following is a typical protocol.

HSS is thawed and cleared by centrifugation as above. All antisera are treated with PMSF and Triton X-100 as above. Subsequent procedures are carried out on ice. HSS (20 μl) containing 100,000–500,000 dpm of protein radioactivity is incubated with 10 μl primary anti-serum for 15–24 hr in a 1.5-ml polypropylene centrifuge tube. Secondary antiserum (40 μl) is added, and the incubation is continued for 2 hr. PBS-T (1 ml) is added, the sample mixed, and centrifuged for 4 min at 10,000 g. The supernatant is discarded, and the pellet is washed three more times as described for direct immunoprecipitation.

Comments

1. The choice of secondary antiserum will depend on the species of primary antiserum. With rabbit primary antisera we have used goat anti-rabbit γ globulin and for goat primary antisera we have used sheep anti-goat γ globulin. While higher antibody titers are almost always obtained by raising the secondary antiserum oneself, secondary antisera from commercial sources are quite acceptable for most purposes. High-quality goat anti-rabbit γ globulin and sheep anti-goat γ globulin are available from Cooper Biomedical, Inc., Malvern, PA.

2. Titration with primary antiserum should be done as described above for the direct immunoprecipitation assay. In this case, however, one must first titrate the secondary antiserum with the primary antiserum as antigen. This is most easily accomplished by performing a classical precipitin reaction with a fixed volume of secondary antiserum under the exact conditions that are to be used in the immunoprecipitation assay. When the equivalence for the secondary antiserum is established, one can proceed to titrate a fixed volume of HSS with the primary antiserum while the secondary antiserum is held in excess. Figure 2B shows a typical result in which avian liver HSS containing [³H]leucine-labeled proteins was titrated with rabbit anti-chicken apoA-I. The second antibody in this case was goat anti-rabbit γ globulin. For purposes of routine assay, each HSS sample is assayed in duplicate at three concentrations.

3. For the indirect immunoprecipitation assay, one may use protein A as an alternative to the secondary antibody. Protein A is a 42,000-Da protein present in the cell wall of *Staphylococcus aureus*. Since protein A binds to the Fc portion of immunoglobulins with high affinity, it may be used to immobilize a primary antibody in an indirect immunoprecipitation

[10] M. L. Blue, A. A. Protter, and D. L. Williams, *J. Biol. Chem.* **255**, 10048 (1980).

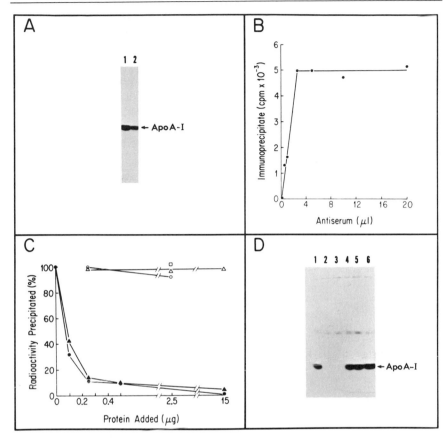

FIG. 2. Specificity of the apoA-I immunoprecipitation assay. (A) Purified chicken apoA-I was analyzed by SDS–10% PAGE (lane 1, 24 μg; lane 2, 8 μg). The gel was stained with Coomassie Blue. (B) Chicken liver HSS (5 μl) containing 1.2×10^5 cpm ^{3}H-labeled protein was immunoprecipitated with the indicated volume of rabbit anti-apoA-I serum in the indirect immunoprecipitation assay. Goat anti-rabbit γ globulin was the secondary reagent. Data points show immunoprecipitate radioactivity after correction for nonspecific precipitation with normal rabbit serum. (C) Liver extract (10 μl) was reacted with 8 μl of anti-apoA-I serum in the presence of the indicated amounts of the following proteins: ●, Purified apoA-I; ▲, HDL; △, chicken immunoglobulin; ○, purified apoB; □, chicken serum albumin. (D) Immunoprecipitates containing [^{3}H]apoA-I formed in the presence of competitors as in (C) were analyzed by SDS–10% polyacrylamide gel electrophoresis followed by fluorography. The competitors were (lanes 1–6, respectively): none, 15 μg of purified apoA-I, 15 μg of HDL, 2.5 μg of chicken serum albumin, 2.5 μg of chicken immunoglobulin, and 15 μg of chicken immunoglobulin. The arrow indicates the mobility of plasma apoA-I.

assay. Protein A can be used in two forms. In the first the formalin-fixed bacteria are used directly as an immunoabsorbant. Ivarie and Jones[11] have presented a detailed discussion and protocol for immunoprecipitations with protein A in this mode. The formalin-fixed bacteria are available from a variety of commercial sources. Protein A also is available in purified form covalently coupled to agarose (Sigma Chemical Company, P-9269) or Sepharose (Pharmacia). One should note that protein A does not bind all classes of immunoglobulins from all species.[12] The majority of IgG subclasses in rabbit, guinea pig, and human sera bind to protein A whereas the reactive fraction is much smaller in goat, sheep, mouse, and rat sera. Avian immunoglobulins show almost no reactivity with protein A.

Specificity and Quantitation in Immunoprecipitations

Specificity

Whether an immunoprecipitation assay is used in a quantitative or qualitative fashion, it is essential to establish the specificity of the assay. Immunoprecipitation of pulse-labeled proteins generally is much more sensitive than other methods of evaluating antibody specificity. As a result, an antibody that is judged highly specific by most immunological criteria may detect pulse-labeled proteins other than the desired antigen. It is important, therefore, to confirm that the assay specifically measures the protein of interest. This is generally done through an analysis of the immunoprecipitated proteins by SDS–PAGE combined with competition or peptide-mapping techniques. Procedures for the limited proteolysis mapping[13] of immunoprecipitated proteins directly or after SDS–PAGE have been described.[6,10,14] An example of a competition analysis is illustrated in Fig. 2. Figure 2B shows that titration of a radiolabeled liver extract with anti-apoA-I yielded a saturation curve with the maximum value equal to 4% of the [^{3}H]leucine incorporated into hepatic protein. Since these data were obtained by direct counting of immunoprecipitate radioactivity, the conclusion that apoA-I was measured rests completely on the specificity of the antiserum. Analysis of the immunoprecipitate by SDS–10% PAGE (Fig. 2D, lane 1) showed directly that the radioactivity was present in a protein with the same electrophoretic mobility as apoA-I.

[11] R. D. Ivarie and P. P. Jones, *Anal. Biochem.* **97,** 24 (1979).

[12] J. J. Langone, *Adv. Immunol.* **32,** 157 (1982).

[13] D. W. Cleveland, S. G. Fischer, M. W. Kirschner, and U. K. Laemmli, *J. Biol. Chem.* **252,** 1102 (1977).

[14] S.-Y. Wang, D. E. Smith, and D. L. Williams, *Biochemistry* **22,** 6206 (1983).

As a further test of specificity the immunoprecipitation was carried out in the presence of increasing quantities of various competitors. As shown in Fig. 2C, apoA-I and HDL completely eliminated [^{3}H]apoA-I from the immunoprecipitate while apoB, serum albumin, and immunoglobulin were ineffective. SDS–PAGE of the immunoprecipitates (Fig. 2D) confirmed that only the apoA-I band was eliminated by competition with authentic apoA-I (lane 2) or HDL (lane 3). These data provide a rigorous demonstration of the specificity of the immunoprecipitation assay.

In the example shown in Fig. 2, apoA-I represents 4% of newly synthesized hepatic protein and is the predominant protein detected when the immunoprecipitate is analyzed by SDS–PAGE. Careful inspection of the fluorogram (Fig. 2D), however, also shows several minor bands that are uneffected by the competition. These bands are proteins that are nonspecifically adsorbed to the immunoprecipitate. The nonspecific bands also are precipitated when control antiserum is substituted for anti-apoA-I.[1] When constructing an immunoprecipitation assay it is important to recognize several things about nonspecific bands. First, they are not representative of the spectrum of newly synthesized proteins. Instead, these bands represent a subset of proteins that bind to immunoprecipitates. Second, although some bands are found in almost all tissues, the pattern of nonspecific bands is different in various tissues. Third, the pattern of nonspecific bands is not the same with different methods of immunoprecipitation. Indirect immunoprecipitation with protein A as the secondary reagent yields different nonspecific bands than with antibody as the secondary reagent. Fourth, the intensity of prominence of the nonspecific bands relative to the specific antigen is determined by the abundance of the specific antigen among the newly synthesized proteins.

Several of these points are well illustrated by the SDS–10% PAGE profiles in Fig. 3 which show indirect immunoprecipitates of newly synthesized apoE from monkey liver. In this tissue apoE represents approximately 0.4% of protein synthesis or an order of magnitude less than apoA-I synthesis in chicken liver. While the apoE band is readily detected by anti-apoE (lanes 1,3,5), there are numerous prominent nonspecific bands that are also present in both anti-apoE and control antibody samples (lanes 2,4,6). The nonspecific bands are not eliminated by competition with authentic apoE (data not shown). Note that if apoE were only 0.04% of protein synthesis, the prominence of the apoE band would be 10-fold less, but the nonspecific bands would show the same intensity. In addition, nonspecific bands are different when the secondary reagent is sheep anti-goat γ globulin (lanes 1,2) as compared to formalin fixed *S. aureus* (lanes 3,4) as compared to protein A–agarose (lanes 5,6). Chang-

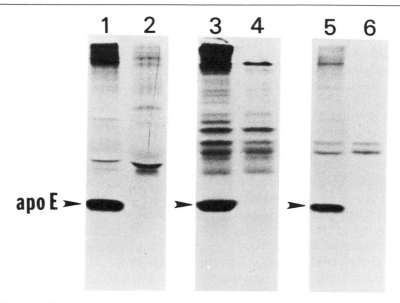

FIG. 3. Indirect immunoprecipitation of monkey apoE with different secondary reagents. Monkey liver [^{35}S]methionine-labeled HSS was reacted with goat anti-apoE (lanes 1, 3, 5) or preimmune serum (lanes 2, 4, 6) in the indirect immunoprecipitation assay. The secondary reagents were sheep anti-goat γ globulin (lanes 1, 2), formalin-fixed *S. aureus* (lanes 3, 4), and protein A-agarose (lanes 5, 6). Washed immunoprecipitates were analyzed by SDS–10% polyacrylamide slab gel electrophoresis followed by fluorography. The arrow indicates the mobility of monkey apoE.

ing the secondary reagent may be useful should the desired antigen comigrate with a nonspecific band. Thus, the ability to detect and quantitate newly synthesized apoE depends on the abundance of apoE synthesis and whether or not a nonspecific band happens to comigrate with apoE in this electrophoretic system.

Quantitation

Nonspecific Immunoprecipitation. Nonspecific immunoprecipitation is largely a function of the mass of the immunoprecipitate. With direct immunoprecipitation, nonspecific immunoprecipitation is usually in the range of 0.05–0.2% of the input protein radioactivity in the assay. In this case direct counting of the immunoprecipitate as described below is sufficient for the measurement of a specific antigen that represents greater than 0.5% of total protein radioactivity. Correction for nonspecific immunoprecipitation is made by carrying out parallel reactions with an unrelated antigen–antibody pair in the same HSS. For example, bovine albumin and anti-bovine albumin were used to correct for nonspecific immunoprecipitation when measuring avian apoB in liver HSS.[3] This control immunoprecipitation is constructed such that the mass of the immunoprecipitate is the same as that produced with the specific antigen. With indirect immunoprecipitations, nonspecific immunoprecipitation is usually in the range of 0.5–1% of the input protein radioactivity in the assay. Correction for nonspecific immunoprecipitation is made with parallel reactions using preimmune primary antibody or an unrelated antibody that will not react with protein in the HSS. The amount of control antibody or antiserum used is adjusted to give the same mass of immunoprecipitate as with the specific antiserum.

When the specific antigen is not in great enough abundance to permit direct counting of the immunoprecipitate, a secondary method for resolving the specific antigen from the nonspecific background is required. This is generally done with SDS–PAGE. Figure 3, for example, shows that the nonspecific background (lanes 2,4,6) is quite substantial in comparison to the apoE radioactivity (apoE band, lanes 1,3,5). In this case, resolution of apoE from the nonspecific bands is necessary to accurately quantitate apoE. ApoE radioactivity is determined by direct counting of the apoE band excised from the gel as described below. Nonspecific or background radioactivity is determined by counting the apoE gel region from the SDS–PAGE of the control immunoprecipitate. In the case of apoE this method permits the measurement of apoE synthesis at levels greater than or equal to 0.02% of total protein synthesis. The limit of detection for any specific antigen, however, will be determined by the nonspecific proteins that happen to comigrate with the specific antigen.

Efficiency of Immunoprecipitation. The efficiency of immunoprecipitation can be determined in several ways. In the case of avian apoB,[3] the efficiency was determined by SDS–PAGE of the immunoprecipitate and the HSS. This is only possible, however, when the specific antigen is resolved completely from other HSS proteins by SDS–PAGE. A second method is to add a tracer amount of the specific antigen to the HSS prior to immunoprecipitation. This can be done with a radiolabeled antigen added to a mock reaction in which the HSS proteins are not labeled. Another method is to use a double-label approach in which the tracer antigen is labeled with a different isotope than the proteins of the HSS. Apolipoproteins labeled with ^{3}H or ^{14}C by reductive methylation[15] have

[15] G. E. Means and R. E. Feeney, "Chemical Modification of Proteins." Holden-Day, San Francisco, 1971.

proved satisfactory for this purpose. When the immunoprecipitation assay is carefully optimized, immunoprecipitation efficiency is usually 90% or greater.

Data Interpretation. Apolipoprotein synthesis can be determined as an absolute rate or as a relative rate of synthesis. With the absolute rate method, amino acid pool specific activities are determined such that incorporation of precursor into the specific apolipoprotein can be expressed as molecules synthesized per unit time. With the relative rate method, the incorporation of radiolabeled precursor into the specific apolipoprotein is expressed as a percentage of precursor incorporation into total protein.

The advantages of the relative rate method are that it requires only two measurements (the specific antigen and total protein radioactivity), is easily applied to large numbers of samples, and controls for the same variables (specific activities of amino acid pools and pool equilibration kinetics) that must be determined directly with the absolute rate method. The relative rate method, for example, controls for precursor availability by normalization to precursor utilization for total protein synthesis. In this fashion, nutritional or hormonal alterations in precursor transport or variations in tissue slice geometry that influence the absolute level of precursor uptake are internally controlled since these parameters will influence precursor incorporation into all proteins. The relative rate of synthesis of a specific apolipoprotein reflects its abundance among newly synthesized proteins or the frequency at which the protein is synthesized. Since most cells make about 10^4 different proteins, the relative rate of synthesis is informative as to whether a specific protein is a common or rare product of a cell. In the avian liver, for example, serum albumin is the most abundant protein product and represents 5–10% of total protein synthesis.[16] In contrast, apoB represents 1–2% of avian liver protein synthesis.[3,4] These proteins are very abundant products and reflect a significant commitment of the protein synthetic capacity of the tissue. Most mammalian apolipoproteins appear to be in the range of 0.1–1% of hepatic protein synthesis and are considered moderately abundant products.

With the absolute rate method it is necessary to document amino acid pool-specific activities and equilibration kinetics in each experimental situation. The absolute rate method has the advantage, however, that it permits one to estimate the actual amount of a specific protein made by a tissue. This information may be of importance in assessing the physiological significance or quantitative contribution of a particular tissue to apolipoprotein production. One should note, however, that cellular heteroge-

[16] D. L. Williams, S.-Y. Wang, and H. Klett, *Proc. Natl. Acad. Sci. U.S.A.* **75**, 5974 (1978).

neity of a tissue complicates the interpretation of data with the relative rate or absolute rate method since the experimentally derived value is the average of all cell types.

Procedures. The following procedures are used for the quantitative determination of apolipoprotein synthesis by immunoprecipitation as outlined above. For established procedures that are presented elsewhere, only relevant modifications are described.

1. *Direct counting of immunoprecipitates.* After the last wash, the immunoprecipitate is smeared on the tube walls by vigorous vortexing. Protosol (0.5 ml) (New England Nuclear) is added to cover the immunoprecipitate, and the tube is incubated 12–18 hr at room temperature or until the precipitate is completely solubilized. The sample is transferred to a counting vial, and the tube is rinsed into the counting vial with several washes of scintillation fluid. Samples are counted in a cocktail containing 3.8 g Omnifluor (New England Nuclear) per liter of toluene.

2. *SDS–PAGE of immunoprecipitates.* The immunoprecipitate is solubilized by the addition of 50–75 μl of 0.03 M Tris-HCl, pH 6.8, 3% SDS, 0.01 M EDTA, 2.5% β-mercaptoethanol, 5% glycerol. This solution (lacking tracking dye) is added to the drained immunoprecipitate without mixing, and the sample is placed at 37°. It is important not to vortex the sample or much of the precipitate will be lost on the tube walls. The precipitates generally dissolve completely in 1–2 hr. This can be ascertained by gently swirling the tube in front of a bright lamp to visualize small fragments of the pellet. If the sample does not readily dissolve, the process can be accelerated by gentle up-and-down pipetting or by breaking up the pellet with a small glass stirring rod. To the dissolved precipitate Bromophenol Blue is added (1 μl of a 0.2% stock solution), the sample is boiled 3 min, and the sample is transferred to the sample well of an SDS–polyacrylamide slab gel. We have employed the discontinuous buffer system of Laemmli[17] with various acrylamide concentrations depending on the specific apolipoprotein under study. Gel recipes and protocols for slab gel electrophoresis have been described in detail.[17,18] Following electrophoresis, the gel is fixed and stained in 50% trichloroacetic acid containing 0.1% Coomassie Blue and destained by diffusion in 7% acetic acid. If the gel is to be exposed by autoradiogrpahy, it is dried at this point. If the gel is to be exposed by fluorography, it is carried through the procedure as described[19] before exposure to X-ray film. Note that pre-

[17] U. K. Laemmli, *Nature (London)* **227**, 680 (1970).
[18] T. Maniatis, E. F. Fritsch, and J. Sambrook, "Molecular Cloning." Cold Spring Harbor Press, Cold Spring Harbor, New York, 1982.
[19] W. M. Bonner and R. A. Laskey, *Eur. J. Biochem.* **46**, 83 (1974).

flashing of X-ray film is required for quantitative measurements of radio-activity by fluorography.[20]

3. *High-resolution two-dimensional gel analysis.* Two-dimensional gel analysis by the method of O'Farrell[21] provides a sensitive means of comparing the immunoprecipitated protein with the authentic antigen. For this purpose the washed immunoprecipitate is given a final wash with water and dissolved in 25 μl of 0.05 M 2-(N-cyclohexylamine) ethanesulfonic acid, pH 9.5, 2% SDS, 1% dithiothreitol, 10% glycerol, 0.005 M PMSF. The sample is placed at 37° until dissolved, boiled for 5 min, and applied to an isoelectric focusing tube gel prepared and prefocused as described.[21] The remainder of the procedure is as described.[21] The sample buffer employed here[22] provides better solubilization of the immunoprecipitate than that originally described. Although deamidation artifacts due to the high pH of this buffer are a potential problem, such artifacts have not been observed with avian apoA-I[1] or human apoE.[2] To facilitate identification and comparison of apolipoproteins resolved by two-dimensional gel analysis, authentic apolipoproteins can be added to the washed immunoprecipitate and resolved on the same gel. If the gel is exposed by autoradiography, the stained gel and the X-ray film can be superimposed for comparison. With fluorography there is considerable shrinkage of the gel and the Coomassie Blue stain elutes from the gel during the work-up. This problem is eliminated with silver staining[23] of proteins which is stable to dimethylsulfoxide extraction during fluorography. The dried gel and the fluorograph can be superimposed for comparison.

4. *Scintillation counting of gel slices.* Quantitation of immunoprecipitated apolipoprotein radioactivity is done by scintillation counting of gel slices. The dried gel is marked in several places with radioactive ink (^{14}C, 1 μCi/ml) before exposure to X-ray film to facilitate alignment of the film with the gel after exposure. After exposure, the desired protein band is located on the X-ray film, and a tracing of the band and the gel alignment marks is made on a thin plastic sheet. The band dimensions are approximately 0.3–0.4 cm × the width of the gel lane. Care is taken to ensure that the band dimensions are the same in each lane. An identical area is marked in the control antibody lanes at the position of the apolipoprotein. The band outlines are cut from the plastic sheet with a razor blade or scalpel to prepare a template. The template is aligned with the gel via the alignment marks, and the band outlines are marked directly on the dried

[20] R. A. Laskey and A. D. Mills, *Eur. J. Biochem.* **56,** 335 (1975).

[21] P. H. O'Farrell, *J. Biol. Chem.* **250,** 4007 (1975).

[22] N. G. Anderson, N. L. Anderson, and S. L. Tollaksen, "Operation of the Isodalt System." Argonne National Laboratory, Publ. ANL-BIM-79-2, 1979.

[23] C. R. Merrill, M. L. Dunau, and D. Goldman, *Anal. Biochem.* **110,** 201 (1981).

gel with a fine-tip blue marker. Each band is numbered with the marker. The band is cut from the dried gel with scissors and transferred to a glass scintillation vial. Hydrogen peroxide (30%) (0.25 ml) is added to the sample, and, after several minutes, the paper on which the gel was dried is removed with forceps. The vial is sealed and placed at 37° overnight at an angle such that the H_2O_2 remains in contact with the gel slice. Protosol (1 ml) is added, and the sample is shaken for 6 hr at room temperature. To the sample is added scintillation fluid (10 ml) containing 3.8 g Omnifluor, 25 ml Protosol, and 3.5 ml water per liter of toluene. The sample is shaken for 12–18 hr at room temperature before counting. The gel slice swells to a volume of several cm^3 and is completely clear. As judged from measurements of ^{35}S-labeled protein polymerized into 10% polyacrylamide gels, this procedure yields quantitative recovery of radioactivity.

5. *Measurement of isotope incorporation into total protein.* For the measurement of radiolabeled amino acid incorporation into HSS protein, 10 μl HSS is added to 0.5 ml bovine serum albumin (200 μg) in a 13 × 100 mm glass tube. Trichloroacetic acid is added to 7.5%, the corresponding unlabeled amino acid is added to 0.001 M, and the sample is heated at 90° for 15 min to hydrolyze aminoacyl tRNA. The sample is cooled on ice for 15 min, filtered over a glass fiber filter (Whatman G F/C), washed four times with 20 ml 7.5% trichloroacetic acid, and the filter is dried under an infrared lamp to remove trichloroacetic acid. The filter is placed in a vial with 3–10 ml of the Protosol-containing scintillation cocktail described above (Procedure 4) and counted after sitting overnight to solubilize protein from the filter.

Acknowledgments

Excellent technical assistance was provided by Penelope Strockbine. This research was supported by Grants AM 18171 and HL 32868 from the National Institutes of Health and Grant 83-849 from the American Heart Association. P.A.D. is a predoctoral trainee in Pharmacological Sciences (NIH GM 07518).

[17] Lipoprotein Synthesis and Secretion by Cultured Hepatocytes

By ROGER A. DAVIS

Introduction

Although the important role of the liver in the synthesis and secretion of plasma lipoproteins has been established several years ago, relatively little is known about its regulation. In part, due to the rapid metabolism of lipoproteins that takes place upon secretion into plasma, analyzing plasma lipoproteins obtained *in vivo* in regard to biosynthesis is difficult. An isolated *in vivo* model eliminates the myriad of rapid catabolic reactions that occur *in vivo* and can yield data that can be more easily interpreted. However, for these data to be valuable, physiologically relevant biosynthetic processes must be preserved. The recent development of the culture hepatocyte models provides an valuable experimental model with which to examine questions not easily addressed using other models. Recently, several laboratories have used this model to obtain important new concepts in regard to the synthesis and secretion of lipoproteins.[1-4] The obvious advantage of this model is the ability to examine the synthetic processes that are confined to the hepatocyte, without interference from extrahepatic metabolic processes. In this report we will concentrate on methodological aspects of the adult rat hepatocyte system. Because there are advantages to individual culture systems, readers are urged to examine the material contained in references 1-4 to aid in their choice of the system most appropriate for their research. Before committing one's resources to a hepatocyte culture model, it is important to consider the limitations of the model imposed by the differences in the anatomy and physiology between an isolated cultured hepatocyte and the *in vivo* liver.

The parenchymal cell *in vivo* has three distinct surface membranes (i.e., cell–cell surfaces, bile–canalicular surfaces, and blood plasma–sinusoidal surfaces). Each is segregated from the other. The segregation of bile from blood plasma affords the liver at least two separate secretion pathways. Whereas secretion of serum proteins such as lipoproteins occurs across the sinusoidal surface, secretion of bile components occurs

[1] D. M. Tarlow, P. A. Watkins, R. E. Reed, R. S. Miller, E. E. Zwergel, and M. D. Lane, *J. Cell. Biol.* **73**, 332 (1977).
[2] R. A. Davis, S. C. Engelhorn, S. H. Pangburn, D. B. Weinstein, and D. Steinberg, *J. Biol. Chem.* **254**, 2010 (1979).
[3] J. Bell-Quint and T. Forte, *Biochim. Biophys. Acta* **663**, 83 (1981).
[4] W. Patsch, S. Franz, and G. Schonfeld, *J. Clin. Invest.* **71**, 1161 (1983).

across the canalicular surface. This segregation of the biliary system affords an excretory route for metabolized toxins, drugs, and biochemical wastes via the feces.

Hepatic parenchymal cells are situated on double-cell plates. The highly fenestrated hepatic endothelium offers little restriction to the passage of high-molecular-weight substances. Thus, the hepatocyte, with little or no restriction at the endothelial boundary, comes in contact with plasma macromolecules at high concentrations. Furthermore, the liver is the first tissue that receives intestinally derived nutrients via the portal blood. It is likely that the rich supply of nutrients the hepatocyte obtains from the portal blood enables it to maintain a rather high anabolic rate. In regard to culturing conditions, the isolated hepatocyte is, therefore, likely to require a rich source of nutrients in order to maintain its anabolic function.

Another unique characteristic of the liver cell is its capacity for regeneration. Within 3 days of partial (75%) hepatectomy, the adult rat liver will return to its previous size. The ability to induce hepatocyte proliferation in culture has been described.[5] However, the rate of cell division in this model is not as great as it is after partial hepatectomy in vivo. The low proliferation rate of adult primary cultured hepatocytes has impeded attempts to obtain cloned mutations in the VLDL assembly/secretion pathway.

Plating isolated cells as a monolayer might be expected to alter many of the anatomical features upon which the hepatic functions depend. However, it is remarkable that even on a plastic culture dish, hepatocytes maintain a polar morphology and reassociate bile canalicular like junctions. Moreover, many of the functions attributed to the liver in vivo are also expressed by cultured cells. In addition to lipoprotein synthesis and secretion,[1–4] cultured hepatocytes demonstrate the production of most hepatic derived serum proteins,[6] drug metabolism,[7] matrix protein synthesis and secretion,[8] bile acid synthesis,[9] bile acid uptake and conjugation,[10] and uptake of chylomicron remnants[11] and disialyted proteins.[12]

[5] M. E. Kaigh and A. M. Prince, Proc. Natl. Acad. Sci. U.S.A. **68,** 2396 (1971).

[6] H. L. Leffert, K. S. Koch, B. Rubalcava, S. Sell, T. Moran, and R. Boorstein, Natl. Cancer Inst. Monogr. **48,** 87 (1978).

[7] R. P. Evarts, E. Marsden, and S. S. Thorgeirsson, Biochem. Pharmacol. **33,** 565 (1984).

[8] D. M. Bissell, Fed. Proc., Fed. Am. Soc. Exp. Biol. **40,** 2469 (1981).

[9] R. A. Davis, P. M. Hyde, J.-C. Kuan, M. M. Malone-McNeal, and J. Archambault-Schexnayder, J. Biol. Chem. **258,** 3662 (1983).

[10] B. F. Scharschmidt and J. E. Stephens, Proc. Natl. Acad. Sci. U.S.A. **78,** 986 (1981).

[11] C.-H. Floren and A. Nilsson, Biochem. Biophys. Res. Commun. **74,** 520 (1977).

[12] A. D. Attie, R. C. Pittman, and D. Steinberg, Proc. Natl. Acad. Sci. U.S.A. **77,** 5923 (1980).

Materials

1. Liver perfusion apparatus (Fig. 1). All materials can be obtained from any scientific supply company
2. One rat. The rat can be any size from 150–350 g. We have had best results with male rats that weigh between 170 and 250 g
3. A 20-gauge intravenous catheter (Delmed, Canton, MA)

Solutions

1. Hanks' Ca^{2+}-free buffer (i.e., KCl—400 mg/liter, KH_2PO_4— 60 mg/liter, NaCl—8 g/liter, $NaHCO_3$—1.008 g/liter, $NaHPO_4 \cdot H_2O$—53 mg/liter, $MgSO_4 \cdot 7H_2O$—246 mg/liter, HEPES—2.38 g/liter, dextrose—2.97 g/liter, and Phenol Red—5 mg/liter). The buffer is adjusted to a pH of 7.8 and then sterilized by passage through a 0.45-μm filter.
2. Collagenase (Sigma, type IIa): It is our experience that the preparation of the collagenase is the single most important factor that determines the quality and quantity of cells obtained. There are marked differences among collagenase preparations from the same supplier. We order 500-mg "test" batches of collagenase (keeping 5–10 g reserved for us). When we obtain a good yield of cells in regard to quality and quantity we okay the delivery of the rest of the collagenase and store it in 200-mg aliquots at $-70°$. Normally, we add 1 mg of collagenase to 1 ml of Ca^{2+}-free Hanks' buffer. A single liver requires 100 ml of collagenase solution.
3. Dulbecco's modified Eagle's medium (formula #78-5433; i.e., arginine free, with L-glutamine), from Grand Island Biological Company: Before use, we add L-ornithine—420 mg/5 liters, $NaHCO_3$—10.1 g/5 liters, HEPES—11.9 g/5 liters, and the appropriate amount of glucose for a total concentration of 5–40 mM, depending upon the experimental design. The medium is sterilized by filtration and stored in 500-ml sterile bottles at 4°. We use the medium within a month of preparation.
4. Insulin solution (1 mg/ml): Sigma porcine insulin is dissolved in DME containing 0.05% fatty acid-free bovine serum albumin (also obtained from Sigma). The pH is adjusted to 5.0, the solution is sterile filtered and stored at $-20°$.
5. Plating medium contains 20% sterile calf serum, insulin (1 μg/ml), and penicillin–streptomycin, 5000 U and 5000 μg/ml, respectively (Irvine Scientific, Irvine, CA).
6. Feeding medium contains everything as in (4), but without calf serum.

Procedure (See Fig. 1)

It is important to begin with the idea that preparing intact, well-functioning hepatocytes is a bit of an art. It requires good sterile surgical techniques, laboratory skills, patience, and practice. Also, it is a good idea to begin using reagents (especially collagenase) that have been used successfully by others to prepare hepatocytes. This allows one to differentiate faulty technique from inadequate reagents.

1. Turn on a heating block and begin gassing the water bath of the perfusion apparatus with 95% oxygen/5% CO_2. The Hanks' buffer should reach 37° by passing through the water bath, which is at 40°. Within 5 min the water bath has reached equilibrium in regard to oxygen saturation. Add 2 ml of sterile 100 mM $CaCl_2$ solution to the lower reservoir.

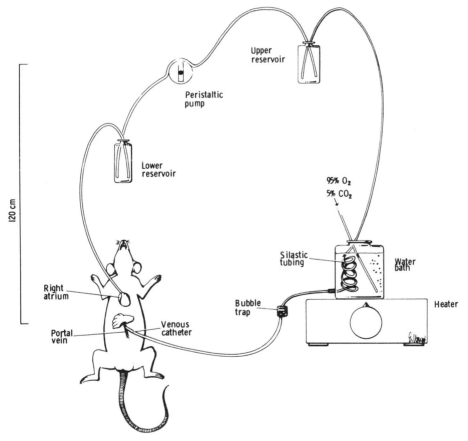

FIG. 1. Schematic representation of rat liver perfusion.

2. Anesthesize the rat using ether. It is important to maintain the respiration rate as close to normal as possible. Place rat on surgical box with rubber bands on front paws (an old polystyrene tube rack works well). Soak entire body with 50% ethanol.

3. Open the abdominal cavity via a midline incision (being careful not to open the chest cavity). The intestines are moved to the right side, exposing the hepatic portal region. A test tube is then placed under the back. This allows unobstructed access to the portal vein. Three pieces of surgical 00 silk thread are placed around the portal vein. One is placed on the end distal to the liver and is tied to restrict the blood flow. Into the now-dilated portal vein is inserted the intravenous catheter which has the Hanks' buffer flowing at a rate of one drop/3 sec. Once in the portal vein, the needle is removed from the catheter (being careful not to puncture the vein or letting the catheter come out of the vein). The catheter is then tied securely in place using the two remaining sutures. The male connector to the perfusion apparatus is attached to the catheter. From the time the portal vein is punctured, it is important to maintain a flow of the Hanks' buffer through the portal vein. Do not hesitate attaching the perfusion tube to the catheter after pulling the needle out of the catheter! Do not let blood clot in the catheter!

4. Once the Hanks' buffer is flowing into the vein via the catheter, quickly open the chest cavity and cut the right atrium. A polyethylene cannula (PE 50) is placed into the atrium and should enter the hepatic vein area via the vena cava. (Be careful not to push the catheter into the liver!) The catheter is then tied tightly (00 silk) in place (to prevent the perfusion medium from escaping from around the catheter).

5. Hanks' buffer (total volume = 75 ml) is perfused at a rate of 10 ml/min. During this time the liver is rapidly blanched (within 1 min). Be careful not to let small gas bubbles occlude the perfusion tubing or the circulatory system! A major reason for poor digestion of the liver is obstructed perfusion due to blood clots and air emboli. If the liver is not clearly blanched and does not swell evenly, it is a good idea to start again with a new rat, or be satisfied with a poor yield of cells.

6. Just before the Hanks' buffer has been depleted add 75 ml of collagenase solution to the upper reservoir. It is a good idea to allow an air bubble to enter the upper reservoir tubing in order to determine when the collagenase solution has entered the liver. (The air bubble will be trapped by the bubble trap which is placed proximal to the liver.) Once the collagenase solution has entered the liver begin collecting the perfusate by placing the catheter coming out of the atrium into the lower reservoir. The peristaltic pump is then turned on. The rate of return of the perfusate to

the upper reservoir should be fast enough not to allow the reservoir to be depleted.

7. After 10–15 min (depending upon the particular liver and the collagenase used) the liver should become digested, evidenced by a loss of its rigidity and a freeing of the cells. This is a judgment which will require some practice. We test the liver by gently touching it with the blunt side of sterile forceps. Do not allow surgical instruments (sterilized by soaking in 50% ethanol) to drip the ethanol on the liver. Rinse first with sterile Hanks' solution. If the liver capsule offers no resistance and cells are movable under the capsule, the liver is well digested. If the liver is poorly digested after a total of 15 min, start another rat. Either increase the collagenase concentration in the perfusion medium or try a different preparation of collagenase.

8. After the liver is judged to be well digested, cut it out and place in a sterile Petri dish. The liver capsule is cut up sufficiently to allow the cells to be gently shaken off. The liver is then scraped into a sterile culture flask and 20 ml of the collagenase solution is added. The flask is then shaken (without allowing foaming to occur) using a water bath shaker for 10 min at 37°. The individual liver cells are then drawn off using a sterile culture pipet (10 ml) with gentle suction. Be careful not to draw up connective tissue. The cells are added to large round-bottom sterile centrifuge tubes (50 ml volume) which contain 15 ml of plating medium (precooled to 4°). Two centrifuge tubes are sufficient for each hepatocyte preparation. Be careful not to damage cells while pipetting—let the cells drain from the pipet via gravity.

9. Hepatocytes are separated from other cell types by low-speed centrifugation (2 min at 500 rpm using a refrigerated IEC centrifuge at 4°). The supernatant is carefully poured off. The cell pellet (usually about 2 ml/tube) is dispersed by tapping and rotating the tubes. Plating medium (20 ml) is then added and the tubes are gently rotated until all of the cells are dispersed. It is important to break up the cell pellet before adding the plating medium or the cells will not disperse properly. The cells and the plating medium are combined into one centrifuge tube. The centrifugation–washing procedure is repeated two more times.

10. The final cell pellet is brought up in 15 ml of plating medium and placed on ice. Cells are counted using a hemocytometer in the presence of 0.4% Trypan Blue (i.e., 0.1 ml cells, 0.1 ml 4% Trypan Blue, and 0.8 ml of plating medium). Greater than 90% of the cells should exclude the dye during the first 5 min. More importantly, cells should look round and there should be minimal contamination with broken cells and connective tissue. It is our experience that the best morphological criterion of cell viability is

the ability to rapidly attach to the culture dish and spread out. This should occur within 2 hr of plating. After counting, the cells should be brought to a concentration of 1.3×10^6 cells/ml. As a reference for plating density, cells are plated on a 60-mm-diameter culture dish with 3 ml of suspended cells. This changes proportionately depending on the size of plating dishes to be used. It is important during plating to constantly (but gently) swirl the bottle containing the suspended cells because they settle to the bottom quite rapidly. Also, before placing the culture dishes in the incubator, the cells should be evenly dispersed on the culture dish with gentle swirling. Even plating of cells is essential and can be judged by using an inverted phase microscope.

11. After 4 hr, the culture medium is changed to the serum-free feeding medium and the experiment is begun. It is always a good idea to observe the cells at various times after plating to get an idea of their quality. (Figure 2 shows a good preparation of hepatocytes cultured for 24 hr.) It is important when adding various chemicals to the culture medium to periodically examine the basic morphology of the cells via phase contrast microscopy.

Experimental Procedures

Cultured hepatocytes can be used to study several aspects of VLDL assembly/secretion. We will describe different experimental protocols that have been used to study hepatic lipoprotein synthesis.

Lipogenesis

Using the appropriate methods, a single 60-mm plate containing 3.9×10^6 cells can be used to study lipogenesis and lipoprotein secretion. Both mass production as well as *de novo* synthesis (using radiolabeled precursors) can be quantitated. We have used [^{3}H]glycerol to measure glycerolipid synthesis and both [^{14}C]acetate and ^{3}H$_2$O to measure the synthesis of all lipid classes.[13] The advantage of using ^{3}H$_2$O is that many of the problems associated with differences in intracellular substrate pool sizes are avoided. Moreover, since ^{3}H$_2$O is not used up, its incorporation into lipids is a reliable measure of *de novo* synthesis.

For labeling studies using [^{3}H]glycerol or [^{14}C]acetate the following procedure is used. Four hours after plating the cells, the medium is

---

[13] R. A. Davis, M. M. McNeal, and R. L. Moses, *J. Biol. Chem.* **257**, 2634 (1982).

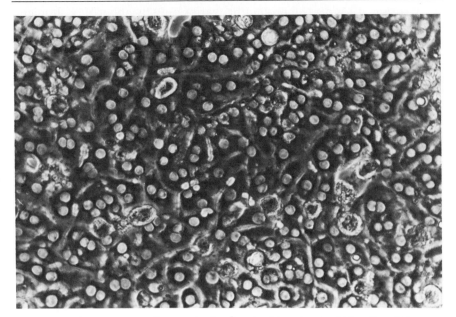

FIG. 2. Phase contrast micrograph of cultured rat hepatocytes 24 hr after plating.

changed to the serum-free feeding medium. The appropriate amount of [2-^{3}H]glycerol (10 μCi, final concentration 0.2 nM) or [^{14}C]acetate (3 μCi, final concentration 0.1 mM) is added. After the desired time of incubation (depending on the questions examined) at 37° in a 5% CO_2 incubator, the dishes are removed. Once the culture dishes are removed from the incubator, it is important to harvest the cells and medium as quickly as possible. The culture medium is drawn off using Pasteur pipets. The medium is centrifuged at 4° (2000 rpm) to sediment broken cells and debris. The organic solvents used for extracting the lipids is added to the supernatant. We usually use $CHCl_3$: MeOH (2 : 1, v/v).

The cells are scraped from the dish into a centrifuge tube using a rubber policeman and phosphate-buffered saline (3 ml). The plates are rinsed further with an additional 3 ml of phosphate-buffered saline. The cells are centrifuged at 4° (2000 rpm) and the supernatant is discarded. The pellet is brought up into 1 ml distilled water and disrupted with a sonicator (cell disruptor, Kontes Glass Co., Vineland, NJ). A portion of the cells are assayed for protein content (Lowry assay) and the remaining are extracted with the organic solvent.

The rate of triacylglycerol (mass) secretion by hepatocytes from chow-fed rats is quite reproducible by several different laboratories. Although different strains and suppliers of rats were used, among different labs this rate, 1 μg/mg cell protein/hr, varies only by about 50%. It might be a good idea to use this rate as one criterion of function when setting up the hepatocyte culture system.

The incorporation of the label into lipids is determined following thin-layer chromatography separation.[2] We usually bring the lipid extracts up in 0.3 ml CHCl$_3$ and spot half on a silica gel G plate. The remaining lipid can be used for quantitating lipid mass. We can determine both the relative rate of *de novo* synthesis and the mass content of phospholipid, free and esterified cholesterol, and triacylglycerol using these methods.[2,13]

There are several important differences in techniques when ^{3}H$_2$O is used. First, because the label is in a volatile form, care is needed to ensure that the radioactivity is confined. For this reason we plate 7.8×10^6 cells onto 25 cm^2 tissue culture flasks that have sealable caps. Four hours after plating the medium is changed to the serum-free feeding medium which contains ^{3}H$_2$O (1 mCi/ml, 2 ml/dish). The flasks are gassed with 95% O$_2$, 5% CO$_2$, and tightly sealed. The dishes are placed on a plastic tray and then put into a plastic bag and sealed with tape. The procedures are performed in a hood that has a good exhaust system. The bag is then placed in the incubator.

Harvesting a ^{3}H$_2$O experiment is roughly the same as described above except it is performed in the exhaust hood. The bag containing the culture flasks is opened in the exhaust hood. The medium is drawn off and placed in screw-capped glass test tubes. The cells are stripped off of the dishes using trypsin.[10] The lipid extraction procedure is the same as that described above. The water layers are combined and placed in sealed bottles for disposal by the radiation safety department. Assays (mass and radioactivity incorporation) are performed as described above.

When it is desired to examine the secretion of lipids in certain lipoprotein density classes, the culture medium can be subjected to ultracentrifugation using standard techniques. We also have separated lipoproteins secreted into the culture medium via size using agarose gel permeation column chromatography. We have found that VLDL is the major lipoprotein secreted into the culture medium. We have not been able to isolate significant amounts of lipoproteins in the "LDL" (i.e., $d = 1.02–1.05$ g/ml) density class from a medium of hepatocytes obtained from chow-fed rats.[2] Under normal conditions, there is almost no triacylglycerol lipase activity expressed by our hepatocyte culture model.[2] It is likely that the isolation of LDL in the culture medium, as reported by Bell-Quint et al.,[3] might indicate the presence of a lipase in their culture model.

Isolation of a $d = 1.05–1.21$ g/ml "HDL" fraction yields a significant amount of protein and lipid. A protein having a molecular weight similar to apoA-I is the predominant protein present in this fraction. However, we believe that this is not a normal secretory product of the cultured hepatocyte. If we first wash the cells with serum-free culture medium and then isolate the "HDL" fraction, >90% of the "A-I" protein disappears. We believe that a large amount of this "A-I" protein is due to a component in the calf serum-containing medium that has associated with the cell surface. More work needs to be done on the question of "HDL" secretion by cultured hepatocytes.

Apolipoprotein Synthesis

The cultured hepatocyte model provides many obvious advantages for studying apolipoprotein synthesis. Although a single plate of cells represents less than 1% of the total liver, the availability of high specific radioactivity amino acid precursors provides adequate sensitivity for determining relative rates of *de novo* apolipoprotein synthesis. We have used several different radioactive precursors. The advantage of using [^{35}S]methionine is that it produces definitive fluorograms in a relatively short time. Fluorograms are easy, accurate, and inexpensive to assay in regard to electrophoretic separation of proteins. There are other advantages offered by ^{3}H-labeled amino acid mixture. ^{3}H-labeled amino acid mixture is biosynthetically incorporated into most residues irrespective of the amino acid sequence. The increased incorporation of several amino acid residues (obtained from the ^{3}H-labeled amino acid mixture) allows one to use a lower specific radioactivity. The ^{3}H-labeled amino acid mixture can be added to the complete feeding medium. In other words, the ^{3}H-labeled amino acid precursors are available at substrate concentrations, rather than as a radioactive pulse. Therefore, problems associated with differences in amino acid pool sizes are to a large degree avoided.

The procedures for using both isotopes are similar. Four hours after plating, the medium is changed to the serum-free feeding medium (^{3}H-labeled amino acid mixture) or to methionine-free, serum-free feeding medium ([^{35}S]methionine). The labeled amino acid precursor is added. The amount added depends upon the length of the incubation and the experimental design. For most studies we add 200 μCi/2 ml/60-mm culture dish of ^{3}H-labeled amino acid mixture or 100 μCi/2 ml/60-mm culture dish of [^{35}S]methionine. The incorporation of ^{3}H-labeled amino acid mixture into apolipoproteins is linear for at least 24 hr, whereas the incorporation of [^{35}S]methionine into apolipoproteins is linear for at least 8 hr (we have not done time course experiments longer than these times). After the appropriate time of incubation in a CO_2 incubator, the dishes are removed and harvested. Apolipoproteins are isolated either by ultracentrifugation or by immunoprecipitation.

If it is desired to examine the apolipoprotein content of specific density classes of lipoproteins, the culture is drawn off as described above (for the lipid biosynthetic studies). The medium is brought to the proper density by adding the appropriate amount of KBr. EDTA (0.01%) and NaN_3 (0.01%) are added preservatives. The lipoproteins are isolated by ultracentrifugation. The individual lipoprotein fractions are dialyzed against ammonium bicarbonate (5 mM), and NaN_3 and EDTA (both 0.01%), pH 7.4, for 48 hr (four changes of 4 liters each). The fractions are then lyophilized. The residues are brought up in sample buffer containing 8 M urea, 2% SDS, 10% glycerol, 5 mM mercaptoethanol, and 10 mM Tris/glycine, pH 8.3, and boiled for 10 min. The apolipoproteins are separated by SDS/PAGE using a 1–20% linear gradient of acrylamide and the running buffer system of Laemmli.[14]

Immunoprecipitation is quicker and easier to perform than is the procedure detailed above. After incubating hepatocytes with the desired labeled amino acid precursors, the medium is drawn off as described above. The cells are immediately lysed by adding boiling buffer B.[15] The medium is diluted 2-fold with buffer B and boiled for 5 min. Additional buffer B (without SDS) is added to bring the final SDS concentration to 0.15%. Antibody (purified IgG fraction) is added. (It is important to first determine the proper amount of antibody to add in order to obtain quantitative recoveries.) After mixing for 16 hr at room temperature using a shaking platform, an appropriate amount of protein A–Sepharose is added. After two more hours of mixing, the beads are centrifuged in a microfuge (5 min). The pellet is washed six times with buffer D.[15] The beads are brought up in 25 μl of the SDS/PAGE sample buffer and subjected to electrophoresis as described above. The SDS/PAGE gels are examined by fluorography ([^{35}S]methionine) and by cutting the individual apolipoprotein bands out of the gels, depolymerizing in H_2O_2, digesting with a tissue solubilizer and assaying by scintillation counting.

Additional Procedures

The amount of lipoproteins secreted by a single plate of hepatocytes is sufficient to perform many more analyses than those described above. Mass measurement of apolipoproteins can be achieved using several immunochemical techniques. Lipoprotein particles can be negatively stained and viewed by electron microscopy. With the development of

[14] U. K. Laemmli, *Nature (London)* **227**, 680 (1970).
[15] J. R. Faust, K. L. Luskey, D. J. Chin, J. L. Goldstein, and M. S. Brown, *Proc. Natl. Acad. Sci. U.S.A.* **227**, 5205 (1982).

methods which can detect proteins and lipids in a sensitive and precise manner, determining the lipid and apolipoprotein composition and structure of lipoproteins secreted by cultured rat hepatocytes should not present major problems to the investigator.

Conclusion

The cultured hepatocyte system can be a valuable model for investigation of lipoprotein synthesis and secretion. Many experiments not possible with other models can be performed using this system. For some questions, data obtained from the hepatocyte model may be difficult to interpret in regard to the *in vivo* state. For this reason hypotheses derived from data obtained using the hepatocyte culture system should be examined further *in vivo*.

There is no reason to assume that differences between cultured hepatocytes and the liver *in vivo* exist in regard to the intrahepatic processes which regulate lipoprotein synthesis and secretion. However, the environmental, hormonal, and nutritional conditions in culture are markedly different than those which occur *in vivo*. Understanding the basis of these differences is by itself valuable to our knowledge of physiology. Depending upon the questions examined, these differences either can be used to aid the investigator or may be prohibitive in affording physiologically relevant conclusions.

[18] Studies of Nascent Lipoproteins in Isolated Hepatic Microsomes and Golgi Cell Fractions

By Debendranath Banerjee

Plasma lipoproteins are lipid–protein aggregates which are responsible for the transport of lipids in blood. The four major classes of lipoproteins have been described in detail by Gotto *et al.* in Chapter [1], Volume 128. These are chylomicrons ($d \leq 0.95$ g/ml; >100 nm in diameter), very-low-density lipoproteins (VLDL; $d = 0.95–1.006$ g/ml; $30–80$ nm in diameter), low-density lipoproteins (LDL; $d = 1.006–1.063$ g/ml; 28 nm in diameter), and high-density lipoproteins (HDL; $d = 1.063–1.21$ g/ml; $8–12$ nm in diameter). Each lipoprotein class consists of a layer of phospholipids, free cholesterol, and apolipoproteins surrounding a neutral lipid core composed mainly of cholesterol esters and triglycerides (see reviews).[1-4] The apolipoproteins and some of the lipid components of all the plasma lipoproteins are synthesized in the hepatocytes and in the absorptive cells of the small intestine.[5,6] Two of the classes of lipoproteins, VLDL and HDL, are secreted by the hepatocytes. Available evidence suggests that most of LDL is generated by the catabolism of VLDL in the plasma.[5] Morphological,[7] biochemical,[8,9] and cell fractionation[10] studies indicate that many of the intracellular organelles and processes involved in the synthesis and secretion of plasma proteins, such as albumin and transferin, also are involved in the synthesis and secretion of VLDL. There is, however, little detailed information of the intracellular sequence of events which leads to the formation of lipoproteins.

Almost all published procedures for the isolation of lipoproteins from intracellular organelles involve the isolation of partially enriched Golgi cell fractions with subsequent isolation of VLDL particles from their contents by sonication.[11-13] These VLDL particles may represent a mixture of nascent VLDL and of VLDL and chylomicron remnants that are taken up by the hepatocytes. Our approach to studying the sequence of lipoprotein assembly by liver cells has been to radiolabel either the protein or lipid moieties *in vivo* by the administration of appropriate radioactive precursors, and then at various time intervals, at which nascent lipoproteins are within the hepatocytes, to fractionate the liver into its intracellular organelles and to isolate the newly synthesized lipoproteins. In this chapter I review some of the procedures that have been used for the isolation and characterization of nascent lipoproteins from the subcellular organelles involved in plasma protein secretion obtained from liver.

[1] R. L. Jackson, J. D. Morrisett, and A. M. Gotto, Jr., *Physiol. Rev.* **56**, 259 (1976).
[2] L. C. Smith, H. J. Pownell, and A. M. Gotto, Jr., *Annu. Rev. Biochem.* **47**, 751 (1978).
[3] G. S. Getz and R. V. Hays, *in* "The Biochemistry of Atherosclerosis" (A. M. Scanu, R. W. Wissler, and G. S. Getz, eds.), p. 151. Dekker, New York, 1979.
[4] M. J. Chapman, *J. Lipid Res.* **21**, 789 (1980).
[5] R. L. Hamilton, *in* "Plasma Protein Secretion by the Liver" (H. Glaumann, T. Peters, Jr., and C. Redman, eds.), p. 357. Academic Press, New York, 1983.
[6] P. H. Green and R. M. Glickman, *J. Lipid Res.* **22**, 1153 (1981).
[7] A. Claude, *J. Cell Biol.* **47**, 745 (1970).
[8] J. J. Bungenberg and J. B. Marsh, *J. Biol. Chem.* **243**, 192 (1968).
[9] A. C. Nestruck and D. Rubinstein, *Can. J. Biochem.* **54**, 617 (1976).
[10] H. Glaumann, A. Bergstrand, and J. L. E. Ericsson, *J. Cell Biol.* **64**, 356 (1975).
[11] R. W. Mahley, R. L. Hamilton, and V. S. LeQuire, *J. Lipid Res.* **10**, 433 (1967).
[12] R. W. Mahley, T. P. Berson, V. S. LeQuire, R. I. Levy, H. G. Windmueller, and W. V. Brown, *Science* **168**, 380 (1970).
[13] M. J. Chapman, G. L. Mills, and C. E. Taylour, *Biochem. J.* **128**, 779 (1972).

The emphasis is on techniques that have been useful in our studies of HDL synthesis and secretion by the liver of young chickens.

In Vivo Labeling of Apolipoprotein and Lipid Moieties

Young chickens (5–10 days old) are used in our studies. However, the procedures described are applicable to other small experimental animals. Nascent proteins and/or lipids are labeled by administration of radioactive amino acids and/or lipid precursors into whole animals. An essential amino acid, L-leucine, which is not easily metabolized by the liver, is commonly used to radiolabel proteins. Lipids are labeled by administration of radioactive choline, fatty acids such as palmitic acid and oleic acid or glycerol. About 10–20 μCi/100 g body weight is usually administered intravenously in any one of the following veins: femoral, portal, or saphenous of rats and wing or jugular veins of chickens. More than 98% of the injected radioactivity is cleared from the blood within 2–5 min after the intravenous administration of radioactive precursors.

The length of time required for a protein to be secreted from the cell is usually called the intracellular transit time and it is measured by determining the time of appearance of radioactive proteins in the blood.[14] For most plasma proteins there is a lag of at least 15 min after the intravenous administration of radioactive amino acids and the appearance of radioactive plasma proteins in the blood and thereafter the radioactive plasma proteins continue to rise linearly for 40–60 min. The radioactive curve will intersect the time scale, if extended toward the abscissa, and the point of intersection constitutes the intracellular transit time. This transit time varies for different plasma proteins. In young chickens the intracellular transit time of apolipoprotein A-I, a major protein component of HDL, is 16 to 18 min.

Preparation of Microsomes and Golgi Cell Fractions

Solutions. The solutions are prepared using deionized double distilled water, filtered through a 0.45-μm Millipore filter and stored in the cold.

1. Homogenization solution: 0.25 M sucrose in water neutralized to pH 7.0
2. Sucrose: 2.1, 1.15, 0.85, and 0.6 M prepared in double distilled water

[14] C. M. Redman and D. Banerjee, *in* "Plasma Protein Secretion by the Liver" (H. Glaumann, T. Peters, Jr., and C. Redman, eds.), p. 285. Academic Press, London, 1983.

Methodology. The majority of published procedures for the isolation of rough microsomes, smooth microsomes, and the Golgi apparatus from rat liver utilize discontinuous sucrose gradient centrifugation.[15–18] We have adapted this well-established procedure to the fractionation of chicken liver. Young chickens are decapitated using a guillotine (Harvard Apparatus Co., Natick, MA). After opening the body cavity, the vena cava is cut above and below the liver and blood is allowed to drain from the liver. The liver is excised and placed in ice-cold 0.25 M sucrose. All subsequent steps are carried out with the solutions immersed in ice. The liver lobes are blotted on filter paper (Whatman #4), weighed, and placed into 5 vol (w/v) of homogenization solution. The liver is minced using a pair of scissors and transferred to a 50-ml Potter-Elvehjem glass–Teflon-type tissue grinder (size C tissue grinders, A. H. Thomas, Philadelphia, PA). The minced tissue is homogenized with four or five vertical strokes of the Teflon pestle. The clearance is 0.008 in. and the pestle is driven at 1000–1500 rpm by a motor equipped with speed control and tachometer (Talboys Engineering Corp., Emerson, NJ). The homogenization procedure requires about 5–10 sec. The homogenate is filtered through two to four layers of cheesecloth (Johnson & Johnson Products, Inc., New Brunswick, NJ).

The filtrate is centrifuged at 9500 rpm for 10 min (Sorvall SS-34 rotor; RC-2B centrifuge; 10,800 g). The pellet containing cell debris, nuclei, and mitochondria is discarded. The supernatant fraction containing microsomes, Golgi complex elements, and soluble proteins is centrifuged at 40,000 rpm for 90 min (Beckman, 60 Ti rotor; L2-75B centrifuge; 105,000 g_{av}). The supernatant fraction contains soluble cytoplasmic proteins. The pellet, which contains microsomes and the Golgi apparatus, is resuspended in a small volume of 0.25 M sucrose using three to four strokes of a loosely fitting motor-driven Teflon pestle and is diluted with an equal volume of 2.1 M sucrose to obtain a final concentration of 1 g (wet wt) tissue equivalent per 1.5 ml of suspension and a refractive index of 1.392, which is equivalent to that of 1.21 M sucrose.

The procedure we have adopted for the isolation of Golgi complex from the microsomal suspension is very similar to that described by Ehrenreich *et al.,*[19] but contains a modification which results in better

[15] M. R. Adelman, G. Blobel, and D. D. Sabatini, this series, Vol. 19, p. 201.

[16] D. J. Morre, this series, Vol. 3, p. 130.

[17] D. E. Leelavathi, L. W. Estes, D. S. Feingold, and B. Lombardi, *Biochim. Biophys. Acta* **211,** 124 (1970).

[18] B. Fleischer and S. Fleischer, *Biochim. Biophys. Acta* **219,** 301 (1970).

[19] H. J. Ehrenreich, J. J. M. Bergeron, P. Siekevitz, and G. E. Palade, *J. Cell Biol.* **59,** 45 (1973).

separation of the more dense Golgi fractions from the smooth endoplasmic reticulum.[20] Ten-milliliter aliquots of the adjusted total microsomal fraction are loaded in the bottom of nitrocellulose tubes (SW-27 rotor, Beckman Instrument Co.) and discontinuous gradients, made by layering 8.5 ml each of 1.15, 0.86, 0.6 M sucrose, are constructed over it. Finally 0.25 M sucrose is layered on the top and this is also used to balance the tubes and then they are centrifuged at 25,000 rpm for 3.5 hr (Beckman SW 27 rotor; L5-50B centrifuge; 82,500 g_{av}). The membranous materials that float on the 0.25 M/0.6 M, 0.6 M/0.85 M, and 0.85 M/1.15 M sucrose interfaces are collected and termed light, intermediate, and heavy Golgi complex fractions.

The remainder of the materials in the bottom portion of the gradient is collected and sufficient 2.1 M sucrose is added to obtain a refractive index of 1.396, which is equivalent to that of 1.3 M sucrose. Rough and smooth microsomes are then fractionated from this mixture by the procedure of Ragland et al.[21] A 20-ml portion of the adjusted fractions is layered over 10 ml of 2 M sucrose and overlaid with 5 ml of 0.6 M sucrose in a polycarbonate 60 Ti rotor tube. This discontinuous gradient is centrifuged at 35,000 rpm for 18 hr. The smooth microsome fraction is collected from the 1.3/0.6 M sucrose interface and the rough microsomal fraction is collected from the boundary between the 1.3 and 2 M sucrose layer.

Criteria of Purity. The organelles are characterized by both morphological and biochemical methods

Morphological—electron microscopy.[22] Fixatives:

1. 4% formaldehyde–glutaraldehyde mixture in 0.2 M sodium cacodylate–HCl buffer, pH 7.4: This fixative is prepared by placing 1 g of paraformaldehyde powder (Fisher Chem. Co.) and 5 ml of distilled water in an Erlenmeyer flask. The flask is warmed on a hot plate (60–70°) for a few minutes. When water condensation appears in the neck of the flask, one to three drops of 1 N NaOH is added and stirred to clear the solution. This solution is then cooled in an ice bucket and 2 ml of 50% glutaraldehyde (w/v) and 6.25 ml of 0.8 M sodium cacodylate buffer, pH 7.4, is added. The final volume is adjusted to 25 ml with distilled water.

2. 2% osmium tetroxide prepared in 0.1 M cacodylate buffer, pH 7.4.

Procedures: Microsomal and Golgi cell fractions should be fixed immediately after isolation from the sucrose gradient by mixing the sample with an equal volume of a freshly prepared 4% formaldehyde–glutaralde-

[20] C. M. Redman, D. Banerjee, K. Howell, and G. E. Palade, *J. Cell Biol.* **66**, 42 (1975).
[21] W. L. Ragland, T. K. Shires, and H. C. Pitot, *Biochem. J.* **121**, 271 (1971).
[22] A. M. Glauert, "Practical Methods in Electron Microscopy," Vol. 3, Part I. North Holland Publ., Amsterdam, 1975.

hyde mixture. After 30 min fixation in an ice bucket the cell fractions are collected by centrifugation at 25,000 rpm for 20 min (Beckman SW-50.1 rotor; L2-75B centrifuge; 37,500 g_{av}). The pellets obtained are postfixed with 2% osmium tetroxide for 60 min, stained en bloc with 2% aqueous uranyl acetate, dehydrated with graded alcohol followed by propylene oxide, infiltrated, and embedded in Epon. Thin sections are cut through the entire depth of one pellet, stained with uranyl acetate and lead citrate, and examined in a Philips 410 microscope.

Biochemical. The yield and the purity of the cell fractions are further established by determining the amount of RNA and protein and the enzymatic activities of glucose-6-phosphatase and N-acetylglucosamine galactosyltransferase. Glucose-6-phosphatase is determined by the procedure of Swanson[23] and UDP-galactose : N-acetylglucosamine galactosyltransferase activity is determined by the method of Babad and Hassid[24] as outlined by Fleischer.[25] RNA is determined by the orcinol method[26] and protein is measured as described by Lowry *et al.*[27] using bovine serum albumin as standard. Most of the galactosyltransferase activity is recovered in the Golgi cell fractions and glucose-6-phosphatase is localized in both the rough and smooth microsomal fractions. The ratio of RNA to protein should be markedly increased in the rough microsomal fractions relative to that of Golgi cell fractions.

Contamination of the microsome and Golgi preparation with other membranous organelles can be estimated using other enzyme assays, such as Mg^{2+}-stimulated ATPase[28] for plasma membrane, succinate cytochrome c reductase[28] for mitochondria, and acid phosphatase[29] or aryl sulfatase[30] for lysosomes.

Subfractionation of Microsomes and Golgi Cell Fractions

There are three major groups of secretory serum proteins found in the lumen of the microsomal and Golgi vesicles. They are serum albumin,[31]

[23] M. A. Swanson, *J. Biol. Chem.* **154,** 647 (1950).

[24] H. Babad and W. Z. Hassid, this series, Vol. 8, p. 346.

[25] B. Fleischer, this series, Vol. 31, p. 180.

[26] W. Mejbaum, *Z. Phys. Chem.* **258,** 117 (1939).

[27] O. H. Lowry, N. J. Rosebrough, A. L. Farr, and R. J. Randall, *J. Biol. Chem.* **193,** 265 (1951); see also this series Vol. 3, p. 448.

[28] S. Fleischer and B. Fleischer, this series, Vol. 10, p. 406.

[29] C. B. deDuve, B. C. Pressman, R. Gianetto, R. Wattiaux, and F. Appelmans, *Biochem. J.* **60,** 604 (1955).

[30] A. B. Roy, *Biochem. J.* **53,** 12 (1953).

[31] T. Peters, Jr., B. Fleischer, and S. Fleischer, *J. Biol. Chem.* **246,** 240 (1971).

serum glycoproteins,[32,33] and serum lipoproteins, particularly VLDL.[8-10] A variety of procedures have been used to release these secretory proteins; they include disruption by mechanical force, such as sonication[34] and french press[19]; treatment with 0.15 M Tris–HCl, pH 8.0, followed by resuspension and incubation at 30° for 15 min[35]; and hydrophobic washing[36] and treatment with low concentrations of detergents.[37] The methodology for the use of low concentration of detergents for release of plasma proteins from microsomal vesicles has been described in detail by Kreibich and Sabatini in Volume 19 of this series.[38] Ehrenreich et al. have also summarized a variety of procedures to release the content of Golgi vesicles.[19]

Recently Fujiko et al.[39] and Howell and Palade[40] released the contents by treating microsomal and Golgi cell fractions, obtained from rat liver, with ice-cold Na_2CO_3, pH 11.3. The organelles are suspended in ice-cold 100 mM Na_2CO_3, pH 11.3, to a final protein concentration of 0.02 to 1 mg/ml. The suspensions are then incubated at 0° for 30 min and centrifuged at 4° for 60 min at 40,000 rpm (Beckman 50 Ti rotor; 106,500 g_{av}). The supernatant fractions contain the "contents" of the microsomes and Golgi apparatus.

In our laboratory, chicken liver microsomes and Golgi cell fractions are disrupted as detailed by Fleischer in Volume 31 of this series.[25] The cell fractions (1–2 mg protein/ml) are suspended in 2 vol of 2 M NaCl and 7 vol of 0.2 M NaHCO$_3$. The mixture is loaded in a Parr Cell Disruption Bomb Model No. 4635 (Parr Instruments Co., Moline, IL), compressed at 1800 psi of N_2 for 15 min and released dropwise from the chamber. The disrupted mixture is adjusted to 0.25 M sucrose and the membranes were collected as a pellet by centrifugation at 35,000 rpm (Beckman 60 Ti rotor; L5-50B centrifuge; 87,000 g_{av}) for 60 min. The supernatants, which contain the cisternal contents, are dialyzed against 10 mM Tris–HCl, pH 7.4, containing 154 mM NaCl, 1 mM EDTA, and 0.01% NaN_3 and used for flotation of lipoproteins. In microsomes, 15% of the total proteins and 60–70% of the apolipoprotein A-I are released by nitrogen cavitation. In the

[32] C. M. Redman and M. G. Cherian, J. Cell Biol. 52, 231 (1972).
[33] H. Schachter, I. Jabbal, R. L. Hudgin, L. Pinteric, E. J. McGuire, and S. Roseman, J. Biol. Chem. 245, 1090 (1970).
[34] P. R. Dallman, G. Dallner, A. Bergstrand, and L. Ernster, J. Cell Biol. 41, 357 (1969).
[35] P. O. Sandberg and H. Glaumann, Exp. Mol. Pathol. 31, 1 (1980).
[36] J. E. Elder and D. J. Morre, J. Biol. Chem. 251, 5054 (1976).
[37] G. Kreibich, P. Debey, and D. D. Sabatini, J. Cell Biol. 58, 436 (1973).
[38] G. Kreibich and D. D. Sabatini, this series, Vol. 19, p. 215.
[39] Y. Fujiki, A. L. Hubbard, S. Fowler, and P. B. Lazarow, J. Cell Biol. 93, 97 (1982).
[40] K. E. Howell and G. E. Palade, J. Cell Biol. 92, 822 (1982).

Golgi complex cell fraction 29% of the total proteins and 76% of the apolipoproteins are released.[41]

Isolation of Apolipoproteins from Cell Fractions

Solubilization of the Organelles

The microsomal fractions and the Golgi cell fractions are resuspended in 0.01 M sodium phosphate buffer, pH 7.4, containing 0.15 M NaCl, and sodium deoxycholate and Triton X-100 are added to a final concentrations of 0.5% each. At this detergent concentration, ~90% of the total acid-insoluble protein radioactivity sequestered within or associated with the microsomes and Golgi cell fractions is released. The samples are then cleared of nonsolubilized material by centrifugation at 40,000 rpm (Beckman 50 Ti rotor; L5-50B centrifuge; 105,000 g_{av}) for 60 min. The clear supernatants are used for immunochemical isolation of apolipoproteins.

Immunochemical Isolation of Apolipoproteins

Preparation of Antisera. Antisera are raised in New Zealand white rabbits (1500 g body weight). A week prior to immunization the rabbits are bled (20–25 ml) and the preimmune sera are saved. Purified lipoproteins, namely HDL, VLDL, and LDL, can be used as antigens. If purified rooster plasma HDL–apolipoprotein A-I is used it is dialyzed against 0.154 M saline. One-milliliter aliquots (0.6 mg to 1 mg/ml protein) are emulsified with 1 ml of Freund's complete adjuvant (Miles Laboratories, IL); 0.2-ml portions are injected intradermally in the back near the shoulders. A booster injection of 0.3 to 0.5 mg of apolipoproteins emulsified with Freund's incomplete adjuvant is given subcutaneously in subsequent weeks and, if necessary, repeated for an additional 2–3 weeks. The rabbits are first bled 2 months after the primary immunization. Preimmune and immune sera are obtained by allowing the blood to clot for several hours at room temperature; the clot is retracted at 4° followed by centrifugation to remove the residual red blood cells.

Isolation of Immunoglobulin G. The antisera are dialyzed against 0.015 M potassium phosphate, pH 8.0, and applied to a DEAE–cellulose column (1 ml serum/5 ml of DEAE–cellulose) equilibrated with the same buffer.[42] The void volume is collected, titered, and stored at −20° in small aliquots as crude immunoglobulin (IgG) fraction.

---

[41] D. Banerjee and C. M. Redman, *J. Cell Biol.* **96,** 651 (1983).
[42] E. A. Peterson and H. A. Sober, *J. Am. Chem. Soc.* **78,** 751 (1956).

Isolation of Radioactive Nascent HDL Apolipoprotein. Aliquots (0.24–0.5 g tissue equivalent) of detergent-soluble supernatant obtained from the microsomes and Golgi apparatus are pipetted into three separate conical centrifuge tubes. To one tube add 100 μl of IgG obtained from preimmune serum. To the remaining tubes add 100 μl of anti-HDL apolipoprotein IgG which is at a titer sufficient to quantitatively precipitate 15 μg of apolipoprotein A-I. To each tube add 0.1 ml of 10× diluted rooster serum and incubate at 37° for 60 min, and at 4° for 18 hr. The antigen–antibody precipitate is then processed according to the procedure of Taylor and Schimke.[43] The reaction mixture is layered on a discontinuous sucrose gradient consisting of 1 ml 1 M sucrose and 0.5 ml 0.5 M sucrose containing 0.01 M leucine, 1% Triton X-100, and 1% sodium deoxycholate and centrifuged at 10,000 rpm (Sorvall HB-4 rotor; RC-2B centrifuge; 15,000 g) for 10 min. The pellet containing the antibody–antigen complex is washed three times with 0.01 M phosphate buffer, pH 7.4, containing 0.154 M NaCl, 1% Triton X-100, 1% sodium deoxycholate, and 0.01 M appropriate unlabeled amino acid. The antibody–antigen complex is solubilized and analyzed by SDS–PAGE.[44,45] The gels are stained with Coomassie Blue and prepared for fluorography as described by Bonner and Laskey.[46]

Isolation of Nascent Lipoprotein Particles from Organelles

Ultracentrifugal Flotation

All centrifugations are performed at 8–10° in a Beckman L5-50B ultracentrifuge using either a 60 Ti or SW-41 rotor (Beckman Instruments, Inc., Palo Alto, CA). The density of all solutions is determined by refractive index measurements using a refractometer. All materials that float at densities ≤1.063 g/ml are considered to be in the VLDL–LDL class; those that float at densities between 1.063 and 1.21 g/ml are considered the HDL class, and those that remain heavier than 1.21 g/ml are considered to be other plasma proteins.[47]

Serum proteins and proteins present in the cisternal contents of the microsomes and the Golgi cell fractions are dialyzed against 10 mM Tris-HCl, pH 7.4, containing 154 mM NaCl, 1 mM EDTA, and 0.01% NaN_3.

[43] J. M. Taylor and R. T. Schimke, *J. Biol. Chem.* **248,** 7661 (1973).
[44] K. Weber and M. Osborn, *J. Biol. Chem.* **244,** 4406 (1969).
[45] U. K. Laemmli, *Nature (London)* **227,** 680 (1970).
[46] W. M. Bonner and R. A. Laskey, *Eur. J. Biochem.* **46,** 83 (1974).
[47] R. J. Havel, H. A. Eder, and J. H. Bragdon, *J. Clin. Invest.* **34,** 1345 (1955).

The density of the dialyzed serum, the microsomal content, and the Golgi content subfractions are then increased by addition of solid KBr to 1.063 g/ml. The suspensions are next loaded into ultracentrifuge tubes, centrifuged at 35,000 rpm (60 Ti rotor; L5-50B centrifuge; 87,000 g_{av}) for 20 hr, and the top 2-ml fractions collected are termed as VLDL–LDL particles. The density of the remainder of the sample is then increased to 1.21 g/ml by further addition of solid KBr, reloaded into centrifuge tubes, centrifuged at 40,000 rpm (60 Ti rotor; L5-50B centrifuge; 102,000 g_{av}) for 20 hr and the top 2-ml fractions collected are HDL particles. All lipoprotein classes are then further purified by recentrifugation at their upper density limits.

Methods have also been described for separating serum lipoproteins by a single ultracentrifugation in a discontinuous gradient.[48,49] In this method the serum samples are adjusted to a density of 1.25 g/ml with solid KBr. The density-adjusted serum is loaded into an ultracentrifuge tube and overlaid with 1.5 ml of salt solutions of densities equal to 1.10, 1.063, 1.02, and 1.006 g/ml. The gradients are centrifuged in a swinging bucket rotor (SW-41; Beckman Instrument Co., Palo Alto, CA) and the lipoprotein classes band at their respective densities. Both of these procedures have also been used successfully to float different classes of lipoproteins from the rat liver Golgi content.[50]

Biochemical Characterization

The VLDL–LDL and HDL classes, recovered from the serum and from the microsomal and Golgi content, are either dialyzed against 1 mM EDTA or concentrated after removing KBr using a PM 10 membrane attached in an Amicon ultrafiltration cell (Amicon Corp., Lexington, MA).

For protein analysis, the lipoproteins are delipidated by extracting with 5 vol of ice-cold diethyl ether : ethanol (2 : 3, v/v) followed by three extractions with diethyl ether. The lipid-free proteins are collected by centrifugation and analyzed by SDS–PAGE[44,45] and fluorography.[46]

For lipid analysis, the lipoproteins are extracted with chloroform–methanol (2 : 1, v/v) by the method of Folch et al.[51] The extracts are dried in vacuo, weighed, redissolved in chloroform, and separated into phospholipids, cholesterol, cholesterol esters, and triglycerides by thin-layer

[48] T. G. Redgrave, D. C. K. Roberts, and C. E. West, Anal. Biochem. 65, 42 (1975).
[49] J. R. Foreman, J. B. Karlin, C. Edelstein, D. J. Julin, A. H. Rubenstein, and A. M. Scanu, J. Lipid. Res. 18, 759 (1977).
[50] K. E. Howell and G. E. Palade, J. Cell Biol. 92, 833 (1982).
[51] J. Folch, M. Lees, and G. H. Sloan Stanley, J. Biol. Chem. 226, 497 (1957).

chromatography on 250-μm layers of Silica gel GF Uniplate (Analtech, Inc., Newark, DE) using hexane–diethylether–acetic acid (80 : 20 : 1, v/v) as a solvent system.[52]

The free cholesterol and cholesterol esters are scraped off the plates and quantitated according to the method of Searcy et al.[53] The phospholipids are recovered from the thin-layer plates and analyzed for phosphorus following the procedure of Bartlett.[54] The triglyceride content is obtained by subtracting the total amount of cholesterol, cholesterol esters, and phospholipids from the total lipid values previously determined by weighing the samples.

The phospholipids are further separated by two-dimensional thin-layer chromatography on 250-μm layers of Silica gel GF Uniplate (Analtech, Inc., Newark, DE). The solvent system used is choroform : methanol : 14.8 N ammonium hydroxide (65 : 25 : 5, v/v)[55] for the first dimension and chloroform : acetone : methanol : acetic acid : H_2O (50 : 20 : 10 : 10 : 5, v/v) for the second dimension.[56]

Morphological Characterization

The morphology of nascent lipoprotein classes can be studied in the electron microscope after negative staining or thin sectioning of the final pellet. For thin sectioning, lipoproteins are first fixed, pelleted, and then prepared by the same procedures that are used for blocks of cell fractions or tissue. However, by far the most widely used technique for examining the morphology of lipoprotein particles is negative staining.[57]

A drop of freshly prepared dialyzed lipoprotein fraction is deposited on a carbon-reinforced formvar-coated grid, held in fine jeweler's tweezers secured with a rubber band. After a few seconds the drop is removed and replaced by a drop of 2% aqueous solution of phosphotungstic acid (PTA) which has been neutralized to pH 6.5–6.7 with KOH. The sample is allowed to stand so that the lipoprotein particles can settle and bind to the surface of the grid. The drop is removed within a minute and the excess fluid is removed by touching the edge of the grid with a triangular wedge of Whatman No. 1 filter paper. Care must be taken to remove all fluids that might have traveled between the arms of the tweezer by

[52] V. Skipski and M. Barclay, this series, Vol. 14, p. 530.

[53] R. L. Searcy, L. M. Bergquist, and R. C. Jung, J. Lipid. Res. 1, 349 (1960).

[54] G. R. Bartlett, J. Biol. Chem. 234, 466 (1959).

[55] G. Rouser, S. Fleischer, and A. Yamamoto, Lipids 5, 494 (1970).

[56] T. Kremmer, M. H. Wisher, and W. H. Evans. Biochim. Biophys. Acta 455, 655 (1976).

[57] R. J. Haschmeyer and R. J. Myers, in "Principles and Techniques of Electron Microscopy" (M. A. Hayat, ed.), Vol. 2, p. 99. Van Nostrand-Reinhold, Princeton, New Jersey, 1970.

capillary action. The grids are allowed to dry thoroughly (5 to 10 min) and then they are examined in an electron microscope operating at 80 kV. The entire grid is surveyed at low magnification and the areas where the particles are well spread are chosen for detailed examination and photography. If the material is poorly distributed on the surface of the grid, fresh grids should be used.

Intracellular Localization

By Immunoelectron Microscopy

Treatment of a liver specimen with an antibody prepared against specific proteins permits localization of the antigen if peroxidase, an enzyme found in plants (horseradish), is covalently conjugated to IgG molecules.[58] The location of the peroxidase is revealed by incubating the specimens in a solution of H_2O_2 and an electron donor such as 3,3′-diaminobenzidine (DAB).[59] An oxidized polymer of DAB precipitates at its site of formation and is rendered electron opaque during fixation with osmium tetroxide. Alexander et al.[60] used this approach and localized apolipoprotein B in rat hepatocytes. Antisera against apoLDL and apoVLDL proteins have been raised in New Zealand white rabbits and the immunoglobulins (IgG) are precipitated from the serum at 4° with saturated solution of ammonium sulfate, pH 6.5, and purified on DEAE–cellulose (Whatman DE-52, W & R, Balston, Ltd., Springfield Mills, Kent, England) by a batch elution.[61] The purified IgG is then digested with papain,[62] applied to a carboxymethyl cellulose column (Cellex CM, Bio-Rad Laboratories, Richmond, CA) and the FAB fragments are eluted with step gradients of acetate buffers of increasing ionic strength.[63] Horseradish peroxidase (Worthington) is conjugated to the FAB fractions by the two-step method of Avrameas and Ternynck.[64]

For electron microscopic localization of apolipoprotein B, rats are starved overnight and anesthetized by intraperitoneal injection of 0.3 ml sodium methohexital (Eli Lilly & Co., Indianapolis, IN) per 100 g body weight. Blood is flushed from the liver by cutting the abdominal vena cava

[58] L. A. Sternberger, "Immunocytochemistry." Wiley, New York, 1979.
[59] R. C. Graham and M. J. Karnovsky, J. Histochem. Cytochem. 14, 291 (1966).
[60] C. A. Alexander, R. L. Hamilton, and R. J. Havel, J. Cell Biol. 69, 241 (1976).
[61] D. R. Stanworth, Nature (London) 188, 156 (1960).
[62] R. R. Porter, Biochem. J. 73, 119 (1959).
[63] D. H. Campbell, J. S. Garvey, N. E. Cremer, and D. H. Sussdorf, "Methods in Immunology," 2nd Ed. Benjamin, New York, 1970.
[64] S. Avrameas and T. Ternynck, Immunochemistry 8, 1175 (1971).

after forcing 10–15 ml 0.135 M phosphate buffer through a canula inserted into the portal vein with a 16-gauge catheter (Angiocath intravenous placement unit, Deseret Pharmaceutical Co., Sandy, UT) connected to a Holter peristaltic pump (Series 1100, Extracorporeal Medical Specialties, Inc., King of Prussia, PA). The drained livers are then perfused for 5–10 min with 1 or 4% formaldehyde freshly prepared from paraformaldehyde (Matheson, Coleman and Bell, Manufacturing Chemists, Los Angeles, CA) in 0.135 M phosphate buffer, pH 7. The excess fixatives are removed by flushing the liver with the same buffer for another 3–10 min. Portions of the liver are removed and 40-μM-thick sections are prepared using a Smith-Farquhar tissue chopper (TC-2, Ivan, Sorvall Inc., Norwalk, CT). The sections are rinsed in 0.135 M phosphate buffer and incubated for 18–24 hr at 8° with FAB conjugates prepared from either anti-LDL IgG or anti-VLDL IgG. As a control, identical sections should also be incubated either in a conjugate made from FAB of an unimmunized rabbit or in a 5 mg/ml solution of peroxidase in 0.135 M phosphate buffer. After incubation, the sections are rinsed in cold 0.135 M phosphate buffer with several changes and postfixed for 60 min in 2% glutaraldehyde in the same buffer and rinsed again with several changes of the same buffer. The postfixed slices are preincubated for 15 min in a solution of DAB (0.5 mg/ml) in 0.05 M Tris-HCl buffer, pH 7.5–7.6, containing 0.25 M sucrose. Finally, the sections are incubated for another 5 min in a fresh solution of DAB sucrose which is supplemented with 0.01% hydrogen peroxide. The sections are rinsed in 0.05 M Tris-HCl buffer followed by 0.135 M phosphate buffer and fixed overnight in 3% osmium tetroxide and 0.135 M phosphate buffer, dehydrated in acetone, and embedded in Epon. A 600- to 800-Å-thick section is cut and each ribbon viewed in the electron microscope (Siemens Elmiskop 101).

By Electron Microscopic Autoradiography

Stein and Stein, using [³H]palmitic acid, [2-³H]glycerol, and choline [ME-³H]chloride, have examined the site of lipid synthesis in rat liver by means of autoradiography, a technique which localizes radioactive isotopes in tissue sections.[65,66] Liver samples are taken, while the animal is under anesthesia, from 2 min to 4 hr after administration of labeled precursor into the tail vein of rats. Liver samples are fixed in glutaraldehyde in cacodylate buffer, followed by five washes in 0.25 M sucrose in cacodylate buffer, and postfixed in 2.5% osmium tetroxide for 1 hr at 0°. The

---

[65] O. Stein and Y. Stein, *J. Cell Biol.* **33**, 319 (1967).
[66] O. Stein and Y. Stein, *J. Cell Biol.* **40**, 461 (1969).

postfixed specimens are dehydrated at 0° with two changes of 70% ethanol (5 min each), two changes of 95% ethanol (5 min each), followed by three changes of pure Epon for 1 hr each. The tissues are transferred to the complete Epon mixture and left overnight at 4° followed by embedding.

Silver-to-gold interference colored thin sections are cut, picked up on thin meshed grids, and placed in rows on a glass microscope slide previously coated with collodion, with identifying marks on the back of the slide opposite each grid. Autoradiography is carried out according to the method of Caro and van Tubergen.[67] Emulsions (Ilford K5 or L4), diluted 1 : 1 with water at 43–45°, are cooled to near 20°. A 4-cm-diameter platinum loop is dipped into the emulsion and carefully withdrawn, so that a thin film of emulsion remains stretched on the loop. The loop is touched to the surface of the slide containing the grids causing the film of emulsion to drop from the loop and coat the grids. The coated sections are exposed at 4° in a light-tight box in the presence of a desiccant for 2–12 weeks. They are developed in Microdol-X and fixed in Kodak acid fixer for 5 min each at room temperature. The grids are then washed briefly, stained with lead citrate and released by cutting around them through the emulsion. They are viewed with either a Siemens or RCA EM3G electron microscope at 80 or 100 kV.

Applications and Limitations

In vivo studies, in which radiolabeled precursors of the protein and/or lipid moieties are used, are valuable in determining the types of apolipoproteins and/or lipids made by the liver. When these types of experiments are combined with kinetic studies, cell fractionation, electron microscopy, and isolation of defined lipoproteins it is possible to track the intracellular movement of nascent lipoproteins. The use of monospecific antibodies directed against specific apolipoproteins also permits the measurement of the amount of nascent apolipoproteins in each of the hepatic organelles and buoyant density flotation of lipoproteins from cellular organelles allows the researcher to identify and study the morphology and the composition of different classes of nascent lipoproteins, which can then be compared with those obtained from serum.

One of the major limitations of the *in vivo* labeling technique is obtaining sufficient incorporation of radioactive precursors into the isolated nascent lipoproteins. Cell fractionation procedures for liver have been well defined but they required mechanical disruption of liver which often results in a poor yield of organelles and subsequently of their constituent

[67] L. G. Caro and R. P. Van Tubergen, *J. Cell Biol.* **15,** 173 (1962).

radioactive lipoproteins. It is difficult therefore, to fully characterize the intracellular lipoprotein particles. The potential for cross contamination within the lipoprotein classes during buoyant density flotation is significant. It must be remembered that classification of lipoproteins is based on the buoyant density obtained from serum samples. Intracellular nascent lipoproteins, which may not be fully assembled, may float at different, or disparate, buoyant densities.

The immunocytochemical technique is useful in the localization of apolipoproteins but provides a static picture and does not differentiate between nascent and recycled apolipoproteins.

The electron microscope autoradiographic technique may be useful in the identification of sites of nascent lipid synthesis, but this technique lacks specificity and does not differentiate between nascent lipids that are associated with cellular membranes and those that are associated with lipoproteins.

Despite the limitations which have been enumerated, the combination of kinetic studies, cell fractionation, and electron microscopy is an effective method in eliciting information of the intracellular sequence of events which leads to the formation of nascent lipoproteins.

Acknowledgment

This research is supported by a research Grant HL-09011 from the National Institute of Health. The author is grateful to Dr. C. M. Redman for his continuous encouragement and support on many aspects of this work.

[19] Morphological Localization of Apolipoproteins and Their mRNA by Immunocytochemistry and in Situ Nucleic Acid Hybridization

By CHIN-TARNG LIN and LAWRENCE CHAN

Introduction

Immunocytochemical techniques are powerful tools for the localization of specific antigens in various organs and tissues. These techniques can be used to observe the localization of specific antigens at the light microscopic and the ultrastructural levels. When used together with biochemical methods, they can be invaluable in the elucidation of the details of synthesis, transport, processing, secretion, as well as receptor binding and internalization, all dynamic processes involving lipoprotein metabolism. For biosynthetic aspects of lipoprotein metabolism, the technique of *in situ* nucleic acid hybridization on tissue sections is also important. This method has been used successfully for the localization of the mRNAs for specific apolipoproteins. In this chapter, we shall describe the use of immunocytochemistry and *in situ* nucleic acid hybridization in the tissue localization of apolipoproteins and their mRNA. Much of the experimental detail was developed in the authors' laboratories involving a model system.[1-3] However, we have also drawn on our experience with antigens other than apolipoproteins.[4-6]

Immunocytochemistry

Preparation of Antigen and Antibodies

The use of polyclonal antisera for immunolocalization of specific proteins requires the highest degree of purity of the antigen. Occasionally, mixtures of proteins have been used as the immunogen. In such instances, careful adsorption of the antisera or further purification with the antigen by affinity chromatography is necessary. However, the use of monoclonal antibodies has obviated this requirement and they can be obtained from relatively impure antigens.

The purification of specific apolipoproteins is described in detail in Chapters [10]–[15], Volume 128, and the production of polyclonal and monoclonal antisera to purified apolipoproteins is described in Chapter [31], Volume 128.

Preparation of Tissues

Tissue Fixation. For immunocytochemical localization of apolipoproteins in the liver or other organs, an optimal fixation condition is essential for preservation of both apolipoprotein antigenicity and tissue morphology. In general, perfusion fixation is superior to immersion fixation. For perfusion fixation, the fixative is perfused into the animal circulatory

[1] C. T. Lin and L. Chan, *Endocrinology* **107**, 70 (1980).
[2] C. T. Lin and L. Chan, *Differentiation* **18**, 105 (1981).
[3] C. T. Lin and L. Chan, *Histochemistry* **76**, 127 (1982).
[4] C. T. Lin, J. P. Chang, and J. P. Chen, *J. Histochem. Cytochem.* **23**, 624 (1975).
[5] C. T. Lin, S. K. Tsai, Z. Y. Lin, T. C. Teng, and W. S. J. Lin, *J. Ultrastruct. Res.* **73**, 310 (1980).
[6] C. T. Lin, L. R. Dedman, B. R. Brinkley, and A. R. Means, *J. Cell Biol.* **85**, 473 (1980).

system through the left ventricle of the heart or in a retrograde manner through the abdominal aorta. The aortic perfusion method may yield better results than the cardiac perfusion because it is easier to handle and the circulation can be maintained throughout the whole procedure before the fixative is administered. In certain circumstances, immersion fixation may be the only practical method; for example, in fixing the tissues from a large animal or human being or when other organs from the same experimental animal are used for other purposes.

The choice of fixative depends on the nature of the antigen and the required degree of preservation of tissue structure. In order to determine the optimal conditions of fixation, it is necessary to use different concentrations of paraformaldehyde and glutaraldehyde and to check the tissue morphology and the resulting immunoreactivity of the antigen in the tissue. For a number of apolipoproteins we have found the following method satisfactory for immunostaining. The initial fixative contains a 4% paraformaldehyde and 0.1% glutaraldehyde mixture. The concentration of glutaraldehyde can be varied from 0 to 1% until the optimal fixative condition is found. The fixative is prepared in 0.04 M sodium phosphate buffer, pH 7.4, containing 0.01% $CaCl_2$ for light microscopic immunohistochemistry and in 0.1 M phosphate buffer, pH 7.4, containing 8.5% sucrose and 0.01% $CaCl_2$ for immunoelectron microscopy. The liver or other organs are removed once the animal has been perfused with 200–600 ml of fixative in 2.5 hr. The tissues should be dissected immediately and immersed overnight in a weaker fixative, such as 2% paraformaldehyde, in order to remove excess glutaraldehyde.

Tissue Sectioning. For tissue sectioning, the following three methods are satisfactory: (1) paraffin block sectioning, (2) cryostat sectioning, and (3) vibratome sectioning. Since most apolipoproteins in the tissue are stable to the paraffin embedding processes, one can easily use paraffin block sectioning (5 μm) for immunostaining.[1,2] Cryostat sectioning is also a good method for light microscopic immunohistochemistry since the antigenicity of certain antigens is preserved better than by the use of paraffin sectioning, especially if used with cryoprotectants such as 20% sucrose or 10% dimethylsulfoxide.[7] For immunoelectron microscopy, cryostat sectioning is not suitable, since the ice crystal formation during tissue freezing results in an inferior morphology. Vibratome sectioning is easier to handle, but difficult to obtain a thin section (< 20 μm). Usually, 50-μm sections can be obtained without difficulty. Although the thickness is somewhat higher than ideal for immunohistochemistry, it is quite suit-

[7] C. T. Lin, H. Z. Li, and J. Y. Wu, *Brain Res.* **270**, 273 (1983).

able for EM immunocytochemistry. For both light and EM localization in one tissue section, vibratome sectioning is the best choice.[8-10]

Immunostaining Procedures

Light Microscopy. There are four methods that can be used for localization of apolipoprotein.

1. *Indirect immunohistochemistry using peroxidase-labeled second antibody*[1,2]

Reagents

Normal serum
Primary antiserum
PBS: 0.04 M phosphate buffer, pH 7.4, 0.15 M NaCl
Peroxidase substrate: 0.05% 3,3'-diaminobenzidine-4-HCl in 200 mM Tris-HCl, pH 7.2, and 0.01% H_2O_2. Filter substrate before adding H_2O_2
Peroxidase-labeled second antibody (from commercial sources, such as Miles Laboratory)

Protocol

1. Tissue sections from paraffin blocks: Deparaffinize by xylene and rehydrate with decreasing concentrations of ethanol to 50%. Tissue sections from cryostat sectioning: Mount the cryostat sections on a clean slide. Air dry for about 5 min. The sections are then dipped in the PBS.
2. Incubate in 1% H_2O_2 in absolute methanol at 23° for 10 min in order to inhibit endogenous peroxidase activity.
3. Wash in PBS twice, 10 min each.
4. Incubate with 0.05% Pronase in distilled H_2O at 23° for 10 min to reduce background staining. Wash in PBS twice, 10 min each.
5. Immerse section for 10 min in normal serum at 1:40 dilution in buffer or in 0.1 mg/ml of its IgG. (The normal serum or its IgG fraction is obtained from the same animal species which is used to produce the second antibody.)
6. Incubate with primary antiserum for 60 min (1:40 dilution) or overnight (1:200 to 1:800 dilution) after blotting of excess normal serum from sections (do not wash).
7. Wash twice in PBS, 10 min each.
8. Incubate with peroxidase-labeled second antibody at 1:40 dilution for 1 hr.

[8] C. T. Lin, *J. Histochem. Cytochem.* **28,** 339 (1980).
[9] C. T. Lin, J. Garbern, and J. Y. Wu, *J. Histochem. Cytochem.* **30,** 853 (1982).
[10] C. T. Lin, C. C. Su, W. Palmer, and L. Chan, *Tissue Cell* **15,** 259 (1983).

9. Wash twice in PBS, 10 min each.
10. Refix in 2.5% glutaraldehyde for 5 min.
11. Wash twice in PBS, 10 min each.
12. Incubate with peroxidase substrate for 5 min.
13. Wash twice in PBS, 10 min each.
14. Counterstain in 0.01% OsO_4 for 30–60 sec.
15. Rinse in distilled H_2O, three times.
16. Dehydrate and mount with permount.

The concentration of the primary antiserum and peroxidase-conjugated second antibody to be used are determined by their titer. Usually, a dilution at 1 : 40 to 1 : 200 in PBS should be tried. The peroxidase substract solution should be prepared immediately before use. During the staining procedure, the section should not be allowed to dry out. To achieve this, a humidified chamber is used for all incubations.

2. *Indirect immunohistochemistry using peroxidase–anti-peroxidase (PAP) method.*[11] The PAP method uses a preformed immune complex (peroxidase–anti-peroxidase complex) to replace the peroxidase-conjugated second antibody. In this method, the peroxidase-conjugated second antibody is replaced by ordinary second antibody (for example, if the primary antibody is produced in the goat, the second antibody may be produced in the rabbit using goat IgG as antigen), followed by incubation with PAP complex (which in this case is produced in the goat). [The PAP complex can be obtained either by immunization of horseradish peroxidase (HRP) to the goat and incubation of goat antiserum against HRP with HRP molecules to form PAP immune complex, or from any commercial source, such as Sternberger-Meyer Co.] In this procedure, since each IgG molecule has two antigen-binding sites, the first antigen-binding site of the second antibody (rabbit antibody against goat IgG) binds the Fc portion of the primary antibody (goat anti-apolipoprotein), while the second binding site binds on the Fc portions of goat IgG in PAP complex. Since each PAP complex contains two or more HRP molecules, each tissue antigen molecule is identified by at least two or more HRP molecules. Thus, a more intense reaction product is obtained in the tissue section.

Reagents

Primary antiserum
Second antiserum
PBS
PAP complex

[11] L. A. Sternberger, "Immunocytochemistry." Wiley, New York, 1979.

Protocol

  1–5. Same as the peroxidase-labeled second antibody method.

    6. Incubate with primary antiserum for 60 min (1 : 80 dilution) after blotting of excess normal serum from sections (do not wash).

    7. Wash twice in PBS, 10 min each.

    8. Incubate with second antibody (such as rabbit antiserum against goat IgG, 1 : 40 to 1 : 80 dilution) for 1 hr at room temperature.

    9. Wash in PBS twice, 10 min each.

  10. Incubate with PAP complex for 15 min (dilution at 1 : 40 to 1 : 200) at room temperature.

  11. Wash twice, in PBS, 10 min each.

  12. Same as the peroxidase-labeled second antibody method (steps 10–16).

3. *Indirect immunohistochemistry using avidin–biotin peroxidase complex (ABC).*[12,13] The "ABC" technique employs primary antibody, biotinylated secondary antibody, and a preformed avidin–biotinylated HRP complex. Each avidin molecule has four available binding sites for biotin. The affinity of these binding sites is extraordinarily high (10^{15} M^{-1}). Most proteins can be coupled to several molecules of biotin. Hence, biotinylated HRP forms a stable macromolecular complex with avidin. Each specific antigen can thus be juxtapositioned to approximately 20–25 HRP molecules,[14] resulting in a very strong reaction product formation in the tissue section.

Reagents

  Primary antiserum
  PBS
  Biotinylated second antibody
  Biotinylated peroxidase
  Avidin

Protocol

  1–5. Same as the peroxidase-labeled second antibody method.

    6. Incubate with primary antiserum at 1 : 100 to 1 : 200 dilution for 60 to 120 min after removal of excess normal serum from sections (do not wash).

    7. Wash twice in PBS, 10 min each.

[12] S. M. Hsu, L. Raine, and H. Fanger, *Am. J. Clin. Pathol.* **75,** 734 (1981).
[13] S. M. Hsu, L. Raine, and H. Fanger, *J. Histochem. Cytochem.* **29,** 577 (1981).
[14] S. M. Hsu, personal communication.

8. Incubate with biotinylated second antibody at 1 : 200 dilution for 30–60 min at room temperature.
9. Wash twice in PBS, 10 min each.
10. Incubate with ABC solution for 30 min at room temperature. The avidin–biotinylated HRP complex (ABC solution) should be prepared 30 min before use. The proportion of avidin molecule to biotinylated HRP molecule should be 5 : 2. (For example, 1 ml of buffer solution contains 10 μg of avidin and 4 μg of biotinylated HRP.)
11. Wash in PBS twice, 10 min each.
12. Follow steps 10 to 16 in the peroxidase-labeled second antibody method.

4. *Immunofluorescence method.*[15] This is an older method for localization of tissue antigen. There are several disadvantages of this method: (1) the section cannot be kept for more than 1 year; (2) any endogenous fluorescent material in the tissue section cannot be readily blocked; (3) the method requires a special fluorescence microscope; and (4) the fluorescent labels are not electron dense and cannot be used for electron microscopic study. The major advantage is that the tissue section does not have to be incubated with the enzyme substrate, thus the whole procedure is considerably shorter than the immunoperoxidase method.

Reagents

Normal serum
Primary antiserum
PBS
Fluorescein-conjugated second antibody

Protocol

1. Tissue sections from paraffin blocks: Deparaffinize by xylene and rehydrate with decreasing concentrations of ethanol to 50%. Tissue sections from cryostat sectioning: Mount the cryostat sections on a clean slide. Air dry for about 5 min. The sections are then dipped in the PBS.
2. Immerse in PBS for 10 min.
3. Immerse the sections in normal serum at 1 : 40 dilution in PBS or in 0.1 mg/ml of its IgG fraction for 10 min. The normal serum or its IgG fraction is obtained from the same animal species which is used to prepare the second antibody.

---

[15] B. R. Brinkley, G. M. Fuller, and D. P. Highfield, *Proc. Natl. Acad. Sci. U.S.A.* **72,** 4981 (1975).

4. Incubate with fluorescein-conjugated second antibody (use 1 : 40 dilution; it can be obtained from commercial sources) for 1 hr at room temperature or 30 min at 37° after the excess normal serum is removed.
5. Wash in PBS twice, 10 min each.
6. Mount with 50% glycerine in PBS and store at 4°.
7. Observe under the fluorescence microscope.

Immunoelectron Microscopy. For electron microscopic immunocyto-chemical study, two different techniques have been used: the preembedding staining method and the postembedding method. The preembedding method has been widely used for EM localization of a large number of proteins including apolipoproteins.[3,16–18] Under optimal experimental conditions, both good morphology and strong staining are preserved. We have used this technique to localize apolipoproteins in the avian liver,[3] as well as a number of other antigens.[4–6,8–10,16–18] The postembedding method is especially suitable for detection of high concentrations of antigens, such as proteins in secretory granules (e.g., neuropeptides and apolipoproteins). It is generally less sensitive for antigens present in very low concentrations, such as proteins in the early part of the synthetic pathway, due to loss of protein antigenicity during tissue-embedding procedures.

There are two experimental approaches in preembedding immuno-peroxidase staining. One is the direct method, which uses IgG peroxidase conjugate or Fab–peroxidase conjugate to stain the tissue directly. The other approach is the indirect immunoperoxidase method, similar to that described for light microscopy. The morphological preservation is generally better in the direct method because of the shorter tissue preparation time. The detailed procedure, including the preparation of IgG–peroxidase conjugate and Fab–peroxidase conjugate are described in the following:

1. *Preembedding staining method*

Direct immunoperoxidase method using Fab–peroxidase conjugate[3–6,8,18] (see Fig. 1): First, purify the IgG fraction from the antiserum. Then, digest it with papain, and use an antigen–Sepharose-4B affinity chromatography to purify the monospecific Fab fragments against the

[16] V. Chan-Palay, C. T. Lin, S. Palay, M. Yamamoto, and J. Y. Wu, *Proc. Natl. Acad. Sci. U.S.A.* **79**, 2695 (1982).
[17] C. T. Lin, K. Mukai, and C. Y. Lee, *Cell Tissue Res.* **224**, 647 (1982).
[18] C. T. Lin and L. H. Chen, *Lab. Invest.* **48**, 718 (1983).

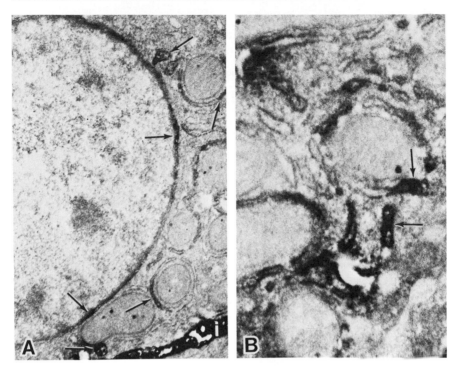

FIG. 1. The cockerel liver was fixed with 4% paraformaldehyde and stained with Fab–peroxidase conjugate against apoVLDL-II. (A) Reaction product is found in the cisternae of the nuclear envelope and rough ER and in some secretory vesicles (arrows). i, Intercellular space. ×16,300. (B) A high magnification taken from a hepatocyte showing reaction product on the membrane and in the cisternae of RER (arrows). ×29,300.

specific antigen. Finally, conjugate the Fab fragments to HRP either by glutaraldehyde[19] or the periodate-activated HRP method.[20,21]

Isolation of IgG fraction from the antiserum: Since ammonium sulfate precipitation of IgG from the antiserum may cause some damage to the antibody-binding activity, the ion exchange chromatography method is recommended. QAE-Sephadex A-50 chromatography is quite effective in the purification of the IgG fraction. Since the lipid components in the serum may interfere with the ion exchange chromatography, we use aerosil-380[22] to remove the lipid component.

[19] S. Avrameas and T. Ternynck, *Immunochemistry* **8,** 1175 (1971).
[20] P. K. Nakane and A. Kawaoi, *J. Histochem. Cytochem.* **22,** 1084 (1974).
[21] M. B. Wilson and P. K. Nakane, in "Immunofluorescence and Related Staining Techniques" (W. Knapp, K. Holubar, and G. Wick, eds.), p. 215. Elsevier, 1978.
[22] J. Jousta and H. Lundgren, *Protides Biol. Fluids Proc. Colloq., Brugge* **17,** 511 (1969).

Reagents

Aerosil-380 (Degussa Comp., Frankfurt, West Germany)
Buffer A: Ethylenediamine–acetic acid buffer, pH 7.0, ionic strength
   0.1
QAE-Sephadex A-50
Buffer B: 0.05 M Tris–saline, pH 8.0

Protocol

1. Mix 50 ml of antiserum with 1 g of aerosil-380. Stir at room temperature for 4 hr.
2. Centrifuge at 12,000 rpm for 30 min.
3. Remove the precipitate from the lipid–aerosil-380 complex.
4. Dilute the supernatant 1 : 1 with buffer A and apply to the QAE-Sephadex A-50 column, which has been previously equilibrated with buffer A.
5. Elute the IgG fraction directly from the column with buffer A. The first main peak is collected, concentrated, and dialyzed against buffer B overnight. The yield of IgG fraction from QAE-Sephadex A-50 column is about 85%.

Preparation of monospecific Fab fragments against an apolipoprotein:

Reagents

CNBr-activated Sepharose-4B (Pharmacia)
0.1 M NaHCO$_3$, pH 11, + 0.5 M NaCl
1 M ethanolamine
0.1 M acetate buffer + 1 M NaCl, pH 4
0.1 M borate buffer + 1 M NaCl, pH 8
Reducing solution: 2 mM EDTA, 30 mM cysteine, pH 7.5
Papain (Sigma Chemical Company)
Buffer B: 0.05 M Tris–saline, pH 8.0
0.03 M iodoacetamide
Sephadex G-100 gel
0.2 M glycine · HCl buffer, pH 2.7
0.2 M Tris · HCl buffer, pH 8.0

Protocol

1. The antigen is conjugated to CNBr-activated Sepharose-4B (Pharmacia) according to the coupling procedures described in the brochure provided by the company with some modifications. Briefly, swell 1 g of CNBr-activated Sepharose-4B gel and wash for 15 min

with 200 ml of 10^{-3} N HCl solution. Dissolve 10 mg of antigen in 5 ml 0.1 M NaHCO$_3$ buffer solution, pH 11, containing 0.5 M NaCl. Mix with the CNBr-activated Sepharose-4B in a test tube. Rock the mixture end over end for 2 hr at room temperature. Wash the antigen–Sepharose-4B complex to remove unbound apolipoprotein with coupling buffer. Determine the unbound apolipoprotein concentration. By subtraction, we can obtain the amount of bound protein. Block the remaining active groups on the column with 1 M ethanolamine at pH 8 for 1–2 hr. Wash three times to remove noncovalently adsorbed protein, each cycle consisting of a wash at pH 4 (0.1 M acetate buffer containing 1 M NaCl) followed by a wash at pH 8 (0.1 M borate buffer containing 1 M NaCl).

2. Digest the IgG fraction with papain according to Porter[23] with some modifications. First, reduce the IgG fraction (10 mg/ml in buffer B) by adding reducing solution. Incubate at 37° for 60 min. Then add papain 2% (w/w) (substrate-to-enzyme ratio of 50 : 1), pH 7.5, and incubate at 37° for 4 hr.

3. Stop the digestion by addition of 0.03 M iodoacetamide at room temperature for 10 min. Separate the undigested IgG from the reaction mixture by Sephadex G-100 gel filtration in buffer B.

4. The monospecific Fab fragment is obtained by passing the reaction mixture, which includes Fab and Fc fragments, through the antigen–Sepharose-4B affinity column prepared in step 1.

5. Remove the Fc and nonspecific Fab fragments by extensive washes with buffer B. Elute the specific Fab fragments bound on the affinity column with 0.2 M glycine–HCl buffer, pH 2.7. Neutralize the eluate immediately with 0.2 M Tris-HCl buffer, pH 8.0. The Fab fragments are concentrated on Amicon ultrafiltration cell and are ready for conjugation.

Conjugation of Fab fragments and HRP: The Fab fragments are conjugated to HRP (Sigma type VI) which has been activated by periodate according to Wilson and Nakane.[21] The conjugation procedure is performed as follows:

Reagents

Horseradish peroxidase (HRP) (Sigma type VI)
NaIO$_4$, 38.5 mg/ml in distilled H$_2$O
0.001 M sodium acetate buffer, pH 4.2
0.2 and 0.01 M sodium carbonate buffer, pH 9.5
NaBH$_4$, 4 mg/ml in distilled H$_2$O

[23] R. R. Porter, *Biochem. J.* **73**, 119 (1959).

PBS: 0.04 M phosphate buffer, pH 7.4, + 0.15 M NaCl
Sephacryl S-200
Sephadex G-100

Protocol

1. Dissolve 4 mg of HRP in 1 ml of distilled water. While stirring, add 50 μl of 38.5 mg/ml of $NaIO_4$ (in distilled water). After stirring for 20 min at room temperature, dialyze the activated HRP against 0.001 M sodium acetate buffer, pH 4.2, for 16 hr.
2. To the dialyzed activated HRP, add 20 μl of 0.2 M sodium carbonate buffer, pH 9.5, followed immediately by 10 mg of goat Fab fragments against the apolipoprotein in 0.01 M sodium carbonate buffer, pH 9.5.
3. Stir for 2 hr at room temperature. Add 100 μl of freshly prepared $NaBH_4$ (4 mg in 1 ml distilled H_2O) and incubate for another 2 hr at 4° without stirring.
4. The conjugation mixture is then dialyzed against PBS at 4° overnight.
5. Further purify the Fab–peroxidase conjugate on a Sephacryl S-200 or Sephadex G-100 column. Collect the peak containing OD_{280} (for protein) and OD_{403} (for peroxidase). It can be concentrated on an Amicon ultrafiltration cell using YM10 ultrafiltration membrane. Store the conjugate in the presence of one-to-one protein concentration of bovine serum albumin (BSA) to Fab–peroxidase conjugate at −79° until use. The BSA protects antigen-binding activity of the Fab fragment.

The procedure for immunoelectron microscopy using Fab–peroxidase conjugate is as follows:

Reagents

Fab–peroxidase conjugate
0.025 M hydroxylamine
PBSC: 0.1 M phosphate buffer, pH 7.4, containing 8.5% sucrose and 0.01% $CaCl_2$
2.5% glutaraldehyde in PBSC
Saturated uranyl acetate in 50% ethanol
Spurr's medium
Peroxidase substrate: 0.05% 3,3'-diaminobenzidine-4-HCl in 0.2 M Tris-HCl buffer, pH 7.2, containing 8.5% sucrose plus 0.01% H_2O_2
0.05% Aquasil (Pierce Chemical Company)

Protocol

1. Allow 50-μm tissue sections (from a vibratome) to float free for 30 min in 0.025 M hydroxylamine in PBCS in order to block excess aldehyde group.
2. Wash in PBSC twice, 10 min each.
3. Incubate in the specific Fab–peroxidase conjugate ($\sim$0.08 mg/ml) for 60 min at room temperature during shaking.
4. Wash in PBSC, 10 min each.
5. Refix in 2.5% glutaraldehyde in PBSC for 30 min.
6. Wash in PBSC twice, 10 min each.
7. Incubate for 5 min with peroxidase substrate. (The substrate has to be prepared before use and to be filtered before addition of H_2O_2.)
8. Wash in PBSC twice, 10 min each.
9. Postfix in 1% OsO_4 in PBSC for 60 min.
10. Rinse in distilled H_2O three times.
11. Prestain in saturated uranyl acetate in 50% ethanol for 10 min.
12. Dehydrate with ethanol.
13. Embed in Spurr's medium[24] using silanized glasses to press the sections. The silanized glasses are prepared from a water solution containing 0.05% Aquasil (Pierce Chemical Company).
14. Allow Spurr's medium to polymerize at 68° for 12 hr.
15. Separate the glasses by a razor blade and observe the tissue section under a light microscope.

The stained areas are cut and mounted on the blank blocks for thin sections. The advantage of this method is that the penetration of Fab–peroxidase conjugate is deeper than the IgG molecule and the incubation time is shorter than the indirect method, resulting in better morphological preservation and stronger antigen binding, especially at the ultrastructural level.

Direct immunoperoxidase technique using IgG–peroxidase conjugate[3]: For preparation of IgG–peroxidase conjugate, the IgG fraction is isolated from the serum as described above and is applied directly to the antigen–Sepharose-4B affinity column. The monospecific IgG fraction against the antigen is then conjugated to HRP as described for conjugation of Fab fragments to HRP. The IgG–peroxidase conjugate is further purified by Sephacryl S-200 gel filtration. It is protected by the same concentration of bovine serum albumin and stored at −79°. Since IgG–perox-

---

[24] A. R. Spurr, *J. Ultrastruct. Res.* **26**, 31 (1969).

idase conjugate is larger than the Fab–peroxidase conjugate, the tissue penetration of the former is not as good as the latter. In this case, addition of 0.02% saponin (a mild detergent) in the IgG–peroxidase solution during tissue incubation will allow the IgG–peroxidase conjugate to penetrate better the membrane barrier. The whole immunostaining procedure is similar to the Fab–peroxidase method except for the addition of saponin in the incubation solution as mentioned above.

The immunostaining procedure is as follows:

Reagents

Normal serum
IgG–peroxidase conjugate
0.02% saponin
PBSC: 0.1 M phosphate buffer, pH 7.4, containing 8.5% sucrose and 0.01% CaCl$_2$

Protocol

1–2. Same as Fab–peroxidase method.
3. Immerse the sections for 10 min in normal serum (1 : 40 dilution). The normal serum is obtained from the animal from which the primary antibody was obtained.
4. Remove normal serum solution and incubate with IgG–peroxidase conjugate against the specific antigen (0.1 mg/ml) containing 0.02% saponin for 60 min at room temperature during shaking.
5. Wash in PBSC twice, 10 min each.
6. The following steps are the same as steps 5 to 15 in the Fab–peroxidase method.

Indirect immunoperoxidase technique using peroxidase-labeled second antibody[9,16,17]: The basic principle is similar to the light microscopic method. This technique is recommended only when a small amount of primary antibody is available. The procedure is as follows:

Reagents

PBSC: 0.01 M phosphate buffer, pH 7.4, containing 8.5% sucrose and 0.01% CaCl$_2$
Peroxidase-labeled second antibody
0.02% saponin

Peroxidase substrate: 0.05% 3,3'-diaminobenzidine-4-HCl in 0.2 M Tris-HCl buffer, pH 7.2, containing 8.5% sucrose plus 0.01% H_2O_2
Saturated uranyl acetate in 50% ethanol
Spurr's medium

Protocol

1–2. Same as the Fab–peroxidase technique.

3. Immerse the sections for 10 min in normal serum at 1 : 40 dilution. The normal serum is obtained from the animal species which is immunized to produce second antibody.

4. Remove normal serum and incubate for 60 min with the primary antibody (1 : 40 dilution) or overnight (1 : 200 dilution) which contains 0.02% saponin at room temperature during shaking.

5. Wash in PBSC twice, 10 min each.

6. Incubate with peroxidase-labeled second antibody (1 : 40 dilution) for 60 min at room temperature during shaking.

7. Wash in PBSC twice, 10 min each.

8. Refix in 2.5% glutaraldehyde in PBSC for 30 min.

9. Wash in PBSC twice, 10 min each.

10. Incubate with peroxidase substrate for 5 min. (The substrate is prepared before use and should be filtered before addition of H_2O_2.)

11. Wash in PBSC twice, 10 min each.

12. Postfix in 1% OsO_4 in PBSC for 60 min.

13. Rinse in distilled water, three times.

14. Prestain in saturated uranyl acetate in 50% cthanol for 10 min.

15. Dehydrate in ethanol and embed in Spurr's medium using silanized glasses to press the sections and polymerize at 68° for 12 hr.

Indirect immunoperoxidase technique using peroxidase anti-peroxidase and avidin–biotin peroxidase complex[11–13]: These two methods are not recommended for ultrastructural immunostaining, especially for intracellular antigens. The advantages of the PAP and ABC methods are their sensitivity through amplification of the peroxidase molecules. The high degree of amplification of the peroxidase reaction product unfortunately also results in varying degrees of loss of resolution.

Indirect immunoperoxidase technique using peroxidase anti-perox-larger than HRP (ferritin $M_r \cong 400,000$; HRP $M_r \cong 40,000$), this molecule penetrates tissue sectins very poorly. However, under optimal conditions for some types of studies (e.g., localization of calmodulin in cultured cells[25]), highly informative results can be obtained. The procedure for indirect immunoferritin staining is as follows:

Reagents

PBSC: 0.1 M phosphate buffer, pH 7.4, containing 8.5% sucrose and 0.01% $CaCl_2$

0.025 M hydroxylamine

Affinity purified specific IgG

0.02% saponin

Ferritin-conjugated second antibody (Miles Biochemicals)

Saturated uranyl acetate in 50% ethanol

Spurr's medium

Protocol

1. Allow 50-μm sections obtained from a vibratome to float free for 30 min in 0.025 M hydroxylamine in PBSC.
2. Wash in PBSC twice, 10 min each.
3. Incubate with affinity purified IgG of the specific antibody (0.1 mg/ml containing 0.02% saponin for 60 min.
4. Wash in PBSC twice, 10 min each.
5. Incubate with ferritin-conjugated rabbit IgG against goat IgG (1:40 dilution) containing 0.02% saponin for 60 min at room temperature.
6. Wash in PBSC twice, 10 min each.
7. Refix in 2.5% glutaraldehyde in PBSC for 60 min.
8. Wash in PBSC twice, 10 min each.
9. Postfix in 1% OsO_4 in PBSC for 60 min.
10. Rinse in distilled water, three times.
11. Prestain in saturated uranyl acetate in 50% ethanol for 15 min.
12. Dehydrate and embed in Spurr's medium for thin sectioning.

2. *Postembedding method*

In the localization of certain neuropeptides and secretory proteins in fairly high concentrations, the postembedding method has been used for immunoelectron microscopic study, for example, peptide hormones in the pituitary have been localized by this method.[26,27] Using this technique, the proteins in the secretory vesicles in high concentration are detected, but those in the rough ER or Golgi apparatus are not. Recently, a low-temper-

[25] M. C. Willingham, J. Wehland, C. B. Klee, N. D. Richert, A. V. Rutherford, and I. H. Pastan, *J. Histochem. Cyotchem.* **31**, 445 (1983).

[26] G. V. Childs and G. Unabia, *J. Histochem. Cytochem.* **30**, 713 (1982).

[27] G. V. Childs and G. Unabia, *J. Histochem. Cytochem.* **30**, 1320 (1982).

ature polymerized resin–Lowicryl K4M has been introduced for embedding the tissue for postembedding staining.[28] The results are better than with other embedding materials, but the method still needs further improvement to achieve optimal results. Generally speaking, the immunoperoxidase method, the immunoferritin method, and the immunogold method can be applied to thin sections. An extensive review of these methods is available.[29]

Immunoperoxidase method[26,27]:

Reagents

Spurr's medium
Normal rabbit serum
Primary antiserum or IgG
PBSC: 0.1 M phosphate buffer, pH 7.4, containing 8.5% sucrose and
    0.01% $CaCl_2$
Biotinylated antiserum
ABC: Avidin–biotin–peroxidase complex
PAP: Peroxidase–anti-peroxidase complex
Peroxidase substrate: 0.05% 3,3'-diaminobenzidine-4-HCl in 0.2 M
    Tris-HCl, pH 7.2, plus 0.01% H_2O_2. Prepare the substrate before
    use and filter before adding H_2O_2.

Protocol

1. After perfusion fixation and immersion in weaker fixative overnight, cut the tissue fragments into $0.1 \times 0.1 \times 0.1$ cm pieces and dehydrate in ethanol and embed in Spurr's medium. The block embedded in the Spurr's medium is polymerized at 68° for 12 hr. Mount the thin sections, which are about 800–900 Å in thickness, on nickel grids. The latter have been coated with formval or carbon film.
2. Immerse the grids in 2% sodium borohydride in distilled water for 5 min or in distilled water saturated with sodium metaperiodate for 30–60 min.
3. Wash in PBSC three times, 10 min each.
4. Immerse in normal rabbit serum (1 : 40 dilution) for 10 min.
5. Remove excess solution and immerse in goat IgG (0.1 mg/ml) or goat antiserum against the specific antigen (1 : 40 dilution) for 120 min.
6. Wash in PBSC twice, 10 min each.

[28] E. Carlemalm and E. Kellenberger, *EMBO J.* **1,** 63 (1982).
[29] A. N. Van Den Pol, *J. Exp. Physiol.* **69,** 1 (1984).

7. Incubate in rabbit anti-goat IgG (1 : 40 dilution) for 60 min or in biotinylated rabbit IgG against goat IgG (1 : 40 dilution) for 60 min.
8. Wash in PBSC twice, 10 min each.
9. Incubate in peroxidase–anti-peroxidase complex (prepared in goat; 1 : 100 dilution) or in avidin–biotin–peroxidase complex (avidin, 10 μg, + biotinylated HRP, 4 μg, per milliliter of PBSC) for 30–60 min.
10. Wash in PBSC three times, 10 min each.
11. Incubate for 5 min in peroxidase substrate.
12. Wash in PBSC three times, 10 min each.
13. Postfix in 1% OsO_4 in PBSC for 2 min.
14. Wash in distilled H_2O three times, 10 min each.
15. Stain with saturated uranyl acetate in 50% ethanol for 3 min.
16. Rinse in distilled H_2O, five times.
17. Stain in lead citrate for 3 min.
18. Rinse in distilled H_2O, 5 min.
19. Observe under the electron microscope.

Immunogold method[29,30]:

Reagents

Gold-labeled IgG (available commercially, e.g., Janssen Pharmaceutica, Belgium)
Gold-labeled protein A (available commercially, e.g., Janssen Pharmaceutica, Belgium)

Protocol

1–6. Same as the immunoperoxidase method.
7. Incubate in gold-labeled rabbit IgG against goat IgG (at 1 : 40 dilution) or in gold-labeled protein A (at 1 : 40 dilution) for 60 min.
8. Wash three times in PBSC, 10 min each.
9. Counterstain with uranyl acetate and lead citrate as conventional method (same as steps 15–18 in immunoperoxidase method).
10. Observe under the electron microscope.

[30] M. Cartin, M. Ballak, C. L. Lu, N. G. Seidah, and M. Chretien, *J. Histochem. Cytochem.* **31,** 479 (1983).

Control experiments for immunostaining: Both light and electron microscopic immunostainings require rigorous controls. Several control experiments should be included to verify the specificity of the staining: (1) use of preimmune serum in place of the specific antiserum; (2) use of specific antiserum which has been previously adsorbed with excess pure antigen; (3) use of antibodies against antigens which are known to be absent in that particular tissue or cell; (4) omission of the first antibody; (5) use of peroxidase substrate only.

In Situ Nucleic Acid Hybridization

Preparation of [^{3}H]cDNA Probe (See Chapter [43], Volume 128)

Tissue Preparation. Tissue fixation. mRNAs are generally very sensitive to glutaraldehyde fixative. Even when the concentration of glutaraldehyde is only 0.02%, the results of hybridization seem to be very poor. However, in our experience, 4% paraformaldehyde is a good fixative for preservation of hybridizable mRNA. Perfusion fixation seems to be better than immersion fixation, though the latter is acceptable for certain tissues, for example, the frozen sections of the pancreas after brief immersion fixation have been used for hybridization with pancreatic cDNAs.[31] In our experiments with avian liver, we perfuse the avian liver at first with normal saline in order to get rid of all erythrocytes, followed by perfusion with 200 ml of 4% paraformaldehyde in 0.1 M phosphate buffer, pH 7.4, containing 8.5% sucrose and 0.002% $CaCl_2$. In our other experiments, the livers are cut into tissue fragments and immersed in the same fixative at 4° overnight. Those livers which have been perfused with fixative are also cut into tissue fragments (0.1 × 0.4 × 0.6 cm) and immersed in the same fixative overnight at 4°.

Tissue sectioning. After fixation, the tissue fragments are washed in phosphate buffer for 60 min with several changes of buffer, followed by dehydration and embedding in paraffin block as in conventional histological methods. Cut tissues into 5-μm sections and mount them on the pretreated glasses which have been treated with 10^{-3} N HCl in 75% ethanol and coated with 0.02% gelatin and air dried. Then incubate the glass sections at 57° overnight and deparaffinize them. Frozen sections can also be used after tissue fixation.

Hybridization procedure (see Figs. 2 and 3)

---

[31] J. D. Harding, R. J. MacDonald, A. E. Przybyla, J. M. Chirgwin, R. L. Pictet, and W. J. Rutter, *J. Biol. Chem.* **252,** 7391 (1977).

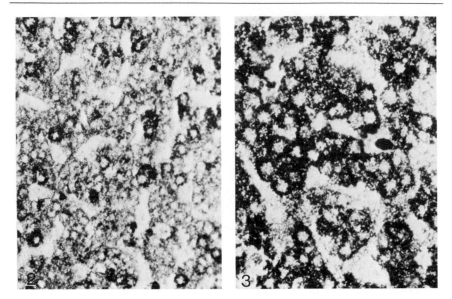

FIG. 2. A cockerel was treated with one dose of diethylstilbestrol. The liver was perfused with normal saline and fixed with 4% paraformaldehyde. The liver fragment was dehydrated and embedded in paraffin block. The 5-μm paraffin section was hybridized with [^{3}H]cDNA of chicken apoVLDL-II gene for 48 hr. Silver grains are shown clearly in certain hepatocytes. $\times$360.

FIG. 3. A cockerel had been treated with 14 daily doses of diethylstilbestrol. The liver was perfused with normal saline and fixed with 4% paraformaldehyde. The hybridization condition is similar to Fig. 2. Silver grains are shown clearly in many hepatocytes. $\times$500.

Reagents

Hybridization buffer: 50% (v/v) deionized formamide, 0.01 M Tris-HCl (pH 7.5), 1 mM EDTA, 0.6 M NaCl, 0.02% Ficoll, 0.02% poly(vinylpyrrolidone), 1 mg/ml of bovine serum albumin, 0.1 mg/ml of sonicated denatured calf thymus DNA, 1 mg/ml of bovine RNA, 1 mg/ml of yeast tRNA, and 0.1 mg/ml of poly(A) (Brahic and Haase buffer).[32] In addition, we have modified this formula by addition of 0.05 mg/ml of poly(C)

[^{3}H]cDNA: 2 $\times$ 10^8 dpm/μg

Unlabeled *E. coli* DNA (20 μg/ml)

Buffer A: 50% formamide, 10 mM Tris-HCl, pH 7.4, 1 mM EDTA, and 0.1 M NaCl

Buffer B: 33% formamide, 10 mM Tris-HCl, pH 7.4, 1 mM EDTA, and 0.1 M NaCl

[32] M. Brahie and A. T. Haase, *Proc. Natl. Acad. Sci. U.S.A.* **75,** 6125 (1978).

Buffer C: 0.3 M NaCl, 0.03 M sodium citrate, pH 7.0
Chloroform
0.3 M ammonium acetate in 70% ethanol
0.3 M ammonium acetate in 90% ethanol

Protocol

1. After deparaffinization, coat the sections with hybridization buffer for 5 min at room temperature in a wet chamber.
2. Remove excess buffer. Add one drop of unlabeled *E. coli* DNA (20 μg/ml) in 10 mM Tris, 1 mM EDTA (pH 7.5) onto the section, and incubate for 5 min in a wet chamber.
3. Remove excess solution. Add two drops (100 μl) of the hybridization medium and incubate in the wet chamber at room temperature for 48 hr. The hybridization medium contains 2×10^8 dpm/μg of [^{3}H]cDNA in 0.5 ml of hybridization buffer. Preheat the medium to 100° for 5 min and quick chill at 4° immediately before use.
4. Rinse in pure cholorform solution twice, 5 min each.
5. Wash in buffer A twice, 5 min each.
6. Wash in 2 liters of buffer B for 48 hr with stirring at 4°. Change buffer at 24 hr.
7. Wash in buffer C twice, 5 min each, at room temperature.
8. Dehydrate in 70% ethanol containing 0.3 M ammonium acetate for 5 min, then in 90% ethanol containing 0.3 M ammonium acetate.
9. Coat with Kodak fine-grain autoradiographic stripping film (AR10) or Ilford emulsion (L4).
10. Dry in a ventilated black box overnight at room temperature.
11. Store in a light-proof box at 4° for 4–12 weeks.
12. Develop in Kodak D-19 developer for 3 min.
13. Wash in stop bath for 30 sec.
14. Fix in rapid fixer for 5 min.
15. Wash in running water for 10 min.
16. Air dry and observe under the light microscope.

Control Experiments for in Situ Hybridization

At least two control experiments should be routinely performed: (1) The tissue sections from an untreated species which has no cross-hybridizing mRNA should be included and processed in the same manner as the experimental sections; (2) the specimen for *in situ* hybridization should be incubated with the hybridization mixture without the radioactive DNA probe. If the control sections show significant amounts of silver grains compared to the experimental one, the experiment should be repeated.

Pitfalls in Techniques and Interpretation

Immunocytochemistry

In all immunocytochemical experiments, control sections should always be included. Occasionally the control sections may also show some spurious immunostaining. This may be due to (1) the control serum containing some specific antibody against some unknown cross-reacting tissue antigen, and (2) more importantly, the peroxidase-labeled second antibody cross-reacting with local tissue immunoglobulins, resulting in some staining in the control sections. Whenever the control sections show any staining, the whole experiment should be repeated. Since peroxidase and its substrate have some nonspecific affinity to the damaged tissues, dead cells, and nucleic acids, an interpretation of the results from severely damaged tissue (or tissues showing severe postmortem change) is usually difficult. These nonspecific stainings and some background staining can be minimized by incubation for more prolonged periods with serial dilutions of primary antibody. Similarly, further dilutions of the peroxidase-conjugated second antibody and PAP concentration may also reduce background staining. In electron microscopic immunocytochemistry, diaminobenzidine (DAB) reaction product may travel some distance by diffusion, thus caution should be exercised in the interpretation in such experiments. One way of minimizing the diffusion artifact is to decrease the concentration of DAB and H_2O_2 or to reduce the incubation time.

In Situ Nucleic Acid Hybridization

Proper tissue preparation is critical for *in situ* nucleic acid hybridization experiments. In our experience, the best results have been obtained by perfusion fixation. Furthermore, certain cDNAs show affinity to the surfaces of erythrocytes, resulting in silver grains appearing on the erythrocytes. Also the effective concentration of the cDNA can be decreased if there is a lot of blood in the section. Therefore, an attempt should be made to remove all erythrocytes in the tissue section.

To lower the background silver grains in some sections, two precautions should be taken. First, after hybridization, the sections should be thoroughly washed. The slides should be incubated in a large beaker containing 2 liters of washing buffer with agitation and the buffer changed several times. Second, after coating with stripping film or emulsion, the slides should be dried completely before storing them in the light-proof box in the refrigerator. If the emulsion is not dried and the slides are stored in the refrigerator, a very strong background of silver grains will occur.

Conclusion

Different apolipoproteins and their mRNAs can be readily identified in the tissue section morphologically by both immunocytochemistry and *in situ* nucleic acid hybridization, if one follows carefully the procedures described in this chapter. The diffusion effect of DAB reaction product is a major drawback for the immunoperoxidase method. Techniques to minimize diffusion in this procedure are being investigated in various laboratories. Methods for the immunolocalization of more than one antigen on the same tissue section are also under intense study. Furthermore, the sensitivity of the *in situ* nucleic acid hybridization has been improved to the extent that as few as 1–10 molecules of RNA per cell can be detected. The future application of these refinements in morphological localization of biomolecules will be useful in our understanding of the biosynthesis as well as cellular metabolism of lipoprotein.

Acknowledgments

Experiments performed in the authors' laboratories described in this chapter were supported in part by grants from the National Institutes of Health, HL-27341 and IIL-16512. We thank Dr. Daniel Medina, Baylor College of Medicine, for a critical review of the manuscript.

[20] The Role of Apolipoprotein Processing in Receptor Recognition of VLDL

By SANDRA H. GIANTURCO and WILLIAM A. BRADLEY

Introduction

The metabolism of lipoproteins is determined in large part by their component apolipoproteins which target the particles to specific cell surface receptors or enzymes. Lipoproteins coexist in plasma with proteases. Conceivably these proteases could specifically modify *in vivo* critical apolipoproteins and thereby affect the metabolic fate of the lipoproteins. Indeed, a specific, single bond cleavage *in vitro* of a critical apolipoprotein (apo) E molecule by thrombin or by proteases associated with very-low-density lipoprotein (VLDL) abolishes or greatly diminishes the binding of hypertriglyceridemic (HTG) VLDL S_f 100–400 to the LDL receptor[1,2] while enhancing binding to the β-VLDL receptor.[3] Such seemingly specific proteolysis which is associated with profound functional changes may be a form of "processing" involved in the metabolic routing of lipoproteins. This chapter will focus on the effects of processing of apoE and apoB in VLDL on the interaction of the lipoprotein with the LDL receptor *in vitro*.

Very-low-density lipoproteins (VLDL) ($d < 1.006$) are heterogeneous in size, composition, density, cellular interactions, and metabolic fate. There is a continuum of apoB-containing lipoproteins from the largest, lightest VLDL, which are triglyceride rich and contain in addition to apoB many moles of apoC peptides and at least 1 mol of apoE, to low-density lipoproteins (LDL), which are cholesteryl ester rich and contain apoB as the only apoprotein. Upon hydrolysis and removal of triglyceride by lipoprotein lipase and loss of water-soluble, transferable surface apolipoproteins and phospholipid, most large VLDL are converted into small VLDL. Intermediate-density lipoproteins (IDL) are derived primarily from small VLDL and LDL in turn are derived primarily from IDL. Metabolic studies demonstrate that not all large VLDL are converted into LDL, however. In hypertriglyceridemia, as much as two-thirds of total VLDL are not catabolized to LDL but are lost directly from the plasma compartment.[4] Cell culture studies have provided potential cellular mechanisms to explain the direct loss of VLDL from plasma.[5–9]

Large VLDL (S_f 60–400) from hypertriglyceridemic (HTG) subjects,

[1] W. A. Bradley, E. B. Gilliam, A. M. Gotto, Jr., and S. H. Gianturco, *Biochem. Biophys. Res. Commun.* **109,** 1360 (1982).
[2] S. H. Gianturco, A. M. Gotto, Jr., S-L. C. Hwang, J. B. Karlin, A. H. Y. Lin, S. C. Prasad, and W. A. Bradley, *J. Biol. Chem.* **258,** 4526 (1983).
[3] S. H. Gianturco, A. M. Gotto, Jr., and W. A. Bradley, *Adv. Exp. Med.* **183,** 47 (1985).
[4] M. F. Reardon, N. H. Fidge, and P. J. Nestel, *J. Clin. Invest.* **61,** 850 (1978).
[5] S. H. Gianturco, A. M. Gotto, Jr., R. L. Jackson, J. R., Patsch, H. D. Sybers, O. D. Taunton, D. L. Yeshurun, and L. C. Smith, *J. Clin. Invest.* **61,** 320 (1978).
[6] A. Poyser and P. J. Nestel, *Artery* **6,** 122 (1979).
[7] S. H. Gianturco, F. B. Brown, A. M. Gotto, Jr., and W. A. Bradley, *J. Lipid Res.* **23,** 984 (1982).
[8] S. H. Gianturco, W. A. Bradley, A. M. Goto, Jr., J. D. Morrisett, and D. L. Peavy, *J. Clin. Invest.* **70,** 168 (1982).
[9] B. J. Van Lenten, A. M. Fogelman, M. M. Hokom, L. Benson, M. E. Haberland, and P. A. Edwards, *J. Biol. Chem.* **258,** 5151 (1983).

in sharp contrast to VLDL S_f 60–400 from normal subjects, bind to and are degraded via two genetically distinct cell surface receptors: the classic LDL receptor[5,7] and the β-VLDL receptor present on macrophages.[8,9] These receptor-mediated pathways, originally described by Goldstein and Brown and colleagues,[10,11] afford at least two potential direct cellular catabolic routes for VLDL S_f 60–400 in hypertriglyceridemic subjects which are not available for VLDL S_f 60–400 in normal subjects. Uptake by these receptor-mediated routes could account for the observed direct loss of VLDL apoB from plasma in hypertriglyceridemic subjects.

In contrast to the functional distinctions between large VLDL from normal and hypertriglyceridemic subjects, a population of small VLDL S_f 20–60 from both normal and hypertriglyceridemic subjects bind to the LDL receptor.[7,12] This provides a cellular catabolic route for direct loss of small VLDL from plasma in normal as well as hypertriglyceridemic subjects, which appears to occur *in vivo*.

Although the LDL receptor can recognize either apoE or apoB in small particles, apoB of large VLDL S_f 60–400 is not recognized by the LDL receptor[2,7] or by certain antisera against apoB.[13] ApoE of a particular conformation is absolutely required for the binding to and uptake of large S_f 100–400 VLDL by the LDL receptor.[2,7] This conformation of apoE is generally present in S_f 100–400 VLDL from hypertriglyceridemic subjects but not in the corresponding VLDL from normal subjects.[3,14] In hypertriglyceridemic subjects, in contrast to normal subjects, most of the apoE in plasma is associated with VLDL rather than HDL.[15] Analogous to this transfer *in vivo*, exogenous apoE can be incorporated into normal S_f 60–400 VLDL *in vitro*, creating the LDL receptor recognition domain.[7]

Thus we hypothesize that the apoE in HTG-VLDL S_f 100–400 involved in binding to the LDL receptor is that which transfers into the VLDL in hypertriglyceridemic subjects but not normal subjects. This apoE is accessible not only to the LDL receptor but to thrombin as well: thrombin cleaves only the apoE in HTG-VLDL which binds to the receptor, thereby abolishing its binding to the LDL receptor, leaving other apoE molecules intact, indicating that these resistant apoE molecules are

[10] J. L. Goldstein and M. S. Brown, *Annu. Rev. Biochem.* **46,** 897 (1977).

[11] M. S. Brown and J. L. Goldstein, *Annu. Rev. Biochem.* **52,** 223 (1983).

[12] S. H. Gianturco, C. J. Packard, J. Shepherd, L. C. Smith, A. L. Catapano, H. D. Sybers, and A. M. Gotto, Jr., *Lipids* **15,** 456 (1980).

[13] G. Schonfeld, W. Patsch, B. Pfleger, J. L. Witztum, and S. W. Weidman, *J. Clin. Invest.* **64,** 1288 (1979).

[14] W. A. Bradley, S-L. C. Hwang, J. B. Karlin, A. H. Y. Lin, S. C. Prasad, A. M. Gotto, Jr., and S. H. Gianturco, *J. Biol. Chem.* **259,** 14728 (1984).

[15] C. B. Blum, L. Aron, and R. Sciacca, *J. Clin. Invest.* **66,** 1240 (1980).

in a conformation inaccessible to both thrombin and the LDL receptor.[2] In contrast to the case with HTG-VLDL, all of the apoE of normal VLDL is not only inaccessible to the LDL receptor[7] but to thrombin as well.[3] We have suggested that this inaccessible conformation of apoE, which is present in both normal and HTG-VLDL (1 to 2 mol/particle), is that present in nascent, hepatic VLDL ("endogenous" apoE) and the accessible conformation present in HTG-VLDL is that which transfers into the VLDL after it reaches the plasma compartment (1–4 mol of "exogenous" apoE).

Thrombin hydrolyzes isolated apoE or the accessible apoE in HTG-VLDL initially and primarily at one out of eight potential cleavage sites (the Arg–Ala bond at amino acid residues 191–192), producing two major fragments which we have termed E-12 and E-22.[1,2] E-22 is the N-terminal two-thirds and contains a domain which appears to be important in binding to the LDL receptor.[16] E-12 is the C-terminal one-third which contains a primary lipid binding region of the molecule. Even without removal of the E-22 fragment, thrombin cleavage abolishes or significantly reduces the binding of HTG-VLDL S_f 100–400 to the LDL receptor and, consequently, its suppression of HMG-CoA reductase activity.[2] Reincorporation studies confirm that this loss of biological activity was indeed due to the cleavage of apoE: reincorporation of intact apoE into thrombin-inactivated HTG-VLDL S_f 100–400 (~1 mol/mole particle) restores efficient receptor-mediated reductase suppression. Addition of the E-12 fragment failed and the E-22 fragment was on a molar basis only 50% as effective as intact apoE in restoring the ability of thrombin-inactivated HTG-VLDL S_f 100–400 to suppress reductase activity. Thus the E-12 domain is crucial in maintaining the intact molecule in the appropriate conformation for receptor recognition in VLDL. Intact apoE of the appropriate conformation, at least 1 mol/mole VLDL, is required for uptake of the largest HTG-VLDL subclass by the LDL receptor-mediated pathway. Although both proteases hydrolyze apoB, thrombin produced no change in the immunoreactivity of apoB of HTG-VLDL S_f 100–400 and neither trypsin nor thrombin affected the ability of LDL to suppress HMG-CoA reductase.[2] Retention of biological activity of LDL after proteolysis is well documented.[17,18] Thus apoB-mediated binding, in contrast to apoE-mediated binding, is not affected by proteolysis and the inactivation of HTG-VLDL S_f 100–400 by thrombin was due to its effect on apoE and not due to an effect on apoB.

[16] R. W. Mahley and T. L. Innerarity, *Biochim. Biophys. Acta* **737,** 197 (1983).
[17] G. A. Coetzee, W. Gevers, and D. R. Van der Westhauyzen, *Artery* **7,** 1 (1980).
[18] K.-S. Hahm, M. J. Tikkanen, R. Darger, T. G. Cole, J. M. Davie, and G. Schonfeld, *J. Lipid Res.* **24,** 877 (1983).

Thus the binding determinants for large lipoproteins are clearly different from those of small lipoproteins: apoE of the thrombin-accessible conformation is necessary and apoB insufficient for uptake of VLDL S_f 100–400 whereas apoB alone is sufficient for receptor binding of LDL.

These studies raised the question, where in the metabolic continuum from HTG-VLDL S_f 100–400 to LDL does this switch in receptor binding determinants occur and, therefore, where does the potential of routing lipoproteins by proteolysis end? To answer these questions we took advantage of the abilities of thrombin and trypsin to inactivate apoE but not apoB as a ligand for the LDL receptor.[2,14] These studies demonstrate that apoE of the thrombin-accessible conformation is required for the receptor binding of HTG-VLDL S_f 60–100 and a small population of HTG-VLDL S_f 20–60, as was the case for HTG-VLDL S_f 100–400, but not for the receptor binding of most of HTG-VLDL S_f 20–60, IDL, or LDL.[14] The shift from apoE to apoB as the primary LDL receptor binding determinant in the hypertriglyceridemic VLDL S_f 100–400 to LDL series therefore occurs precisely within the region in the metabolic cascade where one first observes binding, uptake, and degradation of lipoproteins from subjects with normal plasma triglycerides: within the smallest VLDL subclass of S_f 20–60.[7,12,14] These conclusions have been confirmed using monoclonal antibodies against apoE and apoB.[19]

The apoB in the larger VLDL (S_f 60–400) appears to be in an inappropriate conformation for receptor binding. With the decrease in surface area that occurs upon lypolysis, the conformation of apoB changes and such a change appears to be required before apoB can interact with the LDL receptor.[13,18–20] The inability of apoB to mediate uptake of large VLDL by the LDL receptor pathway could serve to protect cells from the potentially toxic effects of uptake of triglyceride-rich particles. The production of large amounts of intracellular fatty acids upon lysosomal hydrolysis may be responsible for the observed toxicity of HTG-VLDL S_f 100–400 but not normal VLDL S_f 100–400 in endothelial cells and may cause endothelial injury if it occurs *in vivo*.[21]

Thus apoB normally is the physiologically important ligand for the LDL receptor. Even though native IDL and VLDL S_f 20–60 contain apoE, and even LDL contain traces of apoE, our studies indicate that apoB normally mediates binding of most of the VLDL S_f 20–60, IDL, and

[19] E. S. Krul, M. S. Tikkanen, T. G. Cole, J. M. Davie, and G. Schonfeld, *J. Clin. Invest.* **75,** 361 (1985).

[20] A. L. Catapano, S. H. Gianturco, P. K. J. Kinnunen, S. Eisenberg, A. M. Gotto, Jr., and L. C. Smith, *J. Biol. Chem.* **245,** 1007 (1979).

[21] S. H. Gianturco, S. G. Eskin, L. T. Navarro, C. J. Lahart, L. C. Smith, and A. M. Gotto, Jr. *Biochim. Biophys. Acta* **618,** 143 (1980).

LDL to the LDL receptor, as originally proposed by Brown and Goldstein.[22]

We suggest that apoE only plays a role in the binding of lipoproteins to the LDL receptor in abnormal, pathologic situations such as hypertriglyceridemia, abetalipoproteinemia, or hypobetalipoproteinemia. The redundancy in receptor recogition could be beneficial in hypo- or abetalipoproteinemia, in which HDL containing apoE rather than LDL could deliver cholesterol to cells *in vivo,* as has been demonstrated *in vitro.*[23] The ability of apoE, in contrast to apoB, to bind to the receptor in large as well as small particles could be beneficial in hypertriglyceridemia, in which the abnormal LDL receptor-mediated uptake of large HTG-VLDL may be an alternate route for degrading the excess VLDL that are not lipolized efficiently.

Thus proteolytic processing could play a role in diverting large HTG-VLDL away from the LDL receptor as seen *in vitro* by the thrombin inactivation of apoE as a ligand for the LDL receptor. By contrast, the binding of IDL and LDL and most of the small VLDL (S_f 20–60) to the LDL receptor is unaffected not only by thrombin cleavage but even by extensive digestion with trypsin, in which only small apoB fragments remain.[14] The observation that thrombin-treated VLDL but not thrombin-treated LDL showed enhanced uptake by macrophages[3] suggests that such processing *in vivo* could divert VLDL (but not LDL) away from cells possessing LDL receptors to macrophages for disposal. Processing could therefore provide an alternate cellular catabolic route for the triglyceride-rich lipoproteins without disturbing the normal catabolic route of LDL, which is important in maintaining cellular cholesterol homeostasis throughout the body.

Lipoprotein Preparation

VLDL Isolation and Subfractionation

Principle. Because of their extreme heterogeneity, metabolic and cell culture studies using unfractionated VLDL are often ambiguous and difficult to interpret. Total VLDL have traditionally been designated as the $d < 1.006$ fraction of plasma. Using this criterion alone, a great percentage of the protein in the "total VLDL" isolated in a single ultracentrifugal step is not really VLDL protein at all, particularly in VLDL from normal

[22] M. S. Brown and J. L. Goldstein, *Proc. Natl. Acad. Sci. U.S.A.* **71,** 788 (1974).
[23] T. L. Innerarity, T. P. Bersot, K. S. Arnold, K. H. Weisgraber, P. A. Davis, T. M. Forte, and R. W. Mahley, *Metabolism* **33,** 186 (1984).

subjects. In hypertriglyceridemic subjects an average of only 57 ±13% (n = 6) of the protein in the d < 1.006 fraction is VLDL protein; the remainder of the protein (43 ±12%) consists primarily of IDL, but other plasma proteins and even LDL contaminate this fraction.[5,14]

Rate zonal ultracentrifugation demonstrates that the d < 1.006 fraction obtained by flotation or "total VLDL" isolated by gel filtration chromatography from normal plasma contain IDL (Fig. 1).[5] The presence of IDL explains why "total VLDL" from normal subjects, in contrast to purified VLDL S_f 60–400, can cause receptor-mediated reductase suppression.[5,7,12] "Total VLDL" so isolated from normal subjects actually con-

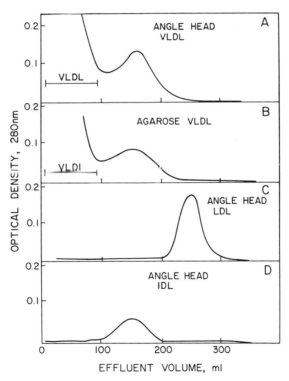

FIG. 1. Contamination of d < 1.006 VLDL by IDL as determined by rate zonal ultracentrifugation. Normal VLDL isolated at d < 1.006 in an angle head rotor (A) and by gel filtration over BioGel A5m (B) contain IDL. (C) shows the relative position of LDL (angle head rotor, d = 1.02–1.063), and (D) shows the relative position of IDL (angle head rotor d = 1.006–1.02) when these lipoproteins are recentrifuged in the zonal rotor. Fractions in the barred area of (A) and (B) are VLDL. Only the first 400 ml of the total rotor volume (665 ml) was collected. Reproduced from *J. Clin. Invest.* **61,** 320 (1978) by copyright permission of The American Society for Clinical Investigation.

tain very few triglyceride-rich lipoproteins (S_f 60–400). And, among the purified VLDL subclasses, only approximately one-third of the particles is S_f 60–400 and two-thirds is S_f 20–60 in normal subjects.[14] In general, the opposite is true of HTG-VLDL subclasses: two-thirds is VLDL S_f 60–400 while one-third is S_f 20–60.[14] Thus if one compares compositions of total HTG-VLDL vs total normal VLDL, even after eliminating contaminating IDL, one is given the false impression that HTG-VLDL particles are enriched in triglyceride in comparison to normal VLDL. If one compares the total compositions of more homogeneous subfractions of VLDL, however, it is apparent that normal VLDL subclasses are qualitatively quite similar to HTG-VLDL subclasses in major components (see the table), but the distribution of particles is different. When VLDL are subfractionated and freed of contaminating lipoproteins by either rate zonal ultracentrifugation[5] or cumulative flotation,[7,12] normal VLDL S_f 60–400 do not bind to the LDL receptor whereas HTG-VLDL S_f 60–400 do. This great functional distinction is due to relatively subtle differences in apoE content (as little as 2% of the protein mass) and conformation rather than gross differences in components.[2,14] These studies emphasize the need in cell and metabolic studies to fractionate total VLDL ($d < 1.006$ fraction) due to its great heterogeneity not only in VLDL subclasses but due to contamination by other classes of lipoprotein, especially IDL.

Initial VLDL Isolation

Reagents

EDTA, pH 7.4, 0.1 M, 1 ml/100 ml plasma
NaN$_3$, 1 M, 0.1 ml/100 ml plasma
Trasylol (FBA Pharmaceuticals), 10,000 KIU/ml, 0.5 ml/100 ml plasma
Phenylmethylsulfonyl fluoride (PMSF) (Sigma), 10 mM in dioxane, 0.1 ml/100 ml plasma, added drop by drop

Blood for the isolation of plasma lipoproteins and lipoprotein-deficient serum ($d > 1.21$) is obtained by venipuncture (500 ml) or by plasmapheresis and collected in citrated bags by a trained technician in a hospital blood donor center. All individuals have fasted for 12 to 14 hours prior to donation. For the isolation of normal VLDL, LDL, and lipoprotein-deficient serum, healthy adult males with normal lipid values donate blood. Diagnoses of hyperlipoproteinemia types are based on commonly used criteria.[24]

---

[24] D. S. Fredrickson, J. L. Goldstein, and M. S. Brown, in "The Metabolic Basis of Inherited Diseases" (J. G. Stanbury, M. F. Wyngaarden, and D. S. Fredrickson, eds.), 4th Ed., p. 604. McGraw Hill, New York, 1978.

COMPOSITIONS OF NORMAL AND HTG-VLDL SUBCLASSES

VLDL fraction	Percentage weight of component[a]				
	Protein	Phospholipid	Cholesterol	Cholesteryl ester	Triglyceride
VLDL S_f 100–400					
HTG type 4	5.0 ± 1.0	15.5 ± 1.4	5.8 ± 1.1	8.4 ± 1.9	65.4 ± 3.4
(N = 4)					
Normal (N = 5)	5.9 ± 0.7	18.5 ± 1.9	5.9 ± 1.4	6.1 ± 1.6	63.6 ± 2.6
VLDL S_f 60–100					
HTG type 4	7.2 ± 1.2	18.8 ± 2.7	6.7 ± 0.3	11.9 ± 4.9	55.3 ± 5.1
Normal	8.7 ± 1.4	21.8 ± 3.3	5.8 ± 1.0	8.2 ± 2.9	55.6 ± 3.9
VLDL S_f 20–60					
HTG type 4	10.8 ± 1.4	20.6 ± 2.3	7.0 ± 0.8	19.6 ± 5.5	42.0 ± 6.6
Normal	11.4 ± 4.6	23.8 ± 3.3	7.5 ± 1.8	15.9 ± 4.5	41.3 ± 2.6

[a] The total weight of each particle was taken as the sum of the amounts of cholesterol, cholesteryl ester, triglyceride, phospholipid, and total Lowry protein. The values given are the means ± SD.

Because of the presence in plasma of enzymes such as lecithin cholesterol acyl transferase, transfer proteins, and proteases which can affect VLDL structure and function, VLDL is isolated only from fresh plasma, and the initial centrifugation is begun no more than 3 hr after the plasma is obtained. Prior to preparation the unopened bag is held at 4°. Then the plasma volume is determined, aliquots removed for lipid analyses, and the plasma is placed in a large beaker with a magnetic stirring bar. EDTA, NaN$_3$, PMSF, and Trasylol are then added to the plasma while it stirs. PMSF should be added dropwise and under a fume hood because of the toxicity of dioxane.

After stirring for 5 min, the plasma is loaded into Beckman QuickSeal tubes using a syringe fitted with a large-bore needle, avoiding bubbles. The tubes are then sealed with the Beckman QuickSeal tube sealing apparatus, placed in a 60 Ti or 50.2 Ti rotor and centrifuged for 18–22 hr at 14° and 45,000 rpm. The top $d < 1.006$ fraction, which contains the impure VLDL, is then isolated by slicing the tubes with a tube slicer (Beckman) beneath the VLDL. The top fraction (~10 ml) is aspirated using a plastic syringe fitted with a large-bore needle through a hole made by snipping off the sealed filling port with sharp scissors. The $d < 1.006$ fraction is stored in plastic at 4° and is fractionated into relatively homogeneous VLDL subclasses by the cumulative flotation method of Lindgren et al.[25]

[25] F. T. Lindgren, L. C. Jensen, and F. T. Hatch, in "Blood Lipids and Lipoproteins" (G. J. Nelson, ed.), p. 181. Wiley (Interscience), New York, 1972.

VLDL Subfractionation by Cumulative Flotation. Total VLDL ($d <$ 1.006) can be separated into three (3) subclasses of S_f 100–400, S_f 60–100, and S_f 20–60 by Lindgren's method.[25] Twelve milliliters of sample can be processed at one time, either plasma or the $d <$ 1.006 fraction of plasma. The $d <$ 1.006 fraction from normal but not hypertriglyceridemic plasma can be concentrated prior to this step. We concentrate this fraction at 4° by placing it in dialysis tubing which is then covered with dry Sephadex G-75 (Pharmacia). This is conveniently done by placing the filled dialysis bag in a bed of dry Sephadex in a large plastic weighing boat in a refrigerator. As the Sephadex absorbs water it is removed by peeling the wet layer off the bag (wear clean, disposable gloves) and replaced with dry Sephadex.

Reagents

NaCl
Ethylenediaminetetraacetic acid (EDTA), disodium salt
NaBr
NaOH, 1 N

A. Stock solutions
   1. Density = 1.006 g/ml
      11.40 g of NaCl and 0.1 g of EDTA, disodium salt, are added to a 1000-ml volumetric flask; 500 ml of H_2O and 1 ml 1 N NaOH are added, and the solids are dissolved by mixing. The flask is filled to volume and 3 ml additional H_2O are added; NaCl concentration = 0.195 M.
   2. Density = 1.182 g/ml
      24.98 g of NaBr is added to 100.0 ml of the above $d =$ 1.006 solution; 0.195 M NaCl and 2.44 M NaBr.
B. Working solutions
   VLDL, $d =$ 1.065
   1. $d =$ 1.006 = stock solution no. 1
   2. $d =$ 1.0117 = 100 ml of 1 + 3.35 ml of stock no. 2
   3. $d =$ 1.0197 = 100 ml of 1 + 8.44 ml of stock no. 2
   4. $d =$ 1.0271 = 100 ml of 1 + 13.62 ml of stock no. 2
   5. $d =$ 1.0336 = 100 ml of 1 + 18.6 ml of stock no. 2
   6. $d =$ 1.0464 = 100 ml of 1 + 29.8 ml of stock no. 2

Rotor: SW40.1 or SW41 (Beckman)
Ultra-clear tubes (Beckman)

Procedure

Sample Preparation. Have all solutions at approximately 23° and perform the centrifugations at this temperature. Adjust the sample to $d =$

1.065 g/ml: 12-ml sample (plasma or $d < 1.006$ fraction of plasma) $+1.014$ g NaCl; swirl to dissolve. Try to avoid bubbles.

Gradient Preparation. Slant the tube and carefully add each of the following solutions dropwise from a pipet held against the side of the tube just above the fluid:

1. 0.5 ml stock solution 2 ($d = 1.182$) to form a cushion on the bottom
2. Two (2)-ml sample containing NaCl ($d = 1.065$ g/ml)
3. One (1)-ml working solution 6 ($d = 1.0464$ g/ml)
4. One (1)-ml working solution 5 ($d = 1.0336$ g/ml)
5. Two (2)-ml working solution 4 ($d = 1.0271$ g/ml)
6. Two (2)-ml working solution 3 ($d = 1.0197$ g/ml)
7. Two (2)-ml working solution 2 ($d = 1.0117$ g/ml)
8. Two (2)-ml working solution 1 ($d = 1.006$ g/ml)

Note: There is usually not enough room in the tubes for the full 2 ml of the last solution ($d = 1.006$). Leave about $\frac{1}{8}$-in. space at the top of each tube so that the tubes can be pulled with tweezers into view after the first spin without disturbing the VLDL at the top of the tube.

We use a Beckman ultracentrifuge for these separations, set at the fastest acceleration rate, with the brake on slow, the temperature at 23°, and the speed 35,000 rpm. For VLDL (S_f 100–400), spin for 128 min full speed (35,000 rpm). After the spin, carefully remove the top 0.5 ml from each tube with a Pasteur pipet, preferably plastic. The pipet tip can be broken so that the surface film is more readily aspirated (the tip should not dip beneath the surface or VLDL floating on top will not be removed and will contaminate the next fraction). This is the S_f 100–400 VLDL fraction. Then immediately spin the tubes again for 86 min at full speed (35,000 rpm). Again carefully collect the top 0.5 ml from each tube for the S_f 60–100 VLDL fraction. Check to be sure the tubes are still balanced and immediately spin the tubes again for 18 hr, 22 min at 35,000 rpm. The top 0.5-ml fraction contains the S_f 20–60 VLDL fraction. All VLDL samples are stored in plastic, not glass.

LDL Isolation

LDL, $d = 1.03$–1.05, is isolated from normal plasma by sequential ultracentrifugation using solid KBr to adjust the density.[26] Each standard spin is at 14°, and 45,000 rpm for 18–22 hr in a 60 Ti or 50.2 Ti rotor. Either before or after VLDL are removed from plasma by ultracentrifugation at plasma density ($d = 1.006$), the bottom fraction is adjusted to $d = 1.03$ using KBr and the formula

$$X = [V(d_f - d_i)]/[(1 - \bar{V})d_f]$$

[26] R. J. Havel, H. A. Eder, and J. J. Bragdon, *J. Clin. Invest.* **34,** 1345 (1955).

In this formula, X = grams KBr; d_i = initial density; V = volume in milliliters; d_f = final density; $\bar{V}$ = 0.312, the partial specific volume of KBr.[27]

After the standard spin in polycarbonate tubes or QuickSeal tubes (14°, 45,000 rpm, 18–22 hr in a 60 Ti rotor), the top fraction containing the IDL (and VLDL if not previously removed) is removed with a Pasteur pipet or by tube slicing. The bottom fraction is adjusted to d = 1.05 to float the LDL under the standard spin conditions. The LDL is then washed at each density to provide a fraction that is as free as possible of soluble peptides from contaminating lipoproteins because these complicate receptor binding studies. First, the LDL (≤10 ml) is overlayed with a KBr solution of d = 1.025 (~20 ml) in polycarbonate ultracentrifuge tubes (Beckman). After spinning, the relatively clear top portion is removed with a Pasteur pipet and discarded, leaving the orange LDL at the bottom of the tube. This LDL is removed with a pipet and adjusted to d = 1.055 with KBr, placed in polycarbonate tubes, and overlayered with a KBr solution of d = 1.05. After the standard spin, the orange LDL is at the top of the tube and is removed with a Pasteur pipet.

The LDL are dialyzed at 4° extensively against 0.15 M NaCl containing 20 mM Tris-HCl, 1 mM EDTA, and 1 mM NaN$_3$, pH 7.4. All lipoproteins are sterilized by passage over 0.45-μm Millex filters into sterile plastic tubes and stored at 4°. Protein content is determined by a modification[28] of Lowry's method.[29]

Lipoprotein-Deficient Serum Preparation

The d > 1.21 fraction of normal plasma is used to prepare lipoprotein-deficient serum (LPDS). Plasma from normal individuals is adjusted to d = 1.21 with solid KBr and centrifuged in QuickSeal tubes in a 60 Ti or a 50.2 Ti rotor at 45,000 rpm, 14°, for 40 hr. The lipoprotein fraction is removed after slicing the tube and the bottom fraction is dialyzed extensively against 0.15 M NaCl, 20 mM Tris-HCl, pH 7.4, 0.3 mM EDTA, pH 7.4 (buffer A) at 4°. This is converted into serum by initiating clotting with thrombin (Parke-Davis), 20 U/ml in a beaker for 15 min at room temperature.[30] The clot is removed with applicator sticks. The LPDS is sterilized (Corning 200-ml filter units fitted with a 0.45-μm fitter) and stored at −20° in sterile plastic tubes.

[27] C. M. Radding and D. Steinberg, *J. Clin. Invest.* **39**, 1560 (1960).
[28] A. Helenius and K. Simons, *Biochemistry* **10**, 2542 (1971).
[29] O. H. Lowry, N. J. Rosebrough, A. L. Farr, and R. J. Randall, *J. Biol. Chem.* **193**, 265 (1951).
[30] M. S. Brown, S. E. Dana, and J. L. Goldstein, *J. Biol. Chem.* **249**, 789 (1974).

Cell Studies: Interactions of VLDL with the LDL Receptor

Principle

Cultured human skin fibroblasts are the most convenient model system for studying interactions of lipoproteins with the LDL receptor and these methods are based on those of Goldstein et al.[31] Fibroblasts are easily grown, they do not contain other lipoprotein receptors (such as the β-VLDL receptor) which also recognize VLDL, and there are a number of well-defined mutants available commercially which are useful experimental controls. Normal fibroblasts derived from newborn foreskin retain a predictable LDL pathway longer in culture than do fibroblasts derived from adult skin biopsies. For instance, newborn fibroblasts can be used for at least 10 transfers (split ratio of 1 : 4), whereas adult-derived cells are unreliable as early as the third transfer (unpublished observation).

Several criteria must be met to demonstrate that a lipoprotein binds specifically to the LDL receptor. The lipoprotein must exhibit (1) saturable binding, uptake, and degradation; (2) competition with specific LDL binding, uptake, and degradation; and (3) regulation of an intracellular event known to be mediated by the LDL receptor pathway [reductase suppression or, if the lipoprotein is cholesterol rich, acyl : cholesterol acyltransferase (ACAT) activation]. Binding studies alone are insufficient to demonstrate specific receptor binding; a physiological endpoint must also be monitored. As originally pointed out by Dana, Brown, and Goldstein, LDL can bind to glass beads with apparent high affinity, competability, and therefore apparent specificity.[32] Thus one must be sure that an intracellular endpoint is also regulated appropriately by the lipoprotein in question.

Binding studies with iodinated VLDL are complicated by several additional factors not affecting studies with LDL.[7] Even more caveats than those for studying LDL binding must be applied to studies of VLDL/receptor interaction. This is due both to the large size and nonspecific stickiness of VLDL and to the presence in VLDL of soluble, transferable peptides; both the VLDL and the soluble peptides have a high affinity for negatively charged surfaces such as glass, plastic, and cell membranes. The soluble apolipoproteins, the apoC peptides and apoE, transfer between lipoproteins. Thus when iodinated VLDL are mixed with unlabeled lipoproteins, the labeled peptides in VLDL can transfer to originally unlabeled lipoproteins. Moreover, both VLDL and the soluble apolipopro-

---

[31] J. L. Goldstein, S. K. Basu, and M. S. Brown, this series, Vol. 98, p. 241.
[32] S. E. Dana, M. S. Brown, and J. L. Goldstein, Biochem. Biophys. Res. Commun. 74, 1369 (1978).

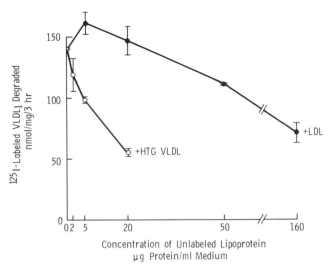

FIG. 2. Relative effects of hypertriglyceridemic VLDL and LDL on the LDL receptor-mediated degradation of ^{125}I-labeled hypertriglyceridemic VLDL. The fibroblasts were grown and incubated as described, then incubated with ^{125}I-labeled VLDL (1.2 μg of protein/ml, S_f 100–400 VLDL from a primary type 4 subject) alone or with the indicated concentrations of unlabeled homologous VLDL or LDL. Trichloroacetic acid-soluble, noniodide, nonlipid radioactivity in the medium represents ^{125}I-labeled VLDL degradation. Each point is the average of values from duplicate dishes; variations between dishes are indicated by bars. Normal LDL (●); hypertriglyceridemic S_f 100–400 VLDL (○). Reproduced with permission from *J. Lipid Res.* **23,** 984 (1982).

teins adsorb both to plastic tissue culture dishes and to cell surfaces.[7,33] Thus binding studies with labeled VLDL can be meaningless without appropriate controls, including assessment of an intracellular receptor-dependent endpoint in addition to blanks run in dishes without cells for each point. Even with the appropriate controls, binding studies with labeled VLDL are often ambiguous and difficult to interpret. For example, in the competition experiment shown in Fig. 2, it looks as though iodinated HTG-VLDL S_f 100–400 and LDL are binding to separate sites, since low levels of LDL do not compete with the labeled HTG-VLDL. In fact, it looks as though the LDL actually enhanced the degradation of the HTG-VLDL, and only at high levels of LDL does one observe competition. The complementary experiment, with the label on LDL rather than HTG-VLDL, indicates that the HTG-VLDL compete as well or better than LDL itself for specific LDL binding, uptake, and degradation (Fig. 3). Gel filtration experiments indicate that labeled soluble peptides from

[33] L. Holmquist, *J. Lipid Res.* **22,** 357 (1982).

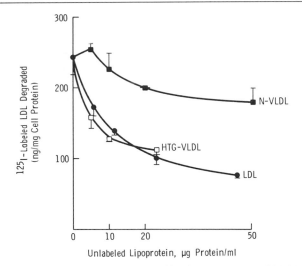

FIG. 3. Competitive effects of hypertriglyceridemic VLDL, normal VLDL, and LDL on the degradation of ^{125}I-labeled LDL. ^{125}I-labeled LDL (5 μg protein/ml) alone and with unlabeled lipoproteins were incubated for 4 hr with normal fibroblasts preincubated in LPDS-containing medium for 36 hr. Trichloroacetic acid-soluble, noniodide, nonlipid radio-activity in the medium represents ^{125}I-labeled LDL degradation. Each point represents the mean ± SD of values from three dishes, expressed as nanograms ^{125}I-labeled LDL acid-soluble material formed per milligram of total cell protein. Normal LDL (●); hypertriglyceridemic VLDL S_f 60–400 (□); normal VLDL S_f 60–400 (■). Reproduced with permission from *J. Lipid Res.* **23**, 984 (1982).

the labeled HTG-VLDL S_f 100–400 transfer to the originally unlabeled LDL (unpublished observation). The LDL and associated labeled peptides can then be taken up and degraded via the LDL receptors on the cell not occupied by HTG-VLDL (most receptors are not occupied by ^{125}I-labeled HTG-VLDL S_f 100–400 at the low levels used in these competition experiments).[7]

Direct binding studies must demonstrate saturability. Thus normal VLDL S_f > 60 can bind to both plastic tissue culture dishes and to cells, both normal and receptor negative, but this binding is not high affinity and therefore is a linear function of concentration rather than saturable at low levels which are sufficient to demonstrate saturability with HTG-VLDL.[7]

Even competition experiments with ^{125}I-labeled LDL are complicated because of the nonspecific adsorption of both normal and hypertriglyceridemic VLDL to negatively charged surfaces. Thus, even some preparations of normal VLDL S_f > 60 which do not bind specifically to the LDL receptor can compete with the binding of labeled LDL to the receptor. These preparations also compete with LDL for binding to empty tissue culture dishes.[7] Thus competition with LDL binding, although a necessary criterion to indicate specific interaction with the LDL receptor, is alone insufficient to prove that the VLDL are specifically competing for the receptor per se in the same way which unlabeled LDL compete. The affinity of VLDL preparations for negatively charged surfaces varies, but in general, HTG-VLDL have higher affinities than do normal VLDL. Because of the large size of VLDL relative to LDL, binding of VLDL to cell surfaces near LDL receptors can hinder LDL binding without actually binding to the receptor itself. If VLDL do not compete with high-affinity LDL binding, obviously they do not bind to the LDL receptor. If, however, VLDL compete with specific LDL binding, the additional criteria must be met before one can be sure that the VLDL are binding with high affinity to the LDL receptor itself. Normal VLDL $S_f > 60$, even those that exhibit some ability to compete with the surface binding of LDL, do not compete effectively with the high-affinity specific degradation of low levels of labeled LDL (Fig. 3). LDL that do get past the "screen" of nonspecifically bound large VLDL on the cell surface are rapidly internalized and degraded. Thus it is advantageous to perform competition experiments at 37° as well as at 4° so that degradation can be monitored. This also lessens the complications due to lipoprotein–plastic interactions. Moreover, recent studies by Hoeg et al. indicate that the LDL receptor is poorly detected at 4°;[34] this may be due to an effect of temperature on the receptor or on apoB or on both.

Cells

All cultures of human fibroblasts are established from the newborn preputial specimens (for normal cells) or purchased from the Human Genetic Mutant Cell Repository for either normal or LDL receptor mutants (Camden, NJ). Monolayer cultures are maintained in a humidified incubator under 5% CO_2 at 37° in 100-mm plastic tissue culture dishes (Falcon). Each dish contains 10 ml of complete medium consisting of Dulbecco's modified Eagle's medium (Grand Island Biological Co., Grand Island, NY, Cat. No. H21), supplemented with 10% (v/v) Nu-Serum (Collaborative Research), 2 mM glutamine, 36 mM NaHCO$_3$, 100 U/ml potassium penicillin G, and 100 μg/ml streptomycin sulfate. Cells are used between the third and seventh passage. The split ratio for cells in each passage is 1 : 4.

For experiments, confluent cell monolayers are dissociated from stock plates with trypsin (100 μg/ml; Grand Island Biological Co.) and the cells

[34] J. M. Hoeg, S. J. Demosky, Jr., E. J. Shaefer, T. E. Starzl, and H. B. Brewer, Jr., J. Clin. Invest. **73**, 429 (1984).

combined in a large sterilized beaker containing a stirring bar. Approximately 10^5 cells are then dispersed in 5 ml of complete medium into 60-mm plastic tissue culture dishes and placed in a 37° humidified incubator under 5% CO_2 for this and subsequent incubations of cells. When the cells are ~75% confluent (2 to 3 days), the medium is removed; the cells are washed with 3 ml of saline and placed on 3 ml of medium containing 5% human lipoprotein-deficient serum (LPDS) and preincubated for 24 hr for HMG-CoA reductase suppression studies or 36 hr for binding studies.

Intracellular Endpoint: Suppression of HMG-CoA Reductase Activity

Principle. HMG-CoA reductase suppression by a lipoprotein requires first that the lipoprotein bind to the LDL receptor and be internalized and hydrolyzed in the lysosomes. Reductase suppression in normal, but not receptor-negative, fibroblasts by a lipoprotein is a sensitive indicator that the lipoprotein was taken up and degraded by the LDL receptor pathway.

The intracellular endpoint of choice to indicate receptor-mediated uptake (or lack of uptake) of VLDL is HMG-CoA reductase suppression rather than activation of ACAT. This is because the latter is activated only after accumulation of a certain "threshold" level of cholesterol, whereas extremely low levels of cholesterol can rapidly suppress reductase activity. This is especially important in experiments to show loss of receptor interaction. Only if excess cholesterol is delivered does one obtain ACAT activation. The levels of lipoprotein required for rapid activation of ACAT are approximately 5- to 10-fold higher during the same time frame as is required for HMG-CoA reductase suppression.

Reductase suppression can be a more sensitive endpoint than binding assays to assess receptor function because one can take advantage of the effects of cumulative uptake. Even low levels of binding that cannot be detected by binding assays can be detected using reductase as an endpoint.[35] A 16-hr incubation for reductase suppression magnifies the effects seen in a 3- to 4-hr binding or reductase suppression experiment. This is especially valuable when testing for a loss of binding determinants upon proteolysis of the lipoprotein. If there is no suppression after a 16-hr incubation with a lipoprotein, one is more certain that there is no low-level receptor interaction not detectable by binding assays. Moreover, one is then assured that one is measuring a cellular receptor-dependent process and one needn't be concerned about nonspecific binding to plates or to nonreceptor membrane surfaces.

To test the effects of lipoproteins on HMG-CoA reductase activity,

[35] J. L. Goldstein, S. E. Dana, G. Y. Brunschede, and J. S. Brown, *Proc. Natl. Acad. Sci. U.S.A.* **72**, 1092 (1975).

the cells are first placed on medium containing 5% human lipoprotein-deficient serum (LPDS) and incubated for 24 hr, to allow HMG-CoA reductase activity to increase to unsuppressed levels.[36] Several different concentrations (0.5–15 μg protein/ml) of each lipoprotein to be tested are then added to duplicate dishes and incubated for 5 to 16 hr. (The effects of several levels of normal LDL on HMG-CoA reductase are included as a control in each experiment.) The medium is removed, the cell monolayers washed twice at room temperature with 2 ml of saline, and the cells scraped with a rubber policeman into 2.0 ml of 0.15 M NaCl, 50 mM Tris-HCl, pH 7.4. The cells are sedimented by centrifugation, the supernatant is discarded, and the cell pellets stored at −80°. The cells are then extracted and assayed for HMG-CoA reductase activity, following the method of Brown et al.[36] except that ether extraction is omitted.[37] VLDL subclasses that suppress reductase in normal but not in mutant receptor-negative fibroblasts interact with the LDL cell surface receptor.

Procedure. Thawed cell pellets are incubated for 10 min at 37° with 0.1 ml of 50 mM K$_2$HPO$_4$, pH 7.4, containing 5 mM dithiothreitol, 1 mM EDTA, 0.15 M KCl, and 0.25% Kyro EOB (Proctor & Gamble) or 0.25% Zwittergent (Calbiochem-Behring Corp.). After centrifugation for 1 min in a Beckman microfuge at room temperature, the clear supernatants are assayed for HMG-CoA reductase activity in duplicate. Cell extracts (20 μl) are incubated at 37° in a final volume of 35 μl with 3 mM NADP, 22 glucose 6-phosphate, 14 mM Tris-HCl, (pH 7.5), 5 mM dithiothreitol, 0.15 U of glucose-6-phosphate dehydrogenase (type XV, sulfate free; Sigma), and 43 μM DL-3-hydroxy-3-methyl [3-^{14}C]glutaryl-CoA.[5] Extraction buffer is incubated as a control for background. After 2 hr, 10 μl of 2.5 N HCl containing 60 mM mevalonolactone as carrier is added. After 20 min at 37°, 15 μl of the acidified reaction mixture is streaked on plastic-backed silica thin-layer chromatograms (Eastman) which have been cut to a size of 2 × 6.5 cm. The origin is 1 cm from the bottom. The chromatograms are air dried and developed in weighing jars (Kontes) in acetone–benzene (1 : 1, v/v) until the solvent front moves to the top of the chromatogram, 5.5 cm above the origin. Areas of the chromatogram containing mevalonolactone (R_f 0.63) and residual HMG-CoA (origin) are cut out and counted in a liquid scintillation counter. To calculate the picomoles of mevalonate formed during the incubations, the fraction of total radioactivity (residual substrate plus product) which appears in the mevalonolac-

---

[36] M. S. Brown, S. E. Dana, and J. L. Goldstein, *Proc. Natl. Acad. Sci. U.S.A.* **70**, 2162 (1973).

[37] D. J. Shapiro, J. L. Nordstrom, J. J. Mitschelen, V. W. Rodwell, and R. T. Schimke, *Biochim. Biophys. Acta* **370**, 369 (1974).

tone region (adjusted for background) is multiplied by the picomoles of HMG-CoA initially present in the assay. Cell protein content of the extracts is determined by the method of Lowry[29] or Bradford.[38]

Binding Studies

These procedures are based on the methods developed by Brown and Goldstein which are detailed in this series.[31] Only the differences in our procedure are presented here.

Iodination

Reagents

Glycine-NaOH buffer, 1 M, pH 10

Iodine monochloride (Fisher): The initial two dilutions of ICl (poisonous) are made under a fume hood in glass tubes within an hour of the iodination; the third (last) dilution is made immediately prior to use. Dilute 10 μl of stock ICl into 1 ml of 2 M NaCl; vortex (dilution #1); 0.1 ml of dilution #1 is added to 0.9 ml of 2 M NaCl and vortexed (dilution #2). This should be protected from light with aluminum foil. The final dilution is 0.1 ml of dilution #2 into 1.9 ml of 2 M NaCl

[[125I]Na (for protein iodination) (Amersham/Searle): 10 mCi in 0.1 N NaOH

Buffer A (0.15 M NaCl, 20 mM Tris-Cl, 0.3 mM EDTA, pH 7.4)

LDL at a concentration of 2.5 to 5 mg/ml or VLDL at a concentration of 1 to 3 mg protein/ml in a 12 × 75 mm polystyrene tube

P-2 column (P-2 from Bio-Rad supported by glass wool in a 10-ml plastic pipet fitted with a short (2–3 cm) piece of tubing at the opening, equilibrated with buffer A)

Procedure. This is based on Bilheimer's modification of McFarland's procedure.[39] All steps are conducted at room temperature behind a 1-in.-thick Plexiglas shield in a fume hood approved for use with volatile radioactivity. Two sets of disposable gloves are worn. Immediately before iodination, 0.2 ml of the 1 M glycine-NaOH is added to 1 ml of lipoprotein solution (dropwise with mixing by tapping the tube) and the final ICl dilution is made. [[125I]Na (0.5 mCi) is added to the lipoprotein and immediately vortexed for 2 sec, after which 50 μl of the diluted ICl solution is added and vortexed for 5 sec. This mixture is immediately applied to the

[38] M. M. Bradford, *Anal. Biochem.* **72,** 248 (1976).
[39] D. W. Bilheimer, S. Eisenberg, and R. I. Levy, *Biochim. Biophys. Acta* **250,** 212 (1973).

column and eluted with buffer A to separate most of the free iodine from the labeled lipoprotein. Fractions of approximately 1 ml are collected, and the lipoprotein, which is visible, is collected after approximately 7 ml have emerged. Five-microliter aliquots of each fraction are counted. The fractions containing the labeled lipoprotein (about 2 ml) are combined and dialyzed extensively against buffer A ($\geq$ three changes of 4 liters each). The lipoproteins are removed from dialysis on the same day that the binding studies are performed, 1 to 2 days after iodination. The lipoproteins are sterilized by passage over a 0.45-μm Millex filter and specific activities are determined. If the lipoproteins are used subsequently (up to 2 weeks after iodination), they are redialyzed and resterilized immediately prior to use. Specific activities between 50 and 100 cpm/ng protein are preferred for VLDL; for LDL, specific activities up to 200 cpm/ng are good. VLDL at higher specific activities tend to be damaged. In all binding studies, the labeled lipoprotein is compared with homologous unlabeled lipoprotein for effects on HMG-CoA reductase activity to be sure the labeling procedure has not changed the biological activity of the lipoprotein. In addition, the amount of label extractable into organic solvents should be less than 5%.

Direct Binding Studies. Direct binding studies at 4° with labeled VLDL will reveal whether the binding is high affinity and saturable (as for HTG-VLDL) or low affinity and linear (protease-inactivated HTG-VLDL or normal VLDL S_f 60–400). At 4° binding of HTG-VLDL S_f 100–400 saturates at levels below 15 μg protein/ml with one-half maximal binding at 1 to 2 μg protein/ml.[14]

Procedure: Fibroblasts are grown to ~75% confluency in complete medium. Blank dishes (without cells) are incubated under identical conditions. Fibroblasts and a like number of no-cell controls are preincubated in lipoprotein-deficient medium for 36 hr prior to the binding studies. Binding of ^{125}I-labeled VLDL is measured as the cell-associated radioactivity in duplicate dishes less the amount "bound" in the blank dishes. The dishes are precooled for 30 min with 10 mM N-2-hydroxyethylpiperazine-N'-2-ethanesulfonic acid (HEPES) added. The dishes are then incubated with six different levels of ^{125}I-labeled VLDL alone (0.1–15 μg VLDL protein/ml) and with excess unlabeled homologous VLDL ($\geq$100 μg protein/ml) for 3 hr at 4°. The dishes are then washed five times with chilled albumin-containing buffer (0.15 M NaCl, 50 mM Tris, pH 7.4, 2 mg bovine serum albumin/ml) and once without albumin.[31] The cells are dissolved in 0.1 N NaOH (1 ml) and counted in a gamma counter. Each value is corrected by subtracting the amount "bound" in control dishes which contain no cells. These blanks are essential for each point. Specific binding is calculated by subtracting the amount of ^{125}I-labeled VLDL bound by cells in the presence of excess unlabeled homologous VLDL (these plots are linear) from the amount bound in the absence of unlabeled lipoproteins (curvilinear for saturable, high-affinity binding of HTG-VLDL but linear for low-affinity binding of normal VLDL S_f 60–400 or protease-inactivated HTG-VLDL $S_f > 60$).[2,7] Cell protein contents are determined by the method of Bradford.[38]

Competitive Binding Studies. Competitive binding studies with iodinated VLDL are difficult to interpret because of (1) the ambiguities introduced by labeled, transferable peptides present in the VLDL (Fig. 1), and (2) the high degree of nonspecific binding of VLDL. Because of the latter, classic competition curves are obtained when ^{125}I-labeled VLDL are competed against unlabeled VLDL and LDL even in empty tissue culture dishes, indicating how inconclusive these experiments can be. To determine competitiveness for the LDL receptor, it is preferable to compete against ^{125}I-labeled LDL. The LDL should be of a narrow density cut and thoroughly washed to eliminate contaminating lipoproteins which contain the transferable peptides that so confound studies with labeled VLDL. It is preferable to perform the studies at 37° so that the effects on LDL degradation can be monitored in addition to binding studies only. This eliminates false impressions referred to above caused by the nonspecific hindrance of binding (but not degradation) that can occur with some preparations of large VLDL that do not bind with high affinity to the receptor.

Procedure. Duplicate dishes of cells preincubated in medium containing LPDS for 36 hr are incubated with ^{125}I-labeled LDL at 5 µg protein/ml with and without several levels of unlabeled VLDL and LDL (2 to 50 µg protein/ml) at 37° for 4 to 6 hr so that specific degradation as well as cell uptake can be determined. No cell controls (^{125}I-labeled LDL ± unlabeled LDL) are included to ensure that there is no significant contribution due to nonspecific binding to plastic.

At the end of the incubation, the dishes of cells are placed on a metal tray embedded in crushed ice and the medium removed and processed to determine degradation exactly as described previously.[31] The cells are washed as described above and previously[31] to determine the amount of ^{125}I-labeled LDL bound and internalized by the cells.

Protease Processing of ApoE and ApoB of VLDL

During the isolation of human apolipoprotein E from plasma VLDL, we had often noticed that the apolipoprotein was degraded into smaller fragments. One specific fragment which we always found associated with the VLDL proved to be, upon characterization, the carboxy terminal one-third of apoE (fragment 192–299). This suggested that a specific degradation process was involved. We also showed that the delipidated, purified apolipoprotein, when treated with purified human α-thrombin, yielded the identical fragment plus the N-terminal two-thirds of apoE.[1]

As described above, α-thrombin treatment of HTG · VLDL generates the same fragments from apoE while in its native conformation in the phospholipid matrix of large triglyceride-rich VLDL from hypertriglyceridemic subjects.[2] The following section describes the methods and structural results of processing VLDL with human α-thrombin and trypsin.

Protease Incubations.[2] Lipoproteins are incubated with trypsin (Worthington $3\times$ crystallized, 1% by weight) or with α-thrombin (200 U/ mg apolipoprotein) in 0.15 M NaCl, 20 mM Tris, pH 7.4, 10 mM CaCl$_2$ (buffer), or with buffer alone (control). Purified human α-thrombin was a generous gift from Dr. John W. Fenton. Bovine thrombin (Parke-Davis) also inactivates HTG-VLDL; the high protein content of bovine thrombin interferes with immunochemical blotting of the E-22 fragment described below, however. After incubation at 37° for 2 hr, aliquots of the protease-treated and control lipoproteins are then recentrifuged through a discontinuous salt gradient (d = 1.05–1.006 g/ml) to reisolate VLDL subclasses; IDL or LDL are reisolated at d = 1.03 g/ml or d = 1.05, respectively. Centrifugation conditions are identical to those used for the original isolations. The lipoproteins are then tested for their effects on HMG-CoA reductase activity in normal fibroblasts as a sensitive intracellular indicator of receptor-mediated uptake or by competition with uptake and degradation of ^{125}I-labeled LDL, or analyzed on SDS–PAGE as described below.

Identification and Characterization of ApoE and ApoB and the Fragments of Proteolytic Processing

ApoE: Sodium dodecyl sulfate–12% polyacrylamide slab gels are prepared as detailed previously.[2,40] Samples (100–300 μg) are solubilized in 20–30 μl buffer containing 1% SDS, 0.06 M Tris-HCl, pH 6.8, 10% glycerol, 0.001% Bromophenol Blue, and 5% β-mercaptoethanol and then boiled for 3 min prior to loading. ApoE and apoE thrombin fragments are prepared[1] and identified by an immunochemical blot transfer technique[41] using horseradish peroxidase-conjugated second antibody as seen in Fig. 4.[2] Several features of thrombin cleavage of apoE on HTG-VLDL are revealed by immunochemical blotting. After thrombin digestion, only a portion of apoE is actually cleaved (~1 mol/mole VLDL) with much of the apoE remaining intact. That apoE which is hydrolyzed yields frag-

---

[40] U. K. Laemmli, *Nature (London)* **227**, 680 (1970).
[41] H. Twobin, T. Staehelin, and J. Gordon, *Proc. Natl. Acad. Sci. U.S.A.* **76**, 4350 (1979).

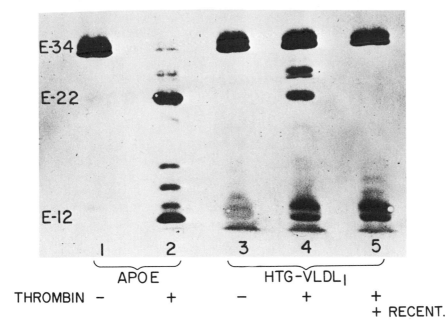

FIG. 4. Immunological identification of apoE fragments in thrombin-inactivated (TI)HTG-VLDL S_f 100–400. HTG-VLDL S_f 100–400 were incubated ± human α-thrombin (50 U/mg) for 2 hr at 37° and tested for reductase suppression ± recentrifugation. The samples were extracted and electrophoresed on sodium dodecyl sulfate–polyacrylamide gel electrophoresis (12%) before transfer to nitrocellulose paper for immunological detection of apoE and apoB fragments using antisera against apoE. Lane 1, purified total apoE; lane 2, isolated apoE incubated with thrombin (50 U/mg, 2 hr at 37°); lane 3, HTG-VLDL S_f 100–400 (starting material); lane 4, HTG-VLDL S_f 100–400 incubated with thrombin (50 U/mg, 2 hr at 37°; TI HTG-VLDL S_f 100–400); lane 5, TI HTG-VLDL S_f 100–400 reisolated by flotation through a salt gradient. Reproduced from *J. Biol. Chem.* **258**, 4526 (1983).

ments identical to those generated by proteolysis of the delipidated apoE. Finally, after reisolation of the thrombin-treated lipoprotein particles by cumulative flotation, the E-22 fragment is lost but the smaller E-12 fragment remains associated with the lipoprotein. Trypsin completely degrades the apoE of all lipoproteins. Neither intact apoE nor apoE fragments are detected by immunochemical blots after trypsinization or by radioimmunoassay.[2,14]

ApoB: Samples are prepared and electrophoresed on 3% SDS–polyacrylamide gels as described by Kane *et al.*[42] As shown in Fig. 5, thrombin converts most of the apoB-100 of IDL and HTG-VLDL S_f 100–400

[42] J. P. Kane, D. A. Hardman, and H. E. Paulus, *Proc. Natl. Acad. Sci. U.S.A.* **77**, 2465 (1980).

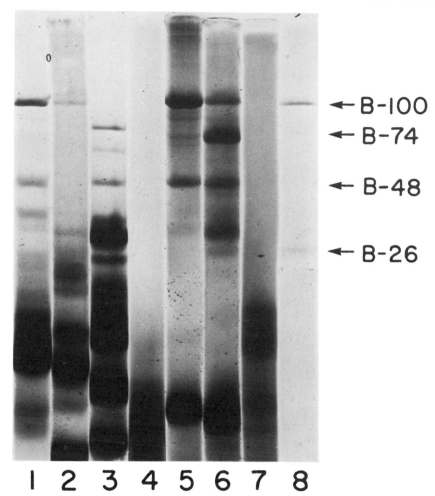

← B-100

← B-74

← B-48

← B-26

1 2 3 4 5 6 7 8

FIG. 5. Effects of thrombin and trypsin on apoB of HTG-VLDL S_f 100–400 and IDL. Aliquots of HTG-VLDL S_f 100–400 and IDL were incubated with thrombin or trypsin as described in the text, extracted, and electrophoresed on 3% polyacrylamide disc gels as described by Kane.[42] Gel 1, plasma chylomicrons from a type 5 subject; gel 2, IDL; gel 3, IDL incubated with thrombin; gel 4, IDL incubated with trypsin; gel 5, HTG-VLDL S_f 100–400; gel 6, HTG-VLDL S_f 100–400 incubated with thrombin; gel 7, HTG-VLDL S_f 100–400 incubated with trypsin; gel 8, LDL. Reproduced from *J. Biol. Chem.* **259,** 14728 (1984).

into species with faster mobility. Two major products have electrophoretic mobilities similar to species found in some LDL preparations as described by Kane (B-74 and B-26)[42] and a third species is similar to apoB-48 of chylomicrons. Analysis on 3% disk gels of HTG-VLDL S_f 100–400, HTG-VLDL S_f 60–100, HTG-VLDL S_f 20–60, and IDL before and after treatment with thrombin indicates in each case patterns similar to those shown in Fig. 5.[14] Under these incubation conditions most of the apoB-100 of IDL and LDL, but not all of the apoB-100 of the VLDL subclasses, are processed by thrombin. Thus thrombin affects the structure of apoB in each lipoprotein even though it has no effect on the ability of S_f 0–20 lipoproteins (IDL and LDL) to bind to the LDL receptor.[14] Thrombin processing of apoB does not affect apoB-mediated binding to the LDL receptor although radioimmunoassay indicates that thrombin can—in IDL and LDL but not VLDL—affect epitope expression using polyclonal antisera.[2]

Trypsin degrades apoB more extensively than does thrombin. After trypsinization of an apoB-containing lipoprotein, no B species with an apparent M_r over 100,000 is seen on 3% SDS–PAGE, as shown in lanes 4 and 7 of Fig. 5 for IDL and HTG-VLDL S_f 100–400. In spite of this extensive degradation of apoB, LDL and IDL still bind to the LDL receptor.[14]

One final note about the processing of apoB. Freshly isolated HTG-VLDL S_f 60–400 but not normal S_f 60–400 VLDL contain species which migrate like B-48 and occasionally B-74 and other large fragments, even though the lipoproteins are isolated from plasma containing protease inhibitors, including thrombin-specific inhibitors.[14] This suggests the possibility that proteolytic processing may occur in vivo or during isolation. We have found that proteases are tightly associated with isolated, highly purified VLDL subclasses and can produce thrombin-like cleavage of apoE in the isolated HTG-VLDL, resulting in the expected loss of LDL receptor-binding determinants.[1] Kane suggested the possibility that B-74 and B-26 are proteolytic fragments of apoB-100.[42] Intestinal apoB (with B-48 migration), however, has a different amino acid composition[42] and is genetically distinct[43] and thus does not appear to be a product of proteolysis of apoB-100. The thrombolytic conversion in vitro of apoB-100 into an apoB species with mobility similar to that of intestinal apoB reinforces Kane's admonition that relying on relative electrophoretic mobility alone as a marker for intestinal apoB may be misleading.[42]

[43] M. J. Malloy, J. P. Kane, D. A. Hardman, R. L. Hamilton, and K. B. Dalal, J. Clin. Invest. 67, 1441 (1981).

Acknowledgments

This work was supported in part by a SCOR Grant in Arteriosclerosis HL 27341. We thank Dr. John W. Fenton for his generous gift of human α-thrombin. S.H.G. is an Established Investigator of the American Heart Association. We thank Antonio M. Gotto, Jr. for his encouragement. In addition, we thank Alice H. Y. Lin, S.-L. C. Hwang, and F. Brown for expert technical assistance.

Section III

Metabolism of Plasma Lipoproteins

A. *In Vivo* Metabolism of Lipoproteins
Articles 21 through 29

B. Cellular and Tissue Metabolism of Plasma Lipoproteins
Articles 30 through 40

C. Enzymes of Lipoprotein Metabolism
Articles 41 through 48

D. Methods
Articles 49 through 51

[21] Plasma Lipoprotein Conversions

By SHLOMO EISENBERG

Introduction

Studies carried out during the last 15 years have demonstrated that all plasma lipoproteins are metabolically related. Volwiler *et al.*[1] and Gitlin *et al.*[2] reported already in the 1950s that after the injection of labeled VLDL in humans radioactivity appears in LDL and postulated that part of the LDL arises in plasma from VLDL. Further research on this process, however, was delayed by about 15 years, until the nature of the plasma apolipoproteins began to be elucidated. Simple considerations led investigators to reinitiate research on the plasma lipoprotein conversion process. Most important was the finding that different lipoproteins share the same apolipoproteins.[3] ApoB-100 is the main protein of plasma LDL and contributes about 40% of the protein moiety of VLDL. An analogous apolipoprotein, apoB-48, is essential for the formation and/or secretion of chylomicrons but is usually not detected in LDL. Substantial amounts of the C apolipoproteins are present in chylomicrons, VLDL, HDL, and trace amounts in LDL. ApoA-I and apoA-II are the major HDL proteins, but when secreted from intestinal absorptive cells are mostly present in chylomicrons. ApoE is apparently a major protein of secreted hepatic VLDL but in the circulation is associated to a large extent with a specific HDL population designated HDL_1. The realization that such distributions of apolipoproteins cannot be arbitrary initiated many of the investigations discussed in the present, and other chapters of this book. The available data suggest two major pathways of lipoprotein interconversion: the "core pathway" and the "surface pathway." In both, the triglyceride-rich lipoproteins chylomicrons and VLDL serve as the primary substrates for the interconversion process. The "core pathway" describes a chain of metabolic events resulting in conversion of core lipid constituents (cholesteryl esters and triglycerides), together with specific surface molecules between large-sized and small-sized apoB-containing lipoproteins. Along this pathway VLDL serves as the precursor in plasma for IDL and later LDL, while chylomicrons are precursors of specific remnant particles

[1] W. Volwiler, P. D. Goldsworthy, M. P. MacMartin, P. A. Wood, I. R. Mackay, and K. Fremont-Smith, *J. Clin. Invest.* **34,** 1126 (1955).
[2] D. Gitlin, D. G. Cornwell, D. Nakasato, J. L. Oncley, W. L. Hughes, Jr., and C. A. Janeway, *J. Clin. Invest.* **37,** 172 (1958).
[3] B. Shore and V. Shore, *Biochemistry* **8,** 4510 (1969).

METHODS IN ENZYMOLOGY, VOL. 129

that are catabolized predominantly by cells in the liver. The "surface pathway" describes events that result in rearrangement of the surface domain of the shrinking triglyceride-rich lipoproteins. Redundant lipid and proteins present at the lipoprotein surface constitute a major source of HDL precursors[4] or become associated with existing HDL particles and determine the nature of HDL subpopulation distribution.[5] The surface pathway therefore is an obligatory counterpart of the core pathway. It is the purpose of this text to critically discuss the methods involved with the plasma lipoprotein metabolic cycles and to evaluate the data obtained by these methods.

Methods

A prerequisite for most studies on lipoprotein metabolism in the vascular bed is the ability to mark the lipoprotein at a given moment (usually referred to as time "zero") with a stable marker that can be traced at subsequent times. Usually, radioactive isotopes are used. Theoretically, other markers (e.g., fluorescence derivatives and nonradioactive isotopes) can be as effective and techniques to use such markers will possibly be developed in the future. In general, lipoproteins are labeled either *in vivo* or *in vitro* and their metabolic fate is being investigated either in the whole animal (*in vivo* studies), or in isolated organs or test tubes (*in vitro* studies). The different techniques and experimental procedures are in essence complementary. In fact, many if not all facets of the plasma interconversion process have been fully evaluated only when a combination of techniques was applied to the problem. In this section the basic techniques and their advantages and disadvantages are described.

Labeled Lipids

In Vivo Labeling Procedures. In vivo labeling of lipoproteins is achieved after injection of labeled lipid molecules into the plasma, or into a space that allows rapid transfer of the molecule to the plasma compartment (e.g., peritoneum) or after feeding an absorbable lipid molecule. It is usually assumed that with the plasma administration route the labeled lipid is incorporated into lipoproteins of hepatic origin, while the oral route labels predominantly intestinal lipoproteins. The following radioactive lipid precursors are commonly used: fatty acids, glycerol, choles-

---

[4] T. Chajek and S. Eisenberg, *J. Clin. Invest.* **61,** 1654 (1978).
[5] J. R. Patsch, A. M. Gotto, T. Olivecrona, and S. Eisenberg, *Proc. Natl. Acad. Sci. U.S.A.* **75,** 4519 (1978).

terol, and phosphorus. With these precursors, it is possible to label lipo-protein triglycerides, lipoprotein cholesterol (free and ester), and lipoprotein phospholipids (predominantly phosphatidylcholine). The available lipid separation techniques allow determination of fatty acids, tri-, di-, and monoglycerides, free and esterified cholesterol, and most phospholipids (phosphatidylcholine, lysophosphatidylcholine, sphingo-myelin, phosphatidylethanolamine, and lysophosphatidylethanolamine). For certain investigations, it is possible to label the lipoproteins simulta-neously with several different isotopes, e.g., ^{3}H-labeled fatty acids, [^{14}C]cholesterol, and ^{32}P-labeled phospholipids.[4] For *in vivo* turnover studies, it is customary to inject a radioactive precursor at time zero and to follow the intraplasma fate of newly synthesized lipid molecules that contain the injected precursor. This technique is used predominantly to study VLDL-triglyceride metabolism by injection of labeled fatty acids[6] or glycerol.[7] In humans, about 30 min is needed before the labeled precur-sor appears in the plasma triglycerides. This delay represents the time needed for the molecule to reach the liver and become incorporated into lipoprotein triglycerides and the period of intracellular processing of the lipoprotein particle. After labeled triglycerides appear in the plasma, the specific activity in VLDL increases rapidly, levels off, and then decreases slowly. While the turnover study is a relatively simple procedure, the analysis of the data is extremely complicated. Processes such as reutiliza-tion of the precursor, multiple synthetic pools in the liver, and multiple catabolic compartments in and out of the plasma must be considered. Multicompartmental analysis seems to be best fitted for proper interpreta-tion of the data.[8] This technique has infrequently been applied to study lipoprotein cholesterol or phospholipid metabolism. Free cholesterol and phospholipids exchange rapidly among lipoproteins and between lipopro-teins and cell membranes. That makes interpretation of data extremely complicated, the main exception being metabolic processes that occur within minutes of the administration of a labeled lipoprotein.

Lipoproteins labeled biosynthetically in their lipid moieties have been frequently used in *in vitro* studies, during organ perfusion experiments, or for reinjection in animals and humans.[9] In these experiments, plasma is obtained from animals injected with radioactive lipids and the biosyntheti-

[6] S. J. Friedberg, R. F. Klein, D. L. Trout, M. D. Bogdanoff, and E. Estes, Jr., *J. Clin. Invest.* **40**, 1846 (1961).
[7] J. W. Farquhar, R. C. Gross, R. M. Wagner, and G. M. Reaven, *J. Lipid Res.* **6**, 119 (1965).
[8] D. M. Shames, A. Frank, D. Steinberg, and M. Berman, *J. Clin. Invest.* **49**, 2298 (1970).
[9] R. J. Havel and J. P. Kane, *Fed. Proc., Fed. Am. Soc. Exp. Biol.* **34**, 2250 (1975).

cally labeled lipoproteins are isolated and used for further experiments. The main advantage of the procedure is that the labeled lipids are incorporated into the lipoprotein particles exactly at the same location and orientation as the unlabeled lipids. This approach therefore assures the native state of the labeled compound. The main disadvantage is the damage to lipoproteins that always occurs during the isolation steps.

In Vitro Labeling Procedures. Several different methods for labeling lipoproteins with radioactive lipids by *in vitro* procedures have been suggested. Some of these methods use physical exchange of lipids between inert surfaces (e.g., glass or celite beads) and lipoproteins. This method has been employed mainly for free cholesterol[10] and less frequently for phospholipids.[11] To label lipoproteins with the hydrophobic lipid molecules, triglycerides, and cholesteryl esters, other methods must be used. Addition of trace amounts of radioactive triglycerides in organic solvents followed by evaporation of the solvent has been suggested.[12] HDL can be labeled with cholesteryl ester by the following procedure: the lipoprotein is first labeled with free cholesterol and then incubated with LCAT or an LCAT source. Excess free cholesterol is removed by exchange to lower density lipoproteins or erythrocytes. The main advantages of *in vitro* labeling procedures are simplicity and high efficiency of utilization of radioactive materials. For example, to label lipoproteins with cholesterol, efficiencies up to 80–100% can be achieved, compared with 1–5% when injected in an animal. The main disadvantages are the uncertainty of location and orientation of the labeled molecule and the need to manipulate the lipoprotein, which invariably causes further damage to the already damaged isolated particles.

An exciting new approach to label lipoproteins with radioactive triglycerides, cholesteryl esters, and phospholipids has been developed recently and is based on the use of lipid transfer proteins (see Chapters [48] and [49], this volume). With this procedure the desirable lipid molecule is first introduced into a lipoprotein,[13] intralipid,[14] or lipid–protein complex[15] by the LCAT reaction or by sonication. The labeled particle is then incubated with unlabeled lipoproteins in the presence of lipid transfer proteins and the transfer reaction is allowed to proceed for a desirable period.

[10] J. Avigan, *J. Biol. Chem.* **234,** 787 (1959).
[11] O. Minari and D. B. Zilversmit, *J. Lipid Res.* **4,** 424 (1963).
[12] C. J. Fielding, *Biochim. Biophys. Acta* **573,** 255 (1979).
[13] R. Deckelbaum, S. Eisenberg, Y. Oschry, and T. Olivecrona, *J. Biol. Chem.* **257,** 6509 (1982).
[14] Y. Stein, Y. Dabach, G. Hollander, G. Halperin, and O. Stein, *Biochim. Biophys. Acta* **752,** 98 (1983).
[15] O. Stein, G. Halperin, and Y. Stein, *Biochim. Biophys. Acta* **620,** 247 (1980).

While with this procedure it is also necessary to incubate and reisolate the lipoprotein, the reaction apparently introduces the labeled lipid at its physiological environment. The yield of the reaction may be as high as 60–70%. As discussed elsewhere in this book, such procedures proved to be of importance to introduce cholesteryl–ether analogs into a variety of rat plasma lipoproteins (Chapter [49], this volume).

Labeled Proteins

In Vivo Procedures. Only in a few studies was endogenous labeling of lipoprotein–apolipoproteins with labeled amino acids attempted. [^{75}Se]-Selenomethionine[16] and [^{3}H]leucine[17] were injected intravenously in humans and apolipoprotein-specific activities were followed. Multicompartmental models are required for analysis of the data.[17] In addition to complexity of analysis, the technique involves administration of very large amounts of radioactivity to the subject.

In Vitro Procedures. Methods of radioiodination of apolipoproteins are the most commonly used procedures to achieve labeling of lipoprotein proteins. Although several methods for iodination of proteins are available, modifications of the McFarlane's iodine monochloride method[18] are by far the commonest.[19] The method first requires isolation of lipoproteins free of contamination with other lipoproteins and apolipoproteins. For optimal results, the lipoproteins should be used at protein concentrations of 10 mg/ml or above. This, however, is not mandatory, and lower protein concentrations have been used successfully. The ratio of iodide to protein in the iodinated preparation should be kept below 1.0, to avoid radiation damage. We have used glycine buffer, pH 10.0, to minimize labeling of lipids.[19] Efficiencies of iodination are highest for HDL (60–70%), intermediate for LDL (25–30%), and lowest for VLDL (10–15%) and chylomicrons (<10%). These values can conveniently be used to calculate the amount of ICl injected to the iodination mixture.

Iodination of chylomicrons, VLDL, and LDL is usually performed on the whole lipoprotein. Because apoB is not soluble in aqueous solutions and cannot be reincorporated into lipoprotein particles, the use of whole lipoproteins is mandatory. Labeling of VLDL or chylomicrons with other apolipoproteins can be achieved by incorporation of labeled apolipoproteins into the lipoproteins. According to this procedure, labeled apolipo-

[16] R. P. Eaton and D. M. Kipnis, *Diabetes* **21**, 744 (1972).
[17] R. D. Phair, M. G. Hammond, J. A. Bowden, M. Fried, W. R. Fisher, and M. Berman, *Fed. Proc., Fed. Am. Soc. Exp. Biol.* **34**, 2263 (1975).
[18] A. S. McFarlane, *Nature (London)* **182**, 53 (1958).
[19] D. W. Bilheimer, S. Eisenberg, and R. I. Levy, *Biochim. Biophys. Acta* **260**, 212 (1972).

proteins (apoC,[4] apoE[20]) are prepared, allowed to associate with the lipoprotein, and unassociated apolipoproteins are removed. To label apoC (or apoA-I, see below), we always iodinated the whole lipoprotein[21] and then prepared purified apolipoproteins by established procedures. The reason for this sequence is our belief that by this procedure we prevent attachment of iodine molecules to tyrosines that are present in close proximity to lipid-binding domains of the protein. Such attachments may interfere with the lipoprotein-binding properties of the protein. Several procedures have been developed to label HDL with specific iodinated proteins. We used the approach described above, i.e., iodination of HDL preparations followed by isolation of labeled proteins and then either reassociation of proteins with lipoproteins or direct injection of the apolipoproteins into the plasma.[21] Shepherd and associates used a different procedure.[22] These investigators iodinated purified apolipoprotein preparations and exchanged the apolipoproteins back into HDL. Of interest, only part of the HDL–apoA-I but all of the apoA-II was exchangeable with the unassociated apolipoproteins. The plasma decay of HDL labeled with exchanged [^{125}I]apoA-I is about 20% faster than that of labeled whole HDL. That presumably reflects technical errors, at least in part. Schaefer *et al.* used a slight modification of Shepherd's method, and have injected the subjects with the labeled proteins.[23] In these studies, apoA-I decay is 10–20% faster than apoA-II. In spite of the slight differences obtained with the various procedures, the results are probably comparable.[24]

More recently, attempts were made to label lipoproteins by covalent linkage of other molecules. Sucrose,[25] glucose,[26] and tyramine cellobiose[27] were studied in detail. One advantage of such labels is the inability of cells to metabolize the marker molecule (sucrose or tyramine cellobiose) and the possibility to determine accumulation of protein degradation products in the cells. With glucosylation of LDL, however, antibodies against the lipoprotein were produced.[26] It is thus doubtful whether these labels can be used for human studies.

[20] F. van't Hooft and R. J. Havel, *J. Biol. Chem.* **256,** 3963 (1981).
[21] S. Eisenberg and D. Rachmilewitz, *Biochim. Biophys. Acta* **326,** 391 (1973).
[22] J. Shepherd, A. M. Gotto, Jr., O. D. Taunton, M. J. Caslake, and E. Farish, *Biochim. Biophys. Acta* **489,** 486 (1977).
[23] E. J. Shaefer, L. A. Zech, L. L. Jenkins, T. J. Bronzert, E. A. Rubalcaba, F. T. Lindgren, R. L. Aamodt, and H. B. Brewer, Jr., *J. Lipid Res.* **23,** 850 (1982).
[24] S. Eisenberg, *J. Lipid Res.* **25,** 1017 (1984).
[25] R. C. Pittman, S. R. Green, A. D. Attie, and D. Steinberg, *J. Biol. Chem.* **254,** 6876 (1979).
[26] Y. A. Kesaniemi, J. L. Witztum, and U. P. Steinbrecher, *J. Clin. Invest.* **71,** 950 (1983).
[27] R. C. Pittman, T. E. Carew, C. K. Glass, S. R. Green, C. A. Taylor, and A. D. Attie, *Biochem. J.* **212,** 791 (1983).

In Vivo Techniques to Study Lipoprotein Metabolism

By far the *in vivo* lipoprotein turnover techniques provided the most important data on the intraplasma conversion of lipoproteins. This technique is described in detail in other chapters of this book and therefore is described here only as related to the lipoprotein interconversion process. The turnover technique is based on follow-up of injected lipid or protein precursors (endogenous labeling) or of injected labeled lipoprotein (exogenous labeling). Ideally, multiple time points and follow-up of disappearance and appearance of the labeled molecule from and into multiple lipoprotein fractions are necessary. In practice, several time points are chosen during the first few hours (or first few days) after the injection of the label, followed by daily plasma samples. Centrifugation is customarily used to separate lipoprotein fractions. That last technique obviously lacks proper discrimination of physiologically important lipoprotein families. One example is VLDL. As defined in the ultracentrifuge, VLDL consists of lipoproteins present in fasting plasma that are of density less than 1.006 g/ml. VLDL, however, contains lipoproteins of different apoB molecules (B-48 and B-100), lipoproteins of intestinal and hepatic origin, and particles with a wide range of size, density, weight, and composition. Similarly, LDL and HDL can be separated into populations of distinctly different properties. Yet, in many if not all turnover studies one or more of these lipoproteins is treated as a single lipoprotein class.

Another concern is the native state of injected exogenously labeled lipoprotein. Preparation of labeled lipoproteins includes isolation procedures, iodination, dialysis (or other methods to separate free and protein-bound iodide), and concentration steps, if necessary. Undoubtedly, each of these steps can, and probably does, damage the lipoprotein. This is particularly true for the triglyceride-rich lipoproteins, chylomicrons and VLDL. In many studies, for example, attempts were made to quantitate conversions within VLDL populations and between VLDL and IDL or LDL. Incomplete conversion at least in part may reflect inability of damaged particles to follow the native metabolic pathways. Similarly, rates of lipolysis of VLDL particles may change by the preparation procedures. Because a prerequisite for interpretation of turnover data is identity between tracer and tracee, the above considerations must call for caution when conclusions are drawn on the basis of the turnover kinetics. An example of artifacts induced by preparation procedures has been recently reported by us.[28] We observed that concentration of lipoproteins by ultrafiltration in Amicon systems frequently caused considerable damage to

---

[28] S. Eisenberg, Y. Oschry, and J. Zimmerman, *J. Lipid Res.* **25**, 121 (1984).

the lipoproteins and induced a rapid decay of the lipoprotein from rat plasma. Concentration by "reverse dialysis," in contrast, seemed adequate. Some of these difficulties may be solved by using "biologically screened" preparations. With this method, an animal is injected with the labeled lipoprotein and the injected preparation is allowed to circulate for several hours. It is assumed that during that period damaged particles are rapidly degraded, predominantly by reticuloendothelial cells. The animal is then bled and the plasma that contains the biologically screened particles is injected in another animal. With this technique, initial rapid decay of HDL or LDL injected into rats is largely eliminated.[28] The procedure, however, cannot be employed in humans and is of course inadequate for lipoproteins with very rapid metabolic rates, e.g., chylomicrons and VLDL.

The methodology to analyze turnover data is discussed elsewhere in this book (Chapter [22], this volume). At this point it is sufficient to mention that very different quantitative results are obtained when the same data are analyzed by different methods. That, however, usually has only a small effect on qualitative or semiquantitative conclusions. It appears that data derived from several related but different techniques will eventually be required for generation of definite conclusions. For example, if similar conclusions are derived from endogenous VLDL triglyceride turnovers and exogenous VLDL-apoB turnover studies, the data are much more meaningful than those obtained with either method alone.

In Vitro Techniques to Study Lipoprotein Metabolism

This approach should be reserved to evaluate specific facets of the lipoprotein conversion process. Normal metabolism depends on the activity of enzymes, lipid transfer proteins, and cellular receptors against the different lipoprotein populations. Lipoprotein lipase, the hepatic triglyceride hydrolase, the LCAT system, the lipid transfer and phospholipid transfer protein(s), and the "remnant," B-VLDL, B, E and E receptors all participate in lipoprotein metabolism. To study the effects of a single or a combination of these reactions on lipoproteins, controlled experimental systems must be constructed. When constructing an *in vitro* system for studies that may have relevance to physiology, several guidelines have to be employed. *First,* least damaged lipoproteins must be used. Artificial lipid emulsions or lipid : protein complexes are frequently used as substrates in enzymatic studies. While the results of such studies are invaluable for evaluation of the properties of the enzyme (or transfer protein), they seldom are relevant for elucidation of a metabolic event. In our studies, freshly prepared, biosynthetically labeled lipoproteins were always used when we wished to mimic physiological events *in vitro*. *Second,* in the circulation lipoproteins are always exposed to multiple reactions. It is important in *in vitro* studies to attempt to complement the system with as many reactants as possible. For example, the activity of an enzyme against one lipoprotein may change considerably if another lipoprotein is added to the system. In some instances, that may reflect greater affinity of the enzyme toward a certain lipoprotein, although it is active on all lipoproteins. Similarly, it is important in some instances to include "contaminants" in the system. Lipoprotein-deficient plasma fraction or even whole plasma should be added wherever possible. *Third,* it should be realized that in many instances one reaction is affected by another, or that activity of one reactant is necessary for that of another. Thus, metabolic cycles are being formed. The test tube conditions must allow continuous operation of the cycle, or it should be taken into consideration that the cycle has been interrupted. An example that clarifies this situation is the activity of the lipid transfer proteins and of lipases on the cholesteryl ester-rich lipoproteins, LDL and HDL. By inducing replacement of cholesteryl ester molecules by triglycerides, the lipid transfer reaction provides LDL and HDL with substrate for the lipases. Lipase activity depletes the lipoprotein of the transferred triglycerides, decreases the triglyceride-to-cholesteryl ester molar ratio, and thereby allows further exchange of cholesteryl esters for triglycerides. Operation of the cycle provides the only logical explanation for the presence of small-sized triglyceride-rich LDL and HDL in hypertriglyceridemic human plasma, while neither lipase activity nor the lipid transfer reaction alone could explain the observed phenomena.

Intraplasma Conversions of Lipoproteins

The "Core Pathway": Origin of Remnants, Intermediates, and Low-Density Lipoproteins

Endothelial bound lipases (lipoprotein lipase and the hepatic lipase) play a key role in this process. Both enzymes hydrolyze triglycerides and phosphoglycerides in all lipoproteins; lipoprotein lipase, however, possesses a high affinity toward triglyceride-rich lipoproteins while the hepatic lipase, toward high-density lipoproteins.[29] Interconversion of chylomicrons and VLDL begins as soon as the lipoproteins interact with lipases. This process may start already in the intestine and the liver,

---

[29] R. J. Deckelbaum, S. Eisenberg, E. Levy, T. Olivecrona, and G. Bengtsson-Olivecrona, *Arteriosclerosis* **4,** 563a (1984).

before the lipoproteins enter the blood stream. Cells in either tissue (hepatocytes, macrophages) can produce hepatic or lipoprotein lipase that may interact with newly secreted VLDL or chylomicrons. Triglyceride hydrolysis becomes very prominent once VLDL and chylomicrons enter the blood stream. In normolipemic humans, 50% of the chylomicron triglycerides are hydrolyzed within 5–15 min.[30] The circulating half-life time of VLDL triglycerides is longer and depending on the activity of the lipase system may vary between 30 min and 4 hr. Immediate structural consequences of triglyceride hydrolysis are decreased size, increased density, and changed composition of either chylomicron or VLDL particles. The metabolic consequences of lipase activity, however, differ between the two particles and are discussed separately.

The VLDL–IDL–LDL Cascade. Numerous studies have demonstrated that along the VLDL–IDL–LDL density range smaller and denser lipoprotein populations are formed in the plasma compartment from larger and lighter particles. The most convincing evidence is derived from *in vivo* studies using ^{125}I-labeled VLDL and tracing the apoB moiety of the lipoprotein.[19,31,32] Precursor–product relationship is readily demonstrable within the VLDL density range,[31–33] between VLDL and IDL and IDL and LDL.[19,31,32] Crucial for understanding of the VLDL–LDL conversion process was the discovery (based on compositional analysis of VLDL density subfractions and LDL) that along this pathway *one and only one product particle is produced from each precursor particle.*[31,34] This finding proves that all of the apoB moiety present in precursor particles is retained in the products, and provides an exceedingly important tool to investigate in detail the mechanism of the interconversion process.

Comparing product and precursor particles allows evaluation of processes that take place during the conversion process, once it is realized that a 1 : 1 relationship is kept between the two. For example, to form one LDL ($M_r = 2.2 \times 10^6$ Da) from one VLDL (16×10^6 Da, S_f 60–100), it is necessary to remove all the apoC and apoE molecules, 98% (9670 molecules) of triglycerides, 80% (3460 molecules) of phospholipids, and 90% (2250 molecules) of free cholesterol (the behavior of cholesteryl esters is more complex and is discussed below). The molecular mechanisms responsible for rearrangement of particles along the VLDL–IDL–LDL cas-

[30] S. M. Grundy and H. Y. I. Mok, *Metabolism* **25**, 1225 (1976).
[31] S. Eisenberg, D. W. Bilheimer, F. T. Lindgren, and R. I. Levy, *Biochim. Biophys. Acta* **326**, 361 (1973).
[32] M. Berman, S. Eisenberg, M. Hall, R. I. Levy, D. W. Bilheimer, R. D. Phair, and R. H. Goebel, *J. Lipid Res.* **19**, 38 (1978).
[33] M. F. Reardon, N. H. Fidge, and P. J. Nestel, *J. Clin Invest.* **61**, 850 (1978).
[34] S. Eisenberg and D. Rachmilewitz, *J. Lipid Res.* **16**, 341 (1975).

cade were studied mainly in *in vitro* experimental systems. Triglycerides are hydrolyzed by lipoprotein lipases, and IDL- or LDL-like particles can be produced *in vitro* with this enzyme alone.[4,34–36] Incubation of VLDL with bovine milk lipoprotein lipase can induce hydrolysis of more than 98% of the triglycerides.[35,36] With the perfused rat heart (endothelial bound lipase) we achieved hydrolysis of more than 90% of VLDL triglycerides.[4] Yet, these observations also offer an excellent example for the caution that is needed for interpretation of *in vitro* data. Although the studies demonstrate the *ability* of lipoprotein lipase to hydrolyze VLDL triglycerides almost to completion, it is very possible that *in vivo* the hepatic lipase contributes significantly to this process. In particular, hepatic lipase activity may be important for the final part of the conversion process, i.e., of IDL to LDL. IDL contains only a few apoC molecules and even fewer (if any) apoC-II. Thus, IDL is predicted to be a poor substrate for lipoprotein lipase but an excellent substrate for the hepatic lipase. Indeed, accumulation of IDL and small VLDL particles has been demonstrated in rats and monkeys injected with antibodies against the hepatic lipase.[37,38] Thus, *in vivo*, both enzymes hydrolyze triglycerides along the VLDL–IDL cascade. As suggested originally by us,[39] and later by others,[38] lipoprotein lipase activity appears to predominate early in the conversion pathway while hepatic lipase predominates at later steps.

Rearrangement of the surface coat of lipolyzed VLDL is a mandatory process along the VLDL–LDL conversion pathway. A substantial amount of phosphoglycerides (mostly phosphatidylcholine) is hydrolyzed by lipoprotein and hepatic lipases. This reaction is responsible for the decreasing lecithin to sphingomyelin molar ratio in lipoproteins along the VLDL–IDL–LDL cascade.[31,34–36,40] The remaining excess phospholipids leave the lipolyzed particle as intact molecules. Free cholesterol, apoC, and apoE molecules leave the lipolyzed particles in proportion to the degree of triglyceride hydrolysis.[4,35] Hence, deletion of surface components is proportional to the reduction of the particle's core volume. Indeed, volume-to-surface relationships agree extremely well with experimental data obtained during simultaneous measurements of triglyceride,

[35] S. Eisenberg and T. Olivecrona, *J. Lipid Res.* **20**, 614 (1979).
[36] R. J. Deckelbaum, S. Eisenberg, M. Fainaru, Y. Barenholz, and T. Olivecrona, *J. Biol. Chem.* **254**, 6079 (1979).
[37] J. Glosser, O. Schrecker, and H. Greten, *J. Lipid Res.* **22**, 437 (1981).
[38] I. J. Goldberg, N. A. Le, J. R. Paterniti, Jr., H. N. Ginsberg, F. T. Lindgren, and W. V. Brown, *J. Clin. Invest.* **70**, 1184 (1982).
[39] S. Eisenberg and R. I. Levy, *Adv. Lipid Res.* **13**, 1 (1975).
[40] Y. Oschry, T. Olivecrona, R. J. Deckelbaum, and S. Eisenberg, *J. Lipid Res.* **26**, 158 (1985).

phospholipid, free cholesterol, and apoC content of lipolyzed VLDL particles.[4,35] This almost perfect coordination between the hydrolysis of core lipids and deletion of surface constituents preserves the general structure of the lipoproteins along the cascade.[41,42] Yet, the composition of both the core and surface domains changes continuously. In the core, cholesteryl esters gradually become the major lipid class, whereas at the surface the ratio between lecithin and sphingomyelin decreases and apoB becomes the predominant apolipoprotein.

The molecular mechanism responsible for exclusion of molecules from the surface of lipolyzed lipoproteins is not clear. When HDL is present, phospholipids, free cholesterol, apoC, and apoE molecules are transferred to the HDL. In the absence of HDL, vesicular, spherical, and discoidal structures accumulate (see below). Whether molecules leave the lipolyzed particles as "surface fragments" or as free molecules that reassociate in the water phase is also unclear. Obviously, more research and perhaps better methodology are needed to elucidate this facet of the conversion process.

The behavior of cholesteryl esters along the VLDL → IDL → LDL cascade is more complex. In addition to the lipase-induced effects, the content of cholesteryl esters reflects the activity of lipid transfer reactions. In species whose plasma contains cholesteryl ester/triglyceride transfer activity (e.g., humans, rabbits), cholesteryl esters are transferred from HDL and LDL to VLDL and chylomicrons and triglycerides are transferred in the opposite direction (see Chapter [48], this volume). Therefore, along the cascade, the particles acquire substantial amounts of cholesteryl esters depending on rates of cholesterol esterification, activity of the lipid transfer proteins, and the time of exposure of the triglyceride-rich lipoproteins to the transfer reaction. That last factor is crucial in understanding the lipoprotein system in subjects with hypertriglyceridemia, especially when the circulating lifetime of triglyceride-rich lipoproteins is prolonged. In such subjects, the VLDL becomes abnormally enriched with cholesteryl esters, while both LDL and HDL are enriched with triglycerides.[43] The impact of cholesteryl ester redistribution on the metabolism of plasma cholesterol and the conversion of lipoproteins is beyond the scope of the present text and is discussed by us elsewhere.[40,43] It should, however, be mentioned that this process is possibly responsible

[41] B. W. Shen, A. M. Scanu, and F. J. Kezdy, *Proc. Natl. Acad. Sci. U.S.A.* **74**, 837 (1977).

[42] S. Eisenberg, *in* "Lipoprotein Metabolism" (H. Greten, ed.), p. 32. Springer-Verlag, Berlin, 1976.

[43] S. Eisenberg, D. Gavish, Y. Oschry, M. Fainaru, and R. J. Deckelbaum, *J. Clin. Invest.* **74**, 470 (1984).

for the incomplete conversion of VLDL to LDL in hypertriglyceridemic states.[32] Absence of cholesteryl ester/triglyceride transfer reactions in rats represents the opposite situation where paucity of cholesteryl esters in VLDL is associated with the production of triglyceride-rich LDL.[44]

An exceedingly important feature of the conversion process is the change of biological reactivity along the VLDL–IDL–LDL cascade. While large VLDL particles do not seem to interact with cell receptors (at least in normolipemic subjects), the smaller VLDL, IDL, and obviously LDL do so avidly. We believe that the changed reactivity of lipoproteins with cell receptors (mainly the B, E receptor) reflects the altered apolipoprotein profile at the surface of the particles.[42,45] Specifically, we pointed out that with the deletion of apoC molecules and the conservation of apoB, the concentration of these apolipoproteins at the surface of lipoprotein particles changes along the cascade: apoC concentration (molecules/surface area) decreases, whereas apoB concentration increases. Because of surface-to-volume considerations, these two changes accelerate at S_f rates of 60, a point in the cascade that coincides with the changed biological reactivity of the lipoproteins. That hypothesis however, has not been tested experimentally, and other possible alterations of the lipoproteins, such as decreased diameter and increased radius of curvature, must be considered. Change of size alone may induce unfolding of receptor binding domains in apoB.

The Chylomicron–Remnant Cascade. In principle, events along the chylomicron–remnant cascade closely resemble those described above for VLDL. Undoubtedly, lipoprotein lipase plays a key role in chylomicron lipolysis and the particle undergoes surface and core alterations similar to those described above. Hepatic lipase apparently does not play any role in chylomicron triglyceride hydrolysis, at least initially, as this enzyme is present in the plasma of patients with type I hyperlipoproteinemia.[46] Whether the hepatic lipase may interact with partially lipolyzed chylomicrons is not known. Another difference between the chylomicrons and VLDL cascades is the apolipoprotein profile of the two. ApoA-I is a predominant protein constituent of nascent chylomicrons, and is displaced rapidly from the particles at the early stages of their lipolysis. ApoE appears to become associated with chylomicrons and is retained at the surface of the particle along the chylomicron–remnant cascade. The behavior of apoC, phospholipids, free cholesterol, and cholesteryl esters is apparently identical to that described for VLDL.

[44] Y. Oschry and S. Eisenberg, *J. Lipid Res.* **23,** 1099 (1982).
[45] S. Eisenberg, *Ann. N.Y. Acad. Sci.* **348,** 30 (1980).
[46] R. M. Krauss, R. I. Levy, and D. S. Fredrickson, *J. Clin. Invest.* **54,** 1107 (1974).

The hallmark of the chylomicron–remnant cascade is the rapid interaction of the lipolyzed lipoproteins with specific receptors in the liver. ApoE-3 or E-4 are essential for that interaction,[47] and in their absence (e.g., apoE-2 homozygosity), lipoproteins of chylomicron origin can be identified even in LDL.[48] The identity of the determinants in chylomicron remnants that are responsible for their interaction with the hepatic E receptor is not clear. ApoE is also present in VLDL and is an important protein constituent of HDL_1.[44] Yet, HDL_1 is not catabolized rapidly in the rat,[28] an animal species with extremely avid mechanism for degradation of chylomicron remnants. Perhaps the combination of B-48 and apoE is necessary. In the absence of the liver (e.g., in the functionally hepatectomized rat[49] or *in vitro*[50]), cholesteryl ester-rich IDL-like or even LDL-like particles are produced from chylomicrons, and their metabolic fate is determined at least in part on their dimension.[50] In intact animals, however, it is doubtful whether chylomicrons contribute significantly to either IDL or LDL.

The "Surface Pathway": Origin of HDL and Effects on HDL Subpopulations

Lipid and protein constituents freed from the surface of lipolyzed triglyceride-rich lipoproteins constitute a major source of plasma HDL. These constituents contribute to the plasma pool of HDL precursors and associate with existing HDL particles. Because apolipoproteins, phospholipids, and free cholesterol possess potent detergent properties, their packaging to spherical lipoprotein particles may prevent severe damage to cells and cell membranes.

Formation of HDL Precursors. In vitro experiments[4,35] and observations in patients with LCAT deficiency[51] or Tangier disease[52] have demonstrated accumulation of vesicular, spherical, or discoidal high-density lipoprotein structures during lipolysis of triglyceride-rich lipoproteins. While the *in vitro* experiments are open to criticism, the studies in the patients were carried out during clearance of alimentary chylomicronemia

[47] E. Windler, Y.-S. Chao, and R. J. Havel, *J. Biol. Chem.* **255,** 8303 (1980).
[48] M. S. Meng, R. E. Gregg, E. J. Schaefer, J. M. Hoeg, and H. B. Brewer, Jr., *J. Lipid Res.* **24,** 803 (1983).
[49] T. G. Redgrave, *J. Clin. Invest.* **49,** 465 (1970).
[50] T. Chajek-Shaul, S. Eisenberg, Y. Oschry, and T. Olivecrona, *J. Lipid Res.* **24,** 831 (1983).
[51] J. A. Glomset, K. R. Norum, A. V. Nichols, W. C. King, C. D. Mitchell, K. R. Applegate, E. L. Gong, and E. Gjone, *Scand. J. Clin. Lab. Invest.* **35,** (Suppl. 142), 1 (1975).
[52] P. N. Herbert, T. Forte, R. J. Heinen, and D. S. Fredrickson, *N. Engl. J. Med.* **299,** 519 (1978).

and therefore provide conclusive evidence that such structures are indeed generated as part of the physiological process of fat transport. Because of the similarities between the high-density discoidal structures found in *in vitro* experiments and the nascent discoidal high-density lipoproteins of hepatic and intestinal origin, we[45,53] and others[54] have suggested that surface remnants contribute directly to the plasma pool of HDL precursors. All HDL precursors contain apolipoproteins, phospholipids, and free cholesterol. The nature of the apolipoproteins, however, varies according to the source of the precursors: apoE is the predominant protein in particles of hepatic origin, apoA-I (? and A-II) in intestinal particles, and apoC in surface remnants generated during VLDL (? and chylomicron) lipolysis. Because plasma HDL contains mainly apoA-I and A-II, rearrangement of the apolipoproteins must take place with the formation of spherical HDL. It is significant to note that in Tangier disease or in conditions where apoA-I is either absent or is present in small amounts, other apolipoproteins cannot substitute for apoA-I and HDL itself is either absent or is present in minimal amounts. We suggested elsewhere that availability of apolipoproteins and of phospholipids determines the amounts of HDL precursors that are formed in plasma or even in the liver and intestine.[24] The surface coat of lipolyzed triglyceride-rich lipoproteins is a major source of all constituents necessary for the formation of HDL precursors. Yet, it is important to realize that cells and cell membranes can also contribute phospholipids for this process, if apolipoproteins are available. Indeed, HDL is present in the plasma of subjects with abetalipoproteinemia, a condition where neither chylomicrons nor VLDL is produced. In such patients, lipolysis does not occur and alternate routes of HDL formation must operate. That example alone demonstrates the complexity of intravascular metabolic events and indicates that while lipolysis is potentially the major source of plasma HDL precursors, other processes contribute and in extreme situations may predominate.

Transformation of discoidal HDL precursors to spherical lipoproteins depends on cholesterol esterification by the LCAT system. It has been suggested that esterification occurs at the surface of the disks and that the hydrophobic cholesteryl ester molecules are then displaced to the hydrophobic domain of the phospholipid bilayer.[55] As mentioned above, however, formation of spherical HDL necessitates adequate supply of

[53] S. Eisenberg, T. Chajek, and R. J. Deckelbaum, *Pharmacol. Res. Commun.* **10,** 729 (1978).

[54] A. R. Tall and D. M. Small, *N. Engl. J. Med.* **299,** 1232 (1978).

[55] R. L. Hamilton, M. C. Williams, C. J. Fielding, and R. J. Havel, *J. Clin. Invest.* **58,** 667 (1976).

apoA-I, and apoA-I replaces at this stage all other apolipoproteins present in the precursors.

Effects on HDL Subpopulations. The second pathway that participates in the intraplasma metabolism of surface lipids and apolipoproteins is their assimilation by existing HDL particles. Distribution of molecules generated from the surface of lipolyzed triglyceride-rich particles to other lipoproteins had been documented on numerous occasions. In many studies, the effects of heparin injection in humans or experimental animals has been determined, while more recently controlled *in vitro* incubation systems were employed. All studies demonstrated increase of HDL phospholipid, cholesterol, apoC, or apoE content during the course of lipolysis. Definite proof that these constituents are transferred to *existing* particles has, however, been obtained in only a few investigations. This reflects the methodology used in different studies. Standard density centrifugation, for example, cannot differentiate between two lipoprotein populations of similar densities or the presence of a single lipoprotein complex that contains both existing *and* transferred molecules. Other methods, however, are adequate for that question. Analytical ultracentrifugation, for example, was used to determine changes induced in lipoproteins after the injection of heparin in human subjects.[31] A remarkable shift of HDL_3 toward HDL_2 was observed with an increased HDL mass from 165 to 215 mg/dl. More recently, we used zonal centrifugation and *in vitro* lipolysis systems to investigate mechanism(s) of transfer of surface constituents from human VLDL to HDL_3.[5] With the experimental techniques employed we were able to show that phospholipids, free cholesterol, and apoC molecules are indeed transferred to existing HDL_3 particles. Because the ratio of protein to lipids of the transferred molecules is lower than that in HDL_3, the density of the complex has shifted toward HDL_2. The capacity of HDL_3 to accept phospholipids was limited, and when the phospholipid content increased by 50–100%, excess molecules were found with the plasma protein fraction. Another observation was the finding that apoC molecules transferred to HDL_3 displace apoA-I from the particle. As much as 50% of the apoA-I could be removed from the lipoprotein and were recovered in the plasma protein fraction (unpublished). Finally, in these systems where cholesterol esterification has not taken place, the change of HDL_3 was reversible. Apparently, an increase of core volume is mandatory for true conversion of HDL_3 to HDL_2 (see below).

Other HDL populations can also serve as acceptors for molecules freed from the surface of lipolyzed triglyceride-rich lipoproteins. We showed that HDL_2 is as potent as HDL_3, even when the two are present together in the lipolysis system (unpublished). In an attempt to quantitate the lipolysis-dependent transfer of surface molecules to HDL, we investigated distribution of free cholesterol from biosynthetically labeled rat plasma VLDL to various cellular and lipoprotein acceptors.[56] This study was conducted *in vitro* and in the perfused rat heart. More than 75% and possibly all of the lipolysis-generated free cholesterol remained with the plasma lipoproteins, and none was transferred to blood cells. LDL, however, seemed to be the preferred acceptor for free cholesterol (unpublished). Yet, LDL apparently serves a role of an intermediate in the process of metabolism of surface molecules, while HDL is the final particle.

HDL Conversions

HDL exists in plasma as three distinct populations—HDL_1, HDL_2, and HDL_3. Another HDL population—HDL_4—has also been described.[57] HDL populations differ in density, size, and composition; HDL_1 is the lightest and HDL_4 the heaviest. Data obtained in the last few years indicated that conversions between subpopulations take place in plasma, and the pathways that regulate HDL subpopulation distribution have been elucidated in part.

Unlike very low, intermediate, and low-density lipoproteins, HDL conversions are apparently reversible. Increased size and decreased density presumably reflect supply of free cholesterol and phospholipids followed by cholesterol esterification, while cholesterol ester/triglyceride transfer reactions, followed by hydrolysis of the transferred triglycerides, causes decreased size and increased density of the HDL. For convenience, these processes are described separately. However, it is important to realize that HDL particles are exposed simultaneously to all different reactions and the final density, size, and composition of particles reflects a dynamic equilibrium of many metabolic activities.

Conversion of HDL_3 to HDL_2. This process begins with supply of free cholesterol and phosphoglycerides to HDL_3. The contribution of molecules transferred from the surface of lipolyzed triglyceride-rich lipoproteins has been discussed above. A second source, especially for free cholesterol, is cells and cell membranes, which under unusual situations (e.g., abetalipoproteinemia) may constitute the major or even the only source of the lipids. In most other situations, however, lipolysis appears to predominate as it can be calculated that the amount of phosphatidyl-

[56] B. P. Perret, S. Eisenberg, T. Chajek-Shaul, R. Deckelbaum, and T. Olivecrona, *Eur. J. Clin. Invest.* **13**, 419 (1983).

[57] R. J. Deckelbaum, S. Eisenberg, Y. Oschry, M. Cooper, and C. Blum, *J. Lipid Res.* **23**, 1274 (1982).

choline generated in a day from lipolysis is approximately the same as that consumed by the LCAT reaction.[24] This consideration alone is sufficient to explain the strong relationship that exists between HDL and, in particular, HDL_2 levels and the activity of the lipoprotein lipase system. Yet, lipolysis supplies only the lipid precursors necessary for $HDL_3 \rightarrow HDL_2$ conversion. As discussed in the previous section, cholesterol esterification by LCAT is mandatory for the formation of HDL_2 particles from HDL_3 (see also refs. 5 and 24). *In vivo,* such conversion was recently demonstrated in our laboratory after the injection of human HDL_3 into rats (unpublished).

HDL_4 is an HDL population of very high density and very small diameter (65–70 Å) found in the plasma of patients with abetalipoproteinemia.[57] We postulated that HDL_4 is a tiny particle of intestinal origin that possibly represents nascent spherical particle. If HDL_4 is a normal constituent of the HDL system, then a rapid conversion of HDL_4 to HDL_3 is expected to occur by the same mechanism described for HDL_3 to HDL_2 conversion. More experiments, however, are needed before any definite conclusion on the HDL_4 system can be drawn.

Conversion of HDL_2 to HDL_1. HDL_1 is a lipoprotein of low density and high diameter that is present in relatively large amounts in rat[44] and smaller amounts in human[58] plasma. ApoE is a major protein constituent of HDL_1.[44,58] Because the cholesteryl ester fatty acid profile of HDL_1 in the rat (an animal species devoid of cholesteryl ester transfer activity) indicates LCAT origin of the esters, we suggested that HDL_1 is formed in the plasma compartment.[44] That suggestion was confirmed in a subsequent study when we demonstrated conversion of rat HDL_2 (labeled with [^{3}H]cholesteryl esters) to HDL_1.[28] Our assumption is that in the absence of lipid transfer, LCAT-derived cholesteryl esters accumulate in HDL_2 and some particles are converted to HDL_1. During this process, however, apoA-I molecules must be replaced by apoE molecules. As discussed elsewhere,[24] that possibly reflects the differential affinity of apoA-I and apoE toward particles of different diameters and radii of curvature.

Conversion of HDL_2 to HDL_3. To produce an HDL_3 particle from HDL_2 it is necessary to remove from the HDL_2 about two-thirds of the cholesteryl ester molecules and an appropriate amount of surface molecules. We suggested that replacement of cholesteryl esters by triglycerides followed by hydrolysis of the transferred triglycerides provides the basis for deletion of core molecules from HDL_2. This suggestion was originally based on our observations in human subjects with abetalipoproteinemia[57] and in rats.[44] In these two instances, LCAT-derived cholesteryl

[58] K. H. Weisgraber and R. W. Mahley, *J. Lipid Res.* **21,** 316 (1980).

esters remain with HDL: in abetalipoproteinemia because of lack of acceptors, and in rats, lack of transfer activity. The HDL system in these two examples is compatible with our hypothesis, and exhibits predominance of larger sized and lower density particle populations. Conversion of HDL_2 to HDL_3-like particles was reproduced by us while using *in vitro* incubation mixtures, and either abetalipoproteinemic serum[57] or normal human plasma HDL_2.[59] As expected, initiation of the lipid transfer reaction caused replacement of cholesteryl esters by triglycerides but has not induced a change of density of HDL_2. (In fact, a slight decrease of density was observed.) Addition of lipases, however, caused hydrolysis of most triglycerides and a substantial reduction in size and density of the HDL_2. Of the two lipases, the hepatic lipase was more active, especially when VLDL has been added to the system.[29,59] That conclusion agrees well with the suggestion that increased hepatic lipase activities are associated with decreased levels of HDL_2 and increased levels of HDL_3.[60]

Conclusions

During their circulating life time, lipoproteins are exposed to the activities of enzymes and lipid transfer proteins and are affected by equilibrium states determined by mass actions. Lipoprotein lipase, hepatic lipase, and LCAT exert their effects by direct action on lipid molecules. Lipid transfer proteins induce compositional changes at the core (cholesteryl ester and triglyceride) and surface (phospholipids) domains of the lipoproteins. Laws of mass action determine distributions of apolipoproteins and free cholesterol molecules between lipoproteins. Each activity changes the physical and chemical properties of lipoproteins and thereby alters the behavior of the particle in other reactions. Presence of proteins that activate enzymatic reactions and differential affinities of lipoproteins toward enzymes, transfer proteins, and receptors determine their metabolic fate. Above all, each lipoprotein particle is exposed simultaneously to multiple metabolic reactions and its biological behavior can be fully understood when all are considered. Because metabolic cycles are composed of more than one lipoprotein and more than one metabolic activity, the plasma lipoproteins represent one integrative system with multiple and complex relationships between its individual components. These considerations lead us to believe that only an integrative experimental approach that investigates simultaneously several lipoproteins and several reactions can

[59] R. J. Deckelbaum, S. Eisenberg, Y. Oschry, E. Granot, I. Sharon, and G. Bengtsson-Olivecrona, *J. Biol. Chem.* **261,** in press (1986).
[60] T. Kuusi, P. Saarinen, and E. A. Nikkila, *Atherosclerosis* **36,** 589 (1980).

properly be applied to studies of lipoprotein metabolism. The currently available methodology and experimental procedures, however, do not allow such studies. *In vivo* studies in intact animals or humans are too complex to enable clear delineation of individual metabolic pathways. The *in vitro* approach on the other hand, does not allow the multiple interactions that occur *in vivo*. Perhaps for these reasons, "the law of uncertainty" must be employed to most, if not all, metabolic studies. The present text, therefore, represents at best an up-to-date approximation of the process of fat transport in lipoproteins. The next approximation, especially if assisted by more advanced methodology, will undoubtedly be better and different from the one presented here.

[22] The Methodology of Compartmental Modeling as Applied to the Investigation of Lipoprotein Metabolism

By LOREN A. ZECH, RAYMOND C. BOSTON, and DAVID M. FOSTER

Introduction

While there are many methodologies associated with the study of lipoprotein metabolism the mathematical construct of compartmental models has resulted in the formation and testing of a formal, uniform theory of lipoprotein metabolism. This theory has all the requirements of a theoretical science (hypotheses, nonobservables, predictions) which sets it apart from the experimental science of lipoprotein metabolism. As in most sciences the present is built upon the shoulders of past results, and can therefore be understood in a superior way if the historical progression is recognized and understood. We therefore propose to examine this progression in the remaining paragraphs of this chapter.

The most important concept in the study of regulated metabolic systems is the concept of the steady state of the system. In general the metabolic pools of a physiologic system are related to one another through the nonlinear relationship of a chemical or physical reaction which, if there is any net flow of reactants, is not at equilibrium.[1] For a pool which is not infinite to exist when there is flow into, there must be flow out of the pool. It is this quiescent state of the pool size in the presence of flow through the pool which defines the steady state. If this

---

[1] It has been said that only those of us resting in *Boot Hill* are in equilibrium.

quiescence in flow, either in transport units of mass or volume per unit time or in flux units of mass or volume per unit area, is achieved then the rate constants as well as the fractional rate constants which describe the system remain constant. Because of this fact there is information concerning the quiescence of the physiologic system under study contained in the measurement of mass or concentration over the time period of the study. The amount of information contained in the measurement of the pool size in a physiologic system is independent of any tracer study which might also be performed.[2] Because of the increased difficulty associated with an interpretation of the results of studies performed under non-steady-state conditions, most but not all of the tracer studies of the lipoprotein system have been performed in the steady state.

Each compartment in a compartmental model delineating the theory of lipoprotein metabolism represents a single, homogeneous, well-mixed pool in the analogy with experimental data. Mass can flow in and out of each compartment and the direction of flow is designated by an arrow in the model. Arrows pointing to a compartment which do not originate at another compartment represent flow into the model from the environment. The environment is considered to be constant or changing at a prescribed rate in the theory. The topology of each model is determined by the number of compartments, the connectivity of the compartments, and the measurement or calculation of tracer or tracee in each group of compartments. Associated with each compartment are numbers representing the time material spends in the compartment (residence time) and the amount of material in the compartment (tracer or tracee).

We will describe the essence of modeling and the establishment of a theory of lipoprotein metabolism from the point of view that models are developed to separate the details of a metabolic system into its constituent parts, and then compare these details across a wide spectrum of experimental results. From the historical progression it can be seen that many particulars of a model were included in the form of a hypothesis or set of hypotheses which, while necessary to explain the analogy between theory and experimental results, were not generally accepted at the outset but proved to be a basis for subsequent development. There remain, however, some concepts which are completely hypothetical and are only accepted because the same hypothesis has been necessary in the explanation of data from a wide variety of sources over an extended period of time. An example of such a theoretical construct is the "delipidation chain."

As everyone is aware, many lipids are not soluble in aqueous solution.

[2] M. Berman and R. Schoenfeld, *J. Appl. Phys.* **27,** 1361 (1956).

The physiologic system transports these compounds in plasma by forming lipoprotein particles. These macromolecular complexes when viewed with an electron microscope are found to be approximately spherical. They are constructed such that the more polar constituents are on the lipoprotein surface (S) forming a boundary layer which interacts directly with the aqueous environment. The less polar moieties make up a core (C) which is relatively isolated from bulk plasma.

Human plasma lipoproteins are a polydisperse collection of macromolecular complexes which range in size from 7 to 160 nm and contain the moieties shown in the following tabulation.

Moiety	Location	Polar
Triglyceride	C	−
Cholesterol	S	+
Cholesterol ester	C	−
Phospholipid	S	+
Apolipoproteins (10 or more)	S	+
Sphingomyelin	S	+

Theory of Classification

Since the early work of lipidologists, lipoproteins have been classified by methods of separation with the tacit assumption that these separation techniques, which were based on physicochemical properties, also had physiological significance. The two classical methods of separation are based on flotation in a prescribed density solution (very-low-density lipoproteins or VLDL, low-density lipoproteins or LDL, high-density lipoproteins or HDL), and charge in a prescribed buffer solution (pre-β migrating lipoproteins, β-migrating lipoproteins, and α-migrating lipoproteins). While the density of lipoprotein particles is dependent on the makeup of the particle and all molecules contained therein, it is most sensitive to changes in triglyceride content when indexed by the remaining lipids. The second separation method is sensitive to changes in charge and therefore sensitive to both protein makeup and surface lipid (cholesterol and phospholipid) content. Separation based on charge at a specified pH is most sensitive to protein makeup.

Although the lipids make up a major fraction of each class of lipoproteins, the major structure–function information (e.g., receptor recognition, cofactor function) is contained in the protein constituents. This structure–function concept is recognized in a third classification system which divides the lipoproteins into families based on their apolipoprotein makeup. While this method of classification attempted to move to a more physiological classification, it still fails in that it is completely dependent on a major manipulation of the lipoprotein sample outside the physiological system. Whereas this system of classification was heralded as an advance in the 1970s, it does not have the sensitivity of the fourth classification system, which was begun in the 1950s, and is a direct result of the formation of a theory of lipoprotein metabolism.

This fourth classification system of lipoproteins, which is yet largely unrecognized as a classification method, is based on kinetic interactions with the physiological system. This kinetic classification transcends the other three systems in that the kinetics of each fractional product based on a particular method of separation can be labeled and examined through its interaction in the physiological system. In addition, it possesses the ability in principle to support phylogenetic studies. A major capability of this method of classification, in addition to the ability to detect transient populations of lipoprotein particles, is that it has a framework in which each moiety of the lipoprotein can be examined within the boundaries of the complete system. It is this framework which allows the incorporation of information from all methods of separation upon which other classifications are based. This classification then is dependent on only the physiologic–molecular interactions and provides a methodology for the formulation of precursor–product relationships based on the convolution and superposition integrals.

Development of Metabolic Theory

Following the study of the metabolism of free fatty acids (FFA), the first lipoprotein moiety to be studied by several groups in a variety of species including dogs, rabbits, rats, and man, a theoretical model was described in terms of simplified system of compartments (Fig. 1). The evidence from the study by Fredrickson and Gordon[3] that linoleic acid undergoes much more recycling into plasma seemingly without a comparable effect on its oxidation, raised the possibility that this fatty acid may be in equilibrium with another compartment not concerned with oxidation. This fact, in addition to the evidence in man that (1) FFA's fluctuate rapidly and markedly from time to time in normal individuals (controlled?), (2) FFA's rise after injection of fat but also rise on fasting, (3) carbohydrate feeding produces a fall in FFA's, (4) epinephrine injection causes an increase in FFA's and (5) FFA's are the blood lipid fraction primarily concerned with the supply of fats to tissue for oxidation metabo-

[3] D. S. Fredrickson and R. S. Gordon, Jr., *J. Clin. Invest.* **37**, 1504 (1958).

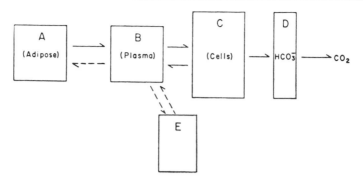

FIG. 1. Hypothetical diagram for plasma FFA metabolism.[3] Reproduced from *The Journal of Clinical Investigation*, 1958, **37,** by copyright permission of The American Society for Clinical Investigation.

lism, were utilized in the model for free fatty acid metabolism. Calculation of transport from plasma suggested that a large portion of the radiolabeled material recycled one or more times in plasma before oxidation, indicating that the transport from plasma may be considerably greater than the utilization of these molecules for energy.

This was further documented in a model in 1959 by Sigfrid Laurell.[4] In the rat labeled FFA's were followed from plasma into liver and adipose tissue. Following a delay, a portion of the labeled FFA's returned to the plasma incorporated in triglycerides. These pathways were examined under conditions of starvation and glucose infusion and differences in transport as a function of caloric intake were assessed. Laurell stated that the quantitative pattern in this complex transport system was still vague and a number of alternative transport passages were possible. This applied in particular to plasma triglyceride whose elimination other than via plasma FFA had not been examined. In other words, was the lipoprotein particle taken up as a particle containing intact triglyceride?

In 1961 Friedberg *et al.*[5] began a systematic study of the decay of plasma triglycerides. Following the injection of precursor free fatty acids, the appearance and decay of plasma triglyceride were determined. Using an equation and arguments to suggest that plasma triglyceride was a single pool, a model was proposed. Using a deconvolution technique, it was determined that the input function consisted of a delay and a single compartment. Even though the triglyceride curve was not monoexponential at the later time points (24-hr studies) this "tail" was neglected.

[4] S. Laurell, *Acta. Physiol. Scand.* **47,** 218 (1959).
[5] S. J. Freedberg, R. F. Klein, D. L. Trout, M. D. Bogdonoff, and E. H. Estes, Jr., *J. Clin. Invest.* **40,** 1846 (1961).

Also, in the early 1960s Baker and Schotz[6] examined the specific activities and total radioactivity of liver and serum triglycerides following the intravenous injection of [1-^{14}C]palmitate in rats. Using these data, a multicompartmental model of plasma and liver triglyceride metabolism was developed. This model was consistent with observed physiological and biochemical behavior of triglyceride and the observed data from their study.

Figure 2 contains a representative sample of serum and liver triglyceride specific activities following the injection of [1-^{14}C]palmitate. A simpler first model was solved by hand to determine the rate constant for the secretion of hepatic triglyceride into plasma (0.16 mg/min/100 g of body weight). To extend the model, however, it was necessary to use a computer program [a forerunner of Simulation, Analysis, And Modeling (SAAM)] and modeling techniques which had been developed by Mones Berman.

Using Berman's techniques, it became clear that a number of models gave equal resolution of the data and this lead the investigators to state, "further experiments should allow construction of a more correct model." The final calculations of triglyceride (TG) synthesis from plasma FFA's, of total liver TG turnover, or of hepatic TG secretion into the circulation were, however, insensitive to the choice of model.

Over the same period of time another group was performing studies of triglyceride metabolism in man using radiolabeled glycerol. Farquhar et al.[7] developed a two-compartment, nonrecycling catenary model to explain the synthesis and catabolism of VLDL-TG in man. They suggested that there were differences in rat and human triglyceride metabolism, and these differences explained the difference between the model of Baker and Schotz and their model. On comparing the VLDL-TG curves using [1-^{14}C] and [2-^{3}H]glycerol the differences in recycling were secondary to the labeled precursor in that labeled glycerol recycled much less than labeled palmitate. In man only 5% of the newly synthesized hepatic triglyceride is secreted into plasma; in the rat 95% of this pool was secreted into plasma. In the rat, 35% of the plasma triglyceride recycled through the liver without intervening hydrolysis, yet no recycling of intact triglyceride was necessary in man. These differences were considered to be due to assumptions in model building, biochemical methods, or species. It was these differences which formed the basis of further experimental design.

Even though by 1969 it could be stated that plasma transport of fatty

[6] N. Baker and M. C. Schotz, J. Lipid Res. 5, 188 (1964).
[7] J. W. Farquhar, R. C. Gross, R. M. Wagner, and G. M. Reaven, J. Lipid Res. 6, 119 (1965).

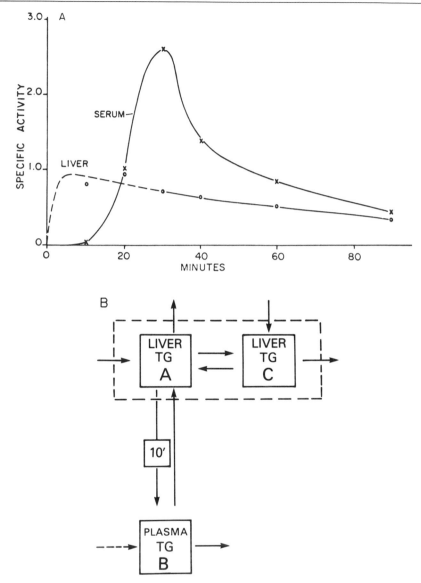

FIG. 2. (A) Rat liver and serum TG specific activity following [1-^{14}C]palmitate injection. (B) A simple model for TG metabolism.[6]

acids in the free form (FFA) and in the form of glycerides (TGFA) has been extensively studied in animals and man, it was clear from the modeling that the kinetics of fatty acids were poorly understood. Eaton *et al.*[8] conducted a study in which the kinetics of the precursor FFA's and the kinetics of the product triglycerides were examined simultaneously. Following these studies, the inverse study was performed with precursor very-low-density lipoprotein TG and product FFA's. From these data a two-compartment model was developed for FFA's. This model was shown to hold for both single injection and constant infusion of FFA's. Following the verification of the free fatty acid mode, it was used to provide a driving function for the plasma triglyceride model. This compartmental model proved to be valid when combinations of single injections and infusions of FFA's were used separately and in combination. This model also proved valid for data collected in subjects with normal and abnormal lipid metabolism.

Following the above results, the inverse experiments (consisting of the injection of lipoproteins labeled in the triglyceride fraction and collection of free fatty acid data) were conducted. A second model was developed to explain this set of data.

The models were combined into a general model, Fig. 3, and this model was tested using the simultaneous injection of radiolabeled FFA and radiolabeled lipoproteins. Using the model, the investigators were able to establish that (1) 20–30% of FFA's leave the plasma compartment in normals and quickly return to the plasma pool without change, (2) the rate of irreversible delivery of FFA's from plasma to tissues averaged 358 μEq/min in normals, and (3) a large portion of the lipoprotein triglyceride fatty acid underwent direct conversion to FFA's. Furthermore, the model explained the kinetics in many abnormal subjects by only changing the rate constants of the model. In other abnormal subjects, particularly individuals with type III hyperlipoproteinemia, the model predicted a liver pool of FFA's or triglyceride several orders of magnitude larger than physically possible. This problem remained unsolved until many years later when the theory of apoB metabolism was developed.

The above series of experiments and analysis concerned itself with the data collected in the first 4 hr after the injection of the labeled material. To further verify the general model, the next task was the examination of data for a longer period of time. Even though earlier investigators, such as Farquhar *et al.*, had conducted studies for as long as 72 hr, they did not construct a model which would account for data much past the first 4–12

[8] R. P. Eaton, M. Berman, and D. Steinberg, *J. Clin. Invest.* **48,** 1560 (1969).

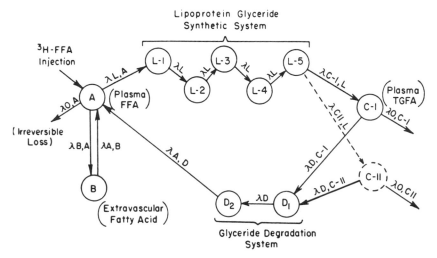

FIG. 3. General compartmental model incorporating all features.[8] Reproduced from *The Journal of Clinical Investigation,* 1969, **48,** by copyright permission of The American Society for Clinical Investigation.

hr. The analysis of triglyceride data in man continued using data collected over 24 hr in 1970 by Shames *et al.*[9] Because of the apparent tail on the curve, the general model developed in 1969 was extended; however, it was clear from this analysis that up to seven different models could account for the data. The investigators were able to discard four of the models based on the data in the tail of the curve; however, the remaining three models could not be separated. Further, the authors suggested experiments which would distinguish between these explanations of the data and the physiological consequences of accepting or rejecting each particular model.

By 1973, despite the fact that several models were available to explain the data yielding different results for triglyceride transport rate, investigators had examined the system under a large number of experimental conditions and for various periods of time. At this point in the development physiologists had decided to use a method of calculation without considering the influence the methodology had on the interpretation of the experimental outcome. Even though several direct experimental methods had been developed to measure the splanchnic bed secretion rate of triglycerides, these methods were not applicable to large population studies.

[9] D. M. Shames, A. Frank, D. Steinberg, and M. Berman, *J. Clin. Invest.* **49,** 2298 (1970).

In 1973 a review of triglyceride turnover in man appeared[10] and the conclusions concerning the theory as reflected in the model were as follows:

> "1. The various models and techniques that have been used to measure plasma triglyceride kinetics have been reviewed. In fasting man, the mass of TGFA that is transported in plasma VLDL has been variously estimated at between 15 and 100 μmol/ min. Even higher values have been reported with nontracer techniques in which the size of the triglyceride pool has been altered. The deficiencies of some of these models have been discussed.
> 2. The effect of dietary carbohydrate on triglyceride turnover has been even more difficult to evaluate for the following reasons: (a) most models assume a single precursor, usually plasma FFA whereas, with carbohydrate-rich diets, additional precursors might contribute significantly; (b) hepatic lipogenesis from glucose can provide newly synthesized fatty acids for TGFA formation even in the postabsorptive state, and (c) stored hepatic fatty acids may be an additional source of TGFA."

These conclusions taken from a set of eight[10] were based on the work of many investigators; however, following the 1970 report of Shames *et al.*[9] no new developments of the theory as reflected in the model had been advanced.

Apolipoprotein Metabolic Theory

Having come this far in the development of a kinetic model for triglyceride metabolism, efforts in the area of theoretical development through model building were shifted toward the protein moiety of lipoproteins. The first lipoprotein turnovers were reported in 1958 by Gitlin *et al.*[11] These investigators separated the lipoproteins into three fractions by precipitation and ultracentrifugation, radioiodinated the three fractions, and reinjected the subjects to be examined. Even though the catabolism of the different fractions was compared for normal adults and nephrotic children, no model was proposed. These investigators did notice that the S_f 10–100 lipoproteins were converted to S_f 3–9 lipoproteins, thus establishing the precursor–product relationship between VLDL and LDL for the protein moiety. Even though several groups subsequently performed lipoprotein turnovers by labeling the protein moiety, it was not until 1975[12] that the first model (Fig. 4) appeared for apoB and apoC metabolism. This model for apoB and apoC metabolism, which included the "delipidation chain," became the working model for the next series of

[10] P. J. Nestel, *Prog. Biochem. Pharmacol.* **8**, 125 (1973).
[11] D. Gitlin, D. G. Cornwell, D. Nakasato, J. L. Oncley, W. L. Hughes, Jr., and C. A. Janeway, *J. Clin. Invest.* **37**, 172 (1958).
[12] R. D. Phair, M. G. Hammond, J. A. Bowden, M. Fried, W. R. Fisher, and M. Berman, *Fed. Proc.* **34**, 2263 (1975).

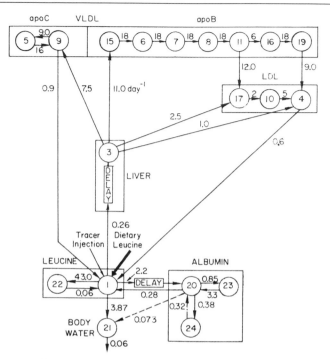

FIG. 4. Modified model including all subsystems.[12]

experiments which were subjected to quantitative analysis. Beginning in the 1970s, a group at NIH performed VLDL and LDL turnovers in subjects with various types of hyperlipidemia and in 1976[13] a model was presented for the metabolism of apoB and apoC in subjects with type III hyperlipoproteinemia, as shown in Fig. 5. This model differed from previous models in that two kinetic components of VLDL were detected. The usual delipidation chain was needed; however, a slowly turning over component of VLDL which was metabolized in parallel with the VLDL in the delipidation chain was necessary to account for the data. This species of lipoprotein made up more than 30% of the VLDL isolated from type III subjects, while it made up less than 1% of the VLDL isolated from normal subjects.

Subsequent studies in normals, type I, type II, and type IV subjects were subjected to analysis and presented in 1978.[14] Other findings include

[13] R. D. Phair, M. Hall, III, D. Bilheimer, R. I. Levy, R. Goebel, and M. Berman, *Proc. Summer Comput. Simul. Conf.* **July** (1976).

[14] M. Berman, M. Hall, III, R. I. Levy, S. Eisenberg, D. W. Bilheimer, R. D. Phair, and R. Goebel, *J. Lipid Res.* **19,** 38 (1978).

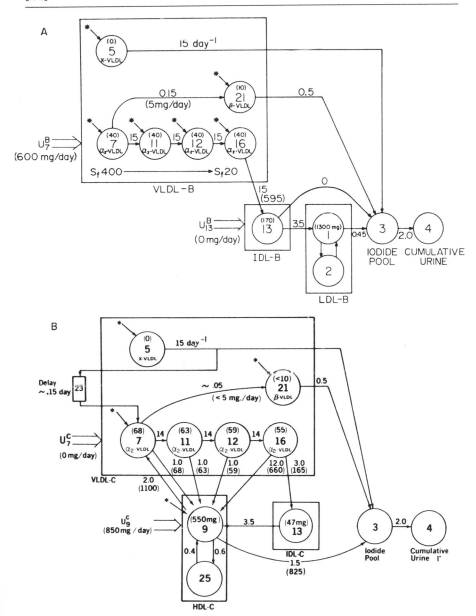

FIG. 5. (A) ApoB model. (B) ApoC model. The values next to the arrows are rate constants in units of day^{-1}. The values in parentheses are calculated or measured steady state transports.[14]

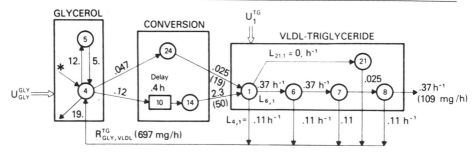

FIG. 6. Detailed compartmental model for subsystems. Fractional transports rates (L_{ij}) in hr^{-1}, transport rates (R_{ij}) in mg/hr.[15] Reproduced from *The Journal of Clinical Investigation*, 1979, **63**, by copyright permission of The American Society for Clinical Investigation.

(1) a pathway for the direct degradation of IDL without going to LDL, which was present in all phenotypes except normals, and (2) a synthesis pathway directly into IDL for type III subjects and one type I subject.

The next major advance in the modeling of lipoprotein metabolism was published in 1979 when the model for VLDL-triglyceride metabolism was again reexamined.[15] Because of the previous apoB model, a delipidation chain for VLDL-TG was established and proved to be necessary in fitting the new data, as shown in Fig. 6.

In addition, an isotopic difference in the metabolism of [^{14}C]glycerol and [^{3}H]glycerol used as a precursor to label VLDL-TG resulted in establishing one of the seven models of the previous 1970 analysis[9] as the better expression of the theory of lipoprotein metabolism.

Additional modeling has also been done for the LDL apolipoprotein B, establishing the fact that the theory is consistent with the existence of two and perhaps three kinetically distinct lipoprotein moieties within the LDL density fraction.[16,17] Using the [^{3}H]leucine-labeled apoB it was established that the hypothesis of secretion into and catabolism directly out of the IDL density fraction represented a large metabolic pathway in subjects with dyslipoproteinemia.[18]

Following the study of the apoB and apoC metabolic pathways, the next logical step was the examination of the HDL apolipoproteins. Even though α-lipoproteins had been studied in the original lipoprotein turn-

[15] L. A. Zech, S. M. Grundy, D. Steinberg, and M. Berman, *J. Clin. Invest.* **63**, 1262 (1979).

[16] M. Berman, S. M. Grundy, and B. V. Howard, "Lipoprotein Kinetics and Modeling." Academic Press, New York, 1982.

[17] R. P. Eaton, R. C. Allen, and D. S. Schade, *J. Lipid Res.* **23**, 738 (1982).

[18] W. R. Fisher, L. A. Zech, P. Bardalaye, G. Warmke, and M. Berman. *J. Lipid Res.* **21**, 760 (1980).

over of 1958,[11] no modeling effort had been extended toward an HDL model until the mid-1970s. Blum *et al.*[19] published the first compartmental model fo HDL metabolism using radiolabeled HDL. These investigators suggested that apoA-I and apoA-II had identical fractional catabolic rates and therefore could be represented by an HDL model. Subsequently, several sets of data indicate this may not be the case and recently independent models[20] for apoA-I and apoA-II have been developed, as shown in Fig. 7.

Cholesterol Metabolic Theory

A third major category of modeling to be reviewed deals with cholesterol metabolism. Cholesterol, both free and esterified, has also been subject to considerable analysis using compartmental modeling methods.

Approaches to the study of cholesterol metabolism in man have generally involved the determination of whole body cholesterol synthesis and turnover, intestinal cholesterol absorption, and bile acid synthesis and turnover. The first theory of cholesterol metabolism was examined using models in 1962 by Avigan *et al.*[21] Subsequently these two compartmental models were generalized into a single two-compartment model by Goodman and Nobel.[22] Over a period of 12 to 14 years studies were conducted and contrasted between the various methods of measuring cholesterol metabolism and between subjects with various lipid and lipoprotein abnormalities. In 1973, using long duration studies,[23] the two-compartment model was extended to a three-compartment model and various comparisons were again carried out.

The theory of cholesterol metabolism implied in compartmental models has been used to drive bile acid synthesis where in such an application it is assumed that the rapidly miscible pool contains both the plasma and liver subpools.[24] Even though considerable effort has been exerted toward the analysis of bile acids and their recirculation including several compartmental models, the description of the quantitative aspects

[19] C. B. Blum, R. I. Levy, S. Eisenberg, M. Hall, III, R. H. Goebel, and M. Berman, *J. Clin. Invest.* **60,** 795 (1977).

[20] L. A. Zech, E. J. Schaefer, T. J. Bronzert, R. L. Aamodt, and H. B. Brewer, Jr., *J. Lipid Res.* **24,** 60 (1983).

[21] J. Avigan, D. Steinberg, and M. Berman, *J. Lipid Res.* **3,** 216 (1982).

[22] D. S. Goodman and R. P. Noble, *J. Clin. Invest.* **47,** 231 (1968).

[23] D. S. Goodman, R. P. Noble, and R. B. Dell, *Circulation* **45,** 11 (1972).

[24] S. H. Quarfordt and M. F. Greenfield, *J. Clin. Invest.* **52,** 1937 (1973).

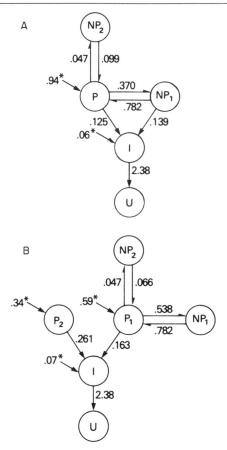

FIG. 7. (A) ApoA-II compartmental model; (B) apoA-I compartmental model.[20]

of cholesterol metabolism was not well addressed until the reexamination by Schwartz *et al.*[25] Given the almost proved experimental fact that in rats the rapidly miscible pool containing the plasma cholesterol compartment was not the precursor pool for bile acid metabolism and the information that newly synthesized cholesterol appeared to be the precursor, Schwartz *et al.* examined this relationship in man. Their model, as shown in Fig. 8, developed from experimental data obtained in patients with total bile fistula, indicates that more than 30% of the cholesterol precursor for bile acid and biliary cholesterol comes from newly synthesized cholesterol. Later data indicate that the total bile fistula plays a small role even

[25] C. C. Schwartz, M. Berman, Z. R. Vlahcevic, L. G. Halloran, D. H. Gregory, and L. Swell, *J. Clin. Invest.* **61**, 408 (1978).

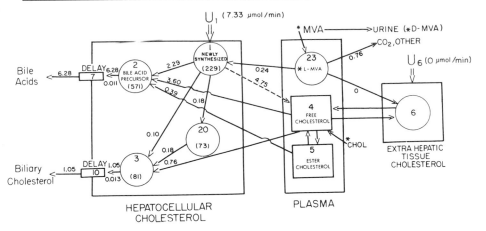

FIG. 8. Model for cholesterol metabolism; rate constants (min⁻¹) are below arrows, transport rates (μmol/min) are below arrows.[25] Reproduced from *The Journal of Clinical Investigation*, 1978, **61**, by copyright permission of The American Society for Clinical Investigation.

though the bile acids modulate fat absorption. These investigators have continued to examine in detail the plasma components in this model and recently[26,27] have published extensions of their original model. The plasma-free cholesterol model, as shown in Fig. 9A, and the HDL cholesterol ester model, Fig. 9B, have overcome the complexities of the system using multicompartmental kinetic analysis to describe a theory that both quantitatively and qualitatively describes HDL free and esterified cholesterol metabolism in man.

Summary

The statements to this point only give a cursory review of the beginning (20 years) of the kinetic approach to the classification of lipoproteins and subsystems which are involved in their synthesis and metabolism. At the present time the following partial list of theoretical findings expressed through model building can be made for the lipid and lipoprotein field:

1. A cascade process of delipidation for VLDL exists and the rate of this process is decreased in subjects with hyperlipoproteinemia.

2. ApoC recycles between VLDL and HDL in response to the dynamics of the delipidation cascade.

[26] C. C. Schwartz, Z. R. Vlahcevic, M. Berman, J. G. Meadows, R. M. Nisman, and L. Swell, *J. Clin. Invest.* **70**, 105 (1982).
[27] C. C. Schwartz, M. Berman, Z. R. Vlahcevic and L. Swell, *J. Clin. Invest.* **70**, 863 (1982).

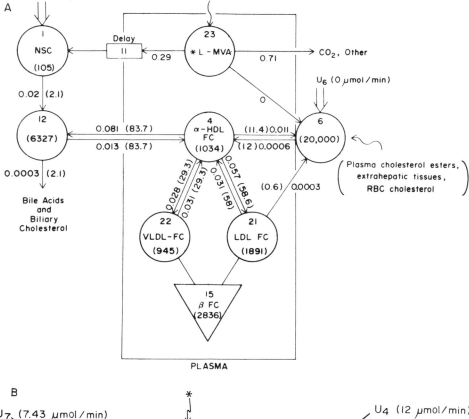

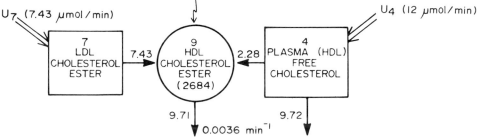

FIG. 9. (A) Free cholesterol model.[26] (B) HDL cholesterol ester model.[27] Rate constants in (min⁻¹), transport rates in (μmol/min). Reproduced from *The Journal of Clinical Investigation,* 1982, **70,** by copyright permission of The American Society for Clinical Investigation.

3. New synthesis of apoB first appears with newly synthesized VLDL. VLDL apoB synthesis decreased in hyperlipidemic states examined.

4. Multiple synthesis pathways exist for the triglyceride moiety of VLDL as determined by the transit time of a precursor through the conversion pathway. Thus the isotopic precursor methods now yield the same results as the more invasive techniques.

5. Kinetic heterogeneity of VLDL, IDL, LDL, and HDL has been established. Kinetic heterogeneity for apoA-I, apoA-II, apoB, apoC-2 and apoC-3 have been established.

6. The presence of direct pathways for IDL and LDL synthesis have been established.

7. Irreversible loss of apoC from HDL has been predicted using models.

8. Synthesis of apoC has been found to be invariant in hyperlipoproteinemic states studied.

9. Hepatic cholesterol compartments associated with the formation of bile acids and biliary cholesterol derive a majority of their cholesterol from newly synthesized and lipoprotein-free cholesterol.

10. More than 85% of the free cholesterol in the β-lipoproteins cycles directly through the HDL.

11. Free cholesterol recycles between HDL and tissue pools and between HDL and β-lipoproteins.

12. Only 20% of the esterified cholesterol in HDL is newly synthesized from plasma HDL free cholesterol at any point in time. The remaining 80% is obtained from LDL cholesterol ester.

13. Apolipoproteins A-I and A-II turn over at different rates, and each has more than one pathway of metabolism which are not common.

Even with this list the exact magnitude of work necessary to carry out a complete characterization of the lipid and lipoprotein system is not clear. The fact that very little quantitative (theoretical) consideration has been given to the dynamical control and regulation of the known lipoprotein system and the fact that not even the end point kinetics of lipoprotein moieties such as apoD, apoH, apoB-100, apoB-48, apoA-IV, phospholipids, and various exchange proteins are established indicate that only a small portion of the task has been completed.

Conclusion

It is clear that a common thread throughout this chapter is the fact that the theory and hence the models become larger with each new pathway.

It probably appears to the casual observer that techniques exist for the continued expansion of the analysis through the use of compartmental models. However, this is not the case or was it ever the case over the past quarter decade as a theory developed for lipoprotein metabolism. The tools and techniques for the realization, identification, and analysis of a large system (such as the system of lipid and lipoprotein metabolism) are developed as the need presents itself. Even though some techniques can be borrowed from other disciplines they usually need to be modified according to the requirements of the specific topic under investigation.

Furthermore, in the process of following the historical progression of the development of the theory of lipoprotein metabolism as represented by mathematical models, it becomes clear that the power of the methodology is derived from its ability to integrate many facts into a general theory. This provides a common framework for the development of a classification of lipoproteins based on the kinetics, and in the future, the dynamics of the interactions of these lipoproteins with the physiologic system under study.

While the mathematical details have not been considered in the above discussion, they are important. These details are not as important to the building of models of large integrated systems such as the lipoprotein metabolic system, however, as the powerful technique of using the compartmental model as the test bed for diverse experimental outcomes. When used in this capacity the model allows the separation of these outcomes into a set which are alike and represent the physiologic system and a set which differ and represent the biophysical and mathematical methodology.

[23] Kinetic Analysis Using Specific Radioactivity Data

By NGOC-ANH LE, RAJASEKHAR RAMAKRISHNAN, RALPH B. DELL, HENRY N. GINSBERG, and W. VIRGIL BROWN

Introduction

In kinetic turnover studies a tracer amount of radiolabeled material is injected intravenously as a bolus. The biologic half-life and/or the *in vivo* metabolic fate of the material of interest can then be determined by analyzing the decay curve which describes the disappearance of the injected radioactivity. A general discussion of the use of compartmental analysis to derive metabolic parameters from the decay curves is presented in Chapter [22], this volume.[1]

The data available for analysis may be in the form of either total radioactivity (TR) or specific radioactivity (specific activity, SA). An advantage of study protocols which permit the direct determination of SA data is that quantitative recovery of the material is not required. In other words, the SA will theoretically be the same whether 5 or 95% of the starting material is recovered following the procedure. Furthermore, while TR reflects only the fate of the injected tracer, SA data reflect the simultaneous fate of both the injected tracer and the unlabeled tracee.

Available programs for compartmental analysis, however, are primarily designed for the analysis of TR data. Thus even if SA data are available experimentally a common practice has been to multiply these SA data by an estimate of the appropriate pool size to obtain the TR data used in the analysis. In this chapter we will define a number of relationships characteristic only of SA data which can be used to derive information regarding the tracee. The application of these relationships in the elucidation of pathways for the metabolism of apolipoproteins will be illustrated for two systems: (1) the precursor–product relationship between VLDL and IDL apoB, and (2) the instantaneous equilibration of apoC-III in plasma.

Relationship in Precursor–Product Systems

Kinetic Equations

Let us assume that a dose of radioactivity D_1 is injected into pool 1 of the two-pool system shown in Fig. 1. In the present example, pool 1 would represent VLDL apoB, which is converted to IDL apoB represented by pool 2. We will now proceed to relate the parameters of the SA decay curves directly to the conversion process from pool 1 to pool 2.

If we denote by $y_1(t)$ the amount of apoB radioactivity present in VLDL (pool 1) and $y_2(t)$ the apoB radioactivity in IDL (pool 2) at various time t, the kinetic equations for the transfer of radioactivity in this system can be written as:

$$\frac{d}{dt} y_1 = -(k_{01} + k_{21})y_1 \qquad y_1(0) = D_1$$

$$\frac{d}{dt} y_2 = k_{21}y_1 - k_{02}y_2 \qquad y_2(0) = 0$$

(1)

---

[1] L. A. Zech, R. C. Boston, and D. M. Foster, this volume, [22].

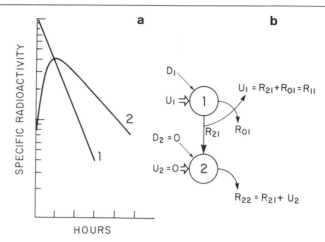

FIG. 1. A simple precursor–product relationship. (a) The precursor pool, pool 1, displays a simple monoexponential decay following the bolus injection of a dose of radioactivity, D_1. (b) Of the total flux U_1 through pool 1, an amount R_{01} is directly removed from the system and a flux R_{21} is converted to the product pool, pool 2. Assuming that there is no initial injection of radioactivity directly into pool 2 and that all of the flux in pool 2 is derived from pool 1, then the specific activity in pool 2 will reach a maximum value exactly as it crosses the specific activity curve corresponding to the precursor pool, pool 1. Furthermore, the areas under the specific activity curves for pools 1 and 2 will be exactly equal to one another.

In other words, the rate of change of radioactivity in pool 1 is proportional to the amount of radioactivity y_1. The fractional rate constant represents the combined effect of two metabolic processes : (1) k_{01} for direct irreversible removal from pool 1 and (2) k_{21} for conversion to the product pool, pool 2. The rate of change of radioactivity in pool 2 is the net result of the input from pool 1, $k_{21}y_1$, and the irreversible loss from pool 2, $k_{02}y_2$. At the beginning of the turnover study, the injected dose D_1 is entirely in pool 1 and there is no radioactivity present in pool 2.

If M_1 and M_2 represent the mass of apoB in VLDL and IDL, pool 1 and 2, respectively, we can represent the total radioactivity in Eq. (1) as the product of the mass by the corresponding specific radioactivity, i.e.,

$$\frac{d}{dt} M_1 S_1 = -(k_{01} + k_{21}) M_1 S_1 \qquad S_1^0 = \frac{y_1(0)}{M_1} = \frac{D_1}{M_1}$$

$$\frac{d}{dt} M_2 S_2 = k_{21} M_1 S_1 - k_{02} M_2 S_2 \qquad S_2^0 = 0 \tag{2}$$

where S_1 and S_2 are the apoB SA in VLDL and IDL, respectively.

In terms of the flux of apoB (mass per unit time) transferred between pools we have the following equations:

$$\frac{d}{dt} M_1 S_1 = -R_{11} S_1$$

$$\frac{d}{dt} M_2 S_2 = R_{21} S_1 - R_{22} S_2 \tag{3}$$

where $R_{11} = (k_{01} + k_{21}) M_1$ is the total flux of apoB through VLDL, pool 1
$R_{21} = k_{21} M_1$ is the flux of apoB transferred from VLDL to IDL
$R_{22} = k_{02} M_2$ is the total flux of apoB through IDL, pool 2

Since the system is assumed to be in steady state, there should be no net change in the apoB mass within each pool, we can divide the rate equations through by the respective mass to obtain:

$$\frac{d}{dt} S_1 = -\frac{R_{11}}{M_1} S_1 = -L_{11} S_1$$

$$\frac{d}{dt} S_2 = \frac{R_{21}}{M_2} S_1 - \frac{R_{22}}{M_2} S_2 = L_{21} S_1 - L_{22} S_2 \tag{4}$$

the rate constants L_{ij} will be referred to as fractional flux rate constants to differentiate them from the fractional rate constants k_{ij} commonly used in rate equations relating total radioactivity data.

The experimental SA decay curves can then be described by sums of exponentials:

$$S_1(t) = S_1^0 e^{-L_{11} t}$$

$$S_2(t) = \frac{L_{21}}{L_{22} - L_{11}} (e^{-L_{11} t} - e^{-L_{22} t}) \tag{5}$$

Thus, using a sum of exponential terms to approximate the experimental SA data corresponding to pools 1 and 2, we can obtain estimates for the fractional flux rate constants L_{11}, L_{21}, and L_{22}. Furthermore, by examining the values estimated for L_{21} and L_{22}, certain conclusions regarding the source of flux in the product pool, pool 2, can be made without a priori knowledge of the pool sizes in either compartment.

If $L_{21} < L_{22}$, this would indicate that there is direct input of material into pool 2 from a source other than pool 1. If $L_{21} = L_{22}$, this would indicate that all of the flux through pool 2 is derived entirely from pool 1.

It should be pointed out that if TR data, defined as the product of the SA data by the appropriate pool size, were analyzed only estimates of the fraction of the radioactivity initially in pool 1 which is converted to pool 2 could be obtained. Information regarding direct input into the product pool would be obtained only if the analysis indicated that the flux through pool 2 is greater than that through pool 1. These flux estimates, however, will be dependent on the accuracy of the determinations of the pool sizes.

Areas Under the Specific Activity Curves

Another relationship of the fluxes which is unique to SA data can also be derived from the areas under the SA decay curves. If we denote by A_1 the area under the SA curve for VLDL apoB (pool 1, in this example) and A_2 the area under the IDL apoB SA curve as described in Eq. (5), then,

$$A_1 = \frac{S_1^0}{L_{11}}$$

$$A_2 = \frac{L_{21}}{L_{11}L_{22}} S_1^0 = \frac{L_{21}}{L_{22}} A_1$$

(6)

Thus for the simple precursor–product system depicted in Fig. 1, a number of relationships must be satisfied by the areas under the experimental SA curves and the steady-state fluxes, including the following:

If pool 1 receives a dose D_1 of radioactivity and has a total flux R_{11} of material then the area under the SA curve corresponding to pool 1 will be

$$A_1 = \frac{D_1}{R_{11}}$$

With respect to pool 2, if a flux R_{21} of material from pool 1 is converted to pool 2 which receives a total flux R_{22} through it, then the area under the SA curve corresponding to pool 2 will be

$$A_2 = (R_{21}/R_{22})A_1 \quad \text{or} \quad A_2/A_1 = R_{21}/R_{22}$$

In other words,

$$\frac{\text{area under the product SA curve}}{\text{area under the precursor SA curve}}$$

$$= \frac{\text{flux from the precursor to the product}}{\text{total flux through the product}}$$

If all of the flux through the product is derived from the precursor, i.e., there is no direct input of material into pool 2 from sources other than pool 1 ($R_{21} = R_{22}$), the ratio will be exactly equal to 1. If there is contribution into pool 2 from sources other than pool 1 ($R_{22} > R_{21}$), the ratio will be less than 1. Two important concepts relating to the ratio of the areas under the SA curves must be emphasized: (1) this ratio of the areas does not depend on the fraction of the flux through pool 1 which is actually converted to pool 2. Thus the same ratio of the areas could be associated with a system in which there was conversion of 90% of the flux in pool 1 to pool 2 as well as with a system in which only 10% of the flux was converted; (2) this ratio of the areas can never exceed 1.

Conversion of ApoB from VLDL to IDL

Following the bolus injection of radioiodinated VLDL, the apoB SA curve displays a typical initial flattening period followed by a biexponential decay. It is generally accepted that the slower terminal component of the VLDL apoB SA curve represents the direct catabolism of a subpopulation of VLDL particles without conversion to IDL and LDL.[2–5]

The simplest approach to characterize the conversion of apoB from VLDL to IDL can be outlined as follows:

1. Approximate the VLDL apoB SA curve by a sum of two exponential terms either by using a nonlinear regression program or by graphical method using Matthews' curve-peeling procedure.[6]

$$B_1 e^{-\beta_1 t} + B_2 e^{-\beta_2 t} \quad (\beta_1 > \beta_2)$$

2. Estimate the area under the fast component of the VLDL upon decay curve. This will correspond to the fraction of the VLDL apoB flux which serves as precursor for IDL apoB. Let this area be denoted as $A_V = B_1/\beta_1$.

3. Approximate the IDL apoB SA curve to a sum of exponential terms. In general, at least three exponential terms will be required to describe the decay curve during the first 48 hr.

$$C_1 e^{-\gamma_1 t} + C_2 e^{-\gamma_2 t} - C_3 e^{-\gamma_3 t}$$

4. Estimate the area A_I under the IDL apoB SA curve,

$$A_I = C_1/\gamma_1 + C_2/\gamma_2 - C_3/\gamma_3$$

[2] M. Berman, M. Hall, III, R. I. Levy, S. Eisenberg, D. W. Bilheimer, R. D. Phair, and R. H. Goebel, *J. Lipid Res.* **19**, 38 (1978).
[3] M. Berman, *Prog. Biomed. Pharmacol.* **15**, 67 (1979).
[4] N.-A. Le, H. N. Ginsberg, and W. V. Brown, *in* "Lipoprotein Kinetics and Modeling" (M. Berman, S. M. Grundy, and B. Howard, eds.), p. 121. Academic Press, New York, 1982.
[5] N.-A. Le, H. N. Ginsberg, R. Ramakrishnan, R. B. Dell, and W. V. Brown, Submitted (1985).
[6] C. M. E. Matthews, *Phys. Med. Biol.* **2**, 36 (1957).

Depending on the relationship between the two areas estimated from the precursor and product SA curves, three cases could be considered:

Case I. If $A_V > A_I$, there is direct input or unlabeled apoB into IDL independent of plasma VLDL. This extent of independent apoB influx may be estimated from the ratio under the apoB SA curves:

$$U_2/R_{22} = 1 - (A_I/A_V)$$

where U_2 is the direct influx and R_{22} is the total apoB flux through IDL.

Case II. If $A_V = A_I$, there is a simple precursor–product between VLDL and IDL apoB and all of the flux in IDL is derived from VLDL. Note again that this does not necessarily imply that all of the VLDL apoB flux is converted to IDL.

Case III. If $A_V < A_I$, there is a theoretical contradiction in the analysis. Under the present assumption of simple precursor–product between VLDL and IDL apoB, it is impossible for the IDL fraction to have an area under the SA curve which is greater than the area under the VLDL apoB SA curve. In most apoB turnover studies completed by our group to date, however, it appears that the area A_I is significantly greater than the area A_V with a ratio of the areas ranging from 1 to 4.5.[5,7] Even greater ratios of the areas have been reported by Reardon *et al.*,[7] who used both exponential components of the VLDL apoB decay curve in the computation of the area of the precursor. Two possible explanations may be proposed:

1. The larger area under the IDL apoB SA curve is due to contamination of the injected radiolabeled VLDL preparation, which results in a portion of the injected dose appearing instantaneously in the IDL density range. In this situation we would expect the difference in the areas be proportional to the dose injected as IDL, i.e., to the initial apoB SA observed in IDL. Available data would suggest that at most, only 5–10% of the injected radioactivity is recovered in the IDL density fraction 5 min after the bolus injection. Most of this radioactivity, however, is associated with apolipoproteins other than apoB. Thus the measured apoB SA in IDL at 5 min is usually low and not sufficient to account for the large difference in the areas observed.

2. The larger area under the IDL apoB SA curve may be indicative of kinetic heterogeneity of apoB within the VLDL density range. The simplest type of heterogeneity would be the situation in which plasma VLDL exist as distinct subpopulations of particles. These subpopulations are metabolically distinct in that one subpopulation can be converted to IDL in a single step while another subpopulation must undergo several transformations within the VLDL density range before becoming IDL. Since

[7] M. Reardon, N. H. Fidge, and P. J. Nestel, *J. Clin. Invest.* **61**, 850 (1978).

all of the subpopulations in the density less than 1.006 g/ml would be initially isolated and radiolabeled, there will be direct injection of radiolabeled VLDL apoB into each of the subpopulations. In such a system, the radioactivity in the IDL density fraction will represent the cumulative contribution of apoB radioactivity injected into each and every VLDL subpopulation. The SA for apoB in each VLDL subpopulation, on the other hand, will reflect only the radioactivity directly injected into it and that derived from the preceding subpopulations. Thus it can be shown that when VLDL is characterized by a series of pools connected in a cascade process,[2,3] The IDL apoB SA curve will have an area which is greater than that for the entire VLDL apoB cascade process.[5] The theoretical basis of the larger area for the IDL apoB SA curve when the VLDL apoB is represented by a cascade process is discussed in detail elsewhere.[5]

Stepwise Conversion of ApoB from VLDL to IDL

The conversion of apoB from VLDL to IDL has been analyzed using any of a number of approaches, including (1) deconvolution method using total apoB radioactivity data,[8] (2) difference method using a portion of the VLDL and IDL apoB SA curves,[7] (3) a simple two-pool model for VLDL with an extravascular delay pool between VLDL and IDL,[4] and (4) a four-pool cascade model for VLDL apoB.[2,3] Except for the method by difference which does not actually fit the experimental curves, the other three approaches can produce theoretical curves which closely approximate the VLDL and IDL apoB TR decay curves. *With respect to the areas under the apoB SA curves, however, none of these analyses can satisfy the experimental data.* From the analysis of SA data, the two-pool model for VLDL apoB[4] would result in equality of the areas under the apoB SA curves for VLDL and IDL. The 4-pool cascade model,[2,3] on the other hand, would predict an area for IDL apoB SA which will be exactly 1.6 times the area of the apoB SA curve corresponding to the VLDL cascade process.

The simplest modification of the four-pool VLDL cascade process which can account for a wide range of ratios between the areas for VLDL and IDL apoB SA curves is to allow for a variable distribution of the total VLDL apoB mass among the four VLDL subpopulations. The basic assumption of this system is that newly secreted VLDL particles are associated with subpopulation 1 which will require, on the average, four metabolic steps to be converted to an IDL particle. The percentage of the total VLDL apoB mass which is associated with this subpopulation of newly

[8] A. H. Kissebah, S. Alfarsi, P. W. Adams, and V. Wynn, *Diabetologia* 12, 501 (1976).

secreted VLDL will depend directly on the relationship between the areas under the SA curves obtained experimentally for VLDL and IDL. Based on our own experimental SA data in normolipidemic and hypertriglyceridemic subjects, from 5 to as much as 45% of the VLDL apoB mass may be associated with the subpopulation of newly secreted VLDL particles. This distribution scheme for the VLDL apoB mass is a theoretical concept and further investigation would be needed to determine the physicochemical characteristics of these VLDL subpopulations.

Although experimental data at this time is not available to further define the subpopulations proposed in our model for the conversion of VLDL apoB to IDL, several statements concerning the apoB flux from VLDL to IDL can be made, including the following:

Direct Loss from Each Subpopulation. With influx of newly secreted VLDL only into the first subpopulation, direct loss of apoB flux from each subpopulation within the VLDL cascade will increase the ratio of the areas, i.e., the area under the IDL apoB SA curve will be greater.

Direct Secretion into Each Subpopulation. With efflux of VLDL apoB only out of the last subpopulation, any direct input of newly synthesized apoB into the various subpopulations within the cascade will decrease the ratio of the areas. In some situations, the area under the IDL apoB SA curve could be less than that for the VLDL apoB cascade.

Direct Conversion of Each VLDL Subpopulation to IDL. Direct conversion of apoB from any of the VLDL subpopulations to IDL will not affect the area under the IDL apoB SA curve to any significant degree. This is an important concept which illustrates the difference between the traditional analysis based on total radioactivity data and analysis using SA data. The area under the SA curve for a product is a weighted average of the areas of all contributing precursors. In this cascade process, since the last subpopulation reflects the apoB radioactivities injected into all of the subpopulations, it will have the largest area. Thus, while direct conversion of VLDL apoB from the first subpopulation to IDL is expected to increase the apoB radioactivity in IDL during the early time points, the larger area under the IDL apoB SA curve must still be explained by the large area generated in the last subpopulation by the cascades process. In fact, it can be shown that, if all of the subpopulations within the VLDL cascade can be directly converted to IDL, the ratio of the areas under the SA curves will be exactly 1.

Relationship in Equilibrating Systems: Metabolism of ApoC-III

Another class of metabolic systems in which specific radioactivity data can provide additional insight into the underlying kinetics is systems in which there is rapid equilibration among metabolically distinct pools. An example of such exchange systems is the dynamic equilibration of apoC among different VLDL particles as well as with lipoprotein particles of other density classes. This exchange of apoC radioactivity was first described by Bilheimer and co-workers.[9,10] Combining *in vitro* observations with *in vivo* data in animals, these investigators concluded that the equilibration of the C-apolipoproteins was instantaneous. They had no data, however, to demonstrate that this equilibration was complete.

If the equilibration of apoC is supposed to be complete, the bolus injection of a dose D as apoC radioactivity should result in a uniform SA for plasma apoC, which depend only on the dose injected and the apoC pool size in plasma. In other words, the same initial SA will be obtained for apoC in VLDL, IDL, LDL, and HDL. Divergence in the SA curves after the initial time would reflect differences in the metabolism of apoC associated with the various lipoprotein classes.

We recently described an immunoaffinity method for the direct determination of the SA of apoC-III in radiolabeled lipoproteins.[11,12] Following the injection of radiolabeled VLDL in a group of subjects we have noted the significant difference in the apoC-III SA in VLDL and HDL. Over a 75-hr period the two SA curves for VLDL and HDL apoC-III were parallel, as demonstrated elsewhere in this volume.[12] Thus, if the experimental SA data for the two curves were multiplied by the appropriate apoC-III pool sizes and expressed as a percentage of the injected dose, then the decay curves for VLDL and HDL apo C-III would be indistinguishable. However, by considering the specific activity data it must be concluded that the variability in the SA between VLDL and HDL apoC-III among subjects is indicative of the lack of complete equilibration of apoC-III in plasma. Experimental protocols can then be designed to specifically characterize the nature of this equilibration processes.[12,13] These studies have demonstrated that portions of the apoC-III pools in both VLDL and HDL did not participate in the equilibration process *in vitro*.

The lack of complete equilibration of apoC-III between VLDL and HDL can also be used to define a relationship between the exchangeable pools of apoC-III in VLDL and HDL. In particular, from the injection of

[9] D. W. Bilheimer, S. Eisenberg, and R. I. Levy, *Biochim. Biophys. Acta* **260,** 361 (1972).
[10] S. Eisenberg, D. W. Bilheimer, and R. I. Levy, *Biochim. Biophys. Acta* **326,** 361 (1973).
[11] P. R. Bukberg, N.-A. Le, J. C. Gibson, L. Goldman, H. N. Ginsberg, and W. V. Brown, *J. Lipid Res.* **24,** 1251 (1983).
[12] N.-A. Le, P. R. Bukberg, H. N. Ginsberg, J. C. Gibson, and W. V. Brown, This volume [27].
[13] P. R. Bukberg, N.-A. Le, H. N. Ginsberg, J. C. Gibson, and W. V. Brown, *J. Lipid Res.* **26,** 1047 (1985).

radiolabeled VLDL, it is expected that the SA of apoC-III in the whole VLDL fraction will be the same as the SA in the equilibrating and non-equilibrating components of VLDL apoC-III prior to equilibration, that is,

$$S_V = S_{Ve} = S_{Vn}$$

where S_V, S_{Ve}, and S_{Vn} correspond to the apoC-III SA in whole, equili-brating, and nonequilibrating pools of VLDL apoC-III, respectively.

Prior to equilibration, no apoC-III radioactivity is expected in HDL, or,

$$S_H = S_{He} = S_{Hn} = 0$$

with S_H, S_{He}, and S_{Hn} denoting the apoC-III SA in whole HDL, the equilibrating, and nonequilibrating pools in HDL, respectively.

After the instantaneous equilibration, the SA in the nonequilibrating pools of VLDL and HDL should not be affected by the *in vitro* incuba-tion. The SA for apoC-III in the exchangeable pools of VLDL and HDL will reach a new equilibrium value which reflects f_{PV}, the fraction of the exchangeable mass of apoC-III which is associated with VLDL, i.e.,

$$S_{He} = S_{Ve} = f_{PV} S_{Ve}$$

The overall apoC-III SA in each lipoprotein class is a weighted average of the SA in the equilibrating and nonequilibrating pool.

$$S_V = S_{Ve} f_V + S_{Vn}(1 - f_V)$$

$$\text{where } f_V = \frac{\text{mass of equilibrating VLDL apoC-III}}{\text{total mass of apoC-III in VLDL}}$$

$$S_H = S_{He} f_H \qquad \text{where } f_H = \frac{\text{mass of equilibrating HDL apoC-III}}{\text{total mass of apoC-III in HDL}}$$

The ratio R of the apoC-III SA between VLDL and HDL as obtained from the decay curves is defined as

$$R = S_V / S_H$$

which is the ratio of the SA of VLDL apoC-III to that of HDL apoC-III observed *in vivo* following the tracer injection of radioiodinated VLDL.

$$R = \left(\frac{S_{Ve} f_V + S_{Vn}(1 - f_V)}{f_{PV} S_{Ve}} \right)$$

or,

$$f_H = - \left(\frac{1 - f_{PV}}{R f_{PV}} \right) f_V + \frac{1}{R f_{PV}}$$

In other words, for a given ratio of the apoC-III specific activity between VLDL and HDL observed *in vivo* following the injection of radioiodinated VLDL, the fraction of apoC-III in VLDL which can participate in the equilibration process is negatively correlated to the fraction of equilibrating apoC-III in HDL. Any kinetic system proposed to explain the metabolism of apoC-III in man must address this relationship.

Acknowledgments

This work was initiated under a postdoctoral training program from NHLBI (HL-07343). Dr. Le is currently supported by a New Investigator Award from NHLBI (HL-27170).

[24] Metabolism of the Apolipoprotein B-Containing Lipoproteins

By George Steiner, Mary E. Poapst, Steven L. Shumak, and David M. Foster

Introduction

This chapter will describe, in general terms, the approaches that can be taken to study the kinetics of apoB-containing lipoproteins *in vivo*. These lipoproteins fall into the classes generally called chylomicrons, VLDL, IDL, and LDL. Unlike glucose, a lipoprotein is not a single molecule in solution in the plasma. Therefore, when describing its metabolism one must consider its various components. One focus here will be on apoB itself, as this is the key apolipoprotein in this family of lipoproteins, and as it appears to stay constant in amount with the particle during its lifetime in the circulation. The other focus will be on the triglyceride of these lipoproteins. This will be of particular importance because it is now recognized that, although the metabolism of the lipoprotein's protein and lipid moieties are related, they may still be independent of each other.[1-3] For example, triglyceride in the VLDL particle is hydrolyzed by the repetitive interactions with lipoprotein lipase (LPL). The result is a succession of VLDL particles of increasing density. Furthermore, triglyc-

[1] J. Melish, N. A. Le, H. Ginsberg, D. Steinberg, and W. V. Brown, *Am. J. Physiol.* **239**, E354 (1980).
[2] G. Steiner and M. F. Reardon, *in* "Lipoprotein Kinetics and Modeling" (M. Berman, S. M. Grundy and B. V. Howard, eds.), p. 237. Academic Press, New York, 1982.
[3] G. Steiner and M. F. Reardon, *Metabolism* **32**, 342 (1983).

eride may enter and leave already existing VLDL particles in the circulation. By contrast, the apoB moiety is not affected by LPL but remains associated with the particle throughout the lipolytic cascade. On the other hand, remnant particles containing apoB and triglyceride may be removed from the circulation as single particles. The main orientation of the chapter will be to the metabolism of these lipoproteins in humans and the chapter will not deal to any great extent with animal data.

This chapter is divided into two parts. The first deals with the kinetics of the apoB-containing particles. In this, various approaches to the analysis of tracer kinetic data will be given focusing on (1) type of analysis and (2) caveats which must be considered in the interpretation of the data. These will point in the direction of the experimental methodology, which is the second part of the chapter. There, current techniques to study the kinetics of the apoB-containing particles are discussed together with areas where future development is needed to address questions raised by the data analysis.

Although this is not a chapter on lipoprotein physiology, some general background is essential to permit one to understand the approaches that will be discussed. The apoB-containing lipoproteins in general are secreted into the circulation as large triglyceride-rich particles. Those made by the intestine and transporting dietary triglyceride are chylomicrons. Those made primarily, but not exclusively, by the liver and transporting endogenous triglyceride are VLDL. Each is delipidated in a reaction catalyzed by lipoprotein lipase, the activity of which is rate limiting. This results in the generation of a chylomicron remnant or a VLDL remnant, often also called IDL, by the ultracentrifugal nomenclature. In most individuals, all of IDL is further delipidated, again in a reaction limited by lipoprotein lipase,[4] to produce the end product, LDL. There are a few, infrequent exceptions to this pathway. In one situation, type III hyperlipoproteinemia, some IDL is produced directly and not as a remnant of VLDL.[5] Another is seen in those who have a particularly high rate of "VLDL–IDL" turnover. They may catabolize some of their IDL directly, not via LDL.[6,7] Finally, those who have familial hypercholesterolemia may produce some of their LDL directly, not via VLDL and IDL.[8] There has recently been some suggestion that this direct production of

[4] M. R. Reardon, H. Sakai, and G. Steiner, *Arteriosclerosis* **2**, 396 (1982).
[5] M. F. Reardon, M. E. Poapst, and G. Steiner, *Metabolism* **31**, 421 (1982).
[6] M. F. Reardon, N. H. Fidge, and P. J. Nestel, *J. Clin. Invest.* **61**, 850 (1978).
[7] M. F. Reardon and G. Steiner, *in* "Lipoprotein Kinetics and Modeling" (M. Berman, S. M. Grundy, and B. V. Howard, eds.), p. 99. Academic Press, New York, 1982.
[8] E. D. Janus, A. Nicoll, R. Wootton, P. R. Turner, P. J. Magill, and B. Lewis, *Eur. J. Clin. Invest.* **10**, 149 (1980).

LDL may be a more general phenomenon and may reflect input of LDL particles from a pool that is not traced by many of the radioactive tracer methodologies that are currently employed. As noted below, these methodologies may be based on the definition of lipoproteins by arbitrary density cuts as opposed to functional classes.

The brief description above should indicate that the major way to consider one class of lipoproteins, as distinct from another, is by their metabolic or functional characteristics. However, most of our current methods to separate lipoproteins (e.g., ultracentrifugation, electrophoresis) from each other do not depend on their metabolic properties. Rather, they are based on physical characteristics of the lipoproteins. Such methods frequently yield subfractions of lipoproteins that are still metabolically very heterogeneous. One example of this is demonstrated in the confusion that results from the previous operational definition of LDL as lipoproteins falling within the S_f range 0–20. More recently, it is recognized that VLDL remnants fall within the range S_f 12–60 and that LDL falls within the range S_f 0–12. Hence, LDL as previously defined (S_f 0–20) contains a portion (S_f 12–20) that is functionally more appropriately considered to be a part of the VLDL remnant class. An example of the problems that were created by the former definition of LDL was the description of a triglyceride-rich LDL in patients with renal failure.[9] Recently, an increase in remnants has been described in such individuals.[10] It is likely that the earlier described increase in triglyceride-rich LDL probably reflected an increase in the S_f 12–20 lipoproteins. It is also noteworthy that even in the Framingham study LDL had been defined as the S_f 0–20 fraction.[11] This may necessitate a reevaluation of some of its conclusions with respect to LDL and to the atherogenic risk of the triglyceride-rich lipoproteins. The recent use of methods such as the heparin–Sepharose columns to separate particles that differ in their apoE content may prove to be a start in separating particles on a more functional basis.[12] Another approach to separate the particles functionally has been to sample from selected anatomical sites, such as the lymphatics draining the intestine.[13–15] However, even here one may be faced with particles

[9] L. S. Ibels, L. A. Simons, J. O. King, P. F. Williams, F. C. Neale, and J. H. Stewart, Q. J. Med. **44**, 601 (1975).

[10] P. J. Nestle, N. H. Fidge, and M. H. Tan, N. Eng. J. Med. **307**, 329 (1982).

[11] W. B. Kannel, W. P. Castelli, T. Gordon, and P. M. McNamara, Ann. Internal. Med. **74**, 1 (1971).

[12] F. A. Shelburne, and S. H. Quarford, J. Clin. Invest. **60**, 944 (1977).

[13] G. Steiner and W. Ilse, Can. J. Biochem. **59**, 637 (1981).

[14] G. Steiner, M. E. Poapst, and J. K. Davidson Diabetes **24**, 263 (1975).

[15] R. M. Glickman and K. Kirsch, J. Clin. Invest. **52**, 2910 (1973).

derived from lymph draining both liver and intestine and therefore one must still be cautious in drawing conclusions. This could be the case in studies of patients with chyluria.[16] Furthermore, even when one is certain that lymph lipoproteins are derived only from the intestine, they may still have all of the characteristics of chylomicrons except that they may contain triglyceride that is entirely endogenous in origin.[14]

In the absence of completely satisfactory ways to separate metabolically distinct lipoprotein species, it will obviously be necessary to be cautious about some of the conclusions that are drawn from current studies. However, one may take another approach to study metabolically distinct lipoproteins, even without separating them one from another. That is to trace the metabolic behavior of a constituent that is unique to one functional class of lipoproteins. Specific questions can be asked which can be answered, in part at least, by tracer studies. This technique, however, does not remove the caution needed in the interpretation of such studies. As discussed elsewhere in this volume (Chapters [22] and [23], these data can be used with other experimental data to develop an integrated kinetic model but, as noted below, these too are not free from caveats. Such an approach has been used to trace the fate of the lipid in intestinally derived particles by the use of retinol.[17] After absorption from the intestine, it is incorporated into chylomicrons and remains with them as they are catabolized to remnants and taken up by the liver. That which is taken up by the liver may reappear in the circulation bound to retinol-binding protein. It has been assumed that retinol does not reappear with VLDL. Recently, it has also become possible to trace the protein component of intestinally derived particles. This is because it has been recognized that apoB exists in a higher and a lower molecular weight form. In humans these two forms are called B-100 and B-48, respectively. This is more completely described elsewhere in this book (Kane et al.). At this point, it is important to recognize that, in man, it appears that B-48 is found exclusively in intestinal particles. There have already been some studies of the kinetics of apoB-100 and apoB-48.[18,19] However, it must be recognized that the quantitative interpretation of such studies depends upon the determination of the radioactivity or specific activities of each form of apoB. This, in turn, requires a knowledge of the efficiency with

[16] P. H. R. Green, R. M. Glickman, C. D. Saudek, C. B. Blum, and A. R. Tall, J. Clin. Invest. 64, 233 (1979).

[17] W. R. Hazzard and E. L. Bierman, Metabolism 25, 777 (1976).

[18] A. F. H. Stalenhoef, M. J. Malloy, J. P. Kane, and R. J. Havel, Proc. Natl. Acad. Sci. U.S.A. 81, 1839 (1984).

[19] P. J. Nestle, T. Billington, and N. H. Fidge, Biochim. Biophys. Acta 751, 422 (1983).

which both forms of apoB are labeled, and requires accurate quantitation of each form.

One other aspect of the lipoprotein's metabolism that may make the interpretation of kinetic data more difficult lies in the fact that many of the components of the lipoprotein may move into, and out of, the particle during its lifetime in the circulation. The triglyceride, cholesterol, and phospholipid may exchange between lipoproteins in a process that requires a transfer protein.[20,21] In addition, newly made triglyceride may enter directly into preexisting circulating VLDL particles, in a process that appears to be different from this exchange.[22] Although such processes may introduce complexities into some of the models used to describe the turnover of triglyceride, this need not always be so. If one is concerned only with the total entry of triglyceride into all of the triglyceride-rich lipoproteins, and not into specific subfraction, many of the apparent complexities will no longer have to be considered. The apolipoproteins may also exchange between lipoproteins. However, this does not appear to be the case for apoB. This characteristic allows apoB's metabolism to reflect that of the whole particle.

Each of the approaches to the measurement of lipoprotein kinetics has limitations and it is virtually impossible to determine which, if any, is absolutely correct. One way to deal with this problem at present is to model data obtained simultaneously by more than one approach, each of which is based on different assumptions. Another way is to make predictions from a particular approach that has been used, or from a model which fits existing data, and then to test that prediction by obtaining a new set of data.

Definitions

Before describing the various techniques used to study the kinetics of the apoB containing particles, it is necessary to define certain terms used in tracer studies. These are given below, and have been taken from several sources.

Tracee: The substance of interest to be traced.

Tracer: A labeled substance which is (1) detectable by an observer, (2) identical with the tracee in all physical, chemical, and biological properties, and (3) negligible in quantity when compared with the tracee.

[20] J. J. Albers, J. H. Tollefson, C. H. Clen, and A. Stienmetz, *Arteriosclerosis* **4**, 49 (1984).
[21] G. J. Hopkins and P. J. Barter, *Metabolism* **31**, 78 (1982).
[22] D. Streja, M. A. Kallai, and G. Steiner, *Metabolism* **26**, 1333 (1977).

Pool or *Compartment:* An amount of material which is (1) homogeneous, (2) well mixed, and (3) distinct from other material in the system. Material can flow into or out from a compartment. A compartment in this sense is not usually a physiological space and may not necessarily conform to anatomic boundaries.

Pool Size: The amount of tracee in a specific pool (units: mass).

Fractional Catabolic Rate: The fraction of tracee lost irreversibly from a pool per unit time (units: pool/time). It is equal to the reciprocal of the residence time, the average time a particle spends in the pool, or to the pool size divided by the turnover rate.

Turnover Rate: The rate at which tracee moves through a pool (units: mass/time). Also called synthetic rate, production rate, or absolute catabolic rate.[23]

Steady State: Occurs when the system under study has constant pool sizes because the rates at which tracee enters and leaves each pool are equal.

Radioactivity: Disintegrations per minute of a particular substance of interest. In tracer kinetic studies this is often expressed per unit volume. It can also be expressed in terms of fraction or injected dose per unit volume.

Specific Activity: Disintegrations per minute per unit mass of a particular substance of interest.

Precursor–Product: These are terms used to describe two substances. The precursor is the source of the product. If a precursor is labeled (tracer), then some label must appear in the product. Under certain specific conditions, if specific activity curves of the two substances conform to the characteristics described by Zilversmit,[24] the precursor may be concluded to be the only precursor of a given product.

A number of "models" have been devised and tested for the study of turnover of apoB-containing lipoproteins using tracer methodology. The object in all such studies is to infer metabolic properties of the tracee from the behavior of the tracer, under specified physiological conditions, in one or a number of pools. A (mathematical) "model" is used to describe these inferences mathematically. For detailed discussions of the mathematical models used in turnover studies, the reader is referred to other chapters as well as to several standard works in the area.[25,26]

[23] M. Berman, *Prog. Biochem. Pharmacol.* **15,** 67 (1978).
[24] D. B. Zilversmit, *Am. J. Med.* **29,** 832 (1960).
[25] R. H. Shipley and R. E. Clarke, "Tracer Methods For In Vivo Kinetics." Academic Press, New York, 1972.
[26] M. Berman, S. Grundy, and B. Howard, eds., "Lipoprotein Kinetics and Modeling." Academic Press, New York, 1982.

The Kinetics of the ApoB Moiety

Many laboratories have developed methods to study the kinetics of chylomicrons and lipoprotein particles of the VLDL–IDL–LDL family. Many have pointed out that the amount of apoB per particle is constant as a lipoprotein procedes through the catabolic cascade from VLDL to IDL to LDL.[27] Furthermore, in contrast to other apolipoproteins, apoB does not exchange or transfer between lipoproteins. Redgrave and Carlson have suggested that there is no significant difference in the amount of apoB per VLDL particle in normal or hypertriglyceridemic individuals.[28] All of these features make apoB an ideal component through which to follow the metabolism of VLDL, IDL, and LDL particles.

There is no single method of choice to study the metabolism of VLDL, IDL, and LDL. In general, there are two methods of labeling used. One isolates the particle of interest and labels it directly; in this case, the particle is labeled according to efficiency of the labeling procedure and mass of the moieties involved. The other makes use of a radiolabeled precursor for protein such as [^{3}H]leucine. In this case, radioactivity is distributed into various proteins in proportion to their synthesis from the precursor. In the former situation, label is usually introduced as a bolus which allows the resulting data to be analyzed by relatively standard methods. The latter does not introduce the label into the moiety of interest as a bolus, and this must be taken into account in the analysis of the data.

A number of theoretical approaches have been developed for the analysis of turnover data. Certain assumptions are inherent in each. In what follows, some of these assumptions and limitations will be discussed within the context of various ways to analyze turnover data. Limitations in the experimental methodology will also be pointed out. In the subsequent section, current methodology will be summarized with particular attention paid to the caveats associated with kinetic analysis.

Exogenous Labeling of Apolipoproteins with Radioiodine

This technique, the most common in use for labeling apoB-containing lipoproteins, depends on the sequential isolation, iodination, purification, and reinjection of lipoproteins from one subject into himself or into another. The use of both ^{131}I and ^{125}I, each labeling different lipoprotein fractions or subfractions, permits the simultaneous examination of the metabolism of two lipoprotein fractions. Radioiodination of lipoproteins

[27] G. Schonfeld, *Prog. Cardiovasc. Dis.* **26,** 89 (1983).
[28] T. Redgrave and L. Carlson, *J. Lipid. Res.* **20,** 217 (1979).

invariably introduces label into the non-B apolipoproteins and lipid of the lipoprotein in addition to labeling the B apolipoprotein. However, by isolating the apoB with the methods discussed in the following section, one can ensure that only apoB radioactivity or specific activity is measured.

Although investigators have employed heterologous lipoproteins, for instance in the study of types II and II hyperlipoproteinemia,[29,30] the possibility that different individuals have structurally and metabolically different apoB must always be borne in mind. Langer et al.[31] have shown that electrophoretic, immunologic, and flotation characteristics of the iodinated lipoprotein are identical to the same nonlabeled lipoprotein for the same individual (i.e., autologous lipoproteins) when isolated using ultracentrifugal techniques. Likewise, Eisenberg and colleagues[32] and other groups[33-35] have shown that iodination of a lipoprotein does not alter its properties on cellulose acetate electrophoresis, immunoelectrophoresis, gel chromatography or heparin affinity chromatography.

The problem is really 2-fold. First, does labeling itself produce a particle that is functionally different from the tracer? The above indicates in the case of autologous studies that this may not be a problem, although there is always the possibility of some other, as yet unidentified, difference between the tracer and tracee. Second, in studies in which the labeled lipoprotein from one individual is injected into another or in which the labeled lipoprotein has been modified, the labeled material is probably different from the tracee. In such cases, the labeled lipoproteins may not be valid tracers and conclusions relating to metabolic mechanisms may be suspect and hence should be made with care.

When labeled apoB is used to trace the metabolism of a lipoprotein particle, the *tracer* is the apoB-labeled lipoprotein, not just the apolipoprotein. To use tracer kinetic analysis, tracer and tracee must be comparable both with respect to the apolipoprotein, and with respect to the lipoprotein particle composition. Hence, in preparing the tracer, one must be careful to preserve the whole lipoprotein's properties, not just those of

[29] C. Malmendier and M. Berman *J. Lip. Res.* **19**, 978 (1978).

[30] A. Chait, J. D. Brunzell, J. J. Albers, and W. R. Hazzard, *Lancet* **1**, 1176 (1977).

[31] T. Langer, W. Stober, and R. I. Levy, *J. Clin. Invest.* **51**, 1528 (1972).

[32] S. Eisenberg, D. Bilheimer, R. I. Levy, and F. T. Lindgren, *Biochim. Biophys. Acta.* **326**, 361 (1973).

[33] A. H. Kissebah, S. Alfarsi, D. J. Evans, and P. W. Adams, *J. Clin. Invest.* **71**, 655 (1982).

[34] G. Sigurdsson, A. Nicoll, and B. Lewis, *J. Clin. Invest.* **56**, 1481 (1975).

[35] P. J. Nestel, T. Billington, N. Tada, P. Nugent, and N. H. Fidge, *Metabolism* **32**, 810 (1983).

the apolipoprotein. For example, Herbert et al.[36] have demonstrated that ultracentrifugation of VLDL changes its apolipoprotein composition. Although the change involves loss of VLDL apolipoproteins other than apoB, such an effect could cause significant alterations in the metabolism of the VLDL particle. Consequently, turnover data gathered with a tracer that has been "damaged" by ultracentrifugation might be misleading. Methods avoiding excessive ultracentrifugation by employing manganese–heparin precipitation to isolate apoB-containing lipoproteins have been shown by Calvert and James[37] to denature the lipoproteins. Hence, material isolated using this technique would fail to meet the definition of tracer, and any conclusions based on them would probably be incorrect. A similar situation probably occurs in the specific circumstances where glucosylated-LDL (glu-LDL) is introduced into a subject who possesses antibodies to glu-LDL. In this case, the antibody-bound lipoprotein would be very rapidly cleared from the circulation.[38] Finally, with regard to denaturation, the potential for radiation-induced lipoprotein damage also exists. This is usually avoided by adding albumin to the lipoprotein preparation. In terms of both state of the art experimental and data analysis methodologies, these examples are of crucial importance. The theoretical methodology exists to describe apoB metabolism in more detail, but the experimental methodology is only now becoming available. When material that is labeled is not a true tracer, then one is analyzing artifacts and most probably will reach a conclusion that may be supported by the *data* but the data are not valid in the true sense of the word.

When ultracentrifugation is used to isolate the lipoproteins for labeling, or the plasma sample for subsequent analysis, it must be recognized that a given density fraction may be metabolically heterogeneous.[22,39–41] The distribution and composition of particles within a given fraction need not be identical at all times in a single individual. For example, particles isolated in $d < 1.006$ g/ml following an overnight fast, a protocol frequently used in lipoprotein turnover studies, will certainly be different than the same particles isolated 3 hr after a fat load. Thus, it is possible

[36] P. N. Herbert, T. M. Forte, R. W. Shulman, M. J. LaPianna, E. L. Gong, R. I. Levy, D. S. Fredrickson, and A. V. Nichols, *Prep. Biochem.* **5,** 93 (1975).

[37] G. D. Calvert and H. M. James, *Clin. Sci.* **56,** 71 (1979).

[38] U. P. Steinbrecher, J. L. Witztum, Y. A. Kesaniemi, and R. L. Elam, *J. Clin. Invest.* **71,** 960 (1983).

[39] R. P. Eaton, R. C. Allen, and D. S. Schade, *J. Lipid Res.* **23,** 738 (1982).

[40] W. R. Fisher, *Metabolism* **32,** 283 (1983).

[41] W. R. Fisher, L. A. Zech, P. Bardalye, G. Warmke, and M. Berman, *J. Lipid Res.* **21,** 760 (1980).

that a lipoprotein isolated and labeled at one given time may not be a valid tracer at all times during a turnover study. Although the investigator must be aware of this theoretical possibility, it is not known to what extent it actually affects the kinetic data. Similarly, a given density fraction may contain lipoproteins whose apoB may be either apoB-48 or apoB-100. There are no a priori grounds to assume that apoB-48 and apoB-100 iodinate equally. In the specific instance where lipoproteins from a diabetic subject are iodinated, the presence of glucosylated lipoprotein introduces yet another heterogeneity. Witztum *et al.* have shown that LDL from diabetics is glucosylated and that glucosylation of LDL to an extent seen *in vivo* in diabetics alters its catabolism.[42] Moreover, glucosylated lipoproteins have not yet been shown to iodinate to the same extent as nonglucosylated lipoproteins.

The metabolic heterogeneity of a given lipoprotein fraction may introduce yet another potential error. It is conceivable that within a given fraction there is a small pool of a rapidly turning over apoB-containing lipoprotein. If that is so, when a tracer is prepared by exogenous labeling techniques for turnover studies, very little of it would be labeled. If the turnover of this small subfraction is rapid, it could serve as a quantitatively significant precursor for a second lipoprotein. The conclusion from studies with exogenously labeled precursors would be that the product receives input from a nonlabeled pool. This could be interpreted, kinetically, as "independent input" into the second pool, i.e., synthesis of product independently from precursor. If this were the situation, then a kinetic model based on the data would be physiologically invalid. This serves to underline the point that the kinetic models are hypothetical bases which may be used to predict the behavior of metabolites. However, these predictions must be experimentally tested and if they do not hold up, then a new model must be considered.

Exogenous Labeling of Apolipoproteins with ³H or ¹⁴C

Radioiodinated lipoproteins can undergo radiation-induced denaturation. Because of this, as well as the desire to label tyrosine-deficient apolipoproteins, an alternative method of exogenous labeling was developed. That method, reductive methylation, introduced ³H- or ¹⁴C-labeled methyl groups into the ε-amino groups of lysine in LDL apoB. Care was taken to produce minimal methylation. Electrophoretic, analytic ultracentrifugal, and immunologic studies demonstrated that methylation of LDL did not change its physical–chemical characteristics, and preliminary

---

[42] J. L. Witztum, E. M. Mahoney, M. J. Branks, M. Fisher, R. Elam, and D. Steinberg, *Diabetes* **31**, 283 (1982).

turnover studies showed that it did not differ metabolically from native LDL.[43] It should be noted that in contrast to minimally methylated LDL, heavily methylated LDL does differ metabolically from native LDL. These metabolic differences have led to the suggestion that heavily methylated LDL may be cleared from human plasma by receptor-independent pathways.[38,44] However, such heavily methylated LDL does not have a counterpart in plasma. Hence, it does not meet the strict definition of a tracer, a caveat that must be remembered when drawing conclusions from any "tracer studies" that use it.

Endogenous Labeling of Apolipoproteins

These procedures utilize a radioactive amino acid as precursor for apolipoproteins. Two methods of endogenously labeling apoB are currently in use. Fisher and colleagues[41] employ [³H]leucine while Eaton and co-workers[39,45,46] use [⁷⁵Se]selenomethionine. It is assumed that the labeled amino acid mixes equally with all of the apolipoprotein's precursor pools and that the plasma amino acid specific activity is an accurate measure of the immediate precursor's specific activity. Thus, it is assumed that the radioactivity-specific activity of the immediate precursor is equal to, or derivable from, the precursor radioactivity-specific activity in plasma. This requires a model to describe the metabolism of the labeled amino acid and its conversion into apoB. With endogenous labeling of apoB, there is no risk of altering the tracer lipoprotein by any isolation or purification technique. Unlike exogenous labeling, endogenous labeling allows for all lipoproteins to be labeled in proportion to their synthetic rates and not to their plasma concentration. In this way, small rapidly turning over populations of lipoproteins may be followed better. Since either amino acid will be incorporated into the protein during its synthesis, labeling should not be affected by any posttranslational alteration such as glucosylation.

The major disadvantages of endogenous labeling are (1) the amount of label required to get enough radioactivity in the moiety of interest, and (2) the need for a description of the metabolism of the labeled moiety and the conversion pathway(s) to the moiety of interest.

To date, published results obtained by both [³H]leucine and [⁷⁵Se]selenomethionine methods are comparable to those obtained using radioiodinated lipoproteins. Nevertheless, great potential exists for the en-

[43] V. K. Murthy, T. C. Monchesky, and G. Steiner, *J. Lipid Res.* **16**, 1 (1975).
[44] H. R. Slater, C. J. Packard, and J. Shepherd, *J. Lipid Res.* **23**, 92 (1982).
[45] R. P. Eaton, S. Crespin, and D. M. Kipnis, *Diabetes* **25**, 679 (1976).
[46] R. P. Eaton, R. C. Allen, and D. S. Schade, *J. Lipid Res.* **24**, 1291 (1983).

dogenous labeling method to clarify the impact, if any, of rapidly turning over pools of lipoproteins on precursor–product relationships. The methodology also exists to perform double label studies combining endogenous and exogenous labeling. Simultaneous analysis of the data can address problems such as (1) kinetic heterogeneity, (2) existence of various pools, (3) precursor–product relationships, and (4) the problem of "metabolic channeling" as described by Fisher.[47]

Behavior of Chemically Modified ApoB

ApoB may be chemically modified by a number of processes including glucosylation, malondialdehyde alteration, reductive methylation, and cyclohexanedione treatment.[38,48] The products formed by any of these treatments are not identical to native apoB. In particular, apoB so altered is not recognized by the high-affinity receptor for LDL. Thus, LDL containing only chemically modified apoB should be catabolized only by receptor-independent processes. Chemically modified, labeled LDL may therefore be thought of as a tracer for that fraction of native LDL that undergoes receptor-independent catabolism.[38,48–50]

1,2-Cyclohexanedione (CHD) treatment of human LDL blocks the arginyl residues of the apo-LDL. The ultracentrifugal behavior of CHD–LDL is the same as that of native LDL. However, its electrophoretic mobility is increased and the possibility of further, more subtle structural alterations in the CHD–LDL remains.[48] Furthermore, all the caveats concerning exogenous labeling of lipoproteins exist for CHD–LDL. In addition, LDL is both radioiodinated and CHD treated by the authors cited and thus subject to possibility of radiation damage. As well, CHD modification of LDL is spontaneously reversible[38] and thus might lead to overestimation of receptor-independent catabolism.

Glucosylation of lysyl residues of apoB leaves the lipid composition and cholesterol/protein ratio of the lipoprotein unchanged.[38] Compared to native LDL, there is a slight decrease in the migration of glu-LDL on SDS–PAGE.[38] Exposure of LDL to the conditions required for glucosylation, i.e., cyanoborohydride incubation, does not in itself cause an alteration in the kinetic properties of LDL.[51] As with CHD–LDL, glu-LDL in

[47] W. R. Fisher, in "Lipoprotein Kinetics and Modeling" (M. Berman, S. Grundy, and B. V. Howard, eds.), p. 43. Academic Press, New York, 1982.
[48] H. R. Slater, C. J. Packard, and J. Shepherd, J. Lipid Res. 23, 92 (1982).
[49] H. N. Ginsberg, I. J. Goldberg, P. Wang-Iverson, E. Gitler, N. A. Le, H. S. Gilbert, and W. V. Brown, Arteriosclerosis 3, 233 (1983).
[50] J. Shepherd, S. Bicker, A. R. Lorrimer, and C. J. Packard J. Lipid Res. 20, 999 (1979).
[51] Y. A. Kesaniemi, J. L. Witztum, and U. P. Steinbrecher, J. Clin. Invest. 71, 950 (1983).

the studies cited was prepared by modification of exogenously radioiodinated lipoproteins. Consequently, its use as a tracer suffers from the theoretical disadvantages already mentioned.

One must realize that except for glucosylated LDL and malondialdehyde-altered LDL, chemically modified apoB tracers are not found *in vivo,* and cannot, therefore, be identical to any tracee. The use of any form of chemically modified apoB as a tracer of receptor-independent catabolism presupposes first that modified LDL is not degraded by the high-affinity receptor, and second that modified LDL and native LDL are degraded equally via the nonreceptor pathway. Although modified and native LDL have been shown to have identical removal rates from the plasma of homozygous familial hypercholesterolemic subjects (i.e., individuals lacking LDL receptors), this second assumption has yet to be completely validated.[49]

Conclusion

What does this all mean for the investigator wishing to study the metabolism of the apoB-containing particles? It means that since there is no ultimate method of labeling apoB, the investigator must be fully aware of the restrictions of the various methodologies to prevent reaching conclusions that are not supported by the data.

First, the specific question(s) relating to apoB metabolism under study must be carefully stated, and then one of the various labeling methodologies chosen. Because of the assumptions required to apply any tracer kinetic analysis, care must be taken in the preparation of the tracer whether it be exogenously or endogenously labeled.

Second, a method of analyzing the data must be made. The method chosen depends upon the information desired. If one is interested only in estimating the fractional catabolic rate, for example, then one may choose to calculate the area under the radioactivity or specific activity curve following a bolus injection of labeled material. If, on the other hand, one is interested in the amount of material transferred from one moiety to another, then one must deal with questions such as (1) kinetic heterogeneity of the substances under study, (2) complex conversion pathways, and (3) the models that are available to analyze the data. This third question is important for the following reason. The type of analysis chosen generates conclusions that can be compared with conclusions reached by analyzing the results from other laboratories that have used the same methodologies. In this way, one has additional confidence that the conclusions are indeed supported by the data. Finally, special care must be taken when modified lipoprotein particles are used, or when material isolated from one individual is injected into another. The reason is that these particles may not be true tracers, and investigators must be cautioned that the parameters estimated may not relate to "normal" metabolic pathways at all. If one is using a double label study in which either normal and modified particles, or autologous and isologous material, are labeled with ^{125}I and ^{131}I, respectively, then differences in the decay curves point to differences in the characteristics of the particles that are more readily recognized by biological systems than by physical or chemical systems. However, for the reasons noted above, conclusions concerning the "foreign" particles should be made with caution.

The Kinetics of the Triglyceride Moiety

The lipoprotein particle is not necessarily synthesized and catabolized as one complete entity. Not only can the apolipoproteins, other than apoB, exchange between the lipoprotein, so can the lipids.[20,21,27] Furthermore, at least newly made triglyceride can enter directly into preexisting plasma VLDL.[2,22] Hence, the metabolism of the lipoprotein must be studied in more than one way. One, as described above, is to study the metabolism of the particle by tracing a component, such as apoB, that remains with the particle during its life in the circulation. The other is to study the metabolism of its other individual components. Ideally, one should study the metabolism of apoB simultaneously with that of another constituent since that tends to define more precisely the behavior of the other component (subject to the hypothesis that apoB traces the metabolism of the particle per se). That approach has already indicated that triglyceride does not follow the same precursor–product pattern of metabolism in VLDL and IDL as that seen for the particle (apoB) itself.[3]

This section is related to the measurement of the kinetics of its triglyceride. Several different approaches have been taken to do this. These have included radioisotopic methods in which the glycerol[52–54] or the fatty acids[55,56] of the triglyceride have been labeled; transplanchnic balance studies[57]; estimates of the body's capacity to remove triglyceride under

[52] L. A. Zech, S. M. Grundy, D. Steinberg, and M. Berman, *J. Clin. Invest.* **63**, 1262 (1979).

[53] G. Steiner and T. Murase, *Fed. Proc.* **34**, 2258 (1975).

[54] J. W. Farquhar, R. C. Gross, R. M. Wagner, and G. M. Reaven, *J. Lipid Res.* **6**, 119 (1965).

[55] A. H. Kissebah, S. Alfarsi, and P. W. Adams, *Metabolism* **30**, 856 (1981).

[56] S. H. Quarford, A. Frank, D. M. Shames, M. Berman, and D. Steinberg, *J. Clin. Invest.* **49**, 2281 (1970).

[57] R. J. Havel, J. P. Kane, E. O. Balasse, N. Segel, and L. V. Basso, *J. Clin. Invest.* **49**, 2017 (1970).

steady state conditions, the "lipolytic rate" procedure[53,58]; and, in animals, the measurement of the rate of increase of plasma triglyceride when its removal is completely blocked by Triton WR 1339.[59,60]

Of these methods, the one most commonly used in humans is that employing [2-³H]glycerol. In that approach a bolus iv injection of [2-³H]glycerol is given in order to attempt what is essentially a "pulse-labeling" of endogenous triglyceride. That produces a biologically labeled VLDL–triglyceride. The kinetics of the triglyceride are then calculated from the curve describing the decline of this VLDL–triglyceride activity versus time. The decline of radioactivity or specific activity usually consists of two phases. The earlier and more rapid is generally monoexponential over the first 12 hr of the study.[52–54] In order to calculate the fractional catabolic rate (FCR), two approaches have been taken. The simpler is to utilize the monoexponential decline of specific activity.[22,53,54] The other, more complex, approach is to use multicompartmental analysis of the entire curve.[52]

The simpler method of analysis, that of estimating the FCR from the initial rapid decline of radioactivity or specific activity, may not satisfy the basic assumption upon which this method is based, namely that the VLDL–triglyceride specific activity decline after a bolus injection of labeled glycerol is identical to that after a bolus injection of biologically prelabeled VLDL–triglyceride. Frequent early plasma samples taken during a triglyceride turnover study reveal an initial rapid upswing of incorporation of radioactivity into VLDL–triglyceride followed by a plateau which differs in length among different individuals. This shows that there are complex kinetics in the conversion of labeled glycerol into labeled VLDL–triglyceride. The question then becomes one of whether one can use the initial rapid decline as an estimate of VLDL–triglyceride FCR, and if one uses this as an estimate, is one likely to reach erroneous conclusions concerning VLDL–triglyceride metabolism? The only way to test this directly is to compare the VLDL–triglyceride FCR obtained following injection of radioactive glycerol to the VLDL–triglyceride FCR obtained by simultaneously injecting VLDL, the triglyceride of which has been prelabeled biologically in its glycerol moiety. This has only been done in one study in humans, and that showed the two FCR values to be the same.[54] Other "reinjection studies" have suggested that the decline of specific activity of prelabeled VLDL does not follow the same curve as that of endogenously labeled VLDL triglyceride. However, those studies

[58] D. Porte, Jr. and E. L. Bierman, J. Lab. Clin. Med. 73, 631 (1969).
[59] J. D. Bagdade, E. Yee, J. Albers, and O. J. Pykalisto, Metabolism 25, 533 (1976).
[60] G. Steiner, F. J. Haynes, G. Yoshino, and M. Vranic, Am. J. Physiol. 246, E187 (1984).

either were done in species other than humans or used radioactive fatty acids to label the triglyceride.

Zech et al.[52] addressed the question of comparing the FCR of VLDL–triglyceride calculated from the initial rapid decay of VLDL–triglyceride radioactivity to that calculated from a multicompartmental model they developed. That model undertook to account both for glycerol kinetics and for a conversion pathway from glycerol to VLDL–triglyceride. The allowances made for each of these must still be directly tested experimentally. They observed that in some individuals, the estimate using the two methods was quite close while in others it was quite different. The multicompartmental model predicted a higher FCR, particularly in those with low plasma triglyceride concentrations, than that predicted from the rapid decline. The higher FCR resulted in a higher VLDL–triglyceride turnover rate. Hence, a comparison of results among normolipidemics and hyper- and hypolipidemics from the various laboratories using different methodologies has, at times, been difficult.

What does this mean for the individual studying VLDL–triglyceride metabolism using glycerol turnover studies? As with the case for apoB metabolism, it depends upon the questions being asked. It is not clear which estimate of the VLDL–triglyceride FCR is closer to the true one. There are assumptions which are not satisfied in the simpler method of analysis, but there are assumptions underlying the various multicompartmental models as well. It is the goal of those performing experiments and developing the models to unravel the truth, or to get close to it.

For the individual experimentalist, once a specific question is asked, a specific protocol should be adhered to. As discussed in the next section, for VLDL-TG turnover studies, the selection of the protocol itself can be as difficult as the selection of the data analysis. It is important, then, to stick with whatever analytical method chosen, for this will give the experimentalist the basis of comparison of normal and perturbed states, and comparison of values among laboratories using the same data analytical methodology. The qualitative results will probably be correct, but the quantitative ones may be subject to question no matter which technique is used.

Methods

Subject Management

Since the nutritional state of the subjects prior to and during lipoprotein turnover studies has a significant effect on the data generated and their interpretation, careful dietary management is important. Prior to the study, the dietary regimen customarily consists of a balanced, weight-maintaining diet that provides approximately 40% of calories as fat, 40% as carbohydrate, and 20% as protein. The length of the prestudy dietary control may range from days to weeks, although a 2-week period is the average reported in the literature. Subjects are normally fasted for 12 to 16 hr prior to the turnover study. During turnover studies that last up to 48 hr, a hypocaloric "fat-free" diet is usually instituted. This regimen provides the same amount of carbohydrate and protein as did the prestudy diet. However, less than 1% of its calories are derived from fat. Such a diet is assumed to minimize the possibility of chylomicron or chylomicron remnant contamination of the VLDL fraction, thereby limiting study to the metabolism of lipoproteins that transport endogenous lipid. Subjects consuming this diet have been found to maintain their weight and to maintain constant plasma apoB and triglyceride concentrations for the 48-hr period.[5,6] Since a prolonged hypocaloric dietary regimen would result in weight loss and disruption of normal lipoprotein metabolic patterns, this regimen is altered for investigations longer than 48 hr in duration. Subjects are either fasted or fed a fat-free diet for the initial 48 hr of the sampling period, and then returned to the previous 40% fat, 40% carbohydrate, 20% protein diet for the remainder of the study.

In terms of turnover study, the point of diet management is 2-fold. First, the material isolated for exogenous labeling should be a true tracer. Hence, if there are differences in the physicochemical characteristics between the tracer and tracee, one must assume the differences have a negligible effect on the kinetic data. Second, the subjects must be in a steady state in order to use linear kinetics. Since this rarely occurs, diet management is used to achieve a "dynamic steady state." Here the assumption is that those phenomena which exercise control over the metabolism of the substance under study change little during the course of the study. An example of dealing with this can be found in the various protocols used in VLDL–triglyceride turnover studies.[5,6,52,61]

The turnover study itself is initiated usually by a bolus of tracer which is either exogenously labeled tracee material or precursor for endogenously labeled tracer, although oral ingestion of isotopes is sometimes used. Blood samples are drawn for analysis at timed intervals thereafter. Subjects are usually hospitalized for a period of observation and dietary equilibrium prior to the study period and remain in hospital for at least the first days of the study when frequent sampling and strict dietary control are required.

[61] S. M. Grundy, H. Y. T. Mok, L. Zech, D. Steinberg, and M. A. Berman, *J. Clin. Invest.* **63**, 1274 (1979).

Isolation of ApoB-Containing Lipoproteins for Kinetic Studies

Traditionally, plasma lipoproteins have been defined and isolated on the basis of ultracentrifugation criteria. These procedures are conventionally, though not exclusively, used for the isolation of lipoproteins for kinetic studies, and are described in detail in Chapters [6] and [7], Volume 128. Some limitations of the ultracentrifugation procedures routinely employed for the isolation of lipoprotein fractions have important implications for interpretation of experimental data. It has become increasingly apparent that lipoprotein subfractions, isolated on the basis of physical criteria alone, are not metabolically homogeneous.[19,22,40] For this reason, a variety of techniques have been used in an attempt to isolate subfractions that are more discrete from a metabolic viewpoint. These procedures include discontinuous density gradient ultracentrifugation or chromatographic methods based on particle size,[62] charge,[63] or antigenic determinants.[64] It is important to recognize that modifications of the lipoprotein particle composition occurring as the result of any initial ultracentrifugation may affect the subsequent subfractionation. For instance, vigorous ultracentrifugation, leading to loss of the particles' apoE moieties, could be expected to change the apparent partition properties of the fraction on subsequent heparin affinity chromotography. This, of course, impinges on the question of what is being labeled as tracer.

In studies of lipoprotein kinetics, a significant proportion of the constituents under examination may be found in more than one subfraction. This may necessitate isolation of the subfraction(s) of interest by ultracentrifugation or other fractionation techniques in order to prepare the tracer and/or analyze samples drawn after the tracer has been administered. In the case of apoB, this is essential if one wishes to follow its transit through the VLDL–IDL–LDL chain. It also affects conditions relating to transfer of material down the chain, and *de novo* input into IDL and/or LDL. In other words, the density cut definitions are in a sense arbitrary and, for example, an individual might have large particles that contain only apoB (LDL), which float in the usual IDL density range. How label moves through these various subfractions (i.e., metabolic heterogeneity) will affect the conclusions based on the kinetics. Hence, it may be incorrect to treat pools as kinetically homogeneous when they are biochemically heterogeneous. This particular point is especially critical for the experimen-

[62] T. Sata, R. J. Havel, and A. L. Jones, *J. Lipid Res.* **13,** 757 (1972).

[63] R. J. Havel and J. P. Kane, *Proc. Natl. Acad. Sci. U.S.A.* **70,** 2015 (1973).

[64] R. W. Milne, P. K. Weech, L. Blanchette, J. Davignon, P. Alaupovic, and Y. Marcel, *J. Clin. Invest.* **73,** 816 (1984).

talist and modeler to resolve, and the methodologies are now becoming available.

By contrast, when studying the turnover of total plasma triglyceride, the tracer is biosynthetically labeled within the body so no isolation is needed for tracer preparation. The compartment of interest is the total plasma triglyceride so, again, no subfractionation is needed for analysis. This may, at first glance, suggest that it would be difficult to interpret triglyceride turnover. However, if one is interested in the total rate of triglyceride turnover, it does not matter how many input routes there are as long as one samples a pool that reflects them all. This is accomplished by analyzing an unfractionated plasma sample. In fact, the specific activity of triglyceride–glycerol, after the injection of [2-^{3}H]glycerol, is the same in the triglyceride extracted from whole plasma as it is in the S_f 20–400 fraction.[53] This indicates either that the majority of the new triglyceride input is into the S_f 20–400 fraction, or that the exchange of triglyceride between different lipoprotein fractions in humans is extremely rapid. Our previous studies have indicated that newly made triglyceride can directly enter preexisting extracellular VLDL.[2,3,22]

In VLDL–triglyceride turnover studies it may be of interest to know how the VLDL are secreted. Are they secreted only as large, triglyceride-rich particles, or as a spectrum of particles. This question becomes increasingly important when pertubation studies are carried out. One is interested, for example, in how the FCR or turnover rate changes but one would also like to know how the particles are affected. Subfractioning of VLDL can help to answer this question, as can performing double label studies (radioiodinated VLDL and [^{3}H]glycerol). Finally, electron microscopy can be used to see if a pertubation affects particle size distribution. All of these experimental techniques are available; their simultaneous use depends upon the questions being asked.

ApoB Kinetics—Exogenous Labeling

Iodination of Lipoproteins. Lipoproteins for iodination are normally isolated, washed once, and concentrated by conventional ultracentrifugal procedures. With slight variations, the iodination procedure in general use is essentially the iodine monochloride method originally described by MacFarlane, modified for use with lipoproteins by Langer *et al.*,[31] or Bilheimer *et al.*[65]

Efficiency of the iodination procedure averages 15%. All peptide moieties of the lipoprotein particle are labeled. Within VLDL–IDL, 43–60%

[65] D. W. Bilheimer, N. J. Stone, and S. M. Grundy, *J. Clin. Invest.* **64**, 524 (1979).

of the label is bound to apoB. Within LDL, 95% of the label is bound to apoB. Ethanol–ether extractions of the labeled lipoproteins indicate that on average less than 12% of the particle radioactivity is taken up by lipoprotein lipid. Precipitation with 15% TCA in the presence of carrier albumin demonstrates that less than 2% of the total radioactivity in the final preparation is present as free iodide. Electrophoretic, immunologic, and ultracentrifugal properties of the lipoprotein preparation are unchanged by iodination.[31]

The iodinated lipoprotein preparation must be sterile and pyrogen free for reinjection. Sterilization is accomplished by filtration through a 0.22-μm Millipore filter. The Limulus Amebocyte Lysate test is the most sensitive, convenient method currently available for detecting bacterial endotoxin, the leading cause of pyrogenicity. Kits providing reagents for this assay are commercially available from several suppliers (e.g., Pyrogent R, Mallinckrodt; E-toxate, Sigma). Sterile human albumin (3.5 mg/dl lipoprotein preparation) is added to the preparation to prevent radio-destruction of labeled lipoproteins. Labeled lipoproteins are normally reinjected within 96 hr from the time of withdrawal from the subject.

For turnover studies reported in the literature, the doses of radioactivity administered in [125]I- or ^{131}I-labeled lipoproteins range from 20–80 μCi. In order to block thyroid uptake of radioiodine, subjects are given from 300 mg to 1000 mg of KI daily, in one or in divided doses, commencing several days prior to administration of isotope. This is continued for two weeks postinjection.

Determination of Radioiodine Radioactivity and Specific Activity. The kinetic data from an apoB turnover study are expressed either as radioactivity (dpm/unit volume) or specific activity (dpm/unit mass). The determination of radioactivity in any particular fraction requires (1) the isolation of that fraction, (2) the isolation of apoB from that fraction, and (3) the determination of radioactivity of apoB. Specific activity requires these plus (4) determination of apoB mass in the fraction.

Whether one collects their data as radioactivity or specific activity depends upon the experimental protocol, the information desired from the data (e.g., what parameters are to be estimated), and the type of data analysis to be used. An important point should be made here. Most tracer experiments are conducted in the steady state. If this is the case, then the shape of the radioactivity and specific activity curves should be the same (if they are not, then there is either an experimental error or the system is not in the steady state). Hence, for parameters such as the FCR it doesn't matter which curve is used. If one is going to utilize Zilversmit's rule[24] to study precursor–product relationships, then specific activity data are used. Finally, to account for the transfer of apoB among particles that are kinetically heterogeneous and which have complex conversion pathways, radioactivity is used since that and not specific activity is what is transferred. With mass data, any integrated kinetic model can be used to obtain the "steady state solution," i.e., estimates of turnover rates and distribution of mass among pools.

As noted earlier, most of the labeling procedures will not only introduce radioactivity into apoB, but also into other apolipoproteins. In some procedures, radioactivity will be introduced into the lipoprotein lipid as well. Hence, in order to measure apoB activity data, it is not only important to isolate the appropriate lipoprotein subfraction, but it is also essential to ensure that the apoB is separated for analysis from all other lipoprotein components that could be labelled. If one is concerned with apoB specific activity, providing both disintegrations per minute and protein mass are determined on the same apoB precipitate, some losses of apoB are tolerable if it is necessary in order to ensure purity. Any methods used to analyze the apoB specific activity must be sufficiently sensitive to be usable for lipoprotein fractions that are isolated from relatively small plasma samples. The method must also be sufficiently reproducible to assure consistency from sample to sample.

Most methods that are currently used are based on a procedure to isolate and assay apoB from all other lipoprotein components by precipitation with 1,1,3,3-tetramethylurea (TMU) as described by Kane et al.[66] Both the protein mass and the radioactivity may be assayed either directly on the apoB precipitate itself, or estimated indirectly as the difference between total protein mass and protein radioactivity in the lipoprotein sample and those values in the sample supernatant following TMU precipitation of apoB. In both the direct and indirect approaches, the radioactivity is measured by a gamma counter and the protein mass is usually measured by the Lowry method.[67] A variant of the indirect approach makes use of an immunoassay to determine the apoB concentration of the sample first and calculates the apoB radioactivity as the difference between radioactivity in an aliquot of the sample and the supernatant after TMU treatment.[68]

A method to determine the specific activity of the apoB directly on precipitates has been developed by Le et al.[69] The method is designed primarily to isolate and quantitate apoB in lipoproteins samples contain-

[66] J. P. Kane, T. Sata, R. L. Hamilton, and R. J. Havel, J. Clin. Invest. 56, 1622 (1975).
[67] O. H. Lowry, N. J. Rosebrough, A. L. Farr, and R. J. Randall, J. Biol. Chem. 193, 265 (1951).
[68] A. Chait, J. J. Albers, and J. D. Brunzell, Eur. J. Clin. Invest. 10, 17 (1980).
[69] N. A. Le, J. S. Melish, B. C. Roach, H. N. Ginsberg, and W. V. Brown, J. Lipid Res. 19, 578 (1978).

ing from 30–75 μg apoB in a volume of 500 μl, but can be successfully adapted to measure other concentrations of the apolipoprotein. Egusa *et al.* also reported a similar procedure for isolation and quantification of apoB specific activity using 50% isopropanol rather than TMU as the apoB-precipitating agent, followed by washes with 50, then 100% isopropanol.[70] ApoB specific activity is then determined from assays of radioactivity and protein concentration made directly on the pellet.

Some potential problems can arise in the determination of apoB activity. All methods for apoB activity determinations that are based on selective precipitation of apoB from other lipoprotein components share the risk that the precipitated apoB may not be entirely free from contamination by other lipoprotein components or small quantities of plasma albumin. In order to minimize such potential contamination, some workers advocate an additional wash of the precipitate with acetone or 8 M urea to assure complete removal of any adherent apoE. It should be noted, however, that Malloy *et al.* have reported that TMU precipitated apoB-48 is soluble in 8 M urea.[71] The protein determination may also be inaccurate if the precipitated apoB is not completely dissolved in Lowry reagent C, or if there is interference in color development in the Lowry reaction caused by the presence of an excessive amount of NaOH or by residual traces of TMU in the reaction mixture. The chief technical difficulty in these procedures relates to the nature of the precipitate formed. It is fine, not readily visible, and may be easily dislodged from the bottom of the test tube during the aspiration of the supernatants. The apoB precipitate formed on the addition of water to TMU is more dense than the precipitate formed with isopropanol and seems to be more easily handled. On the other hand, the isopropanol precipitate is more readily soluble for protein determinations. Generally, it is particularly difficult to precipitate triglyceride-rich lipoproteins completely from the plasma of hyperlipidemic subjects. Because of the large amount of lipid, the lipoproteins tend to float, and thus are not able to mix properly with the precipitating solution. Essentially identical results have been reported in parallel assays using TMU and isopropanol methods to precipitate selectively apoB.[70]

SDS–polyacrylamide gel electrophoresis has been employed in studies examining the simultaneous metabolism of apoB-48 and apoB-100 after intravenous administration of a labeled TG-rich lipoprotein bolus.[18,19] Either 3.5% polyacrylamide or 2.5–27% gradient polyacrylamide gels containing SDS have been used to identify and separate the

[70] G. Egusa, D. W. Brady, S. M. Grundy, and B. V. Howard, *J. Lipid Res.* **24,** 1261 (1983).
[71] M. J. Malloy, J. P. Kane, D. H. Hardman, and R. L. Hamilton, *J. Clin. Invest.* **67,** 1441 (1980).

two apoB species on the basis of their characteristic mobilities in this medium as described by Kane et al.[72] Following SDS–polyacrylamide electrophoresis, the gels are fixed, stained, and destained essentially as described by Kane et al. If the tracer isotope is a γ emitter, the stained apoB-48 and apoB-100 bands may then be cut from the gels and counted directly. Activity has been expressed either in terms (%) of the injected dose, or in terms of the protein mass of the band as judged from dye uptake of the apoB band compared to that of a known peptide mass. An index of the relative chromogenicities of apoB-100 and apoB-48 has not been clearly established. Application of SDS–PAGE to specific activity determinations for apoB-100 and apoB-48 is subject to the limitation that apoB losses on delipidation, or solubility problems following delipidation, may apply preferentially to one species of apoB or other.

In terms of using apoB radioactivity or specific activity, the methodologies exist for what appear to be relatively pure preparations for subsequent analyses. The methodologies also are being developed to subfractionate samples into apoB-100- and apoB-48-containing particles. As mentioned in the remarks on subject management, what once were viewed as potential problems in this management in terms of isolating true tracers which were then subjected to more traditional analyses, may now lead the way to the next level of understanding apoB metabolism. More detailed information of apoB activity in terms of this subfractionation from particles which can then be "shifted" within a given individual by dietary management can provide data which will result in the next "level" of integrated apoB metabolic models.

Problems do arise, however. These involve how label distributes on the principles. When models are used to estimate turnover rates and mass distribution, assumptions must be made relating initial distribution of label and mass. Methodologies must be developed to address this problem or erroneous steady state estimates may result.

ApoB Kinetics—Endogenous Labeling

Another approach to study apoB kinetics has been to label the apolipoprotein endogenously. This has been accomplished either by intravenously injecting a bolus of ^{75}Se-labeled methionine[73] or [³H]leucine.[74] The rationales for the endogenous labeling approach have been considered above. In both cases the plasma amino acid activity has been measured

---

[72] J. P. Kane, D. A. Hardman, and H. E. Paulus, *Proc. Natl. Acad. Sci. U.S.A.* **77**, 2465 (1980).
[73] R. P. Eaton, S. Crespin, and D. M. Kipnis, *Diabetes* **25**, 679 (1976).
[74] W. R. Fisher, *Metabolism* **32**, 283 (1983).

and assumed to reflect the activity of the apolipoprotein's immediate precursor. The [³H]leucine activity was obtained by quantitating and isolating the amino acid through the use of an amino acid analyzer. The ⁷⁵Se-labeled methionine activity was assumed to be reflected by the radioactivity of the supernatant of a TCA deproteinized aliquot of plasma.

In separating the lipoproteins from which to isolate the apoB, it was necessary to avoid confusion due to the possible presence of any free amino acid tracer. In the case of selenomethionine, this was done by adding [³H]methionine *in vitro* to each plasma and counting the ³H in the apolipoprotein fractions. In the case of [³H]leucine, this was done by dialyzing the plasma sample against a solution that contained nonradioactive leucine.

The lipoproteins themselves were isolated ultracentrifugally by techniques adapted to yield the subfractions of interest. In the case of the [³H]leucine studies, the apoB was isolated after delipidation with TMU and washing with diethylether–ethanol. It was then dissolved in 3% SDS. The protein content of an aliquot of this was measured. Another aliquot was kept in solution with Soluene 350 to permit its radioactivity to be quantitated in a liquid scintillation counter. In the case of ⁷⁵Se-labeled methionine, the radioactivity of which can be quantitated in a gamma counter, a different approach was used. The apoB was isolated by preparative polyacrylamide gel electrophoresis. The segment corresponding to apoB was cut out of the gel and counted. No attempt was made to quantitate the apoB and the counts were related back to a volume of the original plasma.

The advantage of this method of labeling in proportion to how the particles are synthesized is offset by the need for a model describing the metabolism of the precursor amino acid and the small amount of radioactivity or specific activity recovered in the apoB-containing particle under study. However, as described earlier, a protocol combining the two labeling methods could yield considerable information concerning distribution of label, sites of entry of new particles into a proposed model, and turnover rates.

Triglyceride Kinetics

The rationales underlying the various approaches to assess triglyceride kinetics have been considered above. The most commonly used approach in humans is to label the triglyceride's glycerol endogenously by injecting an intravenous bolus of 100–250 μCi of [2-³H]glycerol. The triglyceride is then isolated from plasma samples taken at intervals thereafter. As noted earlier, its activity is the same whether measured in whole plasma or in the S_f 20–400 fraction. The important consideration in isolating the triglyceride for activity determinations is to obtain it free from other plasma or lipoprotein components that could contain radioactivity. These would include free glycerol, and other glycerides such as phospholipids. In order to accomplish this, the raw material is extracted with chloroform : methanol (2 : 1) for 12 hr at room temperature. The extract is then washed by the procedure of Folch *et al.* to remove traces of labeled free glycerol, protein, and other water-soluble contaminants.[75] The washed extract is taken to near dryness at 37° by blowing N_2 over it. It is then redissolved in isopropanol. Polar compounds are removed from isopropanol solution by allowing them to adsorb to Zeolite mixture (2 g/10 ml isopropanol), supplied by Technicon Instruments Corp., Terrytown, NY. The purity of the triglyceride may be verified by thin-layer chromatography. A known aliquot of the isopropanol solution of triglyceride is then transferred to a counting vial, dried, redissolved into an appropriate fluor, and counted in a liquid scintillation counter. The triglyceride mass is quantitated in another aliquot of the isopropanol solution by a procedure compatible with this solvent, such as that of Kessler and Lederer.[76] Others have quantitated the plasma triglyceride directly and, assuming quantitative recovery, have determined the radioactivity of an extract containing the triglyceride. From the data, the activity is determined and the turnover is then calculated.

Correction must be made where radioiodine has also been used to determine the kinetics of apolipoprotein at the same time that [^{3}H]glycerol has been used to determine the kinetics of the triglyceride. This may be done by determining how many counts are contributed by any radioiodine to the total counts detected by the liquid scintillation counter. To do this, one must know how much radioiodine is present in the triglyceride extract. This is determined by gamma counting the extract. One must also know how much this radioiodine may contribute to the apparent "^{3}H" counts. This is obtained by counting a sample containing only radioiodine in both a gamma counter and a liquid scintillation counter in order to calculate the percentage of radioiodine counts appearing as ^{3}H counts. The sample used for this purpose is a lipid extract of some of the radioiodinated lipoprotein tracer that had been prepared for injection to follow apoB kinetics, and contains only the radioiodine label.

[75] J. Folch, M. Lees, and G. H. Sloane Stanley, *J. Biol. Chem.* **266**, 497 (1957).
[76] G. Kessler and H. Lederer, *in* "Automation in Analytical Chemistry" (L. T. Skeggs, ed.), p. 341. Mediad, White Plains, New York, 1966.

Summary

This chapter was designed to describe the approaches one can take to study the metabolism of the apoB-containing particles *in vivo*. The focus has been to blend (1) what is the current tracer kinetics analysis methodology and (2) what are the current experimental protocols being used into a total picture so that the experimentalist wishing to perform such studies may have a better perspective of the strong points and pitfalls of this important experimental tool.

Hence, these points have been summarized from the point of view of what caveats are associated with each methodology. Recognition of these is essential to avoid reaching potentially erroneous conclusions. More important, attention has been focused on the realization that certain methodologies can be chosen depending upon what questions are being asked. Finally, areas where future development is needed in order to proceed to the next level of understanding are pointed out in the context of using tracer kinetic analysis as an integral part of a total experimental design.

Acknowledgment

Dr. Shumak was a Fellow of the Ontario Heart Foundation during the preparation of this chapter.

[25] Metabolism of Apolipoproteins A-I, A-II, and A-IV

By Ernst J. Schaefer and Jose M. Ordovas

Introduction

There has been considerable recent interest in the metabolism of plasma apolipoproteins A-I and A-II, largely because of the findings that decreased plasma levels of high-density lipoprotein cholesterol and apolipoprotein A-I are associated with premature coronary artery disease. Apolipoproteins (apo) A-I and A-II are the major protein constituents of high-density lipoproteins (HDL), and comprise approximately 90% of total HDL protein mass. HDL are found in human plasma in the density range 1.063 to 1.21 g/ml, and consist by weight of approximately 50% protein, 25% phospholipid, 20% cholesterol, and 5% triglyceride. Most of the cholesterol within HDL is in the esterified form. Total HDL levels in human plasma are approximately 300 mg/dl, and HDL has classically been divided into two density classes: HDL_2 (density 1.063–1.125 g/ml), and HDL_3 (density 1.125–1.21 g/ml). The apoA-I : apoA-II weight ratio appears to be considerably higher in HDL_2 than in HDL_3. A negative correlation between HDL_2 and very low-density lipoproteins has been noted. HDL_2 has been further subfractionated into HDL_{2b}, density 1.063–1.10 g/ml, and HDL_{2a}, density 1.10–1.125 g/ml. Females have considerably higher HDL_2 levels than do males, and population studies suggest that fluctuations in HDL levels are largely due to changes in HDL_2 concentrations.

HDL particles containing both apoA-I and apoA-II have been isolated, and these particles appear to be the major lipoprotein species within HDL (see Chapter [8], this volume). Other apolipoproteins within the HDL density region include apoB, apoC-I, apoC-II, apoC-III, apoD, apoE, and apoF. An HDL species containing predominantly apoA-I has been reported within the HDL_2 density region. ApoA-I and apoA-II are also found in lymph chylomicron, as well as in newly synthesized HDL from the intestine. In addition, these proteins can be found in trace amounts in plasma very-low-density lipoproteins (VLDL). Some apoA-I and apoA-II is also found in the 1.21 g/ml infranant region following isolation of lipoproteins by ultracentrifugation.

Both apoA-I and apoA-II are synthesized as preproproteins in the liver and the intestine, and these proteins appear to be catabolized primarily by the liver and kidneys. ProapoA-I (apoA-I_2 isoprotein) is found in lymph chylomicrons as well as in plasma lipoproteins. It is converted to the mature forms of apoA-I by cleavage of a six-amino-acid prosegment resulting in the conversion of proapoA-I to mature forms of apoA-I (apoA-I isoproteins 3, 4, and 5). Mature apoA-I is a single polypeptide of 243 amino acid residues. The protein contains no carbohydrate and has a molecular weight of approximately 28,000. The protein has been reported to activate lecithin : cholesterol acyltransferase, the enzyme responsible for cholesterol esterification in plasma. The mature forms of apoA-I such as apoA-I_3 and apoA-I_4 are converted to apoA-I_5 by deamidation.

ApoA-II has a molecular weight of approximately 17,000, and consists of 2 identical 77-amino-acid peptide chains attached by a single disulfide bond. ApoA-II contains no carbohydrate, and has been reported to enhance the activity of hepatic lipase. Both apoA-I and apoA-II can self-associate in aqueous solutions. These proteins can combine with lecithin to form protein phospholipid complexes with a hydrated density of HDL.

ApoA-IV is a glycoprotein which has a molecular weight of approximately 46,000, and is a constituent of chylomicrons.[1,2] In rat and dog plasma apoA-IV is also a major apolipoprotein of HDL; however, in

[1] J. B. Swaney, H. Reese, and H. A. Eder, *Biochem. Biophys. Res. Commun.* **59,** 513 (1974).
[2] K. H. Weisgraber, T. P. Bersot, and R. W. Mahley, *Biochem. Biophys. Res. Commun.* **85,** 287 (1978).

human plasma this protein is found mainly in the 1.21 g/ml infranate.[1,2] Following injection of radioiodinated apoA-IV reassociated with chylomicrons in rats, there was rapid transfer of apoA-IV radioactivity to the HDL and the lipoprotein-free fraction.[3] The half-life for apoA-IV label was 10 hr, similar to the half-life of apoA-I reported by other investigators in the rat.[3,4] To our knowledge no data are available on plasma kinetics of apoA-IV in man. Therefore, we will not discuss this apolipoprotein in the remainder of the chapter. Nevertheless, apoA-IV's interesting distribution among plasma lipoproteins and its possible relationship to cholesterol ester metabolism suggest that this apolipoprotein may play an important role in lipoprotein metabolism.[5]

The isolation of HDL by sequential flotation ultracentrifugation, density gradient ultracentrifugation, vertical rotor density gradient ultracentrifugation, zonal ultracentrifugation, analytical ultracentrifugation, gel filtration chromatography, or high-performance liquid chromatography is discussed in detail in other chapters. In addition, methods of isolation and characterization of apolipoproteins A-I, A-II, and A-IV are also extensively reviewed in Chapter 10, Volume 128. The focus of this chapter is on methodology for studying the plasma kinetics of apoA-I and apoA-II in man. Various methods of labeling these proteins will be discussed with particular emphasis on radioiodination utilizing isolated lipoproteins or apolipoproteins. In addition, comparisons of radioiodinated apolipoproteins and native apolipoproteins will also be discussed, and methodology for carrying out metabolic ward studies, quantitation of apolipoproteins, and kinetic analysis will be addressed. The results of kinetic studies in normal and dyslipidemic man will also be reviewed, and related to an overall concept of apoA-I and apoA-II metabolism in relationship to HDL and atherosclerosis.

Methodology

Isolation of Lipoproteins

Isolation of lipoproteins for use in kinetics studies is generally carried out by preparative ultracentrifugation. Blood is collected in tubes with a final concentration of ethyldiaminotetraacetic acid disodium salt (Na$_2$ EDTA) of 0.1%, and the plasma separated by a 30-min centrifugation at 3000 rpm at 4°. High-density lipoproteins can be isolated in the density

---

[3] N. Fidge, *Biochim. Biophys. Acta* **619**, 129 (1980).

[4] S. Eisenberg, H. G. Windmueller, and R. I. Levy, *J. Lipid Res.* **14**, 446 (1973).

[5] J. G. Delamatre, C. A. Hoffmeier, A. G. Lacko, and P. S. Roheim, *J. Lipid Res.* **24**, 1578 (1983).

range of 1.063–1.21 g/ml by the addition of potassium bromide to adjust the density. An alternative density region to use is 1.09–1.21 g/ml, in order to minimize isolation of HDL particles containing apolipoprotein apoE as well as the C apolipoproteins. Isolation by ultracentrifugation is carried out in Beckman 60 Ti rotors at 59,000 rpm by sequential spinning. In order to minimize contamination with albumin, an additional ultracentrifugation is carried out at density 1.21 g/ml in the ultracentrifuge. This step also serves to concentrate the HDL sample to a protein concentration of 10 mg/ml or more, which is ideal for iodination. Following ultracentrifugation, HDL is extensively dialyzed against a 100-fold excess volume of 0.01% Na_2EDTA, 0.85% NaCl, pH 7.4, sterile solution with 3 changes of the dialysis fluid. It is important for use in human studies to sterilize the polyallomer ultracentrifuge tubes as well as other laboratory equipment used in the isolation of lipoproteins including the tube slicer and gaskets. This goal can be achieved by gas sterilization.

Isolation of apolipoproteins can be carried out (see Chapter [10], Volume 128) by applying approximately 10 to 20 mg of HDL protein, previously delipidated with chloroform–methanol extraction, to Sephacryl S-200 columns in 6 M urea, 0.1 M Tris, pH 8.2 buffer, utilizing 200 × 2.0 cm glass columns. The portion of the apoA-I peak which is clearly free of apoA-II should be utilized and rechromatographed under identical conditions. The purity of isolated apolipoproteins can be assessed by SDS–polyacrylamide gel electrophoresis. ApoA-II can be isolated in similar fashion. For further description of apolipoprotein isolation, see the chapter by Brewer.

Preparative Isoelectric Focusing

The method of choice for isolating the various isoforms of apolipoproteins A-I is preparative isoelectrofocusing. Agarose-Z, ampholines (pH range 4–6.5) and Coomassie Brilliant Blue G-250 are obtained from LKB (Rockville, MD), and acrylamide, N,N-methylene bisacrylamide, and urea from Bio-rad Laboratories (Richmond, CA). All chemicals are of analytical grade. Up to 500 mg of delipidated HDL are dissolved in 100 ml of 8.6 M urea, 10 mM Tris-HCl, pH 8.2 buffer containing 4% ampholines (pH 4.0–6.5). The agarose gel is prepared by dissolving 2 g of agarose in 100 ml of doubly distilled water in a boiling water bath. The solubilized agarose is then transferred to another water bath at 55° and allowed to cool. At this point, the solution containing delipidated HDL is added to the agarose with constant stirring. This mixture is then poured out on an electrophoresis tray and allowed to gel at 4°. The tray is then placed on the cooling plate of an LKB 2117 Multiphor apparatus, and electrode strips (cathode 0.1 M sodium hydroxide, and anode 0.1 M phosphoric acid) are placed at the end of the gel. The electrophoresis is run for 16 hr at 250 V at 4°. Protein bands are subsequently visualized under ultraviolet light or by staining a test strip of the gel. Sections of the gel containing the bands of interest are then cut, diluted in about the same volume of doubly distilled water, and passed through a 25-gauge needle. The protein is recovered by ultracentrifugation of the agarose in a 60 Ti Beckman rotor at 50,000 rpm for 45 min at 10°. The agarose pellet is washed once with an identical volume of 10 mM Tris-HCl buffer, pH 8.2, and spun again under the same conditions. The supernatant is collected and added to the first extract. The isolated fractions are dialyzed against 10 mM ammonium bicarbonate, lyophilized, and stored at −70° for further analysis. Utilizing this technique, one can obtain significant amounts of the various apoA-I isoforms for use in metabolic studies.

Analytical Isoelectric Focusing and Two-Dimensional Gel Electrophoresis

Purified HDL, apolipoproteins, or apolipoprotein isoforms can be assessed by analytical isoelectric focusing and two-dimensional gel electrophoresis. The HDL fraction is isolated by ultracentrifugation as previously described, dialyzed overnight against 10 mM ammonium bicarbonate, lyophilized, and delipidated with chloroform methanol (2 : 1) twice, followed by a wash with diethyl ether. The apolipoproteins are then dissolved in 10 mM Tris-HCl, pH 8.2, 8.0 M urea, 10 mM DDT solution and the protein concentration measured by the method of Bradford.[6] ApoHDL (20 μg) are loaded onto a 7.5% acrylamide slab gel of 0.75 mm thickness containing 8 M urea and 2% ampholines (4.0–6.5). The electrophoresis is carried out for 16 hr at 250 V at 4°. The upper buffer is 20 mM sodium hydroxide and the lower buffer is 10 mM phosphoric acid. Following electrophoresis, the gels are stained with Coomassie Blue, and scanned with an LKB 2202 Ultroscan laser densitometer.

In order to carry out the second dimension of electrophoresis, gel strips are cut out of the gel after staining and scanning, and incubated for 20 min in a buffer containing 6 mM Tris-HCl, pH 6.8, 2.3% SDS, 10% glycerol, and 5% 2-mercaptoethanol. Gel strips are then loaded on the top of a 4–22.5% gradient slab SDS–polyacrylamide gel prepared as previously described by O'Farrell.[7] The electrophoresis is carried out at 300 V and 15° overnight. The gels are stained with Coomassie Blue.

---

[6] M. M. Bradford, *Anal. Biochem.* **27,** 248 (1976).
[7] P. H. O'Farrell, *J. Biol. Chem.* **250,** 4007 (1975).

Radioiodination of Lipoproteins and Apolipoproteins

The standard method of labeling HDL proteins has been by modification of the iodine monochloride method of McFarlane.[8] Utilizing this method, an aliquot of HDL at a concentration of 10 mg of HDL protein/ml or greater (approximately 0.5 ml volume) is mixed with an equal volume of 1 M glycine buffer, pH 10. Alternatively, the isolated HDL can be dialyzed against 1 M glycine buffer. The high pH of 10 minimizes lipid labeling. Approximately 1 ml of this material is placed in a sterile test tube to which is added approximately 1 μCi of carrier-free [^{125}I]- or [^{131}I]sodium iodide (obtained either from New England Nuclear or Amersham). This procedure is carried out in an iodination hood with adequate ventilation and a filter to pick up the radioactive iodine. Following the addition of the radioactivity, the tube is sealed with parafilm and shaken. Subsequently, iodine monochloride (ICl) is jetted in to cause the actual iodination to precede.

In order to prepare ICl, 150 mg of sodium iodide is placed in 4 ml of 6 N HCl. A solution containing 93 mg of anhydrous $NaIO_3$ in 2 ml distilled water is rapidly injected into this mixture. The volume is then brought to 40 ml with distilled water, 5 ml of a 1 M solution of CCl_4 is added, and the solution is vortexed. Subsequently moist air is bubbled through this solution for 1 hr and the total volume is brought to 45 ml with distilled water. The concentration of this solution is approximately 0.03 N or 4 mg/ml.

An alternative method of preparation of iodine monochloride reagent is to make a solution containing 0.55 g of KI, 0.3567 g KIO_3, and 29.23 g of NaCl in 21 ml of concentrated HCl and distilled water in a total volume of 250 ml. Free iodine is then removed by passing a current of air saturated with water vapor through the solution for a few hours. Completeness of free iodine removal can be tested by extracting a few ml of ICl preparation with CCl_4. If the separated fraction is colorless with only a faint pink blush, then the extraction is complete. The final ICl solution has a concentration of 0.02 M. Iodine monochloride can also be obtained from Sigma.

Other iodination techniques include chlotamine T and lactoperoxidase methods. The advantage of the iodine monochloride method is that the only oxidizing agent present during the iodination procedure is the ICl itself; therefore less protein denaturation occurs than is produced by other oxidizing agents. Generally one should incorporate approximately 0.5 mol of iodine per mole of protein during the iodination procedure. In order to calculate the amount of ICl in milligrams to add, one must multi-

[8] A. S. McFarlane, *Nature (London)* **182**, 53 (1958).

ply the amount of protein in milligrams in the solution to be labeled times 62.5, and divide this product by the product of the molecular weight of the protein and the assumed efficiency of iodination. For practical purposes, one assumes a molecular weight of 30,000 for HDL protein, and an efficiency of iodination of 50%. Therefore, if one has a solution of HDL protein containing 5 mg, one adds approximately 0.02 mg of ICl solution diluted with 1 M glycine buffer, pH 10 (approximately 0.5 ml). This material is drawn up in a small syringe and jetted through parafilm with a 25-guage needle into the test tube containing the radioactivity and the HDL solution. If smaller volumes are used for the iodination procedure, one can directly add the lipoprotein or apolipoprotein solution in a volume of approximately 0.1 to 0.3 ml directly into the iodination vial provided with the isotope (Combi V vial, New England Nuclear). It should be noted that only high specific activity, low pH, carrier-free, radioactive iodine should be used for the iodination. The procedure used for iodination of apolipoproteins is identical to that for lipoproteins except that the lyophilized apolipoproteins are redissolved in 1 M glycine buffer at pH 8.5 instead of pH 10.[9]

Following iodination, the lipoprotein or apolipoprotein solution is extensively dialyzed against a 100-fold excess volume of sterile 0.85% NaCl, 0.01% Na_2EDTA, 0.1 M Tris, pH 7.4 solution. The dialysis solution should be changed at least six times in order to remove all free iodine. The percentage of free iodine present in radiolabeled preparations following dialysis is assessed by precipitation of protein with 20% trichloroacetic acid (TCA) following the addition of 5% bovine serum albumin (BSA). This procedure is carried out by mixing a 1-μl aliquot of the iodinated material, 100 μl of unlabeled ligand, 1 ml of 5% BSA, and 1 ml of 20% TCA, vortexing, and spinning down the protein precipitate (3000 rpm, 20 min). The supernatant should be removed and saved and the pellet washed with distilled water. The amount of radioactivity in the protein pellet, the supernatant, and the wash is determined. The percentage of radioactivity not found in the protein pellet is assumed to be free iodine. In order to test efficiency of iodination, the identical procedure is carried out on a similar aliquot prior to the dialysis step which removed the free iodine. Following the removal of free iodine, the radiolabeled preparations containing iodinated lipoproteins or apolipoproteins are diluted 1 : 10 with sterile 5% human albumin to minimize radiation damage and then subjected to Millipore filtration (0.2-μm filter, Millipore Corp.). The preparation is tested for sterility by standard bacterial culture as well as pyrogenicity by injecting 40% of the human dose directly into each of three

[9] E. J. Schaefer, L. A. Zech, L. L. Jenkins, T. J. Bronzert, E. A. Rubalcaba, F. T. Lindgren, R. L. Aamodt, and H. B. Brewer, Jr., *J. Lipid Res.* **23**, 850 (1982).

rabbits intravenously and checking their rectal temperatures every 3 hr for a 12-hr period. Nonpyrogenic preparations will cause less than a 0.5° rise over this time period.[9]

Preparation of Tracer

Lipoprotein solution diluted 1 : 10 with 5% sterile human serum albumin must be calibrated for radioactivity dose at the time of proposed injection by the radiopharmacy. Doses of 25–50 μCi of either ^{125}I- or ^{131}I-labeled HDL are generally used. For apolipoprotein preparations a number of different procedures have been utilized. In the first method, radioiodinated apolipoproteins have been injected directly into plasma. In the second method, radioiodinated apolipoproteins have been added to a small amount of isolated HDL (approximately 3–5 ml). This HDL has been reisolated by ultracentrifugation and the apoA-I or apoA-II that has been reassociated with the HDL then serves as a tracer. Our current preferred method is to take the radioiodinated apolipoproteins and further purify them by Sephacryl S-200 column chromatography (1.5 × 100 cm column) in 0.01 M Tris, 0.1 M potassium chloride, 0.001 M sodium azide, pH 7.4. This method generally separated two radioiodinated apoA-I peaks and it appears that radioiodination results in some degree of incompetent monomer formation.[10] Therefore the first peak containing the octomeric species should be utilized. This material can be injected directly or it can be incubated with approximately 60 ml of the patient's plasma reisolated at density 1.21 g/ml in the ultracentrifuge, and subjected to Millipore filtration through a 0.2-μm filter. This latter technique is currently our preferred method.

Alternative methods of labeling include the use of [^{14}C]leucine given iv, resulting in endogenous labeling of apolipoproteins. There have been no published data utilizing this technique for apolipoproteins A-I and A-II. Another method that we are currently testing, which appears to be quite promising, is the use of [^{15}N]glycine given orally or intravenously to endogenously labeled apolipoproteins A-I and A-II.[11] The advantage of this technique is that a stable isotope is used rather than a radioactive isotope, and isolation of lipoproteins or apolipoproteins is not required for labeling. The disadvantages of this method include requirement of isolation of apolipoproteins during the course of the metabolic study for specific activity determination, the requirement of gas liquid chromatography–mass spectroscopy, and the need for computer modeling. Nevertheless, this methodology may become the method of choice for future studies.

[10] J. O. Osborne, Jr., E. J. Schaefer, N. Lee, and L. Zech, *J. Biol. Chem.* **259,** 347 (1984).
[11] M. Gersowitz, H. N. Munro, and V. R. Young, *Metabolism* **29,** 1075 (1980).

Metabolic Studies

To prepare subjects for kinetic studies, subjects should be studied on a metabolic ward and given a standardized diet generally containing approximately 15–20% of calories as protein, 40% of calories as carbohydrate, and 40% of calories as fat, with a polyunsaturated : saturated ratio of approximately 0.5.[9] The amount of monounsaturated, saturated, and unsaturated fats should have a ratio of approximately 1 : 2 : 1 to approximate a normal United States diet. An isocaloric diet is recommended (i.e., does not alter the subject's body weight). The amount of exercise that a subject performs during the study should be similar to the general level of activity prior to study. Ten to 14 days of diet should be given prior to injection of isotope. Isotopes should be injected in the fasting state in the morning and the smallest possible dose of radioactivity should be utilized. It is important to start all subjects on a dose of supersaturated potassium iodide (300 mg orally three times daily) prior to and during the study. The dose may have to be reduced if side effects such as a skin rash or marked parotid swelling occur. All subjects should have normal thyroid function, liver, and kidney function tests prior to study and should have a negative pregnancy test if female of reproductive age. Blood samples should be drawn precisely 10 min after injection, at 6 and 12 hr, and then daily for 14 days. We recommend the collection of urine throughout this time period as well in order to quantitate excretion of radioactivity from the body. In addition, if available, and if the patient received ^{131}I, the disappearance of radioactivity from the body should be monitored utilizing a whole body counter. Radioactivity in urine and plasma should be determined from all time points. From the urine radioactivity and the amount of urine excreted during each time period, one can calculate the amount of urinary radioactivity excretion during the course of the study. At 10 min, 6 and 12 hr, and days 1, 4, 7, 10, and 14, larger amounts of blood (20 ml) should be drawn in order to isolate plasma lipoproteins, and determine the distribution of radioactivity among the lipoprotein fractions. These samples can also be used for the quantitation of plasma cholesterol, triglyceride and HDL cholesterol, phospholipid, triglyceride, and protein. During the course of these studies, apolipoprotein A-I and A-II should be quantitated in plasma as well as HDL fractions.[9]

Immunoassay of Apolipoproteins

In our previous studies, we have quantitated apoA-I and apoA-II in all samples by radial immunodiffusion following delipidation with methanol diethyl ether (3 : 7) and resolubilization of 0.05 M sodium barbital, 9 mM sodium azide, pH 8.3 buffer for immunochemical measurement as previously described.[9] More recently, we have preferred to use enzyme-linked immunosorbent assay (ELISA) for the quantitation of plasma apolipoproteins. In this method polystyrene microtiter plates (NUNC Immunoplate I) specifically designed for quantitation were coated overnight with anti-apolipoprotein affinity-purified polyclonal antisera (15 ng/100 μl) in 0.2 M NaHCO$_3$, pH 9.6 buffer. Following washing, 100 μl of standards (purified apolipoproteins and secondary standards) as well as unknown samples to be measured are diluted 1 : 1500 with phosphate-buffered saline, 0.05% Tween 20, 0.5% bovine serum albumin solution (PBST–BSA), are added to wells, and are allowed to incubate for approximately 5 hr at room temperature. It should be noted that samples must be heated for 2 hr at 60° prior to use in this assay. This latter procedure appears to be important to expose all of the antigenic sites. After washing, antibody conjugated to alkaline phosphatase diluted with the PBST–BSA buffer is added to plate wells and allowed to incubate overnight at room temperature. After washing 100 μl of 0.1% p-nitrophenyl phosphate in 0.1 M glycine buffer, pH 10.4, is added to the wells for 20 min, and the reaction stopped with 50 μl of 1 M sodium hydroxide. After 10 min on a microplate shaker, the OD is read at 410 nm on a Dynatech MR600 plate reader. This procedure can be automated, is sensitive, and can be utilized for measuring apolipoprotein A-I and A-II in multiple plasma and HDL samples. Moreover, radioactivity does not interfere with the assay.

Specific radioactivity decay curves within HDL and plasma can be analyzed utilizing least-squares fitting to obtain the area under the specific activity curve.[9] Utilizing this method, one can obtain a residence time which is the reciprocal of the fractional catabolic rate (FCR). Assuming the subject is in a steady state, one can measure a synthesis rate utilizing the formula: synthesis rate (mg/kg · day) = [plasma volume (dl) × plasma apolipoprotein concentration (mg/dl)]/[plasma apolipoprotein residence time (days)] [body weight (kg)]. The plasma volume should be determined utilizing the sample obtained 10 min after injection by standard isotope dilution techniques. An alternative is to do multicompartmental modeling of the specific activity decay curves utilizing the SAAM simulator program.[12] This methodology is discussed in detail in Chapter [22], this volume.

Review of Metabolic Studies

The focus of this review of metabolic studies will be on human studies, since animal data are extremely limited. In the rat, HDL proteins are

[12] L. A. Zech, E. J. Schaefer, T. J. Bronzert, R. L. Aamodt, and H. B. Brewer, Jr., *J. Lipid Res.* **24,** 60 (1983).

catabolized at a significantly faster rate than that observed in humans.[4] In nonhuman primates, apoA-II is in a monomeric form and appears to be catabolized at a faster rate than apoA-I.[13] It should be noted that while apoA-I, apoA-II, other apolipoproteins, and various lipid constituents are all found on particles within the HDL density region, it appears that these constituents are not catabolized as a unit. Available data suggest that HDL constituents are catabolized at different rates.[14] This section will focus on a comparison of various labeling methods and apolipoprotein A-I and A-II kinetics in normal and dyslipidemic subjects (see Tables I–IV).

Comparison of Labeling Techniques

Early workers studying HDL kinetics labeled HDL as a lipoprotein density class with ^{125}I or [131]I.[15–18] Using this method, one not only labels apoA-I and apoA-II but other apolipoproteins within the HDL density region as well. For this reason Shepherd and co-workers undertook to isolate HDL apolipoproteins and label them directly.[19,20] Following isolation of apoA-I and apoA-II by column chromatography, these workers labeled these apolipoproteins by the iodine monochloride method at pH 10 (instead of pH 8.5 as recommended for proteins), incubated them with isolated HDL for 30 min at 37°, and then reisolated the HDL in the ultracentrifuge. The amount of HDL used for reassociating the radiolabeled apolipoproteins was small (approximately 5 ml of concentrated HDL). These investigators noted a significant difference when the specific activity decay of apoA-I labeled on HDL was compared with the decay of labeled apoprotein with subsequent intercalation into HDL by incubation. With intercalation they noted that the apoA-I had a 27% lower residence time than if the apoA-I was labeled on the HDL particle directly.[20] In a further experiment, these investigators demonstrated that this difference did not appear to be due to actual labeling of apoA-I as an apolipoprotein

[13] J. S. Parks and L. L. Rudel, *J. Lipid Res.* **23**, 410 (1982).

[14] E. J. Schaefer, D. W. Anderson, L. A. Zech, F. T. Lindgren, T. J. Bronzert, E. A. Rubalcaba, and H. B. Brewer, Jr., *J. Lipid Res.* **22**, 217 (1981).

[15] D. Gitlin, D. G. Cornwell, D. Nakasato, J. L. Oncley, W. L. Hughes, Jr., and C.A. Janeway, *J. Clin. Invest.* **37**, 172 (1958).

[16] A. Scanu and W. L. Hughes, Jr., *J. Clin. Invest.* **41,**1681 (1962).

[17] R. H. Furman, S. S. Sanbar, P. Alaupovic, R. H. Bradford, and R. P. Howard, *J. Lab. Clin. Med.* **63**, 193 (1964).

[18] C. B. Blum, R. I. Levy, S. Eisenberg, M. Hall, III, R. H. Goebel, and M. Berman, *J. Clin. Invest.* **60**, 795 (1977).

[19] J. Shepherd, J. R. Patsch, C. J. Packard, A. M. Gotto, Jr., and O. D. Taunton, *J. Lipid Res.* **19**, 383 (1978).

[20] J. Shepherd, C. H. Packard, A. M. Gotto, Jr., and O. D. Taunton, *J. Lipid Res.* **19,** 656 (1978).

TABLE I
COMPARISON OF VARIOUS LABELING METHODS

Subjects	Injected tracer	Plasma total residence time (days)			Reference
		HDL	ApoA-I	ApoA-II	
Normals (N = 2)	[131I]HDL		4.16		Shepherd et al.[20]
	[125I]ApoA-I/HDL		3.02		
Normals (N = 2)	[125I]HDL			3.89	Shepherd et al.[20]
	[125I]ApoA-II/HDL			3.89	
Normal (N = 1)	[125I]ApoA-I/HDL		6.25		Shepherd et al.[20]
	[131I]ApoA-I/HDL		5.86		
Normals (N = 2)	[125I]HDL		5.28		Schaefer et al.[9]
	[131I]ApoA-I		5.12		
Normals (N = 2)	[125I]HDL			4.73	Schaefer et al.[9]
	[131I]ApoA-II			4.78	
Normals[a] (N = 4)	[125I]HDL	6.04 ± 0.81			Schaefer et al.[9]
	[I]ApoA-I[b]		5.35 ± 0.69		
	[I]ApoA-II[b]			6.13 ± 0.89	
Normals (N = 2)	[131I]ApoA-I (octomer)		6.61		Osborne et al.[10]
	[125I]ApoA-I (monomer)		6.22		

[a] Radiolabeled HDL injected in a different study than radiolabeled apolipoproteins.
[b] Radiolabeled with either 125I or 131I.

on HDL, but rather due to the intercalation process itself.[20] Such differences were not observed when two different methods for comparing apoA-II labeling were examined (see Table I).

In contrast, Schaefer and co-workers did not see such differences when they compared the apoA-I specific activity decay within the HDL density region of apoA-I radiolabeled as HDL or as isolated apoA-I.[9] They also noted no such differences using these different methodologies for labeling apoA-II. These investigators may not have seen these differences because they allowed the radiolabeled apolipoproteins to associate with an entire plasma volume of native apoA-I- and apoA-II-containing lipoproteins in vivo. In other words, the radiolabeled apolipoproteins were injected directly into plasma rather than reassociating them with a small amount of isolated HDL and reisolating the HDL (see Table I).

More recently Osborne and co-workers examined the physical properties of radioiodinated apoA-I and apoA-II.[10] In these studies, they noted that when radioiodinated apoA-I in 0.1 M Tris buffer was subjected to column chromatography (Sephacryl S-200 or HPLC with a TSK-3000 column) the specific activity of apoA-I was somewhat higher in the monomeric region of apoA-I as compared to the octomeric region. These data suggested that there was a radioiodinated species of apoA-I that did not behave in the same fashion as native apoA-I and was incompetent in terms of its self-association properties.[10] No such differences were observed for apoA-II. These investigators then injected the radioiodinated octomer peak and compared it with the radioiodinated monomer peak of apoA-I. They noted that there was some more rapid, early decay of the specific activity following injection of the monomeric radiolabeled apoA-I, and increased early appearance of radioactivity in the urine. However, when the entire plasma residence time decay was analyzed, the monomer only decayed 5% faster than the octomer. Nevertheless, from these studies it appears that one should subject radioiodinated apoA-I to column chromatography following labeling, and use the octomeric species for kinetic studies.

Optimal methods for labeling apoA-I and apoA-II for kinetic studies remain to be developed. The difficulty of utilizing radioiodinated HDL is the need for subsequent separation apoA-I and apoA-II from plasma at each time point following injection to determine apoA-I and apoA-II specific radioactivity decay. Such a study is extremely labor intensive by column chromatography, and results obtained by polyacrylamide gel electrophoresis in our experience are not very reliable. An alternative approach, of course, is to iodinate apolipoproteins directly, but apparently when one does this for apoA-I one may have some technical problems. Our current recommendation is that for any apolipoprotein other than apoB where one must label the protein directly on the particle, one should isolate and purify the apolipoprotein of interest, label it by the iodine monochloride method, being careful to remove all the free iodine, and for apoA-I one should subject the iodinated apoA-I to column chromatography, remove the octomeric species, and utilize this for tracer studies. One can even do a further screen by allowing the apoA-I to reassociate itself with HDL, but one should use 60 to 100 ml of plasma, and then reisolate the HDL to assure oneself that the radioiodinated species that one injects has normal lipid-binding characteristics.

An alternative approach to use is to endogenously label the HDL apolipoproteins with [^{14}C]leucine as has been carried out for VLDL apoB by Fisher et al.[21] The disadvantages of this technique are the extensive

[21] W. R. Fisher, L. A. Zech, P. Bardalaye, G. Warmke, and M. Berman, J. Lipid Res. 21, 760 (1980).

modeling that is required to analyze the data as well as the large amount of radioactivity that must be utilized. A method that we are currently exploring which appears to be quite promising is utilization of either oral or intravenous [^{15}N]glycine to study apoA-I kinetics, utilizing the method described by Gersowitz et al.[11] The advantage of this technique is that it is relatively rapid, can be carried out over an approximately 72-hr period, and allows for endogenous labeling without having to remove lipoprotein or apoproteins from the body. The disadvantages are that it requires gas liquid chromatography and mass spectroscopy to measure specific activities, and may require significant computer modeling.

Comparison of ApoA-I and ApoA-II Kinetics

Kinetic studies of HDL have been carried out not only to determine the catabolic rate of radiolabeled HDL protein but also the individual protein constituents within HDL. Blum and co-workers were the first investigators to carefully examine the kinetics and apoA-I and apoA-II.[18] In these studies in eight normal individuals (see Table II) HDL (density region 1.09–1.21) was radioiodinated, injected intravenously, and the kinetics of apoA-I and apoA-II were followed for 14 days by isolating HDL

TABLE II
COMPARISON OF APOA-I AND APOA-II CATABOLISM

Subjects	Injected tracer	Residence time (days)		Reference
		ApoA-I	ApoA-II	
Normals (N = 8)	[^{125}I]HDL	4.21 ± 0.51	4.83 ± 0.75[a,b]	Blum et al.[18]
Normals (N = 10)	[^{131}I]ApoA-I/HDL [^{125}I]ApoA-II/HDL	3.41 ± 0.18	4.14 ± 0.63	Shepherd et al.[24]
Normals (N = 3)	[^{125}I]HDL	2.97 ± 0.46	3.40 ± 0.70	Fidge et al.[22]
Dyslipidemic subjects (N = 13)	[^{125}I]HDL	4.13 ± 1.58	4.78 ± 1.87	
Normals (N = 6)	[^{125}I]HDL	5.07 ± 1.53	5.94 ± 1.84[b]	Schaefer et al.[9]
Normals (N = 14)	[I]ApoA-I[c] [I]ApoA-II[c]	4.46 ± 1.04	4.97 ± 1.06[b]	

[a] Data of Blum et al.,[18] derived in Schaefer et al.[9]
[b] Significantly different from residence time for apoA-I.
[c] Labeled as either ^{125}I or ^{131}I.

from plasma obtained at various time points, and subjecting delipidated HDL to column chromatography to separate apoA-I and apoA-II. These workers stated that the terminal slopes of apoA-I and apoA-II specific activity decay curves were similar, and therefore concluded that these apolipoproteins were catabolized as a unit from HDL and one could calculate kinetic parameters utilizing HDL protein only. However, when Schaefer and co-workers reexamined these data, and statistically compared fractional catabolic rate (FCR) or the residence time of apoA-I and apoA-II (1/FCR) utilizing the entire decay curve rather than just the terminal slopes, they noted that apoA-I was catabolized at a significantly faster rate than apoA-II.[9]

Shepherd and co-workers also noted that apoA-I was catabolized at a significantly faster rate than apoA-II, but these workers felt that this was due to the methodological problems that they had in labeling apoA-I in their intercalation procedure as previously described.[20] Subsequently, Fidge and co-workers examined apoA-I and apoA-II kinetics in 3 normal and 13 dyslipidemic subjects and also noted that apoA-I was catabolized at a somewhat faster rate than apoA-II when they examined A-I and A-II specific activity decay curves utilizing polyacrylamide gel electrophoresis separation techniques.[22] However, in their study these differences did not reach statistical significance. Schaefer et al.[9] not only compared the data of Blum and co-workers, but in addition carried out radioiodinated HDL studies in six additional normal subjects, and in these subjects also noted that apoA-I was catabolized at significantly faster rates than apoA-II. Moreover, these investigators simultaneously injected radioiodinated apoA-I and apoA-II into 14 normal subjects utilizing a different radioactive iodine label on each of the apolipoproteins. In these studies it was noted that the mean residence time for apoA-I was approximately 11% faster than that for apoA-II.[9]

The question that remains is whether apoA-I is really catabolized at a faster rate than apoA-II. Is it possible that some radioiodinated species of apoA-I that does not normally self-associate can account for this difference? As previously mentioned, the residence time for labeled apoA-I monomer was approximately 5% faster than that for labeled apoA-I octomer.[10] It should be noted that approximately 40% or less of radiolabeled apoA-I is in the monomeric form.[10] This abnormality of labeling only appears to account for about one-fifth of the enhanced catabolism that is observed for radiolabeled apoA-I as compared to radiolabeled apoA-II.[9,10] Therefore, in our opinion apoA-I is catabolized at a faster rate than

[22] N. Fidge, P. Nestel, T. Ishikawa, M. Reardon, and T. Billington, *Metabolism* **29**, 643 (1980).

apoA-II, although support for this conclusion using other methodology is needed.

Apolipoprotein A-I and A-II Metabolism in Normal Subjects

Kinetic parameters of apolipoproteins A-I and A-II in normal subjects are presented in Table III. Early studies by various investigators revealed that radioiodinated HDL protein had a residence time ranging between 3.4–6.6 days.[15–17] Subsequent studies by Blum and co-workers demonstrated that HDL protein synthesis was approximately 8 mg/kg · day, whereas HDL protein residence time was approximately 5 days.[18] The problem with these synthesis rate determinations was that HDL protein was characterized by doing Lowry analysis on HDL of density region of 1.09–1.21 g/ml, which does not include the entire HDL density region. When Schaefer *et al.*[23] carried out such studies in normal subjects utilizing the entire HDL density region, HDL synthesis rates of approximately 11.5 mg/kg · day were noted. Fidge and co-workers reported synthesis rates of approximately 11–12 mg/kg · day for apoA-I, while Shepherd and co-workers reported even higher synthesis rates for apoA-I of approximately 15 mg/kg · day.[19,20,22,24] It should be noted that these latter synthesis rates are inflated due to the documented 25% enhancement of fractional catabolic rate or residence time utilizing the intercalation process. It appears that normal synthesis rates for apoA-I and apoA-II are 10–15 and 2–3 mg/kg · day, respectively.[9]

It is well known that females have significantly higher levels of plasma HDL constituents than do males. Shepherd *et al.* compared kinetic parameters for HDL apolipoproteins in males and females and noted no significant sex difference between the synthesis rates and residence times for these various apolipoproteins.[24] It should be noted, however, that these investigators were comparing males and females who had apoA-I and apoA-II levels that were not significantly different. Schaefer and co-workers observed that females had a significantly higher synthesis rate of apoA-I and apoA-II as compared to males.[9] Fractional catabolic rates were similar. Moreover, these investigators noted that when premenopausal females were given estrogen at a dose of 0.1 mg of ethinyl estradiol, a significant increase in apoA-I synthesis resulted.[25] These data are

[23] E. J. Schaefer, D. M. Foster, L. L. Jenkins, F. T. Lindgren, M. Berman, R. I. Levy, and H. B. Brewer, Jr., *Lipids* **14,** 511 (1979).
[24] J. Shepherd, C. J. Packard, J. R. Patsch, A. M. Gotto, Jr., and O. D. Taunton, *Eur. J. Clin. Invest.* **8,**115 (1978).
[25] E. J. Schaefer, D. M. Foster, L. A. Zech, H. B. Brewer, Jr., and R. I. Levy, *J. Clin. Endocrinol. Metab.* **57,** 262 (1983).

TABLE III
APOLIPOPROTEIN A-I AND A-II KINETIC PARAMETERS IN NORMAL SUBJECTS[a]

Subjects	Injected tracer	HDL (mg/dl) Chol.	HDL (mg/dl) Protein	Plasma (mg/dl) ApoA-I	Plasma (mg/dl) ApoA-II	Synthesis Rate (mg/kg·day) HDL protein	Synthesis Rate (mg/kg·day) ApoA-I	Synthesis Rate (mg/kg·day) ApoA-II	Residence time (days) HDL protein	Residence time (days) ApoA-I	Residence time (days) ApoA-II	Reference
$N = 2$	[125I]HDL	39	—	—	—	—	—	—	6.62	—	—	Gitlin et al.[15]
$N = 2$	[I]HDL[b]	—	—	—	—	—	—	—	3.40	—	—	Furman et al.[17]
$N = 1$	[125I]HDL	—	—	—	—	—	—	—	4.10	—	—	Scanu and Hughes[16]
Males ($N = 3$)	[125I]HDL	31 ± 8	87 ± 23	—	—	8.71 ± 3.40	—	—	4.47 ± 0.58	—	—	Blum et al.[18]
Females ($N = 5$)	[125I]HDL	39 ± 7	101 ± 28	—	—	7.99 ± 2.29	—	—	5.24 ± 0.53	—	—	
$N = 4$	[I]HDL$_2$[b] [I]HDL$_3$[b]	60 ± 19 —	163 ± 22 —	— —	— —	11.64 ± 0.28 11.30 ± 0.60	— —	— —	5.74 ± 0.91 5.70 ± 1.24	— —	— —	Schaefer et al.[23]
$N = 4$	[125I]HDL	42 ± 6	—	124 ± 12	—	—	11.75 ± 3.16	—	—	2.95 ± 0.38	3.40 ± 0.70	Fidge et al.[22]
Males ($N = 5$)	[I]ApoA-I/HDL[b] [I]ApoA-II/HDL[b]	44 ± 7	—	127 ± 7	28 ± 2	—	15.8 ± 2.7	2.8 ± 0.5	—	3.18 ± 0.07	3.93 ± 0.77	Shepherd et al.[24]
Females ($N = 5$)	[I]ApoA-I/HDL[b] [I]ApoA-II/HDL[b]	59 ± 8	—	136 ± 16	28 ± 3	—	14.8 ± 1.9	2.6 ± 0.5	—	3.66 ± 0.50	4.46 ± 1.23	
Males ($N = 11$)	[I]ApoA-I[b] [I]ApoA-II[b]	43 ± 10	126 ± 14	108 ± 16	23 ± 3	—	11.12 ± 1.92	2.10 ± 0.32	—	4.51 ± 1.07	5.33 ± 1.07	Schaefer et al.[9]
Females ($N = 11$)	[I]ApoA-I[b] [I]ApoA-II[b]	50 ± 12	143 ± 22	124 ± 26	25 ± 2	—	13.58 ± 2.23	2.52 ± 0.46	—	4.18 ± 0.94	4.56 ± 0.79	

[a] Chol., cholesterol.
[b] Labeled with either 125I or 131I.

consistent with the view that the increased plasma levels of apoA-I observed in females as compared to males are due to increased synthesis of this protein, mediated by estrogen.

The plasma concentrations and kinetics of apolipoprotein A-I and A-II are also affected by diet and drug perturbations. Blum and co-workers reported that high carbohydrate diets (80% of calories) decreased HDL protein concentrations and enhanced HDL protein fractional catabolic rate.[18] Nestel and colleagues reported that vegetarians on high carbohydrate (60% of calories), low fat (26%) diets had significantly lower apoA-I plasma concentrations due to enhanced fractional catabolism than did control subjects on diets of normal composition (36–43% fat).[26] Shepherd and co-workers noted that the ingestion of diets high in polyunsaturated fat (P/S ratio 4.0) in normal subjects resulted in a significant decrease in plasma apoA-I concentration due to increased synthesis.[27] Drugs also affect HDL protein metabolism. While niacin increases HDL protein mass, its effect on HDL protein metabolism has not clearly been delineated.[18,28] Estrogen has been reported to increase apoA-I plasma concentration and synthesis rate in premenopausal women.[25]

The kinetics of apolipoproteins A-I and A-II have also been studied in a variety of dyslipidemic subjects. Gitlin and co-workers noted a relatively normal residence time for HDL protein in three nephrotic subjects, while Furman and co-workers noted that hyperlipidemic patients (especially those with hyperchylomicronemia) had an enhanced HDL protein residence time as compared to normal subjects.[15,17] Fidge and co-workers noted that five patients studied with type IV hyperlipoproteinemia had normal synthesis rates but enhanced fractional catabolism of apoA-I and apoA-II.[22] In studies reported by Schaefer et al.[9] an important determinant of plasma apoA-I concentration was apoA-I residence time, which was inversely correlated with plasma triglyceride levels. These investigators have also studied apolipoprotein A-I and A-II kinetics in patients with severe hypertriglyceridemia (three patients with type I hyperlipoproteinemia and three patients with type V hyperlipoproteinemia) (see Table IV). These subjects also had reduced levels of apoA-I due to markedly enhanced fractional catabolic rates or decreased residence times of apoA-I. These data are consistent with the concept that hypertriglyceridemic patients have decreased apoA-I levels due to enhanced fractional catabo-

[26] P. J. Nestel, T. Billington, and B. Smith, *Metabolism* **30**, 941 (1981).

[27] J. Shepherd, C. J. Packard, J. R. Patsch, A. M. Gotto, Jr., and O. D. Taunton, *J. Clin. Invest.* **61**, 1582 (1978).

[28] J. Shepherd, C. R. Packard, J. R. Patsch, A. M. Gotto, Jr., and O. D. Taunton, *J. Clin. Invest.* **63**, 858 (1979).

TABLE IV

APOLIPOPROTEIN A-I AND A-II KINETIC PARAMETERS IN DYSLIPIDEMIC SUBJECTS

Subjects	Injected tracer	HDL (mg/dl)		Plasma (mg/dl)		Synthesis rate (mg/kg-day)			Residence time (days)			Reference
		Chol.	Protein	ApoA-I	ApoA-II	HDL protein	ApoA-I	ApoA-II	HDL protein	ApoA-I	ApoA-II	
Nephrotic (N = 3)	[125I]HDL	29 ± 16	—	—	—	—	—	—	5.18 ± 0.77	—	—	Gitlin et al.[15]
Hyperlipidemia (N = 2)	[125I]HDL	—	—	—	—	—	—	—	2.6	—	—	Furman et al.[17]
Hyperchylomicronemia (N = 2)	[125I]HDL	—	—	—	—	—	—	—	1.8	—	—	
Tangier heterozygotes (N = 2)	[125I]HDL	17	63	—	—	7.93	—	—	3.41	—	—	Schaefer et al.[29]
Tangier homozygotes (N = 2)	[125I]HDL	1	43	1	8	3.66	—	—	0.53	0.18	0.76	
Tangier heterozygotes (N = 3)	[I]ApoA-I[b] [I]ApoA-II[b]	22 ± 9	89 ± 16	73 ± 12	17 ± 3	—	11.62 ± 1.62	2.36 ± 0.24	—	2.80 ± 0.69	3.16 ± 0.81	Schaefer et al.[30]
Tangier homozygotes (N = 3)	[I]ApoA-I[b] [I]ApoA-II[b]	2 ± 1	17 ± 7	2.3 ± 0.6	2.8 ± 0.2	—	4.87 ± 0.56	1.47 ± 0.18	—	0.22 ± 0.05	0.90 ± 0.14	
Tangier homozygote (N = 2)	[125I]HDL infusion	—	—	—	—	—	—	—	—	0.29	0.71	Assman et al.[31]

Group	Tracer									Reference
Deficiency of apolipoproteins A-I/C-III homozygote (N = 1)	HDL infusion	8	0.0059	19	—	—	3.0	—	—	Norum et al.[32]
Familial hypercholesterolemia heterozygotes (N = 6)	[^{125}I]HDL	48 ± 9	129 ± 13	—	11.41 ± 3.24	3.33 ± 1.53	—	5.10 ± 1.12	7.10 ± 1.70	Fidge et al.[22]
Type IV HLP (N = 5)	[^{125}I]HDL	35 ± 8	102 ± 18	—	11.21 ± 1.12	4.68 ± 0.74	—	2.90 ± 0.48	3.08 ± 0.60	
Hyperalphalipoproteinemia (N = 2)	[^{125}I]HDL	99	162	—	11.45	4.70	—	5.85	5.25	
Type I HLP (N = 3)	[I]ApoA-I[b] [I]ApoA-II[b]	13 ± 1	82 ± 5	24 ± 3	13.2 ± 1.15	4.29 ± 0.45	—	2.45 ± 0.31	2.68 ± 0.62	Schaefer (unpublished observations)
Type V HLP (N = 3)	[I]ApoA-I[b]	21 ± 3	97 ± 6	28 ± 4	12.71 ± 1.04	—	—	2.51 ± 0.45	—	
Abetalipoproteinemia (N = 2)	[I]ApoA-I[b] [I]ApoA-II[b]	34	44	23	9.01	1.72	—	2.52	2.43	

[a] HPL, Hyperlipoproteinemia.
[b] Labeled with either ^{125}I or ^{131}I.

lism. The mechanism for this consistent finding has not been elucidated, but may be due to the fact that these subjects have triglyceride-rich HDL.

Schaefer and co-workers studied radioiodinated kinetics in heterozygous and homozygous subjects with Tangier disease (a genetic HDL deficiency state), and observed that HDL protein residence time was significantly enhanced in heterozygotes and accounted for their decreased levels of HDL protein.[29] Moreover, a markedly enhanced residence time for both apoA-I and apoA-II was noted in homozygotes, with apoA-I decaying at a significantly faster rate than apoA-II (see Table IV). It was also noted that homozygotes had a decrease in HDL protein synthesis.[29] In further studies, these investigators injected radioiodinated apoA-I and apoA-II into three Tangier heterozygotes and in three Tangier homozygotes and noted that the primary defect in Tangier disease appeared to be due to hypercatabolism of both apoA-I and apoA-II.[30] In addition, these investigators demonstrated that apoA-I from Tangier subjects was catabolized at a significantly faster rate than normal apoA-I in four normal subjects.[30] This did not appear to be the case for Tangier apoA-II.[30] These findings are consistent with the view that the metabolic defect causing the HDL deficiency in Tangier disease is due to an abnormality of apoA-I.

Both Assman and co-workers and Schaefer *et al.* also demonstrated that Tangier homozygotes had enhanced catabolism of HDL apoA-I and apoA-II following infusion of unlabeled and radiolabeled HDL.[14,31] These studies demonstrated that apoA-I was catabolized at a faster rate than apoA-II and that the hypercatabolism of HDL constituents in Tangier homozygotes was not dependent upon pool size. In addition, it was noted that the HDL_2 was catabolized at a faster rate than HDL_3, and that HDL cholesterol and triglyceride were catabolized at a faster rate than HDL protein and phospholipid.[14] These data suggest that core constituents of HDL may be catabolized more rapidly than surface and components. In addition, the studies in Tangier disease are consistent with the concept that this genetic HDL deficiency state is due to enhanced fractional catabolism of HDL proteins due to an abnormal apoA-I (apoA-I$_{Tangier}$).[30]

Norum and co-workers studied decay of HDL protein (apoA) in a subject with homozygous apolipoprotein A-I and C-III (another genetic HDL deficiency state) and reported an HDL protein residence time of

[29] E. J. Schaefer, C. B. Blum, R. I. Levy, L. L. Jenkins, P. Alaupovic, D. M. Foster, and H. B. Brewer, Jr., *N. Engl. J. Med.* **299,** 905 (1978).
[30] E. J. Schaefer, L. L. Kay, L. A. Zech, and H. B. Brewer, Jr., *J. Clin. Invest.* **70,** 934 (1982).
[31] G. Assmann, A. Capurso, E. Smootz, and U. Wellner, *Atherosclerosis* **30,** 321 (1978).

approximately 3 days, significantly higher than that observed in homozygous Tangier disease (0.5 days).[32] These data are consistent with the concept that the defect in apolipoprotein A-I and C-III deficiency is due to an inability to synthesize these apolipoproteins rather than enhanced catabolism.

Studies carried out by Fidge and co-workers in six subjects with heterozygous familial hypercholesterolemia indicate that these individuals have relatively normal synthesis rate and residence times for apoA-I and apoA-II.[22] Other dyslipidemic patients that have been studied include two patients with hyperalphalipoproteinemia and two subjects with abetalipoproteinemia. In the former subjects studied by Fidge and co-workers, the increased levels of plasma apoA-I were felt to be due to a decreased fractional catabolism.[22] Patients with abetalipoproteinemia have HDL as their sole plasma lipoprotein class. These subjects have a slightly decreased HDL cholesterol and also have decreased levels of apoA-I but relatively normal levels of apoA-II (see Table IV). These subjects were noted not only to have enhanced fractional catabolic rate and a decreased residence time for apoA-I and apoA-II, but a slight decrease in synthesis of these apolipoproteins as well. HDL levels are also reduced in liver disease, and enhanced HDL protein fractional catabolism has been reported in patients with alcoholic hepatitis.[33]

Conclusion

Kinetic studies utilizing radioiodinated apolipoproteins have yielded important insights into lipoprotein metabolism. A major concern in these studies is that the tracer have identical kinetic behavior *in vivo* to native lipoproteins. It is well known that preparative ultracentrifugation alters lipoprotein particles, and iodination may also affect the structure of HDL. It is clear that adding labeled apoA-I to a small amount of HDL (intercalation) will decrease its plasma residence time (see Table I) as compared to apoA-I labeled on HDL.[20] Radioiodination of apoA-I also results in formation of some excess monomeric apoA-I, which does not self-associate as readily as native apoA-I, and is catabolized at a slightly greater fractional rate than octomeric apoA-I.[10] No such difficulties exist when apoA-II is radioiodinated.[10] Optimal methods of studying HDL apolipoprotein kinetics remain to be developed. If one does radioiodinate apoA-I, one

[32] R. A. Norum, J. B. Lakier, S. Goldstein, R. Goldberg, P. Dolphin, A. Angel, and P. Alaupovic, *N. Engl. J. Med.* **306,** 1513 (1982).
[33] P. J. Nestel, N. Tada, and N. Fidge, *Metabolism* **29,** 101 (1980).

should utilize the octomeric form for kinetic studies.[10] Stable isotope methodology with endogenous labeling may solve many of these problems, but then other difficulties with kinetic analysis must be addressed.[11]

Lipoprotein tracer studies have nevertheless given us important knowledge about the kinetics of HDL constituents. Chylomicron apoA-I and apoA-II can serve as precursors for these constituents within HDL.[34,35] Both apoA-I and apoA-II readily exchange within HDL subfractions whether labeled as apolipoproteins or on lipoproteins.[19,20] Despite methodologic problems, apoA-I appears to be catabolized at a greater fractional catabolic rate than apoA-II.[9,22] This finding was also observed following HDL infusion in an HDL-deficient subject.[14] Our own calculations indicate that formation of incompetent apoA-I monomer would only account for a small portion of the rapidly turning over plasma apoA-I compartment necessary to model apoA-I kinetics.[10,12]

Modeling data are most consistent with the concept that there is a plasma population of apoA-I-containing lipoproteins not containing apoA-II, and a lipoprotein species containing both apoA-I and apoA-II, with the latter species being catabolized significantly more slowly than the former.[11] The HDL particle containing mainly apoA-I appears to be within the HDL_2 density region, and may be of great functional significance.[11] It has also been shown that the C apolipoprotein within HDL is catabolized with a residence time of approximately 1 day compared to 4–5 days for apoA-I and apoA-II.[23] Following infusion of HDL into an HDL-deficient subject, different HDL constituents were catabolized at varying rates.[14] These data are consistent with the concept that HDL constituents are not catabolized as a unit.

A major determinant of apoA-I concentration in plasma is apoA-I residence time, which in turn is inversely correlated with plasma triglyceride concentration.[9] Hypertriglyceridemic subjects have markedly enhanced fractional catabolic rate of HDL protein apoA-I, and apoA-II (see Table III).[17,22] These subjects also have an increased triglyceride content within HDL which may cause this hypercatabolism. Enhanced fractional catabolism of apoA-I and apoA-II is also observed in Tangier disease,[14,29,30] but not in familial deficiency of apolipoproteins A-I and C-III.[32] These latter subjects appear to have an inability to synthesize these apolipoproteins. The flux of HDL constituents may be an important determinant of the capacity of HDL to serve as a vehicle to reverse cholesterol transport. This hypothesis may explain why premature coronary artery

[34] E. J. Schaefer, L. L. Jenkins, and H. B. Brewer, Jr., *Biochem. Biophys. Res. Commun.* **86,** 405 (1978).

[35] J. L. Vigne and R. J. Havel, *Can. J. Biochem.* **59,** 613 (1981).

disease incidence is more strikingly increased in homozygous familial apolipoprotein A-I and C-III deficiency where HDL flux is decreased as compared to homozygous Tangier disease where HDL flux or synthesis is reasonably normal.[32,36-38] Future studies will hopefully unravel the precise mechanisms controlling HDL metabolism and elucidate the exact nature of its role in the atherosclerotic process.

[36] E. J. Schaefer, *Arteriosclerosis* **4**, 302 (1984).
[37] E. J. Schaefer, L. A. Zech, D. S. Schwartz, and H. B. Brewer, Jr., *Ann. Intern. Med.* **93**, 261 (1980).
[38] E. J. Schaefer, W. H. Heaton, M. G. Wetzel, and H. B. Brewer, Jr., *Arteriosclerosis* **2**, 16 (1982).

[26] Metabolism of Apolipoprotein C

By Noel H. Fidge and Paul J. Nestel

Introduction

Although the smallest of the apolipoproteins so far identified are the C peptides, they nevertheless exert a major influence on lipid metabolism. Their characteristic rapid exchange and transfer between lipoproteins was one of the earlier observations in this field and provided a clue to their function. A physiological role has now been assigned to some of the C apolipoproteins, all of them having been sequenced and the armory of the most recent, sophisticated biochemical tools (including recombinant DNA techniques) having been employed to examine their synthesis and origin. Inevitably, the question of their metabolism per se arose and the present chapter aims to examine recent methodology used to investigate the regulation of apolipoprotein C formation and removal.

In most species, the apolipoprotein C group resides mainly in the plasma of very-low-density and high-density lipoproteins. In normal man they are almost equally distributed between VLDL and HDL, although of the total apolipoprotein mass, the C apolipoproteins comprise approximately 50 and 5-10% of VLDL and HDL, respectively.[1] A significant proportion is also associated with lymph chylomicrons and although the origin is uncertain the apolipoprotein C group accounts for up to 45% of the total apolipoprotein content of human and rat lymph chylomicrons.

[1] E. J. Schaefer, S. Eisenberg, and R. I. Levy, *J. Lipid Res.* **19**, 667 (1978).

Improvement in protein separation technology over recent years has provided useful methods for the identification and isolation of this heterogeneous group of apolipoproteins, while development of sensitive detection systems, including immunochemical procedures, has enabled a more accurate approach to their quantitation. These improvements have led to more precise methods for determining the metabolism of the C apolipoproteins and one approach, based on the *in vivo* kinetics of radioiodinated C peptides, is described in detail in this chapter. This methodology, based on changes in specific radioactivities of the labeled moiety, allows a simultaneous calculation of turnover rates, identification of exchange or transfer related events, and in most cases a measurement of both synthetic and catabolic rates as well as of pool sizes. Comparing these kinetic parameters in appropriate subjects (or in experimental animals) may provide useful information about the metabolic fate and physiological role of the particular apolipoprotein.

Characterization of the apolipoprotein C group has been described in previous reviews.[1] Briefly, they comprise three readily identifiable groups in human subjects, viz C-I, C-II, and C-III peptides. The C-II and C-III apolipoproteins are also observed in animal species, but some variations in both quantity and polymorphism are known to occur.[2] The C-III apolipoprotein group is numbered according to sialic acid content, which confers different electrophoretic mobility to each isoform, with C-III0 to C-III3 containing, respectively, 0 or 3 mol of sialic acid per mole of protein. A variant of apolipoprotein C-II, C-II2, has been described[3] and minor amounts of peptides have been reported in VLDL.[4]

The physiological effect of C-II apolipoprotein is most striking. The peptide activates one form of triglyceride hydrolase (lipoprotein lipase), which *in vitro* is almost totally inactive in the absence of this cofactor apolipoprotein. Similarly, without cofactor protein, the enzyme activity is minimal *in vivo* as recently shown in patients with apolipoprotein C-II deficiency.[5] In humans and several other species the C-III apolipoproteins are the most abundant of the apolipoprotein C group although their function, in easily recognizable biochemical terms, remains elusive.[1] They have been reported to inhibit lipolytic activity, although the ratios of C-III to C-II apolipoproteins rather than absolute levels of the apolipoprotein C-III group itself, appear to modulate lipoprotein catabolism. C-I apolipo-

[2] M. J. Chapman, *J. Lipid Res.* **21,** 789 (1980).

[3] R. J. Havel, L. Kotite, and J. P. Kane, *Biochem. Med.* **21,** 121 (1979).

[4] A. L. Catapano, R. L. Jackson, E. B. Gilliam, A. M. Gotto, and L. C. Smith, *J. Lipid Res.* **19,** 1047 (1978).

[5] W. C. Breckenridge, J. A. Little, G. Steiner, A. Chow, and M. Poapst, *N. Engl. J. Med.* **298,** 1265 (1978).

protein, present in similar proportions to apolipoprotein C-II in human VLDL, has been sequenced and synthesized and both native and synthetic forms, under certain conditions, activate the enzyme LCAT, but the physiological significance of this activation is also not clear.

The absence of demonstrable functional roles for some C apolipoproteins requires that information be sought using different experimental approaches. A study of the metabolic behavior can, and as discussed below, has, provided clues to function. Before proceeding to a description of methods employed in such investigations, the well-characterized exchange and transfer of the C peptide is quite important to the interpretation of apolipoprotein C metabolism and a brief review follows.

Exchange and Transfer of C Apolipoproteins

Not surprisingly, most studies on human apolipoprotein C metabolism have been carried out *in vitro*. Using ^{125}I-labeled lipoproteins, several groups first demonstrated a transfer of C peptides from VLDL to HDL[1] and later a bidirectional exchange between the lipoproteins. Factors which influenced this exchange were then investigated, including lipolytic activity,[6] specificity of the acceptor, and the status of the peptides themselves, that is, whether they are mobilized per se or in complexes with other apolipoproteins or lipids. Overall, the data with labeled VLDL suggest that lipolysis of triglyceride-rich lipoproteins induces additional transfer beyond that which occurs in the presence of serum (70%) and significantly more than that which occurs with albumin alone (25%). The surface structure of VLDL, however, may be an important factor since lipolysis results in demonstrable changes in composition of the VLDL, paricularly those components involved in the exterior topography of the lipoprotein.[7] Lipoprotein lipase also contains phospholipase activity, which may disrupt phospholipid–apolipoprotein complexes[6] which confer structural stability to the VLDL surface coat. Although HDL is the prime acceptor of VLDL C apolipoproteins, in its absence albumin may also act as acceptor, at least *in vitro*. The C apolipoproteins are most likely removed as a group, in complexes with phospholipids and unesterified cholesterol. These complexes contain the whole complement of C apolipoproteins, at least insofar as C-II and C-III peptides are concerned. This has been shown by measuring the ratios of C-II/C-III apolipoproteins in VLDL and HDL which remain unaltered during artificially produced

---

[6] M. C. Glangeaud, S. Eisenberg, and T. Olivecrona, *Biochim. Biophys. Acta.* **486,** 23 (1977).

[7] T. Chajek and S. Eisenberg, *J. Clin. Invest.* **61,** 1654 (1978).

VLDL lipolysis *in vitro*.[8] The involvement of apolipoprotein C-I in the process is not clear.

Unfortunately, the precise movements of these fast migrating C peptides are not yet fully understood simply because most detailed information to date has resulted from *in vitro* studies, and the movements of the C apolipoproteins under these conditions may not necessarily apply *in vivo*. However, most *in vivo* observations[9,10] also suggest that C apolipoproteins comprise a functional entity and in this form migrate between lipoproteins either during transfers stimulated *in vivo* or *in vitro* by lipolysis. We have previously suggested, on the basis of specific radioactivity data,[9] that the initial events associated with triglyceride-rich lipoprotein entry into plasma are a concomitant transfer of both apolipoprotein C-II and C-III from HDL, the former to stimulate lipolysis and apolipoprotein C-III to prevent premature removal of the triglyceride-rich particles by the liver or other tissues rich in E apolipoprotein receptors. The addition of C apolipoproteins to apolipoprotein E-rich "remnant"-type particles has been shown to prevent recognition and uptake by liver cells, although it is not certain whether C-II and C-III apolipoproteins are equally effective in this process.[11] The extent of these putative interrelationships between apolipoprotein C and other apolipoproteins awaits future metabolic studies combining the simultaneous investigation of other apolipoproteins as well as C peptides.[12] One other interesting aspect to be considered in the interpretation of future metabolic studies is the prospect that the exchange phenomenon observed so far represents yet another example of nature's "conservation" systems, since it appears as if one possible function of HDL is to act as a reservoir pool of apolipoprotein C. As such, the pool would be regulated by various factors, the most likely being the proportions of acceptor molecules such as VLDL or HDL or exogenous chylomicron particles. The involvement of a transfer factor, not unlike the lipid transfer protein (LTP) in this process, may be a concept worth consideration.

Synthesis of C Apolipoproteins

Although a supply of C peptides, including a highly potent activator C-II apolipoprotein, is apparently constantly available in normal plasma,

[8] S. Eisenberg, J. R. Patsch, J. T. Sparrow, A. M. Gotto, Jr., and T. Olivecrona, *J. Biol. Chem.* **254,** 12603 (1979).
[9] M. W. Huff, N. H. Fidge, P. J. Nestel, T. Billington, and B. Watson, *J. Lipid Res.* **22,** 1235 (1981).
[10] R. J. Havel, J. P. Kane, and M. L. Kashyap, *J. Clin. Invest.* **52,** 32 (1973).
[11] F. Shelburne, J. Hanks, W. Meyers, and S. Quarfordt, *J. Clin. Invest.* **65,** 652 (1980).
[12] P. Nestel, M. W. Huff, T. Billington, and N. Fidge, *Biochim. Biophys. Acta.* **712,** 94 (1982).

renewal and removal are major factors contributing to either maintenance, expansion, or diminution of the pool. Metabolic studies carried out under circumstances in which these pools have been changed, such as in various hyperlipoproteinemic states, or artificially contrived, for example by diet, offer excellent opportunities to engage the question. They need to be coupled, however, with a fundamental knowledge of which tissues are involved in C apolipoprotein synthesis so that the nature of the biochemical events and constraints can be integrated with the metabolic data. Communication between these potential extravascular and intravascular pools is another important consideration. Insofar as biosynthesis is concerned, much more detailed information is required and it is anticipated that molecular biology will be a major contributor to this field. Recombinant cDNA approaches are reviewed elsewhere in this series and already tools for quantitating apolipoprotein C production based on tissue mRNA levels are being established.

The synthesis of C apolipoproteins occurs to a major extent in the liver. This has been established using isotopically labeled precursor amino acids in many studies.[13,14] What remains less clear, however, is the role of the intestine in C peptide production. A major contributor of many other plasma apolipoproteins, it appears to secrete only small amounts, if any, of most of the C peptides, although apolipoprotein C-II may be one of the exceptions. Differences which have been found probably result from the various nutritional states of the animals used as well as the techniques employed. Calculations of the relative contributions from gut and liver in rats,[13] using simultaneously administered intraduodenal and intravenous radiolabeled leucine as precursor, showed that the liver produced by far the greater amount of apolipoprotein C. Precise quantitation of apolipoprotein C biosynthesis has been difficult to establish since hepatocytes in tissue culture as well as perfused rat livers synthesize less apolipoprotein C than anticipated. More recent studies with cultured rat liver cells have, however, convincingly demonstrated a hepatic origin of apolipoprotein C.[15] The effect of diet, particularly the proportion of fat, may influence intestinal apolipoprotein C production[14] and these and other studies[16] suggest that possibly C-II apolipoprotein but not C-III is synthesized in this tissue. The absence of C apolipoproteins in polyacrylamide gel patterns of lymph lipoproteins from rats fed ethinyl estradiol (to minimize contribution from plasma lipoproteins) also suggests that little C peptide originates in the intestine.[17]

[13] A. L. Wu and H. G. Windmueller, *J. Biol. Chem.* **254,** 7316 (1979).
[14] P. R. Holt, A. L. Wu, and S. B. Clark, *J. Lipid Res.* **20,** 494 (1979).
[15] J. Bell-Quint, T. Forte, and P. Graham, *Biochem. Biophys. Res. Comm.* **99,** 700 (1981).
[16] N. Fidge and W. Kimpton, *Biochim. Biophys. Acta* **752,** 209 (1983).
[17] B. R. Krause, C. H. Sloop, C. K. Castle, and P. S. Roheim, *J. Lipid Res.* **22,** 610 (1981).

It must be emphasized that studies of tissue origin, and especially quantitation of apolipoprotein production, should be interpreted with caution. There is some evidence that C peptides, as well as other apolipoproteins, may be secreted independently of lipoprotein particles or that circulating apolipoprotein C may be transferred to newly synthesized lipoproteins at the point of secretion. Moreover, these processes need not be mutually exclusive of each other. It is conceivable, for instance, that some lipoproteins like lymph chylomicrons or plasma HDL are secreted with only part of their C apolipoprotein complement, whereas the apolipoprotein C composition seen with freshly isolated particles reflects the apolipoprotein status after prolonged circulation and consequent exchange and transfer. As suggested earlier, the apolipoprotein C content changes rapidly during lipoprotein catabolism or in response to metabolic stimuli and such isolated lipoproteins are unlikely to be representative of the original C peptide composition. Techniques relying on immunochemical identification such as immunofluorescence may not provide reliable data on synthesis either. Increased fluorescence under certain conditions, such as fat feeding, merely implies that the specific antigen is present in higher amounts, and does not differentiate between *de novo* synthesis or accumulation from an extracellular source. Measurement of specific mRNA levels, using probes made available by cloning of C peptide DNA, will provide more reliable information in the future.

To summarize, it is currently held that most apolipoprotein C is produced by the liver, although some apolipoprotein C-II may be produced by the intestine; it is also certainly true that there is rapid equilibration between the plasma and lymph compartments (discussed later) so that newly secreted lipoproteins from both organs have rapid access to these apolipoproteins.

Concentrations of Plasma C Apolipoproteins

Several methods have been used to quantitate plasma levels of C apolipoproteins, the major approach relying on immunochemical tools, especially for measuring absolute levels. Separation and quantitation by electrophoretic methods have mainly been used to estimate relative proportions of the proteins. Although the immunochemical methods are discussed in more detail in Chapter [31], Volume 128, it should briefly be stated that various approaches give similar but not necessarily identical values between laboratories. Radioimmunoassays and electroimmunoassays have been the major tools employed but radial immunodiffusion and enzyme immunoassays have also been used. Various conditions are claimed to produce optimal antigenicity of the C peptides including, for example, treatments with detergents, denaturing agents, or lipoprotein lipase. Polyclonal antibodies will obviously vary widely in their affinity and capacity to bind their respective antigens, so that the increasing acceptance of and production of monoclonal antibodies may benefit this area by providing standard and highly reproducible results through the availability of either the IgG itself or a "bank" of hybridoma cells producing the antibodies. To date, most quantitative data concern the C-II and C-III apolipoproteins although C-I has been quantified by radial immunodiffusion. Normal levels for C-II and C-III peptides vary between 3.7–5.2 mg/ dl and 10.0–15.4 mg/dl, respectively, in human plasma but in some hypertriglyceridemic states (Types IV and V) the concentrations may increase 4- to 5-fold, with extremes of 24.3 and 54.0 mg/dl being reported for apolipoproteins C-II and C-III, respectively.[18] C-I varies between 6.9 and 9.1 mg/dl and is preferentially distributed in the HDL fraction.

Dietary factors influence apolipoprotein C profile. We have observed increases in the proportion of apolipoprotein C-III with high carbohydrate diets and of apolipoprotein C-III2 with high cholesterol diets.[18] On the other hand, when VLDL production is suppressed by dietary ω-3 fatty acids, the proportion of apolipoprotein C-III1 falls.

The various methods have relative advantages and disadvantages, with radioimmunoassay leading electroimmunoassay for sensitivity, with the latter being preferred when speedy results are required and sensitivity is not a problem. The reproducibility in quantitation has been reasonably satisfactory and values reported by groups using either method have been fairly consistent. Finally, the potential use of HPLC for both separating and quantitating apolipoprotein C is being investigated by several laboratories but awaits evidence of reproducibility.

Methods Used for Turnover Studies

We have developed a method which allows a rapid separation and determination of individual apolipoprotein C kinetics in multiple samples, a necessary requirement for *in vivo* investigation in humans or animals. In this approach, isoelectric focusing is combined with densitometric scanning and radioassay of separated apolipoprotein C isoforms.

Another method for determining the specific radioactivity of at least one C apolipoprotein group (C-III) has been described recently[19] and is reported in detail in this series. This method uses immunoaffinity columns to isolate C-III apolipoprotein from VLDL but one disadvantage of the

---

[18] P. Nestel and N. Fidge, *Adv. Lipid Res.* **19**, 55 (1982).
[19] P. R. Bukberg, N. A. Le, H. N. Ginsberg, J. C. Gibson, L. C. Goldman, and W. V. Brown, *J. Lipid Res.* **24**, 1251 (1983).

method is that it does not allow separation of the individual isoforms of C-III apolipoprotein. Future studies may reveal different turnover rates for some apolipoprotein C-III isoforms with important metabolic consequences. The method described in this section has the advantage of measuring the specific radioactivity of all major C apolipoproteins present in each lipoprotein simply by loading the samples onto isoelectric focusing gels.

Separation of C Apolipoprotein. The specific radioactivity (SA) of individual C apolipoproteins in isolated lipoproteins such as VLDL or HDL can be determined,[9] but it is also possible to measure the specific radioactivity of C peptides from the whole $d < 1.21$ g/ml fraction of plasma or lymph[16] and probably of tissue extracts or liver perfusions as well.

If VLDL or chylomicrons are the starting materials, the soluble apolipoprotein fraction is extracted with isopropanol according to Holmquist *et al.*[20] 200- to 300-μl samples of triglyceride-rich lipoprotein at a concentration of 1–2 mg/ml are required. The sample is transferred to small plastic tubes (Eppendorf centrifuge tubes) to which is added an equal volume of pure isopropanol. The apolipoprotein B precipitates after several minutes and is removed by centrifugation. The supernatant (isopropanol : H_2O phase) is transferred to larger plastic (resistant to solvents) or glass tubes, and 2 ml methanol, 3 ml chloroform, and 5 ml diethyl ether are added sequentially. The tubes are stoppered, and the soluble apolipoproteins (mainly C peptides in plasma VLDL, but also significant amounts of A and E apolipoproteins if chylomicrons are present) are next precipitated by chilling at 4° for 1 hr. The precipitates are centrifuged (2000 rpm for 2 min), washed with ether, and dried under N_2. This pellet can be stored at $-20°$ for several weeks or immediately dissolved in 200 μl of 8 M urea in 0.1 M Tris-HCl, pH 8.5, and focused on polyacrylamide gels as described below. All soluble apolipoproteins are extracted in this procedure, since apolipoprotein B is the only protein seen on SDS gels following electrophoresis of the insoluble isopropanol pellet.

Samples of HDL require different treatment. All HDL apolipoprotein is soluble and apolipoprotein C comprises only a small portion of the total protein. At least 2 mg HDL protein (apolipoprotein) is required; the sample is dialyzed against 5 mM NH_4HCO_3, pH 8.0, lyophilized, and then redissolved in 200 μl of the same buffer and delipidated with chloroform, methanol, and diethyl ether as described above. The pellet is washed with ether, dried, and redissolved in 8 M urea in 0.01 M Tris buffer, or stored at $-20°$.

[20] L. Holmquist, K. Carlson, and L. A. Carlson, *Anal. Biochem.* **88,** 457 (1978).

To determine specific radioactivities of individual C apolipoproteins in the total $d < 1.21$ g/ml lipoproteins, plasma is centrifuged at $d = 1.21$, and the whole lipoprotein fraction dialyzed, lyophilized, and delipidated. This pellet is treated as described above. This approach has provided the technique for determining the kinetics of lymph and plasma C apolipoproteins following injection of radioiodinated VLDL into rats.

Separation by Isoelectric Focusing. Preliminary experiments showed that a focusing technique was essential if electrophoresis was the choice for specific radioactivity measurement of C apolipoproteins. Resolution of the apolipoprotein C polymorphs, particularly C-III0 from C-II peptide, cannot be accomplished in SDS- or urea-containing gels. Futhermore, we have found that the presence of ampholines sharply focuses both unlabeled and radioiodinated peptides, which does not always occur with urea gels, possibly because of small charge differences between them.

Polyacrylamide gels (7.5%) containing 6.8 M urea and 2% ampholine (LKB, pH 4–6) are prepared,[9] the urea being deionized through mixed bed resins until the conductivity is below 0.5 μmhos. The gels are placed in the electrophoresis apparatus and cooled by circulating tap water through a glass coil immersed in the lower tank buffer. They are prefocused for 30 min at 1 mA per tube and then samples of VLDL or HDL (50–100 μl) are loaded, together with 20 μl of a 1 : 5 dilution of ampholines in 8 M urea and 20 μl of 80% (w/v) sucrose, on top of the gels. The samples are overlaid with the upper tank buffer (0.02 M NaOH), the lower electrode solution being 0.01 M H$_3$PO$_4$.

A constant-voltage power supply is used to produce a constant voltage of 200 V for 1 hr and then 400 V for the remaining 4 hr. Following staining (with an extract of Coomassie Blue R-250)[9] the gels are destained for 24–48 hr and then scanned at 560 nm in the Pye-Unicam SP 8-400 spectrophotometer with scanning attachment.

Excellent resolution of C apolipoprotein is obtained, whether present in VLDL, HDL, or the whole $d < 1.21$ g/ml lipoprotein fraction, as shown in Fig. 1. The larger total load of apolipoprotein present in HDL or $d < 1.21$ g/ml fractions does not appear to disturb focusing of the apolipoprotein C group. Scanning profiles have confirmed the exact correspondence in migration of C peptides derived from the different samples loaded. In order to convert peak areas into mass (micrograms), standard curves are obtained using purified C-II and C-III apolipoproteins. Unfortunately, apolipoprotein C-I data cannot be obtained since it focuses outside the pH 4–6 range, which is most suitable for resolving the C-II and C-III apolipoproteins.

Sufficient apolipoprotein C-II and C-III1, C-III2, or C-III0 for use as

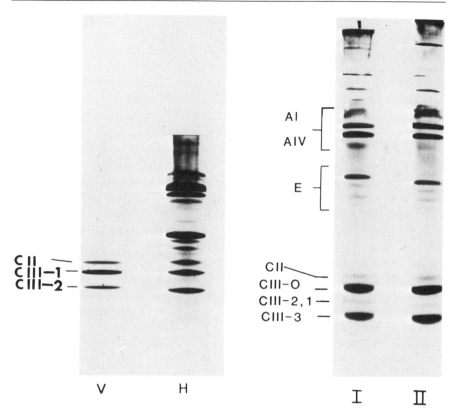

C II
C III-1
C III-2

AI
AIV

E

CII
CIII-O
CIII-2,1
CIII-3

V   H

I   II

FIG. 1. Isoelectric focusing of C apolipoproteins on polyacrylamide gels (pH range, 4–6). Left panel shows a comparison of VLDL (V) and HDL (H) C peptide separation from human subjects; right panel shows separation of rat C apolipoproteins from the $d < 1.21$ g/ml fraction of rat serum (I) and lymph (II).

standards can easily be purified by preparative isoelectric focusing or chromatofocusing. Standard curves are linear in the 5–50 μg range so that 1–2 mg of pure apolipoprotein is sufficient for many experiments. Soluble VLDL apolipoprotein, dissolved in 6 M urea as starting material, is mixed with a gel slurry (Ultradex, LKB, Sweden) and fractionated by flat-bed isoelectric focusing as described in the Pharmacia FBE 3000 bulletin. An improved separation can be obtained by focusing first between pH 2.5 and 5.0, removing the gel section containing apolipoprotein C (located by a stained paper print), and then refocusing the combined gel from two runs between pH 4 and 5. The proteins (in their respective bands) are eluted by adding 6 M urea to liquefy the gel, titrated to pH 8.9 with Tris solution, and then centrifuging to yield a granular gel. Following removal of ampholytes, the protein content is determined and the purity checked by analytical isoelectric focusing.

Chromatofocusing, which separates proteins according to their pI values, is another useful procedure which we have reported elsewhere.[18] This column chromatographic method resolves the C-II and C-III apolipoproteins, but as with all isoelectric focusing techniques, binding of ampholines to peptides is the main problem. We have overcome this by eluting proteins from flat-bed gels or chromatofocusing beads followed by chromatography through Sephadex G-50 (1.5 × 100 cm). This is a useful procedure when precipitation by solvents such as TCA is to be avoided.

Application to Kinetic Studies. As mentioned above, each apolipoprotein C band on IEF gels was associated with a peak of radioactivity with negligible counts between the bands, thus confirming the identity of labeled peptides with native apolipoproteins. In addition, label remaining on top of the gels corresponded closely to the percentage of radioactivity in apolipoprotein B determined by isopropanol precipitation techniques. Therefore, by slicing the bands containing the C-II, C-III0, C-III1, and C-III2 peptides and counting in the Packard auto spectrometer, specific radioactivities can be determined.

Analysis of Specific Radioactivity Data. In order to minimize variations in the concentration of plasma lipoproteins, very hyperlipemic subjects to be injected with ^{125}I-labeled VLDL are placed on a constant fat intake for 48 hr prior to the studies. Labeled VLDL is injected in the morning after an overnight fast and frequent blood samples collected over a 48-hr period during which minimal dietary fat is given.

All studies to date have been characterized by rapid isotopic equilibration between the corresponding C apolipoproteins of VLDL or HDL, with negligible radioactivity present in other plasma fractions. Plotting the data has revealed both monoexponential or biexponential decay curves (see Fig. 2) and in each case the curve parameters are calculated by computer using a nonlinear least-squares technique. Kinetic parameters are therefore calculated from either a one-pool or two-pool model, using conventional techniques. In the former, the parameters are calculated thus[21]:

$$FCR = 0.693/t_{1/2}$$
$$\text{Pool size (mass)} = \text{injected dose/specific activity at } t_0$$
$$\text{Flux} = FCR \times \text{pool size}$$

---

[21] R. H. Shipley and R. E. Clark, "Tracer Methods for In Vivo Kinetics." Academic Press, New York, 1972.

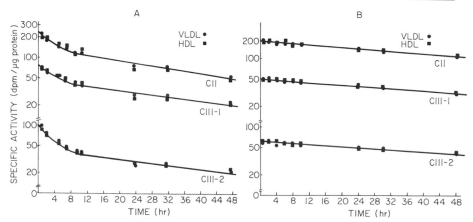

FIG. 2. Examples of specific radioactivity vs time plots of C-II, C-III1, and C-III2 apolipoproteins showing (A) biexponential or (B) monoexponential decay curves, for VLDL (●), or HDL (■). Note that the specific radioactivity of the corresponding C peptides in VLDL or HDL are identical, even at earliest time points, following the injection of ^{125}I-labeled VLDL into these two human subjects.

When the curves could clearly be resolved into two exponentials, kinetic parameters were determined from the two-pool model of Gurpide *et al.*,[22] which describes the relationship between two pools (1 and 2) and allows for independent entry and exit from both pools. Since the origin and magnitude of apolipoprotein C synthesis plus removal from organs is not well established it is possible that both pools may contain tissues which participate in both synthesis and removal. One assumes, however, that direct input into pool 2 and irreversible loss from pool 2 are likely to be much less than from pool 1, which includes plasma, since in most cases a single pool dominated the curve analysis.

Thus, using the two pool model,

Flux (through pool 1) $= R_1 \alpha\beta/\beta C_1 + \alpha C_2$

where R_1 = dose of radioactivity injected
   α, β = rate constants of the exponentials
   C_1, C_2 = intercepts of the two rate constants α and β on the Y-axis of the specific radioactivity curve
Mass (pool 1) $= R_1/(C_1 + C_2)$
   FCR $= (\alpha C_1 + \beta C_2)/(C_1 + C_2)$

Further details of these kinetic parameters can be obtained elsewhere.[9,21]

Interpretation of C Apolipoprotein Kinetics. The technique described above, which allow a quantitative assessment of flux, pool sizes, and

[22] E. Gurpide, J. Mann, and E. Sandberg, *Biochemistry* **3**, 1250 (1964).

removal rates of individual C peptides, provide a more precise description of the kinetics of apolipoprotein C than previously described by simulated models[23] or by using tracer alone to follow redistribution. First, the specific radioactivity data show that radiolabeled apolipoprotein C exchanges rapidly between VLDL and HDL. This rapid redistribution probably occurs mainly within the plasma, but possibly involves a small extravascular pool giving rise to a two-pool system. This was later confirmed in the rat, in which rapid equilibration between the specific radioactivity of apolipoprotein C in both lymph and plasma was demonstrated, suggesting that the biexponential decay phase found in some human subjects may also result from an early rapid equilibration between plasma and lymph, although the presence of other extravascular pools of C peptides cannot be excluded.

Isotopic equilibration between corresponding C peptides in VLDL and in HDL occurred in nearly all normal subjects, though in some hypertriglyceridemic subjects the specific radioactivities within HDL were higher than in VLDL.

Comparison of the C apolipoprotein mass determined from kinetic parameters, with the absolute mass in plasma, measured directly by chemical means, can also assist in the interpretation. For example, we have observed that the removal of radioactive C apolipoproteins is slower in hypertriglyceridemic subjects ($t_{1/2} = 55$ hr vs 28 hr in normal subjects) who also have greatly expanded intravascular pools of all C peptides compared to normals. Since the mass derived from kinetic measurements agreed closely with that determined chemically, we can deduce that most of the C peptide resides within the plasma and due to the characteristic (and prominent) monoexponential decay in these subjects, metabolism appears to have occurred mainly within a single homogeneous pool. Furthermore, all C apolipoproteins studied (apolipoproteins C-II, C-III, and C-III2) had similar fractional removal rates, implying similar sites and modes of catabolism and interrelated functions.

This was true for normal and hypertriglyceridemic subjects and remained so even when metabolism was perturbed by dietary carbohydrate or by infusions of heparin.[12]

The increased mass of C apolipoproteins in hypertriglyceridemia appeared to be due mainly to reduced removal, since the flux (synthesis rates) was not significantly different from normal subjects. However, when normal subjects eat a high carbohydrate diet there is evidence for increased apolipoprotein C production which contributes to the expanded pool. Thus, the increased C apolipoprotein pool in hypertriglyceridemia is

[23] M. Berman, M. Hall, R. I. Levy, S. Eisenberg, D. W. Bilheimer, R. D. Phair, and R. H. Goebel, *J. Lipid Res.* **19**, 38 (1978).

due almost entirely to decreased removal, whereas the expansion in pool size associated with carbohydrate feeding is related to increased production. High levels of plasma triglyceride may therefore reflect an inability to increase apolipoprotein C synthesis whereas carbohydrate-induced hypertriglyceridemia possibly persists only in cases where apolipoprotein C formation fails to match triglyceride synthesis. In other experiments, designed to perturbate apolipoprotein C metabolism by heparin infusion, it was found that, even during extremely rapid catabolism of VLDL particles, the rate of loss of individual apolipoprotein C species was proportional to their initial concentration and their rates of transfer from VLDL to HDL were identical for the three C peptides. These findings are consistent with the previous studies in demonstrating the functional unity of these apolipoproteins. Importantly, during rapid lipolysis induced by heparin, a substantial loss of C apolipoproteins occurred from the plasma compartment. Whereas under normal circumstances the catabolism of triglyceride-rich lipoproteins leads to a quantitative transfer of C peptides to HDL, during the heparin infusions as much as 90% of apolipoprotein C was directly removed with B apolipoprotein from the circulation.[12] Presumably this reflects the uptake of partly catabolized VLDL in the liver. If this process is representative of a normal catabolic event, especially in hypertriglyceridemic subjects, then it represents a potentially serious loss of C peptides and may be a factor in the perpetuation of the hypertriglyceridemia.

Conclusions

The methods described above have proved satisfactory for investigating C apolipoprotein metabolism in man and in experimental animals. We have emphasized that a major advantage of this technique is the opportunity to determine specific radioactivities of all major C peptides except C-I, although this too could be studied if the appropriate pH range used for isoelectric focusing were extended to include the p*I* for this apolipoprotein. A modified approach is to isolate and isotopically label a peptide, e.g., C-II, incorporate it into a native lipoprotein that is then reinjected into a suitable host for the study of the *in vivo* (or *in vitro*) kinetics of the particular apolipoprotein.

The precision of the specific radioactivity values obtained by the focusing technology has been confirmed repeatedly in our laboratory, and is characterized by excellent reproducibility. The finding of almost identical specific radioactivities in VLDL and HDL for an individual C peptide in normolipidemic subjects can be predicted by the demonstrably rapid exchange which characterizes this group of apolipoproteins. The estimation of mass of each C apolipoprotein using this kinetic method has agreed closely with chemical measurements in plasma supporting the predominantly intravascular distribution of C apolipoproteins.

Animal models may be studied by this technique but the success depends on the separation of apolipoprotein C polymorphs, which should be investigated prior to specific radioactivity measurements. Not all animals exhibit a similar distribution of apolipoprotein C. In the pig, for example, those C peptides which migrate on isoelectric focusing in the manner of the human C-III isoforms are not sialylated[24] and are more likely to represent minor amounts of C-II isoforms.

The future offers exciting prospects for understanding the role of C apolipoproteins in lipid metabolism. As briefly mentioned above, quantitative measurement of C peptides by various organs can now be explored using cDNA clones of the individual apolipoprotein C species as probes for measuring their tissue mRNA levels. A gene lesion has recently been identified in association with apolipoprotein A-I–apolipoprotein C-III deficiency, which results in low levels of HDL and premature atherosclerosis.[25] Since these two apolipoproteins are apparently linked in a gene complex to chromosome 11, it will be desirable in future metabolic studies to follow the *in vivo* metabolism of these two apolipoproteins in subjects exhibiting various lipoprotein abnormalities, especially those relating to abnormal HDL levels.

[24] N. H. Fidge, *Biochim. Biophys. Acta* **424**, 253 (1976).
[25] G. A. P. Bruns, S. K. Karathanasis, and J. L. Breslow, *Atheriosclerosis* **4**, 97 (1984).

[27] Direct Determination of Apolipoprotein C-III Specific Activity Using Immunoaffinity Chromatography

By Ngoc-Anh Le, Phillip R. Bukberg, Henry N. Ginsberg, Joyce C. Gibson, and W. Virgil Brown

Introduction*

The C apolipoproteins of human plasma, designated apoC-I, apoC-II, and apoC-III, are small polypeptides of M_r 6000–9000. In fasting plasma, the C apolipoproteins are associated primarily with very-low-density lipo-

* Abbreviations: ApoC-III, apolipoprotein C-III; VLDL, very-low-density lipoproteins; HDL, high-density lipoproteins; LPL, lipoprotein lipase; IgG, immunoglobulin G; TG, triglyceride; FCR, fractional clearance rate; SDS, sodium dodecyl sulfate; EDTA, disodium ethylenediaminetetraacetate; RIA, radioimmunoassay; SA, specific radioactivity.

proteins (VLDL, $d < 1.006$ g/ml) and high-density lipoproteins (HDL, d 1.063–1.21 g/ml), constituting 40–60% of the total protein in the former and 5–10% of that in the latter. *In vitro* studies have demonstrated rapid exchange of labeled C-apolipoproteins between VLDL and HDL, with the final distribution governed by the relative proportions of these lipoproteins in the plasma or incubation mixtures.[1] ApoC-II and apoC-III appear to be involved in the regulation of the rate of triglyceride hydrolysis by lipoprotein lipase[2-4] as well as in the hepatic uptake of lipoprotein particles.[5]

Because of their similar molecular weights and solubility characteristics, purification of the individual C apolipoproteins from the limited amounts of material available in a small plasma sample is extremely difficult. A method for performing turnover studies of apoC which utilizes isoelectric focusing to separate apoC-II from apoC-III has been reported by Huff *et al.*[6] and described in Chapter [26] of this volume. We now present a new method[7,8] for studying the metabolism of apoC-III, which employs immunoaffinity chromatography to isolate this apolipoprotein from lipoprotein fractions in small plasma samples for determination of specific radioactivity.

Materials

Sepharose CL-4B, DEAE-Sephacel, and Sephadex G-100 and G-150 were obtained from Pharmacia Fine Chemicals (Uppsala, Sweden). Bovine albumin (fraction V), ovalbumin, lactoperoxidase, Tween-20, and Triton X-100 were purchased from Sigma Chemical Company (St. Louis, MO). Cyanogen bromide was obtained from Aldrich Chemicals (Milwaukee, WI), polyethylene glycol-6000 from J. T. Baker Chemical Co. (Phillipsburg, NJ), sodium decyl sulfate from Eastman Kodak Co. (Rochester, NY), and aprotinin (Trasylol) from Mobay Chemicals (New York, NY). Carrier-free ^{125}I was purchased from ICN Chemicals and Radioisotopes

[1] S. Eisenberg, D. W. Bilheimer, and R. I. Levy, *Biochim. Biophys. Acta* **280**, 94 (1972).
[2] R. J. Havel, V. G. Shore, B. Shore, and D. M. Bier, *Circ. Res.* **27**, 595 (1970).
[3] W. C. Breckenridge, J. A. Little, G. Steiner, A. Chow, and M. Poapst, *N. Engl. J. Med.* **298**, 1265 (1978).
[4] W. V. Brown and M. L. Baginsky, *Biochem. Biophys. Res. Commun.* **46**, 375 (1972).
[5] F. Shelburne, J. Hanks, W. Meyers, and S. Quarfordt, *J. Clin. Invest.* **65**, 652 (1980).
[6] M. W. Huff, N. H. Fidge, P. J. Nestel, T. Billington, and B. Watson, *J. Lipid Res.* **22**, 1235 (1981).
[7] P. R. Bukberg, N.-A. Le, J. C. Gibson, H. N. Ginsberg, L. Goldman, and W. V. Brown, *J. Lipid Res.* **24**, 1251 (1983).
[8] P. R. Bukberg, N.-A. Le, H. N. Ginsberg, J. C. Gibson, and W. V. Brown, *J. Lipid Res.* **26**, 1047 (1985).

(Irvine, CA). All other reagents were routinely obtained from Fisher Scientific Company (Fair Lawn, NJ).

Buffers

Borate buffers: (1) BB, 0.1 M NaCl, 0.1 M borate at pH 8.0
                (2) BBT1, 0.1 M NaCl, 0.1 M borate, 0.02% Triton X-100, pH 8.0
                (3) BBT2: 0.5 M NaCl, 0.1 M borate, 0.01% Triton X-100, pH 8.0
Phosphate-buffered saline: (1) PBS1: 0.1 M NaCl, 50 mM PO$_4$, pH 7.4
                           (2) PBS2: 0.5 M NaCl, 25 mM PO$_4$, pH 7.4

Preparation of Anti-ApoC-III Antisera

VLDL were isolated from plasma of human donors with type IV hyperlipoproteinemia by ultracentrifugation at d 1.006 g/ml and washed once by reflotation at this density. ApoC-III was purified from delipidated VLDL by a two-step chromatographic procedure involving Sephadex G-150 and DEAE-Sephacel.[9]

Antisera were raised in goats because of the large quantities of IgG required for preparation of the anti-apoC-III affinity columns.

Preparation of Anti-ApoC-III Affinity Columns

Purification of Specific IgG. ApoC-III was covalently bound to CNBr-activated Sepharose CL-4B (50 mg CNBr/ml of gel) at a concentration of 1 mg of apolipoprotein/ml of gel.[10] Coupling efficiency was over 95%.[7] Aliquots of apoC-III–Sepharose (2 ml) were each combined with 12–15 ml of goat antiserum and placed in 50-ml plastic screw-capped tubes. The suspensions were diluted with PBS1 to a total volume of 30 ml, and the tubes incubated for 3 hr at 4° on a mechanical shaker (Eberbach, Ann Arbor, MI). The contents of all tubes were then decanted into a single 1.5 ×10 cm plastic "Econo-column" (Bio-Rad, Richmond, CA). The gel was first washed extensively with PBS1, and then with a higher ionic strength buffer, PBS2, to remove nonspecifically bound protein. Specific IgG was then eluted with 0.2 M glycine at pH 2.5, immediately dialyzed against PBS to minimize denaturation, and stored at −70° until further use.

---

[9] W. V. Brown, R. I. Levy, and D. S. Fredrickson, *J. Biol. Chem.* **244**, 5687 (1969).
[10] P. Cuatrecasas, *Adv. Enzymol.* **36**, 29 (1972).

The binding capacity of the apoC-III–Sepharose was approximately 3 mg IgG/ml of gel. Purity of the IgG eluted from the gel was confirmed by SDS–polyacrylamide gel electrophoresis, and its activity against apoC-III demonstrated by double immunodiffusion. Immunoprecipitation experiments using [^{125}I]apoC-III showed the isolated IgG to have a binding capacity of at least 22.5 μg C-III/mg IgG.

Preparation of Anti-ApoC-III Affinity Column. Specific IgG isolated as above was then covalently bound to cyanogen bromide-activated Sepharose CL-4B (60 mg CNBr/ml of gel) at a concentration of 5.5 mg IgG/ml of gel.[7] The anti-apoC-III–Sepharose was equilibrated in Econo-columns (1 ml of gel per column) with BB. Nonspecific binding sites were saturated prior to initial use of the gel by application of a 0.1% solution of ovalbumin in this buffer. After washing the gel free of residual ovalbumin, ten gel volumes of 0.2M glycine (pH 2.5) were applied, followed by equal volumes of 3M sodium thiocyanate. The gel was then re-equilibrated with BB. Several such elution cycles (3–5 cycles) were carried out until leaching of IgG or ovalbumin was no longer detectable in the eluate.

The efficiency of the coupling reaction was approximately 90%, yielding a final concentration of 5 mg IgG/ml of gel. The capacity of the anti-apoC-III–Sepharose, determined by application of a saturating quantity of apoC-III, extensive washing of the column, and measurement of the apolipoprotein eluted was 80–90 μg apoC-III/ml of gel.

Affinity Chromatography of ApoC-III

Aliquots of lipoproteins (250–400 μg of total protein for VLDL and 1.5–2.0 mg for HDL) were delipidated using acetone followed by two additional extractions with isopropanol. (HDL samples were dialyzed against 1000 vol of 0.9% NaCl with 1 mg/ml EDTA prior to delipidation.) After decanting the second wash of isopropanol, the pellet was resuspended in 1 ml of BBT1. A brief period of sonication in a water bath (Bransonic-52, Branson, Shelton, CT) was usually necessary to break up the delipidated pellet, and the tubes were then allowed to incubate overnight at 4°. The next day, the suspensions were transferred to 1.5-ml microfuge tubes and centrifuged at 10,000 rpm for 20 min (Microfuge-B, Beckman Instruments) to remove visible precipitate and microaggregates.

The supernatants were diluted to 5 ml total volume with BBT1 and transferred to the disposable plastic Econo-columns containing 1 ml of anti-apoC-III–Sepharose. The apolipoprotein mixture was allowed to run slowly through the gel over a period of 60–90 min. Each column was then washed with 10–15 ml of BBT1 buffer, followed by 20–30 ml of BBT2 buffer to remove nonspecifically bound protein. ApoC-III was eluted by addition of 3.0 ml of 0.2 M glycine (pH 2.5) containing 0.01% Triton; after discarding the initial 0.5 ml, the subsequent 2.5 ml of effluent containing the apoC-III was collected. For determination of specific activity (SA), measured aliquots of the apoC-III eluate were counted directly in a Packard autogamma spectrometer (Packard Instruments, Downers Grove, IL). Protein mass was determined on aliquots of the same eluate either by a modification of the Lowry procedure[11] or by radioimmunoassay.[7] For the modified Lowry method, bovine albumin in 0.2 M glycine, 0.01% Triton (pH 2.5) was utilized as standard.

The entire chromatographic procedure was carried out at 4°. After elution of the apoC-III, columns were washed with an additional 10 ml of 0.2 M glycine and with 20 ml of 3 M sodium thiocyanate before being reequilibrated and stored in BB containing 0.02% sodium azide.

Although contamination of the eluates by small amounts of IgG and ovalbumin was evident during the initial use of the columns, leaching of these proteins ceased to be a problem after several elution cycles. When delipidated VLDL was applied to the affinity column, analysis of eluate samples for the presence of apoC-II by specific radioimmunoassay indicated an average contamination of 1.2%. Contamination of the apoC-III eluates by apolipoproteins B and E, assessed by specific radioimmunoassays, ranged from 0.3 to 0.8%. For HDL samples, minimal contamination of the eluate by apoA-I or apoA-II was detected.

Reproducibility of the column method for measurement of the apoC-III specific activity was assessed by delipidation of six separate aliquots of ^{125}I-labeled VLDL at each of two protein loads. Delipidation and analysis of 400 μg of VLDL protein yielded 72.1 ± 6.4 μg apoC-III of SA 242.4 ± 13.5 cpm/μg; 250 μg of VLDL protein yielded 45.4 ± 4.9 μg apoC-III of SA 249.0 ± 7.8 cpm/μg.[7]

In Vivo Studies

Studies of apoC-III turnover following the injection of ^{125}I-labeled VLDL were carried out in six subjects, two with normal lipoprotein levels and four with type IV hyperlipoproteinemia. One subject received simultaneous injections of ^{131}I-labeled VLDL and ^{125}I-labeled HDL. Clinical data on the six subjects are presented in detail elsewhere.[8] Informed consent was obtained from each subject prior to the study.

Following the injection of the tracer, 18 times plasma samples were obtained over a 72-hr period. VLDL were isolated from these samples by ultracentrifugation at d 1.006 g/ml, and HDL at d 1.063–1.21 g/ml, using a

---

[11] A. Bensadoun and D. Weinstein, *Anal. Biochem.* **70**, 241 (1976).

40.3 rotor at 39,000 rpm for 20 hr (VLDL) or 48 hr (HDL) at 10°. ApoC-III SA curves displayed a biphasic decay in all subjects studied (Fig 1). In all studies in which radiolabeled VLDL was the injected tracer, the SA of apoC-III in VLDL were always higher than corresponding values for apoC-III in HDL and remained constant throughout the sampling period. The ratio of the SA in the two normolipidemic subjects (1.35 and 1.96) were slightly higher than the values obtained for individuals with hypertriglyceridemia (1.16–1.26). In the subject who received both ^{131}I-labeled VLDL and ^{125}I-labeled HDL, the ratios of VLDL apoC-III SA to HDL apoC-III SA were 1.19 for the VLDL tracer and 0.55 for the HDL tracer.

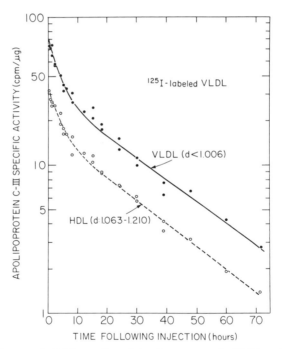

FIG. 1. Apolipoprotein C-III specific activity decay curves in VLDL and HDL following the injection of ^{125}I-labeled VLDL. From each plasma timed sample, VLDL (*d* 1.006 g/ml) and HDL (*d* 1.063–1.210 g/ml) were isolated by sequential ultracentrifugation, delipidated, and then applied to the apoC-III immunoaffinity columns for the separation of apoC-III from other apolipoproteins using separate columns for each sample. The apoC-III bound to the column was eluted and duplicate aliquots were obtained for apoC-III mass and radioactivity. For both VLDL (●) and HDL (○) the duplicate specific activity determinations are presented. The apoC-III specific activity associated with VLDL is higher than that associated with HDL and the two decay curves are parallel over the 72-hr sampling period.

This difference in specific activity for apoC-III between the two lipoprotein fractions was observed in the 5-min sample following the injection of the radiolabeled lipoproteins. In all subjects, the ratio of the SA remained constant throughout the sampling period with standard deviations about the mean being less than 10%. This finding clearly suggested that while the *in vivo* redistribution of apoC-III radioactivity between lipoproteins occurred within minutes, complete equilibration did not occur *in vivo* during several days of observations.

In Vitro Studies

A number of experimental protocols were carried out *in vitro* to further characterize this lack of complete equilibration of apoC-III between VLDL and HDL *in vivo*. In all the studies described below, the concentrations of the two lipoprotein fractions in the incubation mixtures were maintained at levels present in the original plasma. In studies where an increasing mass of unlabeled lipoproteins was used to deplete radioactivity from the exchangeable apoC-III pool, net mass transfer of apoC-III could be detected by radioimmunoassay.

Dynamics of ApoC-III in Sequential Incubation. By exposing radiolabeled lipoproteins to sequential incubations with fresh, unlabeled lipoproteins of a different density class, we expected to deplete selectively apoC-III radioactivity associated with the exchangeable pool.

A tracer amount of autologous ^{125}I-labeled VLDL (less than 10 μg of total protein) was added to several individual VLDL samples and incubated at 37° for 30 min before the addition of the appropriate HDL aliquots (493 μg of VLDL apoC-III versus 88 μg of HDL apoC-III). The incubations were continued at 37° for an additional 3 hr and then allowed to proceed at 4° overnight (16–20 hr). Following incubation, a single ultracentrifugal spin at *d* 1.020 g/ml using a 40.3 rotor was carried out with the VLDL supernatant recovered quantitatively in 2-ml volumetrics. The middle 1 ml was discarded and the 3-ml infranatant representing HDL was saved. Aliquots of some of the VLDL supernatants and HDL infranatants were then delipidated and applied to the immunoaffinity columns for apoC-III specific activity determination (incubation A). The ^{125}I-labeled VLDL supernatants which were not used for SA determination were then combined with fresh unlabeled HDL aliquots for a second incubation (incubation B) and then fractionated as described above. After the second incubation and fractionation, several of the VLDL supernatants and HDL infranatants were used for apoC-III SA determination while some of the VLDL supernatants were subjected to a third incubation. In the third stage (incubation C), ^{125}I-labeled VLDL fractions which had undergone two successive incubations with fresh HDL were allowed to equilibrate further with additional, identical aliquots of fresh unlabeled HDL.

If apoC-III in VLDL and HDL were completely exchangeable the expected apoC-III SA for VLDL and HDL should have been identical with values of 198, 174, and 164 cpm/μg for incubations A, B, and C, respectively (as determined from the total apoC-III radioactivity and the total apoC-III mass). The actual apoC-III SA obtained from the affinity columns were, respectively, 205 vs 155, 187 vs 110, and 177 vs 90 cpm/μg for VLDL vs HDL after incubations A, B, and C. The ratio of HDL apoC-III SA to VLDL apoC-III SA for the three incubations were 0.756, 0.582, and 0.509, respectively, suggesting a decrease in the fraction of the VLDL apoC-III radioactivity which could be transferred to HDL by exchange. Furthermore, for each incubation step, the actual HDL apoC-III SA were significantly lower than the expected SA, suggesting dilution by a large pool of unlabeled apoC-III. The apoC-III SA in VLDL, on the other hand, were only slightly higher than expected, suggesting that the majority of the VLDL apoC-III radioactivity could participate in the equilibration process.

In a parallel experiment with lipoproteins from the same subject, ^{125}I-labeled HDL was incubated with autologous VLDL in proportions identical to that in plasma and the higher apoC-III specific activity was in HDL. The HDL apoC-III specific activities from these incubations were at least 5-fold greater than the values expected with complete equilibration.[8] Thus only a minor fraction of the HDL apoC-III was exchangeable under these conditions.

ApoC-III Labeled by Exchange. Since the higher apoC-III specific activity was consistently associated with the lipoprotein fraction which was radioiodinated, the lack of complete equilibration might have been due to alterations in the binding affinity of apoC-III following radioiodination. To test this hypothesis, ^{125}I-labeled HDL (69 cpm/μg of apoC-III) was first allowed to equilibrate with autologous VLDL at a mass ratio of apoC-III identical to that present in the original plasma. This VLDL fraction, which had acquired apoC-III radioactivity by exchange, was then reincubated with fresh, unlabeled autologous HDL. The incubation conditions were identical to those described in the preceding section.

Only 1.6% of the total radioactivity initially in HDL was transferred to VLDL with a resulting apoC-III specific activity of 2.2 cpm/μg in VLDL. The specific activity for apoC-III in HDL after the incubation was 63.0 cpm/μg. When this exchange-labeled VLDL sample was incubated with an appropriate quantity (i.e., maintaining the mass ratio of apoC-III present in plasma) of fresh unlabeled autologous HDL, the final apoC-III specific activities remained different (1.7 cpm/μg in VLDL compared to 0.9 cpm/μg in HDL). These results indicate that the lack of complete equilibration in the preceding experiments was not due to an artifact of the radioiodination procedure.

Effect of the Method of Lipoprotein Isolation. Another possible source of artifacts which might result in the incomplete equilibration of apoC-III between VLDL and HDL could be the sequential ultracentrifugal protocol utilized. To test this hypothesis, an aliquot of purified apoC-III$_2$, iodinated with ^{125}I using lactoperoxidase,[12] was added to 10 ml of plasma and was incubated at 37° for 3 hr and at 4° overnight. Lipoprotein fractions were then isolated from aliquots of this mixture by three different methods: (1) sequential ultracentrifugation,[13] (2) single-spin discontinuous gradient ultracentrifugation,[14] and (3) molecular sieve chromatography using 4% agarose.[15] The plasma concentrations of apoC-III in the three lipoprotein fractions were 135, 86, and 125 μg/ml for VLDL, IDL + LDL, and HDL, respectively. The apoC-III SA was different in each lipoprotein fraction (14, 20, and 24 cpm/μg for VLDL, IDL + LDL, and HDL, respectively) and these differences in specific activity were similar independent of the methods of lipoprotein fractionation.

Of interest is the fact that with this subject's plasma, the higher apoC-III specific activity obtained by exchange was in HDL (23.9–24.7 cpm/μg compared to 14.0–14.8 cpm/μg for VLDL). This would be compatible with the presence of a major portion of HDL apoC-III in an exchangeable pool in this subject.[8]

Discussion

The basic features of a hypothetical model which could explain the present *in vitro* and *in vivo* data are delineated in Fig. 2. The notations "E" and "NE" are used here to denote the equilibrating and nonequilibrating pools of apoC-III, respectively, within each lipoprotein density fraction. These pools may represent different regions on the same lipoprotein particles and/or different subpopulations of particles within the density class. Upon the injection of radiolabeled VLDL, apoC-III radioactivity associated with the exchangeable pool in VLDL is instantaneously distributed among other exchangeable apoC-III pools in IDL, LDL, and

[12] J. J. Marchalonis, *Biochem. J.* **113**, 299 (1969).

[13] R. J. Havel, H. A. Eder, and J. H. Bragdon, *J. Clin. Invest.* **34**, 1354 (1955).

[14] M. J. Chapman, S. Goldstein, D. Lagrange, and P. M. Kapland, *J. Lipid Res.* **22**, 339 (1981).

[15] J. C. Gibson, A. Rubinstein, P. R. Bukberg, and W. V. Brown, *J. Lipid Res.* **24**, 886 (1983).

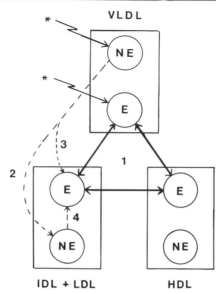

FIG. 2. Basic features of apoC-III metabolism in man. Three lipoprotein classes are considered: VLDL (*d* 1.006), IDL + LDL (*d* 1.006–1.063), and HDL (*d* 1.063–1.210). ApoC-III within each density fraction is distributed between an exchangeable pool (E) and a nonexchangeable pool (NE). After the injection of radiolabeled VLDL, both the E and NE pools of VLDL will have initial radioactivity. Radioactivity in the exchangeable VLDL pool is instantaneously distributed among other exchangeable pools (process 1). All exchangeable apoC-III will consequently have the same specific activity by the time the 5-min sample is drawn. As a result of lipolysis, VLDL are converted to IDL + LDL. The nonexchangeable apoC-III initially in VLDL may remain nonexchangeable (process 2) or become exchangeable (process 3). Process 4 implies that ultimately all of the apoC-III in apoB-containing lipoproteins will become exchangeable. The apoC-III specific activity for each density class will then depend on the fraction of the plasma exchangeable apoC-III pool which is associated with that class and the size of the nonexchangeable pool of apoC-III within that lipoprotein class. From the available data it is not possible to determine whether the exchangeable and nonexchangeable pools of apoC-III are on the same lipoprotein particles or reflect the composition of different lipoprotein subpopulations within each density class. The nonexchangeable apoC-III pool in HDL does not become exchangeable, but does appear to remain associated with the HDL particles as they are irreversibly removed from the circulation.

HDL (process 1). Hydrolysis of VLDL triglycerides by plasma lipases may shift the nonexchangeable VLDL apoC-III pool into the (IDL + LDL) density range either as nonexchangeable (process 2) or as exchangeable apoC-III (process 3). During the final stage of VLDL hydrolysis, most of the nonexchangeable apoC-III in the (IDL + LDL) fraction may become exchangeable (process 4). In both instances, (processes 3 and 4), the exchangeable apoC-III in IDL and LDL would then immediately equilibrate with exchangeable apoC-III in both HDL and nascent VLDL.

Thus, if we assume that nonequilibrating apoC-III in VLDL will ultimately become part of the plasma pool of exchangeable apoC-III as a result of lipolysis, the terminal decay component of the SA curve would reflect the metabolism of plasma exchangeable apoC-III. During this phase, the apoC-III SA in each lipoprotein fraction will depend on the SA of the plasma pool of exchangeable apoC-III and the relative mass contribution of equilibrating and nonequilibrating apoC-III in that fraction. Since for any individual in steady state, the distribution of the two forms of apoC-III is expected to remain constant within each fraction, the ratio of apoC-III SA between VLDL and HDL would remain parallel as observed experimentally in our studies (Fig. 1). In all of our *in vivo* studies, the apoC-III SA in VLDL was greater than that in HDL after injection of radiolabeled VLDL. This would indicate that the majority of apoC-III in VLDL was in exchangeable form, while the apoC-III in HDL was predominantly in a nonequilibrating pool. This is consistent with the majority of our available *in vitro* data.

Our data also imply that with these assumptions, the nonequilibrating apoC-III pool in HDL would never participate in the equilibration process. If the radioactivity injected as nonequilibrating HDL apoC-III (for example, from the ^{125}I-labeled HDL injection in subject 6) participated in the exchange process after some time in the circulation, the HDL and VLDL apoC-III SA curves would have crossed one another, resulting in the same final ratio of SA as observed when the injected tracer in that study was whole VLDL. We did not observe such a crossover in the one study in which kinetic data from a simultaneous injection of ^{125}I-labeled HDL and ^{131}I-labeled VLDL were available.[10] Injection of radioiodinated HDL resulted in a higher HDL apoC-III SA throughout the sampling period, with the ratio of the SA between HDL and VLDL being different from the ratio obtained when ^{131}I-labeled VLDL was injected in the same individual. Furthermore, if the nonequilibrating apoC-III pool of HDL does not participate in the exchange process, then the parallel decay of VLDL and HDL apoC-III following injection of ^{125}I-labeled HDL would suggest that the primary pathway for irreversible removal of plasma apoC-III is via HDL.

The lack of complete equilibration of apoC-III between VLDL and HDL can also be used to define a relationship between the exchangeable pools of apoC-III in VLDL and HDL. In particular, if we can estimate the fraction f_V of the VLDL apoC-III mass and the fraction f_H of the HDL apoC-III mass which participate in the exchange, then the following relationship (see Appendix) must be satisfied:

$$f_H = - \left[\frac{(1 - f_{PV})}{R f_{PV}} \right] f_V + \frac{1}{R f_{PV}}$$

where $f_V = \dfrac{\text{mass of equilibrating VLDL apoC-III}}{\text{total mass of VLDL apoC-III}}$

$f_H = \dfrac{\text{mass of equilibrating HDL apoC-III}}{\text{total mass of HDL apoC-III}}$

$f_{PV} = \dfrac{\text{mass of equilibrating VLDL apoC-III}}{\text{mass of equilibrating apoC-III in plasma}}$

$R = S_V/S_H$ = observed ratio of the SA of VLDL apoC-III to that of HDL apoC-III following the tracer injection of radioiodinated VLDL

In other words, for a given ratio of the apoC-III specific activity between VLDL and HDL observed *in vivo* following the injection of radioiodinated VLDL, the fraction of apoC-III in VLDL which can participate in the equilibration process is negatively correlated to the fraction of equilibrating apoC-III in HDL. Any kinetic system proposed to explain the metabolism of apoC-III in man must address this relationship.

Appendix

From the injection of radiolabeled VLDL, it is expected that the SA of apoC-III in the whole VLDL fraction will be the same as the SA in the equilibrating and nonequilibrating components of VLDL apoC-III prior to equilibration, that is,

$$S_V = S_{Ve} = S_{Vn}$$

where S_V, S_{Ve}, and S_{Vn} correspond to the apoC-III SA in whole, equilibrating, and nonequilibrating pools of VLDL apoC-III, respectively.

Prior to equilibration, no apoC-III radioactivity is expected in HDL, or,

$$S_H = S_{He} = S_{Hn} = 0$$

with S_H, S_{He}, and S_{Hn} denoting the apoC-III SA in whole HDL and in the equilibrating and nonequilibrating pools in HDL, respectively.

After the instantaneous equilibration, the SA in the nonequilibrating pools of VLDL and HDL should not be affected by the *in vitro* incubation,

$$S_{Vn} = S_{Vn} \quad \text{and} \quad S_{Hn} = S_{Hn} = 0$$

The SA in the exchangeable pools of VLDL and HDL will reach a new equilibrium value which reflects f_{PV}, the fraction of the exchangeable mass of apoC-III which is associated with VLDL, i.e.,

$$S_{He} = S_{Ve} = f_{PV} S_{Ve}$$

The overall apoC-III SA in each lipoprotein class is a weighted average of the SA in the equilibrating and nonequilibrating pool.

$$S_V = S_{Ve} f_V + S_{Vn}(1 - f_V)$$

where

$$f_V = \frac{\text{mass of equilibrating VLDL apoC-III}}{\text{total mass of apoC-III in VLDL}}$$

$$S_H = S_{He} f_H$$

where

$$f_H = \frac{\text{mass of equilibrating HDL apoC-III}}{\text{total mass of apoC-III in HDL}}$$

The ratio R of the apoC-III SA between VLDL and HDL as obtained from the decay curves is defined as:

$$R = S_V / S_H$$

[28] Metabolism of Postprandial Lipoproteins

By Alan R. Tall

Overview

Although a significant portion of each day is spent in the postprandial state, there have been relatively few studies of the effects of a meal on lipoprotein metabolism. The absorption and transport of dietary fat lead to significant perturbations of the plasma lipoproteins. The changes reflect both the addition of newly synthesized, triglyceride-rich particles (chylomicrons) to the plasma, and also changes in the other lipoproteins, especially high-density lipoproteins (HDL), probably resulting from the metabolism of chylomicrons. The changes in the $d < 1.006$ fraction (containing chylomicrons) and in the HDL after a fatty meal may be regarded as an important physiological model of the transfer of lipid and apolipoprotein components between the plasma lipoproteins, especially the transfer or exchange of surface and core components between triglyceride-rich lipoproteins and HDL.

Studies of the metabolism of postprandial lipoproteins may shed light on the importance of different particles in the process of atherogenesis. The changes in the plasma lipoproteins that occur in the postprandial state are quite different in different individuals. Otherwise subtle abnormalities of plasma lipoprotein metabolism may be highlighted in the postprandial state. Changes in the postprandial lipoprotein profile in individuals susceptible to atherosclerosis may provide insights into the mechanisms of involvement of apoB-containing particles of HDL in the process of atherogenesis.

This chapter will focus on the metabolism of postprandial lipoproteins in humans, especially the acute response to dietary fat or alcohol. The emphasis will be on the metabolism of chylomicrons and HDL. Studies in animals or *in vitro* will be referred to when they are thought to elucidate human physiology. Methodological aspects of studying postprandial lipoproteins will be discussed. Methodological problems, especially the paucity of nonexchangeable markers of newly synthesized lipoproteins or apolipoproteins, have been a major factor limiting a detailed understanding of the genesis of changes noted in the postprandial plasma lipoprotein profiles.

Study Design

The majority of studies of postprandial lipoproteins in humans have measured sequential changes in the plasma following single or multiple oral fat loads. Triglyceride may be given as corn oil, olive oil, safflower oil, cream, or butter or incorporated into a meal. There are few studies which have compared the acute effects of different types of dietary fat, such as saturated versus polyunsaturated fats. Significant changes in the plasma lipids or apolipoproteins may be observed with doses of 0.75–3.0 g triglycerides/kg body weight. Some subjects experience nausea and diarrhea after ingesting fat alone. When fat is given as part of a meal it is possible that the presence of carbohydrate or protein will alter the subsequent metabolism of triglyceride. There have been few systematic studies of the effects of other dietary components on the response to a triglyceride load. Alcohol has a marked effect on the subsequent metabolism of dietary fat.[1] Alcohol has been given in several forms; there is no information on the effects of the nonalcoholic moieties of different types of alcoholic beverages on postprandial lipoprotein metabolism.

One of the difficulties of studying changes in the lipoproteins after an oral fat load is the variable time between the ingestion of fat and the peak response of the plasma lipoproteins. This variability results primarily from differences in gastric emptying rate. In order to catch the peak changes in plasma lipoproteins it is necessary to take multiple blood samples. The peak in plasma triglycerides usually occurs about 5 to 6 hr after ingestion of 1.5 g/kg of oral fat, and the major alterations in the plasma lipoproteins (such as the HDL) may be observed at 5 to 10 hr. The variability in response necessitates the study of several individuals.

More controlled methods for studying the metabolism of postprandial lipoproteins have involved the intravenous infusion of radiolabeled chylomicrons[2] or intraduodenal fat infusion.[3] These approaches have the advantage that they provide kinetic data for chylomicron removal. Human chylomicrons may be radiolabeled in the protein moiety by the ICl method.[4] Optimally, human chylomicrons can be obtained from thoracic duct lymph. Unfortunately, thoracic duct cannulation is not frequently performed for therapeutic purposes. Human chylomicrons may also be prepared from other sources such as chylous pleural effusions or from the urine of subjects with chyluria. However, the supply of chylomicrons from these sources is irregular and the chylomicrons have probably already undergone extensive alteration due to contact with plasma components.

Grundy and Mok[3] have studied the metabolism of chylomicrons during a constant intraduodenal fat infusion. This method results in a steady rate of fat absorption and constant plasma triglyceride levels and allows the kinetics of triglyceride removal to be determined. The subjects are intubated with a nasogastric tube the night before the study. The next morning the tube is positioned in the region of the ampulla of Vater, under X-ray guidance, then a constant infusion of triglyceride or other lipid (e.g., a monoglyceride/lecithin mixture) is begun. A triple lumen tube may be used to determine the amount of lipid absorbed across a segment of small intestine. In normal subjects, small intestinal triglyceride absorption should be more than 95%. After a 5-hr infusion of triglycerides, the

[1] D. E. Wilson, P. H. Schreibman, A. C. Brewster, and R. A. Arky, *J. Lab. Clin. Med.* **75,** 264 (1970).
[2] P. J. Nestel, *J. Clin. Invest.* **43,** 943 (1964).
[3] S. M. Grundy and H. Y. I. Mok, *Metabolism* **25,** 1225 (1976).
[4] D. W. Bilheimer, S. Eisenberg, and R. I. Levy, *Biochim. Biophys. Acta* **260,** 212 (1972).

plasma triglyceride level becomes constant, and the study is begun. With steady plasma levels the removal rate of triglyceride is equal to the infusion rate (mg/kg/hr), assuming 100% absorption. From the turnover rate and the increment of chylomicron triglyceride, the $t_{1/2}$ of chylomicron triglyceride can be determined as follows: $t_{1/2} = 0.693/K$, where $K =$ input of chylomicron TG (mg/kg/hr)/plasma pool of chylomicron TG. The latter is determined from the plasma triglyceride concentration and the estimated plasma volume ($= 927 + 31.47 \times$ body wt).

The prior state of the patient is an important variable in determining the metabolism of postprandial lipoproteins. In normal subjects plasma triglycerides may continue to fall significantly during the morning after a fast from 11 PM; fasting for a full 14 hr before the onset of the study is advisable. The patient's prior metabolic state has an important influence on the response to a meal. Weight loss, alcohol intake,[1] exercise,[5] and probably a variety of other variables may influence the individual's response to a fatty meal. Ideally, studies should be undertaken in the metabolic ward setting where caloric intake, diet composition, and exercise are all well-controlled variables. However, valuable information has been obtained under less rigorously controlled conditions.

Preexisting hyperlipidemia has a large influence on the changes in plasma lipoproteins after ingestion of a fatty meal. Although several studies have contrasted the responses of broad groups of hyperlipidemic patients with those of normal controls, there is a paucity of information on the response of patients with genetically or otherwise well-defined hyperlipidemias. Even patients who are defined as normolipidemic, on the basis of a plasma triglyceride level of less than 200 mg/dl, have quite different responses depending on their preexisting plasma triglyceride level. This will be discussed in detail below.

Preparation of Lipoproteins

Chylomicrons are generally isolated as plasma lipoproteins of $S_f >$ 400, as described.[6] There is evidence that triglyceride-rich lipoproteins of $S_f < 400$ may make a lesser contribution to the hypertriglyceridemia following a fatty meal,[6] especially during infusion of lecithin.[7] Plasma lipoproteins in the postprandial state may be isolated by procedures reviewed elsewhere in this volume. It is worth noting that during alimentary

[5] J. R. Patsch, J. B. Karlin, L. W. Scott, L. C. Smith, and A. M. Gotto, *Proc. Natl. Acad. Sci. U.S.A.* **80,** 1449 (1983).
[6] T. G. Redgrave and L. A. Carlson, *J. Lipid Res.* **20,** 217 (1979).
[7] F. U. Beil and S. M. Grundy, *J. Lipid Res.* **21,** 525 (1980).

lipemia lipid enrichment of HDL species results in significant alterations of the density of the major subclasses of HDL.[8] It appears that a significant portion of lipid-enriched HDL-3 particles float at d 1.125 g/ml.[8] Thus, the changes in HDL may in general be best defined by methods which make no assumptions about traditional HDL-2 or HDL-3 designations, such as isopycnic density gradient techniques. In the future it is likely that valuable information about changes in HDL subclasses will be obtained by noncentrifugal methods, such as those employing antibody affinity chromatography for isolation of apoA-I-containing particles from plasma.[9]

A major problem in the study of the metabolism of postprandial lipoproteins has been the paucity of nonexchangeable lipid or protein markers to denote the fate of individual lipoprotein particles during lipemia. Although lipid core markers have been employed as a means of following chylomicron metabolism, triglycerides, cholesteryl esters, and retinyl esters are all susceptible to cholesteryl ester transfer protein-mediated exchange with other lipoprotein core lipids.[10] Although these lipids may be used to follow the fate of chylomicron particles in animals which have low lipid transfer protein activity (dogs, pigs, and rats), they are not reliable particle markers in species which have significant cholesteryl ester transfer protein activity (humans, rabbits, and monkeys). Estimates of the extent of core lipid exchange based on incubations of plasma may underestimate physiological exchange rates, since lipolysis enhances the activity of the cholesteryl ester transfer protein.[11] Thus, the redistribution of chylomicron retinyl ester radioactivity into $S_f < 400$ lipoprotein classes in intact animals cannot be taken as evidence that chylomicron remnants contribute to $S_f < 400$ particles. The recent recognition that the low-molecular-weight form of apoB is a nonexchangeable marker for intestinally derived particles[12] has great potential in allowing intestinally derived particles to be traced. For example, this has allowed identification of an intestinally derived subfraction of β-VLDL which is particularly efficacious in delivering cholesterol to macrophages.[13] In future studies mea-

[8] A. R. Tall, C. B. Blum, G. P. Forester, and C. Nelson, *J. Biol. Chem.* **257**, 198 (1982).
[9] J. P. McVicar, S. T. Kunitake, R. L. Hamilton, and J. P. Kane, *Proc. Natl. Acad. Sci. U.S.A.* **81**, 1356 (1984).
[10] R. E. Morton and D. B. Zilversmit, *J. Lipid Res.* **23**, 1058 (1982).
[11] A. R. Tall, D. Sammett, G. Vita, R. J. Deckelbaum, and T. Olivecrona, *J. Biol. Chem.* **259**, 9587 (1984).
[12] J. P. Kane, D. A. Hardman, and H. E. Paulus, *Proc. Natl. Acad. Sci. U.S.A.* **77**, 2465 (1980).
[13] M. R. Fainaru, R. W. Mahley, R. L. Hamilton, and T. L. Innerarity, *J. Lipid Res.* **23**, 702 (1982).

surements of the low-molecular-weight apoB (particularly as specific immunochemical methods become available) should serve to determine if accumulating particles after a fatty meal are intestinally derived or hepatogenous. Also, the characterization of intestinally derived particles should provide insights into their role in atherogenesis.

The lack of a marker for newly synthesized apoA-I has limited the understanding of how apoA-I becomes incorporated into the HDL fraction during lipemia. In humans plasma and HDL apoA-I rise after a fatty meal, but the incremental apoA-I has not been distinguishable from the preexisting apoA-I. A potentially useful marker of newly synthesized apoA-I might be its isoprotein-2, which is the secretory proprotein of apoA-I.[14] The isoprotein-2 has a half-life of several hours in the plasma before being converted to the more basic isoproteins 4 and 5.[15] Analysis of apoA-I isoforms in the HDL fraction after a fatty meal might provide some clue as to how newly synthesized apoA-I becomes incorporated into plasma HDL particles.

Results of Studies

Overview of Chylomicron Metabolism Based on in Vitro Studies and Studies in Experimental Animals

Absorbed dietary fat is packaged into chylomicron particles containing a core of triglyceride and a surface film of phospholipids, cholesterol, apoB, and soluble apolipoproteins (apoA-I, apoA-II, apoA-IV, and apoC-III). The chylomicron triglyceride is rapidly hydrolyzed in peripheral tissues due to the action of lipoprotein lipase, a triglyceride hydrolase located on the surface of the capillary endothelium. Fatty acids and partial glycerides enter the tissues or are bound to albumin. During the process of lipolysis of chylomicrons certain surface materials (phospholipids, apoA-I, and apoA-II) are transferred into the HDL fraction. Conversely, apoE, apoC peptides, and cholesteryl esters are transferred from HDL to chylomicron remnants. Thus, relative to nascent chylomicrons, chylomicron remnants are enriched in cholesteryl esters and apoE and apoC peptides (especially apoC-III), and depleted in apoA-I and apoA-IV; recently we have found that chylomicron and VLDL remnants also bind cholesteryl ester transfer protein. The chylomicron remnant is rapidly

---

[14] V. I. Zannis, D. M. Kurnit, and J. L. Breslow, *J. Biol. Chem.* **257**, 536 (1982).

[15] G. Ghiselli, E. J. Schaefer, S. Law, J. A. Light, and H. B. Brewer, *Clin. Res.* **31**, 500A (1983).

removed from the circulation owing to the action of a hepatic chylomicron remnant or apoE receptor.[16]

Studies of Postprandial Lipoproteins in Humans—Chylomicron Metabolism

Using intravenously injected thoracic duct chylomicrons, Nestel showed that the half-life of chylomicron triglyceride in the circulation was about 5 min.[2] Grundy and Mok[3] obtained similar results using an intraduodenal infusion method. Nestel[2] showed a strong correlation between fasting triglyceride levels and the height of the response of plasma triglyceride to a fatty meal. There was an inverse correlation between the level of fasting plasma triglycerides and the rate of chylomicron metabolism both in hypertriglyceridemia (>200 mg/dl) and also within the normal range of plasma triglycerides (<200 mg/dl). Grundy and Mok[3] speculated that the slower removal of chylomicron triglyceride in subjects with fasting hypertriglyceridemia reflected competition of the chylomicrons with VLDL for removal mechanisms rather than a basic defect in the ability to remove chylomicrons, because with weight loss and a fall in the plasma triglyceride level, most subjects showed a marked improvement in chylomicron clearance. Thus, in many patients with hypertriglyceridemia a marked alimentary lipemic response may reflect a tendency to overproduce VLDL, rather than a primary defect in clearance of chylomicrons. Exceptions to this statement are patients with lipoprotein lipase or apoC-II deficiency and perhaps a minority of other patients who have a predominant clearance defect.[3]

Redgrave and Carlson[6] contrasted the response to lipemia of hypertriglyceridemic and normal subjects. They found that the more pronounced lipemia of hypertriglyceridemic subjects occurred in $S_f > 400$ particles, and was largely due to presence of particles of increased size in these subjects. It is notable that in several studies increases in number or size of triglyceride-rich lipoproteins of both $S_f > 400$ and $S_f < 400$ have been noted during alimentary lipemia.[3,6] It is uncertain if this represents continued degradation of chylomicron remnants to smaller particle size, or competition between chylomicron remnants and hepatogenous VLDL for removal sites, resulting in accumulation of the latter, or increased hepatic VLDL secretion due to the load of dietary fat returning to the liver. It should be possible to determine if the increase in $S_f < 400$ parti-

[16] R. W. Mahley, D. Y. Hui, T. L. Innerarity, and K. H. Weisgraber, *J. Clin. Invest.* **68,** 1197 (1981).

cles is due to hepatic or intestinal particles by determining the distribution of the low-molecular-weight apoB particle in the postprandial VLDL fraction. In patients infused with lecithin the major increase in triglyceride-carrying particles in the plasma occurred in the S_f 20–400 fraction, possibly due to absorption of lecithin over a longer length of small intestine than triglyceride,[7] (distal small intestinal fat absorption tends to result in formation of smaller chylomicrons). Another possibility is that lecithin fatty acid was predominantly absorbed via the portal vein and that the S_f 20–400 particles were, in fact, hepatogenous VLDL. Thus, both the mode of intestinal absorption and the subject's baseline lipid metabolism influence the response of triglyceride-rich lipoproteins during lipemia.

Changes in HDL. Havel *et al.*[17] noted an increase in HDL phospholipids and protein during alimentary lipemia and speculated that these changes might reflect transfer of chylomicron surface materials into the HDL fraction. These workers also showed that during lipemia there is transfer of apoC peptides, including the activator peptide of lipoprotein lipase, apoC-II, from HDL to the chylomicron remnant fraction. More recently, we have performed detailed studies of the changes in HDL subfractions during alimentary lipemia, using primarily density gradient methods to separate HDL subclasses. These studies showed an increase in HDL phospholipids, cholesteryl esters, apoA-I, and apoA-II during lipemia; these changes were thought to reflect at least in part the transfer of these lipids and proteins from chylomicrons into HDL, as demonstrated in the rat.[8] Another potential mechanism could be increased intestinal HDL secretion during lipemia. Analysis of HDL by equilibrium density gradient ultracentrifugation showed the presence of a major and a lesser peak of HDL in fasting plasma, corresponding to the traditional HDL-3 and HDL-2 subclasses, respectively. During lipemia the most pronounced change was an increase in mass and a shift to lower density of the major peak of HDL-3, with similar but less marked changes in the HDL-2 peak. Analyses of molecular compositions suggested that the lipid transferred into HDL was added to preexisting particles, since the particles retained the same basic complement of apoA-I and apoA-II molecules but had additional lipid molecules (especially phospholipids). Although there was an increase in mass of HDL apoA-I and apoA-II the particle contents of apoA-I and apoA-II were unchanged during lipemia, suggesting that apolipoprotein transfer causes interconversion of existing HDL species, or formation of new particles with the same content of apoA-I and apoA-II as existing species.

An interesting recent report[5] has shown that in a group of normolipi-

---

[17] R. J. Havel, J. P. Kane, and M. L. Kashyap, *J. Clin. Invest.* **52**, 32 (1973).

demic healthy young adults there was an inverse relationship between the magnitude of the plasma triglyceridemic response and the levels of *fasting* HDL-2 ($r = -0.86$), HDL–cholesterol ($r = -0.61$), and apoA-I ($r = -0.46$). In two subjects followed for 3 years, when HDL-2 levels rose or fell in response to exercise or the lack of it, the triglyceridemic response decreased or increased, respectively.

Potential Mechanisms of Lipid Transfer Processes during Lipemia

In Vitro Studies of Lipid Exchange Processes. Our recent *in vitro* studies suggest that lipid exchange processes may be stimulated during lipemia.[11] These studies show that lipolysis enhances the cholesteryl ester transfer protein-mediated movement of cholesteryl esters from HDL to triglyceride-rich lipoproteins. The latter were isolated from both fasting and lipemic plasma.[18] The stimulation of cholesteryl ester transfer activity was related to the accumulation of lipolytic products, especially fatty acids in the remnant particle. It appears that small amounts of negatively charged fatty acids enhance the binding of the cholesteryl ester transfer protein to the remnant surface. Under several different circumstances enhancement or reduction of cholesteryl ester transfer activity was related to increased or decreased binding of cholesteryl ester transfer protein to VLDL or VLDL remnants, suggesting that the increased binding resulted in increased cholesteryl ester transfer into the triglyceride-rich particles. There is also theoretical evidence to support the concept that increased affinity of cholesteryl ester transfer protein for acceptor particles relative to HDL should increase cholesteryl ester transfer from HDL to the acceptor particles.[19] The enhancement of cholesteryl ester transfer processes by lipolysis was greatly favored by high VLDL/HDL ratios. Thus, cholesteryl ester transfer processes should be more pronounced in subjects who experience higher levels of plasma triglyceride during lipemia. In summary, these *in vitro* studies suggest that cholesteryl ester transfer from HDL to triglyceride-rich particles during lipemia may be related to activities of lipoprotein or hepatic lipases, and to the magnitude of the rise in plasma triglyceride levels.

In other studies we have shown that the transfer of phospholipids from triglyceride-rich lipoproteins to HDL is enhanced by a partially purified plasma phospholipid transfer protein.[18] This phospholipid transfer protein appears to be a different protein than the lipid transfer protein referred to above.[20] However, the phospholipid transfer protein-mediated transfer of

[18] A. R. Tall, unpublished results, 1984.
[19] P. J. Barter and M. E. Jones, *J. Lipid Res.* **21**, 238 (1980).
[20] A. R. Tall, E. Abreu, and J. Shuman, *J. Biol. Chem.* **258**, 2174 (1983).

phospholipids from VLDL to HDL is also stimulated by lipolysis, suggesting common principles of action.

Mechanisms of Lipid Changes during Lipemia in Vivo. The changes in HDL are thought to result from influx of chylomicron phospholipids, apoA-I, and apoA-II. In incubations of plasma it has been shown that enrichment of HDL with phospholipids is followed by influx of cholesterol from other lipoproteins.[21] The accumulation of phospholipids and cholesterol probably leads to cholesteryl ester formation as a result of activity of lecithin : cholesterol acyltransferase. Within HDL there is a major redistribution of cholesteryl ester mass toward particles of lower density, reflecting in large part the increase in HDL phospholipids.

Different studies have reported that total HDL cholesteryl ester may be increased,[8] unchanged,[17] or lowered[22] during alimentary lipemia. It is possible that the net change in HDL cholesteryl esters is determined by the magnitude of the rise in the plasma triglycerides. Thus, individuals who experience the most profound hypertriglyceridemia may also display the greatest fall in HDL cholesterol during lipemia, owing to lipid transfer protein-mediated exchange of HDL cholesteryl esters with the triglycerides of chylomicron remnants and VLDL. Consistent with this lipid exchange hypothesis is the finding that changes in HDL-2 and HDL-3 cholesteryl esters (which varied from about +15 to −20%) showed a strong negative correlation with changes in triglycerides in the same lipoproteins ($r = -0.85$, -0.88, respectively)[23]; the changes in HDL triglycerides tended to parallel those of plasma triglycerides.

During lipemia there is an approximate 40% increase in lecithin : cholesterol acyltransferase activity, as revealed by incubation of lipemic plasma. The increase in LCAT activity was correlated with the magnitude of rise in the plasma triglycerides.[23] The increase in LCAT activity could reflect the greater availability of substrate (phospholipids and cholesterol) owing to influx of materials into HDL, or the removal of the inhibitory product lipid, cholesteryl esters, owing to lipid exchange processes; however, changes in LCAT activity did not seem to correlate with these processes.[23] An alternative explanation may be related to the recent observations of Barter, which provide evidence that LCAT may act directly on non-HDL lipoproteins.[24] It is conceivable that LCAT binds directly to triglyceride-rich lipoproteins in lipemic plasma, resulting in an increase in LCAT activity. Further support for this concept is derived from the fact that LCAT circulates in a complex with cholesteryl ester transfer pro-

[21] A. R. Tall and P. H. R. Green, *J. Biol. Chem.* **256,** 2035 (1981).
[22] R. M. Kay, S. Rao, C. Arnott, N. E. Miller, and B. Lewis, *Atherosclerosis* **36,** 567 (1980).
[23] H. G. Rose and J. Juliano, *J. Lipid Res.* **20,** 399 (1979).
[24] P. J. Barter, *Biochim. Biophys. Acta* **751,** 261 (1983).

tein.[25] The latter has been shown to bind to triglyceride-rich particles as they undergo lipolysis.[18]

The explanation of the inverse relationship between the magnitude of alimentary lipemia and the levels of fasting HDL-2 and HDL cholesterol levels[5] could be related to some of the above suggestions about lipid transfer and exchange mechanisms. First, the efficiency of clearance of both chylomicrons and VLDL could be related to the quantities of chylomicron phospholipid, cholesterol, and apoA-I transferred into HDL. A second and perhaps more likely explanation is related to the magnitude of cholesteryl ester exchange processes that occur during alimentary hypertriglyceridemia. We have suggested that transfer of cholesteryl esters from HDL to triglyceride-rich lipoproteins may be more pronounced the more plasma triglycerides rise. Since lipemia occurs several times a day, this could have an impact on fasting HDL–cholesterol levels. Also, the transfer of HDL cholesteryl esters into fasting VLDL could be more marked in individuals experiencing greater lipemia, since those who show larger rises in plasma triglycerides during lipemia also have higher fasting plasma triglycerides.[2] A third possibility might be related to the distribution of LCAT activity in lipemic plasma. Individuals with more pronounced hypertriglyceridemia might tend to form cholesteryl esters at the surface of triglyceride-rich particles as they undergo lipolysis, rather than in HDL, with subsequent lipid transfer protein-mediated transfer of cholesteryl ester into the core of the triglyceride-rich lipoprotein. It is evident that these several putative mechanisms may be interrelated.

The inverse correlation between the magnitude of alimentary lipemia and fasting HDL-2 levels may also have its origin in cholesteryl ester–triglyceride exchange processes, or in the distribution of LCAT activity. Thus, exchange of HDL cholesteryl esters with chylomicron triglycerides may result in a triglyceride-enriched HDL-2 particle. *In vitro* the triglyceride-rich HDL-2 can be remodeled into a more dense HDL-3 as a result of lipoprotein lipase activity.[26] Thus, the triglyceride enrichment of HDL-2 that occurs during alimentary lipemia may be an important step in the remodeling of HDL-2 to HDL-3. This remodeling into smaller HDL particles may be more pronounced in individuals who have more marked alimentary hypertriglyceridemia, accounting for their lower mass of HDL-2. Since HDL-2 formation may be dependent on LCAT activity, it is evident that a competitive distribution of LCAT activity between triglyceride-rich lipoproteins and HDL could also affect HDL-2 levels.

[25] P. E. Fielding and C. J. Fielding, *Proc. Natl. Acad. Sci. U.S.A.* **77**, 3327 (1980).
[26] R. J. Deckelbaum, S. Eisenberg, Y. Oschry, M. Cooper, and C. Blum, *J. Lipid Res.* **23**, 1274 (1982).

Further support for a metabolic relationship between triglyceride-rich lipoproteins and HDL-2 levels has been derived from studies of patients with abetalipoproteinemia. In patients with classical abetalipoproteinemia, who lack chylomicrons and VLDL, the major HDL particle is a cholesteryl ester-enriched HDL-2.[26] However, in a patient with normotriglyceridemic abetalipoproteinemia, the major HDL was HDL-3.[27] This patient appeared to make intestinal but not hepatic apoB and thus experienced alimentary lipemia, but lacked hepatogenous VLDL. The difference in HDL in the two types of abetalipoproteinemia implies that the alimentary lipemic process is important in the remodeling of HDL-2 into HDL-3.

Effects of Alcohol

In different studies the acute administration of a single dose of alcohol has led to a variable, small increase in plasma VLDL-triglyceride levels. The changes seem to be more pronounced with larger doses of alcohol and in subjects with higher fasting triglyceride levels.[1,28,29] In a study of normal subjects with mean fasting plasma triglycerides of 40 to 70 mg/dl we found no consistent change in VLDL triglycerides after acute ingestion of 100 g alcohol (as whiskey).[29] However, VLDL phospholipids were markedly increased. Also, HDL-3 and HDL-2a phospholipid levels were increased. These changes in the lipoproteins were associated with a mean 33% decrease in hepatic lipase activity, as determined in postheparin plasma. Given the evidence in experimental animals that hepatic lipase may be involved in the catabolism of HDL and possibly VLDL phospholipids, it appears likely that the changes in the plasma lipoproteins may have reflected in part the inhibition of hepatic lipase activity. It is also possible that increased VLDL flux, without a rise in plasma triglycerides, could lead to increased transfer of VLDL phospholipids into HDL.

Several studies have shown that ingestion of alcohol prior to or with a fat load results in a more pronounced increase in plasma triglyceride level than occurs after ingestion of either alcohol alone or fat alone.[1] The magnitude of this synergistic effect is more pronounced the higher the subject's baseline plasma triglyceride levels. Compared to the effects of fat alone, the synergistic effects of fat and alcohol result in a later and more

[27] M. J. Malloy, J. P. Kane, D. A. Hardman, R. L. Hamilton, and K. B. Dalal, *J. Clin. Invest.* **67,** 1441 (1981).

[28] P. Avogaro and G. Cazzolato, *Metabolism* **24,** 1231 (1975).

[29] C. S. Goldberg, A. R. Tall, and S. Krumholz, *J. Lipid Res.* **25,** 714 (1984).

sustained rise in plasma triglycerides; also, the increment is due to greater levels of $S_f < 400$ particles. It is not known if the increment in $S_f < 400$ triglycerides represents delayed clearance of chylomicron remnants, or greater incorporation of dietary fat and fatty acids into hepatogenous VLDL. Experimental evidence in rats has suggested both greater hepatic secretion of VLDL[30] and also impaired clearance of chylomicron remnants when a fatty meal is given to alcohol-treated animals.[31] Studies in humans in which the low-molecular-weight apoB is quantitated in the $d <$ 1.006 fraction should be able to resolve this question. The combined effects of alcohol and fat on HDL and LDL lipid levels have not been reported.

Significance of the Changes in Postprandial Lipoproteins

The metabolism of chylomicron remnants could have direct relevance to atherogenesis. Zilversmit[32] postulated that chylomicron remnants bearing dietary cholesterol might deposit cholesterol in arterial tissues. Lipoprotein lipase present in arterial tissues could lead to *in situ* remnant formation and movement of remnants or remnant lipids into arterial tissues.[32] There is evidence that perfusion of cholesterol-enriched chylomicrons through the rat heart leads to nonendocytotic transfer of cholesteryl esters into the coronary artery tissue.[33] Tissue culture studies indicate that cholesterol-enriched β-VLDL of intestinal origin (presumably chylomicron remnants) are particularly active in causing cholesterol deposition in macrophages, promoting foam cell formation. Internalization of these particles seems to depend on receptor-mediated endocytosis mediated by the β-VLDL receptor.[13]

Since subjects with low HDL–cholesterol and low HDL-2 levels have a more pronounced lipemia, it is possible that the longer circulation of remnants leads to greater arterial deposition of atherogenic chylomicron remnants.[5] Thus, low fasting HDL–cholesterol levels could be a marker of a propensity to form atherogenic chylomicron remnants.[5]

A more generalized interpretation of the potential significance of the inverse relationship between levels of alimentary lipemia and HDL levels is suggested in this chapter. Subjects with a more pronounced lipemic response may have a tendency to drop their HDL–cholesterol levels dur-

[30] E. Baraona and C. S. Lieber, *J. Clin. Invest.* **49**, 769 (1970).
[31] T. G. Redgrave and G. Marton, *Atherosclerosis* **28**, 69 (1977).
[32] D. B. Zilversmit, *Circulation* **60**, 473 (1979).
[33] C. J. Fielding, *J. Clin. Invest.* **62**, 141 (1978).

ing lipemia and to have lower fasting HDL–cholesterol levels; the more pronounced lipemic response may result from increased VLDL apoB or triglyceride production, on a genetic basis (e.g., in familial combined hyperlipidemia) or on an acquired basis (e.g., in obesity). Thus, the lipemic changes in HDL cholesterol may mark the tendency to overproduce apoB-containing particles. The resulting chylomicron or VLDL remnants or LDL may all be atherogenic.

It should be noted that this hypothesis is based on an opposite interpretation of the significance of the HDL–cholesterol to that made in the reverse cholesterol transport theory. Thus, subjects who show greatest transfer of HDL cholesteryl esters to remnant particles would be at greatest risk for atherogenesis. By contrast, one version of the reverse cholesterol transport theory suggests that individuals with high HDL–cholesterol levels would transport more cholesterol to VLDL or chylomicron remnants, because this step allows a path for exit of cholesterol from the plasma. Further experiments aimed at measuring the transfer of cholesteryl esters between HDL and lipemic particles may be helpful in evaluating these different theories of the relationship between HDL and atherogenesis.

Acknowledgments

Supported by NIH Grants HL 22682 and 21006. Author is an Established Investigator of the A.H.A.

[29] *In Vivo* Metabolism of Apolipoprotein E in Humans

By Richard E. Gregg and H. Bryan Brewer, Jr.

Apolipoprotein E (apoE), a glycoprotein of M_r 34,000, is associated predominantly with very-low-density lipoproteins (VLDL) and high-density lipoproteins (HDL) in human plasma.[1,2] Human apoE is a polymorphic protein[3] with multiple alleles inherited in a codominant fashion at a

[1] C. B. Blum, L. Aron, and R. Sciacca, *J. Clin. Invest.* **66**, 1240 (1980).
[2] R. J. Havel, L. Kotite, J. L. Vigne, J. P. Kane, P. Tun, N. Phillips, and G. C. Chen, *J. Clin. Invest.* **66**, 1351 (1980).
[3] G. Utermann, M. Jaeschke, and J. Menzel, *FEBS Lett.* **56**, 352 (1975).

single genetic locus.[4-10] There are three common alleles in the population that code for the three major isoproteins of apoE present in normal subjects. The alleles are designated ε-2, ε-3, and ε-4 with the corresponding proteins designated apoE$_2$, apoE$_3$, and apoE$_4$.[11] In addition, there are multiple other rare alleles for apoE that code for apoE$_1$,[7,8] additional forms of apoE$_2$[5,6] (that differ in primary amino acid sequence from the common apoE$_2$ isoform), apoE$_5$,[9] and apoE deficiency.[10] ApoE$_1$, apoE$_2$, and apoE deficiency are associated with type III hyperlipoproteinemia while apoE$_4$[12] and apoE$_5$[9] are associated with type V and type II hyperlipoproteinemia, respectively. All of these dyslipoproteinemias are associated with elevations of plasma concentrations of triglyceride-rich lipoprotein particles or remnants of these particles.

ApoE has been demonstrated to bind *in vitro* to the high-affinity apoB,E (LDL) receptor on fibroblasts and to a hepatocyte apoE receptor.[13] ApoE has been proposed to be the recognition site on chylomicrons for the receptor-mediated removal of chylomicron remnants by the liver.[13,14] In addition, apoE has been reported to both inhibit[15,16] and activate[16,17] lipoprotein lipase. Whether one of these functions or some other as yet unrecognized function of apoE is the major physiological role for apoE in lipoprotein metabolism is unknown.

[4] S. C. Rall, Jr., K. H. Weisgraber, and R. W. Mahley, *J. Biol. Chem.* **257**, 4171 (1982).

[5] S. C. Rall, Jr., K. H. Weisgraber, T. L. Innerarity, and R. W. Mahley, *Proc. Natl. Acad. Sci. U.S.A.* **79**, 4696 (1982).

[6] S. C. Rall, Jr., K. H. Weisgraber, T. L. Innerarity, T. P. Bersot, R. W. Mahley, and C. B. Blum, *J. Clin. Invest.* **72**, 1288 (1983).

[7] R. E. Gregg, G. Ghiselli, and H. B. Brewer, Jr., *J. Clin. Endocrinol. Metab.* **57**, 968 (1983).

[8] K. H. Weisgraber, S. C. Rall, Jr., T. L. Innerarity, R. W. Mahley, T. Kuusi, and C. Ehnholm, *J. Clin. Invest.* **73**, 1024 (1984).

[9] T. Yamamura, A. Yamamoto, K. Hiramori, and S. Nambu, *Atherosclerosis* **50**, 159 (1984).

[10] G. Ghiselli, E. J. Schaefer, P. Gascon, and H. B. Brewer, Jr., *Science* **214**, 1239 (1981).

[11] V. I. Zannis, J. L. Breslow, G. Utermann, R. W. Mahley, K. H. Weisgraber, R. J. Havel, J. L. Goldstein, M. S. Brown, G. Schonfeld, W. R. Hazzard, and C. Blum, *J. Lipid Res.* **23**, 911 (1982).

[12] G. Ghiselli, E. J. Schaefer, L. A. Zech, R. E. Gregg, and H. B. Brewer, Jr., *J. Clin. Invest.* **70**, 474 (1982).

[13] R. W. Mahley, D. Y. Hui, T. L. Innerarity, and K. H. Weisgraber, *J. Clin. Invest.* **68**, 1197 (1981).

[14] F. Shelburne, J. Hanks, W. Meyers, and S. Quarfordt, *J. Clin. Invest.* **65**, 652 (1980).

[15] D. Ganesan, H. B. Bass, W. J. McConathy, and P. Alaupovic, *Metabolism* **25**, 1189 (1976).

[16] S. H. Quarfordt, H. Hilderman, M. R. Greenfield, and F. A. Shelburne, *Biochem. Biophys. Res. Commun.* **78**, 302 (1977).

[17] N. Yamada and T. Murase, *Biochem. Biophys. Res. Commun.* **94**, 710 (1980).

Insights into the function and metabolism of apoE may be ascertained by the investigation of the metabolism of apoE or apoE-containing lipoproteins utilizing cellular and molecular biological techniques *in vitro* and by determining the *in vivo* metabolism of apoE and apoE-containing lipoproteins. Methods for evaluating the *in vitro* metabolism of apoE and apoE-containing lipoproteins and the *in vivo* metabolism of lipoprotein particles are discussed elsewhere in this volume. We will describe the methods for investigating the *in vivo* metabolism of apoE in humans. This discussion will be focused on the study of steady-state kinetics of apoE metabolism and will not deal with the acute alterations of apoE distribution and metabolism induced by meals and similar other non-steady-state phenomenon.[18] Methods for radiolabeling apoE, the preparation of the subjects for the study, the techniques for performing kinetic studies, and the analysis of the data will be presented.

Radioiodination of ApoE for Tracer Studies

ApoE may be radiolabeled by many different methods, only a few of which have been extensively investigated. The most common tracers used for labeling have been radioactive iodine isotopes (^{125}I and ^{131}I). The radioactive iodine has been covalently bound to the apolipoprotein utilizing the Bolton–Hunter reagent[19] and by oxidative iodination of tyrosines.[20–22] The Bolton–Hunter reagent method has the advantage that it uses milder conditions for the attachment of the radioactive iodine to the lysine residues. It has the disadvantage that the iodine is attached to a prosthetic group which when bound to the protein forms a large bulky side chain which may be associated with greater disruption of the secondary and tertiary structure of the apolipoprotein than an iodine atom alone. In addition, it has been proposed that the active site of apoE is rich in lysines,[4] and therefore the Bolton–Hunter reagent may be preferentially attached to this site, reducing the interaction of apoE with its putative cell surface receptor. The Bolton–Hunter reagent, however, has been extensively utilized and has been successfully employed in *in vitro* binding assays[13,23] (see Chapter [33] in this volume).

[18] C. Blum, *J. Lipid Res.* **23**, 1308 (1982).
[19] A. E. Bolton and W. M. Hunter, *Biochem. J.* **133**, 529 (1973).
[20] P. Freychet, J. Roth, and D. M. Neville, Jr., *Biochem. Biophys. Res. Commun.* **43**, 400 (1971).
[21] J. J. Marchalonis, *Biochem. J.* **113**, 299 (1969).
[22] A. S. McFarlane, *Nature (London)* **182**, 53 (1958).
[23] T. L. Innerarity, E. J. Friedlander, S. C. Rall, Jr., K. H. Weisgraber, and R. W. Mahley, *J. Biol. Chem.* **258**, 12341 (1983).

We have limited our experience to the oxidative iodination of tyrosines on apoE. We have evaluated the properties of apoE following radioiodination utilizing chloramine-T, lactoperoxidase, and iodine monochloride.[24] The chloramine-T[20] and lactoperoxidase[21] iodination of apoE were performed by standard methods and the iodine monochloride iodination method[22] is described below.

ApoE was isolated by the ultracentrifugal separation of VLDL, delipidation with chloroform–methanol, fractionation by heparin affinity column chromatography, and final purification by Sephacryl S-200 column chromatography.[24] The apoE was stored at $-20°$ in 0.1 N acetic acid at a concentration of approximately 1 mg/ml. For iodination, 20 to 100 μg of apoE was pipetted into a 3-ml glass conical test tube and lyophilized. The lyophilized apoE was redissolved in a freshly prepared and sterile filtered buffer containing 6 M guanidine-HCl and 1 M glycine (pH 8.5) in the small conical test tube such that the final concentration of apoE was 2 μg/μl. One to 5 mCi of carrier-free sodium[^{125}I]iodide or sodium[^{131}I]iodide (NEZ-033L or NEZ-035H, New England Nuclear, Boston, MA) was added and the sample vortexed. Three to 15 μl of 0.33 mM iodine monochloride in 150 mM NaCl was added slowly with a micropipet over 10–15 sec while vortexing the sample. ApoE was iodinated with an efficiency of 35–50% as determined by trichloroacetic acid precipitation with incorporation of approximately 0.5 mol of iodine/mol of apoE.

Characterization of Radiolabeled ApoE

The apoE has been extensively evaluated following iodination. ApoE radioiodinated by the three different techniques was evaluated by two methods: (1) its ability to bind *in vitro* to human lipoprotein particles and (2) its ability to be precipitated by a monospecific polyclonal rabbit anti-apoE antibody.

Radiolabeled apoE was incubated with plasma from a normal fasting subject at a concentration of 1 to 2 μl of the iodination solution/1 ml of plasma for 30 min at 37°, the density of the plasma adjusted to 1.21 g/ml with solid KBr, and the sample centrifuged in a Beckman 60 Ti rotor (Beckman Instruments, Inc., Palo Alto, CA) for 2.2 × 10^8 g-min. The 1.21 g/ml supernatant was separated by tube slicing and dialyzed against 100 vol × 5 of a sterile solution of 150 mM NaCl, 0.1 M Tris-HCl (pH 7.4), and 0.01% EDTA.

---

[24] R. E. Gregg, D. Wilson, E. Rubalcaba, R. Ronan, and H. B. Brewer, Jr., *in* "Proceedings of the Workshop on Apolipoprotein Quantitation" (K. Lippel, ed.), p. 383. DHHS, NIH Publ. No. 83-1266, U.S. Govt. Printing Office, Washington, D.C.

Ten to 20% of apoE iodinated by the chloramine-T and lactoperoxidase methods was associated with lipoproteins while 80 to 90% was in the lipoprotein-free infranatant following ultracentrifugation. In contrast, 40 to 60% of the iodine monochloride-iodinated apoE was associated with lipoproteins and 40 to 60% was in the nonlipoprotein fraction.

The immunoprecipitability of radioiodinated apoE was evaluated as previously described.[24] ApoE iodinated by the iodine monochloride method gave the highest percentage of precipitation (Fig. 1). These results were interpreted as indicating that apoE iodinated by this technique had the least effect on the immunological properties of the apoE. Based on these results, we concluded that the iodine monochloride method of iodination of apoE was superior to the chloramine-T or lactoperoxidase methods for metabolic studies.

Physicochemical and Lipid-Binding Properties of the Iodine Monochloride-Radiolabeled ApoE

The effect of radioiodination on the electrophoretic properties of apoE was evaluated by sodium dodecyl sulfate gel electrophoresis (SDS–PAGE) and isoelectrofocusing.[25] Greater than 90% of the radioactivity comigrated with the apoE band on SDS–PAGE, and the radioactivity associated with the apoE isoforms was directly proportional to the relative quantity of each isoform on isoelectrofocusing.

The distribution of apoE among plasma lipoproteins is determined by the laws of mass action. A similar distribution of radiolabeled apoE and unlabeled apoE with constant specific activities among the plasma lipoproteins would indicate that iodination had no significant effect on the protein- and lipid-binding properties of apoE.

The distribution of radiolabeled apoE$_3$ was initially determined *in vitro* by incubation of radioiodinated apoE with plasma from normal subjects.[25] The plasma lipoproteins were separated by ultracentrifugation, and the apoE radioactivity and apoE mass quantitated. ApoE was present primarily within VLDL and HDL (Table I). The specific activities of each subfraction were determined and normalized with the plasma specific activity being one. The normalized specific activities did not significantly vary from one except for intermediate density lipoproteins ((IDL), which comprised less than 5% of the total radioactivity in plasma (Table II).

The *in vivo* distribution of radiolabeled apoE was also determined following the separation of plasma by gel permeation chromatography.[25]

[25] R. E. Gregg, L. A. Zech, E. J. Schaefer, and H. B. Brewer, Jr., *J. Lipid Res.* **25**, 1167 (1984).

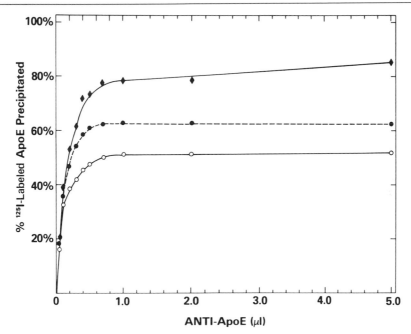

FIG. 1. Comparison of immunoprecipitability of apoE radioiodinated by iodine mono-chloride (◆), chloramine-T (●), and lactoperoxidase (○). ApoE was radioiodinated as de-scribed by these three methods. Radioiodinated apoE was added to each assay tube (10,000 cpm), the indicated amounts of anti-apoE were added, and the immunoprecipitability of apoE radioiodinated by each of the three methods determined.[24]

Plasma for analysis was obtained 10 min following injection of 100 μCi of radioiodinated apoE$_3$ into a normal subject homozygous for the ε-3 allele (see details below for kinetic studies). The chromatographic profiles of radioactivity and apoE concentration as determined by radioimmunoassay were very similar (Fig. 2). Within VLDL the apoE radioactivity and apoE concentration peaks chromatographed with the same K_d as the VLDL triglyceride peak; however, the apoE within HDL preceded the major HDL cholesterol peak consistent with the concept that the major portion of apoE resides on the larger particles within this density range. The specific activities of apoE within VLDL (663 cpm/μg), LDL (671 cpm/μg), and HDL (684 cpm/μg) were similar.

The *in vivo* distribution of radiolabeled apoE was also determined following ultracentrifugation.[25] The distribution of radiolabeled apoE within the plasma lipoproteins in the *in vivo* study was virtually identical to the distribution observed in the *in vitro* study (Table I); the normalized specific activities did not differ from one except for IDL (Table II).

TABLE I

DISTRIBUTION OF RADIOLABELED ApoE$_3$ IN LIPOPROTEIN SUBFRACTIONS

	VLDL	IDL	LDL	IDL + LDL	HDL$_{2b}$	HDL$_{2a+3}$	Total HDL	Infranatant (1.21 g/ml)
10-min *in vivo* sample (n = 12)[b]	23.1 ± 6.0%[a]	4.5 ± 0.7%	11.9 ± 2.1%	14.6 ± 3.3%	24.9 ± 3.5%	17.3 ± 1.8%	41.4 ± 6.5%	20.9 ± 3.6%
In vitro incubation (n = 6)[c]	29.9 ± 3.2%	4.2 ± 1.0%	10.8 ± 0.8%	15.0 ± 1.4%	21.2 ± 3.1%	16.7 ± 4.0%	37.9 ± 4.0%	17.3 ± 0.6%

[a] Plus or minus the SD (from Gregg et al.[25]).

[b] Ten minutes following the injection of radiolabeled apoE$_3$ into normal subjects 5 ml of plasma was obtained and the lipoproteins were fractionated by ultracentrifugation.

[c] Radiolabeled apoE$_3$ (0.2 μCi) was incubated with 5 ml of plasma at 37° for 30 min followed by fractionation of the plasma lipoproteins by ultracentrifugation.

TABLE II

NORMALIZED SPECIFIC ACTIVITIES OF $ApoE_3$ IN LIPOPROTEIN SUBFRACTIONS

	Plasma	VLDL	IDL	LDL	IDL + LDL	HDL_{2b}	HDL_{2a+3}	Total HDL	Infranatant (1.21 g/ml)
In vivo									
ApoE specific activity (n = 12)[b]	1	1.06 ± 0.13[a]	2.06 ± 0.20	1.28 ± 0.17	1.30 ± 0.13	0.92 ± 0.08	1.08 ± 0.10	1.06 ± 0.10	1.13 ± 0.08
In vitro									
ApoE specific activity (n = 6)[c]	1	1.18 ± 0.15	1.60 ± 0.18	1.02 ± 0. 3	1.09 ± 0.11	0.86 ± 0.08	1.02 ± 0.10	0.92 ± 0.09	1.07 ± 0.08

[a] Plus or minus the SEM (from Gregg et al.[25]).

[b] The specific activity normalized to the plasma specific activity 10 min following the injection of radioiodinated $apoE_3$ into $apoE_3$ subjects. The lipoprotein subfractions were isolated by ultracentrifugation.

[c] The specific activity normalized to the plasma specific activity following incubation for 30 min at 37° of radioiodinated $apoE_3$ with $apoE_3$ plasma.

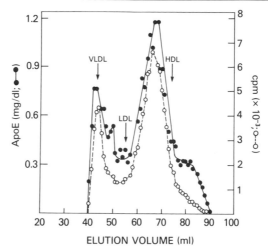

ELUTION VOLUME (ml)

Fig. 2. The profile of apoE mass (●) and apoE radioactivity (○) in normal human plasma separated by 6% agarose column chromatography. Ten minutes following the injection of radioiodinated apoE, 5 ml of plasma was obtained and fractionated by column chromatography on a 1.2×95 cm column of Sepharose CL-6B (6% agarose). The eluting buffer was 0.15 M EDTA, 0.1 M Tris-HCl (pH 7.4), 0.01% EDTA. The flow rate was 18 ml/hr and 1.1-ml fractions were collected. The arrows for VLDL, LDL, and HDL indicate the relative K_d of these lipoproteins.[25]

For *in vivo* kinetic studies in man, apoE was reassociated with plasma lipoproteins prior to injection. Reassociation was employed to ensure that the apoE utilized in the kinetic studies would be in the native state. During iodination guanidine-HCl was employed to prevent aggregation of apoE and to decrease its surface active properties. With removal of the denaturants, apoE undergoes self-association and/or aggregation. To eliminate the formation of irreversible or poorly reversible oligomers, apoE was added to plasma while still in guanidine-HCl. With dilution of the guanidine-HCl apoE associated with lipoproteins in the unaggregated state. In addition, if any irreversible denaturation of apoE occurred during iodination, the altered apoE would be anticipated to have decreased ability to bind to lipoproteins and be removed during isolation of apoE bound to lipoproteins in the 1.21 g/ml supernatant, thus permitting the selection of native, biologically active apoE. The apoE in the 1.21 g/ml infranatant would be a mixture of biologically active apoE that dissociated from lipoproteins during centrifugation and denatured apoE.

The catabolism of apoE radioiodinated as an apolipoprotein and reassociated with a lipoprotein was compared to the catabolism of apoE radiolabeled as a constituent of a lipoprotein.[25] If purified apoE reassociates normally with the lipoproteins, then the catabolism of the apoE iodinated in the two different states should be the same. ApoE and VLDL were isolated from the same subject and radioiodinated. ApoE was radioiodinated with ^{131}I, incubated with the subject's plasma, and the VLDL isolated by ultracentrifugation. VLDL isolated from a separate sample of plasma was radiolabeled with [125]I. The subject was injected with ^{131}I-labeled apoE–VLDL and ^{125}I-labeled VLDL. Since all of the apolipoproteins on the injected VLDL were radiolabeled, it was necessary to isolate apoE to determine its rate of decay. Because of the difficulty in isolating apoE with sufficient radioactivity for quantitation, though, it was possible only to follow the catabolism of the apoE iodinated as a lipoprotein within VLDL. At each sampling time, VLDL was isolated by ultracentrifugation, delipidated, and apoE isolated by SDS–PAGE. The radioactivities associated with ^{125}I-labeled apoE and ^{131}I-labeled apoE were determined in the apoE band cut from the gel and the results presented in Fig. 3. The ^{131}I-labeled apoE labeled as the apolipoprotein decayed at virtually the same rate as ^{125}I-labeled apoE labeled as VLDL, i.e., there was no difference in the rate of catabolism of apoE labeled on a lipoprotein or labeled as an apolipoprotein and then reassociated with lipoproteins. These combined results, therefore, establish that radioiodinated apoE reassociated with plasma lipoproteins is an effective tracer for the analysis of apoE metabolism.

Preparation of Subjects for Kinetic Studies

The investigations that we have performed on the metabolism of apoE have been designed to determine the kinetic parameters in the steady state. The plasma concentration of apoE as well as the relative distribution of apoE among the different types of lipoprotein particles must be in steady state. It is known that apoE exchanges between lipoprotein particles and that the state of alimentation will minimally affect the plasma apoE concentration but will markedly affect its distribution between VLDL and HDL. Compared to fasting plasma, the apoE in postprandial plasma is increased in triglyceride-rich particles and reduced in HDL. To circumvent this problem, a dietary regimen was devised to ensure that apoE is maintained in the steady state.

Initially, the study subjects are placed on a defined isoweight diet containing 42% carbohydrate, 42% fat, 16% protein, 200 mg cholesterol/1000 kcal, and a polyunsaturated-to-saturated fat ratio of 0.1 to 0.3 for 10 days before beginning the apoE metabolic studies. The diet is changed to either liquid formula or solid food of the same nutrient composition given in equal feedings every 6 hr 3 days before the apoE injection and continued for the duration of the study. The coefficient of variation of the apoE

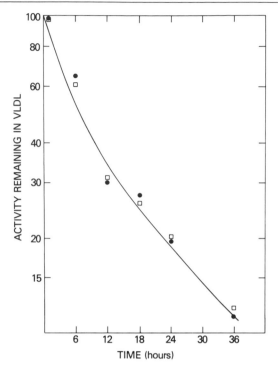

Fig. 3. The decay of apoE from plasma with apoE radioiodinated as an apolipoprotein and as a lipoprotein. ApoE was radioiodinated with ^{131}I, reassociated with lipoproteins, and VLDL isolated by ultracentrifugation. In a separate plasma sample, VLDL was obtained by ultracentrifugation and radioiodinated with [125]I. Both VLDL preparations were injected intravascularly, plasma was obtained, VLDL isolated by ultracentrifugation, and apoE isolated by SDS–PAGE. The catabolism of apoE labeled as the apolipoprotein is indicated by (□) while the catabolism of the lipoprotein-labeled apoE is indicated by (●). The values have been normalized with the amount of radioactivity at the first time point being 100%.[25]

concentration in plasma and in isolated lipoprotein density fractions using this dietary regimen was 10 and 15%, respectively. There was no consistent change in the apoE concentrations over the course of the studies, i.e., the apoE was in steady state in both the plasma and the lipoprotein subfractions.

Protocol for Performing *in Vivo* ApoE Kinetic Studies

Hypolipidemic medications are discontinued 4 weeks before the initiation of the kinetic study. The subjects are placed on the dietary protocol described above 10 days prior to the injection of the radioiodinated apoE. Three days prior to the study the subjects are also started on potassium iodide (300 mg with each meal) to block uptake of radioactive iodine by the thyroid and ferrous gluconate (900 mg/day) in divided doses.

Subjects are injected intravenously with up to 25 μCi of ^{131}I-radiolabeled apoE or 100 μCi of ^{125}I-radiolabeled apoE a half-hour before their morning feeding. Blood samples are obtained 10 and 30 min later, just prior to their next meal. Additional blood samples are drawn at 6, 12, 18, 24, 36, and 48 hr, and then daily through day 7. All samples are drawn just prior to a meal, i.e., after a 6-hr fast, except for the first one, which is drawn after a $5\frac{1}{2}$-hr fast.

Blood samples of 20 ml are drawn in tubes containing EDTA at a final concentration of 0.1%. The blood is placed on ice and plasma is separated by low-speed centrifugation (2000 rpm, 30 min) in a refrigerated centrifuge (4°). Sodium azide and aprotinin (Boehringer-Mannheim, West Germany) are added to plasma at a final concentration of 0.05% and 200 KIU/ml, respectively. The sodium azide is added as an antimicrobial agent while aprotinin is added to inhibit the serine protease cleavage of apoE.[26] One milliliter of plasma is frozen at −20° for apoE quantitation. The plasma lipoproteins are isolated by ultracentrifugation and 1 ml of each subfraction is frozen (−20°) for apoE quantitation. Radioactivities in plasma and the lipoprotein subfractions are quantitated utilizing an automated gamma counter. In addition, the excretion of radioactivity into the urine is quantitated throughout the course of the study, i.e., at each time point when blood is collected, a urine collection is completed and the total radioactivity in that collection quantitated.

Analysis of Data

The residence time (1/fractional catabolic rate) is determined from the area under the plasma decay curve by a multiexponential computer curve-fitting technique.[27] The production rate is calculated from the following formula: production rate = (apoE concentration × plasma volume)/(residence time × weight). The plasma volume can be determined by dividing the total quantity of radioactivity injected by the radioactivity per unit volume determined in the sample obtained 10 min after injection. The specific activity decay curves of apoE in plasma and the lipoprotein subfractions are calculated by dividing the radioactivity at each time point by the mass of apoE quantitated by immunochemical techniques.[24]

---

[26] W. A. Bradley, E. B. Gilliam, A. M. Gotto, Jr., and S. H. Gianturco, *Biochem. Biophys. Res. Commun.* **109**, 1360 (1982).

[27] M. Berman and M. F. Weiss, "SAAM Manual." DHEW, NIH Publ. No. 75-180. U.S. Govt. Printing Office, Washington, D.C.

Using these techniques the apoE concentration and the *in vivo* production rate, residence time, and fractional catabolic rate can be determined. In addition, the radioactivity and specific activity catabolic decay curves for apoE in the lipoprotein subfractions can be obtained. With these data plus the urinary radioactivity excretion data, additional information can be obtained utilizing high-speed digital computer modeling techniques, which will permit the development of a complex multicompartmental model of the whole body *in vivo* metabolism of apoE in humans.[27]

As yet unresolved is the best method to determine the metabolism of apoE in individual lipoprotein density fractions. Possible ways of isolating the lipoprotein fractions to quantitate the apoE mass and radioactivity include lipoprotein electrophoresis, column chromatography, and ultracentrifugation. Lipoprotein electrophoresis can be used to separate different lipoprotein fractions but is time consuming and does not readily permit the separation of large enough samples to quantitate apoE radioactivity and mass. The column chromatographic method has the advantage of being able to separate the lipoprotein fractions on a preparative level for both radioactivity and apoE mass quantitation. In addition, all of the apoE is lipoprotein associated in contrast to ultracentrifugation[1,2,25] (see below). The major disadvantage of the column chromatographic method is that it is time consuming, particularly since 8 to 12 time points from multiple patients are often obtained during a kinetic study and both apoE mass and radioactivity must be quantitated in all column fractions. High-pressure liquid chromatography (HPLC) will be difficult since present columns cannot easily separate the large plasma samples required because of the relatively low specific activity of apoE obtained during human *in vivo* studies. Ultracentrifugation has the advantage of being a time-efficient preparative method for separation of the lipoprotein fractions with uncomplicated quantitation of the apoE mass and radioactivity. Its major disadvantages include (1) 15–25% of the apoE separated by sequential ultracentrifugation is located in the lipoprotein-free fraction,[1,2,25] i.e., not lipoprotein associated, and (2) this method does not permit the resolution of the lipoprotein fractions to the extent possible with column chromatography with respect to the distribution of apoE within individual lipoprotein fractions.

Since there is not a clearly superior method for evaluating apoE metabolism in the lipoprotein fractions, we use a combination of column chromatography and ultracentrifugation. Since the column chromatography method is time consuming, we have utilized this method only on selected samples, depending on the study being performed, while ultracentrifugation has been used to separate the lipoproteins in all timed samples. In addition, we have determined that all of the common genetic forms of apoE, including apoE$_2$, apoE$_3$, and apoE$_4$, are metabolized at the same rate in the lipoprotein-free fraction (1.21 g/ml infranatant) separated by ultracentrifugation in contrast to major differences in their catabolic rates in plasma and lipoprotein fractions.[28] Therefore, the apoE in the lipoprotein-free fraction may not represent an artifact but may be due to the separation by centrifugation of a metabolically distinct pool which column chromatography is unable to resolve.

In Vivo Metabolism of ApoE in Humans

We have investigated the metabolism of apoE in normal and a number of dyslipoproteinemic subjects.[25,29-31] Figure 4 illustrates the plasma decay curve of apoE$_3$ in a normolipidemic apoE$_3$ homozygote.[25] The decay is triexponential and is quite rapid. The residence time was determined to be 0.73 ± 0.18 days ($\pm$ 1 SD, $n = 12$) in normal volunteers. The specific activity decay curve of apoE$_3$ within each lipoprotein subfraction compared to the plasma curve in a normal subject is illustrated in Fig. 5 and is representative of all normal subjects investigated. The following conclusions have been made from these kinetic studies: (1) The decay of radiolabeled apoE$_3$ within the lipoprotein subfractions is most rapid within VLDL. (2) There is a slight delay in the decay of radiolabeled apoE within IDL and LDL, i.e., the specific activity 30 min after injection was equal to or slightly higher than the specific activity 10 min after injection. In addition, the initial slopes of the IDL and LDL decay curves were less than the initial slope of the plasma curve. This result suggests that there is at least a partial precursor–product relationship of apoE between VLDL and IDL/LDL. (3) The decay of radiolabeled apoE in HDL is the slowest of all lipoprotein subfractions. (4) Radiolabeled apoE in the 1.21 g/ml infranatant is removed rapidly and this decay curve of apoE was similar in shape to the decay of apoE in VLDL.

The metabolism of apoE$_2$ and apoE$_4$ has also been investigated in normolipidemic and dyslipidemic subjects.[29-31] In normal subjects, apoE$_2$ was shown to be catabolized slightly, but consistently, more slowly than apoE$_3$ (a residence time of 0.85 days for apoE$_2$ versus 0.73 days for

[28] R. E. Gregg and H. B. Brewer, Jr., unpublished observations.
[29] R. E. Gregg, L. A. Zech, E. J. Schaefer, and H. B. Brewer, Jr., *Science* **211**, 584 (1981).
[30] H. B. Brewer, Jr., L. A. Zech, R. E. Gregg, D. S. Schwartz, and E. J. Schaefer, *Ann. Intern. Med.* **98**, 623 (1983).
[31] R. E. Gregg, R. Ronan, L. A. Zech, G. Ghiselli, E. J. Schaefer, and H. B. Brewer, Jr., *Arteriosclerosis* **2**, 420a (1982).

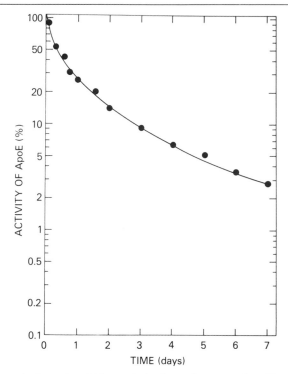

FIG. 4. The catabolism of apoE from whole plasma in a normal apoE$_3$ subject. Radiolabeled apoE$_3$ was reassociated with lipoproteins and injected into a normal subject. Plasma was isolated at various time intervals and the quantity of radioactivity determined in 5 ml of plasma at each time point. The radioactivity was normalized to 1 at the first time point.[25]

apoE$_3$),[29] whereas apoE$_4$ has been catabolized more rapidly (residence time of 0.38 days for apoE$_4$ compared to 0.73 days for apoE$_3$).[31]

Patients with type III hyperlipoproteinemia have elevated plasma apoE concentrations, delayed catabolism of remnants of triglyceride-rich lipoproteins, and the majority are homozygous for apoE$_2$. Analysis of apoE$_2$ metabolism in these subjects have established that apoE$_2$ is catabolized more slowly in type III hyperlipoproteinemic subjects when compared to apoE$_3$ in normals and that there is also overproduction of apoE.[29,30] Therefore, the slowed catabolism of apoE is consistent with the concept that a major metabolic abnormality in type III subjects is a retarded catabolism of remnants or triglyceride-rich lipoprotein particles. In addition, the elevated apoE plasma concentration was due both to delayed catabolism and increased production of apoE.

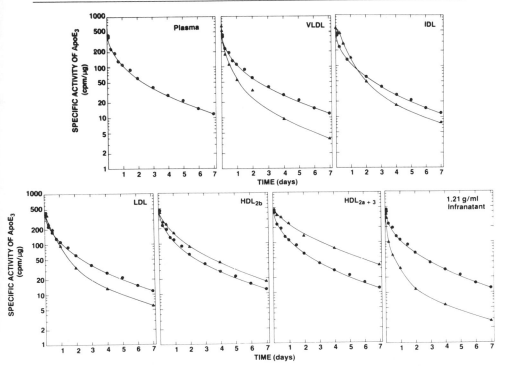

FIG. 5. The specific activity decay curves of apoE, from plasma and lipoprotein subfractions. Following the injection of apoE reassociated with lipoproteins, 5 ml of plasma was obtained at various time intervals and fractionated by ultracentrifugation. The specific activity decay curves of apoE were determined for each lipoprotein subfraction. The plasma-specific activity decay curve (●) is plotted in each panel to facilitate the comparisons with the different lipoprotein subjects (▲).[25]

Summary

A method for the investigation of the *in vivo* metabolism of apoE in humans has been described. In this method, isolated apoE is radioiodinated by the iodine monochloride method and reassociated with lipoproteins. Detailed studies have established that the radiolabeled apoE prepared by this procedure uniformly labels the different plasma lipoprotein pools. The study subjects are studied in steady state on a dietary regimen with multiple feedings. The kinetic results are analyzed by either multiexponential curve fitting or by computer-assisted multicompartmental modeling techniques. Examples of the types of results that may be obtained utilizing these methods are described.

[30] Hepatic Lipoprotein Biosynthesis

By JULIAN B. MARSH

This article discusses some of the methods which have been employed in the study of hepatic biosynthesis and secretion of plasma lipoproteins. It emphasizes the mammalian liver and includes information about studies in the whole animal, perfused liver, slices or pieces of liver, and fresh suspensions of hepatocytes. The techniques of organ culture, hepatocyte culture, and subcellular techniques aimed at localization of biosynthetic intermediates, such as the assembly of VLDL in Golgi vesicles, will not be treated in detail (see Chapters [17] this volume, and [18], Volume 128). The molecular biological approach—the transcription and translation of messenger RNA, and the molecular details of apolipoprotein biosynthesis and posttranslational modification, are likewise beyond the scope of this chapter (see Chapter [2], Volume 128).

The main kinds of questions which can be answered by the techniques under consideration here are those concerned with the nature of the primary hepatic lipoprotein secretory products and the rates at which secretion occurs under a variety of circumstances. Within this framework, the choice of an experimental system always involves a balance between maintenance of physiological conditions and the ease of experimental manipulation, always recognizing that any intact cell study will yield limited information concerning molecular mechanisms. The ultimate objective of most studies of plasma lipoprotein biosynthesis is to relate them to *in vivo* conditions. For example, the amount of mRNA for apoA-I in a given liver should be related to the secretion rate of apoA-I by the liver in the intact animal. The advantages and disadvantages of each of the intact cell systems will be discussed, but it is important to recognize that no *in vitro* technique precisely duplicates that *in vivo*. It may be a mistake to discard a useful *in vitro* technique solely because of an obvious departure from the physiological state.

In contrast to studies of the synthesis and secretion of other plasma proteins (such as albumin) in studies of lipoproteins one must also contend with the lipids. Since the lipids are not covalently attached, and can exchange between lipoprotein particles as well as between lipoproteins and cells, one cannot link the synthesis of the lipid to that of the protein unless the exchange process can be estimated. Free cholesterol exchanges readily; other lipids may require exchange proteins. This problem is more severe in the intact animal than in other systems where exchange proteins may be absent, and it should be kept in mind when choosing the optimal cell system for the study at hand. With lipoproteins, even the proteins don't always stay with a particle. Except for apoB, most of the other apolipoproteins have been found to exchange to some degree between lipoproteins. Whether or not equilibration between lipoprotein classes takes place within the experimental period, the total plasma apolipoprotein must be isolated in order to measure the true synthetic rate. The effects of the isolation procedure on the apolipoprotein content of a given lipoprotein is one of the important problems in the field at present.

Whole Animal Studies

While whole animal studies of lipoprotein biosynthesis may be considered the ultimate biological objective, they are not easy to perform, especially under non-steady-state conditions (precisely those in which there is much interest). They do not often reveal underlying mechanisms. In addition, when the focus is on hepatic synthesis, the contributions of the other tissues must be estimated. Since the liver accounts for about 80–90% of the total synthesis of the protein moiety of the plasma lipoproteins, nonhepatic sources are often neglected. This can cause serious errors for certain apolipoproteins, such as apoA-I, of which more than half is derived from the intestine. While fasting decreases the intestinal contribution, it is not eliminated. Though not of quantitative importance in the body as a whole, the synthesis of apolipoproteins by tissues other than liver and intestine has recently been described.[1]

Turnover Methods

Most studies of lipoprotein synthesis *in vivo* to date have employed the turnover method, in which the lipoprotein is isolated, labeled *in vitro*, and reinjected intravenously. The plasma is then sampled at various time intervals and the resulting curves analyzed mathematically. Good reviews of this methodology have been published[2,3] (see Chapter [22], this volume). In the steady state, the synthetic and catabolic rates are equal. It is also assumed that the labeling itself does not perturb the molecule in such a way as to alter its catabolic rate. Limited radioiodination comes close to fulfilling this condition. The turnover method can also be used with an endogenous amino acid label, though specific activities may be quite low

[1] M.-L. Blue, D. L. Williams, S. Zucker, S. A. Khan, and C. B. Blum, *Proc. Natl. Acad. Sci. U.S.A.* **80**, 283 (1983).
[2] R. Zak, A. F. Martin, and R. Blough, *Physiol. Rev.* **59**, 407 (1979).
[3] M. Berman, *Prog. Biochem. Pharmacol.* **15**, 67 (1979).

due to the large body pool of free amino acids. Recently, Eaton and co-workers[4] have employed [^{75}Se]selenomethionine, which has an isotopic half-life of 120 days and is incorporated into the protein in place of methionine.

Isotope Incorporation Methods

In many whole animal studies, estimates of the synthesis rates of plasma lipoproteins have been made by measuring the incorporation of labeled amino acids into the plasma protein after intravenous injection. Unfortunately, in many of these studies no conclusions can be drawn concerning the actual rate of lipoprotein synthesis because the specific activity of the precursor amino acid was not measured. Although the method is applicable to non-steady-state conditions, the precursor pool size and isotope enrichment must be measured. Even if comparisons between two conditions are of greater interest than the knowledge of the absolute synthetic rate, it is surprising how often measurements of amino acid pool sizes are not made. One way around this difficulty has been to measure the effects of a given perturbation on isotope incorporation into a lipoprotein and a nonlipoprotein (such as plasma albumin) which draw on the same intrahepatic amino acid pool for synthesis.

As outlined by Peters[5] and as originally defined by Zilversmit *et al.*,[6] the slope of the specific activity–time curve of the protein being synthesized is proportional to the difference between the specific activity of the precursor and that of the protein. In spite of the considerable knowledge presently available concerning the details of protein biosynthesis, the specific activity of the amino acid precursor is difficult to estimate. Even if one isolates the aminoacyl-tRNA from the liver there is no assurance that its specific activity reflects that of the true precursor tRNA since some tRNAs may be aminoacylated close to the cell membrane while others may acquire their amino acids after mixing with the intracellular pool. Measurement of hepatic leucyl-tRNA-specific activity after injection of labeled leucine is probably the best approach presently available. If one is limited to sampling the plasma compartment only, the degree to which an amino acid fails to reach isotopic equilibrium with the hepatic pool must be taken into account. Both the single bolus injection and the constant intravenous infusion techniques of isotope administration have been used effectively. In the lipoprotein field, it is striking that several

[4] R. P. Eaton, R. C. Allen, and D. S. Schade, *J. Lipid Res.* **24**, 1291 (1983).

[5] T. Peters, Jr., *in* "Plasma Protein Secretion by the Liver" (H. Galumann, T. Peters, Jr., and C. Redman, eds.), p. 95. Academic Press, New York, 1983.

[6] D. B. Zilversmit, C. Entenman, and M. C. Fishler, *J. Gen. Physiol.* **26**, 325 (1943).

good methods, especially those of McFarlane *et al.*[7] and of Reeve and McKinley,[8] have not been used. The double isotope method devised by Schimke and co-workers,[9] which allows a comparison of the turnover of proteins secreted by the same cells without requiring estimates of precursor and product isotope enrichment, has likewise been neglected.

An example of the kind of data obtainable by turnover methods can be found in the work of Eaton *et al.*,[4] who injected 250 μCi of [^{75}Se]selenomethionine intravenously into subjects fasted for 16 hr and drew blood samples for the next 5–10 days. After isolation of VLDL, IDL, and LDL, apoB-specific activity was estimated by densitometric analysis of stained SDS–PAGE gels. Using the kinetic model of Berman,[3] the turnover rates of apoB of VLDL, IDL, and LDL in five normal individuals averaged 18, 11, and 9 mg/kg body wt/day, respectively. Since the apoB of these lipoproteins in humans does not contain measurable amounts of intestinal apoB, these values represent hepatic synthesis rates.

In Vivo Synthesis of VLDL after Triton WR-1339 Injection

The detergent Triton WR-1339 has been used extensively in experimental animals to measure rates of VLDL triglyceride synthesis, since the original observations of Kellner *et al.*,[10] who found a persistent hyperlipemia in rabbits after intravenous injection of this detergent. The mechanism of this effect is not entirely understood. It may coat the lipoproteins, directly inhibit lipase action, interfere with the access of apoC-II, or do all of these things. Since catabolism is blocked, the rise in plasma concentration can be used to calculate the synthesis rate. In common with the other *in vivo* methods, the intestinal contribution to synthesis must be taken into account. This can be estimated by measuring VLDL accumulation in lymph fistula animals. However, diversion of lymph may not be complete due to collateral vessels and the extent to which removal of lymph affects intestinal synthesis is uncertain. In rats, estimates of 11–19% of plasma triglycerides arising from the intestine have been made by this method.[11] The method cannot, of course, be employed to measure LDL synthesis since the conversion of VLDL to LDL is blocked. Ishikawa and Fidge[12] have shown that changes in apolipoproteins occur in all lipoprotein fractions.

[7] A. S. McFarlane, L. Irons, A. Koz, and E. Regoeczi, *Biochem. J.* **95,** 536 (1965).
[8] E. B. Reeve and J. E. McKinley, *Am. J. Physiol.* **218,** 498 (1970).
[9] J. M. Arias, D. Doyle, and R. T. Schimke, *J. Biol. Chem.* **244,** 3303 (1969).
[10] A. Kellner, J. W. Correll, and A. T. Ludd, *J. Exp. Med.* **93,** 373 (1951).
[11] P. R. Holt and A. A. Dominguez, *Am. J. Physiol.* **238,** G453 (1980).
[12] T. Ishikawa and N. Fidge, *J. Lipid Res.* **20,** 254 (1979).

The dose of Triton has varied between 250 and 800 mg/kg. Since the degree of hyperlipemia is dose dependent, and the detergent/lipoprotein ratio declines with time, higher doses may be advantageous. The rate of plasma lipid accumulation is not linear, declining sharply between 4 and 8 hr. The *recommended technique* follows the work of Reaven et al.[13]:

1. Inject Triton WR-1339 intravenously as a 15% solution in 0.15 M NaCl at a dose of 600 mg/kg.
2. Isolate VLDL by ultracentrifugation and determine the protein and lipid content at 2 and 4 hr. Alternatively, total plasma triglyceride may be measured.

In rats, values of approximately 100 mg of triglyceride secreted/hr/kg body wt have been reported. The corresponding value for VLDL protein is 4.2 mg/kg/hr of which 3.4 mg may be of hepatic origin. This amounts to a hepatic secretion rate of 100 μg/g liver/hr of VLDL protein.

Estimation of the Intestinal Contribution to Lipoprotein Synthesis

In experimental animals, three methods have been used. The first, discussed above, employs the lymph fistula animal. Appropriate controls are necessary since the animal must be restrained and is fed through a duodenal cannula.

In the case of VLDL in rats its intestinal secretion can be measured by feeding orotic acid, which blocks hepatic but not intestinal secretion, of apoB.[14] It also decreases hepatic HDL secretion. Animals must be kept on a low-adenine diet for a week with 1% orotic acid added.

Wu and Windmueller[15] injected [^{3}H]leucine intraduodenally and [^{14}C]leucine intraportally in the rat and measured the isotope ratio in individual apolipoproteins separated by SDS–PAGE. The isotope ratio of plasma albumin was used to correct for hepatic synthesis of labeled protein after the absorption of labeled amino acids into the systemic circulation. By this method, it was estimated that the liver accounted for about 40% of the secretion of plasma apoA-I and A-IV, 84% of apoB, and 90% or more of apoCs.

Liver Perfusion

The direct way to study hepatic lipoprotein biosynthesis is by measuring the output of a perfused liver. Since the early work of Miller et al.[16] in

[13] E. P. Reaven, O. G. Kolterman, and G. M. Reaven, *J. Lipid Res.* **15**, 74 (1974).
[14] H. G. Windmueller and R. I. Levy, *J. Biol. Chem.* **242**, 2246 (1967).
[15] A.-L. Wu and H. G. Windmueller, *J. Biol. Chem.* **254**, 7316 (1979).
[16] L. L. Miller, C. G. Bly, M. L. Watson, and W. F. Bale, *J. Exp. Med.* **94**, 431 (1951).

rats, a wide variety of animal species has been studied, but the rat continues to be the most popular. The following discussion applies specifically to the rat.

Surgical Procedure

Light ether or Nembutal anesthesia is recommended. The abdomen is opened by a midline incision after the skin and subcutaneous tissues are dissected away. The entire abdomen is exposed by grasping the abdominal muscles and making transverse and longitudinal cuts with scissors on either side. The intestinal tract below the stomach is then draped over wet gauze outside of the abdomen. If bile is to be collected, the duct is identified as it runs along the duodenum and ligated distally. One end of the ligature (the author prefers 00 silk) is clamped to the skin distally in such a way as to place the duct under slight tension. After making an angled cut with an iris scissors, the duct is easily cannulated with PE or PP 10 tubing and tied in place. A ligature is next placed loosely around the vena cava above (cephalad to) the renal vessels. Two ligatures are placed around the portal vein, one close to the porta hepatis and one about 1 cm distally. The hemostat holding the bile duct under slight tension by its ligature end is then removed and, after the distal (caudad) ligature is tied, the portal vein is placed under slight tension by clamping the end of this ligature to the skin of the leg or pelvis. An angled cut, no more than one-third of the circumference of the vessel, is made in the portal vein and the portal vein cannula immediately slipped in and tied in place. If an assistant is available, the proximal loose ligature may be used to control bleeding by pulling upward. In the author's experience, the hole in the portal vein is readily seen and entered with the cannula for a second or two before bleeding obscures the field. However, if cannulation is not immediately successful, it is a simple matter to collapse the vein via the proximal ligature, grasp the edge of the hole with iris forceps, and insert the cannula. We favor a 1.5-cm length of polypropylene tubing (2 mm o.d.) tightly nested inside a short length (2–3 mm) of 2.5 mm o.d. polyvinyl tubing, which in turn is nested inside 3 mm o.d. intravenous tubing which can be any convenient length and having a Luer fitting at its other end for easy attachment to a three-way stopcock. Sharp 18- or 19-gauge needles can also be used instead of a cannula but there is always the risk of puncturing the vessel. The needle can be held in place with a bulldog clamp. If the liver is perfused *in situ* needles work well but if the liver is to be dissected from the body they do not. We have not found it necessary to ligate the hepatic artery and prefer to allow arterial blood to reach the liver during the portal vein cannulation. After that is accomplished, a slow flow of oxygenated medium is allowed to enter the liver during the remaining surgery. The ligature around the vena cava above the renal vessels is now tied and the chest opened with scissors after grasping the xiphoid process of the sternum with forceps. Care must be taken not to injure the liver when cutting through the diaphragm and rib cage. The ribs are cut longitudinally on either side, and then a transverse cut is made to remove them. The ribs should be cut to about half of their depth, leaving a sizeable chest cavity intact. After placing a ligature around the vena cava above the diaphragm, the heart can be grasped and a hole cut in the right auricle through which a cannula similar to that in the portal vein can be easily inserted and tied in place. The portal vein inflow rate can then be increased and the perfusion begun. We prefer *in situ* perfusion since there is minimal handling of the liver. The entire procedure (exclusive of bile duct cannulation) can be carried out in less than 4 min. In larger animals, and in rat liver perfusions intended to last longer than 3 hr, we believe it is preferable to remove the liver from the body.

General Considerations

With any intact preparation, the first concern must be its viability and its performance compared to the *in vivo* situation. The following six parameters can be measured:

Oxygen Uptake. Oximeters may be placed in the inflow and outflow stream and these can measure the percentage saturation of the blood with oxygen. Total hemoglobin is readily measured by the cyanmethemoglobin method. A knowledge of the flow rate enables one to calculate the oxygen uptake. If there are no erythrocytes, oxygen electrodes can be placed in the inflow and outflow stream and the PO_2 measured. A simpler method is to have a short length of gum rubber tubing in the inflow and outflow circuits which can be easily punctured with a 26- or 27-gauge needle and a sample of the medium pulled into a syringe without exposure to air. It can then be carried to an oxygen electrode in a different location. Similarly, in the presence of hemoglobin the percentage saturation can be measured by hemolyzing the sample with a detergent without exposure to air and measuring the percentage saturation spectrophotometrically.[17] Alternatively, the pO_2 can be measured and the percentage saturation of the blood calculated from the oxygen dissociation curve. If the red cells are aged, this may be quite different from normal due to changes in diphospho-glycerate content.

Bile Secretion. The simplest overall measurement of liver function during a perfusion is that of bile output. After cannulation of the bile duct, bile is collected and measured by weight.

[17] E. Gordy, D. L. Drabkin, and J. B. Marsh, *J. Biol. Chem.* **227,** 285 (1957).

pH. Secretion of plasma proteins declines below pH 7.3. It can be monitored readily either by taking samples or by placing a glass electrode somewhere in the perfusion circuit. Small amounts of sodium bicarbonate may be added to maintain the pH in the 7.3–7.4 range.

Enzyme Concentrations. Enzyme concentrations in the perfusion medium may be monitored. With liver damage, the amount of transaminases and of lactic dehydrogenase in the medium increases in comparison with normal liver.

Urea Output. This is not often a useful parameter as an absolute measure of hepatic function because it varies with the nitrogen source levels in the medium, but chemical methods for urea[18] are easy to use and, if a consistent urea production is measured under control conditions, urea output is a good index of hepatic function.

Gluconeogenesis. This is not a measurement which is well suited for routine applications, but it does provide a rather stringent method for evaluating hepatic function.

Table I[19–27] gives some normal values for the six parameters of liver function. The values for oxygen uptake warrant some further discussion, in view of reports that effects of insulin in perfused liver can be observed only with relatively high hematocrits.[23] Based on studies in the cat of hepatic portal and arterial blood flow, portal venous pressure, the measurements of oxygen content,[28] mean hepatic blood flow was 26.5 ± 1.9 ml/min/kg body wt, or 1.1 ml/g liver/min, of which 47% was portal. Portal pressure averaged 6.6 mm Hg, or 9 cm of water. These values may differ from those in the rat, but may be a useful approximation of *in vivo* conditions. In rats (setting hepatic portal venous and outflow venous pressure at 10 and 0 mm Hg, respectively) Riedel *et al.,*[20] using pump-perfused livers, found that blood flow increased exponentially as the hematocrit was changed from 80 to 10% while oxygen uptake increased to a maxi-

[18] D. R. Wybenga, J. DiGirogio, and V. L. Pileggi, *Clin. Chem.* **17,** 891 (1971).

[19] C. Soler-Argilaga, R. Infante, J. Polonavski, and J. Caroli, *Biochim. Biophys. Acta* **239,** 154 (1971).

[20] G. L. Riedel, J. L. Scholle, A. P. Shepherd, and W. F. Ward, *Am. J. Physiol.* **245,** G769 (1983).

[21] G. B. Storer, R. P. Trimble, and D. L. Topping, *Biochem. J.* **192,** 219 (1980).

[22] O. O. Thomsen and J. A. Larsen, *Am. J. Physiol.* **245,** G59 (1983).

[23] D. L. Topping, R. P. Trimble, and G. B. Storer, *Biochem. Int.* **3,** 101 (1981).

[24] J. B. Hanks, W. C. Meyers, C. L. Wellman, R. C. Hill, and R. S. Jones, *J. Surg. Res.* **29,** 149 (1980).

[25] T. Ide and J. A. Ontko, *J. Biol. Chem.* **256,** 10247 (1981).

[26] J. B. Marsh, *J. Lipid Res.* **15,** 544 (1974).

[27] R. Hems, B. D. Ross, M. N. Berry, and H. A. Krebs, *Biochem. J.* **101,** 284 (1966).

[28] W. W. Lautt, *Am. J. Physiol.* **232,** H652 (1977).

TABLE I
FUNCTION PARAMETERS IN PERFUSED LIVER[a]

Flow (ml/g/min)	Perfusion pressure (mm Hg)	Hematocrit (%)	Oxygen uptake (ml/g/hr)	Bile output (μl/g/hr)	Medium composition
0.55	—	23	2.8	25	Heparinized rat blood, KRB, 4.5% HSA[19]
0.85	10	20	3.9	—	Canine red blood cells, 3% BSA, KRB[20]
0.95	—	17	3.5	—	Defibrinated blood[21]
0.95	—	37	5.5	—	Defibrinated blood[21]
1.15	11	34	4.8	102	Human red blood cells, KHB, 2.4% BSA[22]
1.23	9	19	2.9	—	Human red blood cells, KHB, 2.4% BSA[22]
1.20	—	19	3.1	—	Defibrinated blood[23]
1.20	—	28	4.4	—	Defibrinated blood[23]
1.20	—	38	6.2	—	Defibrinated blood[23]
1.50	10	10	1.2	24	Canine red blood cells, KRB, 3% BSA[24]
1.80	—	25	—	36	Human red blood cells, KHB, 1.5% BSA[25]
3.00	27	0	3.0	51	KRB[26]
3.50	18	8	3.0	—	Human red blood cells, KHB[27]

[a] These values do not constitute a complete survey of the literature. Low-molecular-weight compounds such as glucose and amino acids were added in some of the studies. Abbreviations: KHB, Krebs–Henseleit bicarbonate medium; KRB, Krebs–Ringer bicarbonate medium; BSA, bovine serum albumin; HSA, human serum albumin.

mum at 20%. At a hematocrit of 20%, blood flow was approximately 0.9 ml/g/min and oxygen uptake was about 3.9 ml/100 g/hr in a medium containing washed canine red cells in Krebs–Ringer bicarbonate medium with 10 mM glucose and 3% bovine plasma albumin. Below 1 ml/g/min, oxygen uptake was flow limited. The higher oxygen uptakes at a hematocrit of 37% observed by Storer et al.[21] may have been due to provision of lactate by the red cells but the explanation for the insulin effects observed is still not clear. From the standpoint of experiments to study lipoprotein secretion, hematocrit effects have not been investigated. In the absence of red cells in the perfusion medium, high flow rates (hence high perfusion pressures) are necessary but satisfactory oxygen uptakes of 3 ml/100 g/hr have been reported (Table I).

Choice of Perfusion Medium

Whole Blood. Though close to physiological, it has several drawbacks for lipoprotein secretion studies. Prevention of blood coagulation can be approached in several ways, each presenting different experimental complications. Defibrinated whole blood is readily prepared by bleeding into a beaker while mixing the blood with a rod or wooden stick. During the formation of fibrin, proteolytic enzymes are activated which may act on apolipoproteins, as in the case of apoE.[29] Anticoagulants which chelate calcium ions can have deleterious effects on hepatic structure and function, while heparin releases hepatic lipase into the medium, causing lipolysis of secreted VLDL. Furthermore, serum is a highly complex medium containing hormones and other compounds which may be modulators of lipoprotein synthesis and secretion. Perhaps the best use of whole blood, or suspensions of red cells in serum, is when information is sought concerning the role of the liver in converting proforms of apolipoproteins to the dominant plasma forms or in studies of amino acid incorporation into secreted lipoproteins.

Red Cell Suspensions in Buffered Medium. This is perhaps the commonest perfusion fluid employed in recirculating perfusions. Rat livers do not seem to mind being perfused with red cells from humans, cows, rabbits, or dogs. A useful precaution, especially with fresh cells, is to dialyze the suspension to remove vasoconstrictors, though this may not be necessary. In our experience with rats, we have not observed lipoprotein binding to red cells. There is always some degree of hemolysis during perfusion with any medium containing red cells, which of course is worse with aged cells. Peristaltic roller pumps, commonly used, can cause hemolysis, which can be lessened by increasing the diameter of the tubing. However, we have not observed effects of liberated red cell proteins or of hemoglobin itself on secreted lipoproteins though no thorough studies of this possibility have been made. Hemoglobin is a useful marker protein of high density and low molecular weight in estimating convective contamination of ultracentrifugally isolated lipoproteins. In dilute solutions, the hemoglobin alpha and beta chains dissociate. Oxygen-carrying fluorocarbons have not been used in lipoprotein secretion studies, and probably should be avoided since they can be taken up by hepatic Kupffer cells with unknown consequences.

Is Albumin Necessary? The usual electrolyte-containing medium is Krebs–Ringer bicarbonate, equilibrated with 95% oxygen and 5% carbon

---

[29] S. H. Gianturco, A. M. Gotto, Jr., S.-H. C. Hwang, J. B. Karlin, A. H. Y. Lin, S. C. Prasad, and W. A. Bradley, *J. Biol. Chem.* **258**, 4526 (1983).

dioxide. To achieve an approximately normal oncotic pressure, albumin is often added to perfusion media, at slightly less than the normal concentration, such as 3%. In perfusions of up to 2 hr, only slight changes in water content of the liver were seen in the complete absence of albumin, and no significant changes in oxygen uptake were observed in our laboratory. Therefore, albumin does not seem to be a necessary component of the perfusion medium. Perhaps the most important reason for including it is the fact that if free fatty acids are to be added as substrates for VLDL synthesis, albumin must be present. Work from Heimberg's laboratory has established the relationship between free fatty acid levels and composition on the synthesis and secretion of VLDL.[30] Under certain circumstances, albumin has been found to actually inhibit VLDL production.[31] Most investigators employ heterologous albumin. Bovine serum albumin is by far the most popular, and rat livers do not seem to mind. However, many commercial BSA preparations are contaminated with apolipoproteins which could affect the interpretation of some experiments. These can be partially removed by treatment with phospholipid vesicles followed by ultracentrifugation at d 1.21.[32] It is also possible to find some commercial preparations which are not contaminated with apoA-I.

Are Amino Acids Necessary? In a recirculating system, hepatic protein catabolism provides amino acids so that the medium concentration reaches some kind of steady state during the perfusion. It would seem wise, however, to add them at the beginning. Eagle's minimal essential amino acid medium containing glutamine is commercially available in concentrated form. Stimulatory effects of amino acid addition on protein synthesis in the perfused liver have been described.

Preperfusion. Removal of blood is necessary if net secretion of lipoproteins is to be measured. A 10-g liver contains about 0.7 ml of blood. Perfusion with 200 ml of red cell-free medium removes more than 99.5%. A far more difficult problem is the removal of preexisting lipoproteins from the space of Disse or otherwise bound to the liver. This binding must be expected because of the existence of hepatic lipoprotein receptors such as the LDL receptor. Bound lipoproteins reveal their presence when the accumulation of lipoproteins in the medium with time is measured, since the washout curve is nonlinear. Inhibition of protein synthesis by cycloheximide also may be used to estimate the amount of preexisting lipoproteins. The washout problem is greater when plasma lipoprotein

[30] C. Soler-Argilaga and M. Heimberg, *J. Lipid Res.* **17**, 605 (1976).
[31] S. Yedgar, D. B. Weinstein, W. Patsch, G. Schonfeld, F. E. Casanata, and D. Steinberg, *J. Biol. Chem.* **257**, 2188 (1982).
[32] M. Fainaru and R. J. Deckelbaum, *FEBS Lett.* **97**, 171 (1979).

levels are elevated and also may be severe in larger animals. For example, in the perfused pig liver,[33] about half of the apoB released over a 5-hr period was in LDL, and the amount in the medium was half as much in the second hour as compared to the first hour of perfusion. VLDL apoB, on the other hand, did not show a similar decline. When labeled amino acids are employed, release of bound lipoproteins during the perfusion results in lower specific activities than expected, taking into account the fact that there is about a 15- to 20-min delay between the time a polypeptide chain is synthesized on a ribosome and the time it appears in the perfusion medium. When preperfusion is carried out in a red cell-free medium, one must employ flow rates high enough (3–4 ml/g/min) to allow adequate oxygen uptake during this period.

Perfusion Techniques

Recirculating Perfusion

Most liver perfusions carried out for the study of lipoprotein synthesis have employed the recirculating mode using red cell suspensions in Krebs–Ringer bicarbonate medium containing glucose, amino acids, and 3% BSA. Volumes of medium are typically 50 to 100 ml at hematocrits of about 20.

The apparatus (see Fig. 1) consists of a pump, an oxygenator, and a reservoir. The use of small-bore silastic tubing placed inside a vessel gassed with oxygen was introduced by Hamilton et al.[34] and constitutes an important technical advance since no foaming occurs and evaporative losses are minimized.

Two general methods of perfusion may be used. Either medium from the reservoir is pumped directly into the liver, or into a second reservoir with an overflow tube, which allows direct gravity perfusion. In the gravity method, changes in flow rate at constant pressure are readily observed. In the direct pumping method, flow is constant but pressure varies. The author has used both methods successfully, with a slight preference for gravity feed because the filter (which should be in every perfusion circuit) can be very simple. A nylon or silk screen of about 150 mesh is suitable. We use a sample applicator supplied with chromatography columns which consists of a plastic cylinder, a nylon mesh screen, and an O-ring to hold it in place. With the direct pumping method, it is

[33] M. Nakaya, B. H. Chung, and O. D. Taunton, J. Biol. Chem. 252, 5258 (1977).
[34] R. L. Hamilton, M. N. Berry, M. C. Williams, and E. M. Severinghaus, J. Lipid Res. 15, 182 (1974).

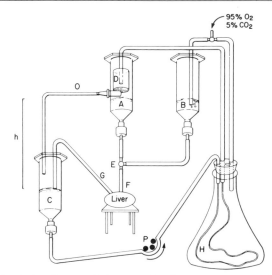

FIG. 1. Recirculating liver perfusion. In this diagram, no means for maintaining the temperature at 37–38° is shown. Three methods can be used. The simplest is to carry out the entire procedure in a constant temperature room. Alternatively, the apparatus can be placed in a constant temperature box. Tubing can be coiled in a constant temperature water bath which can also circulate water through jacketed reservoirs. Reservoir B, attached via a three-way stopcock (E) to the portal vein inflow cannula (F), contains Krebs–Ringer bicarbonate buffer with 0.1% glucose and is used to remove blood from the liver as well as to flush out lipoproteins which may be in the space of Disse or bound to the liver. This is a single-pass mode and high flow rates are required to allow for adequate oxygenation. In the apparatus shown here, the flow is adjusted by changing the height (*h*) of the reservoir. The overflow tubing (O) must be of wide bore. It is connected to a plastic tubing connector inserted through a hole bored in a 50-ml plastic syringe barrel (A) and drives into the overflow reservoir (C). The filter (D) rests conveniently on the wide-bore end of the connector protruding into the lumen of the barrel. We commonly employ 160-mesh nylon bolting cloth (Nytex HC-3-160) of 60-μm thread diameter with 53% open area in the filter. The silastic tubing in the lung (H) has a 1.47 mm i.d., and is 0.25 mm thick (Dow-Corning 602-235) as described by Hamilton *et al.*[34] and is 12–24 ft long, depending on the flow rate employed (G) in the outflow cannula.

important to place a small bubble trap in the line just before the fluid enters the liver.

Single-Pass, or Nonrecirculating Perfusion

This technique can be very useful for certain kinds of studies, such as those designed to investigate binding, rates of hepatic uptake of compounds, and studies of lipoprotein secretion. The main virtue of this method of studying lipoprotein secretion is that the secretory product is removed from the liver, and thereby protected from hepatic catabolism, or the action of secreted enzymes such as lecithin–cholesterol acyltransferase (LCAT). In spite of the generalization that substances secreted by the liver are usually poor substrates for hepatic catabolism, there is no doubt that in the recirculating mode VLDL triglyceride is hydrolyzed to some extent, even if heparin is not present. Another virtue of single-pass perfusion is that the investigator has complete control over the composition of the perfusion medium. The penalty one pays for these considerable advantages is that one must deal with large volumes of medium. This limits the concentration of albumin which can conveniently be present. High flow rates (3–4 ml/g/min) are essential if no red cells are present, which exacerbates the problem of concentrating large volumes of dilute perfusates. For this purpose, we have employed membrane ultrafiltration using 90-mm PM-10 membranes (Amicon Co.) in a thin-channel flow system, which will produce about 200 ml of ultrafiltrate/hr at 5° and a pressure of 20 psi, depending on the protein concentration of the medium. There is some adsorption of lipoproteins to these filters, which can be decreased by a preliminary ultrafiltration of an albumin solution. So far, we have not found evidence for selective adsorption of individual apolipoproteins but a systematic investigation of this possibility has not been carried out in our laboratory. The simplest way to correct for these adsorptive losses is to use labeled iodinated lipoproteins.

Figures 1 and 2 depict typical recirculating and nonrecirculating perfusion systems. Some parameters of hepatic function are given in Table I. Table II[35-43] indicates some values for lipoprotein secretion which have been reported, and Table III[44] shows values for apolipoprotein secretion.

Liver Slices and Liver "Snips"

The slice technique for the study of tissue metabolism was one of the earliest employed in the study of hepatic lipoprotein synthesis. Details of

[35] J. B. Marsh and C. E. Sparks, *J. Clin. Invest.* **64,** 1229 (1979).

[36] J. B. Marsh, *J. Lipid Res.* **17,** 85 (1976).

[37] S.-P. Noel, L. Wong, P. J. Dolphin, L. Dory, and D. Rubinstein, *J. Clin. Invest.* **64,** 674 (1979).

[38] T. E. Felker, R. L. Hamilton, J.-L. Vigne, and R. J. Havel, *Gastroenterology* **83,** 652 (1982).

[39] S. Calandra, E. Gherardi, M. Fainaru, A. Guaitani, and I. Bartosek, *Biochim. Biophys. Acta* **665,** 331 (1981).

[40] L. S. S. Guo, R. L. Hamilton, R. Ostwald, and R. J. Havel, *J. Lipid Res.* **23,** 543 (1982).

[41] C. A. Hornick, T. Kita, R. L. Hamilton, J. P. Kane, and R. J. Havel, *Proc. Natl. Acad. Sci. U.S.A.* **80,** 6096 (1983).

[42] F. L. Johnson, R. W. St. Clair, and L. L. Rudel, *J. Clin. Invest.* **72,** 221 (1983).

[43] L. A. Jones, T. Teramoto, D. J. John, R. B. Goldberg, A. H. Rubenstein, and G. S. Getz, *J. Lipid Res.* **25,** 319 (1984).

[44] J. L. Witztum and G. Schonfeld, *Diabetes* **28,** 509 (1979).

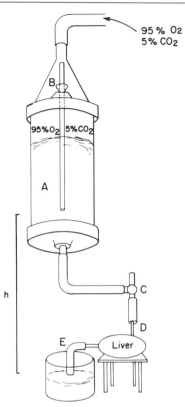

Fig. 2. Nonrecirculating liver perfusion. The apparatus may be placed in a constant temperature box or room. The medium in the reservoir (A) must be equilibrated with the oxygen–carbon dioxide gas mixture, for at least 20–30 min, before tightening the stopper. A slow flow of gas through the funnel (B) prevents dilution of dissolved gas in the reservoir with room air; in short perfusions this may not be necessary. It is also advisable to pass the gas mixture through water prior to its entry into the system. Dual reservoirs may be convenient for some applications; a second reservoir can be attached to the inflow (D) by the stopcock (C). The flow is regulated by changing h, the net pressure height. The contents of the reservoir can be stirred by a magnetic stirrer if red cells are present, and the collecting flask (E) can, of course, be placed in an ice bucket for rapid cooling. As a guide to values of h which will result in the desired flow, we have found that between 25 and 50 cm the increase in flow with increasing h is linear with a slope of 0.9–1.5 ml/min/cm and that 40 cm produces a rate of 26 ± 5 ml/min. These pressures are considerably above the normal pressure of about 15 cm and may produce endothelial damage. If so, it appears not to be reflected in parameters such as transaminase output, oxygen uptake, or bile flow in our experience. Where precise control of flow, independent of hepatic resistance, is required one can employ the silastic tubing lung and deliver the medium directly to the liver by means of a pump. When this is done, it is advisable to insert a bubble trap in the line leading to the liver.

TABLE II

A. Output of Plasma Lipoproteins by Perfused Liver[a]

Reference	VLDL	LDL (μg protein/g/hr)	HDL
Rat			
35	45	35	16
36	39	15	20
37	29	5.0	—
38	19	4.7	7.2
39	18	7.2	15
Guinea pig			
40	15	4.4	1.8
Rabbit			
41	8.6	0.9	2.7
Monkey			
42	6.0	0.9	—

B. Composition of Lipoproteins in Liver Perfusates[b]

	VLDL					LDL					HDL				
	Pr	C	CE	PL	TG	Pr	C	CE	PL	TG	Pr	C	CE	PL	TG
Rat															
35	13	1.7	0.8	35	49	19	1.7	0.9	43	36	57	2.0	1.8	29	8.6
36	20	5.9	1.8	18	54	22	8.6	4.8	17	48	54	9.3	3.5	13	20
37	18	5.7	5.7	13	58	—	17	18	24	37	46	8.6	23	15	7.6
38	14	9.5	1.9	31	46	18	5.5	7.0	53	17	58	3.1	8.9	28	1.6
Guinea pig															
40	8.9	3.2	0.3	16	71	27	5.1	42	17	9.0	45	3.8	19	19	13
Rabbit															
41	7.3	4.2	2.6	13	73	15	7.7	13	25	40	44	7.9	4.4	28	16
Monkey															
42	17	7.3	5.0	3.7	67	35	7.5	6.3	29	22	—	—	—	—	—
43	7.7	7.3	2.2	22	62	18	19	6.7	56	—	58	8.1	1.5	33	—

[a] Fatty acids were not added to the medium in these studies. The strain of rat, dietary status, and duration of perfusion also differed. In refs. 35–37, single-pass perfusions were carried out. VLDL, LDL, and HDL were isolated at $d < 1.006$, $1.006 < d < 1.063$, and $1.063 < d < 1.210$ in most studies except for those in the guinea pig where the sequential upper densities were 1.015, 1.10, and 1.21, respectively.

[b] Abbreviations: Pr, protein; C, cholesterol; CE, cholesteryl esters; PL, phospholipid; TG, triglyceride.

TABLE III
APOLIPOPROTEIN SECRETION BY
PERFUSED LIVER[a]

Reference	ApoA-I[a]	ApoB	ApoC	ApoE
35	5	20	40	40
37	8	13	—	22
38	3	—	—	—
39	—	—	—	12
44	7	—	—	—

[a] These values were calculated from data given in each reference and must be considered approximate since different methods were used and production rates were not linear.
[b] Columns 2–5 in μg/g liver/hr.

the method are given in the article by Peters.[5] The success of hepatocyte suspensions and of cultured hepatocytes has greatly diminished the use of the slice technique, which has some severe drawbacks. It is very difficult to cut and handle slices thinner than 0.5 mm. At this thickness, even in 100% oxygen, the cells in the center of the slice are anoxic, and there is leakage of hydrolytic enzymes from the cut surfaces and probably from the anoxic cells. Adding to the mechanical problems, slices must be shaken at 100 strokes/min or faster to allow adequate oxygenation of the medium by diffusion. In spite of these drawbacks, liver slices can yield valuable information under comparative circumstances. It is easy to compare slices from treated and untreated animals or to add different compounds to the medium bathing slices from the same animal. Of course, this can also be done with hepatocyte suspensions but the use of hydrolytic enzymes during the preparation of the cells, and the variability in perfusing the liver to prepare the cells, may present difficulties. A large number of slices can be rapidly prepared from a single liver and all of the cells normally present are there. In some situations, such as human liver biopsies, it may be better to use slices than to attempt the preparation of hepatocytes. Organ culture techniques may be useful also in that situation. In many circumstances, such as experimental nephrosis, increased lipoprotein synthesis has been demonstrated in slices to the same extent as in perfused liver. Another example of the effective use of slices was a study in which catecholamines were found to be effective inhibitors of VLDL triglyceride secretion by slices of rat liver incubated in whole rat serum from which VLDL had been removed by ultracentrifugation.[45] In

[45] J. B. Marsh and A. Bizzi, *Biochem. Pharmacol.* **21**, 1143 (1972).

this study VLDL triglyceride output was estimated at about one-third of the *in vivo* rate. Another advantage of the slice technique is that it is easy to analyze the liver tissue itself at any time in the course of the experiment.

In 1982, Pollard and Dutton[46] described the preparation of liver "snips" cut 1 mm wide by hand along the thin edges of the lobes of the liver. Each snip, cut 4 mm long, weighs about 0.3 mg and 200 snips can be prepared in about 20 min, while the liver is kept cold. Based on glucuronidation rates, it was estimated that 10% of the cells or less were damaged. Isolated hepatocytes gave glucuronidation rates 30% higher than snips, partly attributed to the presence of nonhepatocytes in the snips. Slices, however, gave about half the glucuronidation rates compared to snips. The main disadvantage of the technique at present would appear to be the relatively limited amount of tissue that can be obtained from one liver by one operator.

Fresh Hepatocytes

Hepatic lipoprotein synthesis can be conveniently studied in suspensions of freshly prepared hepatocytes. Since the work of Berry and Friend,[47] several modifications of the enzymatic digestion method have been published. The methods of Seglen[48] and Berry and Werner[49] have been commonly employed. The following method, the work of Wittman and Jacobus,[50] serves to illustrate the technique and allows a more objective criterion to be used concerning the time at which a given liver can be deemed suitably digested. The liver is prepared for perfusion as described above.

1. The medium used is 130 ml of Hanks medium, pH 7.4, without Ca^{2+}, at 16.6 mM glucose and 8 mM bicarbonate, gassed with 95% oxygen–5% carbon dioxide, and placed in a reservoir for recirculation.

2. The contents of the reservoir are pumped directly to the liver via a portal vein cannula. A manometer consisting of vinyl tubing attached to a meter stick via a T-tube is inserted in the line.

3. The outflow cannula, placed in the vena cava, drains into the reservoir via a flow meter, and the rat board is tilted so that leakage drains via a funnel into the reservoir.

[46] M. R. Pollard and G. J. Dutton, *Biochem. J.* **202,** 469 (1982).
[47] M. N. Berry and D. S. Friend, *J. Cell Biol.* **43,** 506 (1969).
[48] P. O. Seglen, *Methods Cell Biol.* **13,** 29 (1976).
[49] M. N. Berry and H. V. Werner, *In* "Regulation of Hepatic Metabolism" (F. Lundquist and N. Tystrup, eds.), p. 751. Munksgaard, Copenhagen, 1973.
[50] J. S. Wittman, III, and K. E. Jacobus, *Anal. Biochem.* **79,** 58 (1977).

4. At the start, 30 mg of collagenase (Worthington, CLS II) and 25 mg of hyaluronidase (Sigma, type I) in 10 ml of medium are added to the reservoir and perfusion at 20 ml/min is carried out for 5 min, discarding the effluent. Then the perfusion rate is increased to 32 ml/min and terminated when the return flow drops to 24 ml/min and channels are seen on the surface of the liver. There may also be a slight rise in hydrostatic pressure. About 20 min is required. However, with experience, it is probably not necessary to monitor pressure and flow to decide when a liver is ready. The problem relates mainly to the variability of collagenase from batch to batch. Evidently, small noncollagenase protease contamination is important to the process. Many workers experiment with several batches of enzyme and maintain a reserve stock.

5. The liver is removed, cut into pieces, mashed in a Petri dish with the bottom of a beaker, and filtered through silk or gauze. The debris is washed with 30 ml of medium. After incubation for 6 min at 37°, the cells are centrifuged at room temperature at 50 g for 2 min. With this method, Wittman and Jacobus report that 89 ± 3.0% of the cells excluded Trypan Blue when suspended in a 0.025% solution of the dye.

Krul et al.[51] point out that Trypan Blue, while convenient, is not as accurate a measure of cell viability as can be obtained by measuring lactate dehydrogenase in the cells and medium. They also reported that there is an excellent linear correlation between total LDH activity and total cell protein.

The advantages of fresh hepatocyte suspensions are considerable. The drawbacks relate mainly to the damage to the cell membrane which occurs during the digestion procedure, which may be important for receptor studies. Capuzzi et al.[52] found that enzyme activity remains bound to the hepatocyte surface even after repeated washing. They also reported specific cleavage of apoC, though addition of albumin to the medium decreased the extent of apolipoprotein proteolysis.

Several investigators have found that VLDL triglyceride output by fresh hepatocyte suspensions is comparable to that obtainable in liver perfusions, of the order of 0.4 μmol/g liver/hr without added fatty acids, and greatly increased triglyceride output occurs when the free fatty acid concentration in the medium is raised. Krul et al.[51] reported that about 15 μg/g/hr of apoB and apoE were secreted in a hepatocyte suspension incubated for 24 hr. ApoA-I and apoC secretion, however, was much lower. The low rate of apoA-I secretion is in accord with perfusion

[51] E. S. Krul, P. J. Dolphin, and D. Rubinstein, Can. J. Biochem. **59,** 676 (1981).
[52] D. M. Capuzzi, C. E. Sparks, and J. L. DeHoff, Biochem. Biophys. Res. Commun. **90,** 587 (1979).

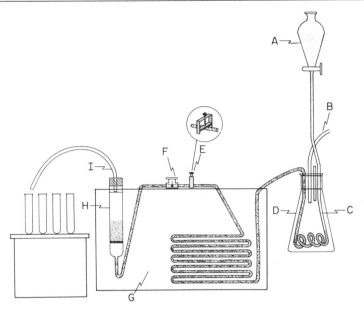

FIG. 3. Apparatus for perfusion of suspended hepatocytes. (A) Reservoir for medium; (B) inflow for 100% O_2; (C) silastic tubing for gas exchange; (D) stoppered flask for oxygenation of medium; (E) clamp; (F) three-way stopcock with inlet for attachment of syringe; (G) 37° water bath; (H) column (50-ml capacity) containing hepatocytes; (I) outlet tubing for effluent. (Reprinted from Capuzzi et al.[53] by permission.)

results, but values for apoC are conflicting. ApoC is labeled quite well in perfusion systems, and our laboratory has reported values for apoC output close to those for apoB (Table III). In view of the fact that some cultured hepatocyte systems show almost no apoC synthesis, it would appear that there may be biological as well as technical problems involved, such as losses during isolation or proteolysis. It would be useful to investigate apoC production using immunoassay procedures, which have not so far been employed. Proteolysis fragments of apoB and apoE can have about the same size as a C apolipoprotein, and SDS–PAGE alone cannot be relied upon to distinguish the two.

One of the disadvantages of the hepatocyte suspension system, which also applies to recirculating perfusions, is the exposure of the secreted lipoprotein to the cells. Although this surely occurs *in vivo*, the secreted lipoproteins enter the systemic circulation before returning to the liver. Capuzzi et al.[53] have devised a perfusion system which circumvents this

[53] D. M. Capuzzi, R. D. Lackman, and G. G. Pietra, *J. Cell. Physiol.* **108**, 185 (1981).

TABLE IV
METHODS OF INVESTIGATING HEPATIC LIPOPROTEIN SYNTHESIS IN
INTACT CELL SYSTEMS

Cell system	Advantages	Disadvantages
Whole animal	Physiological	Limited control of variables; intestinal contribution must be estimated
Isotope incorporation	Applicable to non-steady-state	Precursor isotope enrichment hard to measure
Turnover	Synthesis equals catabolism in steady state	Harder to apply to non-steady-state; *in vitro* labeling may alter molecule
Liver perfusion	Close to physiological	One liver, one experiment
Recirculating	Close to physiological	Hepatic catabolism of secreted products
Single pass	Clearance and binding easily measured; synthesis products removed from liver	Large volumes of medium produced
Liver slice	Easy and rapid preparation of large amounts of tissue, readily aliquoted	Anoxic cells in center of slice; leakage from cut cells
Liver snip	Less damage to cells (than slices)	Limited amount of tissue, longer preparation time
Fresh hepatocytes	Large numbers of cells, easy to aliquot; removal of nonhepatocytes	Cell membrane alteration; proteases not entirely removed
Cultured hepatocytes	Control over culture medium, longer term experiments are possible, cell membrane repaired	No cell growth, limited number of cells, sterility required
Organ (liver) culture	Applicable to biopsies	Mixed cell population, other cell types may predominate

problem, although it has the disadvantage that hepatocytes from fasted rats are difficult to handle because of the decreased density resulting from a lack of glycogen. A diagram of the perfusion system is shown in Fig. 3.

Summary

Methods for the study of hepatic lipoprotein synthesis and secretion have been described, and the advantages of each system discussed. Attention has been focused on intact cell systems. The isolated perfused liver constitutes a standard of comparison for all others, even though it, too, has limitations. Table IV below gives our assessment of some of the

advantages and disadvantages of the techniques for studying hepatic lipoprotein biosynthesis.

Acknowledgments

Thanks are due to Mrs. Elise Winkler for carrying out the experiments relating to pressure and flow in the perfused rat liver. I also thank my departmental colleagues who read and criticized the manuscript. Work in my laboratory was supported in part by grants HL 22633 and HL 07443 from the NIH and by funds from the W. W. Smith Charitable Trust.

[31] High-Density Lipoprotein Formation by the Intestine

By ROBERT M. GLICKMAN and ARTHUR M. MAGUN

In recent years the intestine has been studied as an active site of
lipoprotein biosynthesis. Together, the liver and the intestine are the
major sites of lipoprotein biosynthesis in all species studied. Initial attention focused on the intestine as a site of dietary lipid absorption and
chylomicron formation. These studies demonstrated that the intestinal
epithelial cell actively synthesizes many of the protein and lipid components of chylomicrons during triglyceride absorption. Specifically, triglyceride and phospholipid synthesis as well as cholesteryl ester formation
occur via microsomal enzyme complexes. Specific apolipoprotein synthesis is also required for chylomicron formation. Experimental inhibition of
protein synthesis results in impaired chylomicron secretion[1] and heredi-
tary inability to synthesize apoB (abetalipoproteinemia) results in the
inability to form or secrete chylomicrons.[2] Thus, apolipoprotein synthesis
is an integral component of chylomicron formation. Since the intestine
absorbs large quantities of dietary lipid daily (~ 100 g in man), chylomicron components constitute quantitatively significant amounts of these
constituents which enter the peripheral blood. As detailed in Chapter [25],
this volume, lipoproteins of intestinal origin contribute important apolipoprotein and lipid constituents to plasma lipoproteins. Estimates in the rat
and in man[3,4] indicate that as much as 40–50% of daily synthesis of two

[1] R. M. Glickman and K. Kirsch, *J. Clin. Invest.* **52,** 2910 (1973).
[2] H. B. Salt, O. H. Wolf, J. K. Lloyd, A. S. Fosbrooke, A. H. Cameron, and D. V. Hubble, *Lancet* **II,** 325 (1960).
[3] H. G. Windmuller and A. L. Wu, *J. Biol. Chem.* **256,** 3012 (1981).
[4] P. H. R. Green, R. M. Glickman, C. D. Saudek, C. B. Blum, and A. R. Tall, *J. Clin. Invest.* **64,** 233 (1979).

apolipoproteins, apoA-I and apoA-IV, originates in the intestine. While both of these apolipoproteins are synthesized as chylomicron components in both rat and man, in plasma they are rapidly redistributed in large measure to plasma HDL.[5,6] Thus the intestine is intimately involved in plasma HDL metabolism. This chapter will address the role of the intestine in HDL metabolism and stress those techniques, which, while addressed in other chapters in this volume, have been modified for the study of intestinal lipoproteins. We will specifically review the evidence for the direct synthesis and secretion of HDL particles. Other aspects of intestinal lipoprotein formation have been recently reviewed.[7]

Studies of Mesenteric Lymph HDL

Distribution of ApoA-I in Mesenteric Lymph

The studies which follow have been carried out in the mesenteric lymph fistula rat which is prepared by direct cannulation of the main mesenteric lymph duct.[8] This preparation has the following advantages: (1) Contribution from hepatic lymph is negligible but may be as much as 30% with thoracic duct cannulation. (2) Experiments are carried out 16–24 hr after surgery in conscious, restrained rats with stable lymph flow. (3) Duodenal cannulation permits lipid infusion under steady state conditions monitored by lymph assessment.

We have employed this model to study various parameters of triglyceride absorption and chylomicron secretion. In the course of these studies we determined that the intestine is a major secretory source of apoA-I as assessed by the secretion of apoA-I in mesenteric lymph. Since apoA-I is a major chylomicron apolipoprotein it was assumed that most lymph apoA-I should be associated with triglyceride-rich lipoproteins. It therefore was of interest when lymph was separated into $d < 1.006$ and $d > 1.006$ g/ml fractions and the distribution of apoA-I determined in each fraction. Approximately two-thirds of apoA-I was associated with the more dense fractions of lymph. Further fractionation showed that 60% of this apoA-I was associated with the HDL fraction of lymph.[9] It was of importance to demonstrate that the source of apoA-I in the HDL fraction

[5] A. R. Tall, P. H. R. Green, R. M. Glickman, and J. W. Riley, *J. Clin. Invest.* **64,** 977 (1979).
[6] E. J. Schaefer, L. L. Jenkins, and H. B. Brewer, Jr., *Biochem. Biophys. Res. Commun.* **80,** 405 (1978).
[7] P. H. R. Green and R. M. Glickman, *J. Lipid Res.* **22,** 1153 (1981).
[8] J. L. Bollman, J. C. Cain, and J. H. Grindlay, *J. Lab. Clin. Med.* **33,** 1349 (1948).
[9] R. M. Glickman and P. H. R. Green, *Proc. Natl. Acad. Sci. U.S.A.* **74,** 2569 (1977).

did not derive directly from triglyceride-rich lipoproteins. The following evidence indicated that this was not the case: (1) Chylomicrons when suspended in buffer or in $d > 1.21$ g/ml bottom from lymph, lost only 8% of apoA-I and none was recoverable in the HDL density range. (2) No lipolytic activity was demonstrated in lymph using labeled triglyceride as substrate. (3) As discussed below, biliary diversion which depletes mesenteric lymph of triglyceride-rich lipoproteins results in no significant decrease in lymph apoA-I, strongly suggesting that there are triglyceride-independent mechanisms for apoA-I secretion.

Characterization of Mesenteric Lymph HDL

In view of the finding that a significant proportion of lymph apoA-I was associated with the HDL fraction it was of interest to characterize this fraction both morphologically and chemically. As shown in Fig. 1, using the technique of negative stain electron microscopy (Chapter [26], Volume 128), the HDL fraction d 1.070–1.21 g/ml was examined. It is apparent that morphologically this fraction is polydisperse, containing spherical HDL of varying sizes. In addition, larger forms are apparent which form rouleaux and have a discoidal morphology. Plasma from normal animals contain no discoidal forms and therefore they are unlikely to be filtered from plasma.

In addition to morphological differences from plasma HDL, lymph HDL exhibit marked compositional changes as well. Lymph HDL are relatively enriched in phospholipid and poor in cholesteryl ester. In part, this is compatible with the discoidal morphology described above where there is a paucity of core lipids. The apolipoprotein pattern of lymph HDL as assessed by SDS–polyacrylamide electrophoresis shows that this fraction is enriched in apoA-I and poor in apoE, a characteristic of intestinal lipoproteins.

The compositional data represent the analysis of the whole HDL fraction of mesenteric lymph and clearly tends to blur differences among the various subfractions of HDL particles within this fraction. We therefore attempted to subfractionate mesenteric lymph HDL using a variety of techniques. Initially we studied the distribution of HDL particles using isopycnic density gradient centrifugation (Chapter [7], Volume 128). Figure 2 shows the equilibrium densities of both plasma and lymph HDL. While somewhat broad, the peaks appear unimodal with lymph HDL exhibiting a lower density than plasma HDL. There are no separable discrete peaks representing different particle types. Fractions of the density gradient were examined by negative stain electron microscopy (Fig. 3). The $d < 1.10$ fraction is greatly enriched in disks but also contains

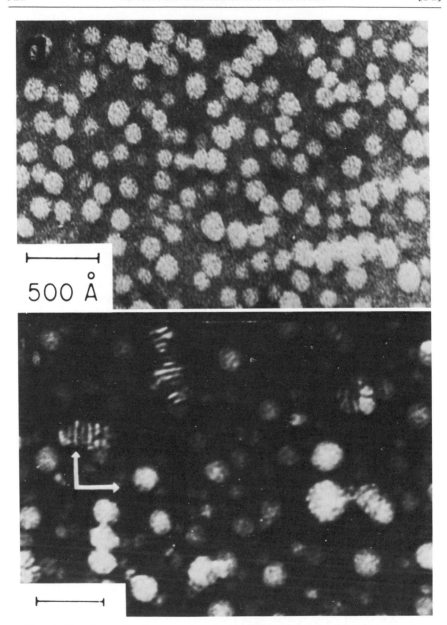

FIG. 1. Negative stain electron micrographs of the 1.07–1.21 g/ml (HDL) fraction of rat plasma (top) and lymph (bottom). The lymph contains disks forming rouleaux. From P. H. R. Green, A. R. Tall, and R. M. Glickman, *J. Clin. Invest.* **61,** 528 (1978).

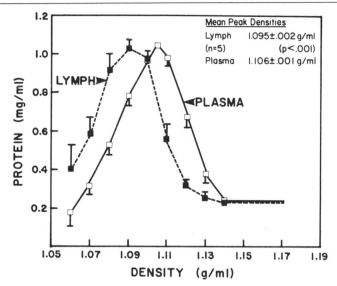

FIG. 2. The protein distribution of rat plasma and lymph HDL fractions on an isopycnic density gradient. From Forester *et al.*[10]

some spherical HDL. The $d > 1.10$ fraction contains few disks and is largely spherical HDL.

Further definition of HDL subfractions in mesenteric lymph was achieved by polyacrylamide gradient gel electrophoresis as described by Anderson and discussed in Chapter [24], Volume 128. Figure 4 shows the pattern of the whole HDL fraction of mesenteric lymph compared with plasma from the same animal. HDL from both sources have particles of overlapping sizes (10–12 nm) although some lymph particles had sizes of 13–14 nm in diameter. In addition, lymph but not plasma contains a population of small particles with an apparent diameter of 7.8 nm. Further evidence for the discrete nature of subfractions of lymph HDL is provided by examining density gradient pools of lymph HDL by gradient gel electrophoresis. As shown in Fig. 5, further definition of a discrete population of larger particles of diameter 13–14 nm is seen in the lymph fraction. These particles were confirmed to be discoidal by electron microscopy. Also the small spherical HDL fraction (d 7.8 nm) is more discrete and can be greatly enriched at $d > 1.13$ g/ml. When examined by electron microscopy this fraction contains a uniform population of small spherical particles.

As shown in Table I, significant compositional differences characterize the density gradient subfractions of mesenteric lymph HDL and are distinctive when compared to similar plasma HDL fractions. Of particular

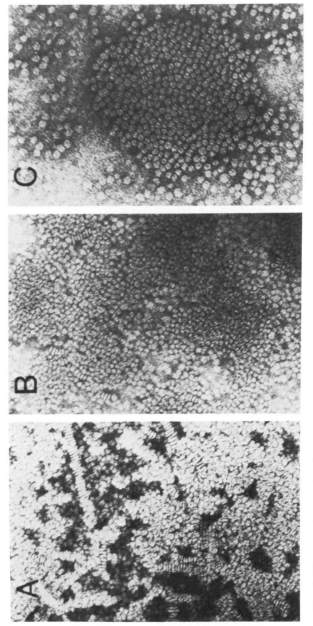

FIG. 3. Electron micrographs of the HDL fractions from rat lymph and plasma. (A) Fractions of $d < 1.10$ g/ml contain disks and spheres. (B) Fractions of $d > 1.10$ contain smaller spherical particles and fewer disks. (C) Plasma HDL. From Forester *et al.*[10]

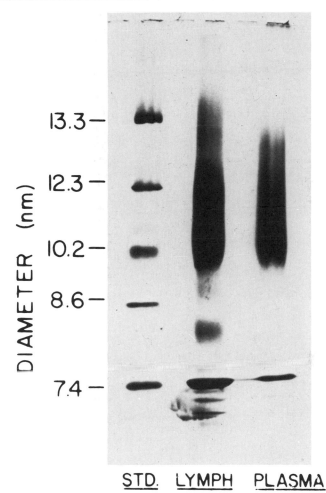

FIG. 4. Polyacrylamide gradient gel electrophoresis of the whole HDL fraction (1.07–1.21 g/ml) of rat lymph and plasma. Lymph contains 7.8-nm particles not seen in plasma. From Forester *et al.*[10]

note is that all lymph HDL fractions are phospholipid rich and cholesteryl ester poor when compared to plasma HDL. The fatty acids of the cholesteryl esters of lymph HDL are enriched in saturated fatty acids when compared to plasma fractions, suggesting ACAT rather than LCAT action in the formation of lymph HDL cholesteryl esters. Also of note is the enrichment of all lymph HDL subfractions with apoA-I, characteristic of intestinal lipoproteins.

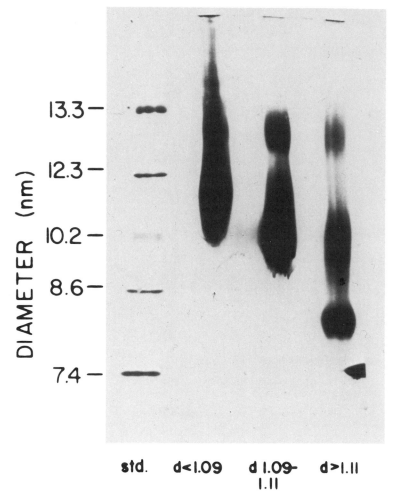

FIG. 5. Polyacrylamide gradient gel electrophoresis of lymph HDL recovered from iso-pycnic density gradient fractions. The $d > 1.11$ g/ml fractions contain the 7.8-nm particle. From Forester *et al.*[10]

Additional studies were carried out to determine whether the core cholesteryl esters of lymph HDL could be newly labeled after intraduodenal infusion of [^{3}H]cholesteryl to lymph fistula rats.[10] Lymph HDL fractions incorporated the isotope to a specific activity of 10–45 times that of plasma fractions, with the highest specific activity in small lymph HDL

[10] G. Forester, A. R. Tall, C. L. Bisgaier, and R. M. Glickman, *J. Biol. Chem.* **258**, 5938 (1983).

TABLE I
COMPOSITION OF RAT MESENTERIC LYMPH AND PLASMA HDL[a]

	Protein/lipid	PL/CE	$\dfrac{\text{ApoA-I}}{\text{ApoE}}$	Fatty acids (%) (18:2 + 20:4)
Plasma HDL				
d 1.07–1.10	0.48	1.8	1.0[b]	64
d 1.10–1.13	0.73	2.0	4.2	60
d 1.13–1.18	1.0	2.8	5.5	32[c]
Lymph HDL				
d 1.07–1.10 (discoidal)	0.53	2.4	5.1	68
d 1.10–1.13	0.69	3.0[d]	4.8	48[b]
d 1.13–1.18 (small spherical particles)	1.5[b]	4.8[b]	5.5	20[b]

[a] Modified from Forester et al.[10]
[b] $p < 0.05$. Significant difference from other fractions.
[c] $p < 0.05$. Significant difference from other plasma fractions.
[d] $p < 0.05$. Significant difference from complete plasma fraction.

of d 1.13–1.18 g/ml (specific activity 45 times greater than plasma). The absence of cholesteryl ester transfer protein in the rat precludes transfer of cholesteryl ester among lymph lipoproteins. Incubation studies of lymph lipoproteins also failed to show transfer. Thus, the above studies have characterized morphologically and chemically distinct forms of mesenteric lymph HDL not normally present in plasma. Evidence is provided that the core cholesteryl ester of these particles is synthesized in the intestine and these particles are of secretory origin.

Effect of Biliary Diversion on Mesenteric Lymph HDL

Further studies were conducted to define the relationship of intestinal lipid flux to the presence and secretory rate of lymph HDL. Since these particles are not major transport vehicles for triglyceride, it was of interest to explore mesenteric lymph HDL metabolism during an experimental condition of interrupted triglyceride flux across the intestine.[11] Biliary diversion is known to severely reduce the entry of triglyceride into the enterocyte with subsequent depletion of the intestinal mucosa of triglyceride and marked reduction in the secretion of lymph triglyceride. Diversion is accomplished via an exteriorized catheter connecting the common bile duct to the duodenum which can divert pancreatobiliary secretions.

[11] H. R. Bearnot, R. M. Glickman, L. Weinberg, P. H. R. Green, and A. R. Tall, *J. Clin. Invest.* **69,** 210 (1982).

The main mesenteric lymphatic duct was also catheterized. After a period of 12–16 hr of postoperative stabilization the biliary catheter was externally divided resulting in acute biliary diversion without further operative risk. After 24 hr of biliary diversion the catheter was reconnected, reconstituting the enterohepatic circulation. Figure 6 shows the secretion of apoA-I and its distribution in lymph during periods of biliary diversion. Of note is that despite the marked reduction in $d < 1.006$ lipoproteins with biliary diversion, apoA-I secretion is largely unaffected, suggesting that lymph apoA-I secretion can be maintained by means other than chylomicron secretion. In bile-diverted animals, 80% of apoA-I in the more dense fractions of lymph resided in the HDL fraction. Negative stain electron microscopy of bile-diverted lymph revealed discoidal as well as spherical particles and was similar to nondiverted lymph with the exception that discoidal particles were somewhat smaller (126 vs 165 Å). Compositional analysis of lymph HDL from bile-diverted animals showed that these particles are phospholipid and apoA-I enriched when compared to plasma. We next wished to determine whether the apoA-I found in the lymph of bile-diverted animals was newly synthesized and what proportion could be filtered from plasma. We chose to infuse radiolabeled HDL

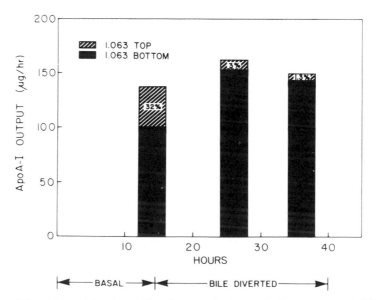

FIG. 6. Lymph apoA-I output in basal state animals and in the same animals following biliary diversion. Though the distribution of apoA-I shifts in the $d < 1.063$ to the $d > 1.063$ particles, the total amount of apoA-I in lymph does not change with bile diversion and consequent loss of chylomicrons.

TABLE II
SPECIFIC ACTIVITY OF LYMPH HDL (ApoA-I) AFTER INTRAVENOUS
INFUSION OF ^{125}I-LABELED HDL (ApoA-I)[a]

| | AI (cpm/μg) | | Lymph |
	Lymph	Serum	Serum
Lymph and bile diverted			
1	341 ± 53	1126 ± 92	0.303
2	80 ± 3	626 ± 12	0.128
3	240 ± 7	940 ± 28	0.256
			Mean = 0.229
Lymph diverted			
1	344 ± 9	945 ± 40	0.360
2	174 ± 28	953 ± 48	0.183
3	104 ± 14	335 ± 23	0.310
4	177 ± 17	597 ± 7	0.280
			Mean = 0.285

[a] From Bearnot et al.[11]

intravenously into bile-diverted, lymph-diverted animals to a constant specific activity and then measure simultaneous specific activities of plasma and lymph. We iodinated purified apoA-I using the chloramine-T method[11] and exchanged this ^{125}I-labeled apoA-I onto plasma HDL by incubation and reflotation of HDL at 1.070–1.21 g/ml; 100% of ^{125}I was TCA precipitable and 90% of counts were found in apoA-I of the HDL. These HDL had a normal half-life ($t_{1/2}$ = 10.5 hr) when infused into donor rats. Table II shows the specific activities in plasma and lymph after a steady state level of radioactivity had been reached in plasma. It can be seen that lymph-specific activities were less than 30% of plasma in both bile-diverted and nondiverted animals. Thus 70% of lymph apoA-I in these animals did not originate from plasma and was secreted into mesenteric lymph.

These data indicate that mesenteric lymph HDL are not dependent on triglyceride secretion and the forms of HDL found in lymph do not derive from coexistent triglyceride rich-lipoproteins in lymph, since these particles are absent from bile-diverted lymph.

Mucosal Aspects of HDL Synthesis

A final series of studies investigated apoA-I synthesis and HDL formation within intestinal epithelial cells. It should be noted that HDL particles have never been visualized within epithelial cells by electron microscopy. The reasons for this could relate to the small size of these particles as well as being obscured by adjacent larger particles in organelles such as the Golgi apparatus. It therefore was of importance to attempt to isolate HDL particles directly from rat intestinal epithelial cells.

Preparation of Dissociated Epithelial Cells and Isolation of Intracellular HDL

Since mucosal scrapings contain variable amounts of interstitial fluid or blood, the starting point for isolation of HDL were washed, dissociated cells. This was carried out by a modification of the technique of Weiser.[12] The entire intestine is removed, clamped at one end, and filled with buffer A (27 mM sodium citrate, 96 mM NaCl, 1.5 mM KCl, 8 mM KH$_2$PO$_4$, 5.6 mM Na$_2$HPO$_4$). The loop is clamped at the other end and placed in a shaking water bath at 37° for 15 min. One clamp is removed and the buffer drained by gravity from the intestine. It is then refilled with buffer B (145 mM NaCl, 4 mM KCl, 5 mM NaH$_2$PO$_4$, 5 mM Na$_2$HPO$_4$, 1.5 mM EDTA, 0.5 mM DTT) and reincubated for 30 min. Dissociated cells are harvested, pelleted by low-speed centrifugation, and washed three times in phosphate-buffered saline. These cells are immunochemically free of albumin, which serves as a marker of extracellular fluid contamination. They are enriched in alkaline phosphatase, a microvillous membrane marker characteristic of mature villous cells. Two approaches were used to isolate cellular lipoproteins. One involved homogenizing cells in isotonic buffers and isolating intact Golgi organelles which were then ruptured to yield content lipoproteins. This was done according to the method of Howell and Palade[13] and is further discussed in Chapter [18], this volume. Content lipoproteins were then separated by density gradient centrifugation to yield an HDL fraction which was then analyzed. In general, yields from this procedure are small, with about 5 μg of HDL protein harvested. We developed an alternative strategy for isolating larger yields of intracellular lipoproteins by homogenization of dissociated cells in a hypotonic buffer, which liberates nascent lipoproteins from cellular organelles with subsequent isolation by gradient centrifugation of lipoprotein fractions.[14] Cells were suspended in ice-cold hypotonic Veronal buffer (37.5 mM sodium diethylbarbiturate, 7.3 mM barbituric acid, pH 8.3). The cells were dispersed in the buffer and then disrupted by nitrogen cavitation at 2000 psi for 15 min. Aliquots were then briefly sonicated on ice for 15 sec (Bran-

[12] M. M. Weiser, J. Biol. Chem. 248, 2536 (1973).
[13] K. F. Howell and G. E. Palade, J. Cell Biol. 92, 822 (1982).
[14] A. M. Magun, T. A. Brasitus, and R. M. Glickman, J. Clin. Invest. 75, 209 (1985).

son, microtip, 3.5 intensity, 50% setting). The solution was then repassed through the nitrogen cavitation apparatus at the same conditions and then centrifuged at 100,000 g for 1 hr. Lipoproteins were isolated from the supernatant by density gradient centrifugation to yield an HDL fraction. A schematic of these two methods of nascent HDL preparation is shown in Table III. Both methods yielded essentially comparable HDL fractions.

Characterization of Intracellular HDL

HDL from either method was subjected to isopycnic density gradient centrifugation and the distribution of apoA-I and apoB determined by radioimmunoassay across the gradient (Fig. 7). A discrete peak for both apolipoproteins is seen at a density of 1.12 g/ml. The material at the top of the gradient is lighter lipoproteins of $d < 1.070$ g/ml. It was of interest to determine whether both apoB and apoA-I resided on the same particle, since apoB is not normally a constituent of HDL.

Immunoelectrophoresis of intracellular HDL against anti-apoB and anti-apoA-I revealed that this fraction formed immunoprecipitation arcs with both antisera. However, these arcs had only partially overlapping mobility, indicating separate particles (Fig. 8). This was further verified by removal of apoB-containing particles either on columns of con-

TABLE III

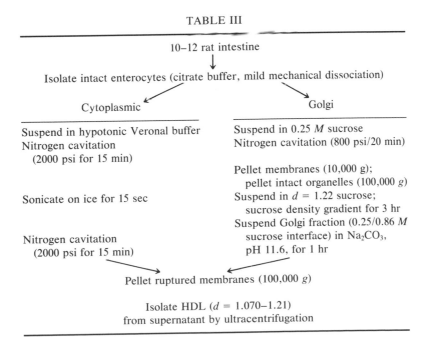

10–12 rat intestine

Isolate intact enterocytes (citrate buffer, mild mechanical dissociation)

Cytoplasmic   Golgi

Cytoplasmic	Golgi

Suspend in hypotonic Veronal buffer
Nitrogen cavitation
  (2000 psi for 15 min)
	Suspend in 0.25 M sucrose
	Nitrogen cavitation (800 psi/20 min)
	
	Pellet membranes (10,000 g);
	  pellet intact organelles (100,000 g)

Sonicate on ice for 15 sec	Suspend in d = 1.22 sucrose;
	  sucrose density gradient for 3 hr
	Suspend Golgi fraction (0.25/0.86 M
	  sucrose interface) in Na₂CO₃,

Nitrogen cavitation (2000 psi for 15 min)	pH 11.6, for 1 hr

Pellet ruptured membranes (100,000 g)

Isolate HDL ($d = 1.070–1.21$)
from supernatant by ultracentrifugation

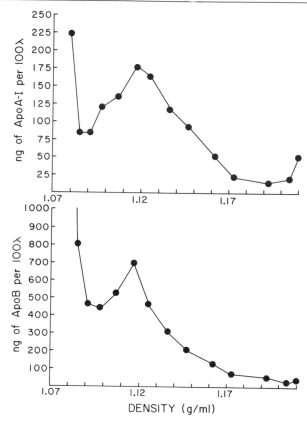

FIG. 7. Distribution of apoA-I (top) and apoB (bottom) as measured by radioimmunoassay of density gradient fractions of HDL isolated from a cytoplasmic intracellular preparation. From Magun *et al.*[14]

canavalin A or by adding an excess of apoB antiserum. When the remaining non-apoB-containing lipoproteins were reexamined by immunoelectrophoresis, only an apoA-I-containing arc was seen in the HDL region. These results confirm that apoA-I-HDL particles not containing apoB can be isolated from enterocytes. Negative stain electron microscopy of these particles revealed small spherical particles with a mean diameter of 8.2 nm. Thus, both the density and size of these particles are remarkably similar to small spherical HDL particles as described in rat mesenteric lymph. The lipid composition of these particles revealed that the major lipids were cholesteryl ester, phospholipid, and triglyceride. Triglyceride comprised 15–20% of the lipid, which is higher than circulating HDL, but has been found in other nascent HDL in comparable amounts.

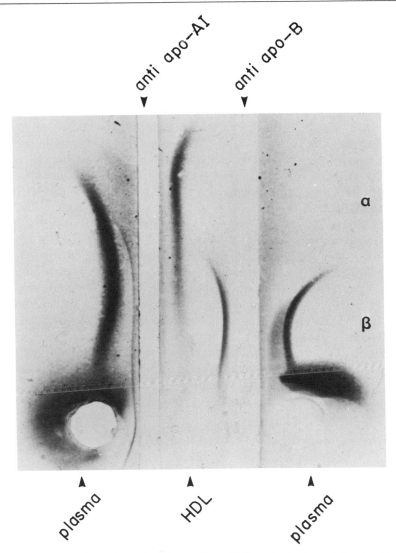

FIG. 8. Immunoelectrophoresis of intracellular HDL against anti-apoA-I and anti-apoB with plasma controls in outer lanes. The precipitin arc in the alpha region of the HDL fraction suggests that there is a particle with alpha mobility containing apoA-I and not containing apoB. From Magun et al.[14]

It is of potential concern that during the isolation of intracellular HDL, fragmentation of triglyceride-rich lipoproteins could have resulted in frag-ments containing apoB or apoA-I. This possibility is unlikely since similar HDL isolations were achieved from the mucosa of bile-ligated animals wherein there is a marked depletion of mucosal triglyceride. In addition, mesenteric lymph chylomicrons were subjected to identical preparative conditions and examined on isopycnic density gradients. No apolipoproteins were found in the density range of intracellular HDL.

Finally, experiments were conducted to determine if the apolipoproteins associated with intracellular HDL were newly synthesized. [³H]Leucine was administered to rats 10 min before sacrifice and HDL isolated. Figure 9 shows the pattern of radioactivity. Three apolipoproteins known to be synthesized by the intestine (apoB, A-I, A-IV) actively incorporated radioactivity. When excess anti-apoB antiserum was used to deplete apoB-containing particles, a clean separation between apoB- and apoA-I-containing particles was achieved. Thus, intracellular HDL contain apoA-I and A-IV.

The above data, taken together, indicate that HDL particles can be synthesized and secreted by the intestine. Evidence has been provided that small spherical HDL particles contain newly synthesized surface and core components and they are actively secreted into lymph. Discoidal HDL, a prominent component of mesenteric lymph, were not recovered from intracellular HDL preparations. It is unlikely that sufficient LCAT exists intracellularly to convert these discoidal to spherical forms. It is possible that discoidal particles are not secreted via Golgi organelles. Alternatively, they could be formed on the basolateral surface of the enterocyte, though direct proof for this is lacking. The finding of discoidal particles in peripheral lymph (Chapter [39], this volume) could support this possibility.

Contribution of the Intestine to HDL Metabolism

The small intestine makes a significant contribution to systemic HDL metabolism via several different mechanisms. As discussed in Chapter [28], this volume, surface components from chylomicrons contribute to plasma HDL through redistribution of phospholipid and apolipoproteins with subsequent remodeling to mature plasma HDL. Surface constituents from hepatic VLDL are similarly processed. The magnitude of this contribution is unknown. However, in the absence of chylomicron or VLDL secretion (as in abetalipoproteinemia), plasma HDL levels are reduced by approximately 50%. The relative contributions of the liver and intestine are unknown.

The data in this chapter suggest an additional contributing mechanism to plasma HDL—direct synthesis and secretion of HDL particles by the intestine. In addition, similar particles have been isolated from rat liver perfusates and from rat hepatocytes. Small, spherical HDL particles have

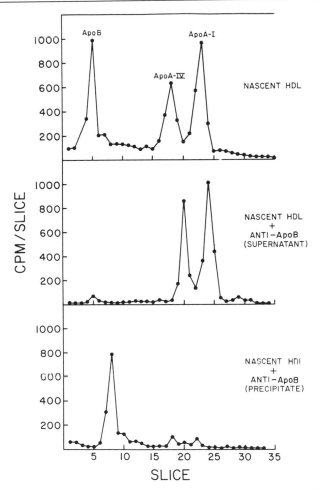

Fig. 9. SDS–polyacrylamide 5.6% gel slices of the HDL fraction prepared from rats that had received [³H]leucine 10 min prior to removal of their intestines. Top: The delipidated HDL fraction shows that there are three newly synthesized apolipoproteins, corresponding in R_f to apoB, apoA-IV, and apoA-I. Middle: ApoB-containing particles were immunoprecipitated with antiserum and the fraction delipidated. The gel reveals that apoA-IV and apoA-I remained in the supernatant. Bottom: The immunoprecipitant contains predominantly apoB. From Magun *et al.*[14]

been described in human plasma. In particular, conditions of LCAT deficiency (genetic or acquired) are associated with prominent smaller HDL.[15] These particles have also been described in abetalipoproteinemia,

[15] T. M. Forte, K. R. Norum, J. A. Glomset, and A. V. Nichols, *J. Clin. Invest.* **50,** 1141 (1971).

eliminating the possibility that they are derived from the catabolism of triglyceride-rich lipoproteins.[16] Recent work has shown that through the action of LCAT these particles can acquire additional core cholesteryl ester and transform into larger HDL.[17] It is unknown what contribution the direct synthesis and secretion of intestinal HDL makes to systemic HDL levels. In addition, factors influencing the synthesis and secretion of these particles are of great interest and the subject of current investigation.

[16] A. M. Scanu, L. P. Aggerbeck, A. W. Kruski, C. T. Lim, and H. J. Kayden, *J. Clin. Invest.* **53**, 440 (1974).
[17] C. Chen, K. Applegate, W. C. King, J. A. Glomset, K. R. Norum, and E. Gjone, *J. Lipid Res.* **25**, 269 (1984).

[32] Phosphorylation–Dephosphorylation of Apolipoprotein B

By ROGER A. DAVIS and ROY A. BORCHARDT

Introduction

There is little doubt of the physiologic importance of apolipoprotein B. In addition to being essential for the secretion of VLDL, apoB plays a fundamental role in targeting cholesterol, as a component of LDL, to tissues via specific high-affinity receptors. Although there are sufficient data demonstrating the physiological and pathophysiologic importance of apoB, little is known about its structure, sequence, and posttranslational modifications. In fact, there still exists some question as to the actual monomeric molecular weight of apoB. Using both SDS–PAGE and molecular permeation chromatography at least two distinct molecular weight forms of apoB are resolvable.[1–5] Kane and co-workers use the centile system of nomenclature.[2] However, since there still exists some doubt as

[1] K. V. Krishnaiah, L. F. Walker, J. Borensztajn, G. Schonfeld, and G. S. Getz, *Proc. Natl. Acad. Sci. U.S.A.* **77**, 3810 (1980).
[2] J. P. Kane, D. A. Hardman, and H. E. Paulus, *Proc. Natl. Acad. Sci. U.S.A.* **77**, 2465 (1980).
[3] C. E. Sparks and J. B. Marsh, *J. Lipid Res.* **22**, 519 (1981).
[4] L. Wu and H. G. Windmueller, *J. Biol. Chem.* **256**, 3615 (1980).
[5] J. Elovson, Y. O. Huang, N. Baker, and R. Kannan, *Proc. Natl. Acad. Sci. U.S.A.* **78**, 157 (1981).

to the exact molecular weights, we use "big B" to designate B_{100} and "little B" to designate B_{48}.

Functional differences exist between the two B forms. Metabolic state differentially affects the synthesis of big and little B. Little B is removed more rapidly from plasma than is big B. In addition, monoclonal antibody studies suggest that an epitope found exclusively on big B is also associated with the part of the molecule which recognizes the "LDL" receptor. These data suggest that there exists differences in the structure as well as the regulation and/or pathways for the metabolism and synthesis of the different apoB forms.

In our attempts to define the biosynthetic differences our attention was directed to the phosphorylation of apoB. All evidence obtained so far suggests that little B is initially secreted by the rat hepatocyte as a phosphoserine-containing protein. Little or no phosphoserine is associated with big B. In this report we describe the methods used to demonstrate apoB phosphorylation.

Materials

1. Rat hepatocyte culture model (description available in this volume)
2. High-titer monospecific rabbit IgG directed against SDS denatured delipidated apoB
3. SDS–PAGE system (1–20% linear gradient of acrylamide)
4. Paper electrophoresis system

Procedure

Antibody Preparation: It is essential when performing these studies to have a high-titer antibody that recognizes apoB in an SDS denatured, delipidated form. Also, it is advantageous if the antibody has a high affinity for protein A. For this reason we used an SDS denatured, delipidated form of purified apoB to immunize rabbits. The antiserum obtained was further purified using protein A–Sepharose affinity chromatography.

1. A $d < 1.21$ g/ml fraction (containing 5 mg of protein) is obtained from rat serum. The sample is dialyzed against 0.1 M phosphate, 5 mM DTT, 10 mM EDTA, 0.01% NaN$_3$. The sample is lyophilized and then extracted according to the procedure of Cardin *et al.*,[6] i.e., three times with ether/ethanol (3 : 1, v/v), once with ether, and once with ethanol. Include 0.02% BHT in solvents to help prevent autooxidation. The protein sample is not allowed to become dry throughout the extraction proce-

---

[6] D. Cardin, K. R. Witt, C. L. Barnhart, and R. L. Jackson, *Biochemistry* **21**, 4503 (1982).

dures, which are performed at 0°. The final protein residue is dissolved in 5 ml of sample buffer (i.e., 0.1 M phosphate, 5 mM DTT, 10 mM NaN$_3$, 0.5 mM PMSF, and 10% SDS; final pH 8.2). Boil the sample if necessary to get the protein into solution.

2. The proteins are then separated using Sepharose CL-6B, according to the procedure of Sparks and Marsh.[3] The sample is loaded onto a 2-m-long, 1 cm i.d. glass column which is pumped via a peristaltic pump connected at the bottom. After loading at a flow rate of 8 ml/hr, the sample is eluted with a buffer which is the same as the sample buffer except it contains 1% SDS (instead of 10%). Samples are collected in 60-drop fractions using a fraction collector. The elution is monitored at 280 nm. The peaks corresponding to apoB (see reference 3) are analyzed using SDS–PAGE. Fractions containing exclusively big or little B are dialyzed further as described[6] and used as antigen for inoculation of rabbits.

3. After at least two injections of the apoB preparation into rabbits, they are bled weekly. The sera from these rabbits are tested frequently for the presence of anti-apoB antibodies. It is a good idea to electroblot the antiserum, in order to assess its specificity. We found that polyclonal rabbit antibody produced in response to either form of apoB was equally effective in reacting with both apoB forms.

4. After production of specific antiserum, the IgG fraction is obtained using protein A–Sepharose affinity chromatography. This ensures that the IgG used will have high affinity for protein A.

Isolation of ^{32}P-Labeled ApoB from Cultured Rat Hepatocytes

Hepatocytes are obtained, plated, and cultured as described in Chapter [17] of this volume, using 60-mm Petri dishes containing 3.9×10^6 cells. After plating for 4 hr, the culture medium is changed to serum-free medium that does not contain phosphate. At this time [^{32}P]orthophosphate (3 mCi/2 ml) is added to the culture medium. The cells are incubated in a CO_2 incubator for the desired time (in most instances, we do a 4-hr labeling). The medium is drawn off with suction, floating cells are removed following centrifugation, and the medium is transferred to test tubes. Hot buffer B (0.5 ml) is added, followed by the addition of 2.5 ml of buffer D with rapid mixing.[7] The solution is then boiled for 5 min. As quickly as possible, 0.5 ml of boiling buffer B is added to the culture dishes. The cells are lysed immediately, as evidenced via phase contrast

---

[7] J. R. Faust, K. L. Luskey, D. J. Chin, J. L. Goldstein, and M. S. Brown, *Proc. Natl. Acad. Sci. U.S.A.* **79**, 5205 (1982).

microscopy. (We have found that dephosphorylation of apoB is effectively prevented by treating cells with this procedure.) The viscous cell lysate is then brought up to a volume of 5 ml by adding 4.5 ml of buffer D that does not contain SDS. The solution is then boiled for 5 min and thoroughly mixed.

Aliquots (0.5 ml) of the cell and medium are placed in 1.5-ml plastic microfuge tubes. An appropriate amount of IgG is added that has been determined to quantitatively bind all of the apoB present. The tubes are then shaken at room temperature for 4 hr. An amount of protein A–Sepharose found to bind all the IgG is added. After an additional 1-hr incubation, the Sepharose is spun down for 5 min in a microfuge. The supernatant is aspirated and 1 ml of buffer D is added and the sample is mixed vigorously. The sample is spun again. The washing with buffer D is repeated five more times. The final pellet is brought up in 25 μl of the sample buffer containing 10 mM Tris · glycine (pH 8.3), 2% SDS, 8 M urea, 10% glycerol, and 5 mM mercaptoethanol. The sample is boiled for 25 min and then cooled before being placed on a 1–20% polyacrylamide linear gradient slab gel. Electrophoresis is performed as described by Laemmli,[8] except we do not use a stacking gel. The samples are electrophoresed for 3–4 hr, until the pyronin Y dye front reaches the bottom of the gel. The gels are immediately stained with Coomassie Brilliant Blue G-250. After an overnight incubation in stain, the gels are destained (10% acetic acid) and shrunk in 2% glycerol, 50% methanol (for 1 hr). The gels are dried onto filter paper. A nice technique we have learned from our colleagues is to make a "radioactive pen" by adding [^{35}S]methionine to a felt marking pen. By marking the filter paper with this pen, we can match up the ink with the autoradiographic marks. This ensures an exact alignment of the autoradiographic bands with the Coomassie-stained protein standards. Also, it provides a permanent identification of the autoradiogram. The dried gels are subjected to autoradiography using Kodak XRP-5 X-ray film and two Dupont Cronex Lightening Plus intensifying screens at −76°. After 1–5 days, the X-ray film is developed.

To quantitate the amount of radioactivity associated with apoB, the areas of the gel corresponding to the apoB Coomassie Blue-stained standards are cut out of the gel and assayed in a scintillation counter.

In Vivo Phosphorylation of ApoB

Precautions. Since ^{32}P is a high-energy emitter, there are precautions one must take to prevent contamination. All work should be done in a

---

[8] U. K. Laemmli, *Nature (London)* **227,** 680 (1970).

radioactive confinement hood. Absorbant paper should cover all areas in contact with the rat. Lead shielding (2 mm) surrounding $\frac{1}{2}$-in. Plexiglas should surround the entire experiment. A sensitive Geiger counter should be available to monitor possible contamination.

Procedure

[^{32}P]Orthophosphate (10 mCi) is injected ip into a 200-g rat. After 4 hr the rat is killed by exsanguination and the liver is removed. The liver is homogenized in buffer D (40 ml). The homogenate is boiled for 15 min and an aliquot is subjected to immunoprecipitation, SDS–PAGE, and autoradiography as described above. A $d < 1.21$ g/ml lipoprotein fraction is obtained by ultracentrifugation of the serum. The $d < 1.21$ g/ml fraction is added to 0.5 ml of boiling buffer B and diluted with buffer D without SDS as described above. An aliquot is subjected to immunoprecipitation, SDS–PAGE, and autoradiography as described above.

Isolation of ^{32}P-Labeled ApoB

There are many analyses that require ^{32}P-labeled apoB to be in a pure water-soluble form. This can be achieved by isolating the apoB previously separated by SDS–PAGE. The gel containing ^{32}P-labeled apoB is sliced out (being careful not to include any protein other than apoB). The gel pieces from 1 or more gels (depending upon the amount of radioactivity needed) is placed in a 5-ml plastic pipet over a cotton plug. Dialysis tubing is placed on the bottom of the tube. Another cotton plug is placed on top of the gel pieces. Electrophoresis sample buffer (100 μl; see above) is layered on top of the top cotton plug. Electrode buffer is then added to fill the top of the tube. The sample is then subjected to electrophoresis for 18 hr, using a standard tube gel apparatus. The apoB is recovered in the dialysis bag. It can be used for phosphoamino acid analysis and peptide mapping.

Phosphoamino Acid Analysis

^{32}P-Labeled apoB is precipitated from the contents of the dialysis bag (see above) using trichloroacetic acid. The precipitate is extracted twice with acetone and once with ethanol. The precipitate is brought up into 50 μl of 6 N HCl and placed in a capillary tube. The ends of the capillary tube are flame sealed, and heated at 110° for 2 hr. Purified standards of phosphoserine, phosphotyrosine, and phosphothreonine are added to the hydrolysate and it is subject to paper electrophoresis. Ninhydrin is used to stain the phosphoamino acid standards. Autoradiography is used to determine the presence of ^{32}P in the areas of the chromatogram paper corresponding to the phosphoamino acid standards.

Dephosphorylation of ^{32}P-Labeled ApoB

It became obvious from our studies of apoB phosphorylation that a significant amount of phosphate can be lost during the isolation of ^{32}P-labeled apoB. Although we have yet to characterize the exact nature of this dephosphorylation, it is likely to be enzymatic in nature. The evidence most strongly supporting this statement is that if the culture medium or cell lysate is boiled in SDS-containing buffers, dephosphorylation is dramatically reduced.

Procedure

^{32}P-labeled apoB is isolated from the culture medium of hepatocytes incubated with [^{32}P]orthophosphate as described above. The washed immunoprecipitate is then subjected to a 1-hr incubation with hydroxylamine at neutral pH. Analysis of the immunoprecipitate by SDS–PAGE shows a complete loss of the autoradiographic band corresponding to little B. These data are consistent with hydroxylamine cleaving the phosphate ester bond.

In regard to physiologic dephosphorylation reactions, we have found that incubation of culture medium, obtained from hepatocytes previously labeled with [^{32}P]orthophosphate, with fresh rat serum for 8 hr resulted in a 70% loss of ^{32}P-labeled apoB. Culture medium (0.5 ml) is incubated at 37° with 0.5 ml fresh rat serum for 8 hr. The reaction solution is treated as described above for immunoprecipitating ^{32}P-labeled apoB. Analysis by SDS–PAGE and autoradiography is also performed as described above. Incubation of culture medium without added rat serum also resulted in a loss of ^{32}P-labeled apoB. However, compared to the results obtained with rat serum, there was a smaller loss of ^{32}P with culture medium only (i.e., 30%). There was no significant reduction in the amount of ^{35}S-labeled apoB subjected to the same treatment. These data suggest that rat serum and culture medium contain processes which can dephosphorylate ^{32}P-labeled apoB.

Conclusions

Phosphorylation of little B is an exciting new development. The finding that phosphoserine accounts for at least 20% of the ^{32}P that is present in little B, whereas similar anlsysis of phosphotidylserine does not yield phosphoserine,[9] strongly argues against the idea that the phosphoserine is due to contamination with phospholipid. Preliminary studies suggest that little B is polyphosphorylated, further suggesting that the phosphorylation is not an adventitious event. However, apoB is an intricate, fastidious protein and one must be cautious when interpreting results of its behavior. It is because of these problems that many are deterred from investigating apoB structure and metabolism. Yet, there is little doubt about the physiologic importance of apoB. Perhaps the phosphorylation of apoB will open a new door toward gaining an understanding of apoB structure and function.

[9] R. A. Davis, G. M. Clinton, R. A. Borchardt, M. M. McNeal, T. Tan, and G. R. Lattier, *J. Biol. Chem.* **259,** 3383 (1984).

[33] Lipoprotein–Receptor Interactions

By THOMAS L. INNERARITY, ROBERT E. PITAS, and ROBERT W. MAHLEY

Lipoprotein receptors function as a major control mechanism for the regulation of lipoprotein catabolism. The classical studies of Goldstein and Brown demonstrated the importance of low-density lipoprotein (LDL) receptors in supplying cells with cholesterol and in regulating plasma LDL levels.[1] Over the past 5 years, a number of other lipoprotein receptors have also been defined. These include the apolipoprotein (apo) E receptor, the high-density lipoprotein (HDL) receptor, the acetyl LDL receptor, and the immunoregulatory receptor.[2] Although each of these lipoprotein receptors appears to be unique, common procedures are employed for their characterization, similar to those used for the characterization of hormone–receptor interactions.

Certain criteria, which have been described in studies of LDL–receptor interactions and hormone–receptor interactions, establish that a lipoprotein is interacting with a cell or that a membrane fraction is binding to a

[1] J. L. Goldstein and M. S. Brown, *Annu. Rev. Biochem.* **46,** 897 (1977).
[2] R. W. Mahley and T. L. Innerarity, *Biochim. Biophys. Acta* **737,** 197 (1983).

receptor.[3] In an examination of lipoprotein–receptor interactions, these criteria (listed below) should be considered.

1. The receptor should demonstrate specificity. Most receptors bind only one ligand or closely related, structural analogues of the ligand. The LDL receptor, also referred to as the apoB,E(LDL) receptor, binds only lipoproteins containing apoB or apoE and no other proteins or apolipoproteins. Although the sequence of the domain of apoE that binds to the apoB,E(LDL) and apoE receptor has been determined,[4] the structure of the receptor binding domain of apoB has not been elucidated. Therefore, it is not known whether the receptor binding domain of apoB and apoE are identical. However, it has been demonstrated that arginine and lysine residues of both apoB- and apoE-containing lipoproteins are involved in mediating lipoprotein–receptor interactions.[2]

2. The ligand should bind to a finite number of specific binding sites. Thus, the specific binding of the ligand to the receptor should be saturable. In practice, the binding of most ligands to cells or tissues has two components. One is "nonspecific" (nonsaturable), low-affinity binding and the other is specific (saturable), high-affinity binding. Saturable binding alone is not sufficient evidence to establish the existence of a receptor.

3. Lipoproteins, like hormones, have been shown to have a very high affinity for their receptors. Values for the equilibrium dissociation constant (K_d) of 10^{-9} to 10^{-10} M are common. In equilibrium studies undertaken at 4°, human LDL bind to fibroblasts with a $K_d = 2.8 \times 10^{-9}$ M, while apoE high-density lipoproteins (HDL) from cholesterol-fed animals (HDL$_c$) bind to the same receptors with an even higher affinity ($K_d = 1.2 \times 10^{-10}$ M).[5] The apoE HDL$_c$ bind to the hepatic apoE receptor with a $K_d = 2.3 \times 10^{-10}$ M,[6] while β-VLDL bind to the β-VLDL receptor on macrophages with a $K_d = 1 \times 10^{-10}$ M.[7] The binding of hormones to their receptors is slower than would be predicted for a simple random collision process, suggesting that a specific alignment of ligand and receptor is necessary for binding to occur. Most hormone–receptor interactions have association rate constants on the order of 10^4 to 10^7 M^{-1} sec^{-1} compared with 10^8 M^{-1} sec^{-1} for a simple diffusion-limited process. Similarly, in their interaction with the apoB,E(LDL) receptor, LDL and apoE HDL$_c$

[3] C. R. Kahn, *J. Cell Biol.* **70,** 261 (1976).

[4] K. H. Weisgraber, T. L. Innerarity, K. J. Harder, R. W. Mahley, R. W. Milne, Y. L. Marcel, and J. T. Sparrow, *J. Biol. Chem.* **258,** 12348 (1983).

[5] R. E. Pitas, T. L. Innerarity, K. S. Arnold, and R. W. Mahley, *Proc. Natl. Acad. Sci. U.S.A.* **76,** 2311 (1979).

[6] D. Y. Hui, T. L. Innerarity, and R. W. Mahley, *J. Biol. Chem.* **259,** 860 (1984).

[7] T. L. Innerarity and R. W. Mahley, in "Drugs Affecting Lipid Metabolism" (R. Fumagalli, D. Kritchevsky, and R. Paoletti, eds.), p. 53. Elsevier, Amsterdam, 1980.

have association rate constants of $5.5 \times 10^4 \, M^{-1} \, sec^{-1}$ and $1.8 \times 10^5 \, M^{-1}$ sec^{-1}, respectively.[5]

4. The binding of the ligand to the receptor should mediate a physiological event, and the number of receptors occupied by the ligand should be related to the magnitude of the biological effect. For example, cholesterol delivery to cells is directly dependent upon the number of LDL receptors occupied by LDL.[1] Moreover, the receptor-mediated delivery of cholesterol regulates HMG–CoA reductase activity and cholesteryl ester formation.

5. For most receptors, the number of receptors per cell is not constant and can be regulated by the absence or presence of the ligand. For example, insulin and epidermal growth factor down regulate or desensitize their respective receptors. The number of apoB,E(LDL) receptors is down regulated by the amount of cholesterol delivered to the cells by lipoprotein internalization. Normal human fibroblasts, which have been cultured in the absence of lipoprotein cholesterol (up regulated), possess 70,000 to 100,000 apoB,E(LDL) receptors per cell, while fibroblasts grown in medium containing lipoproteins have <10,000 receptors.[5]

6. In addition to the criteria explained above, other characteristics may apply. For example, the binding of LDL and apoE HDL_c to the apoB,E(LDL) receptor requires Ca^{2+} or another divalent cation.[1] The binding of a ligand to a specific receptor is dependent upon pH and ionic strength. In addition, specific modifications of either the ligand or the receptor can abolish the binding interaction. For example, when the lysine or arginine residues of LDL or apoE HDL_c are chemically modified,[8,9] the binding of these lipoproteins to either the apoB,E(LDL) receptor or the apoE receptor is abolished. In addition, proteolytic enzymes, e.g., Pronase, have been used to destroy the apoB,E(LDL) and apoE receptors.[1,2]

The properties mentioned above probably will not be exhibited by every receptor, but they are classical criteria for the demonstration of a ligand–receptor interaction and should be assessed when describing a new system.

Many of the techniques used in the study of LDL–receptor interactions have recently been described by their originators in another volume of this series.[10] These include the radioiodination of LDL by the iodine monochloride procedure, preparation of lipoprotein-deficient serum, as-

[8] K. H. Weisgraber, T. L. Innerarity, and R. W. Mahley, *J. Biol. Chem.* **253**, 9053 (1978).
[9] R. W. Mahley, T. L. Innerarity, R. E. Pitas, K. H. Weisgraber, J. H. Brown, and E. Gross, *J. Biol. Chem.* **252**, 7279 (1977).
[10] J. L. Goldstein, S. K. Basu, and M. S. Brown, this series, Vol. 98, p. 241.

says for the binding, internalization, and degradation of ^{125}I-labeled LDL by cells in culture, the assay of HMG–CoA reductase activity in cultured cells, and the assay for cholesteryl ester formation. For the most part, these procedures will not be described again in this chapter, but rather additional procedures will be described with the hope that an investigator using the two chapters together can find most of the techniques needed to study lipoprotein–receptor interactions.

In this chapter, the following procedures are described.

1. Radioiodination of apoE-containing lipoproteins
2. Preparation of apoE · phospholipid complexes
3. Chemical modification of apolipoproteins
4. Cell surface receptor binding
5. Cholesterol, cholesteryl ester, and triacylglycerol mass determination
6. Receptor studies using monoclonal antibodies
7. Preparation and use of fluorescently labeled lipoproteins in receptor-binding experiments

Radioiodination of Lipoproteins by the Bolton–Hunter Procedure

A number of different radioiodination procedures are available for the preparation of biologically active ^{125}I-labeled lipoproteins. The method of choice for human LDL is the iodine monochloride method modified by Bilheimer et al., which is described by Goldstein et al.[10] This procedure can also be used for other lipoproteins that bind to the apoB,E(LDL) receptor. For lipoproteins containing mainly apoE, such as HDL_1 or HDL_c, the Bolton–Hunter procedure[11] is preferred, because apoE iodinates poorly using procedures based on the iodination of tyrosine.

Materials

1. Sodium borate buffer, 0.1 M, pH 8.5
2. Bolton–Hunter reagent (N-succinimidyl-3,4-hydroxy-5-^{125}I]iodophenyl propionate) (Amersham/Searle, Arlington Heights, IL)
3. Saline–EDTA: 0.15 M, NaCl, 0.01% EDTA, pH 7.4
4. N_2 tank

Procedure

The apoE-rich lipoprotein (1–5 mg of protein/ml) is dialyzed at 4° against borate buffer. Bolton–Hunter reagent is supplied in benzene,

[11] T. L. Innerarity, R. E. Pitas, and R. W. Mahley, J. Biol. Chem. **254,** 4186 (1979).

which is evaporated using a gentle stream of nitrogen. This is performed by inserting two hypodermic syringe needles through the seal, one for a nitrogen inlet and the other for an outlet. After all the benzene has been evaporated, a maximum volume of 1.5 ml of the cold lipoprotein is added directly to the dried, radioiodinated ester in the vial at 0° (on ice). The mixture is incubated for 30 min on ice and is occasionally mixed using a syringe with a needle inserted through the seal. The seal is then removed, and the solution is transferred with a Pasteur pipet to a 1-cm-diameter dialysis bag and dialyzed against saline–EDTA for 24–48 hr. When 1 mg of apoE HDL$_c$ protein in 1 ml of borate buffer is iodinated with 1 mCi as described, a specific activity of 600–800 cpm/ng of protein is obtained. Lipoproteins radioiodinated as described are comparable to those radioiodinated by the iodine monochloride procedure as determined by the following criteria: (1) biological activity in tissue culture (the Bolton–Hunter procedure modifies only 0.4% of the total available lysine residues of apoE, and at this low level of modification the iodination has no measurable effect on receptor binding[11]); (2) radioactivity (^{125}I) precipitated by trichloroacetic acid, which is >97% by both procedures; and (3) ^{125}I radioactivity extractable into chloroform–methanol. The actual extent of radioiodination of lipids depends on the species and type of lipoprotein. For example, the Bolton–Hunter procedure radiolabels less lipid in rat VLDL than the iodine monochloride procedure, while the reverse is true for β-VLDL isolated from cholesterol-fed dogs.

Preparation of Lipid · Apolipoprotein Complexes

Although apoE-containing lipoproteins bind avidly to the apo-B,E(LDL) and apoE receptors, purified apoE devoid of lipids does not bind to these receptors.[11] The apoE must be recombined with phospholipid to restore the apolipoprotein's receptor binding activity. We have used a procedure described by Roth *et al.*[12] to recombine the apoE with the phospholipid dimyristoylphosphatidylcholine (DMPC) to produce apoE · DMPC complexes that bind with a very high affinity to apo-B,E(LDL) and apoE receptors. The procedure has been used extensively to test the ability of isolated mutant forms of apoE to bind to receptors.[2]

Materials

1. Dimyristoylphosphatidylcholine (DMPC) (Avanti Polar Lipids, Birmingham, AL, Cat. No. 850345)

[12] R. I. Roth, R. L. Jackson, H. J. Pownall, and A. M. Gotto, Jr., *Biochemistry* **16**, 5030 (1977).

2. Branson sonicator #200, stepped microtip
3. β-Mercaptoethanol
4. Saline–EDTA: 0.15 M NaCl, 0.01% EDTA, pH 7.4
5. Buffered saline: 0.15 M NaCl, 10 mM disodium EDTA, 1 mM Tris-HCL, pH 7.6

Procedure

One milliliter of DMPC (20 mg/ml in chloroform) is placed in a 15-ml glass conical centrifuge tube and dried under a stream of N_2. The residue in the tube is redissolved in 1–2 ml of benzene, shell frozen with dry ice, and lyophilized. Two milliliters of a buffer containing 0.15 M NaCl, 10 mM disodium EDTA, and 1 mM Tris–HCl, pH 7.6, is added to the dry DMPC. The DMPC is allowed to hydrate for at least 30 min at room temperature, sonicated for 30 min, and centrifuged at low speed to remove titanium released from the sonicator tip. The DMPC vesicles in solution, which should be only slightly translucent, are kept at room temperature. An aliquot of apoE, which is isolated as described in article [13], Volume 28 and stored frozen in 0.1 M NH$_4$HCO$_3$, pH 8.1 (0.5–2 mg/ml) until use, is pipetted into a plastic tube. Apolipoproteins E-3 and E-2, especially apoE-2, form intramolecular disulfide bonds that must be reduced. The apoE is reduced by the addition of β-mercaptoethanol (0.5 μl/100 μg of apolipoprotein), and the mixture is incubated for 30 min at room temperature. The DMPC vesicles are added to the protein (37.5 μl of vesicles/100 μg of apolipoprotein), and the mixture is recycled (three times) through the transition temperature of the DMPC (23.5°) by warming to ~30° and cooling to ~15°. These apoE · DMPC disks can be used immediately for receptor studies; however, for quantitative results the lipid · protein complex should be separated from uncomplexed protein and lipid by density gradient centrifugation. A d = 1.006–1.21 g/ml linear KBr salt gradient is prepared in 0.5 × 2-in. polyallomer tubes (Beckman Instruments, Fullerton, CA), which are first treated with polyvinyl alcohol to make them wettable.[13] The apolipoprotein · phospholipid complexes are layered on top of the gradient and centrifuged in an SW-55 rotor at 15° for 20 hr at 55,000 rpm (369,000 g). The tubes are punctured using a Beckman gradient fractionator, and the contents are fractionated into ~10 fractions of 500–600 μl. The majority of the apolipoprotein · phospholipid complex is in the density range of 1.09–1.10. The absorbance of each fraction is monitored at 280 nm to locate the protein peak. (Oxidation products from the β-mercaptoethanol produce a high absor-

[13] L. Holmquist, *J. Lipid Res.* **23**, 1249 (1982).

bance background in the lowest density fractions.) If the apolipoprotein has been radioiodinated, the complex is located by counting aliquots from the tubes. Fractions containing the complexes are pooled and dialyzed against saline–EDTA before use in tissue culture experiments.

Chemical Modification of Apolipoproteins

Selective chemical modification of the lysine or arginine residues of apoB or apoE[8,9] abolishes their ability to bind to the apoB,E(LDL) receptor and the apoE receptor.[2] Several modification procedures for lysine are available,[8] including acetylation, acetoacetylation, carbamylation, and reductive methylation. Cyclohexanedione is used for arginine modification.[9] Each procedure has certain advantages. For example, acetoacetylation of the lysine residues and cyclohexanedione modification of the arginine residues are readily reversible reactions. The reversibility of the reaction is useful in establishing that the procedure has not irreversibly altered or denatured the apolipoprotein. Reductive methylation not only blocks lipoprotein binding to the apoB,E(LDL) and apoE receptors but also abolishes the binding of β-VLDL to lipoprotein receptors on macrophages.[7] On the other hand, acetylation or acetoacetylation promotes the rapid uptake of LDL or β-VLDL by the acetyl LDL receptor on macrophages.[14] Cyclohexanedione modification and reductive methylation have also been used to abolish the *in vivo* receptor-mediated catabolism of LDL and apoE HDL_c.[15]

Modification of Arginine Residues

Materials

1. 1,2-Cyclohexanedione (Aldrich, Milwaukee, WI)
2. Sodium borate buffer, 0.1 M, pH 8.1
3. Saline–EDTA: 0.15 M NaCl, 0.01% EDTA, pH 7.4
4. 1 M hydroxylamine, 0.3 M mannitol, pH 7

Procedure

Cyclohexanedione modification of the arginine residues of apoB- and apoE-containing lipoproteins is performed as follows[9]: Lipoprotein (2–5 mg) protein in 1 ml of saline–EDTA is mixed with 2 ml of 0.15 M 1,2-

[14] M. S. Brown and J. L. Goldstein, *Annu. Rev. Biochem.* **52,** 223 (1983).
[15] R. W. Mahley, T. L. Innerarity, and K. H. Weisgraber, *Ann. N.Y. Acad. Sci.* **348,** 265 (1980).

cyclohexanedione in 0.1 M sodium borate buffer (pH 8.1) and incubated at 37° for 2 hr. The sample is dialyzed against saline–EDTA at 4°. The modified lipoproteins are then filtered through a 0.45-μm Millipore filter.

Cyclohexanedione modification can be reversed to demonstrate that modification has not altered the chemical or biological properties of the lipoprotein. This is accomplished by mixing the modified lipoprotein with an equal volume of a solution containing 1 M hydroxylamine, 0.3 M mannitol, pH 7, and then incubating the mixture at 37° for 16 hr. Before use in tissue culture experiments, the lipoprotein is extensively dialyzed against saline–EDTA, ~12 liters for 48 hr.

Modification of Lysine Residues

Materials

1. For acetoacetylation: diketene (Aldrich), which has been vacuum distilled and stored in 200-μl aliquots at $-20°$; diketene reagent (0.06 μmol/μl), freshly prepared by vortexing 50 μl of diketene in 10 ml of 0.1 M borate buffer, pH 8.5; and 0.2 M sodium carbonate/bicarbonate buffer, pH 9.5
2. For acetylation: saturated sodium acetate solution; acetic anhydride
3. For carbamylation: 0.1 M sodium borate buffer, pH 8.0, and potassium cyanate
4. For reductive methylation: 0.1 M sodium borate buffer, pH 9.0; sodium borohydride; and formaldehyde diluted with H_2O to 7.4%
5. For determination of the extent of modification: 0.01% EDTA, pH 7.4; 1-ml glass ampoules; 6 N HCl; mercaptoacetic acid; dinitrofluorobenzene (DNFB); absolute ethanol; and NaHCO₃
6. Prior to tissue culture experiments, modified lipoproteins are dialyzed against saline–EDTA (0.15 M NaCl, 0.01% EDTA, pH 7.4)

Procedure

Acetoacetylation of Lipoproteins. To block the ability of lipoproteins to interact with apoB,E(LDL) or apoE receptors,[8] the following procedure is used: The lipoprotein is diluted 1 : 1 with 0.2 M borate buffer, pH 8.5, or dialyzed against 0.1 M borate buffer, pH 8.5, at the final protein concentration of 1–5 mg/ml. The diketene reagent is added (1–22 μl/mg of lipoprotein, 0.04–1.29 μmol), and the mixture is allowed to stand for 5 min at room temperature. To hydrolyze acetoacetylated hydroxyl groups, the mixture is dialyzed against a 0.2 M carbonate/bicarbonate buffer, pH 9.5, for 4–6 hr at room temperature and then overnight at 4°. The modified

LDL are then dialyzed for an additional 16–24 hr against saline–EDTA before use in tissue culture experiments.

To acetoacetylate LDL extensively, so that they are maximally recognized by the acetyl LDL receptor on macrophages, a more concentrated diketene reagent is used: A 25-μl aliquot of diketene is added to 1.0 ml of 0.1 M borate buffer, pH 8.5, in a glass test tube and vortexed vigorously to dissolve the diketene. For this more extensive modification, 280 μl of the diketene solution is added per milligram of lipoprotein protein. The dialysis against the carbonate/bicarbonate buffer can be omitted.

The extent of acetoacetylation of lysine residues can be quantitated by amino acid analysis (see below) or by colorimetric determination (540 nm) of the ferric–acetoacetamide complex formed in 3% $FeCl_3$,[16] using acetoacetylglycine as a standard. About 0.43–0.56 μmol of diketene/mg of lipoprotein protein modifies ~20% of the lysine residues of LDL, a level of modification sufficient to abolish apoB,E(LDL) or apoE receptor binding. Acetoacetylation can be reversed with the regeneration of lysine residues by dialysis of the modified lipoprotein against 0.5 M hydroxylamine, pH 7.0, for 16 hr at 37°.

Acetylation of Lipoproteins. To ice-cold lipoprotein, 1 to 7 mg/ml in saline–EDTA, in an ice-water bath, add an equal volume of saturated sodium acetate solution with continuous stirring. Add acetic anhydride in multiple 1.5- to 2-μl amounts over a period of 1 hr until the total mass of anhydride equals one and a half times the mass of the lipoprotein protein. For small amounts of lipoprotein, 4 × 1.5-μl aliquots of anhydride are used. The modified lipoproteins are dialyzed extensively against saline–EDTA.[17]

Carbamylation of Lipoproteins. The lipoprotein (1–5 mg of protein/ml) is dialyzed against 0.1 M sodium borate buffer, pH 8.0, and then a 20-fold by weight (protein) excess of potassium cyanate is added.[8] The time of incubation controls the extent of carbamylation. An incubation time of 90 min results in the modification of ~15% of the lysine residues of LDL, a level sufficient to eliminate >90% of the receptor binding.[8] The reaction is stopped by dialysis against saline–EDTA.

Reductive Methylation of Lipoproteins. The lipoprotein (1–10 mg of protein/ml) is dialyzed against 0.1 M sodium borate buffer, pH 9.0. The lipoprotein protein and all reagents are chilled to 0° and the procedure is performed on ice as follows[18]: 1 mg of sodium borohydride and 5 μl of

[16] A. Marzotto, P. Pajetta, L. Galzigna, and E. Scoffone, *Biochim. Biophys. Acta* **154**, 450 (1968).

[17] S. K. Basu, J. L. Goldstein, R. G. W. Anderson, and M. S. Brown, *Proc. Natl. Acad. Sci. U.S.A.* **73**, 3178 (1976).

[18] G. E. Means and R. E. Feeney, *Biochemistry* **7**, 2192 (1968).

7.4% aqueous formaldehyde are added to 6–10 mg of lipoprotein protein. Additional 5-μl aliquots of 7.4% aqueous formaldehyde are added every 6 min for 30 min. For a more extensive modification of the lipoprotein, another 1 mg of NaBH$_4$ is added after 30 min, and additions of formaldehyde are continued at 6-min intervals for a total of 60 min. Thus, 2 additions of sodium borohydride and 12 additions of formaldehyde are used. The reaction is stopped by dialysis against cold saline–EDTA.

Extent of Modification. The extent of chemical modification can be determined by amino acid analysis.[8,9] The lipoproteins are dialyzed against 0.01% EDTA in water, pH 7.4, and lyophilized, and the lipids are extracted with chloroform : methanol (2 : 1). The dried apolipoproteins (50–100 μg) are sealed under nitrogen in ampoules containing 1 ml of 6 N HCl. The apolipoproteins are hydrolyzed by heating them for 22 hr at 110°. To determine the extent of cyclohexanedione modification, 20 μl of mercaptoacetic acid are added with the HCl, and the number of arginine residues modified is determined by comparing the number of arginine residues in the modified and unmodified samples. Variations of this procedure, used to determine the level of modification by other procedures, are described below.

Because the diketene modification or the acetylation is labile to acid hydrolysis, an indirect procedure is necessary to determine the extent of modification of the lysine residues.[19] The unreacted, free lysyl groups are first converted to acid-stable dinitrophenyl (DNP) derivatives by reaction with dinitrofluorobenzene (DNFB). The apolipoproteins are then hydrolyzed with HCl, which converts the acetoacetyl derivative of lysine back to lysine, while the DNP lysine remains intact. Because DNFB does not modify 100% of the lysine residues (~95% of the LDL residues are modified), duplicate determinations should also be performed on controls. Added to the dried apolipoproteins in glass ignition tubes or 1-ml glass ampoules are 1 ml of absolute ethanol, 50 mg of NaHCO$_3$, and 25 μl of DNFB. The DNFB is allowed to react for 28 hr at 23° in the dark. The ampoules are centrifuged at 2500 rpm for 10 min, the supernatant is removed, and the DNP protein is washed once with distilled absolute ethanol, three times with H$_2$O, and once with acetone. Hydrolysis of the protein is performed as described above.

To determine the extent of carbamylation,[8] follow the DNFB procedure as described above. Carbamylation converts lysine to homocitrilline, which is partially converted back to lysine upon acid hydrolysis. Therefore, after hydrolysis of the DNP-carbamylated apolipoproteins and amino acid analysis, it is necessary to combine the lysine and homocitril-

[19] L. Wofsy and S. J. Singer, *Biochemistry* **2**, 104 (1963).

line residues to determine the total number of lysine residues carbamylated.

Reductive methylation produces mono- and dimethyl lysines that are stable under conditions of acid hydrolysis and are easily quantitated by slightly modifying the amino acid analysis system. The hydrolyzed protein is analyzed by the program recommended for physiological fluids (Beckman), using a lithium citrate buffer. The extent of methylation is determined by measuring the difference between the lysine content in the modified apolipoprotein versus that in the unmodified apolipoprotein.

Assays for Lipoprotein–Receptor Interaction Using Cultured Cells

The elegant studies of Goldstein and Brown have delineated the LDL pathway in cultured fibroblasts.[1] These studies demonstrated that LDL is first bound to the apoB,E(LDL) receptor on the surface of the cell. The LDL–receptor complex is concentrated in coated pits and then internalized into coated vesicles. The coated vesicles lose their coat, become more acidic, and at some point between the cell surface and the lysosomes, the receptor recycles back to the cell surface while the LDL continues to the lysosomes. In the lysosomes, the apoB of the LDL is hydrolyzed to amino acids and the cholesteryl ester is hydrolyzed to free cholesterol. The free amino acids are detected in the tissue culture medium, while the free cholesterol migrates to the cytoplasm where it down regulates the enzyme 3-hydroxy-3-methylglutaryl-CoA reductase (HMG–CoA reductase), the rate-limiting enzyme in cholesterol biosynthesis, thus suppressing cholesterol production. The free cholesterol also activates and serves as a substrate for acyl-CoA : cholesterol acyltransferase (ACAT), which reesterifies the excess cholesterol. The excess cholesterol is stored as cholesteryl ester droplets in the cytoplasm. Assays for the major steps in the LDL receptor pathway have been reported.[10]

Cell Surface Receptor Binding

Reagents

1. Dulbecco's modified Eagle's medium (DMEM) (Gibco, Grand Island, NY, Cat. No. 430-2100)
2. DMEM buffered with 25 mM N-2-hydroxyethylpiperazine-N'-2-ethanesulfonic acid (HEPES), pH 7.4 (instead of bicarbonate), containing 10% lipoprotein-deficient serum (LPDS) (referred to as DMEM–HEPES–LPDS)

3. Phosphate-buffered saline, without calcium and magnesium (referred to as PBS–CMF); 0.1 N NaOH; and PBS–CMF containing 0.2% bovine serum albumin fraction V (BSA)
4. PBS–BSA, PBS–CMF containing 0.2% BSA
5. Fetal bovine serum (Hyclone characterized, Hyclone Laboratories, Logan, UT)
6. Polyphosphate glass (Sigma, St. Louis, MO, Cat. No. P8510)
7. Tissue culture plasticware (Costar, Rochester, NY; Falcon, Pittsburg, PA; Corning, Corning, NY)

Procedure

Fibroblasts are grown in 35-mm plastic Petri dishes or 24-well, 16-mm multiwell plates in DMEM containing 10% fetal bovine serum. The cells are plated at 3.5×10^4 and 9×10^3 cells for 35- and 16-mm Petri dishes, respectively, for use 7 days later. At 48 (or 24) hr before the experiment, the cells are washed once with DMEM containing 5% lipoprotein-deficient serum followed by the addition of DMEM containing 10% lipoprotein-deficient serum. Prior to the experiment, the Petri dishes are placed on a metal tray, which should rest on a large plastic dishpan filled with ice, and cooled for 15 min in a 4° cold room. Radiolabeled lipoproteins at the desired concentration(s), with or without unlabeled competing lipoproteins, are added to DMEM–HEPES–LPDS and are also cooled to 0–4° on ice in a cold room for 20–30 min prior to use. The medium on the cells is replaced with the cold medium (DMEM–HEPES–LPDS) containing the radioactive lipoproteins (0.3 ml for 16-mm dishes, 1.0 ml for 35-mm dishes).

Direct binding studies to determine lipoprotein affinity by subsequent Scatchard analysis of the data are set up using a specific protocol: Duplicate dishes of cells are incubated with ^{125}I-labeled lipoproteins at 7 to 12 different concentrations. The following range of concentrations of radiolabeled lipoproteins is used: human LDL, 0.25–12 μg/ml; type III human β-VLDL, 0.2–10 μg/ml; canine apoE HDL_c, 0.01–1.2 μg/ml; and canine β-VLDL, 0.05–4 μg/ml. A third dish at each concentration of radioiodinated ligand is used to assess nonspecific binding. This is done by incubating the cells with the labeled ligand and a 20- to 100-fold excess (by weight) of unlabeled ligand. Specific binding is the difference between the binding of the ^{125}I-labeled lipoprotein in the absence and presence of excess, unlabeled lipoprotein. At 4°, the receptor binding of LDL and apoE HDL_c reaches equilibrium at 5–6 hr with gentle rocking, but for most purposes, a 3- or 4-hr incubation, which approaches equilibrium, is sufficient. At the end of the incubation period, the cells, which are still on ice, are rapidly washed three times with 1–2 ml of cold PBS–BSA, followed by two 10-min incubations with the same buffer, and finally a rapid wash using cold PBS–CMF.

Because in the 4° binding assay the lipoproteins are bound only to the cell surface receptors and are not significantly internalized, the number of receptor-bound lipoproteins can be measured by quantifying ^{125}I by gamma counting. The cells are dissolved by incubating them with 0.5 ml of 0.1 N NaOH for 10 min. Each dish is covered with a Petri dish lid and rocked manually to allow the NaOH to cover the entire dish. The dissolved cells are removed with a Pasteur pipet into 12 × 75-mm glass tubes. Each dish is washed twice with 0.5 ml of 0.1 N NaOH, and each rinse is added to the tube. The tubes are capped, and the ^{125}I is quantitated by gamma counting. After counting, the tubes are stored at 4° until protein concentrations can be measured. For 35-mm dishes, 200-μl aliquots out of the final 1.5 ml should be used for Lowry protein determinations.

When cells are incubated with ^{125}I-labeled lipoproteins at either 4 or at 37°, lipoproteins bound to receptors on the cell surface can be quantitated by releasing the bound lipoproteins from the receptors. The ^{125}I-labeled LDL can be released from cell surface receptors by incubating the cells with either heparin or dextran sulfate.[10] Lipoproteins containing apoE, such as apoE HDL$_c$ and β-VLDL, bind to LDL receptors with a much higher affinity than LDL, and cannot be released from the receptors by heparin or dextran sulfate. To release these lipoproteins, 1 ml (for 35-mm dishes) of PBS containing 30 mg/ml of polyphosphate is added to the dishes, and they are incubated at 4° with shaking for 60 min. Medium containing the released lipoproteins is transferred quantitatively to 12 × 75-mm tubes for ^{125}I counting.

The receptor binding curves can be linearized by Scatchard analysis either manually[20] or by using an interactive program called "SCAT-DATA" that we have developed for the IBM PC microcomputer. This program converts raw counts per minute (cpm) data obtained from a 4° binding experiment into a Scatchard plot. The user inputs the specific activity of the ^{125}I-labeled lipoproteins, total bound cpm, nonspecific bound cpm, and the cpm's in an aliquot of the medium at each concentration of ^{125}I-labeled lipoproteins used. For each data point, the program calculates the lipoprotein concentration in the medium, the nanograms bound (B), the nanograms free (F), and the ratio of bound over free (B/F). The program also performs linear regression analysis, calculates the maximum amount of ligand bound at receptor saturation (B_{max}), the K_d (μg/ml

[20] G. Scatchard, *Ann. N.Y. Acad. Sci.* **51**, 660 (1949).

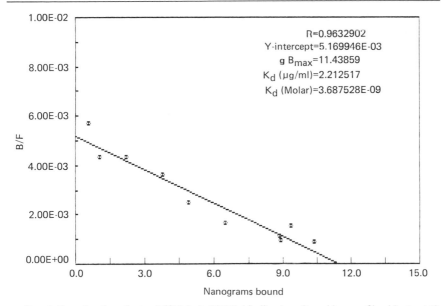

FIG. 1. Scatchard analysis of ^{125}I-labeled LDL binding to cultured human fibroblasts at 4°. *R* is the correlation coefficient.

and *M*), and plots the results. Figure 1 shows the results of "SCAT-DATA" analyses for LDL binding to cultured human fibroblasts. A second graphing program, "MULTIGRF," plots multiple curves on the same graph, using data files set up by "SCATDATA." Both programs are available on request.[21]

Assays for Binding and Internalization and Degradation of
    ^{125}I-Labeled Lipoproteins

Both the procedure for binding and internalization (cellular uptake) and for proteolytic degradation have been described in detail by Goldstein *et al.*[10] When lipoproteins are radioiodinated by the Bolton–Hunter procedure, the assay for proteolytic degradation measures ^{125}I-labeled lysine instead of ^{125}I-labeled tyrosine. Direct comparison of human LDL binding to human fibroblasts using the two iodination procedures yields identical results for binding and internalization. However, degradation values for LDL radioiodinated by the Bolton–Hunter procedure are about 30%

---

[21] To obtain copies of these programs, send a 5 1/4″ double-sided, double density diskette to: Gladstone Foundation Laboratories, Attention: Kay Arnold, P.O. Box 40608, San Francisco, CA 94140.

lower than those for LDL radioiodinated by the iodine monochloride procedure.

Assay for HMG–CoA Reductase

The assay for HMG–CoA reductase in cultured cells, described in detail by Goldstein et al.,[10] can be used for measuring reductase activity in fibroblasts, smooth muscle cells, and macrophages. The K_m (4 μM) of HMG–CoA reductase in macrophages and fibroblasts is similar; however, regulation of the reductase activity differs in the two cell types. In human fibroblasts, the enzyme activity is reduced by incubating the cells with low (10%) levels of fetal bovine serum or low levels of LDL. In mouse peritoneal macrophages, the enzyme activity is not reduced by incubating the cells with 10% fetal bovine serum, or by incubation with 60 μg of canine or human LDL protein/ml. Macrophages have low apoB,E(LDL) receptor activity, and only β-VLDL or lipoproteins taken up by the acetyl LDL receptor are effective in down-regulating the HMG–CoA reductase activity.

Assay for Cholesteryl Ester Formation

Goldstein et al.[10] have described in detail the assay for cholesteryl ester formation in cultured cells grown in 60-mm Petri dishes. This procedure, with a few minor modifications, can be adapted for use with smaller dishes of cells with no loss of precision. If the sodium[1-^{14}C]oleate · albumin complex is prepared with a specific activity of 3000 dpm/nmol, the procedure can be used for cultured cells on 16-mm dishes that have a cell protein content of only 20 μg/dish. For cholesteryl ester synthesis experiments, DMEM without serum is used, because serum promotes cholesterol efflux and thus reduces cholesterol esterification. The cells are incubated with serum-free DMEM (0.3 ml for 16-mm plates) containing 25–250 μg of lipoprotein cholesterol/ml and [^{14}C]oleate · albumin complex. A concentration of 0.2 mM [1-^{14}C]oleate is employed for macrophages and 0.1 mM for cultured fibroblasts. An incubation time of 16 hr is best, although macrophages incubated for 6 hr can produce measurable amounts of cholesteryl [1-^{14}C]oleate. The cells are then washed and extracted as described by Goldstein et al.[10] The substitution of Whatman prechanneled thin-layer chromatography plates (Clifton, NJ, Cat. No. 4855-821) for normal TLC plates allows the spotting and development of 100 samples per experiment. Each dried sample is dissolved in 50–100 μl of chloroform : methanol (2 : 1) for TLC application. The Whatman TLC plates are developed in hexane : diethyl ether : concentrated ammonium hydroxide (80 : 20 : 1).

Cholesterol, Cholesteryl Ester, and Triacylglycerol Mass Determination

We have derived a simple procedure to measure the mass of triacylglycerols, free cholesterol, and cholesteryl ester in a single 35-mm Petri dish of cells. Previous methods have either failed to use an internal standard throughout the entire extraction and analysis procedure, or have used two internal standards: a radioactive internal standard for the extraction procedure and a lipid internal standard for the analysis procedure. In our procedure, the radioisotope internal standard is omitted. The lipid internal standards are added directly into the organic solvents that are used to extract the lipids from the cultured cells. Thus, the losses from all extraction and analysis steps can be calculated directly.

Materials

1. Internal standards: stigmasterol and triheptadecanoin were purchased from Supelco (Bellefonte, PA) and NuChek Prep, Inc. (Elysian, MN), respectively. Stigmasteryl oleate was synthesized by a procedure analogous to that described by Patel *et al.* for the synthesis of cholesteryl esters.[22] Oleic anhydride and 4-pyrrolidinopyridine used in the synthesis were purchased from NuChek Prep and Fluka (Buchs, Switzerland), respectively
2. Solvents: hexane : isopropanol, 3 : 2; hexane : diethyl ether : concentrated ammonium hydroxide, 80 : 20 : 1; chloroform : methanol, 2 : 1
3. Thin-layer chromatography (TLC) plates: Whatman LK5D linear-K silica gel, 20 × 20 cm × 250 μm thick (Clifton, NJ, Cat. No. 4855-821)
4. Methanolic KOH: 0.5 *M* KOH in methanol
5. 7% HCl gas in MeOH: prepared by bubbling HCl gas into anhydrous methanol

Procedure

The internal standards stigmasterol, stigmasteryl oleate, and triheptadecanoin are added to the hexane : isopropanol used to extract lipids from the cells in tissue culture dishes. For 35-mm dishes, 1 ml of hexane : isopropanol (3 : 2), containing 5 μg of stigmasterol, 7.5 μg of stigmasteryl oleate, and 5 μg of triheptadecanoin, is added to the washed cells, and the mixture is incubated for 30 min at room temperature. The solvent is removed and the plates are washed two additional times with hexane : iso-

[22] K. M. Patel, L. A. Sklar, R. Currie, H. J. Pownall, J. D. Morrisett, and J. T. Sparrow, *Lipids* **14**, 816 (1979).

propanol without internal standards. The combined extracts and washes are evaporated to dryness under a stream of dry N_2.

Chloroform (25 μl) is added to the dried extract, and each sample of dissolved lipids is applied to a lane on the TLC plate. The plate is developed in hexane : diethyl ether : concentrated NH_4OH (80 : 20 : 1). The separated neutral lipids are visualized by briefly placing the TLC plate in a chamber of iodine vapors. Areas corresponding to the separated cholesterol, triacylglycerol, and cholesteryl esters are marked, and the iodine is allowed to sublime before proceeding.

The area containing the cholesterol, which also contains the stigmasterol internal standard, is scraped with a razor blade into a glass tube, and 1.0 ml of chloroform : methanol (2 : 1) is added. The mixture is then vortexed, centrifuged at low speed, and the supernatant is removed and saved for analysis.

The triacylglycerol band, which also contains the triheptadecanoin internal standard, is scraped into a glass screw-capped tube and transesterified in 0.5 ml of methanol (containing 7% by weight of anhydrous HCl) by heating the sealed vial at 70° for 2 hr. The reaction mixture is cooled, and 0.5 ml of H_2O and 1 ml of hexane are added. The mixture is then vortexed, and the phases are separated by low-speed centrifugation. The upper hexane layer is removed and saved for analysis of fatty acid methyl esters.

The cholesteryl ester band, which also contains the stigmasteryl oleate internal standard, is scraped into a glass screw-capped tube. The cholesteryl ester is hydrolyzed in 0.5 ml of methanolic KOH by heating it at 80° for 30 min. After cooling, 0.5 ml of H_2O and 1 ml of hexane are added, the mixture is vortexed, and the phases are separated by low-speed centrifugation. The upper hexane layer, containing the cholesterol (and the internal standard stigmasterol), is removed and saved for analysis.

The cholesterol samples are analyzed by gas–liquid chromatography essentially as described by Brown et al.[23] The extracts are evaporated to dryness, one to two drops of chloroform are added, and the samples are analyzed by gas–liquid chromatography on a 6-ft column packed with 3% OV-17 on 100/120 mesh Gas Chrom Q at 275°. The flow rate for the nitrogen carrier gas is 33 ml/min. Quantitation is by reference to the internal standard stigmasterol.

The fatty acid methyl esters from the triacylglycerol are also analyzed by gas–liquid chromatography. For this procedure, a 6-ft column packed with 10% Sp-2330 on 100/120 mesh Chromosorb WAW is used. The car-

[23] M. S. Brown, J. R. Faust, and J. L. Goldstein, J. Clin. Invest. 55, 783 (1975).

rier gas flow rate is 30 ml/min, and the temperature is 165°. Quantitation is by reference to the internal standard methyl heptadecanoate.

Binding Determinants Mapped Using Monoclonal Antibodies

Specific monoclonal antibodies have proved to be a valuable tool in studying lipoprotein–receptor interactions. Monoclonal antibodies have been used to map the receptor binding domain of apoB and apoE.[8,24] It is necessary to use high-affinity antibodies that bind to the apolipoprotein with an affinity similar to or greater than the affinity of the lipoprotein for the receptor. The main assumption made in using monoclonal antibodies for these studies is that the antibody interferes with binding by interacting at or near the functional binding domain of the apolipoprotein. The interference may be steric or the result of an antibody-induced alteration in the conformation of the binding site. A corollary to this is the assumption that an antibody interacting with a noncritical region of the protein will not interfere with binding. Monoclonal antibodies are useful as probes and are likely to meet the criteria described above because they interact with only a small region of the protein: typically, a linear epitope or determinant consisting of a sequence encompassing between 4 and 10 amino acids.

Monoclonal antibodies have also been used to determine which apolipoprotein in a lipoprotein is binding to a receptor. For example, VLDL, β-VLDL, and hypertriglyceridemic VLDL contain apoB and apoE, both of which can bind to the apoB,E(LDL) receptor. By using monoclonal antibodies, it is possible to determine quantitatively the contribution of each of these apolipoproteins in binding to the apoB,E(LDL) receptors.[4,25] It should be emphasized that most monoclonal antibodies to either apoB or apoE do not inhibit the receptor binding of these apolipoproteins, presumably because they do not bind to the receptor binding domain of the apolipoprotein.

Materials

1. Monoclonal antibodies are prepared by standard techniques and characterized by isotyping and subisotyping.[26] To inhibit the receptor binding activity of an apolipoprotein, IgG or a Fab frag-

---

[24] Y. L. Marcel, M. Hogue, R. Theolis, Jr., and R. W. Milne, *J. Biol. Chem.* **257**, 13165 (1982).

[25] D. Y. Hui, T. L. Innerarity, R. W. Milne, Y. L. Marcel, and R. W. Mahley, *J. Biol. Chem.* **259**, 15060 (1984).

[26] B. B. Mishell and S. M. Shiigi, eds., "Selected Methods in Cellular Immunology." Freeman, San Francisco, 1980.

ment of IgG should be used. These specific IgG or Fab fragments can best be isolated by using lipoprotein–Sepharose affinity columns or protein A–Sepharose columns[25]

2. ^{125}I-labeled sheep anti-mouse IgG (Amersham/Searle, Arlington Heights, IL, Cat. No. 1M. 131)
3. Removawell strips and holders (Dynatech Labs, Alexandria, VA, Cat. No. 011-010-6302, 011-010-6601)
4. Glycine, 5 mM, pH 9.2
5. Saline–Tween: 0.15 M NaCl, 0.025% Tween 20
6. PBS containing 1% bovine serum albumin (BSA)
7. PBS containing 0.2% BSA

Procedure

A plate assay is used to determine the apolipoprotein specificity of the antibody and to identify the epitope. Lipoproteins, isolated apolipoproteins, or apolipoprotein fragments are diluted in 5 mM glycine, pH 9.2, to a protein concentration of 2.5–10 μg/ml. A 100-μl aliquot of this solution is adsorbed to plastic Removawells by incubation overnight at room temperature. The addition of protein and all washes are performed using the "dip and flick" procedure. For washing, the Removawell holder is dipped into a wash solution, and the solution is removed by flicking the holder upside down. After the 100 μl of the original protein solution is flicked off, the Removawell is incubated for 1 hr in PBS containing 1% BSA. A range of 8 to 10 concentrations of antibody is then tested starting with 10 μg/ml. The wells are washed five times in saline–Tween, and 100 μl of antibody (diluted in PBS containing 0.2% BSA) is added per well. The antibody is incubated at room temperature (4–24 hr), the plates washed seven times (saline–Tween), and then incubated for 2–4 hr at room temperature with ^{125}I-labeled rabbit anti-mouse IgG in PBS, 0.2% BSA (100 μl; ~50,000 cpm per well). The bound radioactivity is determined following 12 washes (saline–Tween).

Receptor binding studies using monoclonal antibodies are performed at 4° as described in the section on cell surface receptor binding. The antibody, in increasing concentrations, is incubated 1–2 hr at 23° with the ^{125}I-labeled lipoprotein in DMEM containing 10% lipoprotein-deficient serum and then cooled to 4° for 30 min. The mixture is then added to the cells for a 2-hr incubation at 4°. The cells are washed, and the binding activity is determined as described in a preceding section (Cell Surface Receptor Binding). A typical experiment is shown in Fig. 2, where both the IgG and Fab fragments of the apoE monoclonal antibody 1D7, when added in equal immunoreactive amounts, were able to completely inhibit

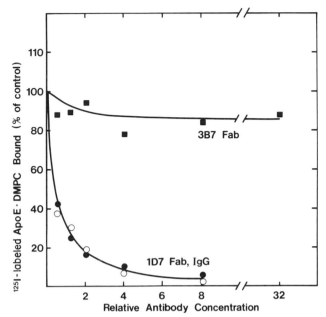

FIG. 2. Ability of 1D7 IgG, and 1D7 Fab fragments, but not 3B7 Fab fragments, to inhibit the binding of ^{125}I-labeled apoE-3 · DMPC to normal human fibroblasts: 1D7 Fab (○), 1D7 IgG (●), 3B7 Fab (■).

the receptor binding activity of ^{125}I-labeled apoE-3 · DMPC. The Fab fragment of the antibody 3B7, which binds to apoE but not at the receptor binding domain, was ineffectual in inhibiting apoE receptor binding activity.

Fluorescently Labeled Lipoproteins Used to Identify Cell Type Responsible for Uptake

The procedures currently used to assess receptor binding activity are appropriate only for pure cell populations. Different procedures are required to study lipoprotein–receptor interactions in mixed cell populations, such as tissue explants or histological specimens of tissues. A method to assess qualitatively the uptake of fluorescently labeled lipoproteins in mixed cell populations has been developed. The fluorescent probe of choice is 1,1′-dioctadecyl-3,3,3′,3′-tetramethylindocarbocyanine perchlorate (DiI).[27] From a practical point of view, DiI is an attractive tool

[27] R. E. Pitas, T. L. Innerarity, J. N. Weinstein, and R. W. Mahley, *Arteriosclerosis* **1,** 177 (1981).

because it is inexpensive, commercially available, easily incorporated into lipoproteins, and is highly fluorescent. From a biological point of view, once the probe is incorporated into the lipoprotein, it does not readily transfer to cell membranes or to other unlabeled lipoproteins and does not affect the receptor binding activity of the lipoproteins. Another important advantage of DiI is that it is retained by cells. When DiI-labeled lipoproteins are internalized through receptor-mediated endocytosis, the apolipoproteins and cholesteryl esters are hydrolyzed in the lysosomes. In contrast to the loss of amino acids and cholesterol from the cells, the fluorescent probe DiI is quantitatively retained for days.[28] This property results in the buildup of intense fluorescence and provides a stable marker for cells that have taken up the lipoprotein. Barak and Webb[29] have suggested that the lipid-soluble DiI is incorporated into lipoproteins with an orientation similar to that of phospholipid, i.e., with the polar head groups on the surface of the lipoprotein.

Using DiI-labeled lipoproteins, it has been possible to demonstrate the specific uptake of LDL, β-VLDL, and apoE HDL$_c$ by fibroblasts and smooth muscle cells, and of acetoacetylated (AcAc) LDL and β-VLDL by macrophages *in vitro* when they are in mixed culture.[27] Using this procedure, it has also been possible to demonstrate that foam cells in explants of atherosclerotic rabbit aorta have receptors for AcAc LDL and β-VLDL.[28] In addition, lipoproteins labeled with DiI and injected intravenously into animals have been used to determine the cell types responsible for the *in vivo* clearance of AcAc LDL and apoE HDL$_c$.[30,31] Hass *et al.*[32] have used this procedure to identify macrophages in sections of rabbit lung. St. Clair[33] has used DiI-labeled lipoproteins to demonstrate receptors for β-VLDL on pigeon macrophages.

Lipoprotein Labeling with the Fluorescent Probe DiI

This procedure can be used to incorporate DiI into LDL, HDL$_c$, β-VLDL, VLDL, and chylomicrons. However, the incorporation of DiI into VLDL and chylomicrons is not as efficient as it is with other lipoproteins.

---

[28] R. E. Pitas, T. L. Innerarity, and R. W. Mahley, *Arteriosclerosis* 3, 2 (1983).

[29] L. S. Barak and W. W. Webb, *J. Cell Biol.* 90, 595 (1981).

[30] R. E. Pitas, J. Boyles, R. W. Mahley, and D. M. Bissell, *J. Cell Biol.* 100, 103 (1985).

[31] H. Funke, J. Boyles, K. H. Weisgraber, E. H. Ludwig, D. Y. Hui, and R. W. Mahley, *Arteriosclerosis* 4, 452 (1984).

[32] A. J. Hass, H. R. Davis, V. M. Elner, and S. Glagov, *J. Histochem. Cytochem.* 31, 1136 (1983).

[33] R. W. St. Clair, *Fed. Proc., Fed. Am. Soc. Exp. Biol.* 42, 2480 (1983).

Materials

1. DiI (1,1′-dioctadecyl-3,3,3′,3′-tetramethylindocarbocyanine per-chlorate; Molecular Probes, Inc., Junction City, OR, Cat. No. D-282) in dimethyl sulfoxide (DMSO) (3mg/ml)
2. Lipoprotein-deficient human serum; $d > 1.21$ g/ml ultracentri-fuged portion of serum
3. Saline–EDTA: 0.15 M NaCl and 0.01% EDTA
4. Filters: 0.8-μm Millex-PF filters (Millipore Corp., Bedford, MA) and 0.4-μm filters (Gelman Acrodiscs, Ann Arbor, MI)

Procedure

The lipoprotein concentration should be between 1 and 2 mg of lipo-protein protein/ml in saline–EDTA. Two milliliters of lipoprotein-defi-cient serum is added for each 1 mg of lipoprotein protein to be labeled with DiI. The solution is then filtered (0.45 μm for LDL and apoE HDL$_c$; 0.8 μm for VLDL, β-VLDL, and chylomicrons) through Millex-PF filters. The solution is gently agitated, and 50 μl of DiI in DMSO (3 mg/ml) is added per milligram of lipoprotein protein. The mixture is incubated for 8–15 hr at 37°.

The lipoproteins are reisolated from the incubation mixture by ultra-centrifugation at a density of 1.063 g/ml for LDL and HDL$_c$ and 1.02 g/ml for VLDL, β-VLDL, and chylomicrons. The density is raised to approxi-mately 1.02 or 1.063 g/ml by adding 0.0199 or 0.0834 g of KBr for each ml of incubation mixture, respectively. For labeling of up to 4 mg of lipopro-tein, lipoproteins are reisolated using a Beckman 50-Ti rotor centrifuged at 49,000 rpm for 16 hr at 4°. After centrifugation, the lipoproteins are isolated by tube slicing, dialyzed against saline–EDTA, and then filtered as described.

It is important to note that lipoproteins must be labeled with DiI prior to acetoacetylation with diketene. Acetoacetylation prior to incubation with DiI inhibits incorporation of the fluorescent probe. In contrast, re-ductive methylation should be performed prior to labeling with the probe, because the procedure for reductive methylation reduces the fluorescence of DiI.

Cellular Binding and Uptake of DiI-Labeled Lipoproteins

The receptor-mediated endocytosis of lipoprotein by cells can be qual-itatively assessed by using DiI-labeled lipoproteins. The cells (fibroblasts, smooth muscle cells, or macrophages) are incubated with the DiI-labeled lipoproteins either at 4 or 37°, washed, and fixed, and the binding (4°) or cellular uptake (37°) is assessed by fluorescence microscopy. For most experiments, incubation at 37° is preferred because intense fluorescence develops due to the accumulation of the fluorescent probe within the cells.

Materials

1. Tissue culture chamber slides (Lab-Tek Products, Naperville, IL); 35-mm tissue culture dishes; and cover slips (22 × 22 mm) for 35-mm dishes
2. Dulbecco's modified Eagle's medium (DMEM); fetal calf serum; lipoprotein-deficient human serum; phosphate-buffered saline (PBS); and bovine serum albumin (BSA)
3. Four percent Formalin in phosphate buffer (0.15 M, pH 7.4) is prepared from depolymerized paraformaldehyde as follows: Paraformaldehyde (40 g/400 ml H_2O) is heated and 1 N NaOH is added dropwise with stirring until the solution clears. The pH of the solution is adjusted to pH 7.4 and 500 ml of phosphate buffer (0.3 M, pH 7.4) is added. The volume is then increased to 1000 ml with water

Procedure

Fibroblasts or smooth muscle cells are plated at a density of 3 × 10⁴ cells in 35-mm tissue culture dishes. If the dishes are difficult to place under the objective lens of the microscope, tissue culture chamber slides can be used. Tissue culture chamber slides allow for the removal of the chamber that holds the tissue culture medium, with the cells remaining adhered to a microscope slide. The cells are maintained in DMEM containing 10% heat-inactivated fetal calf serum. The medium is changed on day 3 of culture. On day 5, the medium is replaced, or the cells are changed to DMEM containing 10% lipoprotein-deficient human serum to induce greater numbers of apoB,E(LDL) receptors.

Mouse peritoneal macrophages are harvested as described,[10] and 0.5 to 1 × 10⁶ cells are plated in 35-mm culture dishes in DMEM. The cells are allowed to adhere for 1.5 hr, and the dishes are washed three times with 1 ml of DMEM without serum. The medium is then replaced with DMEM containing either 10% fetal calf serum or 10% lipoprotein-deficient human serum.

For mixed culture experiments, fibroblasts or smooth muscle cells are grown as described above, and mouse peritoneal macrophages in DMEM are added (2 to 5 × 10⁵ cells/35-mm dish) after the smooth muscle cells or fibroblasts have been cultured for 5 days. After 1.5 hr at 37°, the cells are washed three times with DMEM to remove nonadherent cells, and then DMEM containing either 10% fetal calf serum or 10% lipoprotein-deficient human serum is added. Experiments are conducted 2 days later.

For experiments performed at 4°, a temperature at which binding but not cellular uptake occurs, the cells are cooled for 30 min on a tray of ice in a cold room, and all media and wash solutions are maintained at 4°.

The lipoproteins (5–10 μg of lipoprotein protein/ml of medium) are added to the cells in DMEM containing either 10% fetal calf serum or 10% lipoprotein-deficient human serum, and the cells are incubated for 3–5 hr. The cell monolayers are then washed five times with PBS containing 2 mg/ml of BSA and fixed for 30 min with 4% formalin in phosphate buffer (pH 7.4). Cells incubated at 4° are washed at 4° and fixed for the first 15 min at 4°, followed by a 15-min fixation at room temperature. Cells incubated at 37° are washed and fixed at room temperature. After fixation, the cells are washed with PBS, the buffer is removed, and one drop of 75% glycerol in phosphate buffer is placed on the cells followed by a cover slip. The cover slip can be sealed around the edges with clear nail polish. Visualization of the bound and/or internalized lipoproteins is performed using a fluorescence microscope equipped with excitation and emission filters suitable for use with Rhodamine. The excitation and emission maxima of DiI are approximately 520 and 570 nm, respectively. For black and white prints, Ilford HP-5 film is used.

Binding specificity can be determined by competition with nonfluorescent ligands. Competition with nonfluorescent lipoproteins is performed using a 500-fold excess of the nonfluorescent ligand, as compared to the concentration of the DiI-labeled lipoprotein. In certain cases, materials other than the native ligand can be used. For example, the sulfated polysaccharide fucoidin has been shown[23,28] to compete with ^{125}I-labeled Ac LDL and DiI-labeled AcAc LDL for binding to the acetyl LDL receptor on macrophages. In practice, the specificity of binding of the DiI-labeled AcAc LDL can be assessed in the presence of either a 500-fold excess of AcAc LDL or a 10-fold excess (by weight) of fucoidin.

Acknowledgments

The authors thank Kerry Humphrey and Debbie Morris for manuscript preparation, James Warger and Norma Jean Gagasz for graphic arts, and Russell Levine, Barbara Allen, and Sally Gullatt Seehafer for editorial assistance. We also thank Kay Arnold, who helped to develop many of the procedures described.

[34] Receptor-Independent Low-Density Lipoprotein Catabolism

By J. SHEPHERD and C. J. PACKARD

I. Introduction

Low-density lipoprotein (LDL) contains approximately 70% of the plasma cholesterol and is primarily responsible for transport of the sterol from its site of synthesis in the liver to peripheral tissues. The latter, unable to degrade the steroid nucleus, must control the rate at which it is assimilated and therefore possesses highly regulated mechanisms for this purpose.[1] If they malfunction, a variety of metabolic changes can ensue to the detriment of the organism. So it is not surprising that considerable effort has been expended to develop methods to investigate the catabolic pathways responsible for the degradation of LDL.

LDL is formed in the plasma by the remodeling of triglyceride-rich, very-low-density lipoprotein (VLDL) initially secreted by the liver.[2] From the point of view of its subsequent catabolism, it may be regarded as a complex of cholesteryl ester and a single major protein, apolipoprotein B-100 (apoB-100). Another B protein of smaller molecular weight (apoB-48) is secreted by the intestine in association with chylomicrons but is not found in LDL isolated from the plasma of fasting subjects. ApoB-100 serves as an essential structural component of LDL and also directs its metabolic fate. Recently it has been shown that the plasma LDL concentration shows an equal dependance on the rates of synthesis and catabolism of the B protein.[3] In order to determine the kinetic parameters describing the turnover of the lipoprotein, we must make certain assumptions. The most important of these is that the lipoprotein fraction of d 1.019–1.063 $kg \cdot liter^{-1}$ is metabolically homogeneous. However,

---

[1] J. L. Goldstein and M. S. Brown, *Annu. Rev. Biochem.* **46**, 897 (1977).
[2] J. B. Marsh, article [30], this volume.
[3] C. J. Packard and J. Shepherd, *Atheroscler. Rev.* **11**, 29 (1983).

METHODS IN ENZYMOLOGY, VOL. 129

recent studies[4] have seriously questioned this proposal and it must be acknowledged that the density range encompasses a spectrum of particles differing in size, composition, and charge.

II. General Methods for the Study of LDL Catabolism

The procedure for measuring the parameters of LDL metabolism is well established and has been employed in a number of laboratories. A refinement of the technique, as described later, has allowed the discrimination of receptor-mediated and receptor-independent catabolism but the basic preparation of the tracer and analysis of the results is the same as that initially outlined by Langer et al.[5]

A. Method

LDL of d 1.030–1.050 kg · liter^{-1} is prepared by centrifugation of plasma in a fixed angle rotor (Beckman 40.3) or by rate zonal ultracentrifugation according to Patsch et al.[6] Salt is removed by dialysis against 0.15 M NaCl and 0.01% Na$_2$EDTA, and if necessary the lipoprotein is concentrated by pressure filtration through an XM 100A cellulose membrane (Amicon Corp. Bedford, MA) to a protein concentration of 3–5 mg/ml as determined by the procedure of Lowry et al.[7] Radiolabeling is then performed by mixing 1.0 ml of LDL solution with 0.5 ml of 1.0 M glycine, pH 10.0, and 2.0 mCi of carrier-free Na^{125}I (or Na^{131}I). Finally, an appropriate volume of an iodine monochloride solution (25 mM in 1.0 M NaCl) is added to give an ICl : protein ratio of 2.5 mol : 64,000 Da of protein. This introduces an iodine atom into tyrosine at a level of less than 1 mol iodine/mol of LDL protein. Free and bound radioiodide are separated by passing the iodination mixture over a 1.0 × 10 cm column of Sephadex G-25 (PD10 column, Pharmacia, London, England) and eluting with 0.15 M NaCl and 0.01% Na$_2$EDTA. The efficiency of incorporation is typically 40% and the amount of free iodide which remains in the tracer preparation is less than 1% as determined by paper electrophoresis of the labeled LDL. By employing a pH of 10.0 for the iodination, lipid labeling is minimized and constitutes less than 5% of the total incorporated radioactivity.

The ^{125}I-labeled LDL is sterilized by membrane filtration through a

[4] W. R. Fisher, *Metabolism* **32**, 283 (1983).

[5] T. Langer, W. Strober, and R. I. Levy, *J. Clin. Invest.* **51**, 1528 (1972).

[6] J. R. Patsch, S. Sailer, G. Kostner, F. Sandhofer, A. Holasek, and H. Braunsteiner, *J. Lipid Res.* **15**, 356 (1974).

[7] O. H. Lowry, N. J. Rosebrough, A. L. Farr, and R. J. Randall, *J. Biol. Chem.* **193**, 265 (1951).

0.22-μm filter (Millipore Corp., Bedford, MA) prior to injection of the lipoprotein into the bloodstream of the donor. This can be achieved within 8 hr of taking the blood if the zonal centrifuge isolation procedure is used. Ten minutes after injection, a blood sample is removed to determine the initial level of radioactivity and the subject's plasma volume by isotope dilution. Thereafter, daily samples are obtained to follow the plasma clearance of the lipoprotein. During all studies it is essential to ensure that thyroidal uptake of the radioiodide has been blocked by oral administration of KI (60 mg thrice daily). Under these circumstances, any iodotyrosine released by the degradation of the lipoprotein is rapidly lost into the urine and hence it is sufficient to count the radioactivity in the plasma at the end of the turnover period to determine the clearance rate of LDL protein. The plasma decay curve is composed of two exponentials and is frequently analyzed by the procedure of Matthews.[8] This gives a fractional catabolic rate (FCR) and a measure of the intravascular–extravascular distribution of the tracer. From the serial samples of blood, the apoLDL concentration is estimated and so the circulating pool size of the lipoprotein can be determined and used to calculate the absolute catabolic rate. This parameter, the product of the FCR and apolipoprotein pool size, is often expressed per kilogram of body weight and, under steady state conditions, equals the synthetic rate of the apolipoprotein. These kinetic parameters depend on a number of inherent assumptions, some of which are summarized in Table I and may be oversimplifications. For example, the initial biexponential component of the plasma die-away curve is considered in the procedure of Matthews[8] to represent equilibration of the tracer between intra- and extravascular compartments. However, a more complex kinetic analysis[9] of both urine and plasma data indicates that part of this rapid decline may be due to the loss of a more actively catabolized component in LDL and suggests that there are at least two metabolically distinct species present in the radioactive tracer. As yet these putative LDL subfractions have not been identified or isolated. Investigations such as those performed by Fisher[4] and Krauss and Burke[10] may eventually reveal their character.

B. The Significance of the "Fractional Clearance Rate"

A basic kinetic parameter that is estimated during an LDL turnover is the fractional catabolic rate which mathematically represents the recipro-

[8] C. M. E. Matthews, *Phys. Med. Biol.* **2**, 36 (1957).
[9] R. C. Boston, P. C. Grief, and M. Berman, *in* "Lipoprotein Kinetics and Modeling" (M. Berman, S. M. Grundy, and B. V. Howard, eds.), p. 437. Academic Press, New York, 1982.
[10] R. M. Krauss and D. J. Burke, *J. Lipid Res.* **23**, 97 (1982).

TABLE I

COMMON ASSUMPTIONS IN THE ANALYSES OF ApoLDL METABOLISM

Evidence for	Evidence against
1. Structural homogeneity of tracer Each LDL particle contains the same amount of apoB[a]	LDL is polydisperse in hyperlipidemia[b] LDL contains minor apoproteins[c]
2. Kinetic homogeneity of tracer Homologous and autologous LDL are cleared at the same rate	Simultaneous analyses of urine and plasma indicate the presence of several distinct metabolic species[d]
3. Intravascular catabolism of LDL The liver is the major site of LDL catabolism[e]	Receptors for LDL are found on all tissues[e]
4. Steady-state concentration of tracer Daily measurements show constancy of LDL pool size	
5. Tracer–tracee equivalence Remains to be proved	

[a] D. M. Lee, in "Low Density Lipoproteins" (C. E. Day and R. S. Levy, eds.), p. 3. Plenum, New York, 1976.
[b] W. R. Fisher, *Metabolism* **32**, 283 (1983).
[c] D. M. Lee and P. Alaupovic, *Biochemistry* **9**, 244 (1970).
[d] R. C. Boston, P. C. Greif, and M. Berman, in "Lipoprotein Kinetics and Modeling" (M. Berman, S. M. Grundy, and B. V. Howard, eds.), p. 437. Academic Press, New York, 1982.
[e] T. E. Carew, R. C. Pittman, and D. Steinberg, *J. Biol. Chem.* **257**, 8001 (1982).

cal of the area under the normalized plasma decay curve. It can be viewed as the probability per unit time that a labeled particle will leave the plasma compartment and not return. The fact that this parameter and not the amount of LDL catabolized each day (the absolute catabolic rate) is a better indicator of the metabolic status of an individual became clear from the early studies of Langer et al.[5] and Bilheimer and colleagues,[11] which revealed that the membrane receptor defect of familial hypercholesterolemia was reflected in the fractional rather than the absolute catabolic rate of apoLDL. In fact the latter was substantially higher than normal in these patients. Further evidence that the fractional rate of catabolism did not just mirror the LDL pool size came from studies in which the circulating level of LDL was changed abruptly by plasma exchange[12] midway through

[11] D. W. Bilheimer, N. J. Stone, and S. M. Grundy, *J. Clin. Invest.* **64**, 524 (1979).
[12] G. R. Thompson, A. K. Soutar, F. A. Spengel, A. Jadhav, S. J. P. Gavigan, and N. B. Myant, *Proc. Natl. Acad. Sci. U.S.A.* **78**, 2591 (1981).

a turnover study. There was no automatic rise in the FCR as the pool decreased. Similarly we[13] have observed that acute expansion of the plasma LDL concentration in rabbits given repeated injections of unlabeled lipoprotein does not alter the terminal slope of the clearance curve. Thus the FCR appears to be a valid index of the catabolic efficiency of an individual. Table II briefly summarizes some of the values that this parameter can take in subjects suffering from various hyperlipidemic, endocrine, or hematological disorders. In general, a high FCR is associated with a low pool size but this is not always the case, as noted above. Clearly the catabolic rate of LDL can vary over a wide range. The mechanisms responsible for these changes are currently being elucidated and we now have some understanding of the etiology of the disorders.

One of the most illuminating discoveries was the finding by Goldstein and Brown[1] that there was present on the cell membranes of most tissues a receptor for LDL which facilitated its rapid endocytosis and degradation. This receptor was clearly important in the *in vitro* metabolism of LDL. Measurement of its contribution LDL catabolism *in vivo* became possible with the development of specific chemical modifications which blocked the recognition of LDL by the receptor.

III. Approaches to the Measurement of Receptor-Mediated and Receptor-Independent LDL Catabolism

The following discussion focuses on the theoretical background to the use of chemically modified LDL for the measurement of receptor-independent catabolism of the lipoprotein *in vivo*.

A. Receptor–Lipoprotein Interaction

The LDL receptor, which has been recently purified to homogeneity, is an acidic glycoprotein with a molecular weight of approximately 120,000 Da. It is able to bind LDL with high affinity via forces that appear to be predominantly electrostatic in nature. Evidence for this comes from a number of studies. Polyanions like heparin and polyphosphate and polycations such as platelet factor 4 and polylysine are able to profoundly inhibit the uptake and degradation of LDL by cultured fibroblasts by blocking the interaction of the lipoprotein with the cell surface receptor.[14] The results obtained with polylysines of varying length are particularly informative. It was observed that short (less than 4 residues)

---

[13] H. R. Slater, C. J. Packard, and J. Shepherd, unpublished observations (1982).
[14] M. S. Brown, T. F. Deuel, S. K. Basu, and J. L. Goldstein, *J. Supramol. Struct.* **8,** 223 (1978).

TABLE II

THE FRACTIONAL CATABOLIC RATE OF ApoLDL IN MAN

Subject/disease	apoLDL fractional catabolic rate (pools/day)
Normal[a,b]	0.30–0.45
FH heterozygosity[a,b]	0.16–0.25
FH homozygosity[c,d]	0.11–0.16
Type V hyperlipoproteinemia[d]	0.65
Hypothyroidism[e]	0.11
Myeloproliferative disorders[f]	0.89

[a] C. J. Packard, J. L. H. C. Third, J. Shepherd, A. R. Lorimer, H. G. Morgan, and T. D. V. Lawrie, *Metabolism* **25**, 995 (1976).

[b] T. Langer, W. Strober, and R. I. Levy, *J. Clin. Invest.* **51**, 1528 (1973).

[c] D. W. Bilheimer, N. J. Stone, and S. M. Grundy, *J. Clin. Invest.* **64**, 524 (1979).

[d] G. Sigurdsson, A. Nicoll, and B. Lewis, *Eur. J. Clin. Invest.* **6**, 151 (1976).

[e] G. R. Thompson, A. K. Soutar, F. A. Spengel, A. Jadhav, S. J. P. Gavigan, and N. B. Myant, *Proc. Natl. Acad. Sci. U.S.A.* **78**, 2591 (1981).

[f] H. Ginsberg, H. S. Gilbert, J. C. Gibson, N.-A. Lee, and W. V. Brown, *Ann. Intern. Med.* **96**, 311 (1982).

polypeptides had no inhibitory effect but longer lysine polymers (up to 20 residues) were able to block LDL binding. This suggests that it is the conformation of charges and not just the net total positive charge on the ligand that underlies its attraction to the receptor. This also helps explain why LDL, which itself has a pI of 5.1, can interact with the acidic receptor protein. There must be domains of clustered positively charged amino acid residues which form the focus of the recognition site. Certainly this is the case for the other plasma apolipoprotein, apoE, which is able to bind the receptor. Its primary sequence is known and structure–function studies of a number of human mutants have permitted the location of the recognition site to the center of the protein where there is a region rich in arginyl residues.[15] Further evidence for the functional significance of positively charged groups comes from chemical modification studies. LDL can be treated with a number of agents which react with lysine and arginine present on the exterior aspect of the polypeptide. For

[15] T. L. Innerarity, R. E. Pitas, and R. W. Mahley, article [33], this volume.

example, diketene, cyanate, and acetic anhydride will combine with the ε amino group of lysine while leaving the gross physicochemical properties of the lipoprotein intact.[16] The resultant derivatized particle no longer binds to the receptor and so is not taken up or degraded by fibroblasts. A similar effect can be achieved by modification of LDL with 1,2-cyclohexanedione which blocks arginine residues.[17] Since some of these reactions induce large charge changes in the particle it is possible that, in addition to simple blocking of the ε amino groups, there are also associated conformational perturbations at the receptor recognition site. Indeed, reductive methylation causes a minimal alteration in net apoLDL charge and yet is highly effective in abolishing receptor–lipoprotein interaction.[16] Other amino acid side chains, like those of serine, methionine, and tyrosine, can be modified without influencing the binding of apoLDL to the receptor.[16] So, minor modifications to the protein of LDL can interfere with its metabolism by the specific receptor present on cultured fibroblasts, suggesting that a suitably modified lipoprotein may act as a probe for receptor-independent catabolism *in vivo*. However, some reactions, notably those that replace the positive charge of lysine with a negative one (i.e., acetylation or acetoacetylation), produce lipoprotein preparations that are cleared rapidly from the bloodstream of recipient animals.[18] Within minutes of injection, the radiolabeled, modified tracers appear in phagocytic cells of the liver and spleen, presumably assimilated by these tissues via mechanisms akin to the receptor activity for acetyl LDL seen on cultured macrophages.[18] They are obviously of little value as probes for the receptor-independent pathway since they are recognized as foreign by the recipient. Other modifications, described below, do not behave in this way and permit investigation of receptor-mediated and receptor-independent apoLDL catabolism *in vivo*.

B. The Discrimination of Receptor-Mediated and Receptor-Independent LDL Catabolism

The activity of the LDL receptor pathway *in vivo* can be quantified in man by simultaneously determining the plasma fractional catabolic rates of native and chemically modified, receptor-blocked LDL. An example of the plasma decay curves obtained using native and cyclohexanedione

[16] K. H. Weisgraber, T. L. Innerarity, and R. W. Mahley, *J. Biol. Chem.* **253,** 9053 (1978).
[17] R. W. Mahley, T. L. Innerarity, R. E. Pitas, K. H. Weisgraber, J. H. Brown, and E. Gross, *J. Biol. Chem.* **252,** 7279 (1977).
[18] R. W. Mahley, T. L. Innerarity, and K. H. Weisgraber, *Ann. N.Y. Acad. Sci.* **348,** 265 (1980).

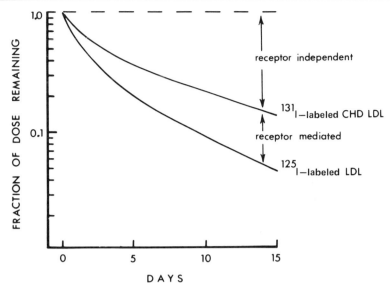

FIG. 1. Measurement of total and receptor-independent catabolism. Typical decay curves for native and 1,2-cylcohexanedione-modified LDL are shown. Since native LDL measures clearance by all mechanisms while the chemically modified LDL represents receptor-independent activity, the difference between the two is a measure of the function of the receptor pathway.

(CHD)-modified LDL in man is shown in Fig. 1. Here blocking of the arginyl groups of LDL leads to a slower plasma removal rate although the decay curve maintains its typical biexponential character, indicating that intravascular–extravascular exchange processes had not been disrupted to any great extent. When the FCRs for the native and modified lipoprotein are calculated it can be seen that receptor-independent catabolism (0.17–0.20 pools/day) accounts for approximately one-half of total catabolism (0.31–0.43 pools/day, $n = 7$) in normal subjects. The other component, the difference between the FCR of native and CHD-modified LDL, represents receptor-mediated catabolism.

Accurate quantitation of the fraction of the LDL pool degraded by the receptor and nonreceptor mechanisms depends greatly on the properties of the modified lipoprotein used to assess the activity of the latter. The features of a satisfactory probe of receptor-independent catabolism are as follows:

1. It should not be recognized by the high-affinity LDL receptor on cell membranes. This presupposes that the extent of modification is such

as to inhibit all receptor–ligand interaction. Cell culture studies have shown that to achieve this it is necessary to modify about 60% of the arginyl residues or 30% of the lysines on apoLDL.[16,17]

2. Its overall chemical and physical properties should not be significantly altered. For example, any change in particle size caused by its aggregation or cleavage would be expected to interfere with its ability to cross endothelial barriers, thereby perturbing its intravascular–extravascular distribution and affecting its exposure to catabolic mechanisms. None of the modifications described in Section IV substantially alter the size or lipid composition of the particle. However, all, to a lesser or greater degree, do affect the net charge carried on the lipoprotein and this, because of the charged nature of cell surfaces and permeability barriers may be expected to influence catabolic mechanisms. Effects of this kind are difficult to monitor *in vivo* but it is noteworthy that the calculated distribution of CHD/LDL between intra- and extravascular compartments (74 ± 6% intravascular) is close to that of the native lipoprotein (68 ± 8% intravascular).

3. It should be recognized as native by all pathways not utilizing the high-affinity receptor. This is one of the most important criteria for the tracer to meet and also the most difficult for which to provide satisfactory evidence. Every tissue in the body possesses, in addition to the high-affinity receptor pathway, nonspecific mechanisms for degrading LDL. It is a matter of concern that chemical treatment of the surface of LDL might not only block its interaction with receptors but also obscure other features essential for the uptake of the particle by nonreceptor routes. Alternatively, new chemical groups may be introduced which activate clearance of the lipoprotein by receptor-independent mechanisms, as is seen to occur following acetylation of LDL. Attempts have been made to establish the extent to which the various modified LDLs conform to this requirement by testing (1) their uptake *in vitro* by cultured macrophage/ scavenger systems and (2) their rates of catabolism in receptor-deficient humans and animals. All probes prepared to date appear to have their own peculiar advantages and disadvantages and often give different answers relative to the contribution of the receptor-independent pathway to total catabolism, probably because of their inability, to a lesser or greater extent, to meet this criterion.

4. It should be chemically and physically stable throughout its lifetime in the circulation. This requirement favors the use of irreversible reactions like reductive methylation with $NaBH_4$/formaldehyde. However, this particular modification was unsuitable for human use, and so in early studies in man, arginine blocking with cyclohexanedione was employed

despite its potential problem of instability. The recent development of stable lysine-modified derivatives of LDL may help resolve this difficulty.

Clearly the precise determination of the contributions made by the two pathways for LDL metabolism is complicated by methodological shortcomings. However, the data that have been acquired using available techniques have provided valuable insight into the factors regulating LDL catabolism *in vivo*, and in particular have elucidated the influence of various dietary regimens and drug therapies on LDL metabolism. It is now obvious that the receptor pathway has a major though not exclusive role in determining the concentration of LDL in plasma, e.g., its underactivity leads to the hypercholesterolemia of FH or hypothyroidism[12] while stimulation causes the decrease in LDL levels seen in cholestyramine therapy. On the other hand, there are situations in which the receptor-independent mechanisms can assume dominance. This is seen in the LDL hypercatabolism that accompanies gross hypertriglyceridemia or chronic myeloproliferative disease.[19,20]

C. Other Methods to Study Receptor-Independent Catabolism

The use of chemically modified LDL as a probe of the nonreceptor pathway has provided valuable information on its properties. But it is necessary to employ other, independent approaches to check the general applicability of the findings. Fortunately there are a number of "model systems" available in which receptor-independent mechanisms of LDL degradation can be examined. These include (1) cultured fibroblasts from FH homozygotes, (2) receptor-suppressed fibroblasts of normal individuals, (3) FH lymphocytes, (4) various cells from the receptor-deficient WHHL (Watanabe heritable hyperlipidemic) rabbit, and (5) cultured macrophages. Also, the pathway can be studied in whole organisms such as receptor-negative FH homozygotes and WHHL animals. The latter is a strain of rabbit inbred from an original male which showed a spontaneous mutation that gave rise to hypercholesterolemia even on a normal chow diet.[21] In the homozygous WHHL rabbit there is a 20-fold increase in the circulating level of LDL as a result of a virtual absence of the high-affinity receptor. With time, the animals develop tendon xanthomata and coronary and aortic atherosclerosis, with histological features similar to those

[19] H. Ginsberg, I. J. Goldberg, P. Wang-Iverson, E. Gitler, N.-A. Le, H. S. Gilbert, and W. V. Brown, *Arteriosclerosis* **3**, 233 (1983).
[20] J. Shepherd, M. J. Caslake, A. R. Lorimer, B. D. Vallance, and C. J. Packard, *Arteriosclerosis* **5**, 162 (1985).
[21] J. L. Goldstein, T. Kita, and M. S. Brown, *N. Engl. J. Med.* **309**, 288 (1983).

seen in the human FH condition. Unlike the cholesterol-fed, wild-type rabbit, hyperlipidemia is not associated with excessive, generalized tissue cholesterol accumulation. In fact, the sterol content of most WHHL tissues is only slightly above normal.[22]

IV. Preparation of Chemically Modified LDL to Probe Receptor-Independent Catabolism

The modified LDL probes which have been used to trace receptor-independent catabolism of the lipoprotein *in vivo* are listed below (Table III). Most involve derivatization of lysine since this residue is readily modified under mild, nondenaturing conditions. However, an alternative approach is to use 1,2-cyclohexanedione, which combines with the guanido group of arginyl residues. As discussed in the preceding section the aim of chemical modification is to abolish the binding of LDL with the high-affinity receptor while leaving intact the multiplicity of other interactions that the lipoprotein undergoes with various tissues in the body. Many of these will recognize some feature of the particles' surface such as the arrangement of amino acid side chains or phospholipid head groups but are likely to be less stringent in their structural requirements for binding than the specific high-affinity receptor. For example, "low-affinity" binding of LDL to hepatic membranes can be competed for by high-density lipoprotein (HDL), which has no major protein in common with the former but probably has a similar phospholipid conformation on its surface.[23]

A. Modification with 1,2-Cyclohexanedione

Bifunctional aldehydes are able to couple selectively to arginyl residues of proteins under mild conditions. Most of these also combine with ε and α amino groups in unwanted side reactions and the products are often intrinsically unstable. Patthy and Smith[24] described the use of 1,2-cyclohexanedione (Fig. 2) as a blocking agent for the arginyl residue of enzymes. In the presence of borate buffer, at pH 8–9 a single product, DHCH–arginine [N^7,N^8-(1,2-dihydroxycyclohex-1,2-ylene)arginine], is formed, which is stable but can be hydrolyzed in the presence of nucleophilic agents. Mahley and co-workers[17] have adapted this procedure to

[22] J. M. Dietschy, T. Kita, K. E. Suckling, J. L. Goldstein, and M. S. Brown, *J. Lipid Res.* **24**, 469 (1984).
[23] D. Y. Hui, T. L. Innerarity, and R. W. Mahley, *J. Biol. Chem.* **256**, 5646 (1981).
[24] L. Patthy and E. L. Smith, *J. Biol. Chem.* **250**, 557 (1975).

TABLE III

FRACTIONAL CLEARANCE RATES OF CHEMICALLY MODIFIED LDL IN MAN

Modified tracer	Fractional catabolic rate (pools/day)	Fractional catabolic rate of control LDL (pools/day)	Subjects	Reference
CHD/LDL[a]	0.19 ± 0.06	0.35 ± 0.06	Normals ($n = 7$)	C. J. Packard, L. McKinney, K. Carr, and J. Shepherd, *J. Clin. Invest.* **72**, 45 (1983)
CHD/LDL	0.15 ± 0.03	0.19 ± 0.03	Heterozygous FH ($n = 5$)	J. Shepherd, C. J. Packard, S. Bicker, T. D. V. Lawrie, and H. G. Morgan, *New Engl. J. Med.* **302**, 1219 (1980)
CHD/LDL	0.081 / 0.077	0.010 / 0.077	Homozygous FH ($n = 2$)	G. R. Thompson, A. K. Soutar, F. A. Spengel, A. Jadhav, S. J. P. Gavigan, and N. B. Myant, *Proc. Natl. Acad. Sci. U.S.A.* **78**, 2591 (1981)
GLC/LDL[b]	0.11 ± 0.03	0.57 ± 0.16	Normals ($n = 4$)	Y. A. Kesaniemi, J. L. Witztum, and U. P. Steinbrecher, *J. Clin. Invest.* **71**, 950 (1983)
GLC/LDL	0.15–0.17	0.29–0.33	Heterozygous FH ($n = 3$)	D. W. Bilheimer, S. M. Grundy, M. S. Brown, and J. L. Goldstein, *Proc. Natl. Acad. Sci. U.S.A.* **80**, 4124 (1983).
HOET/LDL[c]	0.19 ± 0.03	0.37 ± 0.06	Normals ($n = 6$)	H. R. Slater, L. McKinney, C. J. Packard, and J. Shepherd, *Arteriosclerosis* **4**, 604 (1984).

[a] 1,2-Cyclohexanedione-modified LDL.
[b] Glucosylated LDL.
[c] Hydroxyethylated LDL.

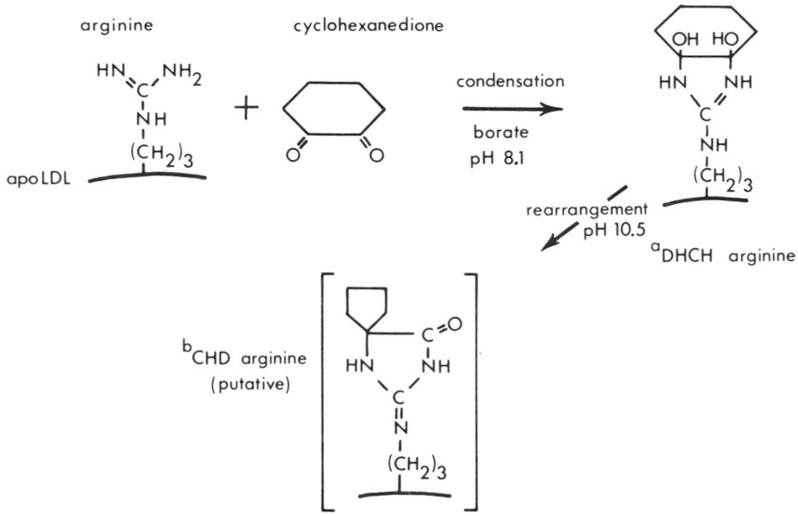

FIG. 2. Chemical reactions of arginyl residues with 1,2-cyclohexanedione. (a) N^7,N^8-(1,2-dehydroxycyclohex-1,2-ylene)arginyl LDL; (b) $N^{5'}$-[4-oxo-1,3-diazaspiro[4.4]non-2-ylidene)arginyl LDL.

modify apoB- and apoE-containing lipoproteins, and the method used in our laboratory follows their technique.

Method

1,2-Cyclohexanedione (CHD) is a hygroscopic reagent which should be stored at 4° in a sealed light-tight container, and renewed at frequent (monthly) intervals. LDL (1.0 ml) at a protein concentration of 3–5 mg/ml in a solution of 0.15 M NaCl/0.01% Na$_2$EDTA is added to 2.0 ml of freshly prepared 0.15 M 1,2-cyclohexanedione in 0.2 M sodium borate buffer, pH 8.1, and the mixture incubated for 2 hr at 35° in a water bath. Usually the modification is performed directly after labeling of the lipoprotein and if this is the case it is important to ensue that no trace of glycine is carried over with the LDL into the solution of CHD. Modified LDL is separated from unbound reagents by passing the mixture over a short (1.0 × 10 cm) column of G-25 Sephadex (PD 10 column, Pharmacia GB) using 0.15 M NaCl/0.01% Na$_2$EDTA as eluant. The lipoprotein is stored at 4° and used within 2 hr of its preparation. In addition to LDL, we have successfully derivatized VLDL and intermediate-density lipoproteins with CHD.

Incubation of LDL with cyclohexanedione results in the loss of about half of the arginine residues in apoB (4–5 out of a total of 9 per 250 amino acid residues), with consequent abolition of the ability of the lipoprotein to compete for the high-affinity receptor on fibroblasts. The size, lipid composition, and flotation properties of LDL remain unaltered and no discernible denaturation occurs under the mild conditions of the reaction. In fact, when the modification is reversed by further incubation of CHD-treated LDL with the nucleophilic agent hydroxylamine, receptor recognition is regenerated almost completely. Thus CHD/LDL has potential as a receptor-independent probe since blocking is achieved with apparently minimal structural perturbation of the lipoprotein although its anodic electrophoretic mobility is significantly accelerated.

On the basis of these findings we[25] investigated the use of CHD/LDL *in vivo*. When injected into the bloodstream of animals or man, it is removed more slowly than native (i.e., radiolabeled but otherwise unmodified) LDL. Its FCR in normolipemic humans is 0.18–0.22 pools/day (Table II) as calculated by the two-pool analysis of Matthews.[8] This represents 50–66% of total catabolism. In five subjects heterozygous for FH, the CHD/LDL catabolic rate was similar (0.14–0.18 pools/day) but now accounted for 80% of overall LDL degradation since the total FCR in these individuals was subnormal. And in two homozygous FH sufferers, CHD-treated and native lipoprotein were cleared at virtually identical rates,[12] consistent with the receptor-deficient status of the subjects. Despite these promising *in vivo* results using the CHD-modified tracer the accuracy of quantitation of the receptor-independent route is open to question because of the finding that prolonged (24 hr) incubation of CHD/LDL in plasma at 37° can restore some of its ability to bind to the receptor of cultured fibroblasts. Awareness of this problem led us and others to examine carefully the metabolic behavior of the probe both *in vitro* and *in vivo*. This was done in a number of ways.

1. CHD-modified LDL was incubated at 37° for 24 hr *in vitro* and its electrophoretic mobility on agarose compared to that of freshly prepared CHD/LDL and native lipoprotein. The incubated material had a reduced mobility in relation to the newly made CHD/LDL but did not revert fully to native. In fact the change that did occur seemed to be limited, as was the restoration of binding ability to fibroblast receptors.[26]

2. CHD-treated LDL was incubated *in vitro* in plasma for 24 hr at 37° or injected into a rabbit and allowed to circulate for the same length of time. Total LDL was then reisolated from the plasma and from the rabbit's blood and the properties of the modified tracer examined in other rabbits given the "screened" LDL and freshly made CHD/LDL labeled with an alternative iodine isotope. No difference was seen in the clear-

[25] J. Shepherd, S. Bicker, A. R. Lorimer, and C. J. Packard, *J. Lipid Res.* **20**, 999 (1979).
[26] H. R. Slater, C. J. Packard, and J. Shepherd, *J. Lipid Res.* **23**, 92 (1982).

ance rate of either preparation when compared to that of CHD/LDL prepared in the standard way and used immediately.[26]

3. LDL was treated with cyclohexanedione as described and the isolated tracer then subjected to a further incubation in 0.1 M phosphate buffer, pH 10.5, for 2 hr at 37°. At this pH, intramolecular rearrangement of DHCH arginine to a more stable, nucleophile-resistant complex is promoted[26] (Fig. 2). When the clearance rate of this material was tested in rabbits it was found to be identical to that of CHD/LDL prepared in the standard way.

4. The altered charge on LDL treated with CHD endows it with resistance to precipitation with heparin–Mn^{2+} when the Mn^{2+} ion is present in limiting concentration. This phenomenon allowed us to check that the difference in charge between native and CHD-modified LDL tracers was maintained during a 14-day turnover in man. Serial blood samples were obtained and treated with the heparin–Mn^{2+} reagent under conditions that gave virtually complete precipitation of the native lipoprotein but not of the modified probe. The difference in precipitability persisted throughout the study, indicating that reversion of CHD/LDL to a particle with native electrical properties had not occurred.[25]

5. Native and CHD-treated LDL were screened in the bloodstream of a homozygous FH individual for 1 week and thereafter reisolated and injected into the circulation of a monkey.[12] In the animal, the native lipoprotein promptly began to be degraded more rapidly than its modified counterpart, providing further evidence for the integrity of the modified tracer *in vivo*.

Taken together this evidence supports the view that CHD/LDL is a suitable probe for receptor-independent catabolism. One possible explanation why the apparent *in vitro* instability is not translated into detectable receptor-mediated degradation *in vivo* may be that the ratio of native to modified lipoprotein in the binding assay is typically 1 : 80, whereas in the bloodstream native LDL particles outnumber the modified lipoprotein by 800 : 1. This partial loss of cyclohexanedione from LDL may be reflected in its ability to compete for receptors in the presence of a small number of native particles, but if the native lipoprotein is present in overwhelming concentration, there is little opportunity for the modified tracer to access the receptor pathway.

B. Reductive Methylation of LDL

In this reaction (Fig. 3), formaldehyde is allowed to form a Schiff base with the ε amino groups of lysine residues on apoLDL. This is reduced with $NaBH_4$ to a monomethyl derivative which undergoes further reac-

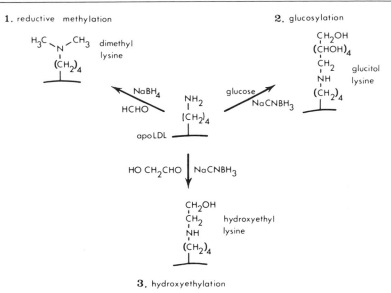

FIG. 3. Reactions of lysyl residues in LDL.

tion to yield the final product, dimethyllysine.[27] Treatment of LDL with these agents is usually performed by the procedure originally described by Weisgraber and colleagues.[16]

Method

LDL at a concentration of 3–5 mg/ml protein in 0.15 M NaCl/0.01% Na$_2$EDTA is diluted (1 vol : 2 vol) with 0.3 M sodium borate buffer, pH 9.0. A 5-ml aliquot of this solution is then added to 1.0 mg of solid NaBH$_4$ in a stoppered glass tube maintained in an ice bath at 0°. Methylation is initiated by the addition of 1.0 μl of 37% aqueous formaldehdye and thereafter at 6-min intervals further 1.0-μl aliquots are dispensed into the LDL solution to obtain the required level of modification. Usually the reaction is terminated after a total of six additions by passing the mixture through a G-25 Sephadex column (PD10 column, Pharmacia) equilibrated with 0.15 M NaCl/0.01% Na$_2$EDTA. By this procedure more than 75% of the lysines in apoB are converted to their dimethyl derivatives. The extent of modification can be readily checked by titration with trinitrobenzenesulfonic acid,[16] which measures the number of patent primary amino groups in the protein.

[27] G. E. Means and R. E. Feeney, *Biochemistry* **7**, 2192 (1968).

Reductive methylation of LDL is irreversible and provides a product which is chemically stable both *in vitro* and *in vivo*. At the level of modification generated by the above procedure no detectable binding to the receptor is left and when injected into rabbits and rats, methylated LDL, like its CHD-modified counterpart, is cleared slowly from the circulation. Unfortunately in several other species, including man,[26] the tracer disappears rapidly from the bloodstream (within minutes less than 20% remains), indicating that it has become recognizably foreign to the recipient. Why this should occur is not clear but it does suggest that reductive methylation induces a larger structural perturbation than the simple introduction of two methyl groups to lysine. Possibly, the disruption of critical hydrogen bonds or of a particular charge configuration leads to extensive conformational change within the protein to a form recognized by scavenger cells. Methylated LDL, however, is not recognized by the "acetyl LDL receptor" of cultured macrophages and so some other mechanism must mediate its rapid plasma clearance.

In those species of animal in which methylated LDL and CHD-treated LDL are catabolized slowly, it is possible to compare the relative properties of the two tracers. Theoretically both should have the same catabolic rate but in rabbits we found that this was not the case.[26] In these animals, the fractional catabolic rate of CHD/LDL (0.80 ± 0.22 pools/day) was consistently faster than that of methylated LDL (0.65 ± 0.25 pools/day). There are two possible explanations for such a finding. Either CHD/LDL is partly recognized by the receptor route or the two tracers are handled differently by receptor-independent catabolic mechanisms. On the basis of our experiments (outlined above) to validate the use of CHD/LDL we feel the second explanation to be more likely since we have observed[28] that when the reticuloendothelial system is inhibited, the difference in the catabolic rates of these two modified LDL tracers is abolished. When rabbits are given an intravenous bolus of ethyl oleate emulsion (see below) to inhibit the activity of phagocytic cells, CHD/LDL and methylated LDL are catabolized at virtually the same rate while the activity of the receptor pathway is left intact[29] under the same conditions. At present there is insufficient evidence to say whether CHD-modified LDL is removed faster, or methylated LDL slower, than normal by macrophages *in vivo*.

As an alternative method of modifying lysine residues on LDL we have examined the potential of carbamylation with cyanate as described by Weisgraber *et al.*[16] This modifies LDL less extensively than methyla-

[28] C. J. Packard, H. R. Slater, and J. Shepherd, *Biochim. Biophys. Acta* **712,** 412 (1982).
[29] H. R. Slater, C. J. Packard, and J. Shepherd, *J. Biol. Chem.* **257,** 307 (1982).

tion but does effectively abolish receptor binding. Since the procedure induces a charge alteration in the particle with neutralization of the amino groups, it seemed possible that this carbamylated derivative might have different properties *in vivo*. However, when its clearance rate was tested in rabbits, we found[30] it to be identical with that of methylated LDL, suggesting that the two tracers are metabolically the same. In the search for other procedures suitable for the preparation of modified human LDL probes, two alternative derivatization techniques have been developed— glucosylation and treatment with 2-hydroxyacetaldehyde.

C. Glucosylation of LDL

Nonenzymatic glucosylation of proteins occurs *in vivo* to an extent that depends on the circulating concentration of the sugar. In diabetics the process leads to the formation of HbA_{1c}, which is used as an index of persistent hyperglycemia. Glucose in its open chain aldehydic form condenses with primary amino groups to form a Schiff's base (Fig. 3). This is reversible but some of the conjugate undergoes rearrangement to a more stable ketoamine (the Amadori rearrangement). Alternatively, *in vitro*, the glucose–lysine adduct can be reduced to glucosylated LDL for use as a tracer *in vivo*.

Method

There are several published procedures[31-33] for preparing glucosylated lipoprotein tracers which are variations of the following.

LDL (approximately 25 mg) at a concentration of 5 mg protein/ml in phosphate-buffered saline (pH 7.4) is coincubated with glucose (at 80 mM[31,33] or 200 mM[32]) and sodium cyanoborohydride (30 mM,[32] 80 mM[33] or 200 mM[31]) under sterile conditions at 37° for a number of days. We[33] chose 7- but other workers have used 2-[32] and 5[31]-day incubations. Thereafter, modified LDL is separated from unused reactants by passage through a column (1.0 × 10.0 cm) of Sephadex G-25 (PD10 columns, Pharmacia) with 0.15 M NaCl/0.01% Na_2EDTA as eluant. The glucosylated lipoprotein is then labeled with radioiodide by the iodine monochloride procedure described in Section II, sterilized, and injected into the donor together with control LDL which has been subjected to the same

30 H. R. Slater, J. Shepherd, and C. J. Packard, unpublished observations (1982).

31 Y. A. Kesaniemi, J. L. Witztum, and U. P. Steinbrecher, *J. Clin. Invest.* **71,** 950 (1983).

32 N. Sasaki and G. L. Cottam, *Biochim. Biophys. Acta* **713,** 199 (1982).

33 H. R. Slater, L. McKinney, J. Shepherd, and C. J. Packard, *Biochim. Biophys. Acta* **792,** 318 (1984).

procedure but without sugar in the incubation mixture. In addition to glucose, we have used ribose and idose to glycosylate LDL.[33] Here, the incubation times required to achieve the necessary lysine blocking were 3 days and 24 hr, respectively.

Under the conditions employed in our laboratory (80 mM glucose, 7-day incubation) more than 50% of the lysines of LDL were modified and the resultant tracer was unable to compete with native LDL for the receptor on cultured fibroblasts.[33] The plasma clearance rate of glucosylated LDL is slower than that of native (or control) LDL in man but the value obtained appears to depend on the method used to prepare the tracer. In the studies of Kesaniemi et al.,[31] the glucosylated lipoprotein was cleared from the bloodstream of normolipemic volunteers in a monoexponential fashion with a fractional catabolic rate of 0.11 pools/day (Table III), one-fifth the clearance value of control (5-day incubated) LDL. That is, much less receptor-independent catabolism occurred than would be expected on the basis of earlier data using a CHD-treated lipoprotein. Furthermore, the monoexponential nature of the glucosylated LDL decay curve led these workers to postulate a role for receptors in intravascular–extravascular compartmental transfer.[31] However, this seems unlikely for two reasons. First, the decay of glucosylated LDL is biexponential in most animals; and second, the plasma decay of native LDL in homozygous FH patients is typically biexponential. It is not obvious why modified LDL prepared using the procedure of Kesaniemi and co-workers[31] should exhibit this metabolic behavior but certainly the low value they obtained for the receptor-independent FCR is in part due to the phenomenon of monoexponential clearance. Our experience with glucosylated LDL was different. Lipoprotein was isolated from healthy donors and divided into two aliquots, which were incubated in the presence and absence of the sugar for 7 days. Both preparations, when labeled and injected, gave biexponential decay curves[34] with a mean FCR of 0.50 pools/day for control and 0.25 for glucosylated LDL. Clearly, the rate of catabolism of the 7-day incubated control material was significantly higher than that of freshly prepared LDL. Such an effect may be caused by the prolonged incubation which alters LDL in a subtle way and leads to accelerated intravascular catabolism of a subfraction of the lipoprotein. Data from a study by Bilheimer et al.[35] is in accord with this idea. They used a much shorter incubation time to modify LDL and reported that the glucosylated lipoprotein was cleared biexponentially from the bloodstream of heterozy-

[34] H. R. Slater, L. McKinney, C. J. Packard, and J. Shepherd, *Arteriosclerosis* **4,** 604 (1984).
[35] D. W. Bilheimer, S. M. Grundy, M. S. Brown, and J. L. Goldstein, *Proc. Natl. Acad. Sci. U.S.A.* **80,** 4124 (1983).

gous FH patients with a plasma FCR of 0.15–0.18 pools/day, a value near to that obtained with CHD-modified lipoprotein (Table III). Thus the shorter preparation time appeared not to produce the problems of the first two studies and may be the most appropriate method to prepare the glucosylated lipoprotein. However, there is another drawback to the widespread use of this modification as a means of probing the receptor-independent pathway. Glucosylated LDL seems to be strongly immunogenic, at least in animals, and this may be the reason why it is cleared rapidly from the plasma of some humans.[36]

D. 2-Hydroxyacetaldehyde Treatment of LDL

2-Hydroxyacetaldehyde reacts with the ε amino groups of apoLDL in a fashion similar to that of formaldehyde and glucose, forming a stable derivative in the presence of reducing agents.

Method

Radiolabeled LDL at a protein concentration of 3–5 mg/ml was incubated for 2.5 hr at 37° in phosphate-buffered saline (pH 7.4) containing 80 mM 2-hydroxyacetaldehyde and 80 mM sodium cyanoborohydride. The modified lipoprotein was desalted over a Sephadex G-25 column equilibrated with 0.15 M NaCl/0.01% Na$_2$EDTA (pH 7.0). It was important to maintain the lipoprotein concentration in the quoted range in order to prevent particle denaturation. Using the procedure, lysine-modified, hydroxyethylated LDL (Fig. 3) could be prepared for injection within 4 hr of isolation of the lipoprotein.

Incubation of LDL with 2-hydroxyacetaldehyde under these conditions resulted in the elimination of 55% of the apolipoprotein lysyl residues as determined by amino acid analysis, and 60% as measured by trinitrobenzenesulfonic acid titration.[33] The reaction could be continued for 10 hr but at 2.5 hr sufficient modification was present to abolish the interaction of LDL with the receptor on fibroblasts. The hydroxyethyl extension to lysine residues neutralized their positive charge and produced a particle of increased anodic electrophoretic mobility. But the size and lipid composition of the LDL particle was unaffected by the treatment. Hydroxyethylated LDL (HOET/LDL), like the CHD-modified or glucosylated lipoprotein, showed no tendency to be rapidly catabolized by cultured macrophages nor did it compete for the "acetyl LDL receptor." The tracer, in vivo, gives a consistently slow clearance with a mean FCR of 0.19 pools/day in normal subjects (Table III). Moreover, the

[36] J. L. Witztum, U. P. Steinbrecher, M. Fisher, and Y. A. Kesaniemi, *Proc. Natl. Acad. Sci. U.S.A.* **80**, 2757 (1983).

conditions of preparation do not in themselves accelerate the initial plasma removal rate of the lipoprotein; that is, control LDL, incubated for 2.5 hr in the absence of 2-hydroxyacetaldehyde, is metabolically identical to native LDL. With hydroxyethylated LDL as a tracer of receptor-independent catabolism, it could be calculated that this process accounts for 50% of overall LDL clearance in healthy humans, a value similar to that obtained using the cyclohexanedione-treated lipoprotein. And indeed both modified tracers were removed at the same rate from the plasma of normal subjects.

V. Characteristics of Receptor-Independent Catabolism *in Vivo*

The following discussion attempts to relate the data obtained with the methods described above to current concepts of the properties of the receptor-independent pathway in cultured cells and animal models.

A. *Receptor-Independent Catabolism in Cultured Fibroblasts and Hepatocytes*

Although they lack the high-affinity LDL receptor, fibroblasts from homozygous FH subjects and hepatocytes from WHHL rabbits are far from inert with regard to LDL catabolism. Both cell lines can degrade large amounts of the lipoprotein under appropriate conditions;[37,38] and in fact WHHL hepatocytes, when exposed to near physiological concentrations of LDL, are able to degrade more lipoprotein than liver parenchymal cells from normal rabbits. This phenomenon may be reflected *in vivo* by the high absolute catabolic rates for apoLDL seen in receptor-deficient animals[39] and humans. Thus the receptor-independent routes have a *high capacity* for lipoprotein removal. Binding studies using liver cells or semi-purified hepatic membranes have revealed that receptor-independent catabolism is mediated through a potentially saturable but *low-affinity* site.[40] The interaction of LDL at this site is relatively resistant to Pronase treatment, is not Ca^{2+} dependent, and appears to be nonspecific since HDL can compete to some extent. The nature of the mechanisms that mediate receptor-independent catabolism in these cells is not yet clear. Bulk fluid

[37] J. L. Goldstein and M. S. Brown, *J. Biol. Chem.* **249**, 5153 (1974).
[38] A. D. Attie, R. C. Pittman, and D. Steinberg, *Hepatology* **2**, 269 (1982).
[39] D. W. Bilheimer, Y. Watanabe, and T. Kita, *Proc. Natl. Acad. Sci. U.S.A.* **79**, 3305 (1982).
[40] T. Kita, M. S. Brown, Y. Watanabe, and J. L. Goldstein, *Proc. Natl. Acad. Sci. U.S.A.* **78**, 2268 (1981).

endocytosis (pinocytosis) is a common property of all tissues but has been calculated[38] to account for only a small fraction of LDL uptake by cultured FH fibroblasts. In addition, the lipoprotein may adsorb to low-affinity binding sites on the cell surface and be ingested during the course of membrane turnover. The LDL-containing endosomes generated during this phenomenon may eventually deliver the lipoprotein to intracellular degradation, but the mechanism seems to be distinct from that mediated by the receptor since (1) the smooth vesicles are not clathrin coated and (2) LDL degradation is not inhibited by colchicine, a drug which disrupts receptor-directed catabolism.[41]

The LDL cholesterol taken in by cells via receptor-independent routes has little of the metabolic impact of sterol delivered by receptor-mediated uptake. Incubation of receptor-deficient fibroblasts or hepatocytes at high concentrations of LDL fails to elicit down regulation of HMG-CoA reductase or stimulate acyl coenzyme A : cholesterol acyl transferase (ACAT) despite the normal or even supranormal sterol delivery to the cells' interior that accompanies the high rate of lipoprotein degradation. This implies that sterol ingested by nonreceptor mechanisms enters a separate metabolic pool from that taken in via the receptor route. Extrapolation to whole organisms suggests that the FH homozygote and WHHL rabbit should exhibit uncontrolled cholesterol synthesis. However, this is not the case. In these receptor-deficient states sterol production and HMG-CoA reductase activity are not significantly different from normal, suggesting that the receptor-independent pathway can achieve some degree of *regulation* over cholesterologenesis *in vivo*. This apparent paradox may be explained by the observation of Ho *et al.*[42] that freshly isolated FH lymphocytes but not cells exposed to lipoprotein-deficient serum (LPDS) down regulate their HMG-CoA reductase levels in response to a high extracellular LDL concentration. These authors suggested that a control mechanism present *in vivo* may be lost in the artificial conditions extant during prolonged exposure to LPDS. Suppression of HMG-CoA reductase in receptor-deficient animals and man may therefore occur when the sterol taken up by pinocytosis and absorptive endocytosis accumulates in the cell to a level sufficient to fill the metabolic pool normally fed by the receptor pathway. This effect may be achieved at the expense of a higher than normal intracellular cholesterol concentration and indeed several tissues of the WHHL rabbit[22] do have increased cellular sterol deposits.

---

[41] R. E. Ostlund, B. Pfleger, and G. Schonfeld, *J. Clin. Invest.* **63,** 75 (1979).

[42] Y. K. Ho, J. R. Faust, D. W. Bilheimer, M. S. Brown, and J. L. Goldstein, *J. Exp. Med.* **145,** 1531 (1977).

B. Receptor-Independent LDL Metabolism in the WHHL Rabbit

The catabolic rate of LDL even in receptor-deficient animals and humans is several times that of albumin, a potential marker for fluid endocytosis *in vivo*. It is possible, therefore, that the low-affinity, high-capacity degradative mechanisms observed in cell culture also operate in whole organisms but the tissues involved in this process and its potential for regulation are largely unknown. A study[43] of LDL metabolism in WHHL rabbits in which the lipoprotein had been labeled with sucrose, a nondegradable and hence cumulative marker of tissue catabolism, has provided some answers. Injection of [14C]sucrose-labeled LDL followed by the examination of tissue levels of the marker after 24 hr revealed that the liver is the single most important organ of LDL clearance, accounting for over half of the degradation that occurred during this time. When uptake was expressed per gram weight, other tissues, including kidney, lung, adrenal glands, and particularly spleen, were also found to be active. Conversely, low LDL catabolic activity was observed in bulk tissues like skeletal muscle, fat, and skin. These results were comparable to data obtained in other species (rat[44] and hamster[45]) using chemically modified LDL to probe the receptor-independent pathway. Again, the liver was the dominant organ but the spleen was very active in removing LDL by nonreceptor mechanisms. So this agreement between the data obtained with native LDL in receptor-deficient rabbits and the receptor-blocked lipoprotein in normal animals gives us some confidence in the answers obtained with the latter probes of the receptor-independent pathway.

C. Role of the Monocyte–Macrophage System in Receptor-Independent LDL Catabolism

The high rate at which the spleen is able to assimilate chemically modified (receptor-blocked) LDL suggests that the reticuloendothelial system may play an important role in receptor-independent catabolism. Perhaps the best evidence for this comes from the observation that macrophages isolated from FH homozygotes are replete with cholesteryl esters,[46] presumably as a result of LDL uptake. However, simple incubation of the lipoprotein with these cells does not result in accumulation of the

[43] R. C. Pittman, T. E. Carew, A. D. Attie, J. T. Witztum, Y. Watanabe, and D. Steinberg, *J. Biol. Chem.* **257**, 7994 (1982).
[44] T. E. Carew, R. C. Pittman, and D. Steinberg, *J. Biol. Chem.* **257**, 8001 (1982).
[45] D. K. Spady, D. W. Bilheimer, and J. M. Dietschy, *Proc. Natl. Acad. Sci. U.S.A.* **80**, 3499 (1983).
[46] J. L. Goldstein and M. S. Brown, *Johns Hopkins Med. J.* **143**, 8 (1978).

sterol and it is now believed that LDL during its circulation in the plasma can be altered over the course of time to a form recognized by phagocytic cells.[47] Two further lines of evidence have indicated a role for monocyte–macrophages in receptor-independent catabolism.

Ethyl Oleate-Induced Suppression of Phagocyte Activity. One of the characteristic features of macrophages is their ability to ingest large particles. Stuart *et al.*,[48] working in the early 1960s, found that phagocytosis *in vivo* could be effectively suppressed by the administration to an animal of fatty acid monoesters. We decided to examine the changes in plasma lipoproteins attendant upon this inhibition of macrophage function in rabbits given ethyl oleate emulsion.

Method

A 10% emulsion of ethyl oleate was prepared by sonicating 1.0 g of the lipid and 0.07 g of Tween 20 (Sigma Chemical Co.) in 10.0 ml of 0.15 M NaCl until a fine, uniform dispersion was achieved. This was injected immediately after preparation into the marginal ear vein of rabbits at a dose of 1.0 ml of lipid/kg body weight. Depending on the experimental protocol, the emulsion was given midway through an LDL turnover to produce an acute effect or 48 and 24 hr prior to measuring LDL kinetics to provide quasi-steady state conditions in which LDL kinetics could be measured. The phagocytic activity present in the animals was monitored by the plasma clearance rate of a colloidal carbon suspension. After one injection of emulsion this fell by 50%, and a second injection produced no further suppression but prolonged the blockade.

Within hours of giving emulsion to rabbits the plasma and LDL cholesterol started to rise, reaching a new plateau value approximately 40% higher than the starting level. Kinetic studies with CHD/LDL indicated that suppression of LDL catabolism via the receptor-independent route was the cause of this rise; clearance via receptors was less affected, if at all. Recently, we[49] have been able to confirm these findings using potent, nonlipid based inhibitors of macrophage function. These compounds, derivatives of muramyl dipeptide (a component of bacterial endotoxin), induce delayed catabolism of LDL and increase its plasma concentration without introducing any lipid to the animal. The conclusion of these experiments was that the reticuloendothelial (macrophage) system made a quantitatively significant contribution to overall LDL removal.

[47] T. Henriksen, E. M. Mahoney, and D. Steinberg, *Atherosclerosis* **3**, 149 (1983).
[48] A. E. Stuart, G. Biozzi, C. Stiffel, B. N. Halpern, and D. Mouton, *Br. J. Exp. Pathol.* **41**, 595 (1960).
[49] J. Shepherd, S. Gibson, and C. J. Packard, *Agents Act. Suppl.* **16**, 147 (1984).

Studies of Myeloproliferative Diseases. Hypocholesterolemia, due mainly to a reduction in LDL levels, is a common finding in patients suffering from myeloproliferative disorders. Ginsberg and his colleagues[19] characterized LDL metabolism in these subjects and found that the decreased levels of the lipoprotein were the result of hypercatabolism, which derived not from the poor nutritional status of the individual but rather from his disease state. Low LDL levels were associated with advanced disease whereas palliative splenectomy or chemotherapy led to a rise in plasma LDL. When the contributions of the receptor and nonreceptor pathways were measured using CHD/LDL it was observed that the increase in LDL catabolism was due specifically to overactivity of the receptor-independent route, again providing evidence for the involvement of monocyte–macrophagelike cells in this pathway.

The potential involvement of the monocyte–macrophage system in receptor-independent LDL catabolism has implications with regard to the regulation of plasma LDL levels. There is a decrease in the FCR of the chemically modified lipoprotein with increasing LDL pool size in humans (Table III) and in cholesterol-fed rabbits,[50] suggesting that some saturation of receptor-independent clearance does occur. Macrophagelike cells may provide this poorly regulated but saturable component of nonreceptor catabolism and operate in addition to pinocytotic and absorptive endocytic pathways which clear a fixed portion of the plasma pool.

VI. Summary

Cultured cells, animal models, and man all possess mechanisms for LDL degradation that do not require the agency of the high-affinity receptor. In healthy individuals their functional significance can be determined using LDL which has been modified by specific chemical reactions designed to inhibit its receptor binding. Although the precise nature of the pathways has not been defined, there is evidence to implicate the monocyte–macrophage system in the process.

Acknowledgments

This work was performed during the tenure of grants from the Medical Research Council (Grant No. G811558SA), the Scottish Home and Health Department (K/MRS/50/C429), and the Scottish Hospitals Endowments Research Trust (Grant No. HERT 673).

[50] H. R. Slater, J. Shepherd, and C. J. Packard, *Biochim. Biophys. Acta* **713,** 435 (1982).

[35] Role of the Liver in Lipoprotein Catabolism

By RICHARD J. HAVEL

The liver accounts for the major fraction of the terminal catabolism of most plasma lipoproteins and may have an important role in lipoprotein interconversions and the nonendocytic uptake of components of lipoproteins as well. Considerable attention has therefore been given in recent years to the mechanisms and regulation of hepatic lipoprotein catabolism and a large number of methods have been used. These include measurements of the clearance of components of isotopically labeled lipoproteins into liver and their breakdown with the cell *in vivo,* similar measurements in isolated, perfused livers or liver cells, and measurements of binding of lipoproteins to hepatic membrane fractions. Although parenchymal cells comprise about 85% of liver cell mass, nonparenchymal cells account for about 35% of the cells in normal livers. Therefore, methods have been devised to evaluate the role of specific hepatic cells in lipoprotein catabolism. In addition to measurements of uptake or binding of lipoprotein components and release of catabolized products, increasing attention is now being given to the intracellular pathways by which lipoproteins are degraded into the liver and the subsequent fate of the undegradable cholesterol moiety.

Because of the variety of methods that have been used, it is impractical to describe them in great detail. Instead, in this chapter I will attempt to summarize the essential features, advantages, limitations, and pitfalls of each of the general methods that have provided useful information about hepatic lipoprotein catabolism. To evaluate the significance of results in this complex field, it is often important to consider and evaluate data obtained by more than one technique.

I. Methods for Quantifying Uptake of Lipoprotein Components by the Liver *in Vivo*

A. *Uptake of Radiolabeled Lipids and Proteins of Lipoproteins after Pulse Injection*

Lipids and protein components of lipoproteins, labeled endogenously by administering radioactive compounds enterally or parenterally, or exogenously by covalent attachment of radioactive atoms or compounds, have been used extensively to determine the role of the liver in lipoprotein

catabolism. The labeled lipoproteins, usually isolated by ultracentrifugation, are administered intravenously as a pulse and the amount of isotope
in liver is measured at selected times thereafter. This simple method is
useful for lipoproteins that are taken up rapidly from the blood,[1] but it is of
relatively little value for those that are removed slowly.

It is now recognized that receptor-mediated endocytosis is a major
mechanism by which lipoproteins are taken up by the liver. Transfer to
lysosomes then ensues in well-defined steps (see Section VII), such that
lipoprotein components begin to undergo lysosomal hydrolysis within 15
min; hydrolysis is complete within 1 hr. Chylomicrons and very-low-
density lipoproteins (VLDL) are rapidly converted to remnant particles
that are taken up by the liver by receptor-mediated endocytosis with high
efficiency. Most injected particles destined for hepatic uptake have entered the endocytic pathway within 15 min. Hepatic uptake can then be
reasonably evaluated at this time after injection, or shortly thereafter.
However, it is important to distinguish entry into the endocytic pathway
from "trapping" of lipoprotein particles within the extracellular compartment of the liver. This compartment is readily accessible to particles with
diameters less than 1000 Å, owing to the size of the numerous fenestrae in
the hepatic endothelium, which the particles traverse to enter the space of
Disse between the endothelium and the surface of the parenchymal cells.
For this purpose, livers are usually flushed via the portal vein with large
volumes of cold saline (~2 ml/g liver).[2] This effectively removes blood
from the liver, but it may not remove particles that have entered the space
of Disse. In principle, inert spherical particles of similar size to the lipoproteins taken up could be used to estimate the extent of such trapping,
but this approach has not been used. Other proteins that enter the extravascular space, such as radioiodinated plasma albumin, have been used,[3]
but they may not reliably reflect trapping of large lipoprotein particles.
For labeled proteins, radioactivity in the flushed liver that is precipitated
by 10% trichloroacetic acid or denaturing solvents, such as ethanol–acetone, 1 : 1, can be used to estimate undegraded protein.[4] For labeled lipids, lipids are extracted, and radioactivity is estimated in the appropriate
compound separated by thin-layer chromatography.[2]

The significance of uptake of tracer must also be evaluated in relation
to its ability to exchange with compounds in liver cells. This is particu-

[1] R. J. Havel, in "Lipids, Lipoproteins and Drugs" (D. Kritchevsky, R. Paoletti, and
W. L. Holmes, eds.), p. 37. Plenum, New York, 1975.
[2] T. Kita, J. L. Goldstein, M. S. Brown, Y. Watanabe, C. A. Hornick, and R. J. Havel,
Proc. Natl. Acad. Sci. U.S.A. **79,** 3623 (1982).
[3] R. J. Havel, J. Felts, and C. Van Duyne, J. Lipid Res. **3,** 297 (1962).
[4] O. Faergeman, T. Sata, J. P. Kane, and R. J. Havel, J. Clin. Invest. **56,** 1396 (1975).

larly the case for cholesterol and phospholipids, which exchange readily with their counterparts in plasma membranes. For this reason, the lipids best suited as tracers are those of lowest polarity that are located mainly in the core of spherical lipoproteins, such as triglycerides, cholesteryl esters, and retinyl esters. However, in many species, these lipids transfer readily among lipoprotein particles by means of specific proteins.[5] In the rat, little transfer occurs, but transfers are rapid and extensive in such species as the rabbit. A number of apolipoproteins also are rapidly exchanged among lipoprotein particles.[5] The exception is apolipoprotein (apo) B. Unfortunately, apoB is not the major protein component of chylomicrons and VLDL, for which the pulse-injection method is most useful. Radioiodination also labels lipoprotein–lipids to a substantial extent in large particles such as chylomicrons and VLDL.[6]

Despite these limitations, the pulse injection method has yielded useful information about hepatic uptake of chylomicrons and VLDL, and even of LDL when its uptake by liver is rapid owing to induction of LDL receptors on parenchymal cells.[7]

B. Use of Hepatectomized Animals

Hepatectomy has been used to evaluate the role of the liver in lipoprotein catabolism by measuring removal rates of labeled lipoproteins from the blood or by measuring levels of lipoprotein components after removal of the liver. Several methods have been employed. In large animals, such as the dog, the liver can be removed totally from the circulation, while maintaining the portal circulation by means of portal–systemic shunting.[8] In small animals, two procedures have been used. In one, a functional abdominal evisceration is performed, removing from the circulation not only the liver, but also the gastrointestinal tract and spleen.[9] A variation of this procedure is to eliminate all blood flow below the diaphragm, so that musculoskeletal structures of the lower part of the body are also eliminated from the circulation.[10] In the second, a partial hepatectomy is performed. In animals such as the rat, in which the liver is divided into distinct lobes, about 70% of the liver mass can be removed readily with maintenance of the portal circulation.[11]

[5] A. M. Gotto, Jr., H. J. Pownall, and R. J. Havel, this series, Vol. 128 [1].

[6] A. F. H. Stalenhoef, M. J. Malloy, J. P. Kane, and R. J. Havel, *Proc. Natl. Acad. Sci. U.S.A.* **81**, 1839 (1984).

[7] C. A. Hornick, A. L. Jones, G. Renaud, G. Hradek, and R. J. Havel, *Am. J. Physiol.* **246**, G187 (1984).

[8] P. Nestel, R. J. Havel, and A. Bezman, *J. Clin. Invest.* **42**, 1313 (1963).

[9] T. G. Redgrave, *J. Clin. Invest.* **49**, 465 (1970).

[10] A. Bezman-Tarcher and D. S. Robinson, *Proc. R. Soc. Lond. Ser. B* **162**, 406 (1965).

[11] G. M. Higgins and R. M. Anderson, *Arch. Pathol.* **12**, 186 (1936).

These procedures have provided information about the production and removal by the liver of chylomicrons and VLDL, and the role of the liver in the catabolism of LDL and components of HDL.[5] After total hepatectomy, animals are often in poor condition with reduced arterial pressure, and flow rates to the remaining organs may be altered differentially. Therefore, inferences concerning altered rates of catabolism of lipoprotein components must be made with caution. Failure to observe effects of partial hepatectomy on removal of lipoprotein components must also be interpreted with caution because of the possibility that the remaining liver may be able to compensate, even in the short term. Differential effects of hepatectomy on removal of components of specific lipoprotein particles, as in the case of chylomicrons, VLDL, and HDL, may provide the most useful information with this method.

C. Use of Lysosomotropic Agents to Inhibit Degradation of Lipoprotein Components after Pulse Injection

Lysosomal degradation of lipids and proteins can be inhibited by compounds that raise intralysosomal pH. Degradation can also be inhibited by specific enzyme inhibitors, such as leupeptin, which inhibits lysosomal cathepsin. The most commonly used lysosomotropic agent for studies *in vivo* has been chloroquine. When given to rats in amounts that do not affect endocytosis of lipoproteins, chloroquine substantially delays, but does not prevent lysosomal degradation.[7] Hence, its use extends the applicability of the pulse-injection method. Leupeptin has been little used to study lipoprotein uptake, but it can be injected into intact animals without notable toxicity in amounts that substantially inhibit lysosomal proteolysis.[12]

D. Use of Slowly Degraded Derivatives or Analogs of Lipoprotein Components

The covalent attachment of certain labeled compounds to lipoprotein–proteins and their use in the study of sites of lipoprotein catabolism is described elsewhere in this volume.[13] These ligands are degraded very slowly by lysosomal enzymes and remain within lysosomes for many hours, so that uptake of lipoproteins that enter the liver slowly can be quantified after pulse injection,[14] or during constant infusion.[15] The latter

---

[12] W. A. Dunn, J. H. Labadie, and N. N. Aronson, *J. Biol. Chem.* **254**, 4191 (1979).

[13] R. C. Pittman and C. A. Taylor, Jr., this volume [36].

[14] R. C. Pittman, T. E. Carew, C. K. Glass, S. R. Green, C. A. Taylor, and A. D. Attie, *Biochem. J.* **212**, 791 (1983).

[15] D. K. Spady, D. W. Bilheimer, and J. M. Dietschy, *Proc. Natl. Acad. Sci. U.S.A.* **80**, 3449 (1983).

method requires maintenance of a steady state, but it can provide information about uptake over a sustained period of time. As described in Section IV, this method can be extended to evaluate the dependence of hepatic uptake upon specific receptors.

Poorly degraded lipid analogs, such as cholesteryl linoleyl ether, which can be incorporated into lipoproteins, are also retained in lysosomes for hours. In rats, this compound, like other nonpolar lipids, is not subject to protein-mediated transfer among lipoprotein particles. It may be incorporated into lipoproteins directly or indirectly by protein-mediated transfer from other lipoproteins.[16]

The behavior of lipoproteins labeled in these ways should be evaluated in comparison with their counterparts labeled with more conventional (preferably endogenous) tracers. For lipoproteins with long residence times in plasma (such as LDL and HDL), slowly degraded tracers have proved to be invaluable in defining the sites of lipoprotein catabolism and the importance of the liver in lipoprotein catabolism.

E. Measurement of Hepatic Extraction of Lipoprotein Components

For lipoprotein components whose single pass extraction by the liver is a few percent or more, uptake can be quantified directly. This method requires sampling of hepatic inflow (both arterial and portal venous) and outflow (hepatic venous). Sampling is best done from indwelling catheters placed in these vessels several days before experiments are performed. In large animals, such as ruminants, dogs, and pigs, indwelling catheters whose distal ends are brought through the skin can be maintained for periods as long as several months. The catheters are usually filled with heparinized saline, which must be removed a few hours before experiments are carried out to obviate release of lipases that affect lipoprotein metabolism. The method is generally applied by giving a constant infusion of labeled lipoprotein together with a second tracer, such as indocyanine green, to measure hepatic blood flow.[17] Measurements of hepatic extraction and plasma flow permit reliable calculation of rates of hepatic uptake. Total rates of removal from the blood can also be calculated, so that the fraction of material removed by the liver can be determined. This method also permits estimation of steady-state release from the liver of labeled components that have been taken up. The method has been particularly valuable to determine the uptake and conversion of chylomicron components (as well as the uptake of albumin-bound free fatty acids) by the liver of animals.

---

[16] G. Halpern, Y. Stein, and O. Stein, this volume [49].
[17] E. N. Bergman, R. J. Havel, B. M. Wolfe, and T. Bohmer, *J. Clin. Invest.* **50**, 1831 (1971).

II. Use of Isolated, Perfused Liver to Evaluate Hepatic Uptake and Catabolism of Lipoproteins

A. General Methodological Aspects

Livers of animals as large as pigs[18] have been perfused successfully in studies of lipoprotein metabolism. The liver is perfused through the portal vein and hepatic venous drainage is collected. Perfusion can be carried out *in situ*[19] or after removal from the animal.[20] Most investigators now use a bicarbonate-buffered medium (such as Krebs–Henseleit buffer) and washed erythrocytes (not necessarily homologous) with or without added plasma albumin (~1%). If albumin is used, it should be dialyzed thoroughly against perfusion buffer. Many preparations of bovine albumin are contaminated with lipoprotein particles containing apoA-I. These can be removed by ultracentrifugation at a density of 1.25 g/ml.[21] For the second method, the bile duct is cannulated first, then the portal vein. Perfusion must be initiated immediately upon cannulating the portal vein to prevent formation of small blood clots. The thorax is then opened and the inferior vena cava is cannulated via the right atrium. The liver is removed and placed on a platform in a warm chamber maintained at 38°. Adequate oxygenation cannot be accomplished in the absence of erythrocytes (at a hematocrit of 18–25%) without using very rapid flow rates that may disrupt the hepatic endothelium. If vasospasm occurs (which may be caused by small clots), it can be minimized by adding papaverine HCl (5 mg/dl) to the perfusate.[22] For recycling perfusions, oxygenation is now usually accomplished with a "lung" of silastic tubing within a jar containing an oxygen–carbon dioxide mixture.[19] This device obviates air–perfusate interfaces that may denature lipoproteins. A blood filter may be placed in the line to remove small clots. Minute flow rates are maintained at 1–1.5 ml/g liver. This yields a perfusion pressure of approximately 12 cm water. Pumps that produced minimal hemolysis and steady flow rates are preferred. Roller-type pumps are commonly employed.

With rat livers, release of lipolytic activity into the perfusate (presumably the heparin-releaseable hepatic lipase) has been observed during the

---

[18] N. Nakaya, B. H. Chung, J. R. Patsch, and D. D. Tauton, *J. Biol. Chem.* **252**, 7530 (1977).

[19] R. L. Hamilton, M. N. Berry, M. C. Williams, and E. M. Severinghaus, *J. Lipid Res.* **15**, 182 (1974).

[20] T. E. Felker, M. Fainaru, R. L. Hamilton, and R. J. Havel, *J. Lipid Res.* **18**, 465 (1977).

[21] L. S. S. Guo, R. L. Hamilton, J. Goerke, J. N. Weinstein, and R. J. Havel, *J. Lipid Res.* **21**, 993 (1980).

[22] C. A. Hornick, T. Kita, R. L. Hamilton, J. P. Kane, and R. J. Havel, *Proc. Natl. Acad. Sci. U.S.A.* **80**, 6096 (1983).

first few minutes of perfusion.[23] If this occurs, the perfusate should be changed and lipoproteins whose uptake is to be studied should then be added to the fresh medium. For single-pass perfusions, the same principles apply as for recycling perfusions. *In situ* perfusion is simpler to perform, but it is unsuitable for perfusions lasting more than 3–4 hr. Small livers are often placed in a simple dish, but larger ones must be placed in a sling or a similar device to prevent vascular obstruction.

Uptake of labeled lipoproteins during recycling perfusions is usually accomplished by sampling small portions of perfusate at intervals. Estimation of initial perfusate concentrations is imprecise because the total perfusate volume (which includes the vascular space of the liver) cannot be measured directly. In principle, a second tracer that is not taken up from the perfusate can be mixed with the added lipoprotein to calculate initial concentrations. Catabolism of radioiodinated lipoprotein–protein is usually estimated from the release of trichloroacetic acid-soluble radioiodine into perfusate and bile. With the McFarlane method of iodination, which labels tyrosyl residues, the soluble products include iodotyrosine and iodide, which can be measured separately.[24] Monoiodotyrosine, 1–2 mg/ml, is added to the perfusate to minimize nonspecific deiodination. Control perfusions through the apparatus in the absence of a liver may be necessary to evaluate nonspecific deiodination.[25] Release of soluble radioiodine begins about 30 min after adding labeled lipoprotein to the perfusate,[25] reflecting the time required for lysosomal catabolism to begin. About 10% of the soluble radioiodine released appears in the bile.

Single-pass perfusions can be used for lipoproteins with appreciable rates of extraction (such as chylomicrons and VLDL). Uptake can be measured by determining the extraction fraction during the steady state, which requires that about 10 vol of perfusate be passed per gram liver.[26] Alternatively, the retention of isotope in the liver can be measured after passing a bolus of labeled lipoprotein through the liver.[27] It is important to measure the content of isotope in the entire liver to obviate inaccuracies owing to inhomogeneous perfusion. With this method, the uptake of different amounts of lipoprotein can be determined, and maximal extraction rates and apparent K_m's can then be estimated.[28]

[23] E. T. T. Windler, Y.-s. Chao, and R. J. Havel, *J. Biol. Chem.* **255**, 5475 (1980).
[24] E. L. Bierman, O. Stein, and Y. Stein, *Circ. Res.* **35**, 135 (1974).
[25] G. Sigurdsson, S.-P. Noel, and R. J. Havel, *J. Lipid Res.* **19**, 628 (1978).
[26] R. J. Havel, Y.-s. Chao, E. E. Windler, L. Kotite, and L. S. S. Guo, *Proc. Natl. Acad. Sci. U.S.A.* **77**, 4349 (1980).
[27] E. T. T. Windler, P. T. Kovanen, Y.-s. Chao, M. S. Brown, R. J. Havel, and J. L. Goldstein, *J. Biol. Chem.* **255**, 10464 (1980).
[28] B. C. Sherrill and J. M. Dietschy, *J. Biol. Chem.* **253**, 1859 (1978).

Measurements of uptake, particularly in single-pass perfusion, may reflect "trapping" in the space of Disse, rather than binding of lipoproteins to cellular receptors or entry of lipoprotein components into cells. Cellular binding can be reversed with appropriate reagents (i.e., heparin or dextran sulfate for the LDL receptor.[29] Endocytosed lipoproteins should remain within cells after separation of liver cells.

Uptake of Specific Lipoproteins and Lipid–Protein Complexes

1. Triglyceride-Rich Lipoproteins and Their Remnants. As discussed above, uptake can be measured in either recycling or single-pass perfusions. The composition of the particles can be modified by producing remnants in functionally hepatectomized animals or by incubation with lipoprotein lipase *in vitro,* by adding specific apolipoproteins, or by harvesting the lipoproteins from animals under different conditions (see Section VI). Chylomicron composition has also been modified by incubation with Pronase to remove all proteins,[30] by incubation with liposomes, which removes most proteins other than apoB-48,[31] or by treatment with phospholipase A_2 to reduce phospholipid content.[32] These maneuvers have been used to define the determinants or receptor-dependent uptake of these lipoproteins by the liver. Competition experiments, in which unlabeled lipoproteins are added in increasing amounts to a small amount of radiolabeled lipoprotein, have also been useful for this purpose.[33] Most investigators have not measured catabolism of components of triglyceride-rich lipoproteins, such as protein, because it has been difficult to label a single protein specifically. This can, however, be accomplished for proteins other than apoB-48, which can be added to the particles by exchange with their unlabeled counterparts or with other surface components (see Section VI).

2. Low-Density Lipoproteins. Uptake and catabolism of homologous radioiodinated LDL have been studied in rat livers during recycling perfusions.[25] Rates of catabolism measured in this way seem to underestimate rates that occur *in vivo.*[34] The reasons for this have not been determined. Uptake is, however, increased appropriately in rats treated with pharmacological amounts of 17-α-ethinyl estradiol,[35] indicating that specific re-

[29] J. L. Goldstein, S. K. Basu, and M. S. Brown, this series, Vol. 98, p. 241.

[30] J. Borenstajn, G. S. Getz, R. J. Padley, and T. J. Kotlar, *Biochem. J.* **240,** 609 (1982).

[31] J.-L. Vigne and R. J. Havel, unpublished observations (1983).

[32] J. Borenstajn and T. J. Kotlar, *Biochem. J.* **200,** 547 (1981).

[33] B. C. Sherrill, T. L. Innerarity, and R. W. Mahley, *J. Biol. Chem.* **255,** 1804 (1980).

[34] R. C. Pittman, A. D. Attie, T. E. Carer, and D. Steinberg, *Biochim. Biophys. Acta* **710,** 7 (1982).

[35] Y.-s Chao, E. Windler, G. C. Chen, and R. J. Havel, *J. Biol. Chem.* **254,** 11360 (1979).

ceptor-mediated endocytosis leading to catabolism does occur in this system.

3. High-Density Lipoproteins. Uptake and catabolism of homologous HDL have also been studied in perfused rat livers.[36] Rates of catabolism of this class of lipoproteins may also be underestimated.[37] HDL specifically labeled with individual apolipoproteins incorporated into the lipoprotein particles by exchange with their unlabeled counterparts have also been used to study uptake in perfused livers.[38]

4. Reconstituted Lipid–Protein Complexes. The uptake of discoidal complexes of radioiodinated apoE with egg phosphatidylcholine by perfused livers has been studied in a single-pass mode.[26] Catabolism of apoE and apoA-I in these complexes has also been measured during recycling perfusions. Like LDL, apoE in such complexes has thus been shown to be bound to hepatic LDL receptors. A short perfusion of labeled complexes in single-pass mode, followed by recycling perfusion, permits clearer analysis of the rate of catabolism.[31]

III. Use of Isolated Hepatic Cells to Evaluate Hepatic Uptake and Catabolism of Lipoproteins

A. General Methodological Aspects

Hepatocytes, separated from liver by perfusion with collagenase,[39] can be used "fresh," after short incubations in suspension, or after culture in primary monolayers. The latter method[40] may be preferable for study of lipoprotein uptake. Surface receptors may be inactive in freshly isolated cells, either because of proteolytic destruction or because disorganization of cellular functions may interfere with such phenomena as recycling of receptors. In primary monolayers, bile "canaliculi" are reconstituted and associated organelles such as lysosomes are reorganized at the bile canalicular pole of the cell 24 to 48 hr after the cells are plated, and microfilaments and microtubules seem to reappear.[41] Unless special conditions are used, however, the viability of the cells decreases and they tend to lose differentiated functions after about 5 days.[42] Hepatoma cells

[36] G. Sigurdsson, S.-P. Noel, and R. J. Havel, *J. Lipid Res.* **20**, 316 (1979).
[37] C. Glass, R. C. Pittman, D. B. Weinstein, and D. Steinberg, *Proc. Natl. Acad. Sci. U.S.A.* **80**, 5435 (1983).
[38] F. van't Hooft and R. J. Havel, *J. Biol. Chem.* **257**, 11001 (1982).
[39] M. N. Berry and D. S. Friend, *J. Cell Biol.* **43**, 506 (1969).
[40] D. M. Bissell, L. E. Hammaker, and U. A. Meyer, *J. Cell Biol.* **54**, 722 (1973).
[41] J.-C. Wanson, P. Drochmans, R. Mosselmans, and M.-E. Ronveaux, *J. Cell Biol.* **74**, 858 (1977).
[42] D. Bernaert, J.-C. Wanson, P. Drochmans, and A. Popowski, *J. Cell Biol.* **74**, 878 (1977).

that retain differentiated functions in culture can also be used. As with other cells, expression of some lipoprotein receptors in hepatocytes and hepatoma cells may be increased by use of lipoprotein-deficient culture medium. The major nonparenchymal cells, endothelial cells and Kupffer cells, separated after perfusion of livers with collagenase, can also be used to study lipoprotein uptake and catabolism (see Section V).

B. Uptake of Specific Lipoprotein Classes

1. Chylomicrons and Chylomicron Remnants. Chylomicron remnant components have been shown to be taken up and degraded by cultured rat hepatocytes in primary monolayers in a saturable manner to a greater extent than lymph chylomicrons.[43,44] Lysosomotropic agents and colchicine decrease the degradation of lipid and protein components of chylomicron remnants incubated with hepatocyte monolayers.[43,44] The nature of the binding site for remnants on cultured hepatocytes has not been established. Competition studies have not been definitive, but C apolipoproteins inhibit uptake of rat lymph chylomicrons and their remnants in rat hepatoma cells in monolayer culture.[45]

2. Other Lipoproteins. Cultured rat and porcine hepatocytes have also been used to study the uptake and degradation of LDL and HDL. At least two classes of binding sites have been identified.[46] One, designated the "lipoprotein-binding site," binds both lipoproteins as well as their reductively methylated counterparts and is unaffected by treatment of the cells with Pronase or preincubation of cells with LDL or HDL. A separate LDL-binding site has been identified whose properties resemble those of the LDL receptor in other cultured cells, such as fibroblasts.[46,47] The characteristics of binding to this site have been difficult to determine because of the large amount of binding to the lipoprotein-binding site, but degradation of LDL incubated with cultured hepatocytes is inhibited by treatment of cells with Pronase or incubation with chloroquine.[46,47]

The lipoprotein-binding site seems also to be present in freshly isolated hepatocytes. Degradation seems poorly coupled to uptake and is only partially inhibited by large concentrations of lysosomotropic agents or protease inhibitors.[48,49]

[43] C.-H. Floren and A. Nilsson, *Biochem. J.* **168,** 483 (1977).
[44] C.-H. Floren and A. Nilsson, *Biochem. J.* **174,** 827 (1978).
[45] N. M. Pattnaik and D. M. Zilversmit, *Biochim. Biophys. Acta* **617,** 335 (1980).
[46] P. S. Bachorik, F. A. Franklin, D. G. Virgil, and P. O. Kwiterovich, Jr., *Biochemistry* **21,** 5675 (1982).
[47] S. H. Pangborn, R. S. Newton, C.-M. Chang, D. M. Weinstein, and D. Steinberg, *J. Biol. Chem.* **256,** 3340 (1981).
[48] L. Ose, I. Roken, K. R. Norum, C. A. Drovon, and T. Berg, *Scand. J. Clin. Lab. Invest.* **41,** 63 (1981).

A similar binding site seems to be present on nonparenchymal liver cells, separated from parenchymal cells by differential centrifugation, but maximal binding of VLDL remnants, as well as LDL and HDL, is higher in nonparenchymal cells.[50]

Acetyl-LDL binds specifically and with high affinity to both of the major nonparenchymal cells (endothelial and Kupffer cells), but binding (per milligram cell protein) is considerably higher in endothelial cells,[51] consistent with observations on the site of uptake of acetyl-LDL in the liver *in vivo* (see Section V). Degradation occurs efficiently and is effectively inhibited by lysosomotropic agents.

Evidence for three classes of lipoprotein-binding sites has been obtained with H-35 rat hepatoma cells, cultured in lipoprotein-deficient medium.[52] One resembles the LDL receptor and another the lipoprotein-binding sites. The third binds chylomicron remnants, which may be mediated by apoE.

IV. Evaluation of Receptor-Dependent Uptake of Lipoproteins by the Liver

A. *Use of Lipoproteins in Which Protein Components Are Covalently Modified*

Lipoproteins with covalently modified arginyl or lysyl residues, which were originally developed to evaluate receptor-dependent and -independent removal of lipoproteins from the blood *in vivo,* can be used to study the contribution of receptor-dependent and -independent mechanisms to the hepatic uptake of lipoproteins, especially when they are labeled with slowly degraded derivatives of lipoprotein components (see Section I). From measurements of the rate of uptake of [^{14}C]sucrose-labeled LDL and similarly labeled methyl-LDL, receptor-dependent mechanisms have been found to account for more than 90% of the hepatic uptake of LDL in the hamster.[15] A similar approach can be applied to other systems. For example, by use of cyclohexanedione-modified LDL, receptor-dependent mechanisms have been found to account for at least 75% of total LDL catabolism in perfused livers of estradiol-treated rats.[35] Reduction in the rate of uptake of cyclohexanedione-modified VLDL remnants by perfused livers has provided evidence that receptor-dependent mechanisms

[49] L. Ose, I. Roken, K. R. Norum, and T. Berg, *Exp. Cell. Res.* **130,** 127 (1980).
[50] T. J. C. van Berkel, J. K. Kruijt, T. van Gent, and A. van Tol, *Biochim. Biophys. Acta* **665,** 22 (1981).
[51] J. F. Nagelkerke, K. P. Barto, and T. J. C. van Berkel, *J. Biol. Chem.* **258,** 12221 (1983).
[52] T. Tamai, W. Patsch, D. Lock, and G. Schonfeld, *J. Lipid Res.* **24,** 1568 (1983).

are important for remnant catabolism in the liver.[53] In lipoproteins such as VLDL remnants, arginyl residues in more than one protein are modified. Therefore, the protein specificity of the receptor interaction cannot be deduced from studies with lipoproteins other than LDL, unless methods to label lipoproteins with a single protein are used.

B. Measurement of Hepatic Lipoprotein-Binding Sites

1. In Whole Cells. As described in Section III, the extent to which lipoprotein receptors are expressed on the surface of isolated hepatocytes and nonparenchymal cells may depend upon the conditions under which the cells are studied. In a number of situations, lipoproteins bind to more than one site, some of which may have little relevance, at least quantitatively, to lipoprotein catabolism *in vivo*. However, covalent modifications of protein moieties provide one method to evaluate the extent to which protein components of lipoproteins bind to sites with properties of described lipoprotein receptors in such systems.

2. In Subcellular Fractions of Liver. Binding of lipoproteins to membrane fractions of liver or liver cells has been used to detect or quantify specific binding sites for lipoproteins. Crude membrane fractions ("microsomes") may contain receptors not only on pieces of plasma membrane, but also on intracellular membranes, so that the extent to which the extent of binding is related to lipoprotein uptake is uncertain. Hence, it is useful to correlate binding to uptake in intact cells or animals.[35,54] Increased binding to plasma membranes, as compared to crude membranes, provides additional evidence of physiological relevance.[55]

With such preparations, binding can be measured by incubating trace amounts of labeled lipoproteins with dispersed membrane fractions and measuring association with the membranes at equilibrium under defined conditions in the presence or absence of a large excess of unlabeled lipoprotein. "Specific binding" is calculated as the amount of labeled lipoprotein displaced by its unlabeled counterpart (see Chapter [33], this volume). From measurements of specific binding at increasing concentrations of labeled ligand, apparent binding affinities and maximal binding can be estimated. Specificity can be evaluated by competition assays with unlabeled lipoproteins (including those with covalently modified proteins), by addition of compounds, such as EDTA, heparin, or suramin that disrupt certain ligand–receptor interactions (see Chapter [33], this

---

[53] B. S. Suri, M. E. Targ, and D. S. Robinson, *FEBS Lett.* **133**, 283 (1981).
[54] P. T. Kovanen, M. S. Brown, and J. L. Goldstein, *J. Biol. Chem.* **254**, 11367 (1979).
[55] M. Carella, and A. D. Cooper, *Proc. Natl. Acad. Sci. U.S.A.* **76**, 338 (1979).

volume), or by enzymatic modification of the binding site with proteases or other hydrolases.

In some cases, the molecular properties of binding sites can be determined by Western blotting. Membranes are solubilized, usually in SDS, and subjected to electrophoresis in polyacrylamide gels. The separated proteins are transferred to nitrocellulose paper, incubated with labeled lipoprotein[56] or specific antibody,[57] and subjected to autoradiography. Western blotting can be adapted to provide a semiquantitative measure of receptor concentration and affinity for ligands. This method obviates the need to measure nonspecific binding. The blotting techniques can therefore provide highly sensitive assays of receptor concentration in membranes, as well as specific identification of binding sites. These methods also permit the same general tests of specificity that are used with suspended membranes.

C. Use of Models with Variable Lipoprotein Receptor Activity

1. Ethinyl Estradiol-Treated Rats. Rats treated with pharmacological amounts of 17-α-ethinyl estradiol for several days express 10 to 20-fold more LDL receptors on their hepatocytes than untreated rats.[54] The livers of these animals provide a useful model to study the consequences of augmented hepatic LDL receptor activity, including effects on lipoprotein concentrations and hepatic cholesterol metabolism. However, these animals have cholestasis,[35,58] so that some of the phenomena observed may not result solely from augmented receptor number. Many of the methods used to study hepatic catabolism of LDL have been evaluated in this model. Evidence of increased receptor number has been observed *in vivo*,[35] in perfused livers[35] and with subcellular fractions of livers and hepatocytes,[54] but not with freshly isolated liver cells.[59] Increased uptake of lipoproteins in perfused livers is limited to those that contain apolipoprotein B-100 and apolipoprotein E and no increase in uptake is observed with cyclohexanedione-modified LDL.[35] Uptake of HDL that contain no apoE is not increased, but HDL levels in treated animals are nonetheless greatly reduced. The estradiol-induced receptor has been found to recognize human apoB-100, but binds it with lower affinity than the apoB of rat LDL.[27]

[56] T. O. Daniel, W. J. Schneider, J. L. Goldstein, and M. S. Brown, *J. Biol. Chem.* **258,** 4606 (1983).

[57] U. Beisiegel, W. J. Schneider, M. S. Brown, and J. L. Goldstein, *J. Biol. Chem.* **257,** 13150 (1982).

[58] R. A. Davis, F. Kern, Jr., R. Showalter, E. Sutherland, M. Sinensky, and F. R. Simon, *Proc. Natl. Acad. Sci. U.S.A.* **75,** 4130 (1978).

[59] J. Belcher and R. J. Havel, unpublished observations.

2. Watanabe Heritable Hyperlipidemic (WHHL) Rabbit. Rabbits homozygous for a mutation which leads to synthesis of an unstable LDL receptor, and virtually no functional hepatic LDL receptors,[60] provide a very useful model to study the overall consequences of LDL receptor deficiency and to evaluate the role of the LDL receptor in the uptake of lipoproteins by the liver. These animals can be readily produced by mating homozygous males with heterozygous females (homozygous females breed poorly). Experiments with these animals have provided the best evidence that the uptake of VLDL remnants by the liver is via LDL receptors and that the formation of LDL and VLDL remnants is an inverse function of hepatic LDL receptor activity.[61] Removal of LDL from the blood of WHHL homozygotes, as in normal rabbits, occurs mainly in hepatocytes by receptor-independent mechanisms.[62] Removal of chylomicron remnants by the liver of WHHL homozygotes occurs at a normal rate, indicating that other mechanisms, presumably related to genetically distinct hepatic receptors, are responsible for the catabolism of this class of lipoproteins.[2] These animals thus provide useful models for studying mechanisms of hepatic uptake and catabolism of lipoproteins that may be obscured or confounded by the normal interaction of the LDL receptor with lipoproteins that contain apolipoprotein E as well as apolipoprotein B-100.

3. Animals Treated with Bile Acid-Binding Resins or Inhibitors of Cholesterol Biosynthesis. In several species, including dogs and rabbits, the activity of hepatic LDL receptors can be increased by administration of bile acid-binding resins (which increase cholesterol catabolism) or by inhibiting the activity of β-hydroxy-β-methylglutaryl-CoA reductase with the fungal metabolites, compactin or mevinolin (which inhibit cholesterol biosynthesis).[63,64] Combined treatment with the two classes of drugs produces a synergistic effect, because the compensatory increase in hepatic cholesterol biosynthesis that accompanies resin treatment is prevented by the reductase inhibitor.

4. Cholesterol-Fed Animals. In a number of species, administration of cholesterol-rich diets causes cholesterol to accumulate in the liver and secretion of cholesterol-rich VLDL. In some cases, reduction of hepatic

[60] M. Huettinger, W. J. Schneider, Y. K. Ho, J. L. Goldstein, and M. S. Brown, *J. Clin. Invest.* **74,** 1017 (1984).

[61] T. Kita, M. S. Brown, D. W. Bilheimer, and J. L. Goldstein, *Proc. Natl. Acad. Sci. U.S.A.* **79,** 5693 (1982).

[62] R. C. Pittman, T. E. Carew, A. D. Attie, J. L. Witztum, Y. Watanabe, and D. Steinberg, *J. Biol. Chem.* **257,** 7994 (1982).

[63] P. T. Kovanen, D. W. Bilheimer, J. L. Goldstein, J. J. Jaramillo, and M. S. Brown, *Proc. Natl. Acad. Sci. U.S.A.* **78,** 1194 (1981).

[64] Y. Chao, T. T. Yamin, and A. W. Alberts, *J. Biol. Chem.* **257,** 3623 (1982).

cholesterol catabolism (by production of hypothyroidism feeding bile acids, or both) may also be required to yield these results. In rabbits and dogs, cholesterol feeding has been found to lead to down regulation of LDL receptors as well.[65,66] The changes in biliary excretion of cholesterol and bile acids, cholesterol biosynthesis, cholesterol content of the liver, hepatic secretion of VLDL, and hepatic LDL receptor number in animals fed cholesterol-rich diets can be used to evaluate the contribution of lipoprotein catabolism to hyperlipoproteinemia in cholesterol-fed animals.

5. *Casein-Fed and Fasted Rabbits*. The hyperlipoproteinemia produced in rabbits by feeding casein-rich diets[64] or by fasting the animals for several days[67] is accompanied by large reductions in the activity of hepatic LDL receptors. These animals thus provide an alternative to WHHL rabbits for study of the effects of reduced receptor number in the liver on hepatic lipoprotein catabolism.

V. Quantification of Role of Parenchymal and Nonparenchymal Cells in Hepatic Lipoprotein Catabolism

A. By Autoradiography after Injection of Labeled Lipoproteins

Lipoproteins labeled with ^{3}H-lipid or by radioiodination can be used to evaluate lipoprotein uptake into parenchymal or nonparenchymal cells by autoradiography of fixed sections of liver at the light microscope level. When lipids are labeled (such as [^{3}H]cholesterol), methods of dehydration of the fixed tissue must minimize solubilization of the tracer.[68] With ^{125}I labeling, labeled lipids can be largely removed during dehydration such that the remaining radioactivity is mainly protein bound.[69] At short intervals after intravenous injection into intact rats, hepatic parenchymal cells have been found to be the primary site of localization of [^{3}H]cholesterol-labeled and ^{125}I-labeled chylomicrons,[68,70] ^{125}I-labeled VLDL,[69,70] LDL (the latter in untreated as well as ethinyl estradiol-treated rats),[71] and

[65] P. T. Kovanen, M. S. Brown, S. K. Basu, D. W. Bilheimer, and J. L. Goldstein, *Proc. Natl. Acad. Sci. U.S.A.* **257**, 3623 (1981).

[66] D. Y. Hui, T. L. Innerarity, and R. W. Mahley, *J. Biol. Chem.* **256**, 5646 (1981).

[67] J. B. Stoudemire, G. Renaud, D. M. Shames, and R. J. Havel, *J. Lipid Res.* **25**, 33 (1984).

[68] O. Stein, Y. Stein, D. S. Goodman, and N. Fidge, *J. Cell Biol.* **43**, 410 (1969).

[69] O. Stein, D. Rachmilewitz, L. Sanger, S. Eisenberg, and Y. Stein, *Biochim. Biophys. Acta* **360**, 205 (1974).

[70] A. L. Jones, G. T. Hradek, C. A. Hornick, G. Renaud, E. E. T. Windler, and R. J. Havel, *J. Lipid Res.* **25**, 1151 (1985).

[71] Y.-s. Chao, A. L. Jones, G. T. Hradek, E. E. T. Windler, and R. J. Havel, *Proc. Natl. Acad. Sci. U.S.A.* **78**, 597 (1981).

HDL.[72] This method depends upon the observation that little degradation of labeled lipid or protein taken up by endocytosis occurs until 20–30 min after injection. Application of the method to lipoprotein components that are slowly removed from the blood requires that large amounts of radioactivity be injected. A major advantage of the method is that labeled components can be visualized in intact liver. Similar studies at the electron microscope level indicate that labeled components of lipoproteins taken up into hepatocytes are processed by a stereotyped pathway consistent with receptor-mediated endocytosis and lysosomal catabolism.

B. By Separating Cells after Injection of Labeled Lipoproteins

Nonparenchymal cells can be separated from parenchymal cells of the liver by two general methods. In one, mixed cell populations, prepared by perfusion of livers with collagenase-containing buffer, are centrifuged at low speed.[39] The larger parenchymal cells sediment much more rapidly than nonparenchymal cells (mainly endothelial and Kupffer cells). In the other, similar mixed cell populations are treated with Pronase, which selectively destroys the parenchymal cells.[73] Alternatively, the liver can be perfused with Pronase directly.[74] The former method permits simultaneous separation of parenchymal and nonparenchymal cell populations, but the nonparenchymal cells are contaminated by some parenchymal cells. They can be purified further by repeated centrifugation.[75] However, the nonparenchymal cells remain contaminated with cytoplasmic fragments ("blebs") of parenchymal cells which can be separated by centrifugal elutriation.[76] The latter method yields virtually uncontaminated nonparenchymal cells. Endothelial cells and Kupffer cells can be separated from each other by centrifugal elutriation.[51,77]

These methods have been used to study the contribution of parenchymal and nonparenchymal cells to the *in vivo* uptake of various lipoprotein components in rats. Cholesteryl esters and retinyl esters of chylomicrons have been found mainly in parenchymal cells,[78,79] but about one-third of protein-labeled chylomicron remnants (with [³H]leucine) was associated

[72] S. Quarfordt, J. Hanks, R. S. Jones, and F. Shelburne, *J. Biol. Chem.* **255,** 2934 (1980).

[73] D. M. Mills and D. Zucker-Franklin, *Am. J. Pathol.* **54,** 147 (1969).

[74] L. Harkes and T. J. C. van Berkel, *FEBS Lett.* **154,** 75 (1983).

[75] T. J. C. van Berkel and A. van Tol, *Biochim. Biophys. Acta* **530,** 299 (1978).

[76] J. F. Nagelkerke, K. P. Barto, and T. J. C. van Berkel, *Exp. Cell Res.* **138,** 183 (1982).

[77] R. Blomhoff, C. A. Drevon, W. Eskild, P. Helgerud, K. R. Norum, and T. Berg, *J. Biol. Chem.* **259,** 8898 (1984).

[78] A. Nilsson and D. B. Zilversmit, *Biochim. Biophys. Acta* **248,** 137 (1971).

[79] R. Blomhoff, P. Helgerud, M. Rasmusson, T. Berg, and K. R. Norum, *Proc. Natl. Acad. Sci. U.S.A.* **79,** 7326 (1982).

with nonparenchymal cells.[80] Kupffer cells accounted for most of the uptake into nonparenchymal cells of lipid-labeled chylomicron remnants added to perfused rat liver.[81] Nonparenchymal cells have also been found to take up a substantial fraction of radioiodinated rat VLDL, LDL, and HDL injected into intact rats, whether the nonparenchymal cells were separated by differential centrifugation or by treatment of mixed cells with Pronase.[82] The same was found for the uptake of human LDL in livers of normal rats, but uptake into parenchymal cells was found to be greatly increased in ethinyl estradiol-treated rats.[74] By contrast, uptake of homologous LDL was found to be mainly into parenchymal cells of the liver of both normal and WHHL rabbits.[62]

This method yields results much more rapidly than autoradiography. A disadvantage is that cells are initially treated with collagenase. This damages or destroys some cells. Lipoproteins released from such cells may be subject to uptake by Kupffer cells, especially if they are altered by the enzyme. Collagenase treatment may also release lipoprotein from cell surfaces, which may then be subject to redistribution before the parenchymal and nonparenchymal cells are separated from each other. The latter may be a particular problem for lipoproteins, such as LDL and HDL, that are taken up slowly by the liver.

The method of cell separation has been used to show major uptake of labeled acetyl-LDL into endothelial cells of rat liver.[51,77] It has also been used to show apparent movement of retinyl esters of chylomicrons, after initial uptake into parenchymal cells, into a population of nonparenchymal cells which may represent the fat-storing cells of Ito.[79]

VI. Role of Apolipoproteins in Hepatic Lipoprotein Catabolism

A. Lipoproteins and Complexes Containing a Single Apolipoprotein

Those lipoproteins that contain a single protein labeled in the intact particle, such as LDL and the lipoprotein obtained from cholesterol-fed dogs which contains only apoE ("apoE-HDL$_c$"), have been particularly useful in evaluating the role of these proteins in hepatic lipoprotein catabolism.[83] Apolipoproteins other than apoB can generally be incorporated readily into lamellar complexes with phospholipids or adsorbed to the

[80] P. H. E. Groot, T. J. C. van Berkel, and A. van Tol, *Metabolism* **80,** 792 (1981).
[81] P. M. Lippiello, J. Dijkstra, M. van Galen, G. Scherphof, and B. M. Waite, *J. Biol. Chem.* **256,** 7454 (1981).
[82] A. van Tol and T. J. C. van Berkel, *Biochim. Biophys. Acta* **619,** 156 (1980).
[83] R. W. Mahley, T. L. Innerarity, and K. H. Weisgraber, *Ann. N.Y. Acad. Sci.* **348,** 265 (1980).

surface of emulsion particles. In this case, the isolated protein is usually radioiodinated and then mixed with phospholipid liposomes to yield defined complexes, such as discoidal particles that resemble "nascent HDL," or with phospholipid-stabilized emulsions. Such specifically labeled particles have been used to define receptor-mediated mechanisms of lipoprotein uptake for which the proteins serve as specific ligands, such as the LDL receptor on hepatocytes.[27] They have also been used to evaluate the effect of point mutations of apoE on protein uptake in perfused livers[26] or binding to receptors in liver membranes.[84]

B. Lipoproteins and Complexes Containing Multiple Apolipoproteins

1. Modifications of Apolipoprotein Composition. The composition of isolated lipoproteins can frequently be modified by incubating them with other lipoproteins or with isolated apolipoproteins. For example, the composition of lymph chylomicrons can be altered substantially by incubating them with HDL or plasma fractions containing HDL.[85] Content of C apolipoproteins is thereby increased substantially, whereas that of A apolipoproteins is reduced. Phospholipid-stabilized triglyceride emulsions containing apolipoprotein mixtures can also be produced in this way. These preparations have been particularly useful to evaluate the inhibitory effect of C apolipoproteins on apoE-mediated uptake of lipoproteins in perfused livers.[86] Such studies may be complicated by the exchange of proteins added to perfusates with those contained on lipoproteins secreted from the liver. Single-pass rather than recycling perfusions may therefore be preferred. Similar considerations hold for studies with isolated liver cells in suspensions or monolayer culture.

2. Labeling with Specific Proteins. Isolated proteins, usually labeled by radioiodination, can be added to lipoproteins, either after isolation or in whole plasma. The latter method has been used to label VLDL and HDL in whole plasma with apoE and HDL in plasma with apoA-I.[87] This method has the advantage that these lipoproteins, specifically labeled with the added protein by exchange, can be separated by gel chromatography, thereby avoiding ultracentrifugal procedures that tend to dissociate these proteins from lipoproteins. As with lipoproteins and complexes

[84] W. J. Schneider, P. T. Kovanen, M. S. Brown, J. L. Goldstein, G. Utermann, W. Weber, R. J. Havel, L. Kotite, J. P. Kane, T. L. Innerarity, and R. W. Mahley, *J. Clin. Invest.* **68,** 1075 (1981).

[85] K. Imaizumi, M. Fainaru, and R. J. Havel, *J. Lipid Res.* **19,** 712 (1978).

[86] R. J. Havel, *in* "Atherosclerosis VI" (F. G. Schettler, ed.), p. 480. Springer-Verlag, Berlin, 1983.

[87] F. van't Hooft and R. J. Havel, *J. Biol. Chem.* **256,** 3963 (1981).

containing a single apolipoprotein, the hepatic uptake of proteins in these complexes can be determined in various systems, but phenomena related to exchange with other lipoproteins must be taken into account in evaluating the results. Similar considerations apply to the analysis of binding experiments. For example, in competitive binding experiments, displacement of a labeled apoprotein in a complex from its binding site by an unlabeled lipoprotein may result from exchange of the labeled protein in the complex with an unlabeled one. The unlabeled protein need not be the same as the labeled one, provided that it can displace the labeled protein.

VII. Evaluation of Pathways of Lipoprotein Degradation in the Liver

A. By Autoradiography after Injection of Labeled Lipoproteins

Autoradiography at the light microscope level is a valuable method to localize the sites of movement of endocytosed lipoproteins within liver cells. It is subject to the limitations related to rates of uptake and intracellular processing described in Sections I and II and large amounts of radioactivity must be injected to permit reasonably short development times ($\sim$0.5 mCi of ^{3}H or ^{125}I for a 300-g rat). Within these limitations, the method has the distinct advantage of avoiding problems related to purity of subcellular fractionation techniques and it can provide direct information about the steps of intracellular transport. The method is technically demanding and quantification of the origin of grains is limited by the resolution of the method ($\sim$850 Å from the site of radioactive decay for ^{125}I).[88] Autoradiographic data obtained after injection of ^{125}I-labeled LDL into estradiol-treated rats[71] have been used to advantage in the purification of organelles of the endocytic pathway in hepatocytes, as described below.

B. By Use of Complexes of Lipoproteins or Antibodies with Colloidal Gold

Proteins, including lipoproteins, can be complexed to uniform colloidal gold particles ranging in diameter from 45 to 150 Å.[89] These gold particles are readily seen in transmission electron micrographs. When used as lipoprotein complexes, they can be seen within coated pits and in endocytic organelles.[90] In this respect they have a distinct advantage over

[88] R. H. Renston, D. G. Maloney, A. L. Jones, G. T. Hradek, K. Y. Wong, and I. D. Goldfine, *Gastroenterology* **78,** 1763 (1980).
[89] J. M. Slot and H. F. Geuze, *J. Cell Biol.* **90,** 533 (1981).
[90] D. A. Handley, C. M. Arbeeny, H. A. Eder, and S. Chien, *J. Cell Biol.* **90,** 778 (1981).

autoradiography. However, gold complexes are taken up readily into Kupffer cells, so that little may be available to bind to lipoprotein receptors on other cells.[91] By inhibiting the phagocytic capacity of Kupffer cells, more lipoprotein may be taken up by specific lipoprotein receptors, which facilitates visualization of the endocytic pathway.[91]

Colloidal gold complexed to specific antibodies or to protein A can be used to identify antigens such as apolipoproteins and lipoprotein receptors within hepatic cells. Complexes of colloidal gold particles of defined size with protein A or species-specific immunoglobulins are available commercially.[92] Application of this method at the electron microscope level requires use of frozen thin sections to permit access of antibodies to proteins within organelles.[93] The method is technically demanding, but can provide unique information. By use of two sizes of colloidal gold particles complexed to asialoorosomucoid and the asialoglycoprotein receptor, both receptor and ligand have been localized in endosomes of hepatocytes and the site of dissociation of the protein from its receptor has been visualized.[94]

C. By Use of Lysosomotropic Agents, Ionophores, and Colchicine

Compounds that delay or interrupt the endocytic pathway, which can be injected *in vivo* or added to perfusates of isolated livers, can provide additional information about pathways of lipoprotein degradation in the liver.[7] They can help in the identification of endocytic organelles, facilitate isolation of organelles, and help to identify morphologically such phenomena as retroendocytosis and receptor recycling.

D. By Subcellular Fractionation Techniques

Unlike most other ligands, lipoproteins can be seen within endocytic organelles, as well as organelles of the secretory pathway of lipoproteins in hepatocytes. When present within these organelles in sufficient number, they can facilitate isolation of the organelle by reducing its density. Radioisotopically labeled lipoproteins can provide ready markers for organelle purification, particularly when purification is guided by morphological data, as described above. Double labeling methods permit organelles of exocytosis and endocytosis to be marked separately. These

[91] G. Renaud, R. L. Hamilton, and R. J. Havel, unpublished data.
[92] Janssen Pharmaceutica, Turnhoutseweg 30, B-2340, Beerse, Belgium.
[93] K. T. Tokayasu and S. J. Singer, *J. Cell Biol.* **71,** 894 (1976).
[94] H. J. Geuze, J. M. Slot, J. A. M. Straus, H. F. Lodish, and A. L. Schwartz, *Cell* **32,** 277 (1983).

principles have been used to advantage in the purification of multivesicular bodies from livers of estradiol-treated rats[95] and they can, in principle, be used to facilitate purification of other endosomal structures. Endocytosed lipoproteins can be isolated from the organelles after the vesicular structures are ruptured by various means (treatment with sodium bicarbonate[96] or passage through a French pressure cell).[97] It is desirable to employ an anti-protease cocktail in processing liver homogenates for such purposes.

E. Biliary Excretion of Labeled Lipoprotein Components

Lipoproteins labeled with unesterified cholesterol have been used to study the pathway of cholesterol from blood plasma to bile.[98] Such experiments must take into account the rate of exchange of cholesterol in specific lipoproteins (without net flux) with cholesterol in plasma membranes and other sites within liver cells. As described in Section I products of protein catabolism also are excreted to some extent in the bile. For unknown reasons, this is particularly the case for proteins labeled by the Bolton–Hunter procedure.[99]

VIII. Role of the Liver in Lipoprotein Interconversions

A. By Measurement of the Concentration of Lipoprotein Components

The release of altered lipoprotein products from the liver can be measured during constant infusion of labeled lipoproteins *in vivo* or through the isolated, perfused liver. This approach has been used to demonstrate the uptake of species of VLDL and their conversion to LDL across the splanchnic region in humans[100] and to study modifications of lipoprotein composition that occur during transsplanchnic passage of lipoproteins.[101]

[95] C. A. Hornick, R. L. Hamilton, E. Spaziani, G. Enders, and R. J. Havel, *J. Cell Biol.* **100,** 1558 (1985).

[96] Y. Fujiki, A. L. Hubbard, S. Fowler, and P. B. Lazarow, *J. Cell Biol.* **93,** 97 (1982).

[97] R. L. Hamilton, *in* "Pharmacological Control of Lipid Metabolism" (R. Paoletti and D. Kritchevsky, eds.), p. 7. Plenum, New York, 1972.

[98] C. C. Schwartz, M. Berman, L. G. Halloran, L. Swell, and Z. R. Vlahcevic, *in* "Lipoprotein Kinetics and Modeling" (M. Berman, S. M. Grundy, and B. V. Howard, eds.), p. 337. Academic Press, New York, 1982.

[99] J. M. Schiff, M. M. Fisher, and B. J. Underdown, *J. Cell Biol.* **98,** 79 (1984).

[100] P. R. Turner, N. E. Miller, C. Cortese, W. Hazzard, J. Coltart, and B. Lewis, *J. Clin. Invest.* **67,** 1678 (1981).

[101] A. Sniderman, D. Thomas, D. Marpole, and B. Teng, *J. Clin. Invest.* **61,** 867 (1978).

B. *By Modifying the Activity of Hepatic Heparin-Releasable Lipase*

Antisera to hepatic lipase have been injected into animals to evaluate the influence of this enzyme upon plasma lipoprotein concentrations[102] and upon the rate of conversion of protein-labeled VLDL to lipoproteins of higher density.[103] Similar studies can be carried out in perfused livers. Alternatively, in this system the lipase can be removed by flushing the liver with perfusate containing heparin.[53] The role of apolipoproteins in lipoprotein interconversions in the liver can, in principle, be studied by the methods used to determine their role in lipoprotein uptake (see Section VI).

[102] T. Murase and H. Itakura, *Atherosclerosis* **39**, 293 (1981).
[103] J. J. Goldberg, N. A. Le, J. P. Paterniti, H. N. Ginsberg, F. T. Lindgren, and W. V. Brown, *J. Clin. Invest.* **70**, 1184 (1982).

[36] Methods for Assessment of Tissue Sites of Lipoprotein Degradation

By RAY C. PITTMAN and CLINTON A. TAYLOR, JR.

Introduction

It is only in the last few years that methods have been available that are generally useful for determining the sites of plasma protein degradation *in vivo*. Application of conventional tracer methodologies to this problem have met with only very limited success because these tracers (usually radioiodine directly incorporated into tyrosine residues) readily escape the cells catabolizing the subject protein. The amount of label, associated with a particular tissue at any time after injection whether as catabolic products, intact protein, or both, bears no necessary relationship to the amount of protein catabolized by that tissue.

The newer methods depend on a common, straightforward principle. The essential idea is to covalently link to the subject protein a tracer ligand. On degradation of the labeled protein, the ligand is not itself degraded but is left behind in the catabolizing cells as a cumulative marker of the amount of degradation that occurs in those cells.

The original ligand chosen and the model for subsequent variant labeling ligands is radioactive sucrose. The bases for this choice further illustrate the principle. Sucrose has been used to measure pinocytosis by cells because it does not interact appreciably with the cell surface and thus is incorporated into the cells only as a fluid-phase solute. The glycoside bond of sucrose is not hydrolyzed by intracellular enzymes of mammalian cells, and the intact molecule cannot cross biologic membranes; thus, it accumulates in cells as a measure of fluid-phase uptake.

Two techniques based on the covalent attachment of radiosucrose have been reported.[1-3] We used an activating ligand, cyanuric chloride, to form a bridge between [14C]sucrose and the protein.[2] Van Zile *et al.* used [3H]raffinose, a reducing trisaccharide containing the sucrose configuration, which they directly attached to the protein by reductive amination.[3] The approach of using sucrose itself has been more extensively characterized and applied to a wider range of proteins. On the basis of the limited studies carried out with the raffinose technique,[4,5] it appears that there may be a greater rate of leakage of this label from tissues in which the label is initially trapped. On these grounds, and because of the greater simplicity of the sucrose–cyanuric chloride technique, only it will be described in detail here. However, the raffinose approach may be the method of choice in at least some applications. This is because the use of cyanuric chloride has the potential to produce protein cross-linking, as discussed further below. Indeed there was such a problem in the studies of apoA-I which required repurification of the labeled protein. The use of the raffinose method may ameliorate this problem (although a cross-linking potential exists in that case also, but probably to a lesser extent). The reader is referred to the primary literature for a description of the [3H]raffinose method.[3]

The major limitation of both the sucrose and raffinose technique is the availability of the radioactive precursors at only limited specific activities, making impractical the study of some proteins that occur in very low concentration in plasma, or the study of tissues of very low catabolic activity. Also pertinent is the difficulty of radioassaying 14C and 3H at low levels in tissues. Consequently, variants of the techniques have been devised which incorporate radioiodine in the labeling ligand. In all cases an acceptor of radioiodine (an activated aromatic ring) was linked to a di- or trisaccharide which still played the role of primary trapping agent.

De Jong and co-workers used as a labeling ligand O-(4-diazo-3,5-

[1] R. C. Pittman and D. Steinberg, *Biochem. Biophys. Res. Commun.* **81**, 1254 (1978).

[2] R. C. Pittman, S. R. Green, A. D. Attie, and D. Steinberg, *J. Biol. Chem.* **254**, 6876 (1979).

[3] J. Van Zile, L. A. Henderson, J. W. Baynes, and S. R. Thorpe, *J. Biol. Chem.* **254**, 3547 (1979).

[4] J. W. Baynes and S. R. Thorpe, *Arch. Biochem. Biophys.* **206**, 372 (1981).

[5] J. W. Baynes, J. Van Zile, L. A. Henderson, and S. R. Thorpe, *Birth Defects: Origin. Article Ser.* **XVI**, 103 (1980).

di[^{125}I]iodobenzoyl)sucrose, which links the aromatic radioiodine acceptor to sucrose via an ester bond.[6] The procedure appears useful for study of proteins of rather rapid decay, but the label leaks too rapidly from tissues degrading the subject protein (perhaps because of the biologic lability of the ester bond) to be applied to proteins that are not rapidly cleared. These authors estimate that the label is retained in tissues with a half-time of about 20 hr. Due to its limited applicability, the method will not be described in detail here.

Also not considered in detail here is a radioiodinated version of the raffinose method.[7] A tyrosine moiety, the iodine acceptor, is attached to raffinose by reductive amination. Cyanuric chloride is then used to link the ligand to the subject protein. The procedure resembles rather closely the tyramine–cellobiose procedure devised in this laboratory which has been more broadly studied and applied. The original advantage of the raffinose method, the presumed lesser chance of cross-linking than in the methods using cyanuric chloride, is lost in this variant. There is the added disadvantage that the carboxyl group of tyrosine adds a negative charge on top of the partial loss of a positive charge resulting from the use of cyanuric chloride. A preliminary report from these same workers of other radioiodinated variants has appeared,[8] but it is too early to evaluate the usefulness of these procedures.

At the present time the method of choice for most applications seems to be the tyramine–cellobiose technique.[9] Tyramine, the radioiodine acceptor, is linked to cellobiose, a nonhydrolyzable reducing disaccharide, by reductive amination. The resulting tyramine–cellobiose adduct (TC) is then attached to the protein using cyanuric chloride. The method has been validated and used in several biological studies,[9–12] which will not be reviewed here.

General and Theoretical Considerations

Two biological requisites must be met for the trapped label approach to be applicable. First, the tracer must be metabolically indistinguishable

[6] A. S. H. DeJong, J. M. W. Bauma, and M. Gruber, *Biochem. J.* **198,** 45 (1981).

[7] J. L. Strabel, J. W. Baynes, and S. R. Thorpe, *A.C.S. Abstr. Pap.* **186,** 33 (1983).

[8] J. L. Strabel, J. W. Baynes, and S. R. Thorpe, *Fed. Proc., Fed. Am. Soc. Exp. Biol.* **43,** 2023 (1984).

[9] R. C. Pittman, T. E. Carew, C. K. Glass, S. R. Green, C. A. Taylor, Jr., and A. D. Attie, *Biochem. J.* **212,** 791 (1983).

[10] C. K. Glass, R. C. Pittman, G. A. Keller, and D. Steinberg, *J. Biol. Chem.* **258,** 7161 (1983).

from the tracee protein, at least through the step of irreversible removal from plasma. It is thought that the most sensitive single index of this is the comparison of the plasma decay kinetics of the tracer to the native protein. The reference kinetics for the native protein may be defined as those obtained using an alternative labeling method appropriate for the subject protein, usually direct iodination. *In vitro* comparisons add a complementary type of evidence. It should be born in mind that the trapped ligands are by necessity fairly large (molecular weight in the range of 500), may change the charge of the protein (use of the TC ligand results in a partial loss of a positive charge at physiological pH), and may cross-link protein. Therefore it is important to document the legitimacy of the labeled protein as a tracer.

Second, the labeling ligand must be adequately trapped in the cells degrading the labeled protein. The TC ligand appears to leak from catabolizing cells at a rate of about 10%/day, somewhat more than the approximately 5%/day observed with the sucrose ligand. However, products that do leak into plasma do not appear to redistribute to other tissues. Catabolic products of LDL labeled with the ^{125}I-labeled TC ligand have been isolated from rabbit liver 24 hr after injection of the tracer and injected into recipient animals; the labeled products were rapidly cleared from plasma, appearing almost exclusively in urine with a modest (5%) trapping in kidney. Thus, uncertainty caused by label leakage is not compounded by redistribution to other tissues. However, tissues may leak at different rates: rat tissues leak the sucrose label at varying rates; cultured macrophages leak the TC ligand at a greater rate than do cultured fibroblasts. Thus a 10% overall leak rate may disguise a greater error in certain cell types.

Leakage from the liver represents a special case. A variable but sometimes very high leakage from this organ has been observed, varying with the labeling ligand (TC leaks more than sucrose), with the subject animal (rabbit liver leaks TC-labeled LDL more than rat liver), and with the subject protein (TC-labeled LDL leaks more than TC-labeled apoA-I in rats). Fortunately this variably high leakage appears to represent exclusively output into bile and not into plasma. Even more fortunately, there is no indication of reabsorption of the label from the gut lumen, even in the case of the sucrose label which one might expect to be subject to hydrolysis by intestinal sucrase. Because the labeling ligand appears to

---

[11] C. K. Glass, R. C. Pittman, D. B. Weinstein, and D. Steinberg, *Proc. Natl. Acad. Sci. U.S.A.* **80,** 5435 (1983).
[12] C. K. Glass, R. C. Pittman, and D. Steinberg, *J. Biol. Chem.* **260,** 744 (1985).

inhibit hydrolysis of protein in its vicinity of attachment, the labeled catabolic products that accumulate in tissues are not the free ligand but the ligand still attached to a short peptide[2,9]; the catabolic products that find their way into bile may be similarly impervious to the action of digestive enzymes so that they pass through the gut intact. Thus, label appearing in bile, gut contents, and feces may be attributed to primary uptake by liver, barring secretion of label by the gut wall which has not been noted to date.

Two further requirements of a theoretical nature must be met to justify the use of the trapped label methodology. A rigorous treatment of these aspects of trapped labels has been published.[13] The important questions involved may be expressed in more intuitive terms by realizing that, for the labeled protein to precisely trace mass flux of the tracee protein, all cells taking up the tracer must be exposed to effectively the same specific activity of the subject protein during tracer clearance. It can be shown that indeed all tissues are exposed to the same *time-integrated* specific activity, *but* only after all of the tracer has been irreversibly cleared (essentially infinite time), and only *provided that* any newly synthesized tracee protein entering the system for the first time enters directly into the initially labeled pool (i.e., the plasma pool). If these conditions are met, the areas under the curves of specific activity versus time will be the same for all exchanging pools, regardless of their number or complexity, and all tissues will have been exposed to the same effective (time-integrated) specific activity during the course of tracer uptake. Thus the tracer will precisely trace the mass flux of traces if tissues are examined after very long times compared to the plasma clearance rates, and if newly synthesized tracee protein enters only into the plasma pool.

In some circumstances these ideal conditions cannot be met. The trapped labels are not perfectly trapped and leak from tissues at significant rates. If one studies a protein of relatively slow clearance rate the leakage may be unacceptably high at times after injection when the tracer is virtually completely cleared. Thus one must balance the uncertainty in the data caused by leakage against the error caused by examining the tissues too soon after injection. If one measures actual catabolic products accumulating in tissues rather than total tissue contents of label, the error resulting from early tissue sampling is somewhat predictable. In that case, catabolism attributable to tissues that are exposed to the subject protein in extravascular pools that exchange slowly with plasma will be underestimated; if total tissue radioactivity is assayed, these tissues will be overes-

---

[13] T. E. Carew and W. F. Beltz, *in* "Lipoprotein Kinetics and Modeling" (M. Berman, S. M. Grundy, and B. V. Howard, eds.), p. 791. Academic Press, New York, 1982.

timated. In most cases the overestimate resulting from using total label content is more severe than the underestimate from using radioactivity in degradation products. Thus, it is usually desirable to measure catabolic products.

Unfortunately, there is no simple, general method available for directly measuring the labeled catabolic products accumulating in tissues. The label trapped in cells is partially precipitated by TCA (evidently because it remains attached to small peptide fragments), and sizing techniques are tedious.

We have suggested a method that can circumvent this problem, illustrated in Fig. 1, which depends on the use of conventionally radioiodinated protein to trace the undegraded protein in tissues. Because the products of catabolism of directly iodinated proteins leak rapidly from the cells and are fairly rapidly excreted into urine, the radioactivity in tissues may be attributed predominantly to intact protein, at least in those tissues of relatively high uptake rate. Thus the subject protein can be directly iodinated with ^{131}I and then labeled with the ^{125}I-labeled TC trapped label; ^{125}I accumulated in a tissue represents degraded and undegraded protein, while ^{131}I represents only undegraded protein. The difference represents degradation products. A problem with this double label method is that products of catabolism carrying the conventional ^{131}I label (iodide and iodotyrosine) which arise in high uptake tissues are not excreted fast enough to avoid some redistribution throughout the extravascular spaces. These redistributed products may make a significant contribution to total ^{131}I radioactivity in very low uptake tissues. Unfortunately these tissues are the very ones that most need the correction for their content of uncatabolized label! The problem can be overcome by measuring only protein-bound conventional label in tissues in order to exclude the catabolic prod-

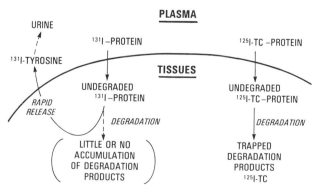

FIG. 1. Basis of the double label approach for measuring the labeled catabolic products accumulating in tissues. For details see text.

ucts arising from high uptake tissues. TCA adequately separates intact protein from catabolic products in the case of directly iodinated proteins. Thus, to accurately determine catabolic products in low uptake tissues using the double label approach outlined above, the tissue is radioassayed for total content of both isotopes; the contribution of TCA-soluble products to total content of the conventional label is determined in a tissue homogenate and subtracted from the total tissue content of that label to give the amount of label attributable to intact protein. This is then used to correct the tissue's total content of the trapped label for that attributable to intact protein. It may appear that this is a rather cumbersome procedure for indirectly measuring the amount of trapped label in catabolic products, but in practice it is much easier than attempting to directly measure such products. It should be remembered that if one is interested only in the major sites of catabolism of the subject protein, the double label method may be applied using the tissue's *total* content of both labels. This is discussed in greater detail elsewhere.[9]

Something should be said about the calculation and expression of experimental results. In those common cases where the ideal situation of allowing virtually total clearance of all tracer cannot be achieved, it would perhaps be ideal if one could determine the amount of the tracer that actually is catabolized and express the labeled catabolic products recovered in a tissue as a percentage of that total catabolism. However, calculation of the amount of tracer actually degraded using the plasma decay data depends on the precise kinetic model applicable to the system, and this is usually not known.

It is our practice to express the data in two ways, both of which assume that we account for all labeled catabolic products that are produced, or at least that we recover a representative sampling of products from all tissues. First, the contribution of a tissue to overall degradation of the subject protein is expressed as the catabolic products recovered in that tissue as a fraction (or percentage) of the total labeled catabolic products recovered from all sources (tissues plus label content of urine, gut contents, and feces). These fractional or percentage contributions of the various organs to total degradation may be converted into fractional rates by incorporating into a calculation the plasma fractional catabolic rate. This is the rate at which the subject protein irreversibly disappears from plasma. The plasma fractional catabolic rate is model independent, and may be calculated from the plasma decay data, as described by Matthews.[14] This whole body rate may be distributed between the tissues according to their individual contributions to whole body catabolism de-

[14] C. M. E. Matthews, *Phys. Med. Biol.* **2**, 36 (1957).

termined as outlined above. This is done by multiplying the plasma fractional catabolic rate by the fraction of whole body catabolism attributable to the organ in question. By this method one can calculate a plasma fractional catabolic rate attributable to each organ, or to 1 g of that organ if the result is divided by the organ weight. These are useful numbers in that they are expressions of rates that are independent of the plasma pool size of the subject protein. Of course, absolute rates may be calculated by simply multiplying these fractional rates by the plasma concentration of the protein.

The [^{14}C]Sucrose–Cyanuric Chloride Method

General Considerations

Cyanuric chloride (2,4,6-trichloro-1,3,5-triazine) was chosen as the linker to join sucrose to the protein because of its high reactivity and its heteropolyfunctional nature.[15,16] In general, displacement of one chloro

Cyanuric Chloride

group from the triazine ring by a nucleophile deactivates the ring toward further chloro group displacement; reaction with an amino group is generally more deactivating than reaction with an oxygen group. Displacement of two chloro groups by amino groups leaves the third chloro group resistant to displacement except under forcing conditions.

Sucrose is activated in a rapid reaction with 2 molar equivalents of cyanuric chloride under basic conditions. The reaction is quenched with excess acetic acid. The activated sucrose is immediately introduced into a solution of the protein to be labeled in a nucleophile-free buffer at near neutral pH. Binding takes place over a few hours at room temperature. The overall yield of [^{14}C]sucrose bound to protein is generally about 20–

[15] A. S. Chaudari and C. T. Bishop, *Can. J. Chem.* **50**, 1987 (1972).
[16] L. T. Hodgins and M. Levy, *J. Chromatogr.* **192**, 381 (1980).

25%. The method has been validated and applied in a number of experiments.[1,17–21]

Possible difficulties with this technique relate to its potential for protein cross-linking: (1) The reaction proceeds with an acceptable yield of [^{14}C]sucrose bound to protein only if excess cyanuric chloride is used; some of the excess cyanuric chloride may survive the activation step and directly cross-link protein. (2) Because sucrose has eight potentially reactive hydroxyl groups (the two primary hydroxyl groups are the most reactive), one molecule of sucrose may react with more than one molecule of cyanuric chloride and thus allow protein cross-linking. (3) Because reaction of cyanuric chloride with sucrose may not deactivate the triazine enough to prevent reaction of all three displaceable chloro groups with nucleophiles, there may be some degree of cross-linking involving only one molecule of the triazine. In practice intermolecular cross-linking has only been a problem with apoA-I. The extent to which this may be a problem with other proteins is not known.

An important virtue of the [^{14}C]sucrose method in application to labeling intact lipoprotein is that less labeling of lipid moieties is observed than is the case with direct radioiodination or with use of the tyramine–cellobiose ligand. Typically, only about 2% of the label of labeled LDL is extracted with lipid solvents in the case of [^{14}C]sucrose, compared to about 6% in the other two cases. The difference has been even more striking in pilot experiments in which VLDL was labeled (about 6 versus about 30%).

Application of the Method

Recrystallization of Cyanuric Chloride. Commercial preparations can be very bad indeed! To recrystallize, cyanuric chloride is dissolved in a minimum volume of hot toluene, rapidly filtered hot, and then allowed to cool and crystallize. The crystallization is repeated and the crystals are collected by filtration and dried. Purity is more important than recovery

[17] R. C. Pittman, A. D. Attie, T. E. Carew, and D. Steinberg, *Proc. Natl. Acad. Sci. U.S.A.* **76,** 5345 (1979).

[18] R. C. Pittman, A. D. Attie, T. E. Carew, and D. Steinberg, *Biochim. Biophys. Acta* **710,** 7 (1982).

[19] S. Yedgar, T. E. Carew, R. C. Pittman, W. F. Beltz, and D. Steinberg, *Am. J. Physiol.* **244,** E101 (1983).

[20] R. C. Pittman, T. E. Carew, A. D. Attie, J. L. Witztum, Y. Watanabe, and D. Steinberg, *J. Biol. Chem.* **257,** 7994 (1982).

[21] T. E. Carew, R. C. Pittman, and D. Steinberg, *J. Biol. Chem.* **257,** 8001 (1982).

because only very small quantities are needed. This very reactive material is stored in the cold in small, tightly capped aliquots in a desiccator.

Activation of [^{14}C]Sucorse and Binding to Protein. [^{14}C]Sucrose has been used rather than [^{3}H]sucrose for two reasons: ^{14}C is more readily radioassayed in tissues than ^{3}H (unless the ^{3}H is oxidized to ^{3}H$_2$O for radioassay); we have not found a source of [^{3}H]sucrose that is dependably free of interfering, nonradioactive contaminants that decrease binding efficiency. We have used [^{14}C]sucrose from Amersham because of its consistently good *chemical* purity.

Proteins are derivatized with sucrose to the extent of less than 0.8 mol sucrose/mole protein in order to ensure that no significant population of molecules is produced which carry more than one sucrose ligand. Because the efficiency of [^{14}C]sucrose utilization is about 25%, we begin with about a 3-fold molar excess of sucrose. Outlined below is a typical sequence for labeling a model 50,000-Da protein:

1. Lyophilize 250 μCi of [^{14}C]sucrose in a microreaction vessel (assuming a commercially available specific activity of 400 Ci/mole, this represents 0.625 μmol of sucrose).

2. Redissolve in 10 μl of water.

3. Add 3 molar equivalents (3 × 0.625 μmol) of NaOH in 15 μl of water.

4. Add 2 molar equivalents (2 × 0.625 μmol) of cyanuric chloride in 20 μl of acetone.

5. Wait 10–15 sec.

6. Add 4 molar equivalents (4 × 0.625 μmol) of acetic acid in 15 μl of water.

7. Immediately add the total reaction mixture to about 0.33 equivalents of the protein (0.33 × 0.625 μmol, or 10 mg of our 50,000-Da model protein). The protein should be at a concentration of 2–5 mg/ml in a nucleophile-free buffer (0.1–0.4 M sodium phosphate buffer, pH 7.2, has been most widely used).

8. Allow binding to proceed at room temperature for at least 3 hr, preferably overnight.

9. Remove unbound sucrose by dialysis, chromatography, or other appropriate procedure. Gel filtration and subsequent dialysis allows a more rapid approach to an acceptably low level of TCA—soluble ^{14}C than does dialysis alone.

Using the above procedure, specific activities of about 500–750 dpm/picomole protein are usually obtained (about 10–15 dpm/ng for our model 50,000-Da protein).

The *I-Tyramine–Cellobiose Method[9,21a]

General Considerations

The scheme for synthesis of the radioiodinated tyramine–cellobiose ligand (TC) is shown in Fig. 2.

Tyramine (an iodine acceptor) and cellobiose (a reducing disaccharide not hydrolyzed by mammalian cells) are joined by reductive amination. The resulting stable TC adduct is purified and stored for routine use. On use, the TC is radioiodinated and then activated with cyanuric chloride for binding to protein.

It was originally thought that the TC adduct would react with cyanuric chloride predominantly by attack of the secondary amino group of TC on the triazine ring,[9] but newer evidence indicates that the phenolic oxygen group is also a good nucleophile toward the triazine ring and contributes importantly to the reaction. The major result of this bifunctionality is some degree of polymerization of the ligand during the activation step. To what extent the ligand actually bound to protein is in polymeric form is not known.

It is recognized that the activation and binding reactions are complex, so that the precise nature of the ligand bound to protein is not well defined. However, these very properties may be what allows the efficient intracellular trapping of the ligand. Variants of the ligand have been synthesized which preclude any degree of polymerization or cross-linking. None of these is as satisfactorily trapped in cells as the TC ligand described here, although satisfactory results are obtained for some proteins with these other ligands. Thus, the TC ligand remains the generally most useful method, at least for proteins of fairly large size. TC labeling does not change the properties of apoA-I, (M_r of about 28,000), but does change the properties of insulin (M_r of about 6000).

The TC labeling method has been validated[9] and applied to studies of apoA-I in rats[10–12] and LDL in rabbits.[22]

Synthesis of the Tyramine–Cellobiose Adduct

Reductive amination of cellobiose by tyramine is carried out using $NaCNBH_3$ as the reducing agent. Because HCN is a by-product, all procedures should be carried out in a fume hood. The synthesis below is an improvement over that previously reported.[9]

[21a] *I, either ^{125}I or ^{131}I used.

[22] T. E. Carew, R. C. Pittman, E. R. Marchand, and D. Steinberg, *Arteriosclerosis* **4,** 214 (1984).

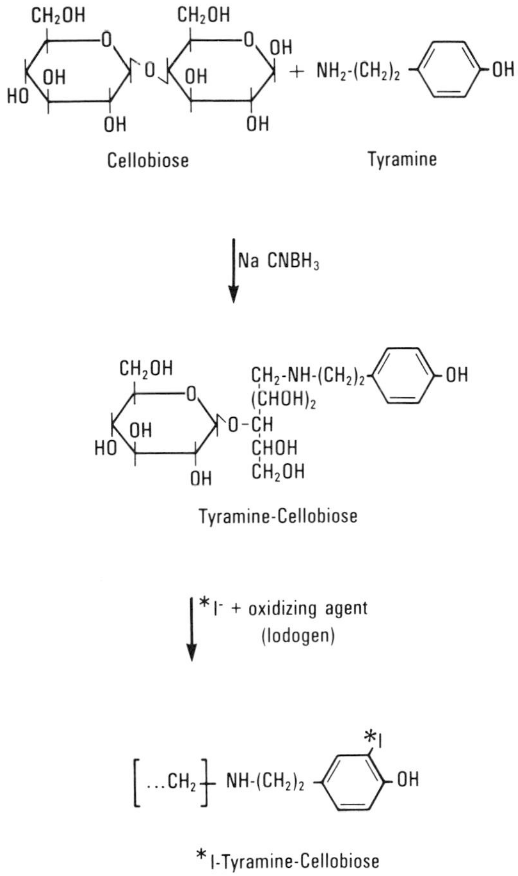

FIG. 2. Scheme for the synthesis of the radioiodinated tyramine–cellobiose ligand (TC).

1. Tyramine (13.7 g, 100 mmol) is added to 1 liter of water in a 2-liter round-bottom flask fitted with a reflux condenser and magnetic stirring bar. Acetic acid, sufficient to dissolve the tyramine, is added. NaCNBH$_3$ (6.3 g, 100 mmol) is added. Cellobiose (13.7 g, 40 mmol) is added. Glacial acetic acid is added to adjust the pH to 5.4.

2. The reaction mixture is refluxed with stirring for 96 hr, with readjustment to pH 5.4 at 24-hr intervals. The solution is then cooled, adjusted to pH 3.5, and refluxed again for 24 hr to destroy remaining NaCNBH$_3$.

3. The volume is reduced to 100 ml using a rotary evaporator. Precipitated material is redissolved by increasing the volume to 250 ml with water.

4. The resulting solution is applied to a 2.6 × 21 cm column of a cation exchange resin in the NH_4^+ form (AG 50W-X8, BioRad). The column is eluted with 2.5 liters of water (eluting unreacted cellobiose) and then with a linear gradient of 2 liters of water and 2 liters of 1 M NH_4OH (eluting the tyramine–cellobiose adduct, which precedes unreacted tyramine from the column). The elution pattern is monitored by TLC on silica gel developed in butanol : acetic acid : water (7 : 1 : 2). The product can be identified as the major component eluting in the gradient which runs on TLC with an r_f of about 0.2–0.3. Tyramine runs with an r_f of 0.4–0.5. The adduct may be viewed by staining for carbohydrate[23] (a spray is made of equal volumes of 2 mg/ml naphthoresorcinal in ethanol and 20% H_2SO_4, and color is developed in 5–10 min at 100°); both the adduct and free tyramine may be viewed with a chlorine–tolidine spray for amines.[24]

5. The fractions containing the tyramine–cellobiose adduct (about 800 ml total) are pooled, reduced in volume using a rotary evaporator, and then dried completely by lyophilization to yield a light tan powder. The yield is about 65%. The material can be stored indefinitely in a desiccator.

Application of the Tyramine–Cellobiose Method

This method was developed in order to provide a higher specific activity label. The major limitation to labeling proteins to high specific activity by this method is the need to deal with chemically manipulatable quantities of reagents which in turn necessitates the handling of relatively large amounts of radioactivity. It is difficult to deal with less than 0.05 μmol of tyramine–cellobiose. In the normal procedure described below (using 0.1 μmol of TC) a specific activity of about 5–10 × 10^4 dpm/picomole protein is readily achieved at an extent of derivatization of 0.7–0.8 mol TC ligand/ mole protein using a total of 10 mCi of radioiodide. The efficiency of binding of the TC ligand to protein is generally about 50%. Thus, in the procedure below about 0.06–0.07 μmol of protein is required.

Radioiodination of the TC Ligand

1. To a 300-μl microreaction vessel is added 10 μg of 1,3,4,6-tetra-chloro-3α,6α-diphenylglycouril (Iodogen, Pierce Chemical Co.) in methylene chloride. The solvent is evaporated by hand warming to leave a thin coating of Iodogen on the walls. The vessel is washed vigorously with water several times to remove any loosely attached Iodogen particles.

[23] E. Stahl, "Thin Layer Chromatography," 2nd Ed., p. 888. Springer-Verlag, New York, 1969.
[24] E. Stahl, "Thin Layer Chromatography," 2nd Ed., p. 863. Springer-Verlag, New York, 1969.

2. To the Iodogen-coated vessel is added 0.1 μmol of tyramine–cellobiose in 10 μl of 0.4 M sodium phosphate buffer, pH 7.4.

3. Carrier-free Na*I is then added in a volume up to 200 μl. In most preparations 5–10 mCi at a concentration of 100 mCi/ml is added (50–100 μl). If higher specific activity is sought commercial preparations of both Na^{125}I and Na^{131}I at about 500 mCi/ml are available. (The 10 μg of Iodogen is sufficient for this greater amount of Na^{125}I.)

4. Iodination is allowed to proceed for 30 min. The solution is then transferred to another vessel containing 5 μl of 0.1 M NaI and 10 μl of 0.1 M NaHSO$_3$ to stop further iodination. Efficiency of iodination is generally >95%. The efficiency may be checked by adding a small aliquot of the iodinated TC to a solution of carrier NaI in acid solution; the iodide is oxidized to iodine with H$_2$O$_2$ and quickly extracted into chloroform. The remaining aqueous phase is then radioassayed for *I-TC. (The CHCl$_3$ is not radioassayed because CHCl$_3$ absorbs *I radiation strongly.)

Activation of Radioiodinated Tyramine–Cellobiose. To the solution of *I-TC prepared above, 0.1 μmol of cyanuric chloride is added in 20 μl of acetone. (Cyanuric chloride is recrystallized as outlined under the sucrose method.) Reaction is allowed to proceed for less than 1 min, at which time protein is added for binding.

Protein Binding. To the solution of *I-TC activated with cyanuric chloride is added 0.3–0.4 μmol of protein at a concentration of 2–5 mg/ml in 0.1–0.4 M sodium phosphate or other well-buffered, nucleophile-free solution. Lower protein concentrations produce a greater wastage of activated ligand, and at high concentrations the protein is more subject to intermolecular cross-linking. Binding is carried out for 3 hr or more at room temperature. The labeled protein is then cleared of unbound label by chromatography and/or dialysis.

Newer Variations on the Method. Because protein cross-linking is a hazard using the potentially trifunctional cyanuric chloride, the bifunctional analogue, 2,4-dichloro-6-methoxy-1,3,5-triazine, may be used. As noted above, oxygen groups on the triazine ring deactivate the remaining chloro groups toward displacement much less than do amino groups; thus this analogue remains bifunctional. However, reaction times must be extended. The drawback is that the ligand linked with this analogue is less stably attached to protein and has been found to leak from cultured cells somewhat faster than the ligand linked with cyanuric chloride.

An analogue of the TC ligand has been synthesized in which the secondary amino group is blocked by methylation; thus reaction is forced exclusively through the phenolic hydroxy group of the ligand. When this is used with the 2,4-dichloro-6-methoxy-1,3,5-triazine just described, cross-linking is positively precluded. However, leakage from tissues makes it useful only for proteins of fairly rapid plasma clearance.

Application to Specific Lipoproteins and Apolipoproteins

Low-Density Lipoprotein

This lipoprotein is the one that has been most studied using the trapped label metholodies, using both sucrose and TC.[2,9,17,18,20–22] In labeling LDL the one major variation from the general procedure outlined above is that reaction of LDL with the activated ligand should be carried out at relatively high pH in order to minimize reaction with lipid moieties. At pH 9.5 and above, lipid labeling is minimal. The usual procedure is to use LDL at 2–5 mg/ml in 0.1 M sodium borate buffer, pH 9.5. Reaction is allowed to proceed for only 1–3 hr to minimize the time of exposure at this elevated pH.

High-Density Lipoprotein

The intact lipoprotein has been directly derivatized with the TC ligand, but the product has not been thoroughly studied. The plasma decay of the resultant labeled HDL is not detectably different from that of HDL directly radioiodinated using the Iodogen method, even though there is no reason to expect that both techniques produce similar patterns of apolipoprotein labeling.

Apolipoprotein A-I

Studies of apoA-I have been carried out by labeling purified apoA-I and then reassociating the labeled apoA-I with HDL.[10–12] As mentioned above, apoA-I and insulin are the only proteins with which we have encountered protein dimerization due to TC labeling. In initial studies of apoA-I labeling this was minimized by carrying out the labeling at high pressure, under which condition the normal reversible self-association of apoA-I is not favored. More recently we have achieved equally good results by labeling in the presence of 4 M guanidine–HCl in sodium phosphate buffer, pH 7.2. In either case, some covalent dimerization and even trimerization still occur due to the labeling procedure. The monomeric labeled apoA-I is isolated from the mixture by gel filtration on Sepharcryl S-200 (Pharmacia). TC-labeled apoA-I may then be reassociated with HDL by overnight incubation at 37° with at least a 10-fold excess of HDL-associated apoA-I. Unbound apoA-I is then removed by floating up the labeled HDL at $d < 1.21$.

Carrying out the in Vivo Experiments

Under some circumstances it may be acceptable to determine the sites of degradation of a plasma protein or lipoproteins simply by injecting the labeled material, waiting until the tracer is assuredly all cleared, and then sampling organs for their label contents. However, in most instances it is wise to determine the plasma decay kinetics to be assured that the tracer behaves normally, and to know that the tracer is adequately cleared. This is particularly important when a protein of relatively slow turnover is studied. As discussed above it is necessary in such cases to balance the requirement that all tracer be irreversibly cleared to obtain accurate results against the rate of label leakage out of tissues.

In the usual procedure, it is most convenient to outfit the subject animal with a venous catheter for tracer introduction and blood sampling. The tracer is introduced through the catheter, which is then thoroughly flushed with normal saline. The catheter is then kept filled with phosphate-buffered saline containing 10^{-3} M EDTA. On each subsequent blood sampling, blood is drawn to clear the catheter of buffer. The sample for analysis is then drawn; the blood used to clear the catheter is reinjected into the animal. Total net blood sampling is kept to under 10% of the blood volume during the time course of the experiment.

Experiments are not always carried out to the theoretically desirable point of near complete clearance of the injected tracer. If one wants to examine low uptake tissues in which trapped blood may make a significant contribution to total tissue radioactivity it is necessary at termination of the experiment to flush the vasculature of residual blood before tissue sampling. In principle, the determination of catabolic products in tissues, either directly or by the double label method described, would obviate the need for this procedure, but tracer concentration in trapped blood may be very high compared to the content of some very low uptake tissues. To clear tissues of as much blood as possible, the animal is exsanguinated via the inferior vena cava. (In small animals the vessel may simply be cut). Before cessation of heart action, the perfusion with normal saline containing 0.01% EDTA is begun through the ascending aorta, usually conveniently reached through the left ventricle, and through the portal vein. Perfusion is carried out beyond the point when the venous effluent runs clear (after 200 ml total in rats, 1–1.5 liters in rabbits). The lungs are separately perfused via the right heart. At harvest of the tissue, discrete organs are harvested *en toto* and multiple aliquots of diffuse tissues such as muscle and adipose tissues are taken. The total body mass of these diffuse organs is approximated from literature values for body composition.[22-27] The gut is washed of its contents, which are preserved for radio-

[25] W. O. Caster, J. Pancelot, A. B. Simon, and W. B. Armstrong, *Proc. Soc. Exp. Biol. Med.* **91,** 122 (1956).

[26] H. B. Latimer and P. B. Sawin, *Anat. Rec.* **129,** 457 (1957).

[27] C. Jelenko, III, A. P. Anderson, T. H. Scott, and M. L. Wheeler, *Am. J. Vet. Res.* **32,** 1637 (1971).

assay. Great care is taken to collect feces and urine during the course of the experiment without cross-contamination because trapped label radioactivity in the feces may be attributed to initial uptake by liver, while trapped label radioactivity in urine represents whole body leakage into plasma.

[37] Reverse Cholesterol Transport

By GEORGE H. ROTHBLAT, MARK BAMBERGER, and
MICHAEL C. PHILLIPS

Introduction

Unesterified cholesterol is essential for cell viability and comes from endogenous and exogenous sources. Synthesis *de novo* or mobilization of cholesteryl ester storage pools can provide the endogenous cholesterol while exogenous cholesterol comes from circulating lipoproteins *in vivo*. The process of sterol removal is an essential component in the maintenance of cellular cholesterol homeostasis. Since peripheral cells do not degrade cholesterol, the only mechanism for removal is by efflux of intact free (unesterified) cholesterol molecules. Such molecules enter the "reverse cholesterol transport" pathway whereby they are delivered to the liver, converted to bile acids, and then excreted from the body (for a review, see Norum *et al.*[1]). On the basis of an inverse correlation between serum high-density lipoprotein (HDL) levels and the incidence of atherosclerosis in the population, the purported lipoprotein vehicle for this transport is HDL. The molecular mechanisms underlying this pathway are under active investigation but are not understood fully at present.

It is apparent that cells can excrete free cholesterol to acceptor particles in the extracellular medium without expenditure of metabolic energy. The process involves a redistribution of free cholesterol molecules between the plasma membrane and phospholipid-containing particles in the medium. The possible equilibration reactions are demonstrated in Fig. 1. Efflux is defined as the rate of movement of free cholesterol from the cells to the medium while influx is the rate of movement in the opposite direction; each process has a characteristic rate constant as indicated in Fig. 1. Obviously, when the lipoprotein particles in the extracellular medium initially contain no free cholesterol, the system is far from equilibrium and

[1] K. R. Norum, T. Berg, P. Helgerud, and C. A. Drevon, *Physiol. Rev.* **63**, 1343 (1983).

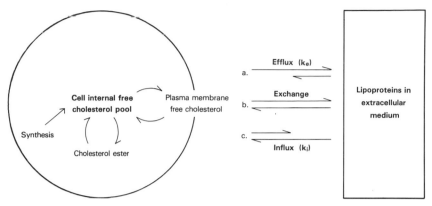

FIG. 1. Free cholesterol equilibria involved in cholesterol flux between lipoproteins in the extracellular medium, and cell free and esterified cholesterol pools. (a) Net movement of plasma membrane free cholesterol out of cells to lipoproteins in the medium. (b) No net cholesterol mass movement. (c) Net movement of cholesterol to cell plasma membrane from lipoproteins in the extracellular medium. k_e and k_i represent the rate constants for efflux and influx, respectively.

essentially unidirectional efflux of cholesterol ensues. When the extracellular particles contain free cholesterol then a bidirectional flux of cholesterol occurs and, if this is unbalanced, a net transfer of cholesterol mass to either the lipoproteins or cells can take place [i.e., net transfer to the medium = (efflux – influx)]. When influx and efflux are equal then cholesterol exchange occurs and there is no change in the cholesterol content of either compartment in the system. Whether or not net transport of cholesterol into or out of cells happens seems to depend upon the physical properties, especially the free cholesterol/phospholipid ratio,[2,3] of the particles in the extracellular medium.

Cells growing in culture provide an ideal system for quantitative investigations of the above free cholesterol transport processes, and a body of literature is developing.[4–6] Efflux or influx of free cholesterol can be followed directly by monitoring the appearance of labeled molecules in each compartment of Fig. 1 and thereby deriving k_e and k_i. Kinetic data on cholesterol efflux are most readily obtained from short-term studies rang-

[2] L. Y. Arbogast, G. H. Rothblat, M. H. Leslie, and R. A. Cooper, *Proc. Natl. Acad. Sci. U.S.A.* **73,** 3680 (1976).

[3] G. H. Rothblat, L. Y. Arbogast, and E. K. Ray, *J. Lipid Res.* **19,** 350 (1978).

[4] G. H. Rothblat and M. C. Phillips, *J. Biol. Chem.* **257,** 4775 (1982).

[5] Y. K. Ho, M. S. Brown, and J. L. Goldstein, *J. Lipid Res.* **21,** 391 (1980).

[6] J. F. Oram, J. J. Albers, M. C. Cheung, and E. L. Bierman, *J. Biol. Chem.* **256,** 8348 (1981).

ing from 1 to 12 hr. In such studies the measurement of the release of labeled cholesterol provides a suitable system which permits rapid, multiple samplings with a minimum of experimental manipulations. A major concern when using the isotopic approach is that the labeled cholesterol is uniformly distributed in all cellular cholesterol pools. In addition, it is necessary to determine the extent to which the labeled cellular cholesterol maintains its initial specific activity throughout the efflux phase of the experiment. In some experimental systems, the synthesis of endogenous cholesterol can significantly dilute the radioactive pool of cholesterol and this decrease in cholesterol specific activity results in an underestimate of cholesterol efflux.

Alternatively, the net transport can be derived indirectly by assaying changes in free cholesterol mass in either the cells or the medium. Direct measurement of the change in cholesterol mass in the incubation medium is limited by the sensitivity of the assay procedure. Even with sensitive techniques, such as GLC and enzymatic cholesterol assays, direct quantitation of released cholesterol is most useful in long-term experiments (i.e., 12 to 24 hr) where the amount of sterol accumulating in the culture medium is sufficiently high to ensure reliable results.

The experimental protocol for conducting cell culture studies designed to quantitate either free cholesterol efflux or cholesteryl ester clearance consists of three phases: labeling, equilibration, and efflux (Fig. 2). These aspects of the problem are dealt with in detail below.

Labeling of Cellular Cholesterol

Cellular cholesterol pools can be labeled using exogenous cholesterol or by supplying labeled precursors (stage I in Fig. 2). In either case, sufficient incubation times (1–4 days) are needed to ensure the equilibration of the label among all subcellular pools. When using the exogenous labeling approach all radiolabeled cholesterol should be repurified prior to use (see below). Repurification is necessary to remove oxidized compounds that, because of their increased polarity, exchange more rapidly than cholesterol. Throughout both the purification and labeling steps, precautions should be taken to reduce the formation of additional oxidized products. Two approaches can be used to incorporate labeled cholesterol into the serum to be used for labeling the cells.

1. Labeled cholesterol, solubilized in an appropriate solvent [e.g., ethanol, dimethyl sulfoxide (DMSO)], can be introduced directly into serum or serum diluted with medium. The solvent should be injected rapidly using a small-bore needle or a Hamilton syringe. Final solvent

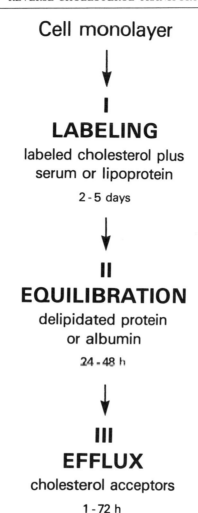

FIG. 2. Protocol for determination of cholesterol efflux from cells in culture.

concentrations should be <0.1% of the final volume of the incubation medium. If higher concentrations of solvent are present the medium should be dialyzed before use. Following filtration through a sterile 0.45-μm filter, the medium can be incubated for 12–24 hr at 37° to ensure equilibration of labeled cholesterol with lipoproteins. A potential problem with this method of adding labeled cholesterol to medium is that the introduction of cholesterol in a solvent may result in the radioactive sterol entering the cell as microcrystals rather than by monomolecular exchange or lipoprotein internalization. An advantage of the procedure is that it can efficiently yield a very high cholesterol-specific activity in the labeling medium. The amount of label used depends on the experimental system. Literature values range from 1–15 μCi/ml medium containing 10% fetal bovine serum. Experiments using low cell densities, cells with slow rates of cellular release, and/or short incubation periods require high cellular cholesterol-specific activity.

2. An alternative method of labeling the incubation medium is by exchange from an inert surface. In this case labeled cholesterol is dried under N_2 onto a flat-bottomed vial (e.g., a scintillation vial) or onto Celite.[7] Serum or serum-containing medium is added and the cholesterol is allowed to exchange into the lipoproteins during a 12- to 24-hr incubation at 37° in the presence of 0.02% (w/v) NaN_3. If Celite is used as the inert carrier, it can subsequently be removed by low-speed centrifugation or filtration. An advantage of this method is that it avoids the possibility of cholesterol microcrystal formation which ensures more uniform labeling of serum lipoproteins. A disadvantage is that the labeling is less efficient, thus resulting in lower specific activities than can be obtained by the direct addition of the cholesterol. In addition, the possibility of generating cholesterol oxidation products is increased.

In both of the procedures described above, serum lecithin–cholesterol acyltransferase (LCAT) should be inhibited prior to the addition of the labeled cholesterol (e.g., by heating or addition of an inhibitor). The presence of active LCAT results in the formation of radioactive cholesteryl esters which decreases the efficiency of cellular labeling and can possibly result in increased levels of labeled cellular cholesteryl ester. In studies designed to specifically quantitate free cholesterol efflux, the labeling medium should be formulated so as to minimize the accumulation of cell cholesteryl ester. As discussed below, the presence of a labeled pool of esterified cholesterol within cells greatly complicates the kinetic analysis of free cholesterol efflux. If the cell system under study is one which responds to the presence of serum by accumulating esterified cholesterol, we have found that dilution of serum to 1% (v/v) and supplementation with delipidized serum protein[8] maintains the growth of most cells while reducing the cellular cholesteryl ester content. Medium is prepared containing 1% fetal bovine serum, delipidized serum protein (5 mg/ml), egg phosphatidylcholine (8 μg/ml), free cholesterol (4 μg/ml), and either [^{14}C]- or [^{3}H]-cholesterol. The phospholipid and cholesterol are dissolved

[7] J. Avigan, *J. Biol. Chem.* **234,** 787 (1959).
[8] G. H. Rothblat, L. Y. Arbogast, L. Ouellette, and B. V. Howard, *In Vitro* **12,** 554 (1976).

in a small volume of ethanol (<0.05%) and added to the sterile protein-containing medium.[4]

Endogenous labeling of cellular cholesterol is accomplished by the addition to the incubation medium of either radiolabeled acetate or mevalonate.[9] The latter precursor is preferred since this compound bypasses HMG-CoA reductase and the majority of the label goes into sterol. Labeling with acetate results in the labeling of all lipid classes which necessitates the separation of the labeled cholesterol from other labeled lipids which are subsequently released during the efflux phase of the experiment.

Depending on the rate of growth of the cells, the labeled medium is added to cell monolayers which are 50–75% confluent. As far as possible, the cells should be maintained at the same density from experiment to experiment since there is evidence that efflux may be influenced by this parameter.[10] Incubation times in the labeled medium can range from 2 to 5 days.

Equilibration of Radioactivity between Cell Cholesterol Pools

After incubation in the labeled medium, we have found it useful to wash the monolayers and incubate for an additional 24 hr in fresh medium containing either delipidized serum protein or albumin (2–5%, w/v). Neither protein promotes significant efflux, and this final 24 hr of incubation allows internalization of surface-bound cholesterol and further equilibration of cellular pools (stage II in Fig. 2).

Throughout the labeling and equilibration period, uninoculated plates as controls should be exposed to the labeled medium. These blank dishes are then exposed to the same free cholesterol acceptors as the experimental monolayers. Cholesterol appearing in the medium represents material previously bound to the plastic and should be subtracted from the cell monolayer values. This "background efflux" is not a serious problem in systems where the efflux is high, but can be a significant factor in experiments exhibiting reduced efflux. An alternative to the blank plate control is to remove the monolayer from the labeling dish using trypsin and to replate the labeled cells onto fresh culture dishes in the presence of delipidized serum protein or albumin. This procedure eliminates the possibility of cholesterol desorption from plastic but can only be used with cells that have a high plating efficiency.

---

[9] J. R. Faust, J. L. Goldstein, and M. S. Brown, *Arch. Biochem. Biophys.* **192**, 86 (1979).
[10] C. J. Fielding and K. Moser, *J. Biol. Chem.* **257**, 10955 (1982).

Preparation of Acceptor Particles

Acceptors for Unidirectional Efflux

A variety of acceptor particles which contain no free cholesterol can be used to induce unidirectional efflux of cholesterol from cells.

Lipoprotein-Deficient Serum (LPDS). Serum from which lipoproteins have been removed has served as an effective promoter of cholesterol efflux. Lipoproteins are generally removed by ultracentrifugation of human or animal sera at $d = 1.21$ g/ml, or in some cases $d = 1.25$ g/ml.[6] LPDS may vary in its efficiency to serve as a free cholesterol acceptor depending on the quantity of phospholipids and apolipoproteins that are present; this is a function of a variety of factors including species, previous treatment of the serum, and the density used for flotation of lipoproteins. Although the composition of LPDS is variable, its ease of preparation and efficiency as a cholesterol acceptor make it a useful preparation for cholesterol efflux studies.

ApoHDL/Phospholipid Complexes. The apolipoproteins from human HDL ($d = 1.063–1.21$ g/ml) can be prepared by extracting the lipids using ether/ethanol following published procedures.[11] The lipid-free apoHDL can be stored at 4° as a solution in saline (0.15 M NaCl, 0.001 M EDTA, 0.02%, w/v, NaN$_3$, pH 8.6).

Complexes of apoHDL with egg PC are prepared by sonicating a mixture of 1 mg/ml of PC and apoHDL (2.5 : 1, w/w) in a Branson Sonifier 350 for at least 30 min at 4°. Following sonication the complexes are centrifuged at 40,000 rpm for 2 hr in a Beckman type-40 rotor to remove titanium and any uncomplexed phospholipids and protein; under these conditions, combination of lipid and protein is essentially complete. Prior to use, the preparation can be dialyzed against minimal essential medium (MEM) plus HEPES and diluted to the required concentrations. Electron microscopic examination shows mostly spherical egg PC/apoHDL complexes.[12] Complexes of apoHDL with rat liver sphingomyelin can be formed by a similar sonication procedure.[13]

As an alternative to the above sonication procedure, apoHDL/phospholipid complexes can be formed using detergent dialysis techniques. Discoidal complexes of egg PC with apoA-I or apoHDL can be formed using sodium cholate.[14] In a typical preparation, apolipoprotein : phospholipid : sodium cholate mass ratios of 1 : 2.5 : 2.5 are used. Chromato-

[11] A. M. Scanu, *J. Lipid Res.* **7**, 295 (1966).
[12] A. W. Kruski and A. M. Scanu, *Chem. Phys. Lipids* **13**, 27 (1974).
[13] O. Stein, J. Vanderhoek, and Y. Stein, *Biochim. Biophys. Acta* **431**, 347 (1976).
[14] C. E. Matz and A. Jonas, *J. Biol. Chem.* **257**, 4535 (1982).

graphically pure egg PC is dried down from chloroform under N_2 and then in a vacuum oven at 40° for 2 hr. The apolipoprotein and cholate are dissolved in a buffer (10 mM Tris, 1 mM EDTA, 1 mM NaN$_3$, 150 mM NaCl, pH 8) and added to the tube containing PC to give a final PC concentration of 5 mg/ml. The PC is dispersed by vortex mixing and incubated overnight at 4°. When the reaction is complete the solution becomes clear. The solution is then dialyzed exhaustively against saline (150 mM NaCl, 1 mM EDTA, 1 mM NaN$_3$, pH 7.4) to remove cholate. In order to size the particles and obtain a homogeneous population, the reaction mixture is passed down a column of Sepharose 4B-CL (90 × 1.5 cm). The column is run at a flow rate of 12 ml/hr using an elution buffer 0.85% NaCl, 0.02% EDTA, NaN$_3$ and 3-ml fractions are collected. The elution profile is monitored by absorbance at 280 nm. The peak fraction and fractions close to the peak are concentrated in an Amicon Model 52 ultrafiltration cell using N_2 at 30 psi and a PM10 filter. Phospholipid and protein determinations are made on the concentrated material; samples can also be taken for electron microscopic examination if desired. The samples are dialyzed overnight before the experiment against the appropriate tissue culture medium and diluted to the requisite concentrations for the cholesterol efflux experiment.

Unilamellar PC Vesicles. Vesicles are prepared by sonication using a modification of the procedure of Barenholz *et al.*[15] Egg PC (chromatographically purified) is dried down from chloroform under N_2, and the lipid is sonicated to clarity under N_2 (for at least 30 min at 4° with a Branson Sonifier 350) in distilled water. Centrifugation at 40,000 rpm for 2 hr in a type-40 Beckman rotor removes any multilamellar particles. By the criteria of negative stain electron microscopy and gel chromatography on Sepharose 2BCL, the vesicles are homogeneous unilamellar spheres with a diameter of 250 Å. Vesicles can be prepared at a concentration of 2 mg/ml in buffer and dialyzed against MEM buffered with 12.5 mM HEPES prior to dilution to required concentrations.

Small unilamellar vesicles of egg phosphatidylcholine can also be prepared by the sodium cholate dialysis procedure described by Brunner *et al.*[16] A film of PC dried on the wall of a tube and a buffer (10 mM Tris, 1 mM EDTA, 1 mM NaN$_3$, pH 8) containing cholate at a molar ratio of PC : cholate = 1 : 10 is added to give a final PC concentration of 7.5 mg/ml. The cholate is separated from the vesicles by subjecting the mixture to gel filtration on Sephadex G-50. Column fractions are collected and the

[15] Y. Barenholz, D. Gibbes, B. J. Litman, J. Gall, T. E. Thompson, and F. D. Carlson, *Biochemistry* **16**, 2806 (1977).

[16] J. Brunner, P. Skrabal, and H. Hauser, *Biochim. Biophys. Acta* **455**, 322 (1976).

peak fractions concentrated and dialyzed against medium prior to the cholesterol efflux experiment.

Sodium Taurocholate/Egg PC Micelles. Bile salt/PC mixed micelles can be obtained following the general procedure described by Mazer *et al.*[17] After dissolving the appropriate amounts of each lipid in methanol, the mixture is dried under N_2. MEM buffered with HEPES is then added and the mixture is vortexed until clear. Typically, stock solutions of the micelles are prepared at a concentration of 3.7 mg/ml of egg PC and 6.3 mg/ml of sodium taurocholate. This stock is diluted to the final concentrations required for the cholesterol efflux experiment by the addition of HEPES-buffered MEM supplemented with 3 mg/ml of sodium taurocholate; this gives a solution containing sodium taurocholate at its critical micelle concentration of 5.5 mM to ensure that the integrity of the mixed micelles is preserved.

Acceptors for Bidirectional Flux

When natural lipoproteins are exposed to cells growing in culture a bidirectional flux of free cholesterol molecules between the lipoproteins and cells is possible (cf. Fig. 1). LDL and HDL fractions isolated by standard ultracentrifugation techniques have been used as acceptor particles to study net changes in the mass of cell cholesterol. In order to examine the flux of cholesterol from the lipoproteins into the cells, it is necessary to prelabel the lipoprotein free cholesterol by introducing radiolabeled molecules from Celite,[7] as described above.

Analytical Procedure

Purification of Labeled Cholesterol

Radiolabeled cholesterol can be purified by thin-layer chromatography (TLC) on silica gel G using diethyl ether. The r_f value for cholesterol in this system is ~0.70. The cholesterol is extracted from the gel with chloroform : methanol (2 : 1, v/v).

Separation of Labeled Cholesterol and Cholesteryl Ester

The distribution of radiolabel between free and esterified cholesterol can be determined by using silica gel G coated on glass plates in a system of petroleum ether : diethyl ether : acetic acid (75 : 25 : 1, v/v). The gel is scraped from the plates into scintillation vials, to which 10 ml of scintilla-

[17] N. A. Mazer, G. B. Benedek, and M. C. Carey, *Biochemistry* **19**, 601 (1980).

tion fluid has been added previously. An easier method which does not sacrifice accuracy is to use fiberglass sheets impregnated with polysilicic acid gel (ITLC-SA, Gelman Sciences, Inc. Ann Arbor, MI) developed in benzene. The cholesterol and cholesteryl ester bands can be cut from the plate and counted directly.

Cholesterol, Phospholipid, and Protein Mass Determinations

Cholesterol mass can be determined by GLC[18] or by enzymatic assay.[19] Phospholipids can be determined by an assay for phosphorus.[20] Protein can be measured by the Lowry method, which incorporates sodium dodecyl sulfate to solubilize hydrophobic proteins.[21]

Efflux of Cholesterol from Cells

Unidirectional Efflux

Cell monolayers containing radiolabeled cholesterol are washed five times and medium containing the acceptors (prewarmed to 37°) is added to replicate plates to initiate efflux (stage III of Fig. 2). For experiments using 1- to 6-hr incubation periods it is not necessary to maintain sterile conditions during washing and sampling, particularly if antibiotics are present. If longer experiments are necessary, the acceptor-containing medium should be sterilized by filtration prior to use. Following filtration, the acceptor concentration should be reassayed (i.e., by a protein or phospholipid determination). Experimental plates are held at 37° and aliquots of incubation medium are removed at designated intervals for measurement of released cholesterol. In this phase of the experiment we have found it advantageous to use medium buffered with HEPES to ensure a constant pH throughout the multiple sampling period. In addition, the plates should be held in a well-humidified incubator to eliminate evaporation. When using cells that grow to high cell densities (i.e., transformed cells), 35-mm dishes containing 2 ml of acceptor can be used. Multiple aliquots (6 × 100 μl) can be removed and directly counted without extraction if appropriate corrections are made for quenching. With lower density cultures it may be necessary to use 60-mm plates and increase the volume of the medium to 5 ml. In this case multiple aliquots of 200 μl or larger can be taken. Since the volume of the incubation medium is being

[18] G. H. Rothblat, *Lipids* **9**, 526 (1974).
[19] J. G. Heider and R. L. Boyett, *J. Lipid Res.* **19**, 514 (1978).
[20] L. Sokoloff and G. H. Rothblat, *Biochim. Biophys. Acta* **280**, 172 (1972).
[21] M. K. Markwell, S. M. Haas, L. Bieber, and N. Tolbert, *Anal. Biochem.* **87**, 206 (1978).

progressively reduced throughout the course of the experiment, appropriate, cumulative corrections have to be made for both the changing volume and the radioactivity removed in each aliquot in order to establish the actual amount of labeled cholesterol released at each time point.

At the end of the efflux period, the remaining medium is removed from the experimental dishes and the cell monolayers are washed five times with fresh medium. The cells are harvested and both cholesterol content and the total remaining radioactivity are determined. From these determinations the cellular cholesterol specific activity can be calculated and compared to similar values established from control cells harvested at zero time. A reduction in specific activity throughout the course of the experiment is a reflection of the extent of dilution of the labeled cholesterol produced by cellular cholesterol synthesis. Extensive cholesterol synthesis may be encountered with some cell types or in long experiments, and such synthesis can lead to a significant underestimate of efflux of cholesterol mass. The release of cellular free cholesterol follows first-order kinetics with respect to cell free cholesterol concentration. A semilog plot of the fraction of cholesterol remaining in the cell versus the time of incubation yields a straight line from which the half-time for cholesterol efflux can be easily estimated (Fig. 3). A divergence from linearity can be an indication of the presence of cellular pools having differing kinetic properties. The half-time for efflux differs widely for various cell types and Fig. 4 presents the half-times for a variety of cell lines exposed to a common cholesterol acceptor.

Bidirectional Flux

The preceding section of this chapter has focused upon the situation in which a unidirectional flow of cholesterol out of the cell is attained. In many instances, however, cholesterol efflux is accompanied by a reciprocal transfer of sterol from particles in the medium, such as lipoproteins, to the plasma membrane. The overall effect on cellular cholesterol levels is determined by the relative rates of these two processes (cf. Fig. 1). A direct determination of influx can be made in an analogous fashion to that described for efflux, by following the movement of radiolabeled cholesterol from the medium into the cell. Thus, two separate experiments can be performed, with either the medium or cellular cholesterol pools being radiolabeled, or, both influx and efflux can be quantitated in a single experiment utilizing cells and medium labeled with different radioisotopes. A protocol used to study the bidirectional movement of cholesterol between HDL and rat hepatoma cells is given below.

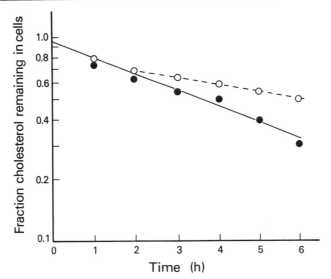

FIG. 3. Efflux of [³H]cholesterol from prelabeled Fu5AH rat hepatoma cells incubated at 36° in minimal essential medium buffered with HEPES and containing 1 mg/ml egg phosphatidylcholine added as unilamellar vesicles (Phillips *et al.*[29]). The efflux data are analyzed in terms of first-order kinetics with all of the cell free cholesterol in a single pool (<5% of the cell cholesterol is esterified). ○, Fraction of counts remaining in cell; ●, fraction of cholesterol mass remaining in cells after correction for the changes in specific activity due to initiation of cellular cholesterol synthesis.

Labeling of Cells. Fu5AH rat hepatoma cells can be prelabeled with [¹⁴C]cholesterol in T-75 flasks by the procedure described above. After 2 days of labeling, at which time the monolayers are confluent, the cells are detached by trypsinization, resuspended in medium containing 5 mg/ml delipidized serum protein, and plated in 35-mm plastic Petri dishes at a concentration which yields 80–90% confluent monolayers after cell attachment. A period of at least 18 hr is allowed for the cells to attach to the Petri dishes. The approximate specific activity of cellular cholesterol at the beginning of the experiment is 10,000 dpm/µg unesterified cholesterol. These conditions of labeling result in little accumulation of cellular cholesteryl ester.

Labeling of HDL with [³H]Cholesterol. Any apoB- and apoE-containing particles are removed from human HDL by passage over heparin–Sepharose[22] before labeling by the Celite procedure described above. Typically, the labeling efficiency is 50%; the specific activity of the HDL

²² S. H. Quarfordt, R. S. Jain, L. Jakoi, S. Robinson, and F. Shelbourne, *Biochem. Biophys. Res. Commun.* **83**, 786 (1978).

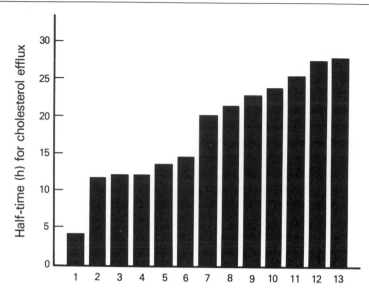

Fig. 4. Comparison of the half-times for cholesterol efflux for a variety of cells. Cells were prelabeled with [³H]cholesterol and incubated with apoHDL/egg PC acceptor particles at a concentration of 1 mg PC/ml MEM for 6 hr. (1) Fu5AH rat hepatoma; (2) mouse P-388 macrophage; (3) mouse L-cells; (4) human Hep G2 liver; (5) human U-937 macrophage; (6) human preputial fibroblast (line GS); (7) human preputial fibroblast (line MLK); (8) human skin fibroblast (GM-2000, LDL-receptor negative); (9) mouse sarcoma 180; (10) mouse J774 macrophage; (11) human preputial fibroblast (line DW); (12) bovine fetal aortic endothelial BFA-39; (13) WIRL rat liver.

free cholesterol after labeling averages 75,000 dpm/μg. To separate Celite from HDL, the vials are spun at 1000 g for 10 min, the supernatant removed, and the lipoproteins passed through a 0.45-μm filter (nonsterile) to remove contaminating Celite particles. HDL is dialyzed against PBS overnight before use to remove the sodium azide.

Incubation of HDL with Cells. In order to initiate the bidirectional flux experiment, cell monolayers prelabeled with [¹⁴C]cholesterol, are washed three times with 3 ml of unsupplemented MEM buffered with HEPES (pH 7.4) at 37°. Medium containing [³H]-labeled HDL is then added and the plates transferred to a 37° humidified incubator. After periods of incubation, the medium is removed from the monolayers to 1.5 ml polypropylene microfuge tubes and immediately put on ice. The monolayers are washed three times with PBS and the cholesterol extracted from the cells by the addition of isopropanol for 15 min at room temperature, with the cover of the Petri dish in place. To quantitate the amount of labeled cholesterol in the cell, 500 μl of the isopropanol extract is placed in a

20-ml glass scintillation vial, the solvent evaporated, and 10 ml of scintillation fluid added.

The radioactive content of the medium can be determined after centrifugation in a Beckman microfuge at 1000 g for 5 min to pellet any cells which may have detached during the course of the experiment. The supernatant (100 μl) is counted. The ^{3}H and ^{14}C in both the cell extracts and the medium can be quantitated by liquid scintillation techniques in a counter using very narrow window settings to minimize leakage. For each experiment, the initial cellular specific activity and medium specific activity are determined.

Analysis of Results. Influx is determined by the appearance of ^{3}H in cell extracts and efflux as the appearance of ^{14}C in the medium. The actual mass of cholesterol that has moved into or out of the cell over time can be estimated from the radiolabel data and the initial specific activities of the cholesterol in the medium and in the cells. No matter how the individual influx and efflux components are quantitated, either separately or in a single experiment, it should be noted that, as the experiment progresses, the specific activity of both the cellular and medium cholesterol pools becomes diluted. Consequently, a certain number of cpm at a later time interval represent a greater mass of cholesterol translocated than an equivalent amount of cpm at a preceding time interval. An estimate of the decrease in specific activity for a particular time interval can be made by taking into account the net loss or gain of cholesterol from the medium or the cell during that time interval and the number of either ^{3}H or ^{14}C cpm which remain in the medium or cell, respectively. The adjusted specific activities can then be used to determine the amount of cholesterol transferred between the cell and medium during the proceeding time interval. This procedure was used to obtain results such as those presented in Fig. 5. A detailed treatment of the kinetics of bidirectional flux in a two-compartment system is available.[23]

An alternative method of studying the bidirectional flux of cholesterol in a cell culture system has been used.[24] In this procedure, the efflux of cholesterol is determined from the amount of [^{3}H]cholesterol in the medium and the initial specific activity of cellular cholesterol while net transport is determined by changes in the cholesterol mass in the medium. Net accumulation of cholesterol by the cells results in a decrease in medium cholesterol content and net efflux causes an increase. Measurement of the typically small changes (<1 μg) in cholesterol content of the medium

[23] R. A. Shipley and R. E. Clark, "Tracer Methods for In Vivo Kinetics," p. 132. Academic Press, New York, 1972.
[24] C. J. Fielding and P. E. Fielding, *Proc. Natl. Acad. Sci. U.S.A.* **78**, 3911 (1981).

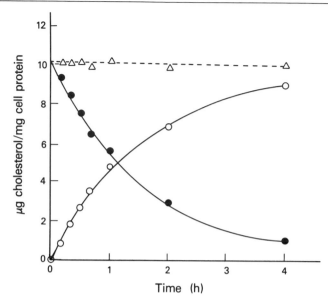

FIG. 5. Bidirectional movement of free cholesterol between HDL and rat hepatoma cells. Human HDL, labeled with [³H] free cholesterol, was incubated with Fu5AH rat hepatoma monolayers containing [¹⁴C]cholesterol at a concentration of 25 μg HDL free cholesterol/ml. The loss of cholesterol from the cell into the medium (efflux, ●) or the micrograms of cholesterol transferred from the medium into the cells (influx, ○) was calculated at each time point taking into account the dilution of specific activity as described in the text. Influx and efflux values, determined from the data of the initial 40 min of the experiment, are 0.081 and 0.079 μg cholesterol/min/mg cell protein, respectively. Net accumulation (△) equals influx minus efflux.

requires that a sensitive assay be employed. Cholesterol influx is estimated indirectly as the difference between net movement and efflux.

Cholesteryl Ester Clearance from Cells

Studies designed to quantitate the clearance of cellular cholesteryl esters can be conducted using modifications of the experimental approach described for free cholesterol efflux (Fig. 2). In contrast to studies on free cholesterol efflux where loading conditions are designed to label cells while maintaining the cell cholesteryl ester content at a minimum, investigation of cholesteryl ester clearance requires incubation of cells under conditions which result in maximum esterified cholesterol deposition. The selection of such conditions will be determined by the cell type under study, and may utilize hyperlipemic serum,[25] cholesterol-rich disper-

[25] G. H. Rothblat, L. Arbogast, D. Kritchevsky, and M. Naftulin, *Lipids* **11**, 97 (1976).

sions,[2] high concentrations of either LDL[26] or acetylated LDL.[5] Addition of appropriate acceptors of cholesterol to the culture medium initiates free cholesterol efflux which in turn results in the hydrolysis and clearance of stored cholesteryl ester pools (cf. Fig. 1). This decrease in cellular cholesteryl ester can be quantitated by either mass or isotopic determinations. The presence of labeled free cholesterol in the culture medium during the loading phase induces labeling of both the cellular free and esterified cholesterol pools which can facilitate the acquisition of kinetic data by radiocounting procedures. However, isotopic analysis can be complicated, particularly if the cellular free and esterified cholesterol pools have different specific activities. We have observed a number of instances where 24 hr was not sufficient time to achieve isotopic equilibration so that the free cholesterol specific activity was higher than the cholesteryl ester specific activity at the initiation of efflux. If this is the case, the continued resynthesis of cholesteryl ester by ACAT from a highly labeled pool of free cholesterol (Fig. 1) results in an underestimation of cholesteryl ester clearance when determined by the isotopic approach. This problem can be eliminated by blocking potential cholesteryl ester synthesis with specific ACAT inhibitors during the efflux/clearance phase of the experiment.[27,28] Since isotopic measurements of cholesteryl ester clearance can yield misleading data, they should always be confirmed by parallel mass determinations. It should be noted that both mass or isotopic determinations can yield overestimates of cholesteryl ester clearance if the data are expressed on the basis of cell protein and no correction is made for continued cell growth. This can be particularly important when studying rapidly growing transformed cells over extended periods.[29] Analysis of the data in terms of the cholesteryl ester mass per culture dish will often reveal situations where cell growth has resulted in a reduction in cholesteryl ester/milligram cell protein in the absence of any significant cholesteryl ester hydrolysis and clearance. Figure 6[30] illustrates the kinetics of cholesteryl ester clearance obtained when human fibroblasts were incubated in the presence of lipoprotein-deficient serum which served as the acceptor of cholesterol. It is apparent that the concentration of the acceptor can markedly influence the rate of clearance of esterified cholesterol. The data currently available are insufficient to es-

[26] R. J. Daniels, L. S. Guertler, T. S. Parker, and D. Steinberg, *J. Biol. Chem.* **256**, 4978 (1981).
[27] A. C. Ross, K. J. Go, J. G. Heider, and G. H. Rothblat, *J. Biol. Chem.* **259**, 815 (1984).
[28] J. L. Goldstein, J. R. Faust, J. H. Dygos, R. J. Chorvat, and M. S. Brown, *Proc. Natl. Acad. Sci. U.S.A.* **75**, 1877 (1978).
[29] M. C. Phillips, L. R. McLean, G. W. Stoudt, and G. H. Rothblat, *Atherosclerosis* **36**, 409 (1980).
[30] R. W. St. Clair and M. A. Leight, *J. Lipid Res.* **24**, 183 (1983).

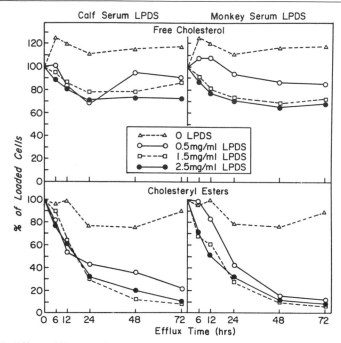

FIG. 6. Effect of lipoprotein-deficient serum (LPDS) concentration on cellular choles-terol efflux (St. Clair *et al.*[30]). Human skin fibroblasts were grown to confluence and incu-bated for 48 hr with basal medium containing LPDS. The cells were then loaded with cholesterol by incubation for 24 hr with 50 μg protein/ml of [³H]cholesterol-labeled hyper-cholesterolemic LDL. After 24 hr, four dishes were harvested and analyzed for free and esterified cholesterol radioactivity. These loaded values average 949,100 ± 15,600 (SEM) and 107,650 ± 3,700 dpm/mg protein for free and esterified cholesterol, respectively, and were set at 100% for the 0-time values on the ordinate. The remaining cells were incubated for up to 72 hr with efflux medium containing 0, 0.5, 1.5, or 2.5 mg protein/ml of LPDS isolated from calf serum or normocholesterolemic rhesus monkey plasma. After the indi-cated efflux period, cells were harvested and analyzed for free and esterified cholesterol radioactivity. Results are the mean of duplicate cultures.

tablish the extent of variation in the rate of clearance of esters from different cell types. Also, it is not known at what point either the hydroly-sis or efflux step becomes rate limiting in different cell systems exposed to various acceptors.

Acknowledgments

Our studies in this area are supported by research Grants HL 22633, HL 07443, and by a grant from the W. W. Smith Charitable Trust.

[38] Receptor-Mediated Transport of Cholesterol between Cultured Cells and High-Density Lipoproteins

By JOHN F. ORAM

High-density lipoprotein appears to play a central role in the transport of cholesterol from peripheral cells to the liver, a pathway termed "reverse cholesterol transport."[1] The obligatory first step in this pathway is the removal of cholesterol from peripheral cells by HDL, a process that may modulate the rate of flux of cholesterol through the entire pathway. Recent evidence suggests that this removal step may be mediated by binding of HDL to a cell surface receptor. This is based on studies demonstrating that cholesterol enrichment of cells in tissue culture leads to an increase in HDL binding to cells by a process that appears to involve synthesis of cellular protein.[2] It was postulated that cells may synthesize specific, cell surface receptors for HDL in response to overaccumulation of cholesterol, and that these receptors may act to facilitate HDL-mediated transport of cholesterol from cells. The physical characteristics of the putative HDL receptor and the mechanism by which it may facilitate cholesterol transport from cells are presently unknown. The purpose of this chapter is to describe experimental protocols that can be used to study cholesterol transport from cultured cells under conditions that may depend on the interaction of HDL with a cell surface receptor. These protocols are divided into three major areas of methodology: (1) modulation of HDL-binding activity, (2) assay of HDL-binding activity, and (3) assay of cholesterol transport.

Modulation of HDL-Binding Activity

Several different types of cultured cells have been shown to undergo marked increases in HDL-binding activity in response to cholesterol loading of cells. These include human skin fibroblasts,[2] arterial smooth muscle cells,[2] bovine vascular endothelial cells,[3] and mouse peritoneal macrophages.[4] Cholesterol loading of cells can be achieved by designing incubation conditions that deliver sterol into cells by both receptor-independent and receptor-dependent processes. These loading procedures offer differ-

[1] J. A. Glomset, *J. Lipid Res.* **9,** 155 (1968).
[2] J. F. Oram, E. A. Brinton, and E. L. Bierman, *J. Clin. Inv.* **72,** 1611 (1983).
[3] E. A. Brinton, R. D. Kenagy, J. F. Oram, and E. L. Bierman, submitted (1986).
[4] T. H. Aulinskas and J. F. Oram, submitted (1986).

ent advantages and limitations, depending on the type of cells used. The major requirement for the choice of methodology is that sterol is delivered to the cell at a fast enough rate that unesterified cholesterol accumulates in cellular pools. To attain this, cultured cells can be treated with media containing either nonlipoprotein cholesterol or lipoprotein cholesterol in the form of LDL or acetylated LDL.

When nonlipoprotein cholesterol dissolved in a solvent such as ethanol is added to culture medium, it enters cells by diffusion across the cell membrane. Thus the rate of delivery is not dependent on the activity of a cell surface receptor, and cells can be enriched with cholesterol several-fold. Moreover, the delivery rate appears to greatly exceed the rate at which cells can esterify the incoming cholesterol, allowing for a massive buildup of unesterified cholesterol. This is important because HDL-binding activity is a direct function of the cell's unesterified, but not the esterified, cholesterol content.[2] The disadvantage of loading cells with nonlipoprotein cholesterol is that the cholesterol source is nonphysiologic, and prolonged incubation may be harmful to cells. The more durable cultured cell types, such as fibroblasts, can be exposed to media containing 50 μg/ml cholesterol for up to 48 hr without appreciable cytotoxicity. Electron micrographs of the treated cells indicate that they contain numerous lipid-filled vacuoles. Light microscopy shows that the cells are covered with granules that appear to be microcrystals of cholesterol or albumin–cholesterol aggregates. Thus, this loading procedure requires that cells be extensively washed prior to assays for HDL binding and cholesterol efflux.

An alternative method to upregulate HDL binding to cells by nonlipoprotein sterols is to use oxygenated sterol derivatives, such as 25-hydroxycholesterol. This derivative is more potent than cholesterol as an activator of HDL binding[2,5] and thus can be used at slightly lower concentrations. However, 25-hydroxycholesterol is toxic to cells and must be used under conditions where cytotoxicity is closely monitored.

The major process for delivery of cholesterol into human peripheral cells is the LDL receptor pathway. Therefore, in order to load cultured cells with cholesterol by a physiological mechanism, LDL can be used as an exogenous sterol substrate. Because LDL sterol is delivered to cells by a receptor-mediated process, this loading procedure is self-limiting. The number of LDL receptors on the cell surface progressively decreases as cells accumulate cholesterol, thus greatly limiting the amount of sterol that can be introduced into cells. However, a modest increase in HDL-binding activity can be attained with treatment of cells with LDL.

[5] J. P. Tauber, D. Goldminz, and D. Gospodarowicz, *Eur. J. Biochem.* **119,** 327 (1981).

Certain types of cultured cells, such as bovine endothelial cells and mouse peritoneal macrophages, have a separate set of receptors that recognize chemically modified LDL particles.[6,7] These "scavenger receptors" do not undergo substantial feedback regulation, and thus cells can amass large amounts of cholesterol when exposed to modified LDL. A convenient method for modification of LDL is to acetylate it in the presence of acetic anhydride.[6] The degree of acetylation can be monitored by agarose gel electrophoresis. The effects of acetylated LDL on cell cholesterol metabolism has been well characterized.[6,7]

Delivery of both LDL and acetylated LDL sterol to intracellular pools requires the enzymatic hydrolysis of cholesteryl esters in lysosomes and transfer of cholesterol to cytosolic compartments and cell membranes. During these processes, a significant proportion of the released cholesterol becomes reesterified by the enzyme acyl-CoA cholesterol acyltransferase (ACAT). This esterification can reduce the cellular unesterified cholesterol content and dampen upregulation of HDL-binding activity. Thus, inhibition of ACAT activity can markedly enhance the lipoprotein-mediated stimulation of HDL binding.[4] The simplest way to achieve this inhibition is to include fatty acid-free albumin in the medium, which sequesters the fatty acids that act as precursors for cholesteryl ester formation. Alternatively, specific inhibitors of the ACAT enzyme can be added during the loading period.

Reagents

Medium A: Dulbecco's minimum essential medium supplemented with albumin (2 mg/ml). The albumin acts as a protein source, a carrier for cholesterol microcrystals, and a sink for trapping cellular fatty acids (to minimize cholesteryl ester formation). Therefore, an albumin preparation should be used that is fatty acid free and is not contaminated significantly with apolipoproteins. Bovine serum albumin lot No. A-6003 from Sigma Chemical Company adequately meets these criteria

Wash buffer: Phosphate-buffered saline containing 1 mg/ml bovine serum albumin (lot No. A-6003, Sigma Chemical Company)

Stock cholesterol or 25-hydroxycholesterol: Prepared by dissolving 10 mg cholesterol or 25-hydroxycholesterol in 1.0 ml 95% ethanol. The stock can be stored in the refrigerator for several weeks. When cold, the cholesterol will crystallize out of solution, but will redissolve upon heating to above 30°

[6] J. L. Goldstein, Y. K. Ho, S. K. Basu, and M. S. Brown, *Proc. Natl. Acad. Sci. U.S.A.* **76,** 333 (1979).
[7] O. Stein and Y. Stein, *Biochim. Biophys. Acta* **620,** 631 (1981).

Lipoprotein-deficient serum: Lipoproteins are removed from plasma by flotation in an ultracentrifuge at d 1.25 g/ml[8]

Low-density lipoprotein (LDL): Isolated from human plasma by sequential ultracentrifugation at d 1.019–1.063 g/ml[8]

Acetylated LDL: Prepared by treatment of LDL with acetic anhydride according to the method of Goldstein et al[6]

Inhibitor of acyl-CoA cholesterol acyltransferase (ACAT): Sandoz ACAT inhibitor 58.035, progesterone, or other well-characterized specific inhibitor of ACAT activity. Stock solutions of inhibitor can be prepared by solubilization in ethanol. Maximum inhibition of ACAT activity in fibroblasts and mouse peritoneal macrophages occurs at concentrations below 1 and 20 μg/ml for Sandoz compound 58.035 and progesterone, respectively

Bovine calf serum

Loading Cells with Nonlipoprotein Cholesterol. Cultured cells, usually human skin fibroblasts or smooth muscle cells, are washed once with wash buffer and then incubated with medium A plus the indicated amount of stock solution of cholesterol or 25-hydroxycholesterol. The standard amount of cholesterol stock added per 1 ml medium is 5 μl, which gives a final cholesterol concentration of 50 μg/ml. With 25-hydroxycholesterol, it is not advisable to exceed a final concentration of 20 μg/ml, or 2 μl/ml stock solution, since cytotoxicity becomes a problem at this concentration. Ethanol concentrations up to 10 μl/ml of media (1%) do not appear to have any deleterious effects on cells or to influence HDL binding or cell cholesterol metabolism.

To prepare the cholesterol-enriched medium, an aliquot of the stock solution is injected into the swirling albumin medium. The medium, stock solution, and injection pipet or syringe are warmed to at least 37° prior to injection. The medium, which appears cloudy, is then added to the cells immediately after preparation. Incubations are usually carried out for 24 to 48 hr prior to measurement of HDL-binding activity.

Loading Cells with LDL Sterol. Cultured cells, usually fibroblasts or smooth muscle cells, are grown to 75 to 90% confluency. Cells are then incubated with medium A containing 10% human lipoprotein-deficient serum for 48 hr with one medium change after 24 hr (to maximize cholesterol efflux from cells and provide fresh growth factors). At this stage, LDL receptors should be induced to near maximum levels. Cells are then incubated with serum-free medium A plus LDL at a concentration of 30 to 50 μg/ml protein. With this protocol, a surge of LDL sterol is delivered

[8] O. de Lalla and J. Gofman, "Methods of Biochemical Analysis," p. 459. Wiley (Interscience), New York, 1954.

intracellularly before feedback regulation of LDL receptor activity shuts down the uptake process. Maximal delivery occurs between 24 and 48 hr of exposure to LDL. Since serum is replaced by albumin in the LDL medium, efflux of the unesterified cholesterol liberated from the incoming LDL is minimized. The albumin also sequesters fatty acids and inhibits cholesteryl ester formation. An ACAT inhibitor can be added to the medium to further enhance unesterified cholesterol accumulation in the cell. Since LDL itself can act as a sink for cholesterol and promote efflux of cholesterol from cells, it is important to use an LDL concentration high enough to saturate LDL receptor sites but not high enough to cause continual depletion of cholesterol from cells.

Loading Cells with Acetylated-LDL Sterols. Both bovine aortic endothelial cells[3] and mouse peritoneal macrophages[4] exhibit increased HDL-binding activity when exposed to media containing acetylated LDL. Since the activity of the scavenger receptor is not modulated by change in the cell cholesterol content, it is not necessary to preincubate cells with medium containing lipoprotein-deficient serum.

Bovine aortic endothelial cells from stock cultures are plated into 35-mm dishes or wells at a density of 1×10^5 and grown to confluency. If cell loss becomes evident during the extensive washing needed for the HDL-binding and cholesterol efflux assays, it may be necessary to coat dishes with 1% gelatin prior to plating. Moreover, with gelatin-treated dishes, cells tend to "sprout" less and have more of a uniform "cobblestone" appearance. To coat the dishes, 1.0 ml of a 1% gelatin solution (distilled water) is added to the dishes, the dishes are gently swirled, and the solution is aspirated. Cells are plated after the dishes are allowed to dry overnight. After cells are grown to confluency (7 to 10 days), the monolayer is washed and incubated with medium A containing 25 to 50 μg/ml acetyl-LDL protein. Adequate cholesterol loading occurs after 24 to 48 hr.

A good experimental model for studying HDL binding to cholesterol-loaded cells is cultured mouse peritoneal macrophages. These cells amass large quantities of sterols via the scavenger receptor pathway when incubated in the presence of acetyl-LDL.[6] Resident macrophages are harvested from the peritoneal cavity of mice after intraperitoneal injection of saline.[9] Approximately 1×10^6 cells suspended in 0.5 ml medium A (albumin free) containing 20% fetal calf serum are plated per one 16-mm culture well. The yield from one unstimulated mouse is approximately 3×10^6 cells, equivalent to three wells. After an overnight incubation, cells

[9] P. J. Edelson and Z. A. Cohn, "In Vitro Methods in Cell-mediated and Tumor Immunity," p. 333. Academic Press, New York, 1976.

are washed with serum-free medium A containing 20–30 μg of acetyl-LDL protein/ml. Substantial cholesterol loading accompanied by increased binding activity occurs within 24 hr of incubation.[4] As described for LDL-treated cells, HDL binding to cells treated with acetylated LDL is further enhanced by including an ACAT inhibitor in the medium.

Assays for HDL Binding

Several different types of cultured cells have been shown to bind ^{125}I-labeled HDL with high affinity when cells are treated with sterol-rich medium. This binding shows specificity for HDL and its apolipoproteins,[2–5] although the precise ligand recognized by the binding sites has yet to be identified. Binding is reversible, in that HDL bound to cells at 4° is released from the cell within minutes upon warming to 37°.[2] The mechanism involved in binding and release of HDL from its binding site is poorly understood.

HDL binding activity is measured at either 4 or 37° by similar assays. The 37° assay is preferred when cells are pretreated with medium containing protein, such as those in serum, that may interfere with subsequent binding of ^{125}I-labeled HDL. At 37°, these interfering proteins should be displaced from the cell surface during the binding assay. Since it is possible that HDL becomes internalized by cells at 37°, the binding incubation is limited to 1 hr or less. For these short times, HDL binding may be in a nonequilibrium state and may not reflect receptor activity. Since cells can bind receptor ligands at 4° but internalize them at very slow rates, the 4° binding assay may provide a more reliable estimate of cell surface binding activity. This assay, however, requires a more extensive washing procedure prior to the binding incubations in order to remove cell surface components that could interfere with HDL binding. When directly compared, both the 37 and 4° assays showed the same relative increase in HDL binding activity when fibroblasts were treated with cholesterol.[2]

Reagents

High-density lipoprotein (HDL): Whole HDL, HDL$_2$, or HDL$_3$ are isolated from human plasma by sequential ultracentrifugation at densities of 1.063 to 1.21, 1.063 to 1.125, or 1.125 to 1.21 g/ml, respectively

Heparin–Sepharose CL-6B (Pharmacia)

Column buffer A, consisting of 15 mM NaCl plus 5 mM imidazole (pH 6.5)

Column buffer B, consisting of 1 M NaCl plus 5 mM imidazole (pH 6.5)

Medium A: Dulbecco minimum essential medium is supplemented with 2 mg/ml essentially fatty acid-free, bovine serum albumin (lot No. A-6003, Sigma Chemical Company)

37° Binding medium: Radiolabeled HDL is added to medium A

4° Binding medium: Dulbecco's minimum essential medium without bicarbonate is neutralized with 1 N NaOH ($\sim$5 μl/ml) and buffered with 20 mM N-2-hydroxyethylpiperazine-N'-2-ethanesulfonic acid (HEPES). Fatty acid-free bovine serum albumin (2 mg/ml) and radiolabeled HDL are added, and the medium is chilled on ice

Wash buffers A and B: Wash buffer A consists of 1 mg/ml fatty acid-free bovine serum albumin in phosphate-buffered saline (pH 7.4). Wash buffer B is the same buffer containing a less expensive preparation of bovine serum albumin (e.g., lot No. A-2153, Sigma Chemical Company) at a concentration of 2 mg/ml

0.1 N NaOH

Preparation of Radiolabeled HDL. In most cases, HDL binding is studied under conditions where cells are cholesterol loaded and the activity of the LDL receptor is suppressed. Therefore, contamination of the radiolabeled HDL preparation with small amounts of apoB and E, two ligands that interact with the LDL receptor, does not interfere with the HDL-binding assay. Under these conditions, the ultracentrifugally isolated HDL₃ fraction can be directly iodinated without further treatment. This fraction usually contains only trace amounts of apoB and E. However, if HDL binding is to be studied under conditions where LDL receptor activity may be enhanced or the HDL preparation contains a large proportion of either or both apoB and E, it is necessary to remove these apolipoproteins prior to iodination. ApoB and E are removed from HDL by passing the preparations through a column containing heparin covalently coupled to Sepharose. Approximately 5 g heparin–Sepharose is suspended in column buffer A at 4° and washed several times with the cold column buffer A by mild centrifugation. After resuspension in $\sim$30 ml of the same buffer, the heparin–Sepharose is added to a column (1.5 cm diameter, bed height of $\sim$10 cm) in a refrigerated room. A 2- to 5-ml aliquot containing 5–15 mg/ml whole HDL or specific subfraction is added to the column and eluted with column buffer A into tubes in a fraction collector set to collect 2 to 4 ml per tube. Tubes with the highest concentration of [apoE]-free HDL are combined and dialyzed against saline/ EDTA or glycine buffer (for iodination). The apoE-, B-containing particles are eluted from the column with column buffer B. The column can be stored for continuous use by reequilibration with column buffer A that contains sodium azide.

Preparations of HDL₃ or [apoE]-free HDL are radioiodinated by the iodine monochloride method described by Bilheimer *et al.*[10] This procedure yields biologically active HDL preparations with specific activities of 2 to 5 × 10⁸ cpm/mg protein and with less than 2 and 5% of the radioactivity in trichloroacetic acid-soluble components and chloroform–methanol-extractable lipids, respectively.

37° Binding Assay. Cultured cells are washed several times at room temperature with either wash buffer A or medium A. Cells are then incubated with 37° binding medium containing either ¹²⁵I-labeled HDL₃ or [apoE]-free ¹²⁵I-labeled HDL (usually at a concentration of 5 μg protein/ml). After 1 hr at 37°, the culture dishes or wells are placed on a stainless steel tray that is sitting on ice in a 4° cold room. After chilling for 10–15 min, cells are washed three times in rapid sequence with ice-cold wash buffer B, followed by two 5-min incubations with the same buffer and two rapid washes with albumin-free wash buffer B. Cells are then dissolved in 0.1 N NaOH at room temperature and aliquots are assayed for ¹²⁵I radioactivity and protein mass.[11]

4° Binding Assay. Cells are treated the same for the 4° binding assay as they are for the 37° assay except for three major differences: (1) cells are washed more extensively prior to chilling, depending on preincubation conditions; (2) binding medium is specially buffered to maintain a constant pH throughout the assay; and (3) the binding incubation is 2 instead of 1 hr. If cells are preincubated with medium that contains proteins that may bind to HDL binding sites, then the cells must be thoroughly washed prior to chilling to 4°. This is because ligands become irreversibly bound to receptors after cells are chilled, blocking subsequent binding of newly introduced ligand. In addition, when cells are treated with nonlipoprotein cholesterol, the 4° binding assay is improved with thorough washing, possibly because the cholesterol-enriched membranes undergo a physical transformation when chilled that affects HDL binding. As a standard protocol that produces optimum binding, cells are washed twice at room temperature with wash buffer A followed by a 1-hr incubation at 37° with medium A. Cells are then washed twice again with wash buffer A and are chilled to 4° while being bathed with the last wash. After chilling for 15 min, the 4° binding medium is added to the dishes.

Assays for Cholesterol Transport

Several different direct and indirect assays can be used to measure cholesterol transport from cultured cells. These include measurement of

[10] D. W. Bilheimer, S. Eisenberg, and R. I. Levy, *Biochim. Biophys. Acta* **260,** 212 (1972).
[11] O. H. Lowry, N. J. Rosebrough, A. L. Farr, and R. J. Randall, *J. Biol. Chem.* **193,** 265 (1951).

(1) efflux of radiolabeled cholesterol, (2) net transport of cholesterol mass, and (3) biochemical processes involved in cellular sterol metabolism (biochemical markers).

Efflux of Radiolabeled Cholesterol

The most sensitive assay for measurement of the rate of cholesterol transport from cells is to first radiolabel cellular cholesterol and then monitor the rate of appearance of radiolabeled cholesterol in the culture medium. To study cholesterol efflux under conditions of upregulated HDL-binding activity, labeled sterol can be introduced into cells via the same delivery process used to induce HDL binding.

Reagents

[³H]Cholesterol in ethanol: For every 10 ml of loading medium, 1 to 5 μl of [³H]cholesterol in toluene (~55 Ci/mmol, New England Nuclear) is dried in a small glass tube under N_2. Fifty microliters of warm (~40–50°C) cholesterol stock solution (10 mg/ml, prepared in 95% EtOH) is added, and the tube is vigorously agitated on a vortex

LDL

Trioleylglycerol stock: 50 mg/ml chloroform (store in freezer)

Cholesteryl linoleate stock: 50 mg/ml chloroform (store in freezer)

[³H]Cholesteryl linoleate: (~100 Ci/mmol, New England Nuclear)

Egg lecithin stock: 30 mg/ml chloroform (store in freezer)

Tris buffer: 0.15 M NaCl, 0.3 mM EDTA, 10 mM Tris (pH 7.6)

Sephadex G-50 (Pharmacia)

Plasma fraction of d >1.21 g/ml, prepared by ultracentrifugation

Silica gel G thin-layer chromatography plates, prescored (Analtech)

TLC standards (mixture A): 1 mg each triolein, cholesteryl oleate, oleate, cholesterol, diolein, monolein in 1 ml chloroform–methanol (2:1, v/v)

Isopropanol (anhydrous)

Hexane

Chloroform

Methanol

Ethyl ether

Acetic acid

Aquasol scintillation fluid (New England Nuclear)

Radiolabeling Cellular Cholesterol. When cells are treated with nonlipoprotein cholesterol to upregulate HDL binding activity, cholesterol efflux can be studied by including radiolabeled cholesterol in the loading medium. An aliquot of [³H]cholesterol in ethanol, warmed to 37°, is injected into medium A, usually at a concentration of 50 μg/ml. Cells are then incubated for 24 to 48 hr with the radiolabeled cholesterol medium.

To measure cholesterol efflux from cells treated with LDL or acetylated LDL, the lipoprotein particles are reconstituted with radiolabeled cholesteryl esters. Labeled sterols can be introduced into LDL by direct addition, where cholesteryl esters dissolved in an organic solvent are added to delipidated LDL, or by enzyme-catalyzed transfer of lipid from microemulsions into LDL. The latter procedure has the advantage of yielding labeled LDL particles that have not undergone harsh physical treatment, such as lipid extraction with organic solvents, freezing, and lyophililization.

The method of choice for reconstitution of LDL with lipid microemulsions is that described by Craig et al.[12] Lipid microemulsions are prepared that contain lipid components at the approximate molar ratio contained in native LDL. To prepare microemulsions with 1.5 mg/ml total lipid, aliquots of stock solution that contain 0.21 mg trioleylglycerol, 0.16 mg cholesterol, 0.71 mg cholesteryl linoleate, 0.42 mg dipalmityl phosphatidylcholine, and 250 μCi [^{3}H]cholesteryl linoleate are combined and dried under N_2. The lipids are solubilized in 60 μl dry isopropanol, heated to 60°, injected with a prewarmed 100-μl Hamilton syringe into 1.0 ml Tris buffer, and the buffer is immediately mixed on a vortex. The isopropanol is then removed by centrifugation through a Sephadex G-50 column. Sephadex G-50 is suspended in Tris buffer and 5 ml is pipetted into a 5-ml plastic syringe plugged with glass wool. The syringe is placed into a plastic centrifuge tube and spun at low speed for 5 min to deplete gel of buffer. The syringe is transferred to another tube, the microemulsion is added to the syringe, and the tube is recentrifuged. The microemulsion is combined with 1.0 mg LDL and 8 ml of $d>1.21$ g/ml fraction of human plasma. This plasma fraction contains lipid transfer proteins that promote the transfer of lipid into LDL. The mixture is incubated at 37° for 24 hr and the LDL is reisolated by ultracentrifugation at $d=1.019$ to 1.063 g/ml. The radiolabeled LDL can be acetylated for introducing label sterols into cells via the scavenger receptor pathway. The procedure yields reconstituted LDL preparations that contain 20 to 30 cpm [H^3]cholesteryl linoleate/ng protein.

Assay Procedure. To measure rates of cholesterol efflux, cells are incubated with efflux medium after being washed at least four times with wash medium at room temperature to remove surface-bound radioactivity. When cells are loaded with [^{3}H]cholesterol, it is advisable to include

[12] I. F. Craig, D. P. Vie, B. C. Sherrill, L. A. Sklar, W. W. Mantulin, A. M. Gotto, Jr., and L. C. Smith, *J. Biol. Chem.* **257,** 330 (1982).

an "equilibration" incubation to further facilitate removal of surface-bound radiolabel. For the equilibration phase, cells are incubated at 37° with medium A for 1 to 2 hr between the second and third wash. Efflux incubations should be less than 4 hr, since the rate of efflux may become nonlinear after longer incubation.

Regardless of the type of radiolabeled sterols used to load cells, only unesterified cholesterol is released from cells to a significant extent. When cells are treated with radiolabeled unesterified cholesterol, efflux is assayed by scintillation counting of an aliquot of the efflux medium. When cell sterols are radiolabeled by treatment of cells with radioactive lipoprotein sterols, medium-unesterified cholesterol must be isolated by thin-layer chromatography prior to counting.

For isolation of medium cholesterol, total lipids are extracted by treatment of 1.0 ml medium with 5.0 ml $CHCl_3$: CH_3OH (2 : 1, v/v) plus 100 μl of TLC standards (mixture A) in glass-stoppered tubes. The tubes are shaken vigorously, the phases separated by mild centrifugation, and the upper water–methanol phase discarded by aspiration. The lower chloroform phase is washed once with 1.0 ml water. The solvent is dried under a stream of air and resuspended in 50 to 100 μl chloroform for thin-layer chromatography.

For isolation of cellular cholesterol, lipids are extracted from cell monolayers by the hexane–isopropanol method described by Goldstein *et al.*[13] Two milliliters of hexane–isopropanol mixture (3 : 2, v/v) are added to cell monolayers previously washed with albumin-free wash medium. Dishes are incubated for 30 min at room temperature. The extract is transferred to a glass-stoppered tube, the monolayers are washed twice with 1 ml hexane–isopropanol, and the washes are combined with the original extract. The remaining monolayer is extracted in 0.1 N NaOH for protein determination.[11] The solvent in each tube is dried under air after the addition of 100 μl/tube of TLC standard (mixture A), and the residue is resolubilized in 50 to 100 μl chloroform for thin-layer chromatography. To separate lipids, the chloroform is spotted on silica gel G thin-layer chromatography plates, and the chromatograms are developed in hexane–ethyl ether–acetic acid (90 : 20 : 1, v/v/v). The esterified and unesterified cholesterol silica spots are scraped into scintillation vials and counted in Aquasol.

Net Transport of Cellular Cholesterol

The procedure described in the previous section is a direct measurement of unidirectional flux of cholesterol from cells. The rate of net trans-

[13] M. S. Brown, Y. K. Ho, and J. L. Goldstein, *J. Biol. Chem.* **255**, 9344 (1980).

port of cholesterol between cells and HDL is determined by the difference between the rate of flux of HDL cholesterol into cells and the rate of flux of cellular cholesterol to HDL particles. Therefore, in order to obtain accurate estimates of net movement of cholesterol from cells, changes in the cholesterol mass in either the efflux medium or cells is measured.

Reagents

Silica gel G thin-layer chromatography plates, prescored (Analtech)
TLC standards (mixture B): 1 mg each triolein, oleate, diolein, mono-
    lein in 1 ml chloroform–methanol (2 : 1, v/v)
Hexane
Isopropanol
Chloroform
Methanol
Ethyl ether
Acetic acid
Ethanolic KOH: 1.0 N KOH in 95% ethanol
Sodium phosphate buffer (50 mM, pH 7.0)
Cholesterol oxidase solution: Enzyme (Calbiochem) is diluted to 0.8
    U/ml with phosphate buffer and stored in small aliquots (100–200
    μl) in the freezer
Peroxidase solution: Enzyme (Boehringer) is diluted fresh to 300
    U/ml with phosphate buffer
p-Hydroxyphenylacetic acid solution: Prepared fresh in water at a
    concentration of 1.5 mg/ml
0.5 N NaOH

Extraction of Medium and Cell Sterols. To measure net transport of cholesterol, cells are washed and incubated as described in the section, Efflux of Radiolabeled Cholesterol. To determine initial rates of transport, incubation times should be limited to 4 hr. To determine the maximum capacity of HDL to promote net transport of cholesterol from cells, the incubations are extended to 24 or 48 hr. Medium and cellular sterols are isolated and separated by thin-layer chromatography according to the procedure described for measurement of efflux of radiolabeled cholesterol, except that esterified and unesterified cholesterol are omitted from the thin-layer chromatography standards mixture that is added to the extracts. (The addition of the glycerides and fatty acid standards aid in the recovery of sterols but do not interfere with the cholesterol mass assay.)

After thin-layer chromatography of the medium and cell lipid extracts, the silica acid spot corresponding to unesterified cholesterol is scraped from the plates into glass tubes. The cholesterol is removed from the silica by suspending the silica in hexane and transferring the hexane to another tube after mild centrifugation. The silica is washed once again with hexane, and the solvent washes are combined. The hexane is dried under a stream of air and the cholesterol is quantified by enzymatic assay.

To measure the esterified cholesterol content of the medium and cell extracts, the esters are first hydrolyzed. The cholesteryl ester silica spot is scraped into a 15-ml glass-stoppered tube. One milliliter of ethanolic KOH is added to the tube, and the stoppered tube is heated at 80° for 1 hr. After the addition of 1.5 ml of H_2O and 4.0 ml of hexane, the tubes are vigorously shaken and the phases are separated by mild centrifugation. An aliquot of the top hexane phase is transferred to a glass tube, dried under air, and the cholesterol is quantified.

Enzymatic Assay for Cholesterol.[14,15] Tubes containing samples and standards of unesterified cholesterol are taken to dryness under air. The standard tubes contain known amounts of cholesterol ranging from 0 to 3 μg/tube in 0.5 μg/tube increments. The dried cholesterol is resolubilized in 20 μl isopropanol. The reaction is started by addition of 0.4 ml reagent mixture prepared fresh the day of the assay. Each 10 ml of reagent mixture consists of 7.9 ml sodium phosphate buffer, 1.0 ml peroxidase solution, 1.0 ml p-hydroxyphenylactic acid solution, and 1.0 ml cholesterol oxidase solution. After 5 min at room temperature, 0.8 ml of 0.5 N NaOH is added to each tube. Samples are then quantified in a spectrofluorometer. For an Aminco spectrofluorometer (SPF-500), excitation wavelength is set at 325 nm, band-pass 2 nm, and emission wavelength is set at 415 nm, band-pass 10 nm.

The presence of phospholipids in the sample interferes with this assay. If samples have not undergone thin-layer chromatography prior to the assay, phospholipids must be removed. To do this, the solvent is dried under air, and 0.2 to 0.5 g of silica gel plus 1.0 ml chloroform is added to each tube. After mixing and centrifugation, the chloroform is transferred to another tube, the silica is washed once with 1 ml chloroform, and the combined solvents are dried under air. The cholesterol is then assayed as described above.

Biochemical Markers

Cultured cells possess several different biochemical processes that are involved in maintenance of cell cholesterol homeostasis that can be used as sensitive intrinsic assays for estimating the relative degree of transport of cholesterol from cells. Since HDL-mediated cholesterol transport from cells is studied under conditions where cells are loaded with cholesterol, the most useful of these biochemical markers is the cholesterol-esterifying

[14] J. G. Heider and R. L. Boyett, *J. Lipid Res.* **19,** 514 (1978).
[15] W. King and M. Culala, unpublished results.

enzyme, acyl-CoA cholesterol acyltransferase (ACAT). The activity of this enzyme is a direct function of the amount of excess cholesterol that accumulates in the cell.[16] Another assay that can be used to estimate changes in cell cholesterol content is the rate of cholesterol synthesis from acetyl units. The rate-limiting enzyme in the pathway of cholesterol synthesis, HMG-CoA reductase, is induced in response to removal of cholesterol from cells.[16] This induction, however, only becomes significant when excess stores of cholesterol are totally depleted and cells enter a negative cholesterol balance.[2] The assay described in this section provides a simultaneous estimate of the relative activity of both of these processes. The assay is based on the principle that when cells are exposed to medium containing radiolabeled oleic acid, not only is the fatty acid incorporated into cholesteryl ester, it is also oxidized by cells at a fast enough rate that radiolabeled acetyl units accumulate in the cell. These acetyl units are then utilized as precursors for cholesterol synthesis. Because these two processes are measured simultaneously, this assay provides an estimate of the degree of cholesterol transport from cells over a wide range of initial cell cholesterol concentrations.

Reagents

Medium A and medium B: Medium A is Dulbecco's minimum essential medium containing 2 mg/ml bovine serum albumin (essentially fatty acid free), and medium B is the same medium without albumin

Radiolabeled oleate stock solution: A tube containing 500 μCi [1-^{14}C]oleic acid (57 mCi/mmol, Amersham) in toluene is warmed to 60° in a water bath and dried under nitrogen. One milliliter of 0.02 N NaOH is added to the tube, and the solution is mixed on a vortex and heated at 70 to 80° until clear. The tube is cooled to touch and injected with a Pasteur pipe into 9.0 ml phosphate-buffered saline containing 15 mg fatty acid-free bovine serum albumin. The mixture is stirred immediately on a vortex, and the original tube is rinsed several times with the fatty acid–albumin solution. The solution is pipetted into 200-μl aliquots which are stored in the freezer

Silica gel G thin-layer chromatography plates, prescored (Analtech)

TLC standards (mixture A): 1 mg each of triolein, cholesteryl oleate, oleate, diolein, monolein in 1 ml chloroform

Hexane

Isopropanol

---

[16] M. S. Brown and J. L. Goldstein, *Science* **191,** 150 (1976).

Chloroform
Methanol
Ethyl ether
Acetic acid
Ethanolic KOH: 1.0 N KOH in 95% ethanol
Aquasol scintillation fluid (New England Nuclear)

Assay Procedure. Cholesterol-loaded cells are incubated at 37° with medium A containing varying concentrations of HDL or specific cholesterol acceptors. After the indicated time (usually 24 to 48 hr), cells are washed rapidly at room temperature with phosphate-buffered saline and incubated at 37° with medium B containing 30 μl [^{14}C]oleate stock/ml. After 1 hr, the dishes are placed on a stainless steel tray on ice for 10 min, and the cells are washed twice with ice-cold phosphate-buffered saline. Each dish then receives 2 ml hexane–isopropanol and is incubated for 30 min at room temperature. The solvent is transferred to glass tubes, the dishes are washed twice with 1 ml hexane–isopropanol, and the washes are combined with the solvent extract. After the addition of 100 μl TLC standards (mixture A), the solvent is dried under air, and the lipids resolubilized in 50 μl chloroform. The chloroform is spotted on silica gel G thin-layer chromatography plates, which are then developed in hexane–ethyl ether–acetic acid (90 : 20 : 1, v/v/v). After exposure to iodine vapor for visualization, the cholesteryl ester and unesterified cholesterol silica spots are scraped into separate glass-stoppered tubes. Two milliliters of 1 *N* KOH in 95% ethanol is added to each tube, and the tubes are stoppered and heated at 80° for 1 hr. This saponification step hydrolyzes the esters in the cholesteryl ester samples and any glycerides that may have contaminated the unesterified cholesterol samples. After cooling, 1.5 ml of water and 4 ml hexane are added to each sample. The tubes are then shaken and centrifuged to separate the phases. An aliquot of each of the hexane phases, which contains unesterified cholesterol, is transferred to a glass scintillation vial, dried under air, and counted in Aquasol. An aliquot of the lower ethanol/water phase of the cholesteryl ester samples, which contains the fatty acyl moiety of the hydrolyzed esters, is also counted in Aquasol. (The lower phases for the samples from the free cholesterol silica spots, which contain radiolabeled acyl groups from contaminating glycerides, are discarded.) The counts in the hexane-phase represent incorporation of ^{14}C-labeled acyl units into cell cholesterol, an index of the rate of sterol synthesis. The counts in the lower water/ethanol phases represent incorporation of radiolabeled oleic acid into cholesterol esters, an index of the rate of cholesteryl ester formation.

[39] Interstitial Fluid (Peripheral Lymph) Lipoproteins

By LADISLAV DORY, CHARLES H. SLOOP, and PAUL S. ROHEIM

Introduction

Importance of Interstitial Fluid Lipoprotein Composition

Lipoproteins play an important role in regulating cellular lipid metabolism in the periphery. In order to fulfill this important function, plasma lipoproteins must come in direct contact with peripheral cell membranes. Studies of the interstitial fluid lipoproteins have provided a unique opportunity to detect and characterize lipoproteins in direct contact with these cells.

Much of what is known to date about the interaction between lipoproteins and peripheral cells is derived from *in vitro* tissue culture studies utilizing plasma lipoproteins as substrates. Plasma lipoproteins have been used because of the general assumption that interstitial fluid lipoproteins are simply filtered plasma lipoproteins, and a practical consideration of the unavailability of significant quantities of interstitial fluid lipoproteins.

Chemical characterization of interstitial fluid lipoproteins may provide insight into the role lipoproteins play in maintaining cholesterol homeostasis of peripheral cells, particularly the mechanisms used to excrete excessive cholesterol, a process usually referred to as reverse cholesterol transport. The well-documented protective effect of high levels of HDL in the development of atherosclerosis may be related to its ability to promote cholesterol efflux from cells, a concept first proposed by Glomset.[1]

The ability of HDL or even specific HDL apolipoproteins to promote cholesterol efflux from cultured cells has been clearly demonstrated in a number of laboratories.[2-4] Despite the potential importance of this process, there is little direct *in vivo* evidence for the interaction of HDL with peripheral cells and a resultant removal of cellular cholesterol by HDL. Such an interaction, as manifested by altered chemical composition, would be difficult to detect when examining plasma HDL because it represents a mixture of particles of hepatic, intestinal, and interstitial origin,

[1] J. A. Glomset, *J. Lipid Res.* **9**, 155 (1968).
[2] O. Stein and Y. Stein, *Biochim. Biophys. Acta* **326**, 232 (1973).
[3] Y. K. Ho, M. S. Brown, and J. L. Goldstein, *J. Lipid Res.* **21**, 391 (1980).
[4] M. C. Phillips, L. R. McLean, G. W. Stoudt, and G. M. Rothblat, *Atherosclerosis* **36**, 409 (1980).

modified to varying degrees by plasma enzymes. Compositional studies of interstitial fluid lipoproteins, on the other hand, can provide evidence for the interaction between plasma lipoproteins and peripheral cells.

Recent tissue culture studies also indicate that plasma lipoproteins may be chemically modified by factors secreted by endothelial cells during their passage into the interstitial space and thus "condition" them for subsequent interaction (such as uptake) with other cells, including tissue macrophage or smooth muscle cells.[5] Chemical analysis of interstitial fluid lipoproteins and comparison to plasma lipoproteins may provide direct evidence for such phenomena.

In summary, studies of interstitial fluid lipoproteins and their metabolism may provide the important link between tissue culture studies and *in vivo* studies of plasma lipoprotein metabolism.

Determinants of Interstitial Fluid Lipoprotein Composition

The lipoprotein composition of the interstitial space is influenced by a number of factors: (1) the plasma lipoprotein composition; (2) the permeability characteristics of the plasma–interstitial fluid barrier; (3) the exclusion properties of the interstitial space; (4) modification of filtered lipoproteins by endothelial cells during passage across the capillary endothelium; (5) modification within the interstitium; and (6) peripiheral synthesis of the apolipoprotein components of lipoproteins leading to peripheral lipoprotein assembly/synthesis.

There is considerable controversy concerning the relative importance of the various physical forces which cause movement of plasma constituents across the capillary endothelium. Diffusion (caused by concentration gradients) and convective transport (due to ultrafiltration) have been proposed. It is likely that both forces contribute to capillary transport, and their relative importance may vary with changes in capillary hemodynamics and characteristics of the transported molecules. Charge of the macromolecules may also play an important role in determining their permeability characteristics.

Tracer substances of various molecular size (radii of 1.5–25 nm) have been used to examine the permeability characteristics of capillary endothelium. Results obtained using dextran fractions could be interpreted by postulating the existence of a relatively large number of capillary endothelial pores with a radius of about 3.4–4.5 nm ("small pore system") and a relatively small number of "leaks" (12 to 35 nm in radius) ("large pore

[5] T. Henriksen, E. M. Mahoney, and D. Steinberg, *Proc. Natl. Acad. Sci. U.S.A.* **78**, 6499 (1981).

system")).[6] Studies using endogenous plasma proteins as molecular tracers have yielded similar conclusions.[7]

The endothelial cells of the capillary wall constitute the principal barrier to movement from plasma to interstitium. The most permeable capillaries are the open sinusoids of liver and spleen. The tight junctional capillaries of the brain are least permeable. The continuous endothelium of capillaries in periphery (skin and skeletal muscle) exhibit intermediate permeability characteristics.

Prenodal Peripheral Lymph as a Model of Interstitial Fluid

Direct sampling of interstitial fluid in reasonable quantities is, at the present time, not realistic. Different approaches have been utilized to circumvent this problem. Inflammatory fluid collected into implanted polyvinyl sponges in rabbits may be one approach.[8] Large changes in endothelial permeability in response to inflammatory stimuli and the resultant edema may, however, extensively change the interstitial fluid lipoprotein distribution and composition. Lymph obtained from cannulated prenodal lymphatics is considered an acceptable model of interstitial fluid, since anatomical studies of the terminal lymphatics have shown them to be in free communication with the interstitial space.[9] Much of the information obtained about various properties of the interstitial fluid has, in fact, been obtained using prenodal lymph. The lipoprotein composition of prenodal peripheral lymph collected from the dorsum of the foot in humans[10] and the hindleg of dogs[11,12] has been described. Prenodal peripheral lymph is characterized by relatively low plasma protein "contamination" and may thus be the best model of interstitial fluid in studies seeking to define differences in chemical characteristics of interstitial fluid and plasma lipoproteins. Lymph collected from specific organs, such as cardiac[13] or lung[14] lymph, has also been used. This lymph contains a relatively high concentration of plasma proteins, including lipoproteins, mak-

[6] G. Grotte, *Acta Chir. Scand. Suppl.* **24,** 1 (1956).

[7] R. D. Carter, W. L. Joyner, and E. M. Rankin, *Microvasc. Res.* **7,** 31 (1974).

[8] T. L. Raymond and S. A. Reynolds, *J. Lipid Res.* **24,** 113 (1983).

[9] F. Huth and D. Bernhardt, *Lymphology* **1,** 15 (1977).

[10] D. Reichl, L. A. Simons, N. B. Myant, T. T. Pflug, and G. L. Mills, *J. Clin. Sci. Mol. Med.* **45,** 313 (1973).

[11] C. H. Sloop, L. Dory, B. R. Krause, C. Castle, and P. S. Roheim, *Atherosclerosis* **49,** 9 (1983).

[12] C. H. Sloop, L. Dory, R. Hamilton, B. R. Krause, and P. S. Roheim, *J. Lipid Res.* **24,** 1429 (1983).

[13] P. Julien, E. Downar, and A. Angel, *Circ. Res.* **49,** 248 (1981).

[14] T. M. Forte, C. E. Cross, R. A. Gunther, and G. C. Kramer, *J. Lipid Res.* **24,** 1358 (1983).

ing it more difficult to discern changes in lipoprotein composition. In our laboratory, we use heartworm-free mongrel dogs as experimental animals. We have found that the cannulation of hindleg lymphatics, relatively abundant lymph flow, and individual cost of animals made the dog the most attractive species for these studies. Furthermore, dog plasma lipoproteins and their metabolism have been extensively characterized, providing a solid framework for subsequent studies.[15,16]

Peripheral Lymph Collection

Surgical Preparation

Dogs are fasted for 14–16 hr before the experiment. They are anesthetized with sodium pentobarbital (30 mg/kg iv) and intubated with a cuffed endotracheal tube, but not mechanically ventilated. The external jugular vein is cannulated (18-gauge catheter placement unit, Jelco Laboratories, Raritan, NJ) to allow continuous intravenous administration of saline (0.2–0.5 ml/min), anesthetic supplement, and blood sampling. Prepopliteal peripheral lymphatics of both hindlimbs are cannulated near the lateral saphenous vein.

Only one lymphatic of each hindlimb is cannulated. All other lymphatics near the cannulation site are ligated to direct most of the lymphatic outflow into the collection cannula.

Procedure for Lymph Collection

There is considerable evidence that the large lymphatics of several species are contractile. Periodic contraction of these structures may be important in the propulsion of central lymph. However, normal peripheral flow in anesthetized dogs requires passive limb movement. Most efficient lymph transport occurs when the animal's foot is flexed at the hock.

The apparatus designed to accomplish this is shown in Figs. 1–3. Particular care has been taken to make the device as adjustable as possible and to minimize venous and lymphatic stasis of the hindlegs. The structures referred to in the following text are numbered in parentheses and, correspondingly, in the figures.

Each leg is moved by a separate DC motor (Dayton Model 4Z143)(1) linked to a 24 : 1 gear reducer (Dayton Model 2Z821A)(2). The gear reducer in turn drives a cam and piston arrangement (3), producing a 5-cm reciprocal displacement of the paw. The tip of the piston is padded with

[15] R. W. Mahley and K. H. Weisgraber, *Circ. Res.* **35**, 713 (1974).
[16] R. W. Mahley, K. H. Weisgraber, and T. Innerarity, *Circ. Res.* **35**, 722 (1974).

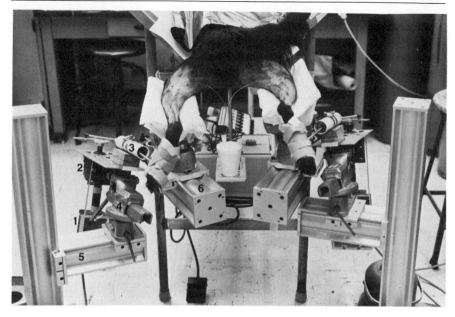

FIG. 1. The apparatus designed to flex the dog's feet at the hock to promote peripheral lymph flow. The numbered structures are described in detail in the text. For a detailed drawing of the adjustable support (5) for the motor-gear reducer–cam and piston arrangement, see Fig. 2. For a detailed drawing of the adjustable system supporting the dog's hindlegs (6), see Fig. 3.

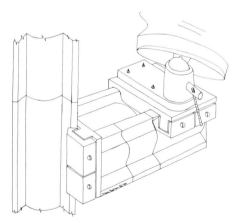

FIG. 2. Detailed drawing of the adjustable support system for the ball and socket machinist's vise holding the unit producing the flexing motion (4) in Fig. 1.

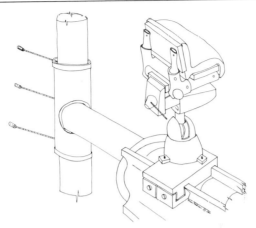

FIG. 3. Detailed drawing of the adjustable support system for the hindleg of the dog (6) in Fig. 1.

closed cell foam (Ensolite) and provided with two strips of "hooked" Velcro. The dog's lower leg is wrapped with several layers of open cell foam attached to "pile" Velcro. This provides a strong attachment when "hooked" Velcro strips of the piston are wound around the dog's leg. The large surface for attachment and extensive foam padding minimize trauma to the foot. Pumping frequency is controlled by adjusting a variable transformer supplying current to a full-wave rectifier on the motor.

Each motor is supported by a large ball and socket machinist's vise which allows the motor, cam, and piston to be tilted to match the natural plane of movement of each leg (4) (detail drawn in Fig. 2). The vise and motor are mounted on cruciform aluminum supports (X95 System, Klinger Scientific, Richmond Hill, NY) (5), which allow adjustment of height and lateral displacement. The dog's hindlegs are supported by foam-covered plastic "V's" attached to a similar adjustable system (6), shown in detail in Fig. 3.

While the hindlegs are supported horizontally, care must be taken to prevent the dog from lying directly on its sternum. We have found that this prevents adequate oxygenation by the animal.

Skeletal Muscle Lymph Collection

Peripheral lymph contains lymph derived mainly from skin and connective tissue. This allows us to study interstitial lipoprotein interactions with fibroblasts and macrophages. Skin, however, represents a small fraction of total body mass. We have chosen to extend our studies to skeletal muscle and develop a model which will allow us to study interstitial lipoprotein interactions in a tissue which constitutes a large fraction of total body mass.

Surgical Procedure and Lymph Collection

Skeletal muscle lymphatics are typically diffuse and rarely coalesce into collecting lymphatics carrying pure skeletal muscle lymph. For this reason, cannulation of skeletal muscle lymphatics is difficult, and for many years the permeability characteristics of skeletal muscle were studied by extraction techniques[17] or by electron microscopic studies using electron dense permeability markers.[18]

Bach and Lewis[19] published a method which allowed collection of "pure" skeletal muscle lymph from rabbits. They noted that the skeletal muscle lymphatics of the rabbit hindlimb joined efferent lymphatics from the popliteal node (draining skin) to form femoral lymphatics. Cutaneous lymph contamination was excluded by ligating the popliteal efferent lymphatics. This made it possible to collect skeletal muscle lymph indirectly by cannulating femoral lymphatics. Skeletal muscle lymph flow is usually slower than simultaneously measured cutaneous lymph flow, and its protein content is greater than that of cutaneous lymph. Thus, skeletal muscle capillaries are slightly more permeable to plasma proteins than are skin capillaries. We have adopted a similar procedure for use in dogs.

Most of the surgical preparation is similar to the one described for prepopliteal lymph collection. The medial and lateral surfaces of both hindlegs are shaved and the dog is placed on its side with both hindlegs on padded articulated supports. Efferent as well as afferent popliteal lymphatics are then isolated and ligated in the perinodal popliteal fat. One of the afferent lymphatics is cannulated in the normal manner to obtain peripheral lymph. This eliminates a major source of nonskeletal muscle lymph from lymphatics draining the skin and connective tissue of the paw. Deep intramuscular injections of sodium fluorescein have shown that lymph drainage from the gracilis muscle is unaffected by this procedure because its lymphatics follow the femoral artery and vein. The dog is then placed on its back and the femoral artery and vein are exposed. Ligation of the connective tissue under and around the femoral vein will cause the femoral lymphatics to distend.

The largest femoral lymphatic is then cannulated at the level of the

[17] C. Crone, *Acta Physiol. Scand.* **58**, 292 (1963).
[18] S. L. Wissig, *Acta Physiol. Scand. Suppl.* **463**, 33 (1979).
[19] C. Bach and G. P. Lewis, *J. Physiol. (London)* **235**, 477 (1973).

cranial femoral artery. Accessory lymph nodes, such as the medial femoral node described by Baum,[20] occur only rarely but must be ligated to prevent contamination of the femoral lymph with postnodal lymph. Connective tissue lymphatics lying near the fibular and dorsal branch of the saphenous vein are also ligated to prevent contamination of the femoral lymph.

Both femoral muscle and peripheral lymph are collected from the dog lying on the surgical table, and each hindleg is attached to piston-like devices to provide passive movement (as described earlier). The femoral lymph collected in this manner may contain small quantities of lymph derived from other deep structures (i.e., connective tissue); however, it should reflect the composition of skeletal muscle interstitial fluid.

Peripheral Lymph Lipoproteins–Examples of Experimental Design

Characterization

Peripheral lymph from both hindlegs is collected into vials placed in a small cup containing ice (see Fig. 1). Each collecting vial contains sufficient anticoagulant (such as EDTA) and sodium azide (optional) to give a final concentration of these agents of 0.01%. Depending on the nature of the experiment, protease inhibitors may also be added. In some experiments it may be important to exclude the lymph collected during the first hour or so, since the slight stasis during the cannulation procedure will result initially in an accelerated lymph flow. After the initial 1-hr period, a steady-state flow is obtained and amounts to 1–3 ml/hr/leg. At the end of the collection period (6–7 hr), the lymph is pooled and an aliquot taken for routine assays including total protein, cholesterol (esters and unesterified), phospholipid, and triglycerides. The remaining lymph is used for sequential ultracentrifugal lipoprotein isolation (described in chapter [6], Volume 128). In order to preserve the unesterified cholesterol/cholesteryl ester ratio, we have routinely adjusted the lymph to contain 1 mM DTNB, an inhibitor of lecithin : cholesterol acyltransferase. As shown below, this may not be necessary because of low levels of this enzyme in peripheral lymph. Blood is also collected from the dog from the cannulated jugular vein (usually at the end of the lymph collection) and treated the same way as the lymph samples. It is very important that lymph and plasma lipoproteins obtained from the same animal be compared.

The initial determinations of total lymph and plasma protein and lipid content allows the calculation of a lymph-to-plasma concentration ratio

[20] H. Baum, *Arch. Wiss. Prakt. Tierheilkd. Bd.* **44,** 521 (1918).

(L/P) of each component. A large L/P ratio may alert the investigator to possible plasma contamination of the collected lymph. Some dogs, in our experience, have a very permeable endothelial barrier. The presence of visible amounts of red blood cells, on the other hand, clearly indicates that the plasma contamination is a result of the surgical procedure.

A rapid estimation of the plasma and peripheral lymph lipoprotein profile may be obtained by agarose electrophoresis of whole plasma or lymph using the initial aliquot. Agarose electrophoresis is carried out essentially as described.[21] The standard procedure allows a 15-μl sample/ well. Because peripheral lymph contains only a fraction of the plasma lipoproteins (approximately 5–20%, depending on the lipoprotein species, dietary status, and individual variability of each dog), it is desirable to apply larger samples. Up to 20-μl samples may be applied if a thicker agarose plate is poured. We use 30 ml of 0.5% agarose in a 0.025 M Tris-Tricine buffer, pH 8.6, containing 0.35% BSA, spread over a 4 × 9 in. Cronar film placed on a glass plate. The plate is run at 200 V for 1½ hr or until the BSA/Evans Blue tracker sample (placed in each of the extreme wells) reaches the 3.5 cm mark from the origin. The gel on the Cronar film is then fixed in a mixture of 95% ethanol/water/glacial acetic acid (370 ml/ 130 ml/40 ml) for 15 min and dried in an oven at 40°. The dried gel may be stained for lipids with Oil Red O.[21] Using this rapid procedure, presence of β-, pre-β-, and α-migrating lipoproteins in both plasma and lymph can be demonstrated. HDL are the predominant lipoproteins of normal plasma and lymph, while increasing hypercholesterolemia is associated with increasing plasma concentrations of β- and pre-β-migrating lipoproteins, a trend also reflected in peripheral lymph. The concentrations of the various apolipoproteins in whole plasma or lymph can be measured by electroimmunoassay, essentially as described by Laurell.[22] The antibodies toward purified apolipoproteins (the purification procedure is described in Chapter [31], Volume 128) can be prepared in rabbits or goats. It is important that the antibody-containing agarose also contain NP-40, a nonionic detergent, in final concentration of 0.05%. We have used, as a standard, dilutions of standard plasma and expressed the concentrations in the samples as a percentage of standard plasma (arbitrary units). Table I demonstrates the concentrations of the various apolipoproteins in plasma and lymph of control and cholesterol-fed dogs. Because the results are presented in arbitrary units (AU), comparisons can only be made within the same apolipoprotein column and not between the different apolipoproteins. The comparison of lymph or plasma samples to a

[21] R. P. Noble, *J. Lipid Res.* **9,** 693 (1968).
[22] C. B. Laurell, *Anal. Biochem.* **15,** 45 (1966).

TABLE I

APOLIPOPROTEIN CONCENTRATIONS (AU) IN WHOLE PLASMA AND LYMPH OF CONTROL AND
CHOLESTEROL-FED DOGS[a]

Group	Compartment	Apolipoprotein concentration (AU)[b]			
		ApoB	ApoE	ApoA-I	ApoA-IV
Control	Plasma	86.9 ± 13.3	46.9 ± 7.2	148.0 ± 12.2	70.6 ± 6.4
(4)	Lymph	6.0 ± 1.5	5.0 ± 0.7	14.5 ± 1.0	10.7 ± 1.3
	Lymph (percentage of plasma)	6.7 ± 1.1	10.9 ± 1.4	10.2 ± 1.5	15.3 ± 1.3
Cholesterol-fed (3)	Plasma	923.6 ± 102.1[c]	271.3 ± 48.5[c]	22.7 ± 7.6[c]	216.0 ± 48.9[c]
	Lymph	20.8 ± 3.5[c]	8.8 ± 1.5	2.9 ± 1.2[c]	25.5 ± 7.7
	Lymph (percentage of plasma)	2.3 ± 0.6[c]	3.7 ± 1.4[c]	9.3 ± 1.7	11.7 ± 2.6

[a] AU, Arbitrary units.
[b] Values are the mean ± SEM.
[c] Significantly different from the corresponding value in control animals, $p < 0.05$. Reprinted from Sloop et al.[11]

standard plasma (stored frozen at $-70°$) was chosen so that the antigen in the test sample and standard be physically in as close a state as possible, an important consideration in electroimmunoassays.[22] The use of known amounts of purified, delipidated apolipoproteins as standards and delipidated samples (with organic solvents or detergents) can be used to obtain absolute amounts of the various apolipoproteins. This may not be important when comparing plasma and lymph concentrations of individual apolipoproteins (concentration units cancel out) but will become very important when the stoichiometry of the apolipoprotein associations on various lipoproteins is investigated.

By sequential ultracentrifugation of plasma and lymph, the following lipoprotein fractions are obtained: VLDL ($d < 1.006$ g/ml); IDL ($d = 1.006-1.03$ g/ml); LDL ($1.030-1.063$ g/ml); and HDL ($1.063-1.21$ g/ml). Table II shows the mass distribution of the various ultracentrifugally isolated lipoproteins in normal and cholesterol-fed dogs. Quantitation of normal dog VLDL and IDL is difficult because of the very low concentration of these lipoproteins in plasma and lymph. When expressed as a percentage of plasma concentrations, peripheral lymph contains an increasing fraction of lipoproteins with increasing density (and decreasing size), reflecting the greater permeability of the endothelial lining for the

TABLE II
DISTRIBUTION OF LIPOPROTEIN–PROTEIN AMONG THE LIPOPROTEIN
FRACTIONS IN PLASMA AND LYMPH OF CONTROL AND
CHOLESTEROL-FED DOGS[a]

		Protein concentration (mg/dl)	
Fraction	Compartment	Control ($n = 4$)	Cholesterol fed ($n = 4$)
VLDL	Plasma	$2.9^b \pm 0.4^c$	95 ± 33
	Lymph	0.3 ± 0.1	3 ± 0.4
	Lymph (percentage of plasma)	10	3
IDL	Plasma	b	92 ± 23
	Lymph	b	2 ± 0.3
	Lymph (percentage of plasma)	—	2
LDL	Plasma	19 ± 4	7 ± 0.7
	Lymph	1 ± 0.2	0.5 ± 0.1
	Lymph (percentage of plasma)	5	7
HDL	Plasma	260 ± 25	47 ± 27
	Lymph	20 ± 1	6 ± 4
	Lymph (percentage of plasma)	8	13

[a] Reprinted from Sloop et al.[12]
[b] Due to the small amount of material available in the VLDL and IDL fraction of control animals, only a combined (VLDL + IDL) fraction was isolated.
[c] Values ± SEM.

smaller lipoproteins. A similar pattern is obtained when the lipoprotein distribution is determined on the basis of cholesterol content.

The apolipoprotein content of each fraction may be determined by electroimmunoassay, as described above, or by resolving the delipidated apolipoproteins of each fraction by SDS–PAGE (according to Laemmli[23]), followed by densitometric scanning.

Important information about the nature of the peripheral lymph lipoproteins may be obtained by electron microscopy (the method is described in Volume 128) using phosphotungstic acid staining on coated copper grids. Figure 4 is an example of negatively stained plasma and

[23] U. K. Laemmli, Nature (London) 227, 680 (1970).

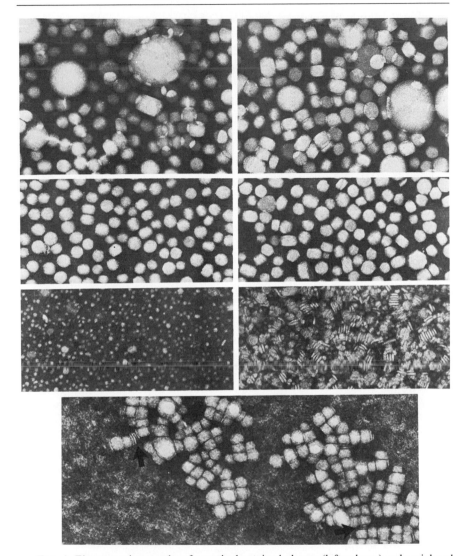

FIG. 4. Electron micrographs of negatively stained plasma (left column) and peripheral lymph (right column) lipoproteins from cholesterol-fed dogs. Top: VLDL ($d < 1.006$ g/ml); middle: IDL ($1.006 < d < 1.030$ g/ml); bottom: HDL ($1.063 < d < 1.21$ g/ml); (LDL not shown). Centered bottom image is of whole uncentrifuged peripheral lymph from cholesterol-fed dogs, demonstrating the presence of most particles in above centrifugal fractions, including discoidal HDL in short stacks of rouleaux (arrows). Note the absence of very large particles seen in VLDL, suggesting that they are artifacts.

peripheral lymph lipoprotein fractions obtained from a cholesterol-fed dog. Of special interest is the observation that a significant proportion of lymph HDL obtained from cholesterol-fed dogs is in the form of disk-shaped particles stacked in rouleau structures. Fewer discoid particles are seen in plasma HDL. The high proportion of discoid particles in lymph HDL is reflected in its chemical composition, shown in Table III. As can be seen, lymph HDL contains a much higher amount (in percentage weight) of unesterified cholesterol and phospholipid at the expense of cholesteryl esters and protein. Thus, both chemical composition and electron microscopy studies support and confirm the conclusion that peripheral lymph HDL differs in a number of aspects from plasma HDL. SDS–PAGE or electroimmunoassay reveals that lymph HDL is also significantly enriched in apoE and apoA-IV when compared to its plasma counterpart. It is obvious that simple filtration does not account for the differences between the two pools of HDL, and it appears likely that the peripheral tissues alter the chemical composition of filtered plasma HDL and possibly participate in the assembly of the discoid HDL, since these are not nearly as abundant in the plasma of the same animal.

The heterogeneity of lymph HDL suggests that further subfractionation should be carried out. We have also found that for a complete recovery of all non-apoB-containing lipoproteins (i.e., HDL), and especially the discoid HDL, the initial density fraction obtained by ultracentrifugation should be wider—1.05–1.21 g/ml. Due to an overlap in densities, this fraction, however, contains significant amounts of apoB-containing LDL. Con A–Sepharose (Pharmacia) affinity chromatography may be used to remove the apoB-containing lipoproteins (LDL) from the HDL.[24] Briefly, the lipoprotein fraction is extensively dialyzed against the equilibrating buffer consisting of 0.05 M Tris-HCl, pH 7.0, containing 1.0 M NaCl and 1 mM CaCl$_2$, MgCl$_2$, and MnCl$_2$. The affinity chromatography is carried out at 4° in a 2.5 × 15 cm column with a flow rate of 0.3 ml min^{-1}. The unretained fraction (HDL mixture) is collected and may be concentrated in dialysis bags against Aquacide. The apoB-containing lipoproteins can be subsequently eluted with 0.2 M 1-O-methyl-α-D-glucopyranoside in the equilibrating buffer.

The Con A–Sepharose column may be recycled several times for repetitive use by the following procedure: the beads are washed in 10 vol of a 0.1 M Tris-HCl buffer, pH 8.5, containing 0.5 M NaCl, followed by a wash (10 vol) with a 0.1 M sodium acetate buffer, pH 4.5, containing 0.5 M NaCl. The material should then be reequilibrated with the running buffer and is ready for use again.

[24] T. Mitamura, *J. Biochem.* **91**, 25 (1982).

TABLE III
CHEMICAL COMPOSITION OF PLASMA AND LYMPH HDL (1.063–1.21 g/ml) OBTAINED
FROM CONTROL AND CHOLESTEROL-FED DOGS[a]

| | Control (n = 3) | | Cholesterol-fed (n = 3) | |
	Plasma	Lymph	Plasma	Lymph
Protein	43.2 ± 0.4[b]	37.1 ± 0.3	40.7 ± 0.9	35.7 ± 1.2
Cholesteryl ester	12.8 ± 1.5	11.0 ± 0.5	19.1 ± 0.8	9.0 ± 0.3
Free cholesterol	4.4 ± 0.4	8.1 ± 0.4	4.2 ± 0.4	15.3 ± 0.4
Phospholipid	39.1 ± 1.1	41.9 ± 1.8	35.2 ± 1.1	40.0 ± 0.8
Triglyceride	0.5 ± 0.1	1.9 ± 0.3	0.8 ± 0.3	ND[c]
Free cholesterol/ester cholesterol	0.36 ± 0.06	0.74 ± 0.07	0.23 ± 0.04	1.70 ± 0.02

[a] Reprinted from Sloop et al.[12] All values are percentage by weight.
[b] Values ± SEM.
[c] ND, Not detectable.

The heterogeneous mixture of HDL obtained after this can be further subfractionated by a number of techniques. Agarose column chromatography was used in our laboratory (Bio-Rad, BioGel A-0.5 M, 1 × 120 cm, equilibrated in 0.15 M NaCl, 0.01% EDTA, azide, pH 7.4) to subfractionate HDL according to size. Two major peaks are obtained. When using peripheral lymph HDL from cholesterol-fed dogs, the void volume peak is composed of almost entirely discoid particles, while the second peak contains spherical HDL.

Often, however, it is more desirable to obtain lipoproteins homogeneous with respect to their apolipoprotein content and, thus, possibly homogeneous with respect to their metabolic fate. For this purpose, we have employed various forms of affinity chromatography. Heparin–Sepharose chromatography (described in Chapters [9, 11], Volume 129) may be used to isolate the apoE-containing lipoproteins. Immunoaffinity chromatography may be used to isolate lipoproteins enriched in any particular apolipoprotein, provided monospecific antibodies are available. An example of an approach to characterize peripheral lymph HDL in our laboratory is summarized in a flow-chart form shown in Fig. 5.

Detailed characterization of homogeneous subfractions of peripheral lymph HDL and the comparison to equivalent subfractions of plasma HDL (if present) will provide important insight about the interaction between peripheral cells and HDL and, thus, about the early events in the process of reverse cholesterol transport. Clearly, similar approaches may be used to characterize other lipoprotein fractions present in peripheral

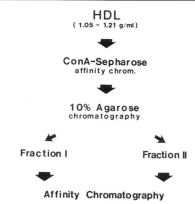

FIG. 5. An example of the fractionation scheme for HDL used in our laboratory. An initially wide density range of HDL is isolated to ensure good recovery ($1.05 < d < 1.21$ g/ml). The contaminating apoB-containing lipoproteins can be adsorbed to Con A–Sepharose and the resulting HDL mixture subfractionated according to size by agarose column chromatography. The subfractions obtained in this step may be further subfractionated, based on their apolipoprotein content, by immunoaffinity chromatography (or, for apoE-containing lipoproteins, heparin–Sepharose chromatography).

lymph, notably the various apoB-containing lipoproteins present especially during cholesterol feeding.

Peripheral Apolipoprotein Synthesis

In addition to the well-established role of the liver and intestine, evidence obtained in a number of laboratories under various *in vitro* conditions strongly points to various peripheral tissues as sites of apolipoprotein synthesis. The ability of a cholesterol-loaded macrophage to synthesize and secrete apoE, as well as excrete cholesterol, has been demonstrated by Basu and co-workers.[25] A number of laboratories have since demonstrated the presence of apoE mRNA in a variety of tissues. Therefore, it appears reasonable to suggest that newly secreted apoE, together with other apolipoproteins filtered from plasma and relatively lipid deficient, may combine with plasma membrane cholesterol and phospholipid (destined for reverse cholesterol transport) to form the discoid particles. The availability of lymph from acutely cannulated hindleg prenodal lymphatics has provided a way to investigate peripheral apolipoprotein synthesis under *in vivo* conditions.

[25] S. K. Basu, Y. K. Ho, M. S. Brown, D. W. Bilheimer, R. G. W. Anderson, and J. L. Goldstein, *J. Biol. Chem.* **257**, 9788 (1982).

Experimental Design. The animal is surgically prepared for lymph collection as already described above. Having established that the cannulation is successful and good lymph flow is obtained, the radioactively labeled amino acid is injected in a small volume (0.1–0.2 ml) at several sites into the skin of the toes of one leg. The injection time is 2–3 min. The skin of the toes of the other leg is injected similarly with an equal volume of saline. Lymph from both legs is collected separately into chilled vials containing appropriate preservatives (EDTA, azide) and enzyme inhibitors (when necessary). Lymph is pooled and tubes are changed at appropriate time intervals. Plasma samples from the cannulated jugular vein are also collected (ca. every 5–10 min or to match the lymph samples) and pooled in the same manner as lymph. This experimental design is based on the assumption that the labeled amino acid enters the interstitial space and, subsequently, the cells. Some of the amino acid may directly enter the blood compartment so that eventual incorporation of labeled amino acid into proteins by the liver and intestine will also be observed. This contamination can be corrected for by determining the protein-associated activity in the lymph collected from the saline-injected leg–it should be contaminated with radioactive plasma proteins to a similar extent as the isotope-injected leg lymph. Having corrected for such factors, comparison of specific activities (SA) of specific proteins from the isotope-injected leg lymph and plasma should clearly establish whether or not peripheral synthesis occurs.

In a typical experiment, over 70% of the total recovered radioactivity was found in the collected lymph of the isotope-injected leg, about 12% was found circulating in the plasma, 18% in the liver, and only about 0.1% in the lymph collected from the saline-injected leg. This demonstrates that the labeled amino acid was preferentially exposed to peripheral tissues of the hindleg and that the bulk of the radioactivity (mostly free amino acid) was drained out in the isotope-injected leg lymph (usually within 30 min of the injection). Much lower levels of radioactivity (several orders of magnitude lower) appear, with a little delay, in the blood. Radioactivity in the saline-injected leg lymph peaks about 1 hr after injection and, like that of plasma, remains low and relatively constant throughout the experiment (up to 4 hr).

The specific activities (SA) of the apolipoprotein in question in lymph or plasma can be evaluated in a number of ways. Each pooled sample may be adjusted, by addition of solid KBr to $d = 1.25$ g/ml, layered with KBr $d = 1.21$ g/ml, and the total $d < 1.21$ g/ml lipoprotein fraction isolated. All or portions of this fraction may be dialyzed against 5 mM ammonium bicarbonate, lyophilized, and delipidated by ethanol/ether mixture

(3 : 1 v/v) at $-70°$. The delipidated apolipoproteins may be resolved by SDS–PAGE and appropriate bands cut out, dissolved overnight in a perchloric acid (0.2 ml) and peroxide (0.4 ml) mixture, and counted. The various apolipoproteins can also be quantitated in the same delipidated samples by electroimmunoassay or radioimmunoassay so that the final results can be expressed in terms of specific activities.

Alternatively, monospecific polyclonal antibodies can be used to immunoprecipitate apolipoproteins from whole plasma or lymph. The immunoprecipitation is carried out in the presence of detergent (such as 1% NP-40). The immunoprecipitates may be washed with saline and dissolved in the SDS–PAGE protein solvent. The solubilized immunoprecipitates can be resolved by SDS–PAGE, the appropriate bands cut out, and further treated as described above. Alternatively, the slab gels may be soaked in Autofluor, EN³HANCE, or similar products, dried, and exposed to X-ray film. The resulting fluorographs may be quantitated by densitometry. In a typical experiment, the specific activity of apoE in peripheral lymph is severalfold higher (orders of magnitude) than the plasma apoE specific activity at the time of peak incorporation. (In one experiment, we found 130,000 cpm/mg apoE in lymph versus 2800 cpm/mg apoE in plasma at 150 min after injection of 1 mCi of [³⁵S]methionine.)

Enzymes in Peripheral Lymph

Lecithin : Cholesterol Acyltransferase (LCAT). LCAT, an enzyme that esterifies free cholesterol into cholesteryl esters in plasma, may play a potentially important role in reverse cholesterol transport through this action. Free cholesterol can freely and rapidly equilibrate between various surfaces, including plasma membranes and lipoprotein surfaces. If a concentration gradient between the surface exists, a net transfer of free cholesterol will result. It is, however, the action of LCAT that will result in relocation of lipoprotein surface cholesterol to lipoprotein core cholesteryl ester. As an ester, the cholesterol cannot equilibrate rapidly with cell plasma membranes any more. The presence of HDL of high unesterified cholesterol content in peripheral lymph indicated the importance of determining the extent that this enzyme is present in the interstitial fluid. LCAT was therefore determined in aliquots of lymph and plasma (0.5 ml) from the same animal, incubated with trace amounts of [¹⁴C]cholesterol at 37° for various times, essentially as previously described[26] (detailed methods of LCAT assay are described in Chapter [45], this volume). The free

[26] L. Dory, C. H. Sloop, L. M. Boquet, R. L. Hamilton, and P. S. Roheim, *Proc. Natl. Acad. Sci. U.S.A.* **80**, 3489 (1983).

cholesterol pool size of the lymph aliquots was, on average, 7.5% of the equivalent plasma free cholesterol pool. At each time interval, the reaction was stopped by the addition of 10 vol of isopropanol. The neutral lipids were separated by thin-layer chromatography, and after visualization by exposure to iodine vapor, the cholesterol and cholesteryl ester spots were scraped into scintillation vials containing 0.5 ml of ethanol and the radioactivity was determined after addition of 10 ml of Aquasol-2. The total free cholesterol pool size at time zero was determined as described below and the results were plotted as the percentage of free cholesterol esterified. For a direct comparison of LCAT activities (independent of the different free cholesterol pool sizes in lymph and plasma), the initial rates were expressed in nmol/hr ml^{-1} of free cholesterol esterified. The results are shown in Fig. 6.

Cholesterol esterification in lymph lagged far behind that of plasma; less than 3% of the lymph free cholesterol pool was esterified, even after 24 hr of incubation. The difference was even more striking when the results were expressed in terms of absolute amount of free cholesterol esterified: the initial rate of cholesterol esterification was 0.3 nmol/hr ml^{-1} in lymph and 40.6 nmol/hr ml^{-1} in plasma, more than 100-fold greater.

Our data suggest that the action of LCAT may not be essential during the initial stages of cholesterol efflux from the peripheral cells (if the formation or reassembly of lymph HDL represents the early stages of

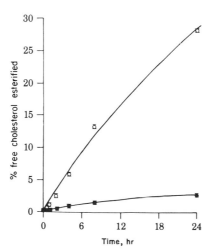

Fig. 6. [^{14}C]Cholesterol esterification in plasma (○) and peripheral lymph (●) of cholesterol-fed dogs, as a function of time. Mean (± SEM) of four experiments is shown. The initial rates of cholesterol esterification were 40.6 nmol/hr/ml in plasma and 0.3 nmol/hr/ml in peripheral lymph. Reprinted from Dory *et al.*[26]

reverse cholesterol transport). This clearly appears to differ from results of tissue culture experiments in which LCAT action was reported to be essential. It is possible that, in tissue culture experiments, promotion of sterol efflux depends on cholesterol esterification in order that the free cholesterol concentration gradient between the cell and medium be maintained. In such a static system, esterification provides an essential sink for the effluent cholesterol and thereby maintains the gradient. Under *in vivo* conditions, however, esterification may not be necessary at the interstitial fluid level because lymph flow ensures constant removal of the free cholesterol-enriched acceptor.

Conclusion

Prenodal peripheral lymph is an accepted model of interstitial fluid. In this chapter, we present techniques which allow collection of prenodal peripheral lymph in quantities sufficient for analytical or metabolic studies of interstitial fluid. The dog as an experimental model offers advantages in ease of surgical procedures, quantities of fluid obtained, and a well-characterized plasma lipoprotein profile. The experimental designs and data, described serve only as examples of the potentially important information that can be obtained from studies of peripheral (or deep muscle) lymph lipoproteins. Such information will provide a link between studies on plasma lipoprotein metabolism and tissue culture studies using peripheral cell cultures (such as macrophage, smooth muscle cells, endothelial cells, etc.) and thus provide an important contribution to the understanding of the function of lipoproteins.

Acknowledgments

The authors wish to thank Ms. Linda Spiess and Ms. Lawrie Lehmann for their skillful preparation of this manuscript, and Mr. Steve Redman for the drawings. This work was supported by NIH Grant #25596.

[40] Lipoproteins and Steroid Hormone-Producing Tissues

By JOHN T. GWYNNE and DARIEN D. MAHAFFEE

Introduction

Studies of lipoprotein metabolism in steroid hormone-producing tissues have substantially increased our understanding of lipoprotein metabolism in general. This discussion will concentrate on the methods used to study the metabolism of lipoprotein cholesterol and protein only since steroidogenic tissue metabolism of other lipoprotein constituents has not yet been systematically examined. It should be noted, however, that tropic hormones stimulate polyphosphoinositide turnover in steroidogenic tissues[1,2] and the role of lipoprotein phospholipids in these pathways deserves further investigation. Six species have been most extensively studied: rat, mouse, pig, cow, rabbit, and man. We will concentrate primarily on methods used in the rat. In most cases, similar methods can be applied *in vitro* to tissue from other species as well. Considerations in adapting these procedures to other species are noted.

General Considerations

Steroidogenic tissues present unique opportunities to study cholesterol and lipoprotein metabolism. First, per unit weight, steroidogenic tissues generally consume more lipoprotein cholesterol, possess more lipoprotein receptors, and degrade more apolipoprotein than other tissue. This concentration of activity has facilitated previous work such as the isolation and characterization of the LDL receptor from the bovine adrenal cortex.[3,4] Despite this concentrated metabolic activity, the adrenal and ovaries or testes normally contribute little to whole body cholesterol and lipoprotein economy because of their small size. In contrast, the term placenta may consume large amounts ($\geq$250 mg/day in woman) of lipoprotein cholesterol. Second, lipoprotein metabolism by steroidogenic tissue is hormonally regulated. This provides an additional means to verify the physiologic significance of an observation and to alter tissue choles-

---

[1] R. V. Farese, M. A. Sabir, and S. L. Vandor, *J. Biol. Chem.* **254,** 6842 (1979).

[2] R. V. Farese, M. A. Sabir, and R. E. Larson, *J. Biol. Chem.* **255,** 7232 (1980).

[3] W. J. Schneider, S. K. Basu, M. J. McPhaul, J. L. Goldstein, and M. S. Brown, *Proc. Natl. Acad. Sci. U.S.A.* **76,** 5577 (1979).

[4] W. J. Schneider, J. L. Goldstein, and M. S. Brown, *J. Biol. Chem.* **255,** 11442 (1980).

terol utilization to advantage. To study lipoprotein metabolism in steroidogenic tissues to greatest advantage requires thorough familiarity with hormonal control of steroidogenesis in the tissue of choice. Third, production of steroid hormones provides an additional end point for studying lipoprotein metabolism not available in other types of tissue. In most, but not all, steroidogenic tissues, lipoproteins are a major source of substrate cholesterol for steroid hormone production. Under selected circumstances steroid hormone production depends largely on available lipoprotein cholesterol. Thus cellular uptake of lipoprotein cholesterol may be monitored not only by measurements of tissue cholesterol content but also indirectly by measurement of steroid hormone release.

Significant disadvantages are also encountered in studying steroidogenic tissue lipoprotein metabolism. First, all steroidogenic glands are composed of mixed cell populations. In the ovary and testes only a small portion of the total tissue mass synthesizes steroid hormones. Thus *in vivo* studies may be difficult to interpret and *in vitro* studies may require laborious cell purification. Second, in rats and other small animals, the small size of steroidogenic glands makes acquisition of large amounts of tissue difficult and expensive. Third, the presence of cholesterol catabolic pathways not found in tissues other than steroidogenic glands and liver, while providing unique opportunities, also makes control and definition of experimental conditions more demanding.

To accurately interpret studies of lipoprotein metabolism by steroid hormone producing tissues, both the lipoprotein and the hormonal milieu of the cells must be accurately known. Moreover, cholesterol and steroid hormone metabolic pathways may interact at several levels. For instance, glucocorticoids, estrogens, and androgen all affect circulating lipoprotein levels. In particular, in the rat, the estrogen, ethinyl estradiol, profoundly inhibits hepatic lipoprotein release and drastically lowers circulating lipoprotein levels.

Studies of lipoprotein metabolism by steroid hormone producing tissues have provided considerable insight into steroidogenic tissue function and metabolism. Of the four steroidogenic tissues, adrenal, ovary, testes, and placenta, the adrenal and ovary have been most extensively studied and are best characterized.[5–8] In each type, steroid output may be stimu-

[5] J. T. Gwynne and J. F. Strauss, III, *Endocr. Rev.* **3**, 299 (1982).

[6] J. F. Strauss, III, L. A. Schuler, M. F. Rosenblum, and T. Tanoka, *Adv. Lipid Res.* **18**, 99 (1981).

[7] J. R. Schreiber and D. B. Weinstein, *in* "Lipoproteins, Receptors, and Cell Function" (A. M. Scanu and A. Spector, eds.). Dekker, New York, 1984.

[8] M. S. Brown, P. T. Kovanen, and J. L. Goldstein, *Recent Prog. Horm. Res.* **35**, 215 (1979).

TABLE I
INHIBITORS OF CHOLESTEROL SIDE CHAIN CLEAVAGE

| Inhibitor | Mechanism of action | Reversible | Usual dose in rats | |
			In vivo	*In vitro*[a]
Aminoglutethimide[b]	Active site analog	Yes	30 mg sc	70 μM
Cycloheximide[c]	Inhibition of protein synthesis	Yes (short term)	10 mg ip	25 μM

[a] H. E. Falke, H. J. Degenhart, G. J. A. Abeln, and H. K. A. Visser, *Mol. Cell. Endocrinol.* **3,** 375 (1975).
[b] R. N. Dexter, L. M. Fishman, R. L. Ney, and G. W. Liddle, *J. Clin. Endocrinol. Metab.* **27,** 473 (1967).
[c] W. W. Davis and L. D. Garren, *J. Biol. Chem.* **243,** 5153 (1968).

lated by specific tropic hormones, in the adrenal ACTH stimulates gluco-corticoid and angiotensin stimulates mineralocorticoid production, while in the ovary and testes FSH and LH stimulate sex steroid production. In addition estradiol, prolactin, insulin, and other growth factors may also stimulate or augment tropic hormone response in the gonads. While hormonal control of ovarian steroidogenesis is not yet completely understood any investigation of ovarian lipoprotein metabolism must define and/or regulate as many hormonal variables as possible. There are no known hormonal stimuli of placenta steroidogenesis. The action of the tissue-specific tropic stimuli can be mimicked by cAMP and its nondegradable analog dibutyryl-cAMP, thus circumventing tropic hormone interactions with cell membrane receptors. Lipoproteins do not directly stimulate steroid hormone release but enhance the action of other stimuli by providing substrate cholesterol. Available evidence neither precludes nor supports additional sites of lipoprotein action. Lipoprotein may also influence growth of steroidogenic tissues.[9]

While the steroid products and hormonal regulation of steroid synthesis differ among the tissues, the cellular pathways of cholesterol and lipoprotein metabolism are similar. Cholesterol is an obligate intermediate in steroid hormone production. In the presence of adequate cholesterol substrate the rate-limiting, hormonally regulated step in steroid hormone production by all steroidogenic tissue is cholesterol side chain cleavage to yield Δ^5-pregnenolone and isocaproic acid. Specific inhibitors of side chain cleavage are available (Table I). Cholesterol side chain cleavage enzyme is located on the inner mitochondrial membrane and is one of a

[9] N. Savion, J. Lui, R. Laherty, and D. Gospodarowicz, *Endocrinology* **109,** 409 (1981).

family of cytochrome P-450 containing mixed function oxidase, essential for steroidogenesis. When cholesterol side chain cleavage is inhibited, tropic hormone stimulation causes accumulation of cholesterol within the mitochondrion.[9] The mechanisms responsible for intracellular cholesterol movement to and within the mitochondrion are not known. One way to assess maximum potential steroid hormone production in the intact cell is to introduce a freely accessible substrate such as 25-OH-cholesterol.[10,11]

There are normally three immediate sources of steroidogenic substrate cholesterol: *de novo* synthesis, release from intracellular cholesterol ester stores, and uptake of lipoprotein cholesterol. The relative contribution of each precursor pool varies among species and must be considered in choosing tissues for experimentation. Hormonal stimulation of steroid production also stimulates mobilization of cholesterol from each pool. As in other tissues, abundant intracellular cholesterol suppresses *de novo* synthesis. The rate of cholesterol synthesis therefore depends on the relative rates of cholesterol uptake, cholesterol ester mobilization, and steroid hormone production but is limited by the intrinsic synthetic potential. Uptake of lipoprotein cholesterol and mobilization of cholesterol esters are more directly responsive to hormone stimulation. The coordination of these pathways and the mechanisms of intracellular cholesterol movement are poorly understood.

Two different pathways for uptake of lipoprotein cholesterol have been identified in steroid hormone producing tissue. These can be conveniently identified as the "LDL pathway" and the "HDL pathway." The general properties of the LDL pathway in steroidogenic tissues are identical to those originally described by Brown and Goldstein in cultured human fibroblasts and have been reviewed by those investigators.[8] The HDL pathway is substantively different and does not require endocytosis nor lysosomal degradation of the HDL particle to provide lipoprotein cholesterol for steroidogenesis or storage.[5] The "LDL pathway" has been identified in all steroidogenic tissues in all species in which it has been sought (Table II). In contrast, the complete "HDL pathway" has been definitely identified only in steroidogenic tissues of the rat and mouse although evidence to suggest its presence in bovine ovarian and porcine ovarian and adrenal tissue has been presented. As can be seen from Table II, some tissues possess characteristic HDL binding activity but HDL does not deliver substrate cholesterol or enhance steroid hormone production. The relationship of HDL binding to HDL enhancement

[10] D. Mahaffee, R. C. Reitz, and R. L. Ney, *J. Biol. Chem.* **249**, 227 (1974).
[11] H. E. Falke, H. J. Degenhart, G. J. A. Abeln, and H. K. A. Visa, *Mol. Cell. Endocrinol.* **4**, 107 (1976).

TABLE II

SPECIES DIFFERENCES IN LIPOPROTEIN METABOLISM BY
STEROIDOGENIC TISSUES

Species	Characteristic receptor activity				Increased steroid hormone production[a]			
	LDL[b]		HDL[c]		LDL		HDL	
	A[d]	O	A	O	A	O	A	O
Rat	+	+	+	+	+	+	+	+
Mouse	+	+	+	+	+	+	+	+
Pig	+	+	+	+	+	+	±	±
Cow	+	+	−	−	+	+	−	±
Man	+	+	+	+	+	+	−	−

[a] A tissue was judged responsive if enhanced steroid production could be demonstrated either *in vivo* or *in vitro*. A tissue was judged responsive to HDL only if available evidence clearly demonstrated that apoE was not required.

[b] LDL receptors were judged to be present if lipoprotein binding *in vivo* or *in vitro* (1) was specific for apoB- and apoE-containing lipoproteins, (2) required calcium, (3) was sensitive to proteolytic digestion, or (4) could be identified by anti-LDL receptor antibodies.

[c] HDL "receptor activity" was judged to be present if HDL binding *in vivo* or *in vitro* was independent of apoE. Binding was judged to be independent of apoE if calcium was not required and/or cyclohexanedione modification of arginine residues in HDL did not abolish binding. Receptor insensitivity to proteolytic digestion was considered further evidence for the presence of the "HDL receptor."

[d] A, Adrenal; O, ovary. Includes responses *in vivo* and/or *in vitro* by granulosa cells and/or luteinized ovarian cells.

of steroidogenesis is unknown. Several lines of evidence now suggest that HDL binding is not required for HDL enhancement of steroidogenesis.

In Vivo Procedures

The above conclusions were derived from a combination of *in vivo* and *in vitro* studies. Several general principles underlie each of these approaches. In both cases the degree and nature of hormonal stimulation as well as the amount and type of lipoproteins to which the tissue is exposed must be defined. Since circulating lipoproteins undergo continuous intravascular modification, they should optimally be monitored throughout the experiment to draw accurate conclusions. Lipoproteins added to *in vitro* culture systems may also be modified and should also optimally be characterized both at the beginning and conclusion of the experiment. The potential for lipoprotein modification *in vitro* is highlighted by recent observations that the human adrenal secretes apoE.[12,13] Two approaches have been used *in vivo* to access the contribution of lipoprotein cholesterol to the steroid substrate pool. First, radiolabeled precursors, as well as native and modified lipids and lipoproteins, have been administered to animals under physiologic conditions or after drug-induced hypolipidemia and/or hormonal stimulation. The use of nondegradable markers,[14] including [³H]cholesterol ethers, [¹⁴C]sucrose-labeled apoB,[15] and [¹²⁵I]tyramine–cellobiose-labeled apoA-I[16] (see also chapter [36], this volume) has been particularly useful in quantitating lipoprotein and cholesterol uptake by steroidogenic glands. Second, unlabeled lipoproteins have been administered to animals pharmacologically rendered hypocholesterolemic. Andersen and Dietschy[17] have successfully applied this approach to quantitate rat adrenal uptake of HDL and LDL cholesterol. Spady and Dietschy have also estimated the contribution of extracellular cholesterol to the steroid substrate pool by substrating the rate of endogenous cholesterol synthesis estimated by incorporation of [³H]water from the rate of steroid hormone production.[18] Finally, note should be made of the use of the Watanabe hereditable hyperlipidemic rabbit[19,20] in estimating the contribution of LDL to steroid hormone production.

Ethinyl estradiol and 4-aminopyrazolopyrimidine have been used to lower circulating lipoprotein cholesterol levels (Table III). Brecher and Hyun[21] have shown that 4-APP treatment leads to depletion of adrenal

[12] L. A. Schuler, L. Scavo, T. M. Kirsch, G. L. Flukinger, and J. F. Strauss, III, *J. Biol. Chem.* **254,** 8662 (1979).

[13] M. L. Blue, D. L. Williams, S. Zucker, A. Khan, and C. B. Blum, *Proc. Natl. Acad. Sci. U.S.A.* **80,** 283 (1983).

[14] Y. Stein, Y. Dabach, G. Hollander, G. Halperin, and O. Stein, *Biochim. Biophys. Acta* **752,** 98 (1983).

[15] R. C. Pittman, A. D. Attie, T. E. Carew, and D. Steinberg, *Proc. Natl. Acad. Sci. U.S.A.* **76,** 5345 (1979).

[16] C. Glass, R. C. Pittman, M. Civen, and P. Steinberg, *J. Biol. Chem.* **260,** 744 (1985).

[17] J. M. Andersen and J. M. Dietschy, *J. Biol. Chem.* **256,** 7362 (1981).

[18] D. K. Spady and J. M. Dietschy, *J. Lipid Res.* **24,** 303 (1983).

[19] R. C. Pittman, T. E. Carew, A. D. Attie, J. L. Witztun, Y. Watanabe, and D. Steinberg, *J. Biol. Chem.* **257,** 7994 (1982).

[20] M. Herettinger, W. J. Schneider, Y. K. Ho, J. L. Goldstein, and M. S. Brown, *J. Clin. Invest.* **74,** 1017 (1984).

[21] P. I. Brecher and Y. Hyun, *Endocrinology* **102,** 1404 (1978).

TABLE III

PHARMACOLOGIC AGENTS USED *in Vivo* TO STUDY LIPOPROTEIN METABOLISM BY
STEROIDOGENIC TISSUE OF THE RAT

Agent	Dose	Reference
A. To lower circulating cholesterol		
4-Aminopyrazolopyrimidine	2 mg/100 g/day ip × 3	*a, b*
17-Ethinyl estradiol	20 mg/100 g/day	*c, d*
B. Adrenal stimulation		
Adrenocorticotropin (ACTH)	20 U in gel sc	*e*
Suppression		
Dexamethasone acetate	1 mg sc qd × 5	*f*
C. Ovary		
To stimulate granulosa cell proliferation		
PMSG	5 IU sc followed (60 hr) by 25 IU hCG (sacrifice 4 hr)	*g*
Diethylstilbestrol (DES)		*h*
To induce luteinization		
Pregnant mare's serum gonadotropin (PMSG)	50 IU sc followed (60 hr) by human chorionic gonadotropin (LCG) 25 IU sc (sacrifice 7 days)	*i*

a J. M. Andersen and J. M. Dietschy, *J. Biol. Chem.* **253,** 9024 (1978).
b P. I. Beecher and Y. Hyun, *Endocrinology* **102,** 1404 (1978).
c Y. Chan, F. F. Windler, G. C. Chen, and R. J. Hand, *J. Biol. Chem.* **254,** 11360 (1979).
d A. H. Veschoor-Klootwyk, L. Verschoor, S. Azhar, and G. M. Reaven, *J. Biol. Chem.* **257,** 7666 (1982).
e D. D. Mahaffee, R. Reitz, and R. L. Ney, *J. Biol. Chem.* **249,** 227 (1974).
f D. K. Spady, S. D. Turley, and J. M. Dietschy, *J. Clin. Invest.* **76,** 1113 (1985).
g J. P. Louvet, S. M. Harman, and G. T. Ross, *Endocrinology* **69,** 1179 (1975).
h M. F. Rosenblum, C. R. Huttler, and J. F. Strauss, III, *Endocrinology* **109,** 1518 (1981).
i M. F. Toaff, H. Schleyer, and J. F. Strauss, III, *Biochim. Biophys. Acta* **617,** 219 (1980).

cholesterol stores, diminished rates of adrenal steroidogenesis, and resultant enhancement of adrenal stimulation by endogenous ACTH. Administration of dexamethasone concurrently with 4-APP suppresses endogenous ACTH release and prevents depletion of adrenal cholesterol ester stores.[22] It is rarely possible to alter independently the hormone and lipoprotein milieu by a single pharmacologic agent. At high doses (>10 mg/100 g/day) 4-APP may also suppress pituitary gonadotropin release. Ethinyl estradiol also suppresses hepatic lipoprotein release in the rat and produces less toxicity than 4-APP. However, its use has obvious disad-

[22] K. Mikami, T. Nishikawa, Y. Saito, Y. Tamura, N. Matsuoka, A. Kumagai, and S. Yoshida, *Endocrinology* **114,** 136 (1984).

TABLE IV
Steroidogenic Cells Used for *in Vitro* Study of Lipoprotein Metabolism

Cell	Major steroid products	References
Adrenal cortex		
Mouse Y-1 tumor (continuous culture)	11β,20α-Dihydroxyprogesterone, 20α-hydroxyprogesterone	*a, b*
Bovine (primary culture)	11-Deoxycortisol, progesterone, 17α-hydroxyprogesterone, 20α-dihydroprogesterone, 17α-hydroxy-20α-dihydroprogesterone	*c–e*
Rat (suspension)	Corticosterone, 18-hydroxydeoxycorticosterone	*f–h*
Snell #494 transplantable rat tumor (suspensions)	Deoxycorticosterone	*i, j*
Human fetal (organ culture) (cell culture)	Cortisol, dehydroisoandrosterone, pregnenolone	*k, l* *m, n*
Ovary		
Granulosa cells		
Rat	Progesterone	*o, p*
Bovine		*q, r*
Porcine		*s, t*
Human		*u, v*
Luteinized rat ovary	Progesterone	*w, x*
Luteinized human granulosa (tissue)	Progesterone	*y*
Testes		
Rat leydig cell	Testosterone	*z, aa*
MA-10 mouse (continuous culture)	Progesterone	*bb*
Human choriocarcinoma cell	Progesterone	*cc*

a J. Koval, *Recent Prog. Horm. Res.* **26,** 623 (1970); *b*J. R. Faust, J. L. Goldstein, and M. S. Brown, *J. Biol. Chem.* **252,** 4861 (1977); *c*D. C. Gospodarowicz, III, P. J. Hornsby and G. N. Gill, *Endocrinology* **100,** 1080 (1977); *d*P. J. Hornsby and G. N. Gill, *Endocrinology* **102,** 926 (1978); *e*P. T. Kovanan, J. R. Faust, M. S. Brown, and J. L. Goldstein, *Endocrinology* **104,** 599 (1979); *f*M. J. O'Hare and A. M. Neville, *J. Endocrinol.* **56,** 529 (1973); *g*M. J. O'Hare and A. M. Neville, *J. Endocrinol.* **56,** 537 (1973); *h*J. T. Gwynne and B. Hess, *J. Biol. Chem.* **255,** 10875 (1980); *i*K. L. Snell and H. L. Stewart, *J. Natl. Cancer Inst.* **22,** 1119 (1979); *j*J. I. Mason and W. F. Robidoux, *Endocrinology* **105,** 1230 (1979); *k*E. R. Simpson, B. R. Carr, C. R. Parker, Jr., L. Milewich, J. C. Porter, and P. C. MacDonald, *J. Clin. Endocrinol. Metab.* **49,** 146 (1979); *l*B. R. Carr, C. R. Parker, Jr., L. Milewich, J. C. Porter, P. C. MacDonald, and E. R. Simpson, *Endocrinology* **106,** 1854 (1980); *m*K. Fujieda, C. Faiman, F. I. Reyes, and J. S. D. Winter, *J. Clin. Endocrinol. Metab.* **53,** 34 (1981); *n*K. Fujieda, C. Faiman, F. I. Reyes, H. Thlives, and J. S. D. Winter, *J. Clin. Endocrinol. Metab.* **53,** 401 (1981); *o*C. P. Channing and Ledruitz, this series, Vol. 39, p. 193; *p*L. A. Schuler, L. Scavo, T. M. Kursch, G. L. Flickinger, and J. F. Strauss, III, *J. Biol. Chem.* **254,** 8662 (1979); *q*N. Savion, J. Lui, R. Laherty, and D. Gospodarowicz, *Endocrinology* **109,** 409 (1981);

vantages in studies of ovary or testes as it effectively suppresses gonadotropin release.

Specific tropic hormones may be used to enhance steroid output and thus increase the demand for substrate cholesterol. Adrenal steroidogenesis can be stimulated *in vivo* by administration of ACTH and suppressed by hypophysectomy or administration of dexamethasone (Table III). In the rat, handling or adverse environmental conditions are powerful stimuli to endogenous ACTH release. Thus treatment with dexamethasone or the use of hypophysectomized rats may be necessary to obtain an adequate control for ACTH effects.

Regulation of ovarian steroidogenesis is somewhat more complex than in the adrenal. Both LH and FSH may stimulate pregnenolone production under appropriate conditions. Estradiol, prolactin, insulin, and other growth factors may also play a role. In the rat ovarian steroidogenesis can be stimulated and luteinization induced by sequential treatment with pregnant mare's serum gonadotropins (PMSG) and human chorionic gonadotropin (hCG) (Table III).

In Vitro Procedure

In vitro systems have also been employed to study lipoprotein metabolism in each of the steroidogenic tissues. Steroidogenic cells in which lipoprotein metabolism has been studied are listed in Table IV. Artifact may arise in such *in vitro* systems from loss of normal glandular architecture and/or changes in structure and function introduced during isolation. Both *in vivo* and *in vitro* approaches offer advantages for examining specific questions. However a clear picture of normal physiology requires a combination of both approaches. Important species differences in lipo-

[t]N. Savion, R. Laherty, G. Lui, and D. Gospodarowicz, *J. Biol. Chem.* **256,** 12817 (1981); [s]J. D. Veldhuis, P. A. Klase, J. F. Strauss, III, and J. M. Hammond, *Endocrinology* **111,** 144 (1982); [t]J. D. Veldhuis and J. T. Gwynne, *Endocrinology* **117,** 1067 (1985); [u]R. W. Tureck and J. F. Strauss, III, *J. Clin. Endocrinol. Metab.* **54,** 367 (1982); [v]R. W. Tureck, A. B. Welburn, J. T. Gwynne, L. G. Paavola, and J. F. Strauss, III, *J. Steroid Biochem.* **19,** 1033 (1983); [w]J. F. Strauss, III and J. L. Flickinger, *Endocrinology* **101,** 883 (1977); [x]M. E. Toaff, H. Schleyer, and J. F. Strauss, III, *Biochim. Biophys. Acta* **617,** 219 (1980); [y]B. R. Carr, R. K. Sadler, D. B. Rochelle, M. A. Stalmach, P. C. MacDonald, and E. R. Simpson, *J. Clin. Endocrinol. Metab.* **52,** 875 (1981); [z]E. H. Charreau, J. C. Calvo, K. Nozu, O. Pragnataro, K. J. Catt, and M. L. Dufau, *J. Biol. Chem.* **256,** 12719 (1981); [aa]M. Behamed, C. Dellamonica, F. Haour, and J. M. Saez, *Biochim. Biophys. Acta* **99,** 1123 (1981); [bb]D. A. Freeman and M. Ascoli, *J. Biol. Chem.* **257,** 14231 (1982); [cc]C. A. Winkel, J. M. Snyder, P. C. MacDonald, and E. R. Simpson, *Endocrinology* **106,** 1054 (1980).

protein metabolism have clearly been documented and care must be exercised in the use of heterologous systems and in accepting the generality of observations confined to only one species. Procedures used in our laboratory for preparation, maintenance, and study of rat adrenocortical cells are described below. Appropriate techniques for preparing ovarian tissues are also cited below. Additional references to studies of lipoprotein metabolism in each of the tissue types cited may be found in Table IV.

Preparation of Adrenocortical Cells

This method is a modification of the procedure initially described by O'Hare and Neville[23] for isolation of rat cells. We have applied the same procedure to adrenal tissue obtained from neonatal and adult swine as well as human adrenocortical fragments with equal success. Adrenal glands are excised and trimmed of surrounding fat. Unless the rats are handled quiescently or endogenous ACTH is suppressed, ACTH stimulation will occur at sacrifice. When cells are being prepared for long-term culture, excised glands are washed thoroughly with phosphate-buffered saline before decapsulation and handled sterilely thereafter. The glands are next decapsulated and minced. Rat adrenal glands may be easily decapsulated by first incising the fibrous capsule and then gently extruding the contents. This is easily accomplished by gently pinching the gland between two fingers, applying pressure starting at the pole opposite the site of capsular incision. Decapsulation affects a partial cell purification since the mineralocorticoid producing granulosa cells remain associated with the capsule. The capsular material may be subsequently digested to obtain a cell preparation enriched in granulosa cells. In the case of larger glands such as porcine or human, the capsular material may be dissected away. The decapsulated cortical tissue is then suspended in 1 ml of digestion medium and finely minced (1–2 mm). Mincing is best accomplished with crossed scalpel blades rather than scissors, which tend to crush the cells adjacent to the cut. Minced tissue fragments are then incubated (60 min, 37°) in digestion medium: MEM supplemented with 10% FCS and 2.5 mg/ml collagenase (type I, Worthington Biochemical Corporation) and 2.5 mg/ml hyaluronidase (type II, Sigma Chemical Company). Digestion medium can alternatively be made up with Ham F-12 or Ringers bicarbonate glucose buffer. Best results are obtained if the medium is supplemented with 10% fetal calf serum (10% FCS) or 1% albumin, possibly providing an alternative substrate for nonspecific proteases contaminating available preparations of collagenase. DNase may be used in place of

---

[23] M. J. O'Hare and A. M. Neville, *J. Endocrinol.* **56,** 529 (1973).

hyaluronidase to prevent cell aggregation which may be promoted by DNA released from damaged cells. We have obtained similar results with either enzyme. Following incubation tissue fragments are allowed to settle and the medium is decanted and fresh medium without digestive enzymes is added. The cells are then released from the tissue matrix by gently pipetting, approximately five times, with a small-bore pipet such as a Pasteur pipet. After allowing the large intact fragments to again settle, the dispersed cells are decanted. This process should be repeated three or four times until the remaining connective tissue has a whitish appearance. Dispersed cells are then recovered from the combined decanted supernatants by low-speed centrifugation (200 g for 10 min) and resuspended in the study medium. Reported ACTH sensitivity for steroid production varies among laboratories from 10^{-12} to 10^{-9} M ACTH required for half-maximal stimulation. ACTH sensitivity can be increased by maintaining cells in cultures for 24 hr.[24] The digestion process largely results in destruction of medullary cells. Nonetheless the remaining cortical cells are structurally and functionally heterogeneous. Two major cell types can be distinguished on the basis of size: a smaller cell comprising, in our hands, 60–70% of the total population, and a larger cell which constitutes the remainder. The total cell yield which ordinarily has a viability greater than 90–95% varies between 0.5 and 1.0 × 10^6 cells per gland. Pretreatment of the animals with ACTH or 4-APP produces a hypertrophied hyperemic gland which nonetheless can be handled as are glands from untreated animals. Greater contamination with erythrocytes occur under these circumstances. Partial purification of dispersed cells from erythrocytes and cell fragments can be achieved by centrifugation over a discontinuous BSA gradient. We have adapted and simplified the method of Miao and Black.[25] Adrenal cells are layered over a discontinuous BSA gradient consisting of 5 ml of 28% BSA overlayered with 10 ml of 4% BSA in a 50-ml Nalgene screw cap tube and the gradient is centrifuged 200 g for 10 min. A band of adrenal cells collects at the interface of the 4/28% gradient; erythrocytes sediment at the bottom and a band of cell fragments and debris collects at the top of the gradient. The band of adrenal cells are aspirated off, diluted with MEM, and sedimented by centrifugation at 200 g for 10 min. The cell pellets are then resuspended in MEM supplemented with 10% FCS or other study medium.

ACTH responsive cells can be maintained in culture for many days. Loosely pelleted cells are resuspended in MEM supplemented with 10% fetal calf serum and plated at a density of 3 to 6 × 10^5 cells/cm^2. At 24 hr

[24] A. S. Leotta and D. T. Krieger, *Endocr. Res. Commun.* **4,** 156 (1977).
[25] P. Miao and V. H. Black, *J. Cell Biol.* **94,** 241 (1982).

cell attachment ranges from 60 to 90%. Inclusion of ACTH during the first 24 hr of culture markedly decreases cell attachment. Utilization of extracellular lipoprotein cholesterol by cultured rat adrenocortical cells can be enhanced by incubating cells in lipoprotein-poor media in the presence of ACTH for 3 days.[26]

Preparation of Ovarian Cells

Two functionally different types of ovarian cell preparations have been employed in studies of lipoprotein metabolism: granulosa cells and dispersed luteinized ovarian cells. Granulosa cells offer the advantage of relative homogeneity compared to cell dispersions prepared from luteinized ovaries. In addition granulosa cells may be obtained without exposure to proteolytic enzymes. However, the two types of preparations differ functionally in steroid product and hormonal regulation. Granulosa cells have been prepared by similar techniques, modifications of the method of Channing and Ledwitz-Rigby[27] from rats, swine, cows, and woman (Table IV). References to methods for maintaining granulosa cells of each species in primary and/or continuous culture have been described and are also noted in Table IV. Proliferation of rat granulosa cells can be enhanced *in vivo* by implantation of diethyl stilbesterol in a slow-release form (Table III). In the rat luteinization can be induced by sequential administration of pregnant mare's serum gonadotropin (PMSG) and human chorionic gonadotropin (hCG) as initially described by Parlow[28] and subsequently adapted by Strauss and colleagues[29] (see Table III). These same investigators have also described methods for preparing dispersed rat luteinized ovarian cells and maintaining them in short-term culture.[30]

[26] J. T. Gwynne and B. Hess, *J. Biol. Chem.* **255,** 10875 (1980).
[27] C. P. Channing and Ledwitz-Rigby, this series, Vol. 39, p. 183.
[28] A. F. Parlow, *Fed. Proc., Fed. Am. Soc. Exp. Biol.* **17,** 402 (1958).
[29] L. A. Schuler, L. Seavo, T. M. Kirsch, J. L. Flickinger, and J. F. Strauss, III, *J. Biol. Chem.* **254,** 8662 (1979).
[30] M. E. Toaff, H. Schleyer, and J. F. Strauss, III, *Biochim. Biophys. Acta* **617,** 219 (1980).

[41] Preparation, Characterization, and Measurement of Lipoprotein Lipase

By PER-HENRIK IVERIUS and ANN-MARGARET ÖSTLUND-LINDQVIST

Introduction

Lipoprotein lipase (EC 3.1.1.3) is secreted by a variety of cell types and plays a key role in the metabolism of the triglyceride-rich lipoproteins, namely, chylomicrons and very-low-density lipoproteins. At its extracellular location on the capillary endothelium in various tissues, the enzyme hydrolyzes the triglyceride component of these particles to glycerol and free fatty acids. The latter are thereby made available for uptake into cells where they are either used as fuel or stored.[1]

The enzyme has attracted considerable clinical interest since various causes of hypertriglyceridemia have been associated with impairment of its function.[2] More recently, lipoprotein lipase (LPL) in adipose tissue has been implicated as a major regulator of fat cell size.[3]

After the introduction of affinity chromatography on heparin–agarose,[4] several purification procedures adopting that technique have appeared and made it possible to purify the enzyme to homogeneity as well as to investigate its chemical and functional properties in more detail.[5-10] This chapter describes protocols for purification of the enzyme from bovine milk[6] and human postheparin plasma.[8] It also presents a versatile assay procedure that can be adopted for tissue extracts, column effluents, cell cultures, postheparin plasma samples, and adipose tissue biopsy specimens.

[1] P. Nilsson-Ehle, A. S. Garfinkel, and M. C. Schotz, *Annu. Rev. Biochem.* **49,** 667 (1980).

[2] J. D. Brunzell, A. Chait, and E. L. Bierman, *Metabolism* **27,** 1109 (1978).

[3] J. D. Brunzell, and M. R. C. Greenwood, in "Biochemical Pharmacology of Metabolic Disease, Vol. I Obesity" (P. B. Curtis-Prior, ed.), p. 175. Elsevier, Amsterdam, 1983.

[4] T. Olivecrona, T. Egelrud, P.-H. Iverius, and U. Lindahl, *Biochem. Biophys. Res. Commun.* **43,** 524 (1971).

[5] T. Egelrud and T. Olivecrona, *J. Biol. Chem.* **247,** 6212 (1972).

[6] P. H. Iverius and A.-M. Östlund-Lindqvist, *J. Biol. Chem.* **251,** 7791 (1976).

[7] P. K. J. Kinnunen, *Med. Biol.* **55,** 187 (1977).

[8] A.-M. Östlund-Lindqvist, *Biochem. J.* **179,** 555 (1979).

[9] L. Wallinder, G. Bengtsson, and T. Olivecrona, *Biochim. Biophys. Acta* **711,** 107 (1982).

[10] S. M. Parkin, B. K. Speake, and D. S. Robinson, *Biochem. J.* **207,** 485 (1982).

Assay

LPL requires the presence of a specific protein cofactor, also called activator, to hydrolyze a triglyceride emulsion at an optimal rate. This factor, apolipoprotein C-II,[11,12] is a normal constituent of the natural substrates (chylomicrons and very low-density lipoproteins). Most published assays for the enzyme consist of a triglyceride emulsion, serum as a source of apolipoprotein C-II, and albumin as an acceptor for free fatty acids.[13] Assays mainly differ with regard to the type of emulsifier used, the relative concentrations of various assay components, and the method used to quantitate the reaction products. Methods using commercial triglyceride emulsions, e.g., Intralipid (Vitrum AB, Sweden), are highly reproducible but less sensitive than those employing radioactive substrates. Larger variability of the radioactive methods is introduced when a new emulsion is prepared for each assay.

Maximal enzyme activity requires a sufficient concentration of triglyceride substrate as well as an optimal concentration of activator.[14] The assay described herein[15] is designed by that principle using a serum activator and radioactive triolein emulsified with lecithin.

Stock Reagents

Phosphatidylcholine (egg), 50 mg/ml: Supplied as a chloroform solution by Serdary Research Lab., London, Ontario (Cat. No. A-31), and stored at $-20°$.

Triolein, 200 mg/ml: Supplied by Sigma Chemical Co., St. Louis, MO (Cat. No. T7502), dissolved in benzene and stored at $-20°$.

Radioactive triolein, 1.25 μCi/ml: Supplied by Amersham, Arlington Heights, IL, as glycerol-tri[1-^{14}C]oleate, 100 μCi/ml (30–60 mCi/mmol), in toluene. Aliquots of 125 μCi are evaporated under a stream of nitrogen, redissolved in heptane (3 ml), and further purified by extraction with 7 ml of 0.05 M NaOH in 50% (v/v) ethanol. The purified material is finally evaporated under nitrogen, dissolved in 100 ml of heptane and stored at $-20°$.

Radioactive oleic acid, 0.5 μCi/ml: Supplied by Amersham (see above) as [1-^{14}C]oleic acid, 100 μCi/ml (>50 mCi/mmol) in toluene. An

[11] J. C. LaRosa, R. I. Levy, R. Herbert, S. E. Lux, and D. S. Fredrickson, *Biochem. Biophys. Res. Commun.* **41**, 57 (1970).

[12] R. J. Havel, C. J. Fielding, T. Olivecrona, V. G. Shore, P. E. Fielding, and T. Egelrud, *Biochemistry* **12**, 1828 (1973).

[13] S. E. Riley and D. S. Robinson, *Biochim. Biophys. Acta* **369**, 371 (1974).

[14] A. M. Östlund-Lindqvist and P. H. Iverius, *Biochem. Biophys. Res. Commun.* **65**, 1447 (1975).

[15] P.-H. Iverius and J. D. Brunzell, *Am. J. Physiol.* **249**, E107 (1985).

aliquot (50 μCi) is evaporated under nitrogen with 50 μmol (14.124 mg) of cold oleic acid, diluted to 100 ml with heptane and stored at $-20°$.

0.223 M Tris-HCl buffer, pH 8.5: Trizma-8.5 (29.289 g), supplied by Sigma Chemical Co., St. Louis, MO, is diluted with water to 1 liter and stored at 4°. At 37° the pH is 8.2.

0.78 M NaCl in Tris buffer (salt correction): 4.558 g NaCl is diluted to 100 ml with Tris buffer (see above) and stored at 4°.

Methanol–chloroform–heptane, 1.41:1.25:1 (v/v/v): Methanol (141 ml) is mixed with 125 ml of chloroform, and 100 ml of heptane.

0.05 M Carbonate–borate buffer, pH 10.5: Anhydrous K_2CO_3 (6.91 g) and boric acid (3.092 g) are dissolved in water, adjusted to pH 10.5 with 5 M KOH, and diluted to 1 liter. Fresh solution is made weekly and stored at 4°.

Albumin, 183.3 mg/ml in Tris buffer: Bovine serum albumin (45.825 g) supplied by Sigma (Cat. No. A4503) is dissolved in Tris buffer (see above), diluted to 250 ml, and stored at $-20°$ in 10-ml aliquots.

Heparin, 1 mg/ml, in Tris or Krebs–Ringer phosphate (KRP) buffer: Sodium heparin (50 mg) supplied by Sigma (Cat. No. H3125) is dissolved in either Tris buffer (see above) or KRP buffer (see below) to a final volume of 50 ml and stored at $-20°$ in 2-ml aliquots.

Pooled human serum. Blood (400 ml per donor) is drawn in 50-ml plastic syringes with 2.5 ml of disodium EDTA (20 mg/ml). The blood from each donor is centrifuged in separate containers, the plasma collected and centrifuged again before pooling. Solid $CaCl_2$ is added to 0.1% (w/v) followed by bovine thrombin to 1 unit/ml (Topical Thrombin, Parke-Davis, Morris Plains, NJ), and the plasma is incubated at 37° for 30 min. After the serum has been recovered from the clot by centrifugation, it is dialyzed against 0.15 M NaCl and heat inactivated at 56° for 30 min in 100-ml glass bottles. Finally, the serum is dialyzed against KRP buffer (see below) and stored in 10-ml aliquots at $-20°$.

Detergent solution, 2 mg/ml sodium deoxycholate–0.08 mg/ml Nonidet P40–0.05 mg/ml heparin–10 mg/ml BSA–0.25 M sucrose in Tris buffer: Sodium deoxycholate (500 mg) and 21.394 g of sucrose are added to 2 ml of Nonidet P40 (1 g/100 ml), 12.5 ml of heparin (1 mg/ml), and 13.7 ml of albumin. Tris buffer (see above) is used for all the component solutions as well as for dilution to 250 ml after warming to 37°. The solution is stored in 5-ml aliquots at $-20°$.

Skim milk standard. Unpasteurized bovine milk is obtained fresh from a diary farm and the cream removed after centrifugation at 4°. The skim milk (500 ml) is stirred for 30 min at 4° with solid trisodium citrate dihydrate (14.7 g) added to 0.1 M, and dialyzed against 0.15 M NaCl–5 mM sodium phosphate buffer (pH 7.4) for 3 × 6 hr with a 10-fold excess of buffer. After the dialysis, glycerol is added to 30% (v/v) and the standard is frozen in 4-ml aliquots with ethanol–dry ice and stored at −70°.

Fresh Reagents

Krebs–Ringer phosphate (KRP) buffer, pH 7.4:
NaCl (45.0 g/liter), 100 ml
KCl (57.4 g/liter), 5 ml
$CaCl_2 \cdot 2H_2O$ (81.4 g/liter), 1.5 ml
$MgSO_4 \cdot 7H_2O$ (191.0 g/liter), 1.0 ml
Distilled water, 432 ml
$NaH_2PO_4 \cdot H_2O$ (13.80 g/liter), 15 ml
$Na_2HPO_4 \cdot$ anhydr (14.196 g/liter), 85 ml
Elution buffer, 25% (v/v) serum–0.05 mg/ml heparin:
Serum, 5 vol
Heparin (1 mg/ml in KRP buffer), 1 vol
KRP buffer, 14 vol
Diluted serum, 25% (v/v):
Serum, 1 vol
KRP buffer, 3 vol

Triglyceride emulsion. Aliquots of phosphatidylcholine (0.12 ml; 6 mg), triolein (0.25 ml; 50 mg), and radioactive triolein (2 ml; 2.5 μCi) are transferred to a 2.2 × 7 cm flat-bottomed glass vial and the solvents evaporated under a stream of nitrogen. After the addition of Tris buffer (1.95 ml), salt correction (1 ml), and albumin (3 ml), the mixture is cooled on ice and sonicated at 50 W with 10 repeated 10-sec bursts interrupted by 10-sec pauses using a $\frac{1}{2}$-in. diameter probe with a flat tip. The emulsion is kept at 0° and used within 2 hr. Each batch of 6 ml is enough for 10 assay incubations. When several batches are required, they are pooled before use.

Sample Preparation

Elution of Adipose Tissue. Weighed adipose tissue (~50 mg), obtained by needle aspiration and cut into pieces of 5–7 mg size, is transferred to 12 × 75 mm glass tubes containing 0.3 ml of elution buffer. The tubes are covered with Parafilm and incubated in a shaking water bath (80 cycles/min) at 37° for exactly 30 min. At the end of the incubation, an aliquot (0.2 ml) of the buffer is removed and assayed for more enzyme activity (see below).

Detergent Extraction of Adipose Tissue. Stock detergent solution is thawed at 37° to ensure complete dissolution of its components. Approximately 50 mg of weighed tissue is homogenized at room temperature with

0.2 ml of detergent solution using an all-glass tissue grinder (Duall, size 20, Kontes, Berkeley, CA). After the addition of another 0.3 ml of detergent solution followed by brief mixing, the homogenate is immediately transferred to centrifuge tubes and spun at 4° and 12,000 g for 15 min. An aliquot (0.2 ml) of the infranatant is assayed for enzyme activity (see below).

Postheparin Plasma. Postheparin blood is collected in Li-heparinate Vacutainer tubes (Becton-Dickinson, Rutherford, NJ), the plasma recovered by centrifugation, frozen with dry ice–ethanol, and stored at −70°. Before analysis, the plasma is thawed rapidly in cold water and an aliquot (0.05 ml) diluted with 0.2 ml of pooled serum, 0.05 ml of heparin (1 mg/ml in KRP buffer), and 0.2 ml of KRP buffer. An aliquot (0.15 ml) of the above dilution is added to an equal volume of inhibiting antiserum or antibody to LPL, appropriately diluted with KRP buffer. A control sample is similarly prepared using diluted nonimmune serum or IgG. The mixtures are incubated at 4° for 2 hr before an aliquot (0.2 ml) is assayed for lipase activity.

Cell Culture Dishes.[16] Culture medium is assayed without further dilution. The cell layer of a 35-mm culture dish is rinsed twice with 1 ml of saline. After careful removal of the second rinse, 0.5 ml of detergent solution is added and the cells are lysed by scraping with a Teflon policeman. The extract is then homogenized and centrifuged as described above for adipose tissue.

Skim Milk Standard. Frozen standard is rapidly thawed in cold water and an aliquot (0.1 ml) gently added to the bottom of a test tube with 2 ml of ice-cold Tris buffer. Immediately before taking an aliquot (0.2 ml) for assay, the standard is mixed with the buffer.

Miscellaneous Enzyme Sources. When needed, a sample is appropriately diluted with ice-cold Tris buffer either to fit the assay range or to reduce the effect of interfering substances, e.g., salt.

Procedure

Assay Mixture. The complete assay mixture (pH 8.2 at 37° and ionic strength 0.16) contains 0.178 M Tris-HCl buffer (ionic strength 0.05), 0.11 M NaCl, 55 mg/ml of albumin, 0.01 mg/ml of heparin, 5% (v/v) of serum, and 5 mg/ml of glycerol-tri[1-^{14}C]oleate (0.05 μCi/mg) emulsified with 0.6 mg/ml of lecithin.

Blank incubations consist of 0.6 ml of triglyceride emulsion, 0.2 ml of Tris buffer, and 0.2 ml of elution buffer.

[16] A. Chait, P.-H. Iverius, and J. D. Brunzell, *J. Clin. Invest.* **69,** 490 (1982).

Aliquots (0.2 ml) of adipose tissue eluates or diluted postheparin plasma samples (see above) are mixed with 0.6 ml of triglyceride emulsion and 0.2 ml of Tris buffer.

Aliquots (0.2 ml) of samples in Tris buffer, i.e., detergent extracts, skim milk standard, or miscellaneous samples are mixed with 0.6 ml of triglyceride emulsion and 0.2 ml of diluted serum.

Incubation. Test tubes containing incomplete assay mixture (0.8 ml) are kept on ice until enzyme (0.2 ml) is added, then covered with Parafilm, and incubated in a 37° shaking water bath (80 cycles/min) for 60 min. Incubations are terminated by the extraction of free fatty acids for counting of radioactivity.

Fatty Acid Extraction.[17] Test tubes (13 × 100 mm) containing 3.25 ml of methanol–chloroform–heptane are prepared in advance. Four aliquots (0.2 ml) from each incubation mixture are transferred to extraction tubes which immediately are agitated vigorously. Carbonate–borate buffer (1.05 ml) is added, the tubes are stoppered and agitated again. Phase separation is accomplished by centrifugation at room temperature for 20 min using a swingout rotor. An aliquot (2 ml) of the aqueous upper phase is then transferred to a scintillation vial containing 4 ml of scintillation liquid (Aquasol, New England Nuclear, Boston, MA) and counted for 10 min in a liquid scintillation spectrometer.

Specific Radioactivity of Triolein. Radioactive triolein solution (20 μl) is transferred in duplicate to empty scintillation vials and the solvent evaporated under a stream of nitrogen. Prior to counting of radioactivity, scintillation liquid (4 ml) and the aqueous phase (2 ml) from a blank extraction are added.

Extraction Recovery. Radioactive oleic acid solution (0.2 ml) is evaporated in two test tubes as well as in two scintillation vials under a stream of nitrogen. The test tubes are processed as blank incubations (see above) which are followed by fatty acid extraction and scintillation counting. The scintillation vials receiving tracer are counted after the addition of scintillation liquid (4 ml) and the aqueous phase from a blank extraction (2 ml).

Comments. Blank incubations are run in triplicate with one replicate used in the determination of triolein specific radioactivity and extraction recovery. It is essential that hydrolysis products of the radioactive triolein are removed by the procedure described above in order to ensure low background radioactivity of the blank.

The procedures used to recover enzyme, e.g., elution of enzyme or detergent extraction of enzyme from adipose tissue, have more variability than the proper assay. Therefore, it is recommended that these proce-

[17] P. Belfrage and M. Vaughan, *J. Lipid Res.* **10**, 341 (1969).

dures are carried out in triplicate to allow computation of the standard error.

Computations

Blank Correction. The gross radioactivity (cpm) of blanks, standards, and unknown samples is computed as the mean of all extractions from a particular sample. Net radioactivity, which is obtained by subtracting the mean of the blanks, is used in all further computations.

Extraction recovery (R): $\qquad R = 5E/F$

where R is the fraction of oleic acid that distributes to 2 ml of the aqueous phase in the extraction tube. F represents the total oleic acid radioactivity (cpm) added to a separate scintillation vial or to a blank incubation (1 ml) and E is the part of that radioactivity extracted from a 0.2-ml aliquot into 2 ml of aqueous phase.

Specific radioactivity (S): $\qquad S = \dfrac{T \times 10 \times 885.43}{(5 \times 10^{-3})(3 \times 10^9)} = 5.903 \times 10^{-4}T$

T is the radioactivity of triolein that was directly added to a scintillation vial and $T \times 10$ represents the radioactivity in 1 ml of complete assay mixture. The remaining figures convert the triolein (M_r 885.43), present at 5 mg/ml, into fatty acid equivalents (nanomoles). Thus the unit for S is cpm/nmol.

Enzyme activity (A): $\qquad A = \dfrac{C}{R \times S \times 0.2 \times 0.2 \times 60} = \dfrac{C}{R \times S \times 2.4}$

The enzyme activity (A), expressed as nanomoles of fatty acid released per minute and per milliliter sample, assumes that the latter was added to the assay mixture as a 0.2-ml aliquot. The parameters R and S are defined above. C is the mean net radioactivity for a particular sample.

Interassay variation is controlled for by multiplying apparent enzyme activity by a correction factor (f):

$$f = A_0/A_1$$

A_1 is the apparent activity of the skim milk standard on a particular assay occasion and A_0 an arbitrarily chosen reference value for the standard.

Postheparin plasma lipolytic activity is obtained by multiplying the apparent activity by 20 (dilution factor). Activity remaining after the incubation with inhibiting antibody to LPL represents hepatic lipase.[18] The

[18] J. K. Huttunen, C. Ehnholm, P. K. J. Kinnunen, and E. A. Nikkilä, *Clin. Chim. Acta* **63**, 335 (1975).

LPL activity is then obtained by subtraction. Calculation of lipase activity in adipose tissue and other sources needs no further explanation.

General Comments

The assay is linear for sample enzyme activities up to 200 nmol/ min · ml. Using a skim milk standard, the interassay coefficient of variation is 5.2% whereas the intraassay variation is only 1.2%. Detergent extraction of adipose tissue reflects total tissue enzyme and yields higher activity than any other method to recover the enzyme. Therefore this procedure should be particularly useful when it is important to distinguish a low activity from zero activity. Elution of adipose tissue with serum and heparin at 37° recovers less than 50% of the activity in a detergent extract. However, the eluted activity is more sensitive to physiological perturbations and is probably also a better indicator of the physiologically active enzyme fraction. Some applications of this assay to human adipose tissue and postheparin plasma will appear elsewhere.[15]

Purification Procedures

Reagents

Heparin-agarose: Heparin is supplied by Sigma (see above). Heparin with low affinity for anti-thrombin is fractionated on anti-thrombin–Sepharose as described by Nordenman and Björk.[19] Conventional heparin and low-affinity heparin are covalently linked to Sepharose CL-6B as described.[6,20,21] After use, the gels are regenerated by alternating washes with 10 bed volumes of 0.1% Triton X-100 in 2 M NaCl and 10 vol of water

C_γ-*aluminum hydroxide gel:* Supplied by Sigma (Cat. No. A8628)
Bovine milk LPL[6]

Skim milk: Bovine milk is obtained fresh from a cow that has been screened beforehand for high enzyme activity. Within 1 hr the milk is chilled to 0° and then centrifuged. The fat cake is pierced with Tygon tubing and the skim milk siphoned off. After another centrifugation to remove traces of fat, the milk (3 liters) is adjusted to 0.1 M with solid trisodium citrate dihydrate (88.2 g)

Heparin–Agarose Chromatography. Conventional heparin–agarose gel (400 ml) is added to the citrate-treated skim milk (3 liters) and kept at

---

[19] B. Nordenman and I. Björk, *Biochemistry* **17,** 3339 (1978).
[20] P.-H. Iverius, *Biochem. J.* **124,** 677 (1971).
[21] A.-M. Östlund-Lindqvist and J. Boberg, *FEBS Lett.* **83,** 231 (1977).

0° for 30 min under frequent stirring. The gel is separated from the skim milk on a sintered glass filter, washed three times with 400 ml of ice-cold 0.5 M NaCl–30% (v/v) glycerol–0.01 M phosphate (pH 7.5), resuspended in the buffer, and packed into a column (7 × 10.5 cm). The column, kept at 4° and run at 174 ml/hr, is washed with one bed volume of buffer and then eluted with a 3-liter linear gradient of 0.5 to 1.5 M NaCl in 30% (v/v) glycerol–0.01 M phosphate, pH 7.5 (Fig. 1).

Adsorption to C_γ Gel. The pooled enzyme fractions (Fig. 1) are diluted with an equal volume of 30% (v/v) glycerol–0.01 M phosphate (pH 7.5) and 20 ml of sedimented C_γ gel in water are added. The suspension is stirred at 4° for 30 min, after which the gel is collected by centrifugation. The gel is washed twice by suspending in 250 ml of 0.5 M NaCl–30% (v/v) glycerol–0.01 M phosphate (pH 7.5) and then centrifuging. The suspension should be accomplished gently with a glass rod or spatula, since more efficient tools like a tissue homogenizer may destroy the enzyme activity. The enzyme is finally eluted from the gel by suspending and centrifuging four times in 50 ml of 1.2 M NaCl–30% (v/v) glycerol–0.01 M phosphate (pH 7.5). The combined eluates (200 ml) are spun at 4° and 12,000 g for 20 min to remove traces of the gel, and thereafter dialyzed for 20 hr at 4° against 3.6 M ammonium sulfate–0.01 M phosphate (pH 6.5). After centrifugation at 4° and 30,000 g for 15 min using one centrifuge tube repeatedly, the white precipitate is dissolved in the buffer for the final fractionation step (see below).

Intervent Dilution Chromatography. The enzyme obtained above after ammonium sulfate precipitation, is dissolved in 3 ml of 0.15 M NaCl–15%

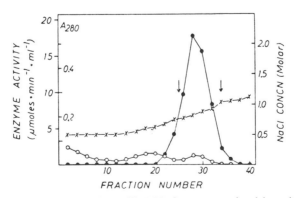

FIG. 1. Elution pattern of bovine milk LPL from conventional heparin–agarose. The fractions (58 ml) were analyzed for enzyme activity (●) by a previously described assay, absorbance at 280 nm (○), and sodium chloride concentration (×) by measuring conductivity. The 10 fractions between the arrows were pooled. The graph is taken from P.-H. Iverius and A.-M. Östlund-Lindqvist, *J. Biol. Chem.* **251**, 7791 (1976).

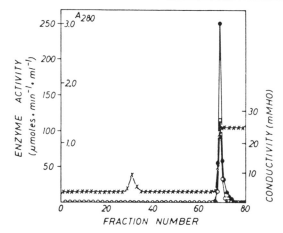

FIG. 2. Intervent dilution chromatography of partially purified bovine milk LPL on conventional heparin–agarose. The fractions (3.5 ml) were analyzed for enzyme activity (●), absorbance at 280 nm (○), and conductivity (×). The three fractions containing the highest enzyme activities were pooled. The graph is taken from P.-H. Iverius and A.-M. Östlund-Lindqvist, *J. Biol. Chem.* **251**, 7791 (1976).

(v/v) glycerol–0.01 *M* phosphate (pH 7.5) and applied to a 1.5 × 85 cm column of conventional heparin–agarose. After washing with one bed volume (150 ml) of the above buffer, the column is eluted with 1.5 *M* NaCl–15% (v/v) glycerol–0.01 *M* phosphate (pH 7.5) at a rate of 14 ml/hr. The enzyme, which emerges as a sharp peak at the front of the elution buffer (Fig. 2), is dialyzed against 3.6 *M* ammonium sulfate–0.01 *M* phosphate (pH 6.5) for 20 hr and centrifuged at 4° and 30,000 *g* for 15 min. The precipitate, which at this stage has a gelatinous appearance, is dissolved in 1 ml of 50% (v/v) glycerol–0.01 *M* phosphate (pH 7.5) and stored at −20°.

Comments. This procedure produces 1.5–2 mg of enzyme with a recovery of 10–13%. The purity, checked by sodium dodecyl sulfate (SDS)–polyacrylamide electrophoresis, may vary between different preparations but is usually better than 95%. Most of the impurities, which almost always have a faster mobility than the enzyme, do react with an affinity-purified antibody to LPL after Western blotting, indicating that they are proteolytic fragments of the enzyme (unpublished results). A recent report confirms this finding and advocates the addition of 1 m*M* phenylmethanesulfonyl fluoride to the milk in order to inhibit proteolysis.[22] Gel electrophoretic methods other than those employing SDS have been un-

---

[22] L. Socorro and R. L. Jackson, *J. Biol. Chem.* **260**, 6324 (1985).

successful for testing of purity, since aggregation of the enzyme prevents it from entering the gel.

The enzyme produced by this particular procedure has not been exposed to detergents and is obtained in the native form at a high concentration. Storage in 50% glycerol at $-20°$ as a liquid solution is extremely convenient, and such preparations have been kept for more than 6 years without substantial loss of activity.

Human Postheparin Plasma LPL[8,21]

Postheparin Plasma. Volunteers are injected with heparin intravenously (100 U/kg). After 20 min, plasma is obtained either by plasmapheresis or as described for the initial steps to make pooled human serum (see "Assay"). The plasma (1000 ml) is immediately chilled to $0°$, mixed with solid NaCl (23.38 g) to a concentration of 0.4 M, and 120 ml of conventional heparin–agarose. After stirring at $0°$ for 30 min, the gel is washed six times with 200 ml of 0.4 M NaCl–30% (v/v) glycerol–0.01 M phosphate (pH 7.5) and packed into a column (2.5 × 25 cm). The column, kept at $4°$ and run at 60 ml/hr, is washed with another 3 bed volumes of the buffer and then eluted with a 700-ml linear gradient of 0.4 to 1.5 M NaCl in 30% (v/v) (glycerol–0.01 M phosphate, pH 7.5 (Fig. 3). The peak fractions of LPL are pooled (200 ml), dialyzed against 3.6 M ammonium sulfate–0.01 M phosphate (pH 6.5), and the precipitate is dissolved in 1 ml of 50% (v/v) glycerol–0.01 M phosphate (pH 7.5) before storing at $-20°$.

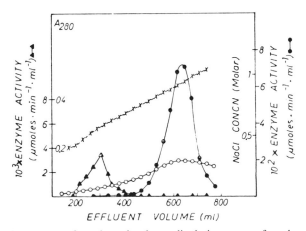

FIG. 3. Elution pattern of postheparin plasma lipolytic enzymes from heparin–agarose. The fractions (5 ml) were analyzed for LPL activity (●) and hepatic lipase activity (▲) by previously described assays, absorbance at 280 nm (○), and sodium chloride concentration (×) by measuring conductivity. The graph is taken from A.-M. Östlund-Lindqvist and J. Boberg, *FEBS Lett.* **83**, 231 (1977).

Affinity Chromatography on Heparin–Agarose with Low Affinity for Anti-Thrombin. Four partially purified LPL preparations, obtained as described above by chromatography on conventional heparin–agarose, are pooled, adjusted to 0.5 M NaCl, and applied to a heparin–agarose column (1.6 × 5 cm) with low affinity for anti-thrombin. The column is washed with 160 ml of 0.5 M NaCl–30% (v/v) glycerol–0.01 M phosphate (pH 7.5) and then eluted with 1.2 M NaCl in the same buffer (Fig. 4).

Adsorption to C_γ Gel. The pooled enzyme fractions (Fig. 4) are diluted with an equal volume of 30% (v/v) glycerol–0.01 M phosphate (pH 7.5) and adsorbed to 2 ml of sedimented C_γ gel. Washing of the gel and elution is carried out as described above under "Bovine Milk LPL" except that the procedure is carried out in a 10-fold smaller scale. The enzyme fraction is finally dialyzed against ammonium sulfate, the precipitate dissolved in 1 ml of 50% (v/v) glycerol–0.01 M phosphate (pH 7.5) and stored at −20°.

Comments. The above procedure is designed for the isolation of both hepatic lipase and LPL from postheparin plasma. By contrast to milk, plasma contains significant amounts of anti-thrombin, which binds to heparin and coelutes with LPL. The purification step on heparin–agarose with low affinity for anti-thrombin is therefore added to remove this protein. Although the resulting lipase fraction (0.1 mg) is free from anti-thrombin, as tested by immunodiffusion, it shows more than one band on SDS electrophoresis and has a specific activity which is approximately half of that of pure bovine enzyme. Nevertheless, this fraction has been

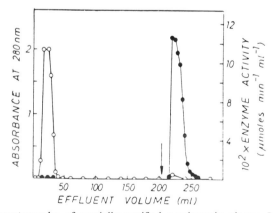

FIG. 4. Chromatography of partially purified postheparin plasma LPL on heparin–agarose with low affinity for anti-thrombin. The fractions (4 ml) were analyzed for LPL activity (●) and absorbance at 280 nm (○). The three fractions of highest enzyme activity were pooled. The graph is taken from A.-M. Östlund-Lindqvist and J. Boberg, *FEBS Lett.* **83**, 231 (1977).

uscful for further purification by preparative SDS electrophoresis. The homogeneous protein resulting from this step has been employed for antibody production and amino acid analysis.[8]

Properties of Purified LPL

There are striking similarities between LPLs from different species. For instance, the human,[21] the bovine,[4] and the rat enzyme[10] all bind to heparin–agarose and elute at a similar salt concentration, are inhibited when assayed in the presence of 1 M NaCl,[23,24] and are all prone to inactivation in the purified state. Furthermore, the human serum activator (apolipoprotein C-II) can also activate the bovine and the rat enzyme, and antibodies raised against the bovine enzyme may cross-react with human, rat, and mouse enzymes.[25] Although not yet substantiated by amino acid sequences, these data suggest that LPL has been well conserved during mammalian evolution. Bovine LPL, which so far is the best characterized enzyme, may therefore serve as a model for lipases from other species.

The bovine lipase is a glycoprotein containing 8.3% carbohydrate including mannose, galactose, glucose, N-acetylglucosamine, and sialic acid. The amino acid composition, which has been published elsewhere,[6] is very similar to that of the human enzyme.[8] The extinction coefficient ($E_{1\,cm}^{1\%}$) is 14.0 at 280 nm.[6]

The molecular weight determined by sedimentation equilibrium ultracentrifugation in guanidine under reducing as well as nonreducing conditions is 48,300.[6] Higher values recorded by SDS–polyacrylamide electrophoresis are probably a result of the carbohydrate moiety causing anomalous behavior of the protein in this method.[26] The native enzyme has a sedimentation coefficient ($S_{20,w}^{0}$) of 5.40 S and a diffusion coefficient ($D_{20,w}^{0}$) of 48.8 $\mu m^2/S$. Using the Svedberg equation, the molecular weight of the native enzyme is 96,900. Thus the native enzyme is a dimer held together by noncovalent interactions.[6] The functional significance of this finding is unclear.

The purified enzyme is notoriously labile and must be handled with care. This may, in large part, be due to a strong tendency to aggregate. Certain maneuvers with the native enzyme, such as concentration by ultrafiltration, electrophoresis under nondenaturing conditions, and sedimentation equilibrium ultracentrifugation, are therefore precluded. How-

[23] P.-H. Iverius, U. Lindahl, T. Egelrud, and T. Olivecrona, *J. Biol. Chem.* **247,** 6610 (1972).
[24] E. D. Korn, *J. Biol. Chem.* **215,** 1 (1955).
[25] T. Olivecrona and G. Bengtsson, *Biochim. Biophys. Acta* **752,** 38 (1983).
[26] J. P. Segrest and R. L. Jackson, this series, Vol. 28B, p. 54.

ever, aggregation and inactivation can be at least partially prevented in the presence of high concentrations of glycerol,[6,27] glycine,[6] and protein.[27] Furthermore, the enzyme is also stabilized by heparin[23] and detergents such as Triton X-100,[7] Nonidet P-40,[6] and sodium deoxycholate.

Acknowledgments

This work was supported by grants from the American Diabetes Association, the National Institutes of Health (AM 02456), and the Swedish Medical Research Council (19X 07193).

[27] J.-S. Twu, A. S. Garfinkel, and M. C. Schotz, *Atherosclerosis* **24**, 119 (1976).

[42] Assays of the *in Vitro* Metabolism of Very-Low-Density Lipoproteins in Whole Plasma by Purified Lipoprotein Lipase

By BYUNG HONG CHUNG and JERE P. SEGREST

Introduction

Very-low-density lipoproteins (VLDL) are the major transport vesicles of triglycerides of hepatic origin, endogenous lipid. VLDL are the spherical particles consisting of triglyceride (60–70% of total mass) and small amounts of cholesteryl ester in the core, and phospholipid, free cholesterol, and apolipoproteins on the surface monolayer film.[1]

The metabolism of VLDL, *in vivo,* occurs on the capillary wall through the activity of lipoprotein lipase.[2] Lipoprotein lipase (triacylglycerol lipase, E.C. 3.1.1.3) is located on the endothelial surface of several extrahepatic tissues[3,4] and is released into the blood stream by heparin.[5] Lipoprotein lipase is also present in high concentration in the bovine raw milk of many mammals.[6]

[1] L. E. Smith, H. J. Pownall, and A. M. Gotto, *Annu. Rev. Biochem.* **47**, 751 (1978).
[2] D. W. Bilheimer, S. Eisenberg, and R. I. Levy, *Biochim. Biophys. Acta* **260**, 212 (1972).
[3] C. F. Chung, G. M. Ousta, A. Bensadoun, and R. D. Rosenberg, *J. Biol. Chem.* **256**, 12893 (1981).
[4] K. Shimada, P. J. Gill, J. E. Silbert, W. H. J. Douglas, and B. L. Fanburg, *J. Clin. Invest.* **68**, 995 (1981).
[5] E. D. Korn, *J. Biol. Chem.* **215**, 1 (1955).
[6] T. Egelrud and T. Olivercrona, *J. Biol. Chem.* **247**, 6212 (1972).

VLDL exposed to lipoprotein lipase lose their triglyceride in the core and become smaller and denser low-density lipoprotein (LDL)-like particles.[7] As the VLDL core shrinks, the excess surface constituents of VLDL are then dissociated from the particles and are largely associated with high-density lipoproteins (HDL).[8,9] The free fatty acids released during lipolysis of VLDL are largely associated with the albumin fractions of plasma.[10,11]

The assays of *in vitro* metabolism of VLDL by lipoprotein lipase thus can be achieved by measuring the extent of hydrolysis of the triglyceride moiety of VLDL as well as by measuring the density conversion of VLDL components into the other density fraction following lipolysis.

In the following section, we describe a method for assaying the *in vitro* metabolism of VLDL in whole plasma by utilizing purified bovine milk lipoprotein lipase.

Principle

carboxyl-^{14}C-labeled triolein is incorporated into the VLDL *in vitro* method, and the VLDL containing [^{14}C]triolein is then reconstituted with the VLDL infranatant plasma fraction. Following incubation of reconstituted plasma with purified bovine milk lipoprotein lipase, the extent of hydrolysis of VLDL–triglyceride is determined by measuring the amount of ^{14}C-labeled free fatty acids (oleic acid) produced. Free ^{14}C-labeled oleic acids produced by hydrolysis are then extracted and adsorbed onto an ion exchange resin. Following washout of residual triolein from the resin, the adsorbed free fatty acids are then displaced from the resin by strong base, and the ^{14}C level is assayed by liquid scintillation counting.

Alternatively, ^{14}C-labeled oleic acids are separated from unhydrolyzed esters into the upper phase of a liquid–liquid partition system, and the ^{14}C radioactivities in the upper phase are determined.

The assays for the transfer of VLDL components or conversion of VLDL into the other density fraction following lipolysis are achieved by the separation of pre- and postlipolyzed plasma into the major lipoprotein density fractions by the short single vertical spin density gradient ultra-

[7] R. J. Dekelbaum, S. Eisenberg, M. Fainaru, Y. Barenholz, and T. Olivercrona, *J. Biol. Chem.* **254**, 6079 (1979).

[8] J. C. LaRosa, R. I. Levy, W. V. Brown, and D. S. Fredrickson, *Am. J. Physiol.* **220**, 785 (1971).

[9] T. ChaJeck and S. Eisenberg, *J. Clin. Invest.* **64**, 162 (1976).

[10] K. A. Mitropoulos, M. D. Avery, M. B. Myant, and G. F. Gibbons, *Biochem. J.* **130**, 363 (1972).

[11] I. Bjorkhemand and J. Gustatsson, *Eur. J. Biochem.* **40**, 667 (1973).

centrifugation method with subsequent analysis of lipid and apolipoprotein moieties of VLDL in each density fraction.

Materials

Biological samples
  Normolipidemic plasma
  Bovine raw milk

Reagents

[^{14}C]Triolein (1-C^{14}), 45–60 mCi/mmol
  (New England Nuclear Corp)
^{14}C-Labeled oleic acids (1-C^{14}), 40–60 mCi/mmol
  (New England Nuclear Corp.)
Unlabeled triolein and oleic acid (Sigma Co.)
Heparin–Sepharose (Pharmacia Co.)
Amberlite resin, IRA-400 in RN(CH$_3$)$_3$Cl form, 16–50 mesh (Sigma Co.)
Protosol (0.5 M quarternary ammonium hydroxide)
  (New England Nuclear Corp.)
LKB-Wallac Internal Standard (LKB Instrument)
Enzymatic cholesterol assay kit (Biodynamics/BMC)
Enzymatic triglyceride assay kit (Biodynamics/BMC)
Fatty acid extraction mixtures
  Isopropyl alcohol:heptane: 1 N sulfuric acid (40:10:1, v/v/v)
    (Dole's mixture)
  Methanol:chloroform:heptane (1.45:1.25:1, v/v/v)
Tris buffer, 0.01 M Tris–0.15 M NaCl, pH 7.4
Tetramethyl urea (Sigma Co.)
^{125}I-labeled Na (New England Nuclear Corp.)
Scintillation cocktail solution
Potassium carbonate borate buffer, 0.1 M, pH 10.5

Equipment

Liquid scintillation counter
Gamma counter
Ultracentrifuge
Vertical rotors (Beckman VTi 80, and Sorvall TV 850)
Gradient fractionator (Isco Co.)
Peristaltic pump (Holter Co.)
Spectrophotometer (Hitachi Co.)
Technicon Autoanalyzer II consisting of proportioning pump, calorimeter, incubator and recorder (Technican Co.)
Tube slicer (Beckman Co.)

Preparation of Enzyme and Substrate

Lipoprotein lipase is isolated from bovine raw milk and is purified by the heparin–Sepharose affinity chromatographic method of Iverius *et al.*[12] Reconstituted plasma containing a known amount of VLDL is used as a substrate for purified bovine milk lipoprotein lipase and is prepared as follows[13]: Fresh normolipidemic plasma obtained from subjects following an overnight fast is placed in a fixed angle ultracentrifuge rotor (Beckman 50.2 Ti) and centrifuged at 49,000 rpm for 18 hr at 7° in a Beckman L8-80 ultracentrifuge. The VLDL floating at the top of the tube are collected by slicing the tube with a tube slicer. The VLDL and VLDL infranatant plasma fractions are then dialyzed overnight against Tris buffer (0.01 M Tris, 0.15 M NaCl, pH 7.4) at 7°. Reconstitution of the plasma is then carried out by mixing the isolated VLDL and VLDL infranatant plasma fractions to give a VLDL–triglyceride level of 150–200 mg/dl. The reconstituted plasma, thus prepared, contains a minimum of 90% of the original plasma constituents. Reconstituted plasma containing [^{125}I]- or [^{14}C]triolein-labeled VLDL is prepared by including a trace amount of the radiolabeled VLDL during the reconstitution processes. ^{125}I-Labeled VLDL is prepared by iodinating the VLDL with ^{125}I-labeled Na using the iodine monochloride method of McFarane.[14] VLDL labeled with [^{14}C]triolein is prepared *in vitro* by a procedure similar to that of Thomas *et al.*[15] for the preparation of plasma radiolabeled with cholesteryl ester. Typically, 20 μCi of [^{14}C]triolein in benzene is placed in a glass tube, and the interior surface of tube is uniformly coated with [^{14}C]triolein by evaporating the solvent under reduced pressure in a Brinkman rotary evaporation. Any residual solvent is then removed by placing the tube in a lyophilizer overnight. Two-milliliter portions of isolated VLDL (4–5 mg VLDL triglyceride) and 1 ml of a lipoprotein-free plasma fraction ($d > 1.230$/ml) are then added to the [^{14}C]triolein-coated tube, and the tube is incubated at room temperature for 4 hr. After the 4-hr incubation VLDL are separated from the free protein fraction by the short single vertical density gradient ultracentrifugation method.[16] By using this procedure, an average of 40–50% of the [^{14}C]triolein is transferred from the wall of the tube into the medium, and more than 98% of the radioactivity in the medium is associated with the VLDL fraction.

Preparation of Dehydrated Hydroxyl-Charged Ion Exchange Resin. Amberlite IRA-400 resin is hydroxylated and dehydrated by the method

[12] P. H. Iverius and A. M. Östland-Lindqvist, *J. Biol. Chem.* **251,** 7791 (1976).
[13] B. H. Chung, J. T. Cone, and J. P. Segrest, *J. Biol. Chem.* **257,** 7472 (1982).
[14] A. S. McFarane, *Nature (London)* **182,** 53 (1958).
[15] M. S. Thomas and L. L. Rudel, *Anal. Biochem.* **130,** 215 (1988).
[16] B. H. Chung, T. Williamson, J. C. Geer, and J. P. Segrest, *J. Lipid Res.* **21,** 284 (1980).

described by Baginsky.[17] Amberlite resin (500 g) is hydroxylated by suspending the resin in 2 liters of 2.5 M NaOH. The mixture is allowed to equilibrate for 4 hr with occasional shaking. The resin is then transferred to a glass funnel which has a coarse-fritted glass filter. The resin is washed with distilled water until the effluent becomes neutral to pH paper, and then the resin is washed with 2 liters of isopropyl alcohol. Finally, the resin is washed with n-heptane until the odor of isopropyl alcohol is no longer detectable. The dehydrated resin is stored in heptane at 4° in a dark, tightly capped container.

Assay Procedures

Enzyme Reactions. A 3-ml aliquot of reconstituted plasma containing trace amounts of [^{14}C]triolein-labeled VLDL is placed in a screw-capped tube, and 0.15 ml of purified bovine milk lipoprotein lipase (specific activity 10–20 mU/μl). The tube is then gassed with a gentle stream of nitrogen gas and incubated at 37° in a thermodynamically controlled water bath.

Determination of Free Fatty Acids Liberated. A 0.1-ml aliquot of sample (in triplicate) is withdrawn and placed in a screw-capped tube following 0, 5, 15, 30, and 60 min of incubation, and the enzyme reaction is then terminated by adding 5 ml of Dole's lipid extraction mixture,[18] isopropyl alcohol–heptane–0.1 N sulfuric acid (40 : 10 : 1, v/v/v). The tubes are shaken vigorously, and 10 min later heptane containing 0.4 μmol of unlabeled oleic acid and 2.9 ml of water are added. The tubes are then shaken for 15 min using a mechanical shaker. The tubes are then placed in an upright position until phase separation of the liquid has been achieved.

It should be noted that temperature has some effect on the relative volume of this two-phase system; thus, all of the above procedures should be carried out at room temperature (24°). A 3-ml portion of the upper phase, which is three-fourths of the total volume, is transferred to a 20-ml glass scintillation vial containing 1 g of hydroxylated Amberlite resin in 1 ml of heptane containing 0.05 μmol of cold triolein. The vials are placed in a vial rack and the rack is mounted firmly on the variable speed rotator and swirled for 10 min. The resin is allowed to settle out, and then the heptane phase is aspirated into a radioactive trap.

The resin is then washed five times with 5-ml aliquots of heptane. After the last heptane wash, the absorbed free fatty acids are released from the resin by treatment with a strong base as described by Khoo *et al.*[19] A 1.5-ml aliquot of 0.5 molar quarternary ammonium hydroxide (Pro-

[17] M. L. Baginski, this series, Vol. 72, p. 325.
[18] V. P. Dole, *J. Clin. Invest.* **35,** 150 (1956).
[19] J. C. Khoo and D. Steinberg, this series, Vol. 25, p. 23.

tosol) is added to each vial, and the vials are incubated in a 65° water bath for 15 min to allow the release of fatty acids from the resin. The levels of displaced ^{14}C-labeled free fatty acids are then assayed by counting the vials following addition of 15 ml of scintillation fluid. In order to determine the recovery of free fatty acids, standard ^{14}C-labeled free fatty acid bound to albumin is run through the entire procedure, i.e., extraction, adsorption, and elution. The resultant recovery value is used to calculate the percentage of recovery. Counting efficiency is estimated by counting LKB internal standard capsules in the presence of resin and Protosol. A typical recovery value for the above assay is approximately 70%, and the counting efficiency ranges from 65–70%. A blank assay tube containing reconstituted plasma without enzyme is run in order to correct for spontaneous lipolysis by residual lipase in the plasma; the blank value is usually less than 2% of the total radioactivity in the sample.

Alternatively, the extent of hydolysis of [^{14}C]triolein-labeled VLDL can be estimated by the simple method of utilizing the partitioning of free fatty acids from unhydrolyzed triacylglycerol described by Nilsson-Ehle and Schotz,[20] which is a modification of the method of Belfraze and Vaughn.[21] Since a high level of protein in the plasma sample interferes with the partitioning of free fatty acids, a 0.05-ml aliquot of sample rather than a 0.1-ml aliquot taken at 0, 5, 15, 30, and 60 min of incubation is diluted to 0.2 ml with Tris buffer, and the enzyme reaction is terminated by adding 3.25 ml of the methanol . chloroform . heptane (1.41 : 1.25 : 1, v/v/v) mixture followed by 1.05 ml of 0.1 M potassium carbonate buffer, pH 10.5. The tubes are shaken vigorously in a vortex mixer for 2 min and centrifuged at 24° for 15 min. Since the partition coefficient of potassium oleate is temperature dependent, the extraction, centrifugation, and all other procedures should be done at room temperature (24°). A 1-ml aliquot of the methanol–water upper phase (total upper phase 2.45 ml) is taken for scintillation counting in 15 ml of scintillation fluid.

Recovery is estimated by determing the level of [^{14}C]oleic acid in the upper methanol–water phase after the standard has been treated in the same manner as the lipolyzed sample. Average recovery in the upper phase is 75% at room temperature. Counting efficiency is determined by using the internal standard method described earlier; counting efficiency usually ranges from 65–75% but should be checked in each assay.

A typical hydrolysis curve of [^{14}C]triolein-labeled VLDL in reconstituted plasma during incubation with purified bovine milk lipoprotein lipase is given in Fig. 1.

[20] P. Nilsson-Ehle and M. E. Schotz, *J. Lipid Res.* **17**, 536 (1976).
[21] P. Belfrage and M. Vaughan, *J. Lipid Res.* **10**, 341 (1969).

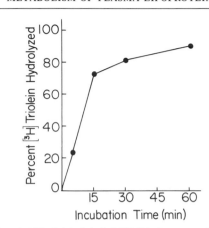

FIG. 1. Hydrolysis of [³H]triolein-labeled VLDL in reconstituted plasma by purified bovine milk lipoprotein lipase.

Kinetic Measurement of the Changes in Plasma Lipoprotein Cholesterol Profile during Lipolysis. For this assay, 6 ml of reconstituted plasma is placed in a capped tube, and 300 μl of purified lipoprotein lipase is added. The tube is then gassed with a gentle stream of nitrogen gas and incubated at 37°. A 1.5-ml aliquot of lipolysis mixture is withdrawn at 0, 15, 30, and 60 min of incubation, and the enzyme reaction is stopped by adjusting the sample density to $d=1.21$ g/ml with desiccated solid KBr (0.49 g). The lipoprotein cholesterol profiles of density-adjusted plasma samples are determined by the vertical autoprofiler method[22] for lipoprotein analysis. This method involves the separation of plasma into the major lipoprotein density fractions by short single vertical spin density gradient ultracentrifugation[16] and continuous on-line analysis of the cholesterol level in the effluent of the gradient sample. Short single vertical spin density gradient ultracentrifugation is performed according to the previously reported procedure[16] with a slight modification. A 1.4-ml portion of the density-adjusted plasma sample ($d=1.21$ g/ml) is pipetted into a 5-ml quick seal ultracentrifuge tube (Beckman) and 3.6 ml of 0.15 M NaCl containing 0.1% EDTA, pH 7.4, is carefully layered over the sample by using a Holter pump. The tubes are immediately placed in a Beckman vertical rotor and centrifuged at 80,000 rpm with a slow acceleration and breaking mode activated to cut off at an ω^2 value of 1.57×10^{11}. This ω^2 value is equivalent to a time setting for 30 min. At the end of the centrifugation, each centrifuge tube containing separated lipoproteins is placed on

[22] B. H. Chung, J. P. Segrest, J. T. Cone, J. Pfau, J. C. Geer, and L. A. Duncan, *J. Lipid Res.* **22,** 1003 (1981).

a gradient fractionator, and the tube is punctured at the bottom. The sample effluent is removed downward by a TAII pump, and the stream is split into two portions by using a Y-shaped connector. Twenty percent of each sample is continually mixed with the enzymatic cholesterol reagent and incubated, and the absorbance of the resulting color is measured in 1.5-mm flow cells of a TAII colorimiter at 505 nm and plotted on a recorder. A typical lipoprotein cholesterol profile obtained employing this procedure is shown in Fig. 2. A similar lipoprotein cholesterol profile can be obtained by manual fractionation of the sample into 25–30 fractions and subsequent assay of the level of cholesterol in each density fraction.

Assay for the Transfer of Other VLDL Constituents into Other Lipoprotein Density Fractions. The transfer of protein, phospholipid, free and esterified cholesterol, and free fatty acid moieties of VLDL liberated during the lipolysis of VLDL to other density fractions can be assayed by the fractionation of pre- and postlipolyzed plasma samples containing [^{125}I]- or [^{14}C]triolein-labeled VLDL into the various density fractions by the short single vertical spin density gradient ultracentrifugation method and subsequent measurement of each VLDL component in the density fractions. Since the assays of each VLDL constituent require a larger sample size, the fractionation of the sample is done in a larger vertical rotor (TV 850).

In a typical assay, 10 ml of reconstituted plasma containing trace amounts of [^{125}I]- or [^{14}C]triolein-labeled VLDL is incubated with 0.5 ml of purified bovine milk lipoprotein lipase (specific activity 10–20 nm/μl) for 60 min at 37°. The enzyme reaction is stopped by adjusting sample density to $d = 1.30$ g/ml with desiccated solid KBr.

A 10-ml portion of the density-adjusted control or lipolyzed sample is pipetted into a 37-ml polyallomer ultracentrifuge tube (Sorvall Co.), and 24 ml of 0.15 M NaCl containing 0.1% EDTA, pH 7.4, is layered over the density-adjusted sample by using a Holter pump. The tubes are then immediately placed in a Sorvall vertical rotor (TV 850) and centrifuged at 50,000 rpm for 150 min. After centrifugation, the tubes are removed from the rotor, the bottoms are punctured, and 27–35 fractions are collected from each tube using a gradient fractionator. The total ^{125}I radioactivity in a 200-μl aliquot of each fraction is counted in a Packard Autogamma counter. Radioactivities associated with the apoB or apoC moieties of apoVLDL are estimated by employing the procedure of Kane *et al.*[23] This procedure consists of adding 100 μl of cold VLDL (2 mg protein/ml) as a carrier to each previously counted 200-μl sample. Tetramethyl urea (300 μl) is then added to each tubes, and the tubes are incubated at 37° for 1 hr

[23] J. P. Kane, T. Sata, R. L. Hamilton, and R. J. Havel, *J. Clin. Invest.* **56**, 1622 (1975).

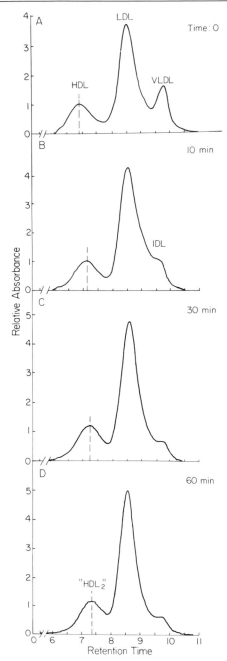

Fig. 2. Kinetics of *in vitro* lipolysis of VLDL in normolipidemic VLDL by purified bovine milk lipoprotein lipase.

and centrifuged at 2000 rpm for 20 min at room temperature. The radioactivity in the clear tetramethyl urea solution (apoC) and in the tetramethyl urea-insoluble precipitates (apoB) is then determined by using an auto-gamma counter. In plasma containing ^{3}H-labeled VLDL the radioactivity in a 100-μl aliquot of each fraction is determined in a Packard liquid scintillation counter after adding 10 ml of liquid scintillation fluid to each vial. The levels of total cholesterol and free cholesterol are determined by an enzymatic assay method.[24,25] Lipids in a 200-μl aliquot of each fraction are extracted with 4 ml of chloroform : methanol (2 : 1, v/v), and the level of phospholipid in the lipid extract is determined by the colorimetric method of Steward.[26] The distribution of ^{125}I-labeled apolipoproteins, ^{3}H-labeled triacylglycerol or free fatty acids, free and esterified cholesterol, and phospholipid in the density gradient fraction of pre- and postlipolyzed plasma obtained by above procedure is given in Figs. 3 and 4.

The net movement of free and esterified cholesterol and phospholipid moieties of VLDL into other density fractions following lipolysis is summarized in the table.

Comment

The *in vitro* lipolytic conversion of VLDL components into the LDL and HDL density fractions in reconstituted normolipidemic plasma can be clearly demonstrated by employing the methods described in this chapter. Although the lipolysis of VLDL and chylomicrons has been shown to occur in a plasma-free incubation system or in systems containing only trace amounts of plasma,[27–30] we have found that the transformation of VLDL into the LDL density fraction or the transfer of VLDL components into the HDL density region cannot be clearly demonstrated unless other plasma fractions, especially HDL, are present. The VLDL infranatant fractions in most normolipidemic plasma, which contain a normal level of HDL (>40 mg% HDL cholesterol), are capable of supporting the lipolytic conversion of VLDL into the LDL density fraction, but the VLDL infranatant fractions from many hypertriglyceridemic plasmas, which contain a low level of HDL (<25 mg% HDL cholesterol), fail to

[24] C. C. Allain, L. S. Poon, C. S. G. Chan, W. Richmond, and P. C. Fu, *Clin. Chem.* **20,** 470 (1974).
[25] G. Bucolo and H. David, *Clin. Chem.* **19,** 476 (1973).
[26] J. C. M. Stewart, *Anal. Biochem.* **104,** 10 (1980).
[27] C. J. Fielding, *Biochim. Biophys. Acta* **205,** 109 (1970).
[28] E. J. Blanchette-Mackie and R. O. Scow, *J. Lipid Res.* **17,** 57 (1976).
[29] R. M. Krauss, R. I. Levy, and D. S. Fredrickson, *J. Clin. Invest.* **54,** 1107 (1974).
[30] J. K. Huttunen and E. A. Nikkila, *Eur. J. Clin. Invest.* **3,** 483 (1973).

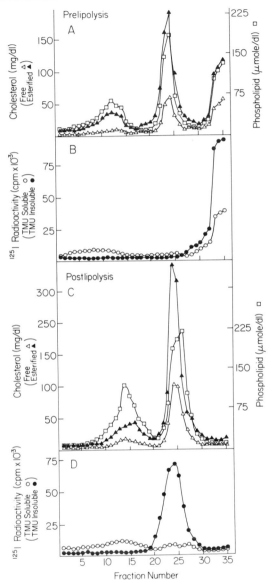

FIG. 3. Changes in the distribution patterns of free and esterified cholesterol, phospholipid, and radioactivity of ^{125}I-labeled VLDL in the density gradient fractions of reconstituted plasma following *in vitro* lipolysis by purified bovine milk lipoprotein lipase of plasma.

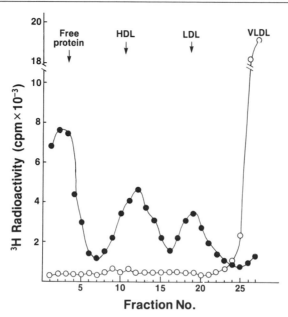

FIG. 4. Distribution of lipolytic products of [³H]triolein-labeled VLDL in the density gradient fractions of pre- and postlipolysis plasma.

MOVEMENT OF FREE AND ESTERIFIED CHOLESTEROL AND PHOSPHOLIPID MOIETIES OF VLDL TO THE OTHER LIPOPROTEIN FRACTIONS FOLLOWING *in Vitro* LIPOLYSIS OF RECONSTITUTED NORMOLIPIDEMIC PLASMA

Lipid moiety	Percentage recovery after lipolysis	Percentage removed		Percentage VLDL and IDL lipid transferred to other lipoprotein fraction[a]	
		VLDL (fraction 32–35)	IDL (fraction 28–31)	LDL (fraction 19–27)	HDL (fraction 1–18)
Free cholesterol	91.0	98.3	71.7	73.1	26.9
Cholesteryl ester	102.6	86.0	26.6	100	−1.4
Phospholipid	109.0	97.5	66.4	57.8	42.2

[a] Percentage transfer was calculated after adjustment to 100% recovery.

support the full lipolytic conversion of VLDL into the LDL density fraction.[31] The VLDL from some hypertriglyceridemic plasma exhibited resistance to lipolytic conversion into the LDL density fraction due to an abnormal enrichment of the core of VLDL with cholesteryl ester, although the hydrolysis of VLDL–triglyceride occurred normally. Thus, before selecting the plasma for this study, the level of HDL as well as the core lipid composition of VLDL should be determined. Since the enrichment of the core of normolipidemic VLDL with cholesteryl ester could occur through the action of core lipid transfer proteins in the free protein fraction of reconstituted plasma, and this process could limit the lipolytic conversion of VLDL into the LDL density fraction, the reconstituted plasma should be kept in a refrigerator to minimize any core lipid transfer reactions or the plasma should be lipolyzed immediately after reconstitution. Since it has been reported that some hypertriglyceridemic plasma contains inhibitors of lipoprotein lipase[32–33] the plasma should be tested for the presence of any inhibitor of lipoprotein lipase before using it in these assays.

[31] B. H. Chung, J. T. Cone, and J. P. Segrest, *J. Biol. Chem.* **257,** 7422 (1982).
[32] J. D. Bruzell, J. J. Alber, A. Chait, S. M. Grundy, E. Groszele, and G. B. McDonald, *J. Lipid Res.* **24,** 147 (1983).
[33] G. A. Crawford, J. F. Mahony, and J. H. Stewart, *Clin. Sci.* **60,** 73 (1981).

[43] Preparation, Characterization, and Measurement of Hepatic Lipase

By CHRISTIAN EHNHOLM and TIMO KUUSI

Hahn[1] originally observed that intravenous injection of heparin into hyperlipidemic dogs caused a rapid clearing of the turbidity of their plasma. It was soon recognized that this clearing was due to the appearance in the bloodstream of lipolytic enzymes released from the tissue sites by heparin.[2] It is at present thought that these enzymes are bound *in situ* to heparin like receptor molecules, presumably to the cell surface glycosaminoglycans. Heparin thus competes with these receptors for the lipase and displaces it from its natural site into the bloodstream.

The "clearing factor" in postheparin plasma was assumed to be identi-

[1] P. F. Hahn, *Science* **98,** 19 (1943).
[2] R. K. Brown, E. Boyle, and C. B. Anfinsen, *J. Biol. Chem.* **204,** 423 (1953).

cal with lipoprotein lipase (LPL), an enzyme found in extrahepatic tissues, i.e., adipose tissue and muscle.[3] The criteria, established by Korn[4] for identification of LPL activity, were as follows: (1) activation by a serum cofactor, (2) alkaline pH optimum, and (3) inhibition by 1 M NaCl and by protamine. Postheparin lipolytic activity, PHLA, was intensively studied by several investigators, and a key role for it in the clearance of plasma triglyceride-rich lipoproteins was suggested. However, no consistent correlation could be found between the serum triglyceride level and PHLA.

Lipase activity was also found in liver preparations, but this did not meet all the criteria set forth for LPL. In addition, the lipase present in liver acetone powders, in contrast to LPL, was more active against artificial fat emulsions than against chylomicrons.[4] More recently, rat liver homogenates have been shown to contain at least three different lipases, viz. an acid lipase with a pH optimum at pH 5.0, an alkaline one with a pH optimum at 8.6, and a heparin-activated lipase with a pH optimum at 7.5. The last lipase is found in the liver membrane fraction.[5] This plasma membrane-associated lipase activity can be released into solution by heparin. It has the same apparent K_m value and the same pH optimum as the lipase present in heparin-containing liver perfusates and can be purified by chromatography on Sepharose containing covalently linked heparin (heparin–Sepharose).[6] However, this lipase activity has characteristics which distinguish it from LPL. Thus, full activity is retained in the presence of 1 M NaCl, the lipase activity is resistant to protamine,[7,8] and the absence of serum in the enzyme assay does not result in decreased activity.[9]

Studies on the lipolytic activity in postheparin plasma indicated that this activity resembled more hepatic than extrahepatic triglyceride lipase activity.[8] Affinity chromatography of postheparin plasma on heparin–Sepharose columns enabled the separation of two different lipases from human postheparin plasma.[10] The first lipase elutes with 0.6–0.8 M and

[3] E. D. Korn, *J. Biol. Chem.* **215,** 15 (1955).

[4] E. D. Korn, *J. Biol. Chem.* **204,** 1 (1955).

[5] W. Guder, L. Weiss, and O. Wieland, *Biochim. Biophys. Acta* **187,** 173 (1969).

[6] G. Assman, R. M. Krauss, D. S. Fredrickson, and R. I. Levy, *J. Biol. Chem.* **248,** 1992 (1973).

[7] J. C. LaRosa, R. I. Levy, H. G. Windmuller, and D. S. Fredrickson, *J. Lipid Res.* **13,** 356 (1972).

[8] R. M. Krauss, R. I. Levy, and D. S. Fredrickson, *J. Clin. Invest.* **54,** 1107 (1974).

[9] H. Greten, B. Walter, and W. V. Brown, *FEBS Lett.* **27,** 306 (1972).

[10] C. Ehnholm, W. Shaw, H. Greten, W. Lengfelder, and W. V. Brown, *in* "Atherosclerosis III" (G. Schettler and A. Waizel, eds.), p. 557. Springer-Verlag, Berlin and New York, 1974.

the second one with 1.2–1.5 M NaCl.[11] The first lipase is not inhibited by high ionic strength, it does not require a serum cofactor for full activity, and it is absent from postheparin plasma of hepatectomized animals.[11–13] It is therefore called hepatic lipase (HL) or salt-resistant lipase, and it is believed to correspond to the lipase activity found in heparin perfusates of liver. The lipase activity eluting with 1.5 M NaCl has the characteristics of LPL.

Hepatectomy experiments implicated liver as the main source of HL in postheparin plasma[14] and the enzyme could be localized, by immunoelectron microscopy, to liver sinusoidal endothelial cells.[15] For some time HL was thought to be derived exclusively from the liver. In 1980, however, Jansen and co-workers demonstrated the presence of HL in adrenals.[16] Subsequently, the enzyme has also been found in ovarian tissue.[17] Despite its wider distribution, the lipase will be referred to as hepatic lipase, HL, throughout this chapter.

Purification of Hepatic Lipase

Hepatic lipase has been purified from various sources, postheparin plasma,[9,10,13,18,19] liver perfusates,[20] and liver homogenates.[21] For obvious reasons, human postheparin plasma is the only readily available source of human hepatic lipase.

Purification of Hepatic Lipase from Postheparin Plasma

Procedures for the purification of HL from several species, i.e., human, pig, rat, and dog PHLA, have been reported and they all are based on one or several of the following characteristics: (1) affinity for the lipid

[11] C. Ehnholm, A. Bensadoun, and W. V. Brown, *J. Clin. Invest.* **52,** 26 (1973).
[12] H. Greten, A. D. Sniderman, J. G. Chandler, D. Steinberg, and W. V. Brown, *FEBS Lett.* **42,** 157 (1974).
[13] C. Ehnholm, A. Bensadoun, and W. V. Brown, *Biochem. J.* **163,** 347 (1977).
[14] C. Ehnholm, T. Schröder, T. Kuusi, B. Bång, P. Kinnunen, K. Kahma, and M. Lempinen, *Biochim. Biophys. Acta* **617,** 141 (1980).
[15] T. Kuusi, E. A. Nikkilä, I. Virtanen, and P. K. J. Kinnunen, *Biochem. J.* **181,** 245 (1979).
[16] H. Jansen, C. Kalkman, J. C. Birkenhäger, and W. C. Hulsmann, *FEBS Lett.* **112,** 30 (1980).
[17] H. Jansen and W. J. de Greef, *Biochem. J.* **196,** 739 (1981).
[18] C. Ehnholm, W. Shaw, H. Greten, and W. V. Brown, *J. Biol. Chem.* **250,** 6756 (1975).
[19] A-M. Östlund-Lindqvist and J. Boberg, *FEBS Lett.* **83,** 231 (1977).
[20] T. Kuusi, T. Schröder, B. Bång, M. Lempinen, and C. Ehnholm, *Biochim. Biophys. Acta* **573,** 443 (1979).
[21] J.-S. Twu, A. S. Garfinkel, and M. C. Schotz, *Biochim. Biophys. Acta* **792,** 330 (1984).

substrate; (2) affinity for heparin; (3) affinity for calcium–phosphate gels; and (4) affinity for lectins.

The purification procedure described here is a combination of several recently described methods.

Assay Methods

The HL catalyzes hydrolysis of tri-, di-, and monoacylglycerol,[22] acyl-CoA thioesters,[23] and phospholipids.[18] To follow enzyme purification we routinely use an assay in which triglyceride lipase activity is recorded with a synthetic [14C]triolein substrate emulsified in the presence of 5% (w/v) gum arabic. However, other of the lipids mentioned above can be used as substrate.

Triacylglycerol Lipase Assay[24]

Materials

Conical glass tubes, 50-ml Pyrex No. 8062

Reagents

Tri-[1-14C]oleylglycerol, The Radiochemical Centre, Amersham, England
Triolein, Sigma No. T-7502
Gum Arabic, Sigma No. G 9752
Albumine bovine, fraction V (BSA), Sigma A-4503
Toluene, heptane, chloroform, and methanol, reagent grade
Trishydroxymethylaminoethane, Sigma 7-9 No. T-1378
Scintillation fluid, ACS, Amersham No. 196 290

Solutions

Nonradioactive triolein in toluene, 20 mg/ml; this solution can be stored well sealed under nitrogen at −20°
0.2 M Tris-HCl buffer, pH 8.4
Gum arabic, 5 g/100 ml, dissolved in 0.2 M Tris-HCl, pH 8.4; the solution is filtered through glasswool and stored frozen in aliquots. The solution should not be refrozen

[22] T. Kuusi, E. A. Nikkilä, M-R. Taskinen, P. Somerharju, and C. Ehnholm, *Clin. Chim. Acta* **122**, 39 (1982).
[23] H. Jansen, M. C. Oerlemans, and W. C. Hulsmann, *Biochem. Biophys. Res. Commun.* **77**, 861 (1977).
[24] J. K. Huttunen, C. Ehnholm, P. K. J. Kinnunen, and E. A. Nikkilä, *Clin. Chim. Acta* **63**, 335 (1975).

10% (w/v) BSA in 0.2 M Tris-HCl pH 8.4; this should be stored in aliquots at $-20°$

Methanol : chloroform : heptane (1.41 : 1.25 : 1.00, v/v/v)

0.14 M K_2CO_3–0.14 M H_3BO_3, pH 10.5

Preparation of Substrate

The triolein emulsion used as substrate must be prepared under strictly standardized conditions. The radioactive tri-[1-^{14}C]oleylglycerol is diluted in toluene at 2 μCi/ml; 5 ml of this solution and 5 ml of nonradioactive triolein (20 mg/ml) are transferred to a 50-ml conical tube and the solvent is evaporated under nitrogen. The triolein is mixed three times with 3 ml of heptane, and the solvent is evaporated under nitrogen between washes. The tubes can be stored under nitrogen at $-20°$.

To make the substrate emulsion, 7.5 ml of 5% gum arabic solution is added to the tube. The microtip of a Branson Sonifer-Cell Disruptor (Branson Instruments Col., Danbury, CT) is centered 0.5 cm below the surface of the solution and the mixture is sonicated in an ice bath at setting 4 for 4 min. After sonication, 5 ml of 10% BSA solution is added and the tube is agitated with a Vortex mixer. This amount of substrate is enough for 75 assays. Although the emulsion retains its properties for at least 48 hr if it is stored at 4° and briefly resonicated before use, a fresh batch should be prepared for each day.

Because two lipolytic enzymes, LPL and HL, are present in PHLA, both activities should be monitored during the purification. A convenient way to do this is to measure total lipolytic activity (LPL + HL) and salt-resistant (HL) activity. In the assay of HL the activity of LPL is inhibited by increasing the NaCl concentration in the assay to 1.0 M and omitting the serum activator. Total lipolytic activity is determined at low ionic strength, 0.1 M NaCl, with normal human serum added as activator. In the assay of HL, 200 μl of substrate, 250 μl of 0.2 M Tris-HCl buffer, pH 8.4, containing 2.0 M NaCl, the enzyme sample, and enough of 0.2 M Tris-HCl buffer to make the volume to 500 μl, are added to each tube. The standard assay for total lipolytic activity is similar except that 50 μl of normal human serum, previously dialyzed against 0.2 M Tris-HCl buffer, pH 8.0, is added and the final NaCl concentration in the incubation is 0.1 M. For blank and total counts similar tubes are prepared omitting the enzyme. The tubes are incubated in a waterbath at $+28°$ for 60 min and thereafter the released free fatty acid radioactivity is determined by a liquid–liquid partitioning system.[25] The reaction is terminated by adding

[25] P. Belfrage and M. Vaughan, J. Lipid Res. 10, 341 (1969).

3.25 ml of the methanol : chloroform : heptane mixture (v/v/v). Thereafter 750 μl of potassium borate–carbonate buffer, pH 10.5, is added and the tubes are vigorously mixed on a Vortex mixer. After centrifugation at 1600 g for 15 min, 1 ml of the upper phase is transferred to a counting vial containing 10 ml of counting solution. From the blank tubes, 1 ml of the upper phase is counted for blank and 1 ml of the lower phase for total counts. The results are expressed as units per milliliter of enzyme source where 1 unit represents 1 μmol of fatty acid hydrolyzed per minute.

Enzyme Purification

Sources of Enzyme. Postheparin plasma is obtained by plasmapheresis from healthy volunteers 10 min after the injection of heparin (Heparin, Medica, Finland), 100 IU/kg body weight.[24] Plasma is stored frozen at $-20°$ until used.

Step 1. Enzyme–Substrate Complex Formation.[26] In a typical experiment 400 ml of postheparin plasma is mixed with 100 ml of 20% triglyceride emulsion (Intralipid, Vitrum, Stockholm, Sweden). The mixture is incubated at 37° for 15 min. After centrifugation at 76,000 g for 60 min, the fat layer at the top of the centrifuge tube is collected. The clear infranatant is again mixed with Intralipid and the incubation and centrifugation repeated. The fat layers are combined and added to 250 ml centrifuge tubes containing 150 ml of cold acetone. After stirring, the tubes are centrifuged at 20.000 g for 10 min. The precipitate is then washed with 150 ml of acetone and twice with 70 ml of ether and finally dried under nitrogen. The dried powder is dissolved in 5 mM sodium barbital buffer, pH 7.4, containing 20% (v/v) of glycerol. The solution is centrifuged at 27,000 g for 15 min. The supernatant is used for further enzyme purification.

Step 2. Heparin–Sepharose Affinity Chromatography. Sepharose CL-4B containing covalently linked heparin can be obtained commerically (Pharmacia Fine Chemicals, Sweden) or can be prepared as described by Iverius.[27,28] The crude enzyme preparation from step 1 is applied on a heparin–Sepharose column (2.5 × 17.5 cm) equilibrated with 5 mM sodium barbital buffer, pH 7.4, containing 20% (v/v) glycerol and 0.15 M NaCl. After washing with 300 ml of 5 mM sodium barbital buffer containing 20% (v/v) glycerol and 0.4 M NaCl, the enzyme is eluted with a linear NaCl gradient (300 + 300 ml, 0.4 ml/min) from 0.4 to 1.5 M NaCl in 5 mM

[26] C. Ehnholm, P. K. J. Kinnunen, and J. K. Huttunen, *FEBS Lett.* **52,** 191 (1975).
[27] P.-H. Iverius, *Biochem. J.* **124,** 677 (1971).
[28] P.-H. Iverius and A-M. Östlund-Lindqvist, *J. Biol. Chem.* **251,** 7791 (1976).

sodium barbital buffer, pH 7.4, containing 20% (v/v) glycerol. Under these conditions the HL elutes as several peaks at salt concentrations between 0.5 and 0.9 M NaCl.[26] Using stepwise elution with 0.9 M NaCl, a more concentrated HL preparation can be obtained. The enzyme preparation is then concentrated by precipitating the protein by dialysis against 3.6 M ammonium sulfate, 0.01 M phosphate buffer, pH 6.5, and dissolving the precipitated protein in 0.01 sodium phosphate buffer, pH 7.5, containing 50% (v/v) glycerol. In this form the enzyme can be stored at $-20°$.

At this stage of purification the enzyme preparation has a specific activity of about 20 U/mg protein and the main contaminant is anti-thrombin III, a protein that also binds to heparin. For further purification several methods are available: (1) adsorption to calcium–phosphate gel[29,30]; (2) affinity chromatography on heparin–Sepharose with low affinity for anti-thrombin[19]; (3) DEAE–Sephacel anion exchange chromatography[31]; (4) gel filtration on Ultrogel AcA34[32]; (5) Concanavalin A–Sepharose affinity chromatography.[18,33]

Calcium–Phosphate Gel Adsorption

The lipase preparation obtained after heparin–Sepharose chromatography is mixed with calcium phosphate gel (Sigma Chemical Co., 7 mg of dry gel/ml of eluate) and the suspension stirred at 4° overnight. After centrifugation at 9000 g for 10 min, the supernatant is removed and the gel is washed with 20 ml of 50 mM NH$_4$OH–NH$_4$Cl buffer, pH 8.3, containing 0.5 mM potassium oleate, 10 mM deoxycholate, and 0.1 M potassium oxalate, and finally with 50 mM NH$_4$OH–NH$_4$Cl buffer containing 0.1 M potassium oxalate. The enzyme is eluted twice with 4 ml of 50 mM sodium citrate–50 mM NH$_4$OH–NH$_4$Cl buffer, pH 8.3, containing 0.5 mM potassium oleate. This step results in a further 1.5 to 2.5-fold purification of HL with the recovery of lipase activity varying between 10 and 40%. However, this purification step results in an HL preparation which is no longer retained by heparin–Sepharose.[26] Therefore it is obvious that the complete separation of lipoprotein lipase and salt-resistant lipase can be achieved only if the heparin–Sepharose chromatography is performed before the calcium phosphate gel treatment.

[29] E. A. Nikkilä, *Biochim. Biophys. Acta* **27**, 612 (1958).
[30] P. E. Fielding, V. G. Shore, and C. J. Fielding, *Biochemistry* **13**, 4318 (1974).
[31] G. Jensen and A. Bensadoun, *Anal. Biochem.* **113**, 246 (1981).
[32] T. Kuusi, P. K. J. Kinnunen, C. Ehnholm, and E. A. Nikkilä, *FEBS Lett.* **98**, 314 (1979).
[33] J. Augustin, H. Freeze, P. Tejada, and W. V. Brown, *J. Biol. Chem.* **253**, 2912 (1978).

Affinity Chromatography on Sepharose Containing Covalently Linked Heparin with Low Affinity for Anti-Thrombin III[19]

As mentioned above, the main contaminant of the HL preparation after the heparin–Sepharose step is anti-thrombin III. Thus a high degree of purification can be obtained using heparin that binds HL but not anti-thrombin.

The HL preparation obtained by conventional heparin–Sepharose affinity chromatography is brought to an NaCl concentration of 0.4 M and applied to a Sepharose column (1.6 × 5 cm) containing covalently linked heparin with low affinity for anti-thrombin. After washing with 200 ml of 0.4 M NaCl and 30% (v/v) glycerol in 0.01 M sodium phosphate buffer, pH 7.5, the enzyme is eluted by increasing the NaCl concentration to 0.8 M. This procedure is efficient and results in a good purification. However, the fractionation of heparin with low affinity for anti-thrombin on anti-thrombin–Sepharose[34] is time consuming.

Anion Exchange Chromatography

The ammonium sulfate precipitate of HL, prepared after heparin–Sepharose affinity chromatography, is resuspended in 50 mM Tris-HCl buffer, pH 7.2, containing 30% (v/v) glycerol and dialyzed against the same buffer for 2 hr. The HL sample is then applied to a 1.6 × 12 cm DEAE Sephacel (Pharmacia, Sweden) column equilibrated with the same buffer. After washing the column with 5 bed volumes of the equilibration buffer and an equal amount of 50 mM Tris-HCl buffer, pH 7.2, without glycerol, HL is eluted with 50 mM Tris-HCl buffer, pH 7.2, containing 0.8 mM NaCl, 0.2% (w/v) Triton N-101 (Sigma), and 100 mM α-methyl-D-mannoside (Sigma). This purification step is efficient but involves the use of detergent which may be undesirable in some experiments.

Gel Filtration[32]

For further purification of the HL, gel filtration can be used. The ammonium sulfate concentrated enzyme preparation is dissolved in a minimal volume of 5 mM sodium barbital buffer, pH 7.4, containing 2.0 M NaCl, 0.1 M D-galactose, 0.1 M α-methyl-D-glycopyranoside, and 0.1% (v/v) Triton X-100. The dissolved protein is dialyzed against the same buffer for 2 hr at +4°, and then subjected to gel filtration on a 1.6 × 91 cm

[34] B. Nordenman and I. Björk, *Biochemistry* **17**, 3339 (1978).

column of Ultrogel AcA34 (LKB, Sweden) equilibrated with the sample buffer. Upon elution at 5 ml/hr, the enzyme emerges as a symmetrical peak with almost complete recovery of activity.

Concanavalin A Affinity Chromatography [18,33]

The HL fractions obtained after heparin–Sepharose chromatography can be applied directly on a 1 × 10 cm concanavalin A–sepharose column (Pharmacia, Sweden) equilibrated with 5 mM sodium barbital buffer, pH 7.0, containing 0.75 M NaCl, 10 mM CaCl$_2$, and 1 mM MgCl$_2$. After washing with 100 ml of 5 mM sodium barbital buffer containing 1.0 M NaCl, the enzyme is eluted with the same buffer containing 1.0 M α-methyl-D-glycopyranoside. This step provides simultaneous concentration and purification of the enzyme.

Purification of HL from Liver Perfusates [31,32]

During HL purification from postheparin plasma the main contaminating protein is anti-thrombin III, a protein which also binds to heparin. By using liver perfusate, which contains very low levels of anti-thrombin as starting material, this contaminant can be avoided. A convenient two-step purification procedure successfully applied to rat,[32] pig,[20] and dog liver[35] perfusates, is as follows: The liver is first perfused with 250 ml of Krebs–Ringer bicarbonate (KRB) buffer to flush out the blood. To release the hepatic lipase into the perfusate, the medium is changed to 50–100 ml KRB containing 40 IU heparin/ml, and the new medium is allowed to recirculate for 5 min. This same medium can be used for consecutive perfusions of several livers. The heparin-containing perfusate is applied to a heparin–Sepharose column (1.2 × 6 cm) equilibrated with 5 mM sodium barbital buffer, pH 7.4, containing 20% (v/v) glycerol and 0.15 M NaCl. The column is first washed with 60 ml of 5 mM sodium barbital buffer, pH 7.4, containing 20% (v/v) glycerol, 0.15 M NaCl, and 0.1% (v/v) Triton X-100, followed by a wash with 60 ml of the same buffer but without Triton X-100. The enzyme is eluted with a linear NaCl gradient from 0.4 to 2.0 M in 5 mM sodium barbital buffer, pH 7.4, containing 20% (v/v) glycerol. Total volume of gradient is 100 ml.

The lipase-containing fractions are combined and the enzyme is concentrated by ammonium sulfate precipitation (see above). The precipitate is dissolved in 2 ml of 5 mM sodium barbital buffer, pH 7.4, containing 0.1 M α-methyl-D-glucopyranoside, 0.1 M D-galactose, 2.0 M NaCl, and 0.1%

---

[35] C. Ehnholm and R. W. Mahley, unpublished observation.

(v/v) Triton X-100. After dialysis against the same buffer for 1 hr at $+4°$, it is subjected to gel filtration on a 1.6×84 cm column of Ultrogel AcA34.

By this method HL from rat livers can be purified to a specific activity of about 65 U/mg protein, using only the heparin–Sepharose affinity chromatography step. From liver perfusates of six rats, about 60 μg protein is obtained. By gel filtration, the specific activity is further increased to about 90 U/mg protein, with almost 100% recovery of lipase activity. The whole purification procedure takes 2–3 days. If further purification of the enzyme is desired, a DEAE–Sephacel step can be added to the purification procedure (see HL Purification from Postheparin Plasma).

Purification of HL from Liver Homogenates

A purification procedure employing liver homogenate as starting material has recently been reported.[21] According to this method, 120 g of rat liver is homogenized for 1 min with a Polytron (Brinkman Instruments, Lucerne, Switzerland) in 1800 ml 5 mM barbital buffer, pH 7.2, containing 20% (v/v) glycerol (buffer A) and 0.5 M NaCl. The homogenate is centrifuged at 35,000 g for 90 min. The supernatant is filtered through glass wool, and the filtrate immediately adsorbed to 150 ml of octyl-Sepharose 4B (Pharmacia, Sweden) equilibrated with the same buffer. After a 2-hr incubation, the slurry is filtered through a sintered-glass funnel. Following washing with 2000 ml of the equilibrating buffer, the gel is packed into a column (2.5×30 cm). The column is washed with an additional 300 ml of the same buffer. About 60 to 70% of the total lipolytic activity binds to the octyl-Sepharose 4B gel. Elution of the HL is carried out with 1500 ml buffer A containing 0.5 M NaCl and 0.4% (v/v) Tween-20. Two peaks with lipolytic activity emerge at this stage. The first lipolytic peak is discarded since it contains most of the contaminating protein. The fractions from the second peak of activity are combined and diluted 1 : 1 with buffer A, and thereafter added to 25 ml of heparin–Sepharose equilibrated with buffer A containing 0.25 M NaCl. After shaking for 1 hr the gel is recovered on a sintered glass funnel, washed with 200 ml of buffer A containing 0.25 M NaCl and packed into a column (1.5×30 cm). The column is then washed with 150 ml of the same buffer. Elution of HL activity is carried out with 100 ml buffer A containing 0.75 M NaCl. The active fractions are pooled and applied to a 0.5×1 cm hydroxyapatite gel column (Sigma Chemical Co., MO) equilibrated with 10 mM sodium phosphate, pH 7.4. The column is washed with 15 ml of equilibrating buffer, followed by 15 ml of 75 mM sodium phosphate, pH 7.4. The HL activity is eluted with 60 ml of 0.18 M sodium phosphate, pH 7.4. The active fractions are combined and applied to a 1-ml heparin–Sepharose column equilibrated with buffer A containing 0.25 M NaCl. After washing with 5 ml of this buffer, the column is eluted by increasing the NaCl concentration to 0.75 M. This enzyme preparation is then subjected to gel filtration on Ultragel AcA34 as described before.

Characterization of Hepatic Lipase

Purity of Enzyme

The purification of hepatic lipase can be followed most conveniently by observing the specific activity in the preparation instead of monitoring the protein by polyacrylamide gel electrophoresis in the presence of sodium dodecyl sulfate (SDS–PAGE). This is due to the fact that the main contaminant in the purification of postheparin plasma HL, anti-thrombin III, has a similar molecular weight to HL upon SDS–PAGE. Hepatic lipase purified from postheparin plasma without low affinity chromatography has a specific activity of 4.52 U/mg protein.[18] The removal of anti-thrombin III from the preparation results in a specific activity of 120 U/mg, which is at present the highest reported for purified human postheparin plasma HL.[19] When studied by SDS–PAGE this preparation contains one major and several minor protein bands. The enzyme purified from canine postheparin hepatic venous blood has a specific activity of 14 U/mg protein.[36] The hepatic lipase in blood-free heparin perfusates of rat livers has a specific activity of 1.7 U/mg without any purification.[32] This starting material, combined with affinity chromatography on heparin–Sepharose and gel filtration on Ultrogel AcA34, results in a preparation with a specific activity of 92.3 U/mg protein, free of anti-thrombin III.[32] This method has been further refined, and with the aid of an additional purification step the specific activity of HL can be increased to 750 U/mg protein.[33] Interestingly, purified bovine milk LPL has a specific activity of 622 U/mg protein. Thus the purest HL preparation has a specific activity similar to that of bovine milk LPL.[37]

Stability

Hepatic lipase isolated by affinity chromatography on heparin–Sepharose from postheparin plasma or from heparin-containing liver perfusates is stable for several months if stored frozen at $-70°$ in 5 mM sodium barbital buffer, pH 7.4, containing 0.75 M NaCl and 20% (v/v)

---

[36] P. H. Frost, V. G. Shore, and R. J. Havel, *Biochim. Biophys. Acta* **712**, 71 (1982).
[37] P.-H. Iverius and A. M. Östlund-Lindqvist, *J. Biol. Chem.* **251**, 7791 (1976).

glycerol.[21,38] It is also stable at $-20°$ when stored in the same buffer but at 50% (v/v) glycerol concentration.[32] At room temperature or even $+4°$ the enzyme is rapidly inactivated with a half-life of about 60 min.[31,39] This loss of activity can be prevented by adding heparin[39] (15 IU/ml) or detergents[21,31] to the enzyme. The effect of different detergents on the stability of hepatic lipase varies. Thus, Triton N-101 seems to be slightly superior to Triton X-100, the most commonly used detergent in the purification of different lipases.[21,31] Of the ionic detergents sodium dodecyl sulfate rapidly inactivates the enzyme activity[40] while deoxycholate has been successfully used for the elution of hepatic lipase from affinity columns.[41] Hepatic lipase activity can be further stabilized by carbohydrates. Thus, the enzyme retains its activity for at least 2 months at $+4°$ if stored in 5 mM sodium barbital buffer, pH 7.4, containing 2.0 M NaCl, 0.1 M D-galactose, 0.1 M α-methyl-D-glucopyranoside and 0.1% (v/v) Triton X-100.[32] It is also stable for long periods of time if stored at $-20°$ as an ammonium sulfate precipitate at 60% saturation. The enzyme activity is rapidly abolished by denaturing agents such as 8 M urea.[31]

Physical Properties

The monomer molecular weight of hepatic lipase purified from human postheparin plasma is 64,000 to 69,000 as determined by SDS–PAGE.[18,33] The HL purified from liver perfusates[31,32] has an apparent minimal molecular weight of 60,000 to 62,000 on SDS–PAGE, whereas the enzyme purified from liver homogenates[21] has a slightly smaller molecular weight, 53,000. As determined by gel filtration in the presence of detergent, the molecular size of HL from rat liver perfusates[31,32] is 180,000, whereas that of HL purified from liver homogenates[21] is slightly larger, 200,000. These data suggest that the active enzyme is an oligomer or an enzyme–detergent complex when gel filtrated in the presence of detergent. In the absence of detergent, the size of the enzyme upon gel filtration is smaller, about 75,000.[18]

Chemical Properties

Binding of HL to concanavalin A and its elution by sugar strongly suggest that it is a glycoprotein.[18] The enzyme purified from postheparin

[38] C. E. Jahn, J. C. Osborne, E. J. Schaefer, and H. B. Brewer, *Eur. J. Biochem.* **131,** 25 (1983).
[39] K. Schoonderwoerd, W. C. Hulsmann, and H. Jansen, *Biochim. Biophys. Acta* **665,** 317 (1981).
[40] M. L. Baginsky and W. V. Brown, *J. Lipid Res.* **18,** 423 (1977).
[41] G. Bengtsson and T. Olivecrona, *FEBS Lett.* **119,** 290 (1980).

plasma has been reported to contain about 8% carbohydrate,[33] but the preparation used in this analysis evidently contained anti-thrombin III as contaminant.[19] A more recent estimation of the carbohydrate content of hepatic lipase is, however, not available.

Amino acid analysis of highly purified human hepatic lipase obtained by chromatography on low-affinity heparin–Sepharose to remove anti-thrombin III revealed that the amino acid composition of HL is significantly different from that of human postheparin plasma LPL.[19]

Hepatic lipase displays heterogeneity upon affinity chromatography on heparin–Sepharose when using delipidated plasma as a starting material.[26] Thus, three separate peaks with salt-resistant activity are eluted with a linear NaCl gradient. All activities are inhibited with anti-HL serum. The reason for this heterogeneity is at present not known. Also in isoelectric focusing HL displays heterogeneity.[18] Thus partially purified HL from human postheparin plasma gives a major activity peak with a pI of 4.1 and two minor peaks at higher pI values. A similar focusing pattern has been observed for highly purified rat HL.[31] This isoelectric heterogeneity might be explained by binding of ampholyte to HL as has been previously reported for LPL.[42] Analytical isoelectric focusing of purified rat or dog HL in 6 M urea-containing gels indicates a very alkaline pI, higher than pH 8.0.[31,36] A similar alkaline shift in focusing properties upon denaturation has been observed for LPL.[42]

Substrate Specificity

Purified hepatic lipase catalyzes the hydrolysis of acyl—ester bonds in various lipids including triglycerides, monoglycerides, phospholipids, and acyl-CoA thiolesters. That the acylester bonds of different lipid classes are hydrolyzed by the same enzyme is concluded from the following observations: constancy of the ratios between different lipolytic activities during purification,[18,43] similar inactivation of the different hydrolytic activities by HL antiserum,[43] by thermal incubation,[18,43] and by diisopropylfluorophosphate.[21] These results suggest the possibility that one catalytic site is responsible for all four lipolytic activities. Competitive inhibition occurs between triglyceride and monoglyceride substrates.[21] This further strengthens this suggestion. Hepatic lipase is inactivated by incubation with benzene boronic acid[44] and diisopropylfluorophosphate,[21] indicating that serine residues may be involved in the catalytic site. The

[42] G. Bengtsson and T. Olivecrona, *in* "Electrofocusing and Isotachophoresis" (R. J. Radola and D. Graesslin, eds.), p. 189. De Gruyter, Berlin, 1977.
[43] G. L. Jensen, B. Daggy, and A. Bensadoun, *Biochim. Biophys. Acta* **710**, 464 (1982).
[44] P. K. J. Kinnunen, J. A. Virtanen, and P. Vainio, *Atheroscl. Rev.* **11**, 65 (1983).

enzyme shows stereospecificity against the *sn*-1-acylester bond of tri-acylglycerol,[45] a specificity similar to that previously reported for LPL.[46] The enzyme has a specificity for the 1-position of phospholipids.[47]

The phospholipase activity measurable in total postheparin plasma correlates closely with the triacylglycerol hydrolase activity of hepatic lipase.[22] The same also applies to the acyl-CoA thiolesterase activity of postheparin plasma.[23] The hydrolysis of triglycerides by postheparin plasma hepatic lipase occurs, however, at higher rates than that of phospholipids.[18,22] This observation is in a good accordance with the almost 20-fold higher V_{max} of triacylglycerol hydrolysis by purified rat hepatic lipase compared to that of dipalmitoylphosphocholine.[21] Hydrolysis of phosphatidylethanolamine occurs more readily than that of phosphatidylcholine.[18] Also, dioleylphosphatidylcholine is hydrolyzed at higher rates than dipalmitoylphosphatidylcholine.[48] Therefore, great caution should be exercised when comparing reaction rates of lipolysis in different studies.

The substrate specificity of purified hepatic lipase can be modified by limited proteolysis.[43,49] The triacylglycerol lipase activity is more sensitive to proteolytic digestion than the monoacylglycerol hydrolase or phospholipase activity. Incubation of purified hepatic lipase with collagenase causes cleavage of a low-molecular-weight fragment from the enzyme. This results in an enzyme preparations devoid of triacylglycerol lipase activity but still capable of hydrolyzing monoglycerides. It has been postulated that this small molecular weight fragment functions in the recognition of triacylglycerol substrates.[49]

A heparin-releasable phospholipase A₁ without triacylglycerol lipase activity has been isolated from rat liver plasma membranes. This enzyme, capable of transacylation reactions between different lipids, was termed monoacylglycerol acyltransferase (MGAT).[50] It was earlier thought to be a separate enzyme. However, more recent studies by Waite *et al.*[51] indicate that the hepatic lipase and MGAT are the same enzyme. MGAT has a slightly lower molecular weight than hepatic lipase and is devoid of triglyceride lipase activity but has monoacylglycerol hydrolase activity.[51]

Hepatic lipase is located on the liver endothelium[15] and, therefore, *in vivo* serum lipoproteins are likely to be substrates for this enzyme. The

[45] B. Åkesson, S. Gronowitz, and B. Herslöf, *FEBS Lett.* **71**, 241 (1976).
[46] F. Paltauf, F. Esfandi, and A. Holasek, *FEBS Lett.* **40**, 119 (1974).
[47] M. Waite and P. Sisson, *J. Biol. Chem.* **249**, 6401 (1974).
[48] W. C. Hulsmann, M. C. Oerlemans, and H. Jansen, *Biochim. Biophys. Acta* **618**, 364 (1980).
[49] T. Kuusi, K. Bry, E. A. Nikkilä, and P. K. J. Kinnunen, *Med. Biol.* **57**, 192 (1979).
[50] M. Waite, P. Sisson, this series, Vol. 71, p. 678.
[51] M. Waite, P. Sisson, and R. El-Maghrabi, *Biochim. Biophys. Acta* **530**, 292 (1978).

enzyme has *in vitro* higher affinity for high-density lipoproteins (HDL) than for triacylglycerol-rich particles, as judged from studies with HDL–Sepharose columns.[41] This is consistent with the observation that human ascites chylomicrons are a poor substrate for hepatic lipase.[52] Neither are β-VLDL isolated from type III hyperlipidemic patients hydrolyzed by HL *in vitro*.[53] Selective hydrolysis of the HDL_2 subfraction lipids by partially purified hepatic lipase has been demonstrated by Shirai *et al.*[54] and Groot *et al.*[55] Thus, incubation *in vitro* of human plasma with purified rat hepatic lipase results in changes in HDL subfraction distribution,[56] supporting the view that the enzyme *in vivo* has a function in the conversion of HDL_2 to HDL_3.[57]

Effect of Serum, Lipoproteins, and Apolipoproteins on HL Activity in Vitro

The effect of apolipoproteins on hepatic lipase activity is highly sensitive to the way in which the substrate used for the measurement of enzyme activity is presented. Thus, different effects of apolipoproteins can be observed depending on whether gum arabic-stabilized or lecithin-stabilized triolein substrates or native lipoproteins are used. Therefore caution should be exercised when interpreting data obtained in artificial systems.

The presence of low concentrations of serum in the assay medium causes varying degrees of activation ranging from 0 to 300% of postheparin plasma HL.[38,58–62] However, higher concentrations of serum invariably induce progressive inhibition of enzyme activity. This inhibition has been attributed to plasma HDL.[58,62] It is evident only at concentrations

[52] N. Yamada, T. Murase, Y. Akanuma, H. Itakura, and K. Kosaka, *Clin. Chim. Acta* **110**, 451 (1981).

[53] C. Ehnholm, R. W. Mahley, D. A. Chappell, K. M. Weisgraber, E. Ludwig, and J. L. Witztum, *Proc. Natl. Acad. Sci. U.S.A.* **81**, 5566 (1984).

[54] K. Shirai, R. L. Barnhart, and R. L. Jackson, *Biochem. Biophys. Res. Commun.* **100**, 591 (1981).

[55] P. H. E. Groot, L. M. Scheek, and H. Jansen, *Biochim. Biophys. Acta* **751**, 393 (1983).

[56] P. H. E. Groot, H. Jansen, and A. van Tol, *FEBS Lett.* **129**, 269 (1981).

[57] T. Kuusi, E. A. Nikkilä, M. J. Tikkanen, M-R. Taskinen, and C. Ehnholm, *in* "Atherosclerosis VI" (G. Schettler, A. M. Gotto, G. Middlehoff, A. J. R. Habenicht, and K. R. Jurutha, eds.), p. 628. Springer-Verlag, Berlin, 1982.

[58] M. Kubo, Y. Matsuzawa, H. Sudo, K. Ishikawa, A. Yamamoto, and S. Tarui, *J. Biochem.* **88**, 905 (1980).

[59] T. Kuusi, thesis, University of Helsinki, 1979.

[60] P. K. J. Kinnunen and C. Ehnholm, *FEBS Lett.* **65**, 354 (1976).

[61] C. Ehnholm, H. Greten, and W. V. Brown, *Biochim. Biophys. Acta* **360**, 68 (1974).

[62] M. Kubo, Y. Matsuzava, H. Sudo, K. Ishikawa, A. Yamamoto, and S. Tarui, *J. Biochem.* **92**, 865 (1982).

above 40 to 60 nmol HDL–cholesterol/μmol triolein,[63,64] stabilized with gum arabic. In contrast even minor quantities of HDL are capable of inhibiting HL activity when lecithin-stabilized triolein or intact lymph chylomicrons are used as substrate.[41,59,63]

Two mechanisms have been postulated to explain the inhibition of HL activity by HDL. Bengtsson and Olivecrona[41] have demonstrated a preferential binding of hepatic lipase to HDL as compared to lecithin-stabilized triolein (Intralipid). This mechanism operates at low HDL concentrations, with no threshold for the inhibition.[41] However, using gum arabic-stabilized triolein as substrate, a threshold for HDL-induced inhibition is evident. Kubo and co-workers demonstrated that purified human apoA-I and apoA-II above a threshold value of about 32 μg protein/μmol substrate triolein both inhibit hepatic lipase activity.[62] Their studies demonstrate that this inhibition depends on the binding of apolipoproteins to the substrate and becomes evident only when the substrate surface is saturated with apoA-I or A-II. Under saturating conditions all available surface is excluded, but below saturation, the remaining free substrate surface is available for the enzyme.[62] This mechanism may explain the inhibition of hepatic lipase activity also noted with other apolipoproteins, such as apoC-I, C-II, C-III[59,60] and the different isoforms of apoE, shown in Fig. 1. Thus, the inhibition of HL activity by purified apolipoproteins seems to be relatively unspecific and due to the common lipid-binding properties of apolipoproteins. This view is supported by the finding that hepatic lipase activity can be inhibited by synthetic lipid-binding model peptides.[59]

Variable effects of apoA-II on the activity of purified HL has been reported using lecithin–glycerol-stabilized triolein as a substrate.[38] These effects probably depend on properties of the substrate as they are critically dependent on the intensity of shaking during incubation. Thus, complete inhibition of HL activity by apoA-II was evident when 26 cycles/min was used, but an almost 3-fold activation was demonstrated at 100 cycles/min. At both shaking velocities apoA-I and C apolipoproteins caused inhibition.[38] These experiments led the authors to postulate that the observed specificity of hepatic lipase for HDL$_2$[54,55] is due to stimulation by apoA-II.[65] This is, however, in contrast to the findings of Jackson and co-workers, who in their *in vitro* studies demonstrated that hydrolysis

[63] T. Kuusi, E. A. Nikkilä, M.-R. Taskinen, and M. J. Tikkanen, *Atherosclerosis* **44,** 237 (1982).
[64] P. N. M. Demacker, P. M. J. Stuyt, A. G. M. Hijmans, and A. van't Laar, *Atherosclerosis* **48,** 101 (1983).
[65] C. E. Jahn, J. C. Osborne, E. J. Schaefer, and H. B. Brewer, Jr., *FEBS Lett.* **131,** 366 (1981).

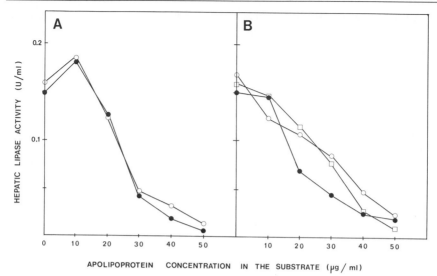

FIG. 1. Effect of purified apolipoproteins A-I and A-II and the different isoforms of apoE on hydrolysis of lecithin-stabilized triolein by hepatic lipase *in vitro*. Ten microliters of hepatic lipase (specific activity 300 U/mg protein) was incubated for 60 min at +28° with 3.0 m*M* acyl-[1-^{14}C]triolein stabilized with 0.3 m*M* lecithin in a total volume of 250 μl of 0.2 *M* Tris-HCl buffer, pH 8.4, containing 0.1 *M* NaCl and 5% (w/v) BSA. The apolipoproteins were added to the substrate before the enzyme at concentrations indicated on the abscissa. Panel A: (○), apoA-I; (●), apoA-II; panel B: (○) apoE-2; (●), apoE-3; (□), apoE-4.

of HDL$_2$ triglycerides by HL was inhibited by purified apoA-II, and that the inhibitory effect was proportional to the amount of apoA-II present in the HDL$_2$ fraction.[66] Also other apolipoproteins inhibited the hydrolysis of HDL$_2$ triglycerides, suggesting that the effect of apolipoproteins depends more on the protein-to-phospholipid ratio in the substrate than on the specific characteristics of an apolipoprotein.

Measurement of Hepatic Lipase

Selective Assay of Postheparin Plasma Hepatic Lipase and Lipoprotein Lipase Activities

Several methods have been published for the selective measurement of HL and LPL, indicating the difficulties encountered in the measurement of these lipolytic activities. The main problems are (1) preparation of

---

[66] M. Shinomiya, N. Sasaki, R. L. Barnhart, K. Shirai, and R. L. Jackson, *Biochim. Biophys. Acta* **713,** 292 (1982).

a reproducible and stable substrate and (2) the development of a selective assay that measures one or the other of the triacylglycerol hydrolase activities. Based on the characteristics of HL and LPL, basically three types of selective assays have been used. The first utilizes the different affinities of HL and LPL for heparin. The second assay relies on the observation that HL is protamine resistant. The third is based on the use of antiserum against purified HL. Additionally, detergents and specially prepared substrates can be used for the selective measurement of HL from postheparin plasma. As LPL has low phospholipase and monoacylglycerol hydrolase activity also, these substrates can be used to monitor HL activity with minor interference from LPL.

The Use of Affinity Chromatography for Specific Assay of HL and LPL[67]

In this method affinity chromatography on heparin–Sepharose columns is used for separation of HL and LPL before assay; 2–4 ml of postheparin plasma is applied to 8 × 20 mm columns of heparin–Sepharose equilibrated with 5 mM sodium barbital buffer, pH 7.4, containing 0.15 M NaCl. After washing the column, the enzymes are eluted stepwise with 0.72 and 1.5 M NaCl, respectively, and assayed, essentially, as described in the previous purification section. The advantage of this method is that the enzyme activities, when assayed, are free from major serum lipoprotein components. In a modification of this method heparin is used for the elution of LPL activity.[68] The lipase activities measured with these methods correlate well with those obtained with the protamine inhibition assays; however, these methods are laborious, difficult to standardize, and require large samples of postheparin plasma.

Determination of Protamine-Sensitive (LPL) and Protamine-Resistant Triglyceride Lipase (HL)[8]

By preincubation of plasma in the presence or absence of protamine, the "protamine-resistant" HL and the "total" lipase activities can be measured in postheparin plasma. From these the LPL activity can be derived. This assay, in our hands, using a gum arabic triolein emulsion as substrate, correlates well with the immunochemical method for assaying lipases[24]; however, discrepancies between the two methods have also been reported.[69]

[67] J. Boberg, J. Augustin, M. L. Baginsky, P. Tejada, and W. V. Brown, J. Lipid Res. **18,** 544 (1977).

[68] C-S. Wang, H. B. Bass, D. Downs, and R. K. Whitmer, Clin. Chem. **27,** 663 (1981).

[69] H. Greten, V. Laible, G. Zipperle, and J. Augustin, Atherosclerosis **26,** 563 (1977).

Immunochemical Measurement of Hepatic Lipase and Lipoprotein
   Lipase Activities[24]

The selective measurement of lipoprotein lipase activity is based on inactivation of hepatic lipase with specific antiserum and subsequent assay of the remaining activity in the presence of serum activator.

In the standard assay 10 μl of postheparin plasma is incubated for 2 hr at 4° with 10 μl of hepatic lipase antiserum (diluted if necessary with normal rabbit serum). Thereafter, 500 μl of the assay mixture containing 3.2 mM acyl-[1-^{14}C]triolein emulsion, 0.2 M Tris-HCl buffer, pH 8.4, 0.1 M NaCl, 2.5% bovine serum albumin, and 50 μl of normal human serum is added, and the tubes incubated at 28° for 60 min. The released free fatty acid radioactivity is determined by the liquid–liquid partitioning system of Belfrage and Vaughan[25] as described previously. In the assay of hepatic lipase the activity of lipoprotein lipase is inhibited by raising the NaCl concentration to 1.0 M and omitting the serum activator. The assay is carried out without preincubation in a reaction mixture of the same composition as that used for lipoprotein lipase, except that the NaCl concentration is 1.0 M and human serum is omitted. Total lipolytic activity toward triolein emulsion is determined under conditions used for lipoprotein lipase but without preincubation with antiserum.

Other Methods

To selectively measure HL and LPL in postheparin plasma, glycerol-based[70] or phosphatidylcholine-stabilized[71] trioleylglycerol emulsions have been used. In a comparison between the assay using a specific phosphatidylcholine-stabilized substrate and the immunochemical assay a satisfactory correlation ($r = 0.92$) was observed; however, the substrate specific assay consistently yielded lower LPL activities.[72] In this assay the purity of the lipids used for substrate preparation is of great importance. Thus the presence of labeled monoacylglycerol deletes the specificity of the "substrate specific" LPL assay. Monoacylglycerol contamination may therefore offer an explanation for discrepancies in reports on LPL activity assays.[73]

The observation that sodium dodecyl sulfate (SDS) completely inactivates HL at concentrations that protect LPL[40] activity is the basis for a

[70] J. E. Corey and D. B. Zilversmit, *J. Lab. Clin. Med.* **89,** 666 (1977).
[71] P. Nilsson-Ehle and R. Ekman, *Artery* **3,** 194 (1977).
[72] C. Ehnholm, E. A. Nikkilä, and P. Nilsson-Ehle, *Clin. Chem.* **30,** 1568 (1984).
[73] P. Nilsson-Ehle, *Clin. Chim. Acta* **14,** 293 (1984).

lipase assay described by Baginsky and Brown.[74] In this assay postheparin plasma is preincubated with 40 mM SDS in 0.2 M Tris-HCl buffer, pH 8.2, for 50 min at 26°. This results in the inactivation of HL while leaving the LPL fully active. HL is determined from the postheparin plasma sample as described for the immunochemical method. This assay has been reported to correlate well ($r = 0.99$) with the immunochemical assay.[74]

As hepatic lipase also hydrolyzes monoacylglycerols, phospholipids, and long chain fatty acid CoA thioesters, these substrate specificities can be used to measure HL activity.

Phospholipase Assay[22]

This assay is based on the hydrolysis of Triton X-100-emulsified phosphatidylglycerol radioactively labeled in the 2-position ([2-³H]PG) and subsequent determination of the released lyso[2-³H]phosphatidylglycerol (lyso-[2-³H]PG) radioactivity. In this modification of the method described by Shaw and Harlan,[75] instead of using phosphatidylethanolamine (PE), we use phosphatidylglycerol (PG) emulsified in Triton X-100 as substrate.

Materials

[2-³H]Glycerol, New England Nuclear Co.
Phosphatidylcholine, grade I, Sigma Chemical Co., MO
Phosphatidylglycerol, grade I, Sigma Chemical Co., MO
Phospholipase D, Boehringer GmbH, Mannheim, FRG
Triton X-100, Sigma Chemical Co., MO

Preparation of Radioactive Phosphatidyl[2-³H]glycerol. [2-³H]PG is prepared from phosphatidylcholine by transphosphatidylation as follows: 30 mg of phosphatidylcholine (PC) is incubated (180 oscillations/min) for 2 hr at room temperature with 1.0 mCi [2-³H]glycerol (specific activity, 200 Ci/mol), and 3 mg phospholipase D in a two-phase emulsion system composed of 1.0 ml diethyl ether and 150 μl of 0.1 mol/liter acetate buffer, pH 5.6, containing 50 mmol/liter CaCl₂. The volume of the aqueous phase is kept minimal to favor transphosphatidylation over hydrolysis but sufficient to form an emulsion with ether. The reaction is terminated by adding 150 μl of 100 mmol/liter ethylenediaminetetraacetate (EDTA). Thereafter the ether phase is evaporated and the lipids extracted by the method of

[74] M. L. Baginsky and W. V. Brown, *J. Lipid Res.* **20**, 548 (1979).
[75] W. A. Shaw and W. R. Harlan, *J. Lipid Res.* **18**, 123 (1977).

Folch *et al.*[76] The [2-³H]PG is separated from PC and phosphatidic acid by preparative thin-layer chromatography on 0.5-mm-thick silicic acid-coated plates (20 × 20 cm) developed with chloroform/acetone/methanol/acetic acid/water (50 : 20 : 10 : 10 : 5, v/v/v/v/v).

Preparation of Substrate. The substrate is prepared by sonication on ice for 3 min with a Branson sonifier microtip of 1.8 μmol [2-³H]PG (specific activity, 0.12 Ci/mol) in 1 ml of 0.14 M glycine buffer, pH 10.0, containing 0.05% (v/v/v) Triton X-100, 10 mM CaCl$_2$, and 1.0 M NaCl to inhibit residual LPL activity. A new substrate should be prepared each day.

The sample (20 μl) is incubated with 25 μl of substrate at 37° for 15 min and the reaction terminated by extracting the formed lysophospholipids by adding 2 ml of chloroform/methanol (1 : 1, v/v). The tube is then mixed vigorously for 10 sec with a Vortex mixer at room temperature. The phases are separated by addition of 2.6 ml of 50 mmol/liter CaCl$_2$, followed by mixing as above and centrifugation for 10 min at 2500 g at room temperature. The upper phase contains lyso[2-³H]PG radioactivity but less than 0.2% of [2-³H]PG radioactivity. The extraction of lysoPG is linear at least up to 16 nmol. One milliliter of the upper phase is counted for radioactivity as described for the triacylglycerol hydrolase assay. The phospholipase activity is expressed as μmol lysoPG formed/ml/min.

Monoacylglycerol Hydrolase Assay

The assay is that described by Tornqvist *et al.,*[77] with the exception that TX-100 is used instead of Nonipol TD-12 as detergent.

Materials

Conical glass tubes, 50-ml Pyrex No. 8062

Reagents

[³H]Glycerol, The Radiochemical Centre, Amersham, England
Oleyl chloride, NuCheck Prep., MN
Monooleylglycerol, Sigma Chemical Co., MO
Albumin bovine, fraction V, Sigma Chemical Co., MO
Triton X-100, Sigma Chemical Co., MO
Trishydroxymethylaminoethane, Sigma 7–9,   Sigma Chemical Co., MO
Scintillation fluid, ACS, Amersham No. 196 290

---

[76] J. Folch, M. Lees, and G. H. Sloane-Stanley, *J. Biol. Chem.* **226,** (1957).
[77] H. Tornqvist, L. Krabisch, and P. Belfrage, *J. Lipid Res.* **14,** 291 (1974).

Solutions

Nonradioactive monooleylglycerol, 20 mg/ml, dissolved in heptane; this solution can be stored well sealed under nitrogen at −20°

Methanol : chloroform : heptane (1.41 : 1.25 : 1, v/v/v)

0.2 *M* Tris-HCl buffer, pH 8.0, containing 0.8% (w/v) Triton X-100

Synthesis of Radioactive Monoacyl [1(3)-³H]glycerol. This is done as described by Tornqvist *et al.*[77] as follows: Sodium mono[³H]glycerate is obtained by reacting [1(3)-³H]glycerol (10 mCi, 1 mmol) with NaOH at 150°. To the glycerol, dissolved in 1 ml of methanol, 1 ml of 1 *M* NaOH is added, and the solvents are evaporated with dry N_2 at 150° followed by another 30 min at 150°. The sodium mono[³H]glycerate is dried in a desiccator overnight. Dry ethyl acetate is thoroughly mixed with the compound in a closed vial, and the glycerate is subsequently acylated at room temprature for 2.5 hr with 400 μl of oleyl chloride. The reaction is interrupted by the addition of water, and the lipids are extracted with diethyl ether, which is evaporated after drying over Na_2SO_4. After the addition of unlabeled monooleylglycerol (0.7 g), the monoleyl[³H]glycerol is purified on a silicic acid column using diethyl ether–hexane mixtures. The compound is characterized by comparison with a pure reference monooleylglycerol on silicic acid thin-layer plates impregnated with boric acid and by the equimolar production of glycerol and oleic acid upon hydrolysis. The monooleyl[³H]glycerol obtained was > 99.5% radiochemically pure by thin-layer chromatography and has a specific activity of about 2.0 μCi/μmol.

Preparation of Substrate. Labeled and unlabeled monooleylglycerols are mixed in a conical glass tube to give 40 μmol of substrate with a specific activity of 0.1–0.3 × 10⁶ cpm/μmol. After removal of solvents under nitrogen, the substrate is washed three times with 3 ml of heptane, evaporating the heptane under nitrogen between washes. To prepare substrate 4.0 ml of 0.8% (w/v) Triton X-100 in 0.2 *M* Tris-HCl, pH 8.0, is added to the tube and the dry lipid is solubilized by sonicating for 2 min (setting 2, Branson Sonifier, type LS 75) at 0°. The resulting solution is sufficient for the assay of at least 37 samples. It can be stored at −20° and used several times after resonication for 15 sec.

For assay of HL monoacylglycerol hydrolase activity 100 μl of substrate, the enzyme preparation, and enough of 0.2 *M* Tris-HCl buffer, pH 8.0, containing 0.8% (w/v) Triton X-100 to make the final volume 200 μl, are added. After a 10-min incubation at 22° the reaction is interrupted by the addition of 3.25 ml of methanol–chloroform–heptane (1.41 : 1.25 : 1, v/v/v) followed by 1.05 ml of 2% (w/v) NaCl to obtain a two-phase system. A 1-ml aliquot of the upper methanol–water phase, which contains all the [³H]glycerol produced and less than 0.3% of the monooleyl-[³H]glycerol, is transferred to a counting vial with 10 ml of counting solution and the radioactivity determined. The results are expressed as units per milliliter of enzyme source where 1 U represents the release of 1 μmol of glycerol/min.

Concluding Remarks

The complete role of HL in lipoprotein metabolism is still controversial. Although originally purified as a triacylglycerol lipase, this enzyme can also hydrolyze several other lipid substrates and the question which is the physiological substrate for HL is far from solved. As has become evident from this chapter, many of the molecular characteristics of HL have not yet been clarified. It is evident that one of the main reasons for the lack of this information involves the problems of purifying and assaying HL. We hope that our presentation will be of help in this respect and lead to an increase in our knowledge on the issue.

Acknowledgments

The authors wish to thank Prof. Jussi K. Huttunen and Prof. Esko A. Nikkilä for fruitful discussions. The work in the authors' laboratory is supported by grants from the Sigrid Juselius and Yrjö Jahnsson Foundation. Major financial support is also provided by the Finnish State Medical Research Council (T.K.). The secretarial aid from Ms. Marita Heinonen is gratefully appreciated.

[44] Mechanism of Action of Lipoprotein Lipase

By LARRY R. MCLEAN, RUDY A. DEMEL, LILIAN SOCORRO, MASAKI SHINOMIYA, and RICHARD L. JACKSON

Introduction

Lipoprotein lipase (LpL, E.C. 3.1.1.34) catalyzes the hydrolysis of plasma tri- and diacylglycerols, phosphatidylcholines, and phosphatidylethanolamines transported in plasma chylomicrons and very-low-density lipoproteins (VLDL).[1-10] LpL does not circulate in the blood but is immobilized to endothelial cell surfaces by its interaction with membrane-associated heparan sulfates. Administration of heparin releases LpL from cell surface binding sites, causing clearing of plasma triglycerides, hence, the name *clearing factor lipase*.[11,12] In the rare autosomal recessive disorder of familial LpL deficiency (type I hyperlipoproteinemia), the near absence of postheparin plasma or adipose tissue LpL activity results in fasting hypertriglyceridemia.

LpL is located primarily in adipose tissue, lung, muscle, and lactating mammary gland. The characteristic properties of the enzyme regardless of tissue source are (1) inhibition of enzyme activity by 1 M NaCl and protamine sulfate; (2) enhancement of enzyme activity by apolipoprotein C-II (apoC-II), a protein constituent of triglyceride-rich lipoproteins and high-density lipoproteins (HDL); and (3) a pH optimum for triacylglycerol substrates of 8–9. Table I gives the range of molecular weights of LpL purified from several sources; the values range from 34 to 78,000. Although the primary physiological substrates for LpL are long chain triacylglycerols and phospholipids, the enzyme also catalyzes the hydrolysis of water-soluble short chain phosphatidylcholines,[13] triacylglycerols,[14] and *p*-nitrophenyl esters.[15-17] The enzyme is specific for the *sn*-1(3) ester bond of triacylglycerols and the *sn*-1 ester bond of phospholipids.

The purpose of this chapter is to describe several of the methods used in our laboratory to determine the mechanism of action of lipopro-

[1] D. Quinn, K. Shirai, and R. L. Jackson, *Prog. Lipid Res.* 22, 35 (1983).

[2] M. Hamosh and P. Hamosh, *Mol. Aspects Med.* 6, 199 (1983).

[3] P. K. J. Kinnunen, J. A. Virtanen, and P. Vainio, *Atheroscler. Rev.* 11, 65 (1983).

[4] A. Cryer, *Int. J. Biochem.* 13, 525 (1981).

[5] P. Nilsson-Ehle, A. S. Garfinkel, and M. C. Schotz, *Annu. Rev. Biochem.* 49, 667 (1980).

[6] D. S. Robinson, *Comp. Biochem.* 18, 51 (1970).

[7] I. Posner, *Atheroscler. Rev.* 9, 123 (1982).

[8] R. L. Jackson, *in* "The Enzymes" (P. D. Boyer, ed.), Vol. 16, p. 141. Academic Press, New York, 1983.

[9] L. C. Smith and H. J. Pownall, *in* "Lipases" (B. Borgstrom and H. L. Brockman, eds.), p. 263. Elsevier, Amsterdam, 1984.

[10] T. Olivecrona and G. Bengtsson, *in* "Lipases" (B. Borgstrom and H. L. Brockman, eds.), p. 205. Elsevier, Amsterdam, 1984.

[11] E. D. Korn, *J. Biol. Chem.* 215, 15 (1955).

[12] P. F. Hahn, *Science* 98, 19 (1943).

[13] M. Shinomiya and R. L. Jackson, *Biochem. Biophys. Res. Commun.* 113, 811 (1983).

[14] D. Rapp and T. Olivecrona, *Eur. J. Biochem.* 91, 379 (1978).

[15] K. Shirai and R. L. Jackson, *J. Biol. Chem.* 257, 1253 (1982).

[16] D. M. Quinn, K. Shirai, R. L. Jackson, and J. A. K. Harmony, *Biochemistry* 21, 6872 (1982).

[17] T. Egelrud and T. Olivecrona, *Biochim. Biophys. Acta* 306, 115 (1973).

TABLE I
MOLECULAR WEIGHT OF LIPOPROTEIN LIPASE OBTAINED FROM
VARIOUS SOURCES

Source	Molecular weight	Method of analysis	Reference
Postheparin plasma			
Human	67,000	SDS–PAGE	a
	77,000	SDS–PAGE	b
Rat	72,600	Sucrose gradient	c
	37,000	SDS–PAGE	d
Heart			
Human	60,000	SDS–PAGE	e
Rat	34,000	Sed. Equil.*	32
Adipose tissue			
Rat	56,000	SDS–PAGE	f
Chicken	60,000	SDS–PAGE	g
Milk			
Bovine	48,300	Sed. Equil.	h
	58,000	SDS–PAGE	i
	41,700	Sed. Equil.	j
	55,000	SDS–PAGE	k, l, m
Human	78,000	SDS–PAGE	o

* Sed. Equil., sedimentation equilibrium measurements.

[a] J. Augustin, H. Freeze, P. Tejada, and V. W. Brown, *J. Biol. Chem.* **253**, 2912 (1978); [b]L. Becht, O. Schrecker, G. Klose, and H. Greten, *Biochim. Biophys. Acta* **620**, 583 (1980); [c]C. J. Fielding, *Biochim. Biophys. Acta* **178**, 499 (1969); [d]P. E. Fielding, V. G. Shore, and C. J. Fielding, *Biochemistry* **13**, 4318 (1974); [e]J.-S. Twu, A. S. Garfinkel, and M. C. Schotz, *Atherosclerosis* **24**, 119 (1976); [f]S. M. Parkin, B. K. Speake, and D. S. Robinson, *Biochem. J.* **207**, 485 (1982); [g]A. H. Cheung, A. Bensadoun, and C. F. Cheng, *Anal. Biochem.* **94**, 346 (1979); [h]P. H. Iverius and A. M. Östlund-Lindqvist, *J. Biol. Chem.* **251**, 7791 (1976); [i]I. Posner, C. S. Wang, and W. J. McConathy, *Arch. Biochem. Biophys.* **226**, 306 (1983); [j]T. Olivecrona, G. Bengtsson, and J. C. Osborne, Jr., *Eur. J. Biochem.* **124**, 629 (1982); [k]P. K. J. Kinnunen, *Med. Biol.* **55**, 187 (1977); [l]L. Socorro and R. L. Jackson, *J. Biol. Chem.* **260**, 6324 (1985); [m]L. Socorro, C. C. Green, and R. L. Jackson, *Prep. Biochem.* **15**, 133 (1985); [o]C. S. Wang, D. Weiser, P. Alaupovic, and W. J. McConathy, *Arch. Biochem. Biophys.* **214**, 26 (1982).

tein lipase purified from bovine milk and the effect of apoC-II on LpL catalysis.

Structural Domains in the LpL System

A schematic representation of the LpL system is shown in Fig. 1. In this model, the LpL molecule has four functional sites: (1) a glycosaminoglycan-binding site which anchors LpL to the endothelial cell surface; (2) a lipid-binding site which interacts with the surface of the lipoprotein interface; (3) an apoC-II-binding site; and (4) an active (catalytic) site.

As shown in Fig. 1, the lipoprotein substrate consists of a neutral lipid core of triacylglycerols and cholesteryl esters and a surface monolayer of lipids and various apolipoproteins. The major lipid constituents of the lipoprotein monolayer are phosphatidylcholine (PC), sphingomyelin, and unesterified cholesterol. In addition, there are small amounts of phosphatidylethanolamine, tri-, di-, and monoacylglycerols, cholesteryl esters, and fatty acids in the surface. ApoC-II, the activator protein for LpL catalysis, is also a component of the lipoprotein interface. The triacylglycerol substrate molecules are distributed between the surface and core

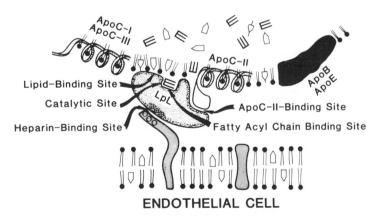

ENDOTHELIAL CELL

FIG. 1. The lipoprotein lipase system. LpL interacts with a heparin sulfate-like molecule at the endothelial cell surface. A lipid-binding site on the enzyme stabilizes the interaction between the enzyme and the surface of a triglyceride-rich lipoprotein. Apolipoproteins C-I, C-II, C-III, B, and E are shown residing in the monolayer surface of the lipoprotein particle; apoC-II is a specific activator for LpL. Triacylglycerol substrates reside primarily in the core of the lipoprotein, with a small percentage in the surface monolayer. The enzyme is shown interacting with a triacylglycerol substrate molecule.

of the lipoprotein particle. Miller and Small[18] have estimated that the concentration of triacylglycerol at the surface of a triglyceride-rich lipoprotein is 2–4%; the maximum solubility of trioleoylglycerol in an egg PC bilayer is <5% by weight.[19]

LpL and Glycosaminoglycan Interactions

LpL binds to heparin immobilized to a matrix of Sepharose. This binding property has been used to facilitate the isolation of the enzyme from a variety of sources.[20] The relative binding affinities of LpL for various classes of immobilized glycosaminoglycans indicate that the enzyme binds most tightly to heparin, followed by heparan sulfate and dermatan sulfate, which are more efficient than chondroitin sulfate.[21] Similarly, LpL is released from cultured endothelial cells more effectively with heparin than with heparan sulfate or dermatan sulfate.[22,23] The removal of heparan sulfate from cell surfaces by a specific platelet endoglucuronidase[22] or heparinase[23] inhibits LpL binding to endothelial cells >90%. The dissociation constant (K_d) for LpL binding to endothelial cell surfaces is $0.14–1.0\ \mu M$ (Table II). Clarke et al.[24] have determined the K_d for the binding of LpL to heparan sulfate in solution and reported a value of $0.16\ \mu M$; the K_d for heparin binding ranged from $0.006–0.044\ \mu M$.[24] These data are consistent with the notion that LpL binds to a cell surface heparan sulfate and is displaced in the presence of circulating heparin to which the enzyme binds more tightly in solution.

Differentiation between Heparin and ApoC-II Binding Sites

That heparin and apoC-II bind to different sites on the LpL molecule was shown by competitive binding studies with immobilized LpL.[25] LpL was coupled to 6-aminohexanoic acid–Sepharose in the presence of 1-ethyl 3-(3-dimethyl aminopropyl)carbodiimide at pH 5.5. Glycine was then added to block remaining reactive sites on the gel. The immobilized LpL retained enzyme activity for trioleoylglycerol and was stimulated

[18] K. W. Miller and D. M. Small, J. Biol. Chem. 258, 13772 (1983).
[19] J. A. Hamilton, K. W. Miller, and D. M. Small, J. Biol. Chem. 258, 12821 (1983).
[20] P.-H. Iverius and A.-M. Östlund-Lindqvist, this volume [41].
[21] G. Bengtsson, T. Olivecrona, M. Hook, J. Riesenfeld, and U. Lindahl, Biochem. J. 189, 625 (1980).
[22] C.-F. Cheng, G. M. Oosta, A. Bensadoun, and R. D. Rosenburg, J. Biol. Chem. 256, 12893 (1981).
[23] K. Shimada, P. J. Gill, J. E. Silbert, W. H. J. Douglas, and B. L. Fanburg, J. Clin. Invest. 68, 995 (1981).
[24] A. R. Clarke, M. Luscombe, and J. J. Holbrook, Biochim. Biophys. Acta 747, 130 (1983).
[25] N. Matsuoka, K. Shirai, and R. L. Jackson, Biochim. Biophys. Acta 620, 308 (1980).

7-fold by apoC-II. The amount of ^{125}I-labeled apoC-II bound to immobilized LpL was not significantly altered in the presence of heparin, nor was the amount of [^{125}I]heparin bound decreased by the addition of apoC-II. Furthermore, LpL that was bound to heparin–Sepharose retained its ability to bind apoC-II; the activator protein did not bind to heparin–Sepharose in the absence of LpL.

Effect of ApoC-II on LpL Binding to Lipid Interfaces

As is shown in Fig. 1, apoC-II is associated with the lipoprotein surface. One possible mechanism by which apoC-II enhances LpL activity is by facilitating the interaction of the enzyme at the lipoprotein interface. However, several lines of evidence support the hypothesis that apoC-II does not increase the binding of LpL to a lipid surface, including the following equilibrium studies.

1. *LpL binding to monolayers in the presence and absence of apoC-II:* A monolayer of 1,2-didecanoylglycerol was spread at surface pressures between 15 and 30 mN/m.[26] ^{125}I-Labeled LpL was injected beneath the monolayer in the presence and absence of apoC-II; hydrolysis was allowed to continue to steady state kinetics and then the film was collected and the interfacial excess of enzyme was determined by radioactive counting. In the presence of apoC-II, the hydrolytic rate was increased. However, less enzyme was at the interface in the presence of apoC-II than in the absence, suggesting that apoC-II does not facilitate the binding of enzyme to the lipid interface.

2. *LpL binding to vesicles in the presence and absence of apoC-II:* LpL (25 μg) and [^{14}C]dipalmitoylphosphatidylcholine (DPPC) vesicles (40 μg) were incubated at 4° in the presence and absence of ^{125}I-labeled apoC-II (20 μg).[27] The lipid–protein mixtures were chromatographed on BioGel A5m at 4° and LpL activity, ^{125}I-labeled apoC-II, and [^{14}C]DPPC were determined. The elution pattern for LpL consisted of two peaks corresponding to lipid-bound and free enzyme. The addition of apoC-II decreased the amount of LpL activity associated with the vesicles; the amount of unbound enzyme correspondingly increased.[27]

3. *Comparisons of K_d for LpL and apoC-II binding to vesicles:* ApoC-II binding. The K_d for apoC-II binding to PC interfaces measured by several techniques ranges from 0.45 to 6.5 μM (Table II). Cardin et al.[28] used a competition assay to determine the binding of apoC-II to lipid. Addition of apoC-II to complexes of dansylated apoC-III/DPPC vesicles

[26] R. L. Jackson, F. Pattus, and G. deHaas, *Biochemistry* **19,** 373 (1980).
[27] K. Shirai, N. Matsuoka, and R. L. Jackson, *Biochim. Biophys. Acta* **665,** 504 (1981).
[28] A. D. Cardin, R. L. Jackson, and J. D. Johnson, *J. Biol. Chem.* **257,** 4987 (1982).

TABLE II

DISSOCIATION CONSTANTS FOR THE LpL SYSTEM

Components	K_d (M)	Molar ratio in complex	Reference
LpL : endothelial cells	0.14	—	22
LpL : endothelial cells	1.0	—	23
LpL : heparin (M_r 7400)	0.044	1 : 1	24
LpL : heparin (M_r 17600)	0.006	1 : 2	24
LpL : heparan sulfate	0.16	1 : 2	24
ApoC-II : DPPC vesicles	6.5	1 : 250	28
ApoC-II : DMPC-ether vesicles	3.0	1 : 20	29
ApoC-II : Trioleoylglycerol/PC particles	0.45–1.1	1 : 85–1 : 120	30
LpL : C_{14}-ether vesicles	0.04	1 : 270	29
LpL : Dansyl-ApoC-II (44–79)			
(with PC vesicles)	0.05	1 : 1	31
(without PC vesicles)	2	1 : 1	31

reduces the fluorescence intensity of the dansyl group as a result of displacement of dansylated apoC-III from the lipid vesicle and a reversal of the phospholipid-induced enhancement in dansyl fluorescence. Analysis of the decrease in fluorescence as a function of apoC-II concentration gives a K_d of 6.5 μM. McLean and Jackson[29] obtained a similar K_d value (3.0 μM) using a direct measurement of the intrinsic fluorescence enhancement of apoC-II when it binds to lipid (Table II). By gel filtration, Tajima et al.[30] reported values ranging from 0.45 to 1.1 μM for trioleoylglycerol particles stabilized by PC.

LpL binding. The K_d for binding of LpL to lipids has been determined by changes in intrinsic tryptophan fluorescence.[29] LpL binding to sonicated vesicles of 1,2-ditetradecyl PC (diC$_{14}$PC ether), a nonhydrolyzable ether PC, causes a 3 nm blue shift and a 40% increase in intrinsic tryptophan fluorescence (Fig. 2, inset). The magnitude of the LpL fluorescence enhancement depends on the concentration of lipid or protein added (Fig. 2). Using an equivalent, independent binding sites model, a K_d value of 0.04 μM was calculated. A comparison of the K_d values for apoC-II binding to a PC interface (0.45 to 6.5 μM) and for LpL (0.04 μM) suggest that LpL binds to a PC interface more strongly than does apoC-II.

4. *LpL binding to apoC-II in the presence and absence of lipid vesicles:* The K_d for LpL and a dansylated fragment of apoC-II in the pres-

[29] L. R. McLean and R. L. Jackson, *Biochemistry* **24**, 4196 (1985).
[30] S. Tajima, S. Yokoyama, and A. Yamamoto, *J. Biol. Chem.* **258**, 10073 (1983).

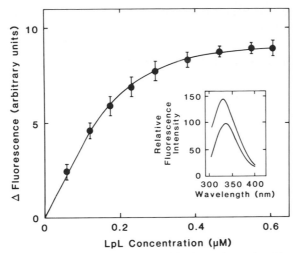

FIG. 2. LpL binding to C_{14}-ether PC vesicles. The inset shows the effect of C_{14}-ether PC vesicles on the fluorescence emission spectrum of LpL. The difference in fluorescence emission of LpL in the absence and presence of PC vesicles (Δ fluorescence) was followed during a titration of 50 μM C_{14}-ether PC vesicles at 25°. (L. R. McLean and R. L. Jackson, unpublished results.)

ence and absence of PC vesicles has been measured by fluorescence energy transfer.[31] In the absence of lipid, a K_d of 2 μM was reported (Table II). Addition of lipid to the system decreased the K_d ~ 40-fold, to 0.05 μM, a value similar to that for LpL–PC interactions (K_d = 0.04 μM), suggesting that the interaction of LpL with lipid enhances the binding of apoC-II to the enzyme. A weak interaction between apoC-II and LpL in the absence of lipid has also been reported by Shirai et al.,[27] who were unable to isolate a complex of LpL and apoC-II by gel chromatography.

Structure of the Active Site

The series of molecular events responsible for LpL catalysis remains speculative primarily because detailed information about the structure of the enzyme is limited. Quinn et al.[1] have reviewed the possible reaction mechanisms and have suggested that LpL exhibits a serine esterase-like catalytic mechanism resembling that for proteases such as α-chymotrypsin and trypsin. In the proposed mechanism,[1] LpL forms an acyl-enzyme complex with the substrate through an active site serine. Evidence that

[31] J. C. Voyta, P. Vainio, P. K. J. Kinnunen, A. M. Gotto, Jr., J. T. Sparrow, and L. C. Smith, *J. Biol. Chem.* **258**, 2934 (1983).

serine is at the active site is based on inhibition studies with the serine protease inhibitors diisopropylfluorophosphate,[32] n-butyl(p-nitrophenyl)carbamide,[33] phenylmethylsulfonyl fluoride (PMSF),[1,16] diethyl-p-nitrophenyl phosphate (E-600)[1] and the transition state analog, α-aminophenylbenzeneboronic acid.[3,34] Studies from this laboratory[1,35] have shown that the half-lives for inhibition of LpL activity by PMSF (1 mM) and E-600 (2 mM) at pH 7.25 and 25° are 15 and 1 min, respectively; the inhibition follows first-order kinetics in both cases consistent with modification of a single amino acid residue. Direct evidence for 1 mol of inhibitor bound per mole LpL has been provided by Socorro et al.[35] using [3H]PMSF. As yet, the existence of histidine in the active site has not been demonstrated by covalent modification of LpL. However, pH rate profiles suggest that a histidine imidazole side chain is in the active site of LpL. Vainio and co-workers[3,34] have also suggested, based on inhibition of LpL activity with various boronic acid derivatives, that LpL forms a tetrahedral complex with an active site serine and a nitrogen group of a histidine imidazole.

Substrates for LpL Catalysis

Lipoprotein Substrates

In this section, we describe four different methods for labeling triglyceride-rich lipoproteins which have been used to assess the mechanism of LpL action. Specifically, these labeled lipoproteins have been used to determine the minimum amino acid sequence requirements of apoC-II for LpL activation, the mechanism of the apoC-II enhancement of LpL catalysis,[36-40] the effect of LpL on the interconversion of plasma HDL subfrac-

[32] J. Chung and A. M. Scanu, J. Biol. Chem. 252, 4202 (1977).

[33] J.-S. Twu, P. Nilsson-Ehle, and M. Schotz, Biochemistry 15, 1904 (1976).

[34] P. Vainio, J. A. Virtanen, and P. K. J. Kinnunen, Biochim. Biophys. Acta 711, 386 (1982).

[35] L. S. Socorro, D. M. Quinn, and R. L. Jackson, unpublished observations (1982).

[36] C. J. Fielding, Biochim. Biophys. Acta 316, 66 (1973).

[37] A. L. Catapano, P. K. J. Kinnunen, W. C. Breckenridge, A. M. Gotto, Jr., R. L. Jackson, J. A. Little, L. C. Smith, and J. T. Sparrow, Biochem. Biophys. Res. Commun. 89, 951 (1979).

[38] K. Shirai, T. J. Fitzharris, M. Shinomiya, H. G. Muntz, J. A. K. Harmony, R. L. Jackson, and D. M. Quinn, J. Lipid Res. 24, 721 (1983).

[39] N. Matsuoka, K. Shirai, J. D. Johnson, M. L. Kashyap, L. S. Srivastava, T. Yamamura, Y. Yamamoto, Y. Saito, A. Kumagai, and R. L. Jackson, Metabolism 30, 818 (1981).

[40] T. J. Fitzharris, D. M. Quinn, E. H. Goh, J. D. Johnson, M. L. Kashyap, L. S. Srivastava, R. L. Jackson, and J. A. K. Harmony, J. Lipid Res. 22, 921 (1981).

tions,[41] and the relative rates of hydrolysis of triacylglycerols in normal and hypertriglyceridemic subjects.[42] To assess the effect of apoC-II on LpL catalysis, VLDL are isolated from two sources which lack apoC-II: (1) guinea pig liver perfusates and (2) patients with apoC-II deficiency.

Preparation of Radiolabeled Guinea Pig Liver Perfusate VLDL.[38,40] Livers (10–12 g) from adult female albino guinea pigs (300–400 g) are perfused in a recirculating system with 65 ml of Krebs–Ringer bicarbonate buffer, pH 7.4, containing 2.15 g (3%) of fatty acid-free bovine serum albumin (BSA) and 65 mg of glucose (0.1%). After 20 min, 10 ml of 0.15 M NaCl, 10 mM Tris-HCl, pH 7.4, containing 0.3 g BSA, 0.15 mmol oleic acid, and 25 μCi [^{3}H]oleic acid are added to the perfusate. After 4 hr, the perfusion is terminated and the perfusate centrifuged for 18 hr at 48,000 rpm in a Beckman type 50 Ti rotor at 4° ($d = 1.006$ g/ml). The top layer is collected, adjusted to a density of 1.019 g/ml by the addition of solid KBr, layered under a solution of KBr ($d = 1.006$ g/ml), and recentrifuged as above. The top layer is removed and dialyzed extensively against 0.15 M NaCl, 50 mM Tris-HCl, pH 7.4. Approximately 90% of the [^{3}H]oleic acid is incorporated into the VLDL–triacylglycerols.

Preparation of Radiolabeled Human VLDL.[39,41] Plasma is obtained from normal human volunteers or patients with apoC-II deficiency[39] by plasmapheresis after 12 hr of fasting. VLDL are isolated at plasma density by ultracentrifugation at 4° in a Beckman type 50 Ti rotor operating at 48,000 rpm for 18 hr. VLDL are removed by aspiration and dialyzed against 0.15 M NaCl, 1 mM EDTA, 10 mM Tris-HCl, pH 7.4, 0.01% NaN$_3$; the VLDL are reisolated by ultracentrifugation. Tri[^{14}C]oleoylglycerol is incorporated into the VLDL by the method of Fielding.[43] Tri[1-^{14}C]oleoylglycerol (100 μCi; 50 mCi/mmol) is dried under ultrapure nitrogen, dissolved in 1 ml of dimethyl sulfoxide, and then added to 4 ml of 10 mM Tris-HCl, pH 7.4, 0.15 M NaCl, 1 mM EDTA. To this solution is added, with rapid stirring, 10 ml of VLDL (5 mg triacylglycerol/ml). After incubation at 37° for 4 hr, the VLDL are dialyzed extensively against 10 mM Tris-HCl, pH 7.4, 0.15 M NaCl, 1 mM EDTA, 0.01% NaN$_3$ at 4°, and then reisolated by ultracentrifugation.

Di[^{14}C]PPC are incorporated into VLDL by the phospholipid-exchange protein (PLEP)-mediated transfer of radiolabeled DPPC from

[41] M.-R. Taskinen, M. L. Kashyap, L. S. Srivastava, M. Ashraf, J. D. Johnson, G. Perisutti, D. Brady, C. J. Glueck, and R. L. Jackson, *Atherosclerosis* **41**, 381 (1982).

[42] M.-R. Taskinen, J. D. Johnson, M. L. Kashyap, K. Shirai, C. J. Glueck, and R. L. Jackson, *J. Lipid Res.* **22**, 382 (1981).

[43] C. J. Fielding, *Biochim. Biophys. Acta* **573**, 255 (1979).

sonicated vesicles.[44] DPPC (20 mg) and di[^{14}C]PPC (30 μCi) are dissolved in chloroform and evaporated *in vacuo*. A buffer of 10 mM Tris-HCl, pH 7.4, 0.15 M NaCl, 1 mM EDTA, 0.01% NaN$_3$ is added to give 6 mg DPPC/ ml. Vesicles are prepared by sonication for 15 min at 42°. To 15 mg of radiolabeled vesicles are added VLDL (7 mg phospholipid), purified PLEP (50 μg) and fatty acid-free BSA (32 mg) in a total volume of 16 ml of 10 mM Tris-HCl, pH 7.4. After incubation at 37° for 2 hr, the radiolabeled VLDL are reisolated by ultracentrifugation. Typically, this procedure has allowed for the preparation of VLDL containing approximately 0.2 μCi of radiolabeled DPPC/mg of VLDL phospholipid.

Incorporation of ApoC-II into VLDL.[39,40] Purified human apoC-II[45] (1 mg/ml in 6 M guanidine-HCl, 10 mM Tris-HCl, pH 8.0) or human plasma HDL ($d = 1.063–1.210$ g/ml) are incubated for 1 hr at 37° (0–50 μg apoC-II) with VLDL (1 mg triglyceride) in 1.0 ml of 10 mM Tris-HCl, pH 7.4, 0.15 M NaCl, 1 mM EDTA buffer. Following incubation, VLDL are reisolated by ultracentrifugation. The amount of apoC-II incorporated into VLDL is measured by radioimmunoassay.[46]

LpL-Catalyzed Hydrolysis of Radiolabeled Lipoproteins.[38–42] The reaction mixtures contain various amounts of VLDL (0.5–3.0 mM triacylglycerol), 30 mg of fatty acid-free BSA/μmol of VLDL triglyceride, and various amounts of apoC-II (0–5 μg/mg VLDL triacylglycerol) in a total volume of 0.25 ml of 0.1 M Tris-HCl, 0.15 M NaCl, at pH 8.5 for trioleoylglycerol hydrolysis or pH 7.0 for DPPC hydrolysis. The amount of LpL added is chosen so that the release of radiolabeled fatty acids is linear for 30 min at 37°. For determination of trioleoylglycerol hydrolysis, lipids are extracted by the method of Belfrage and Vaughan.[47] To each sample is added 3.25 ml of methanol : chloroform : heptane (1.0 : 0.9 : 0.7, v/v/v) and 1.0 ml of 0.14 M potassium borate buffer, pH 10.5. The mixtures are vortexed for 1 min and centrifuged at room temperature for 10 min at 3000 rpm. One milliliter of the upper phase is removed and radioactivity is determined by liquid scintillation counting. To measure radiolabeled DPPC hydrolysis, lipids are extracted by the method of Bligh and Dyer.[48] To each sample is added 0.55 ml of distilled water, 2 ml of methanol, and 1 ml of chloroform. After vortexing for 1 min, 1 ml of chloroform and 1 ml

[44] R. L. Jackson, D. Wilson, and C. J. Glueck, *Biochim. Biophys. Acta* **557**, 79 (1979).
[45] R. L. Jackson, H. N. Baker, E. B. Gilliam, and A. M. Gotto, *Proc. Natl. Acad. Sci. U.S.A.* **74**, 1942 (1977).
[46] M. L. Kashyap, L. S. Srivastava, C. Y. Chen, G. Perisutti, M. Campbell, R. F. Lutmer, and C. J. Glueck, *J. Clin. Invest.* **60**, 171 (1977).
[47] P. Belfrage and M. Vaughan, *J. Lipid Res.* **10**, 341 (1969).
[48] E. G. Bligh and W. J. Dyer, *Can. J. Biochem. Physiol.* **37**, 911 (1959).

of distilled water are added with vortexing. The samples are centrifuged for 10 min at 3000 rpm. The entire lower phase is removed and evaporated under N_2. The lipids are redissolved in 50 μl of chloroform, applied directly to silica gel G plates, and developed in chloroform : methanol : water (70 : 30 : 4, v/v/v). The lipids are visualized with iodine vapor, the spots corresponding to lysoPC and PC are scraped from the plates, and radioactivity is determined by liquid scintillation counting.

Dansyl Phosphatidylethanolamine-Labeled VLDL. Johnson and co-workers[42,49,50] have developed a fluorescent method which provides an accurate, convenient, and sensitive means of monitoring continuously the lipolysis of VLDL by LpL. The assay procedure is based on the hydrolysis of the fluorescent phospholipid, dansyl(5,5'-dimethylaminonaphthalene-1-sulfonyl)phosphatidylethanolamine (DPE), which is incorporated into VLDL by bath sonication. Since there is a linear relationship between the fluorescence intensity increase at 490 nm and the rate of release of [^{14}C]oleic acid, this method has been used to monitor relative lipolytic rates. DPE (Molecular Probes, Inc.) is dissolved in absolute ethanol at 25 mg/ml. Labeling of VLDL is performed by adding DPE to VLDL (2 mg triacylglycerol/ml) in 50 mM Tris-HCl, pH 7.4, 0.15 M NaCl, 0.01% NaN$_3$ to give a molar ratio of DPE to triacylglycerol of 1 : 40. The DPE–lipoprotein mixture is vortexed for 15 sec, bath sonicated at 23° for 6–8 min in a Cole-Parmer (Model 8846-50) bath sonicator, and dialyzed against the Tris-HCl buffer. With these conditions, all of the DPE is incorporated into the lipoprotein particle. The enzyme reaction mixture contains DPE–VLDL (0.1–1.0 mg triacylglycerol) and 30 mg of fatty acid-free bovine serum albumin in 1.0 ml of Tris-HCl buffer. The excitation wavelength is 340 nm and emission at 490 nm is monitored continuously. Temperature is maintained at 24° with a recirculating water bath and measured by a temperature probe. The enzyme reaction is initiated by the addition of LpL. The data are calculated by dividing the fluorescence intensities by the initial fluorescence intensity of the sample prior to addition of LpL. This method has been used to determine the relative rates of LpL catalysis for triglyceride-rich lipoproteins from patients with various forms of hyperlipoproteinemia.[42,51,52]

[49] J. D. Johnson, M.-R. Taskinen, N. Matsuoka, and R. L. Jackson, *J. Biol. Chem.* **255**, 3461 (1980).
[50] J. D. Johnson, M.-R. Taskinen, N. Matsuoka, and R. L. Jackson, *J. Biol. Chem.* **255**, 3466 (1980).
[51] K. Saku, C. Cedres, B. McDonald, B. A. Hynd, B. W. Lui, L. S. Srivastava, and M. L. Kashyap, *Am. J. Med.* **77**, 457 (1984).
[52] A. M. Stoline, K. Saku, B. A. Hynd, and U. L. Kashyap, *Metabolism,* **34**, 30 (1984).

Monolayer Lipid Substrates

As is illustrated in Fig. 1, the interaction of LpL with a triglyceride-rich lipoprotein occurs at a phospholipid interface which contains the triacylglycerol substrate. In this section, two monolayer systems are described, one of which uses pure films of short chain lipids[26,53,54] and the other a mixed lipid system of long chain phospholipids and triacylglycerols.[55–57a-c] Each of these systems has its advantages and disadvantages for studying LpL action. However, both lipid systems allow for an understanding of LpL catalysis at a lipid interface under defined conditions. Monolayer systems also allow one to address questions as to the role of lipid structure, composition, and packing and apolipoprotein requirements for LpL catalysis. In addition, by using ^{125}I-labeled LpL or apoC-II it is possible to determine the amount of enzyme or activator protein at the lipid interface.

Monolayers of Short Chain Lipids. The most common monolayer apparatus for short chain lipids is the constant pressure, zero-order trough method described by Verger and de Haas.[53] This method has been employed to assess LpL action using various short chain fatty acyl glycerols and phospholipids.[26,54] The advantage of the short chain lipids is that the products of catalysis are water soluble, thus negating the use of albumin to remove them from the interface. The apparatus consists of a reaction trough and a reservoir trough separated by a small opening. When the products of LpL catalysis leave the interface of the reaction trough, a surface balance senses the changes in pressure; the surface balance is connected to a drive assembly which forces lipid from the reservoir trough into the reaction trough. Movement of the bar thus allows for measurement of LpL catalysis. The advantage of this system is that the surface pressure (substrate concentration) is maintained constant. This is an important consideration since it is known that LpL activity is dependent on surface pressure.[26,55] The zero-order monolayer method has been used to determine the effect of surface pressure on the LpL-catalyzed

[53] P. Verger and G. H. de Haas, *Annu. Rev. Biophys. Bioeng.* **5,** 77 (1976).

[54] P. Vainio, J. A. Virtanen, P. K. J. Kinnunen, A. M. Gotto, Jr., J. T. Sparrow, F. Pattus, P. Bourgis, and R. Verger, *J. Biol. Chem.* **258,** 5477 (1983).

[55] P. Vainio, J. A. Virtanen, P. K. J. Kinnunen, J. C. Voyta, L. C. Smith, A. M. Gotto, Jr., J. T. Sparrow, F. Pattus, and R. Verger, *Biochemistry* **22,** 2270 (1983).

[56] R. A. Demel, K. Shirai, and R. L. Jackson, *Biochim. Biophys. Acta* **713,** 629 (1982).

[57a] R. A. Demel, P. J. Dings, and R. L. Jackson, *Biochim. Biophys. Acta* **793,** 399 (1984).

[57b] R. A. Demel and R. L. Jackson, *J. Biol. Chem.* **260,** 9589 (1985).

[57c] R. L. Jackson and R. A. Demel, *Biochem. Biophys. Res. Commun.* **128,** 670 (1985).

hydrolysis of 1,2-didecanoylglycerol (dicaprin).[26] The rate of LpL catalysis increases linearly from 11 mN/m to the collapse pressure of the lipid (30 mN/m). The absence of LpL activity at surface pressures <11 mN/m is presumably due to surface denaturation of the enzyme. In the presence of apoC-II, the reaction velocity increases at all surface pressures. The zero-order monolayer method has certain limitations for assessing LpL action. With short chain lipids, the dependency of LpL activity on apoC-II is minimal (<3-fold enhancement of activity at 25 mN/m). In contrast, LpL activation by apoC-II with long chain triacylglycerols and phospholipids in bulk phase systems (see below) is 10- to 60-fold. Another disadvantage is that apoC-II is tensoactive and at high concentrations causes an increase in surface pressure at the lipid–air interface. Since LpL catalysis is measured by changes in surface pressure, the addition of large amounts of apolipoproteins may complicate the interpretation of the kinetic data.

Monolayers of Long Chain Lipids. Because of the limitations of the zero-order trough described above for the study of LpL catalysis of physiologic lipids, we have developed another monolayer method to determine the mechanism of action of LpL at a lipid interface as illustrated in Fig. 3.[56,57a] The interfacial measurements are performed using a 15-ml Teflon trough (5.4 × 5.9 × 0.5 cm); the subphase is stirred with a magnetic bar. ^{14}C-Radiolabeled lipids are spread from chloroform solutions to the desired interfacial surface pressure; pressure is determined with a recording Beckman LM 500 electrobalance equipped with a platinum plate 1.96 cm wide. Surface radioactivity is measured by a gas-flow (helium/1.3% butane) detector (Nuclear Chicago, Model 8731). The Teflon trough contains an extended corner so that additions are made to the subphase without disturbing the interface. The method is based on recording the LpL-mediated loss of radioactivity in the presence of fatty acid-free BSA from the surface and is illustrated in Fig. 3. Egg PC containing 5 mol% tri-[1-^{14}C]oleoylglycerol is spread to a final surface pressure between 20–30 mN/m. To the subphase is then added the following: heparin (20 μg), fatty acid-free BSA (100 μg), LpL (2–20 ng), and the appropriate amount of apoC-II. Since the rate of the LpL-catalyzed hydrolysis of trioleoylglycerol is >100 times that of phospholipid and the triacylglycerol is <5% of the total lipid, there is only a minimal change in surface pressure during LpL catalysis of the triacylglycerol. Thus, it is possible to graphically determine reaction rates by measuring the maximal slope of the kinetic curve (Fig. 3).[56,57a–c]

Precautions. Monolayer methods described above employ extremely small quantities of lipid. Thus, a number of precautions with the equipment and experimental protocols must be taken to maximize reproducibil-

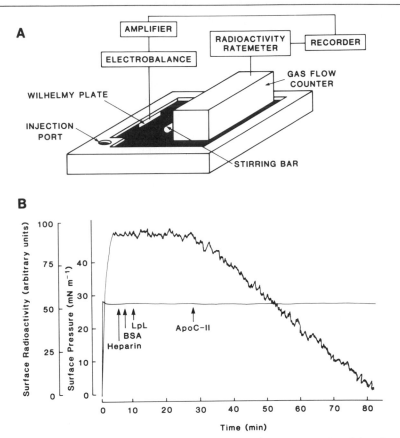

FIG. 3. (A) Scheme for the measurement of surface pressure and radioactivity. See text for details of the system. (B) Desorption of [^{14}C]oleic acid from a monolayer during the LpL-catalyzed hydrolysis of tri[^{14}C]oleoylglycerol in a monolayer of egg PC. At the indicated times, heparin, BSA, LpL, and apoC-II were added to the subphase as described in the text.

ity. The Teflon trough is cleaned after each experiment with soap and gentle brushing, extensive washing with deionized glass-distilled water and ethanol, and is wiped dry with tissue. The platinum plate is cleaned after each experiment in sulfuric acid : H$_2$O$_2$ (1 : 1, v/v) for 30 sec followed by extensive washing with deionized glass-distilled water. Since the packing density of lipids is dependent on temperature, the interfacial measurement is performed in a thermostatically controlled box. The purity of all lipids is routinely checked by thin-layer chromatography in two solvent systems: petroleum ether : diethyl ether : formic acid (60 : 40 : 1.5, v/v/v) and chloroform : methanol : acetic acid : water (90 : 30 : 8 : 2.8, v/v/v/v). Lipids are stored at 4° in chloroform under nitrogen. A decrease in surface pressure after the addition of BSA to the subphase is usually indicative of impurities. Because of the tensoactive properties of BSA, the studies are carried out at surface pressures >20 mN/m; BSA is surface active up to 18 mN/m. The amount of BSA required to maximally remove 0.8 nmol of [^{14}C]oleic acid from the lipid interface is approximately 20 μg. However, 100 μg of fatty acid-free BSA appears to prevent nonspecific binding of LpL to the Teflon surface. Stock solutions of LpL (approximately 1 mg/ml) are diluted with 50% glycerol–10 mM Tris-HCl, 0.15 M NaCl to give a final concentration of 5 μg/ml. Dilutions of LpL are prepared daily. The most common problem that has been experienced studying LpL catalysis is the preparation of the apoC-II solutions. It is imperative that apoC-II be dissolved in 6 M guanidine and added directly to the subphase; dilutions of the apolipoproteins are made in 6 M guanidine, 10 mM Tris-HCl, pH 7.4, 0.15 M NaCl. Presumably, storage of apoC-II in the absence of guanidine results in aggregation of the activator protein.

Model Lipid Substrates

Two classes of model lipid substrates will be considered in this section. First, water-soluble substrates such as *p*-nitrophenyl esters, and second, aggregated substrates such as PC solubilized with Triton X-100 or PC/apolipoprotein complexes. Water-soluble substrates are operationally defined as substrates which are dispersed as monomers at the concentrations used for the enzyme assay. For example, dihexanoyl PC (diC$_6$PC), at concentrations below its critical micellar concentration (cmc), is water soluble; at concentrations >4 times the cmc the predominant structure is a micelle.[58] Water-soluble substrates offer several advantages to the study of LpL action: (1) the products of catalysis are water soluble; albumin or Ca^{2+} is not required in the system to bind the released fatty acids; (2) certain water-soluble products may be continuously monitored as the reaction proceeds (nitrophenyl esters, fluorescent lipids, and thiolester analogs of various lipids); and (3) the enzyme kinetic parameters K_m and V_{max} are more easily interpreted for monomeric water-soluble substrates than for lipids contained in aggregated structures, such as lipoproteins or emulsions.

p-Nitrophenyl Esters.[6,15] A major advantage of *p*-nitrophenyl esters for studying LpL action is that the product of catalysis, *p*-nitrophenoxide, strongly absorbs light at 400 nm, allowing continuous monitoring of the reaction time course. A 15 mM solution of *p*-nitrophenyl butyrate (PNPB) or *p*-nitrophenyl acetate (PNPA), is prepared in neat, dry acetonitrile. A

---

[58] T. T. Allgyer and M. A. Wells, *Biochemistry* **18**, 4354 (1979).

typical reaction mixture contains PNPB or PNPA (0.1–1 mM) in 1.0 ml of 0.1 M sodium phosphate, pH 7.25, 0.15 M NaCl, 1% acetonitrile, containing 10 μg of heparin. The reaction rate is measured continuously at 400 nm using a 1-cm path length cell maintained at 25.0 ± 0.1° by circulating water through the cell holder of the spectrophotometer. Baseline velocities for spontaneous hydrolysis are determined prior to the addition of LpL. After 2 min, LpL (10 μl, 1–10 μg) is added, the velocity is established over a further 2-min period, and then apoC-II (1–20 μg) or DPPC vesicles (5–100 μg) are added. The reaction is followed for a total of 15 min. Initial velocities (v_i) of LpL catalysis are calculated from the slopes of the time courses over <15% reaction using the following equation:

$$v_i = \frac{(\Delta A_{400}/\text{min})(1 \text{ ml}/1000 \text{ ml})}{\varepsilon_{PNP}[\text{mg LpL}]} \times 10^6$$

where v_i = initial velocity (μmol/min/mg LpL), ΔA_{400} = change in A_{400} above background (without LpL), and ε_{PNP} = absorptivity constant of the p-nitrophenolate ion, 14,775 M^{-1} cm^{-1}.

Fluorescent Phosphatidylcholine.[59] The fluorescent phospholipid, 1-acyl-2-[6-(7-nitro-2,1,3 benzoxadiazol-4-yl)amino]caproylphosphatidylcholine (C_6-NBD-PC), has been used to study LpL catalysis. LpL catalyzes the hydrolysis of C_6-NBD-PC in the *sn*-1 position, yielding the fluorescent product lyso-NBD-PC. Associated with LpL catalysis of the C_6-NBD-PC is a 50-fold fluorescence enhancement with no shift in the emission maximum at 540 nm. The increase in fluorescence intensity allows continuous monitoring of LpL catalysis. The reaction mixture contains 10 mM Tris-HCl, pH 7.4, 0.1 M KCl, and 5 × 10^{-8} to 10^{-6} M C_6-NBD-PC (Avanti Biochemical Corp.) with or without apoC-II (2 μg) in a total volume of 1.0 ml. The reaction is initiated by the addition of LpL and is monitored continuously with excitation at 470 nm and emission at 540 nm. With C_6-NBD-PC, apoC-II has no effect on the rate of LpL catalysis.[59] Furthermore, the interfacial activation of hydrolysis of C_6-NBD-PC at the cmc (0.2 μM),[60] which is exhibited by phospholipase A_2, is not evident for LpL.[59]

Thiolester Substrates. The thiolester analogs of short chain, water-soluble phospholipids and di- and triacylglycerol have been used as substrates for various lipolytic enzymes.[61,62] The advantage of these sub-

[59] L. A. Wittenauer, K. Shirai, R. L. Jackson, and J. D. Johnson, *Biochem. Biophys. Res. Commun.* **118,** 894 (1984).

[60] J. W. Nichols and R. E. Pagano, *Biochemistry* **20,** 2783 (1981).

[61] M. Shinomiya, D. E. Epps, and R. L. Jackson, *Biochim. Biophys. Acta* **795,** 212 (1984).

[62] A. J. Slotboom, H. M. Verheij, and G. H. de Haas, *in* "Phospholipids. New Comprehensive Biochemistry" (J. N. Hawthorne and G. B. Ansell, eds.), Vol. 4, p. 359. Elsevier, Amsterdam, 1982.

strates is that enzyme catalysis yields a free sulfhydryl group which in the presence of the chromogenic reagent 5,5′-dithiobis(2-nitrobenzoic acid) (DTNB) produces a yellow color which can be monitored continuously at 412 nm. Shinomiya *et al.*[61] prepared the dithio analog of dihexanoyl-PC (dithioC$_6$PC) and compared the LpL kinetic parameters to the corresponding acyloxyester lipid. The reaction mixture contains 0.1 M Tris-HCl, pH 8.0, 0.15 M NaCl, 0.5 mM dithioC$_6$PC, heparin (15 μg), and 0.5 mM DTNB in a total volume of 0.25 ml. LpL is added to initiate the reaction and the absorbance is monitored continuously at 412 nm. LpL activities are calculated based on a molar extinction coefficient for the thionitrobenzoate anion of 14,000 M^{-1} cm^{-1}. The maximal reaction velocity for the acyloxyester lipid was 40 times greater than the corresponding thio derivative. In addition, a reduced rate of LpL catalysis was observed for 3-butyrylthio-1,2-dibutyryloxypropane as compared to the corresponding tributyrylglycerol ester analog. These results may be the consequence of the reduced ability of sulfur atoms to form hydrogen bonds within a tetrahedral transition state complex with a structure analogous to that observed in several serine proteases.

Short-Chain Phosphatidylcholines.[13] DiC$_6$PC and diheptanoyl-PC (diC$_7$PC) have sufficiently high cmc's to allow comparison of LpL catalytic rates above and below the cmc of the substrate. Such substrates have been used to demonstrate interfacial activation by phospholipase A$_2$.[62] The cmc of diC$_6$PC is 9.3 mM, whereas for diC$_7$PC it is 1.0 mM.[13] The reaction mixture contains 0.1 M Tris-HCl, pH 8.0, heparin (5 μg), and 3.5–60 mM PC in a total volume of 0.1 ml; the reaction is initiated by addition of LpL. Rates of LpL catalysis are determined at 30° for 15–60 min; <20% of the lipid is hydrolyzed. At the appropriate time points, the reaction mixture is applied directly to silica gel G thin-layer chromatography plates. The plates are developed in chloroform : methanol : water (65 : 35 : 5, v/v/v). The lipids are visualized with iodine vapor, the spots corresponding to lysoPC and PC are scraped from the plate, and the lipid content is determined by the method of Bartlett[63] for phospholipid phosphorus. With this system, LpL activity is not enhanced at the cmc, nor does apoC-II alter the rate of LpL-catalyzed hydrolysis.[13]

Phosphatidylcholine Emulsions.[64–66a,b] Triton X-100 and Triton N-101 mixed with PC has been used to compare LpL activities with lipid substrates of various fatty acyl chain lengths. With this system it is also

[63] G. R. Bartlett, *J. Biol. Chem.* **234**, 466 (1959).
[64] M. Shinomiya, L. R. McLean, and R. L. Jackson, *J. Biol. Chem.* **258**, 14178 (1983).
[65] M. Shinomiya, R. L. Jackson, and L. R. McLean, *J. Biol. Chem.* **259**, 8724 (1984).
[66a] M. Shinomiya and R. L. Jackson, *Biochim. Biophys. Acta* **794**, 177 (1984).
[66b] L. R. McLean, A. Subramanian, and R. L. Jackson, *Biophys. J.* **49**, 532a (1986).

possible to study the role of lipid organization in LpL catalysis since the physical form of the substrate can be varied from bilayer structures to micelles by increasing the amount of detergent in the assay mixture.[66-68a,b] At molar ratios of Triton X-100 to dimyristoylphosphatidylcholine (DMPC) of 0 to 1.0, the structure of the substrate is bilayers; at molar ratios >2.0, only micellar structures are formed.[68a,b] Thus, the effect of lipid structure and apoC-II on the LpL-catalyzed hydrolysis of PC can be examined.

Two methods are used to follow the LpL-catalyzed hydrolysis of PC: pH-stat and release of radioactive fatty acid products. Substrates are prepared as follows: unlabeled DMPC (2.5 μmol) with or without the corresponding radiolabeled PC dissolved in chloroform are evaporated to dryness with ultrapure nitrogen and then lyophilized for 15 min. To the dry lipid is added 1.0 ml of 0.1 M Tris-HCl or 0.5 mM Bicine, pH 8.0, 0.15 M NaCl containing various amounts (1.6–7.5 μmol) of Triton X-100; the Bicine buffer is used in the pH-stat experiments. After vortexing, the lipid mixtures are incubated at 37° for 30 min. The assay mixtures contain 0.15 M NaCl, Triton X-100/DMPC substrate (0.16 mM PC), fatty acid-free BSA (8 mg/ml) or 1 mM CaCl$_2$, heparin (1 μg/ml), and apoC-II (0–200 μg/ml) in a total volume of 0.25 ml of 0.1 M Tris-HCl or 10 ml of 0.5 mM Bicine, pH 8.0; LpL is added to initiate the reaction. Temperature is controlled to ±0.1° using a circulating water bath. The pH-stat experiments are performed using a Radiometer automatic titrator with 0.01 M NaOH as titrant. With radiolabeled substrates, the enzyme reactions are terminated by the addition of 3.25 ml of methanol : chloroform : heptane (100 : 88 : 70, v/v/v) and 1.0 ml of 0.14 M potassium borate buffer, pH 10.5, according to the procedure of Belfrage and Vaughan.[47] Released fatty acids are determined by liquid scintillation counting of the upper phase. Initial rates are determined for 10–15% hydrolysis.

Phosphatidylcholine/Apolipoprotein Discoidal Structures.[38] Plasma apolipoproteins are amphipathic molecules whose function is to bind and transport lipid. A common property of apolipoproteins is their ability, under certain defined conditions, to solubilize phospholipid liposomes forming discoidal structures with dimensions approximately 50 × 150 Å. Shirai *et al.*[38] have prepared discoidal structures of DPPC containing various ratios of apoC-II and apoC-III at a constant protein concentration

[67a] D. Lichtenberg, R. J. Robson, and E. A. Dennis, *Biochim. Biophys. Acta* **737**, 285 (1983).
[67b] M. L. Kashyap, L. S. Srivastava, B. A. Hynd, P. S. Gartside, and G. Perisutti, *J. Lipid Res.* **22**, 800 (1981).
[68a] A. A. Ribeiro and E. A. Dennis, *Biochemistry* **14**, 3746 (1975).
[68b] K. Shirai, D. A. Wisner, J. D. Johnson, L. S. Srivastava, and R. L. Jackson, *Biochim. Biophys. Acta* **712**, 10 (1982).

in order to determine the effect of apoC-II on the kinetic parameters of PC hydrolysis by LpL. The maximal rate of catalysis occurred in those structures containing on average one molecule of apoC-II per particle. To prepare apolipoprotein–PC discoidal complexes, DPPC (10 mg) containing 4 μCi of di[^{14}C]PPC is dissolved in chloroform and evaporated to dryness with nitrogen. Lipid-free, lyophilized apoC-II and apoC-III isolated from triglyceride-rich lipoproteins are dissolved in 6 M guanidine-HCl, 10 mM Tris-HCl, pH 8.0, to give 5–10 mg protein/ml and then dialyzed 6 hr at room temperature against 10 mM Tris-HCl, pH 7.4, 0.15 M NaCl, 1 mM EDTA, 0.01% NaN$_3$. After dialysis, apoC-II and apoC-III are mixed in various proportions (final protein concentration 10 mg/3.5 ml of the Tris-HCl buffer) and added to the dry DPPC. The apolipoprotein–DPPC mixtures are incubated at 42° for 24 hr with gentle shaking. The turbid lipid suspensions become clear after ~10 hr, indicating complex formation. The apolipoprotein– DPPC complexes are separated from the free apolipoprotein by ultracentrifugation at d = 1.15 g/ml in KBr for 18 hr at 48,000 rpm and 15°. The content of apoC-II and apoC-III in the isolated complexes is determined by specific radioimmunoassays.[46,67a,b] To determine the rates of LpL hydrolysis, the incubation mixtures contain the apolipoprotein–DPPC complexes (525 μg PC/ml), fatty acid-free BSA (2 mg/ml), and heparin (50 μg/ml) in 0.1 M Tris-HCl, pH 7.4, in a volume of 0.2 ml; LpL is added to initiate the reaction. The incubation is performed at 37° for 60 min; over this time course the rate of fatty acid release is linear. The enzyme reaction is terminated by extraction using the method of Bligh and Dyer.[48] To each sample of 0.2 ml is added 0.6 ml of distilled water, 2 ml of methanol, and 1 ml of chloroform. After vortexing for 1 min, 1 ml of chloroform and 1 ml of distilled water are added with vortexing. The samples are centrifuged for 10 min at 3000 rpm. The entire lower phase is removed and evaporated under N$_2$. The lipids are redissolved in 50 μl of chloroform and applied directly to silica gel G plates which are developed in chloroform : methanol : water (70 : 30 : 4, v/v/v). The lipids are visualized with iodine vapor, the spots corresponding to lysoPC and PC are scraped from the plate, the radioactivity is determined by liquid scintillation counting.

Kinetics of LpL Catalysis

An important question in LpL catalysis is whether a single active site is responsible for the hydrolysis of various LpL substrates. Although the rates of catalysis are markedly different, LpL catalyzes the hydrolysis of di- and triacylglycerols, phosphatidylcholines, phosphatidylethanolamines, and water-soluble esters, such as p-nitrophenylesters. A number of studies strongly support the contention that the same active site is responsible for the hydrolysis of these various lipid substrates. For example, Quinn et al.[16] showed that 1 mM PMSF inhibits the LpL-catalyzed hydrolysis of Triton X-100-emulsified trioleoylglycerol and PNPB to the same extent. In addition, Fab fragments prepared from the γ-globulin fraction of antiserum raised against the bovine milk enzyme inhibit LpL hydrolysis of Triton X-100-emulsified trioleoylglycerol and sonicated vesicles of DMPC to the same extent.[68,68a]

Steady State Kinetics

Quinn et al.[1,16] have recently extended the Verger and de Haas[53] kinetic model for heterogeneous enzyme catalysis to include LpL : apoC-II interactions. How this model can explain several observations in the LpL literature has been reviewed.[1] In this section, we focus on application of the model to the various experimental systems described above. Scheme I

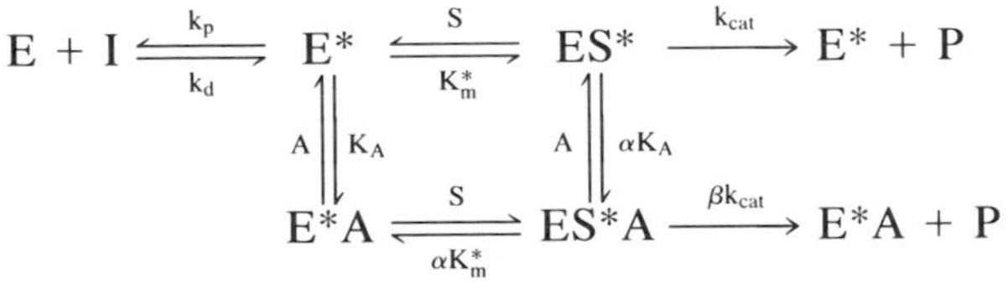

$$E + I \underset{k_d}{\overset{k_p}{\rightleftharpoons}} E^* \underset{K_m^*}{\overset{S}{\rightleftharpoons}} ES^* \xrightarrow{k_{cat}} E^* + P$$

$$A \updownarrow K_A \qquad A \updownarrow \alpha K_A$$

$$E^*A \underset{\alpha K_m^*}{\overset{S}{\rightleftharpoons}} ES^*A \xrightarrow{\beta k_{cat}} E^*A + P$$

SCHEME 1.

outlines the proposed kinetic model.[1] In this model, E = LpL; I = lipid interface; E^* = LpL bound to the lipid interface with active site unoccupied; S = substrate; ES^* = Michaelis complex of LpL and S; P = products of LpL catalysis; A = apoC-II; and K_A = dissociation constant of the LpL–apoC-II complex. Since there are two thermodynamically indistinguishable pathways to ES^*A, α is the factor by which the dissociation constant of ES^*A is changed when LpL and apoC-II interact along either pathway. The factor β accounts for a change in the catalytic rate constant, k_{cat}. In Scheme I, there are two routes to the formation of products: one requires apoC-II (ES^*A) and the other occurs in the absence of activator protein (ES^*). Thus, the proposed scheme describes apoC-II as a nonessential activator of LpL catalysis: in the absence of apoC-II, substrate turnover occurs via ES^*.

Scheme I excludes an effect of apoC-II on LpL binding to the interface. Several lines of evidence support this contention. (1) Fragments of apoC-II which do not bind to lipid activate LpL catalysis as effectively as native apoC-II[37] and bind to LpL molecules in the presence and absence of an interface.[31] (2) The binding data of Table I support the hypothesis that LpL binds more tightly to a lipid interface than does apoC-II. (3) In the LpL-catalyzed hydrolysis of PC in guinea pig VLDL, apoC-II has little effect on K_m^{app}, a term which contains k_d/k_p (see below), the dissociation constant for LpL : lipid complexes. (4) The effect of apoC-II on a variety of substrate structures, including micelles, vesicles, emulsions, monolayers, and monomeric substrates, is independent of the organization of the substrate molecules and depends only on the fatty acyl chain length of the substrate.[64] (5) In the absence of a lipid interface, apoC-II is a noncompetitive inhibitor of the LpL-catalyzed hydrolysis of PNPB with a dissociation constant of the LpL–apoC-II complex of 0.6 μM.[16] In the presence of DPPC vesicles, the binding of LpL to the phospholipid interface decreases the dissociation constant for the LpL–apoC-II complex to 0.13 μM.[38] Thus, the binding of LpL to lipid enhances the binding of the activator protein.

Following LpL binding to the substrate-containing particle interface, the enzyme interacts with the substrate in two dimensions. Therefore, substrate concentrations and K_m^* are expressed in terms of molecules per unit area of interface. The apparent Michaelis constants may be obtained from the initial velocity (v_i) expression[53]:

$$v_i = \frac{V_{max}^{app} S_b}{K_m^{app} + S_b} \tag{1}$$

where the substrate concentration per unit volume, $S_b = S \cdot I/V$, and I/V is the area of interface per unit volume. The subscript b refers to volume (bulk phase) concentrations, and V_{max} and K_m^{app} are experimentally measured parameters. Since S_b is substrate concentration per unit volume, V_{max} and K_m^{app} may be obtained from values of v_i as a function of total substrate concentration. When not all of the substrate molecules are accessible at the interface (e.g., triacylglycerols in lipoproteins), S is reduced by the fraction of inaccessible substrate molecules. The distribution of substrate between core and surface is $S = S_{total} K_s$, where K_s is the equilibrium constant for the distribution. A steady state kinetic derivation[1] based on the kinetic model of Scheme I yields:

$$V_{max}^{app} = \frac{k_{cat} E_T S(1 + \beta A/\alpha K_A)}{K_m^*(1 + A/K_A) + S(1 + A/\alpha K_A)} \tag{2}$$

$$K_m^{app} = \frac{k_d}{k_p} \frac{K_m^* S}{K_m^*(1 + A/K_A) + S(1 + A/\alpha K_A)} \tag{3}$$

where the K_s term is incorporated into S, the concentration of substrate at the interface. Common to V_{max} and K_m^{app} is the term:

$$\theta = \frac{S}{K_m^*(1 + A/K_A) + S(1 + A/\alpha K_A)} \tag{4}$$

which is the fractional occupancy of LpL active sites by substrate molecules.[40]

A summary of the effect of apoC-II on the enzyme kinetic parameters, K_m^{app} and V_{max}, is given in Table III. The major effect of apoC-II on the hydrolysis of apoC-II deficient human or guinea pig VLDL–triacylglycerols is to decrease the K_m^{app} with a limited effect on V_{max}.[38–40] For the LpL-catalyzed hydrolysis of DPPC and PNPB, apoC-II alters V_{max} with little effect on K_m^{app}. ApoC-II has no discernable effect on diC$_6$PC hydrolysis. Although V_{max} for DPPC hydrolysis increases in the presence of apoC-II, it should be emphasized that this does not necessarily imply that k_{cat} is increased. If k_{cat} is unchanged by the addition of apoC-II, then $\beta = 1$. In this case, the effects of apoC-II on K_m^{app} and V_{max} for DPPC and trioleoylglycerol hydrolysis may be predicted by the relative values of the two addends in the denominator of the occupancy factor [Eq. (4)] as described below (cf. refs. 1, 38, 40).

1. For $K_m^* (1 + A/\alpha K_a) <<S(1 + A/K_A)$, the equilibrium for LpL lies toward ES^* and ES^*A. Thus, when $\beta = 1$:

$$V_{max}^{app} = k_{cat}E_T \tag{5}$$

$$K_m^{app} = \frac{k_d}{k_p} \frac{K_m^*}{(1 + A/\alpha K_A)} \tag{6}$$

Therefore, only K_m^{app} is affected by apoC-II (trioleoylglycerol case).

2. For $K_m^* (1 + A/K_A) >>S(1 + A/\alpha K_A)$ and $\beta = 1$:

$$V_{max}^{app} = \frac{k_{cat}E_T S(1 + A/\alpha K_A)}{K_m^*(1 + A/K_A)} \tag{7}$$

$$K_m^{app} = \frac{k_d}{k_p} \frac{S}{K_m^*(1 + A/K_A)} \tag{8}$$

Here, only V_{max} is affected when $A < K_A$ (DPPC case). For the water-soluble substrates PNPA and PNPB, inhibition of LpL fits a noncompetitive kinetic model ($\alpha = 1$ in Scheme I) of homogeneous catalysis. In this system, the K_A in volume units is 0.26–0.83 μM.[16] Equilibrium binding studies using a dansylated apoC-II fragment[31] (Table I) give a similar value of 2 μM in the absence of lipid vesicles and 0.05 μM in the presence of vesicles. For DPPC hydrolysis where the concentration of apoC-II is

TABLE III
ENZYME KINETIC PARAMETERS FOR LpL CATALYSIS

Substrates	pH	K_m^{app} (mM)		V_{max}^{app}/E_t (μmol/min/mg)		Reference
		(−)C-II	(+)C-II	(−)C-II	(+)C-II	
Human VLDL-TG[a]	7.4	7.8	1.0	14	25	39
Guinea pig VLDL-TG	7.4	3.1	0.5	152	150	38
	8.5	2.7	0.6	175	265	
Guinea pig VLDL-DPPC	7.4	0.23	0.31	7.0	16.9	38
	8.5	0.28	0.18	10.2	32.5	
DPPC-apo C-III disks	7.4	0.39	0.40	8.0	53.7	38
	8.5	0.52	0.39	8.2	131	
DiC$_6$PC	7.4	4.0	3.7	5.0	5.1	13
PNPB	7.25	0.9	0.9	6.7	1.6	16

[a] TG, Triacylglycerol.

0.3–1.4 μM, the data are consistent with $K_A > 2\ \mu M$.[38] At pH 8.5 a small increase in V_{max} for trioleoylglycerol hydrolysis is observed with the addition of apoC-II[38] (Table III). These data imply that K_m^* increases from pH 7.4 to 8.5 so that $K_m^* (1 + A/K_A)$ is similar in magnitude to $S(1 + A/\alpha K_A)$ and Eqs. (2) and (3) apply.

Thus, in the simplest mechanism where $\beta = 1$ and $\alpha < 1$, apoC-II reduces the dissociation constant for ES^*A and the enzyme is shifted to ES^*A. The different effects of apoC-II on the apparent Michaelis–Menten constants for PC and trioleoylglycerol are explained by the relative values of K_m^* and S. Since S is reduced by the number of inaccessible substrate molecules for trioleoylglycerol hydrolysis, the data support the notion that the interfacial K_m^* for trioleoylglycerol hydrolysis is much less than that for DPPC hydrolysis.[1,38,40]

Thermodynamic Model of the ApoC-II Effect

Shinomiya et al.[64] have compiled data for the effect of apoC-II on the rate of LpL-catalyzed hydrolysis of a variety of substrates of various acyl chain lengths. Although the rates of hydrolysis of the various substrates could not be predicted based on acyl chain length, the ratio of the rates of hydrolysis in the presence and absence of apoC-II (activation factor) increased with increasing fatty acyl chain length. A quantitative relationship is obtained when data are analyzed at maximal apoC-II concentrations: the logarithim of the activation factor is a linear function of the number of carbon atoms of a single fatty acyl chain of the substrate.

This linear relationship applies to LpL-catalyzed hydrolysis of the water-soluble substrates PNPA,[16] PNPB,[16] and diC$_6$PC,[13] to monomolecular films of 1,2-didecanoyl-sn-glycero-3-phosphoglycerol,[26] trioctoanoyl-glycerol,[32,54] and 1,2-didodecanoyl-sn-glycero-3-phosphoglycerol,[55] to DMPC vesicles[64] and to PC/Triton X-100 emulsions of even number chain lengths from 12 to 18.[64–66] In addition, the activation factor is independent of the ratio of Triton X-100/PC over a range which includes the bilayer to micelle phase transition of this system.[66] Since the linear relationship between the activation factor and fatty acyl chain length is independent of the molecular organization of the substrate, it is unlikely that apoC-II enhances LpL activity by facilitating the binding of the enzyme to the lipid interface. In addition, for Triton/PC substrates the activation factor for apoC-II and for a synthetic peptide of residues 56–79 of apoC-II (which does not bind to lipid) are identical.[66a]

Further insight into the effect of apoC-II was obtained from the dependence of the activation factor on temperature in a Triton X-100/PC system where the predominant substrate organization is micelles. The observed rate constant, k_0, is given by $k_0 = (kT/h) \exp(-\Delta G\ddagger/RT)$ according to the theory of absolute reaction rates.[69,70] The activation factor can be expressed as the ratio of two such Eyring expressions:

$$k_0'/k_0 = \exp(-\Delta\Delta G\ddagger/RT) = \exp(\Delta\Delta S\ddagger/R) \exp(-\Delta\Delta H\ddagger/RT) \quad (9)$$

where $\Delta\Delta G\ddagger = \Delta G\ddagger' - \Delta G\ddagger$, $\Delta\Delta S\ddagger = \Delta S\ddagger' - \Delta S\ddagger$, and $\Delta\Delta H\ddagger = \Delta H\ddagger' - \Delta H\ddagger$ and the primes refer to value measured in the presence of apoC-II. The various thermodynamic functions were measured by fitting Eq. (9) to kinetic data obtained over a range of temperatures. A linear relationship between each of the thermodynamic functions and the fatty acyl chain length of the substrate was observed. As the chain length of the substrate increases, the rate of PC hydrolysis is enhanced to a greater extent by apoC-II. This is the result of a large increase in the activation entropy as a function of chain length which exceeds an unfavorable increase in activation enthalpy. A similar qualitative relationship of increased entropy and enthalpy is observed in the transfer of an amphipathic molecule from a polar to a nonpolar solvent.[71] Analysis of the chain length dependence of the entropy of transfer of a series of n-alkanoic acids ($n = 4$ to 7) from water to hexadecane gives $T\Delta S = 480$ cal/mol per CH$_2$ group at 25°.[72] A

[69] S. Glasstone, K. J. Laidler, and H. Eyring, "The Theory of Rate Processes" p. 189, McGraw-Hill, New York, 1941.
[70] K. J. Laidler and B. F. Peterman, *Methods Enzymol.* **63,** 234 (1979).
[71] H. S. Frank and M. W. Evans, *J. Phys. Chem.* **13,** 507 (1945).
[72] R. Aveyard and R. W. Mitchell, *Trans. Faraday Soc.* **66,** 37 (1970).

similar value is obtained for the effect of apoC-II on the formation of the transition state complex in LpL catalysis, $T\Delta\Delta S\ddagger = 520$ cal/mol per CH_2 group at 25°.[65] Thus, the chain length dependence of the entropy terms for transfer of amphipathic molecules from polar to nonpolar solvents and for the effect of apoC-II on the LpL-catalyzed hydrolysis of PC molecules are comparable. These data suggest a hydrophobic model for apoC-II activation of LpL. In this model, apoC-II changes the reaction pathway such that the fatty acyl chain of the substrate or product is transferred to a more hydrophobic environment in the transition state complex.[65]

An important corollary of this hypothesis is that apoC-II alters the specificity of LpL so that substrates with longer fatty acyl chains are preferred. When 0.5 mol% trioleoylglycerol is incorporated into DPPC vesicles, the addition of apoC-II decreases PNPB hydrolysis with a reciprocal increase in trioleoylglycerol hydrolysis.[38] Thus, apoC-II modulates the specificity of LpL catalysis from short chain esters to the physiological substrates.

Acknowledgments

The authors wish to acknowledge their colleagues who were responsible for much of the work presented in this chapter. These individuals include Drs. Alan Cardin, Moti Kashyap, Judith A. K. Harmony, Kohji Shirai, S. Ranganathan, George Holdsworth, A. Balasubramanium, Norihiro Sasaki, Daniel Quinn, Nobuo Matsuoka, and J. David Johnson, and Ms. Janet Simons who has prepared the manuscript for publication. This research was supported by U.S. Public Health Service Grants P01-HL-22619 and R01-HL-23019. L.R.M. was supported by a Molecular and Cellular Biology Training Grant NIH HL-07527.

[45] Isolation, Characterization, and Assay of Lecithin–Cholesterol Acyltransferase

By John J. Albers, Ching-Hong Chen, and Andras G. Lacko

Introduction

Lecithin–cholesterol acyltransferase (LCAT, EC 2.3.1.4.3) is synthesized by the liver and secreted into the bloodstream, where it is believed to act principally on the surface of high-density lipoproteins (HDL). It catalyzes the hydrolysis of fatty acid from the sn-2 position of phosphatidylcholine and transfers the fatty acid to unesterified cholesterol to form cholesteryl ester. This transesterification reaction is the major source of plasma cholesteryl ester in man. Cholesterol is removed from the HDL surface through the transesterification reaction and the subsequent transfer of cholesteryl ester to acceptor lipoproteins, permitting the transfer of additional unesterified cholesterol to HDL from the surface of cells and other lipoproteins.[1] Through this mechanism, LCAT is postulated to play a key role in the transport of cholesterol from peripheral tissues to the liver, in the interconversion of HDL subclasses, and the maintenance of lipoprotein structure. LCAT along with the lipid transfer proteins (see Chapter [48] by Tollefson and Albers[2]) participate in the redistribution of apolipoproteins and in the removal of surface lipids, particularly unesterified cholesterol and phosphatidylcholine, from the degradation products of very-low-density lipoproteins (VLDL) and chylomicrons. Recently, LCAT has also been shown to catalyze the transfer of fatty acid to lysolecithin to form lecithin in the presence of low-density lipoproteins (LDL)[1] (see Chapter [47] by Subbaiah[3]).

The transesterification mediated by LCAT not only occurs in the plasma of all mammalian species examined, it can also be found in reptiles and amphibians. The proportion of plasma cholesterol esterified by the LCAT reaction, the rate of cholesterol esterification, the fate of the newly formed cholesteryl esters, and the nature of the cholesteryl esters formed, may vary considerably from one animal species to another. For example, a significant proportion of the plasma cholesteryl esters in rats may be formed by the acyl-CoA cholesterol acyltransferase enzyme, and little of the cholesteryl esters formed in association with rat HDL by LCAT appears to be transferred to other lipoproteins, and rat plasma contains a greater proportion of cholesterol arachidonate than is found in human plasma. Thus, the differences in the LCAT reaction that appear to exist among animal species are both quantitative and qualitative in nature.

Isolation of LCAT

The first attempts to isolate LCAT yielded only partially purified preparations even though a variety of chromatographic techniques were used, including hydroxylapatite adsorption, salt precipitation, ultracentrifugation, and affinity chromatography. Albers *et al.* reported the first homoge-

[1] P. V. Subbaiah, J. J. Albers, C.-H. Chen, and J. D. Bagdade, *J. Biol. Chem.* **255,** 9275 (1980).
[2] J. H. Tollefson and J. J. Albers, Chapter 48, this volume.
[3] P. V. Subbaiah, Chapter 47, this volume.

neous preparation of purified LCAT.[4] This procedure used the following steps: (1) isolation of the d 1.21–1.25 g/ml plasma fraction by sequential ultracentrifugation; (2) HDL affinity chromatography; (3) hydroxylapatite chromatography; and (4) anti-apolipoprotein D affinity chromatography to remove apolipoprotein D contaminants. This laboratory has made a number of modifications to the procedure for isolating homogeneous LCAT from human plasma. These modifications gave rise to a 16,500-fold purification with an apparent yield of approximately 13%.[5] The procedure includes isolation of the d 1.21–1.25 fraction by ultracentrifugation, phenyl–Sepharose affinity chromatography instead of HDL–Sepharose affinity chromatography, DEAE–Sepharose chromatography, and hydroxylapatite chromatography.[5] Other workers also have reported having isolated a homogeneous preparation of LCAT with similar yield(s) and degree of purification, using alternative procedures such as molecular sieve chromatography and Affi-Gel Blue affinity chromatography.[6,7] Ultracentrifugation is time consuming and expensive, therefore a number of laboratories have used alternatives to centrifugation for the isolation of LCAT. For example, the procedure of Kitabatake *et al.* involves binding of the enzyme for human plasma on DEAE–Sephadex, followed by treatment with 1-butanol in the presence of ammonium sulfate, DEAE–Sephadex, treatment with dextran sulfate in the presence of calcium ions, and hydroxylapatite chromatography.[8] The final enzyme preparation was purified approximately 30,000-fold over the starting plasma in a yield of 10%. Alternatively, Lacko and Chen reported the removal of LCAT from plasma by dodecylamine agarose chromatography.[9] More recently, Chen and Albers have developed a procedure for preparing of LCAT from plasma that does not require preparative ultracentrifugation.[10] A slightly modified purification procedure of Chen and Albers[10] is as follows.

Large-Scale Preparation of LCAT without Ultracentrifugation

Reagents

Dextran sulfate–Mg^{2+} solution, 10 g/liter dextran sulfate with a molecular weight of 50,000 (SOCHIBO, Boulogne, France) dissolved in 500

[4] J. J. Albers, V. G. Cabana, and Y. D. Barden Stahl, *Biochemistry* **15**, 1084 (1976).

[5] C.-H. Chen and J. J. Albers, *Biochem. Med.* **25**, 215 (1981).

[6] L. Aron, S. Jones, and C. J. Fielding, *J. Biol. Chem.* **253**, 7220 (1978).

[7] J. Chung, D. A. Abano, G. M. Fless, and A. M. Scanu, *J. Biol. Chem.* **254**, 7456 (1979).

[8] K. Kitabatake, U. Piran, Y. Kamio, Y. Doi, and T. Nishida, *Biochim. Biophys. Acta* **573**, 145 (1979).

[9] A. G. Lacko and T. F. Chen, *J. Chromatogr.* **130**, 446 (1977).

[10] C.-H. Chen and J. J. Albers, *Biochim. Biophys. Acta* **834**, 188 (1985).

mM MgCl$_2$. Phenyl–Sepharose Cl-4B, DEAE–Sepharose Cl-6B, hydroxylapatite, and Sephacryl S-200 (Pharmacia Fine Chemicals, Uppsala, Sweden). Affi-Gel Blue from Bio-Rad Laboratories.

Procedure

LCAT is purified from human plasma by the combination of dextran sulfate precipitation and consecutive chromatography on phenyl–Sepharose column, Affi-Gel Blue column, DEAE–Sepharose column, hydroxylapatite column, and Sephacryl S-200 column. All procedures are performed at 4° in a cold room unless specified otherwise. Protein values during chromatographic runs are monitored spectrophotometrically at 280 nm wavelength and protein concentrations are determined by the method of Lowry et al.[11] LCAT activity is assayed as described by Chen and Albers.[12]

1. *Dextran sulfate precipitation.*[13] To 1000 ml of plasma 100 ml of dextran sulfate–Mg^{2+} solution is slowly added with stirring. The mixture is incubated for 10 min at room temperature and then centrifuged at 2500 rpm for 30 min at 4°. The supernatant fluid contains most of the LCAT activity in plasma.

2. *Phenyl–Sepharose chromatography.* To the dextran sulfate supernatant fluid 58 g of solid NaCl is added and dissolved, and the mixture is applied at a flow rate of 60 ml/hr to a phenyl–Sepharose CL-4B column (3.4 × 26 cm) equilibrated with 10 mM Tris buffer containing 140 mM NaCl and 1 mM EDTA, pH 7.4. The column is washed with the same buffer until the optical density at 280 nm is less than 0.01. The column is then eluted with distilled water at a flow rate of 60 ml/hr with 20 ml/fraction. LCAT activity is eluted in fractions 12 through 18.

3. *Affi-Gel Blue chromatography.* The phenyl–Sepharose bound fractions with LCAT activity are pooled, dialyzed against 20 mM sodium phosphate buffer, pH 7.1, and passed through an Affi-Gel Blue column (2.0 × 27 cm) equilibrated with the 20 mM phosphate buffer. The column is eluted with 200 ml of phosphate buffer (20 mM, pH 7.1). Ten-milliliter fractions are collected at a flow rate of 60 ml/hr. LCAT activity is eluted in fractions 5 through 15.

4. *DEAE–Sepharose chromatography.* The post Affi-Gel Blue sample is dialyzed against 1 mM Tris buffer containing 5 mM EDTA and 25 mM NaCl, pH 7.4, and is chromatographed on a DEAE–Sepharose CL-6B column (1.2 × 30 cm) and eluted with 1 liter of a linear NaCl-Tris gradient

[11] O. H. Lowry, N. J. Rosebrough, A. L. Farr, and R. J. Randall, *J. Biol. Chem.* **193,** 265 (1951).
[12] C.-H. Chen and J. J. Albers, *J. Lipid Res.* **23,** 680 (1982).
[13] G. R. Warnick, J. Benderson, and J. J. Albers, *Clin. Chem.* **28,** 1379 (1982).

containing 5 mM EDTA. The initial buffer contains 500 ml of 25 mM NaCl and 1 mM Tris, pH 7.4, and the second buffer 500 ml of 200 mM NaCl, 10 mM Tris, pH 7.4. Twenty-milliliter fractions are collected at a flow rate of 45 ml/hr. LCAT activity is eluted at a NaCl concentration of approximately 120 to 160 mM.

5. *Hydroxylapatite chromatography.* The fractions with LCAT activity from the DEAE–Sepharose eluate are pooled, dialyzed against 4 mM phosphate buffer containing 150 mM NaCl, pH 6.9, and chromatographed on a hydroxylapatite column (3.4 × 10 cm) equilibrated with the same phosphate buffer. LCAT is eluted with 240 ml of a linear phosphate (15 to 60 mM) gradient, pH 6.9, containing 150 mM NaCl. Fractions of 5 ml each are collected at a flow rate of 20 ml/hr. LCAT activity is eluted at a phosphate concentration of approximately 30 to 38 mM.

6. *Sephacryl S-200 gel filtration.* The active fractions from the hydroxylapatite column are pooled and concentrated to 2 ml by ultrafiltration. The concentrated sample is then applied to a Sephacryl S-200 column (1.2 × 90 cm) and eluted with 10 mM Tris, 140 mM NaCl, 1 mM ETDA, pH 7.4. Fractions of 2 ml each are collected at a flow rate of 40 ml/hr. LCAT activity is eluted in fractions 45 through 55.

The final LCAT preparation is purified approximately 20,000-fold with 13% yield. The purified LCAT exhibits a single protein band on SDS–polyacrylamide gel electrophoresis with apparent M_r of 65,000 ± 2000 (n = 5) and exhibits a single protein peak on HPLC size exclusion chromatography with apparent molecular weight of 67,000 ± 2500 (n = 4). A representative purification of LCAT from 1 liter of human plasma is shown in Table I.

TABLE I

PURIFICATION OF HUMAN PLASMA LECITHIN–CHOLESTEROL ACYLTRANSFERASE

	Total volume (ml)	Total protein (mg)	Total activity (U)[a]	Specific activity (U/mg protein)	Yield (%)	Purification (fold)
Whole plasma	1,000	68,670	110,400	1.6	100	1
Dextran sulfate supernatant	1,000	49,253	102,640	2.1	93	1.3
Phenyl–Sepharose eluate	140	557.20	95,050	171	86	106
Affi-Gel Blue fractions	120	146.40	58,560	400	53	248
DEAE–Sepharose eluate	260	26.50	52,320	1,974	47	1,225
Hydroxylapatite eluate	30	0.63	18,710	29,683	17	18,436
Sephacryl S-200 eluate	22	0.44	14,456	32,854	13	20,533

[a] One unit of enzyme catalyzed the esterification of 1 nmol of cholesterol/hr at 37° and pH 7.4.

Characterization of LCAT

Lecithin–cholesterol acyltransferase (LCAT, EC 2.3.1.43) has been one of the most difficult enzymes to characterize from the physical, chemical, and enzymological points of view. The painstakingly slow early development of an efficient purification procedure and the instability of the enzyme precluded rapid progress. For example, the first report of the isolation of a homogeneous LCAT preparation appeared nearly 10 years ago,[4] and yet relatively little is known about the physicochemical properties of the enzyme.

Physical Properties

Molecular Weight. Initial reports, based on SDS–PAGE analysis, estimated a molecular weight in the range of 66,000–68,000.[4,14] These values are now considered to be too high because of the high carbohydrate content of the enzyme. More recent studies, employing sedimentation equilibrium measurements, indicated an M_r of about 59,000.[7] This study employed a value for partial specific volume of 0.710 ml/g, which is appropriate if the carbohydrate content of the enzyme is taken into consideration.

Conformational Properties. Circular dichroism (CD) data on purified LCAT are available from two laboratories.[15,16] The two sets of data agree well concerning the contribution of α-helix (24 and 18%) and the lack of significant change in the overall conformation of the enzyme in the presence of 0.2–0.4 M NaCl. Also, the addition of phosphatidylcholine vesicles does not significantly alter the CD spectrum of the protein. The β-pleated sheet content has been estimated to be 27%. However, using another method of calculation Doi and Nishida have suggested a β-sheet contribution of 53% for the secondary structure.[16]

The environments of the tyrosine and tryptophan residues in LCAT present an intriguing picture. The data from spectrophotometric titrations suggest that essentially all the tyrosine residues are buried until the pH exceeds 11. At this point, ionization of the tyrosine residues occurs with a pK of 12, accompanied by a conformational change.[15] In contrast, from the results of spectrofluorometric studies, about 60% of the tryptophan residues of the enzyme appear to be exposed under nondenaturing conditions and at neutral pH.[17]

[14] J. J. Albers, J. T. Lin, and G. P. Roberts, *Artery* **5**, 61–75 (1979).
[15] K. S. Chong, R. E. Thompson, S. Hara, and A. G. Lacko, *Arch. Biochem. Biophys.* **222**, 553–560 (1983).
[16] Y. Doi and T. Nishida, *J. Biol. Chem.* **258**, 5840 (1983).
[17] K. S. Chong, M. Jahani, S. Hara, and A. G. Lacko, *Can. J. Biochem.* **61**, 875 (1983).

Isoelectric Point and Microheterogeneity. LCAT was found to have multiple isoelectric points.[14] This microheterogeneity of the enzyme was later confirmed in a number of laboratories.[15,18] The isoelectric points of the isoforms are now believed to range from 3.9 to 4.4. Doi and Nishida[19] showed, in addition, that the microheterogeneity is abolished upon extensive digestion of the enzyme with neuraminidase, which is apparently due to the removal of sialic acid residues. The loss of sialic acid is accompanied by an increase in LCAT activity, suggesting that the additional negative charges may be detrimental to the LCAT/substrate interaction.[16]

Other Physical Properties. Stokes radii and frictional coefficient ratio determinations were reported by Doi and Nishida, using sedimentation analysis and gel filtration.[16] The values agreed well between the two types of studies. The native and neuraminidase-treated LCAT samples were found to have Stokes radii of 40 and 36 Å and frictional coefficients of 1.53 and 1.38, respectively. The extinction coefficients ($E_{1\,cm,\,280}^{1\%}$) for the enzyme ranged from 20 to 21.[15,16]

Chemical Properties

Amino Acid Composition. The amino acid and carbohydrate compositions of the enzyme are shown in Table II. An M_r of 45,000 was used to calculate the amino acid composition of LCAT because of the reported apparent M_r of about 60,000 and the approximate carbohydrate content of 25%. These calculations reveal seven methionines and four half-cystines per mole in addition to a relative abundance of acidic (Asp, Glu) and some hydrophobic (Leu) residues.

Carbohydrate Composition. Carbohydrate analyses indicate a carbohydrate content of about 25% with a comparatively high sialic acid content (5%).[7,14,15] The role of the carbohydrate residues in enzyme structure and function is currently unclear, although (as mentioned above) an increase in enzymatic activity and a change in hydrodynamic parameters accompanied the removal of sialic residues from LCAT.[16] It is possible that the sialic residues protect the enzyme against rapid removal by the liver from the circulation.

Specific Properties Related to Catalytic Function

pH Optimum. LCAT has a pH optimum around pH 8.0 for transesterification with both high-density lipoproteins (HDL) and lecithin–choles-

---

[18] G. Utermann, H. J. Menzel, G. Adler, P. Dieker, and W. Weber, *Eur. J. Biochem.* **107,** 225 (1980).

[19] Y. Doi and T. Nishida, this series, Vol. 71, p. 753.

TABLE II

AMINO ACID AND CARBOHYDRATE COMPOSITION OF HUMAN
LECITHIN–CHOLESTEROL ACYLTRANSFERASE

Residue	Chung et al.[7]	Albers et al.[14]	Aron et al.[6]	Doi and Nishida[19]	Chong et al.[15]
Lys	14 (12–15)[a]	15 ± 1[b]	14 – 1[b]	13	14 (14–14)[a]
His	12 (11–13)	13 ± 2	12 ± 1	12	12 (12–13)
Arg	19 (16–20)	19 ± 2	17 ± 2	18	18 (17–20)
Asp	37 (34–40)	43 ± 2	33 ± 1	36	41 (38–45)
Thr	25 (23–27)	27 ± 0	20 ± 1	22	24 (23–25)
Ser	25 (22–28)	28 ± 2	23 ± 2	22	24 (24–25)
Glu	42 (38–45)	46 ± 5	41 ± 2	38	38 (38–39)
Pro	35 (30–39)	36 ± 2	34 ± 1	38	31
Gly	38 (35–42)	44 ± 2	39 ± 2	34	37 (37–18)
Ala	27 (26–28)	28 ± 2	23 ± 1	22	25 (25–26)
Val	27 (25–29)	29 ± 1	21 ± 1	28	28 (27–29)
Met	7 (7–8)	9 ± 1	7 ± 0	8	7 (7–7)
Ile	18 (17–19)	19 ± 1	13 ± 1	17	15 (14–15)
Leu	46 (44–49)	51 ± 5	41 ± 1	48	50 (49–52)
Tyr	13 (12–15)	18 ± 2	17 ± 1	18	19 (19–20)
Phe	14 (13–16)	20 ± 1	17 ± 0	18	19 (18–19)
Cys	4	4 ± 1	—	4	4 (3–4)
Trp	7	—	—	10	10 (10–11)
Hexoses	39				43
Hexosamines	13	14			21
Sialic acid	10				11

[a] Mean and range.
[b] Mean ± SD.

terol liposomes as substrates.[6,20] According to Aron et al.[6] the phospholipase activity of LCAT has a pH optimum of about 8.0, and the acyltransferase activity has a slightly higher pH optimum of 8.5.

Stability. LCAT has been reported to be a fairly unstable enzyme, especially in the highly purified state. Some workers reported a half-life of 3 hr at 4°,[21] whereas others reported a half-life of several weeks.[5] Furukawa and Nishida[22] reported that the stability of the purified enzyme depended on ionic strength. Accordingly, enzyme inactivation at the air–water interface was apparently prevented by buffered media of low ionic strength. Kitabatake et al.[8] reported that their purified LCAT preparations were stable for 4 weeks if they were stored in a 0.4 mM phosphate

[20] M. Jahani and A. G. Lacko, *Biochim. Biophys. Acta* **713**, 504 (1982).
[21] C. J. Fielding and P. E. Fielding, *FEBS Lett.* **15**, 355 (1971).
[22] Y. Furukawa and T. Nishida, *J. Biol. Chem.* **254**, 7213 (1979).

buffer in the presence of 4 mM 2-mercaptoethanol at 4° and under N_2. Jahani and Lacko[20] found that LCAT remained stable indefinitely in the homogeneous state, if it was stored as a precipitate that was generated by dialyzing the enzyme against a solution of saturated $(NH_4)_2SO_4$. The variability in the stability of these "purified" preparations could relate not only to the different conditions used but also to the relative purity of the preparations, because less purified preparations exhibit greater stability than do purified preparations.

LCAT Inhibitors. Numerous compounds have been reported to inhibit LCAT activity including diisopropyl fluorophosphate (DFP) and other organophosphates, dithiobisnitrobenzoic acid (DTNB) and p-hydroxy-mercuribenzoate (PHMB), and other sulfhydryl blocking agents, carnitine, chlorpromazine, and excess amounts of cysteine and reduced glutathione. Much of the information available in the literature deals with the inhibitors that form covalent products with the enzyme. For example, DFP reacts with LCAT rapidly and efficiently; a 10^{-3} M solution effects nearly total inactivation in less than 30 min.[17] Neither the presence of substrate nor pretreatment with DTNB was able to prevent the reaction and subsequent inactivation with DFP.[17] The reaction of sulfhydryl group(s) in LCAT with sulfhydryl blocking agents, such as DTNB, also rapidly destroys the enzyme's activity.[17] Verdery,[23] however, finds that LCAT can also be inhibited by reduced sulfhydryl compounds such as cysteine and reduced glutathione, indicating that perhaps reduction of disulfide bonds and subsequent formation of mixed disulfides with the newly formed SH groups occurred. In fact, LCAT contains four half-cystines, only two of which are reactive with DTNB.[17] Consequently, it is possible that one disulfide bridge is present in the enzyme in addition to the two cysteine residues. As with the reaction with DFP, macromolecular substrates (HDL or proteoliposomes) offered little or no protection against the inhibition of the enzyme by cysteine.[17]

Apolipoprotein Cofactor Requirements. Fielding *et al.*[24] were the first to show that LCAT did not significantly react with lecithin–cholesterol liposomes unless apolipoprotein A-I (apoA-I) was also included in the reaction mixture. Albers *et al.*[14] found that apoA-II, C-II, C-III, and D inhibited the activation of LCAT by apoA-I, presumably by displacement of apoA-I from the vesicle surface. Fielding *et al.*[24] show that HDL apolipoproteins other than apoA-I are inhibitors of the activating effect of apoA-I on enzyme activity. Chung *et al.*[7] reported similar observations

[23] R. B. Verdery, *Biochem. Biophys. Res. Commun.* **98**, 494 (1981).
[24] C. J. Fielding, V. G. Shore, and P. E. Fielding, *Biochem. Biophys. Res. Commun.* **46**, 1493 (1972).

and suggested that the apoA-I/A-II ratio at the vesicle surface controlled the rate of the LCAT reaction. Furukawa and Nishida[22] also suggested that apoA-I interacts with LCAT at the water–vesicle interface in a reversible manner, since in their study apoA-I did not influence the association of the enzyme with lipid vesicles. Investigations using artificial substrates have permitted the precise study of physical–chemical interactions of the enzyme with well-defined lipid–protein complexes. However, the mode of preparation of these artificial substrates varies from one laboratory to another and the findings may occasionally be discordant. For instance, Chen and Albers[25] find that under certain conditions apoA-II can actually enhance apoA-I-activated LCAT activity rather than inhibit it as had been reported.[6,7,14] In the former case with the substrate prepared by the cholate dialysis procedure, it is possible that apoA-II is incorporated into the liposome under conditions whereby apoA-I is not displaced from the surface of the vesicle whereas in the latter cases apoA-I is displaced from the vesicle surface, which would explain the observed inhibition of LCAT activity. Therefore, the interpretation of the studies explaining the specificity of LCAT using artificial substrates may not necessarily have physiological relevance. In fact, studies with intact lipoprotein substrates yielded results that were quite different from those obtained with single bilayer vesicles.[26] Recently, Chen and Albers[27,28] have demonstrated that isolated apoE and apoA-IV activate plasma LCAT when individually incorporated into defined phosphatidylcholine/cholesterol liposomes. ApoA-IV was approximately 25% as efficient as apoA-I for the activation of the enzyme and apoE-30 to 40% as efficient. Similar results have also been obtained by Zorich et al.[29] with apoE and Steinmetz and Uterman with apoA-IV.[30] Furthermore, Chen and Albers[28] have found that patients with Tangiers disease (who have a near-total lack of normal apoA-I) and patients with apoA-I and C-III deficiency (who have only a trace of apoA-I) exhibit normal LCAT activity for the level of enzyme present.[28] It is likely therefore that apolipoproteins other than A-I serve as physiological cofactors for the LCAT reaction. Sotar et al.[31] reported that apoC-I may serve as cofactor for the LCAT reaction although this polypeptide is more efficient in facilitating

[25] C.-H. Chen and J. J. Albers, Circulation 68 (Suppl. III), 232 (1983).
[26] M. Jahani and A. G. Lacko, J. Lipid Res. 22, 1102 (1981).
[27] C.-H. Chen and J. J. Albers, Circulation 70 (Suppl. II), 121 (1984).
[28] C.-H. Chen and J. J. Albers, Biochim. Biophys. Acta 836, 279 (1985).
[29] N. Zorich, A. Jonas, and H. Pownall, J. Biol. Chem. 260, 8831 (1985).
[30] A. Steinmetz and G. Uterman, J. Biol. Chem. 260, 2258 (1985).
[31] A. K. Soutar, C. W. Garner, H. N. Baker, J. T. Sparrow, R. L. Jackson, A. M. Gotto, and L. C. Smith, Biochemistry 14, 3057 (1975).

the transfer of medium chain than that of long chain fatty acids. The role of other apolipoproteins in the LCAT reaction currently remains unclear.

Topography of the Active Site. At this point, the scarcity of published information precludes a detailed postulate for the participation of the functional groups in acyl transfer; however, some observations can be made on the basis of the available data. It has been clearly shown that DFP reacts with the enzyme in a covalent manner[20] and recent data indicate that approximately one residue of DFP is incorporated per mole of LCAT.[20] These observations strongly implicate a role for a serine or a threonine residue, which could be the recipient of fatty acid group, during acyl transfer. It is not certain whether one or two SH groups participate in the catalysis. However, the rapid inactivation of the enzyme by reagents such as DTNB suggests that at least one cysteine residue is involved in the substrate-binding or the acyl transfer step. It is not yet clear whether the -SH and -OH functions are acting in any cooperative manner in the enzyme mechanism.

Assay of LCAT

Measurement of LCAT Mass

Principle. A competitive double-antibody radioimmunoassay was developed to measure LCAT mass in plasma and plasma fractions.[32] Conditions were chosen to permit optimal measurement of LCAT at a 1 : 50 dilution of plasma and precipitate approximately 50% of the ^{125}I-labeled LCAT precipitated by an excess of antibody.

Reagents

Chloramine-T (B.D.H. Chemicals, Gallard-Schlesinger Chem. Mfg., Carle Place, Long Island, NY): 2.5 mg/ml in 0.05 M NaPO$_4$, pH 7.5

Sodium metabisulfite (Na$_2$S$_2$O$_5$): 2.5 mg/ml in 0.05 M NaPO$_4$, pH 7.5

Dialysis buffer: 0.01 M Tris, 0.15 M NaCl, 0.001 M EDTA, pH 7.4

Assay buffer: dialysis buffer + 1% bovine serum albumin (BSA) + 0.02% NaN$_3$

Sample or standard dilution buffer: assay buffer + 0.1% Tween-20

Procedure. LCAT was labeled by the chloramine-T method. The LCAT was dialyzed against 0.05 M NaPO$_4$, pH 7.5, of which 10 μg was placed in a small vial with a microstirring bar, and into which ^{125}I-labeled Na (12 nCi/μg protein) was stirred, followed by chloramine-T (1.5 μg/μg

[32] J. J. Albers, J. L. Adolphson, and C.-H. Chen, *J. Clin. Invest.* **67**, 141 (1981).

protein). The mixture was left stirring for 2 min and then sodium metabisulfite (1.5 μg/μg protein) was added to stop the reaction. The labeled LCAT was then dialyzed for 2 hr against 0.01 M Tris, 0.15 M NaCl, 0.001 M EDTA (dialysis buffer), pH 7.4, after which it was chromatographed on a 0.9 $\times$ 60 cm column of Sephacryl S-200 (Pharmacia, Piscataway, NJ). The fractions were counted, pooled, and again dialyzed against dialysis buffer.

The assay was done in 12 $\times$ 75 mm glass tubes coated with a 1% column coat (California Immuno Diagnostics, San Marcus, CA). The assay buffer was 0.01 M Tris, 0.15 M NaCl, 0.001 M EDTA, 1% bovine serum albumin, 0.01% NaN$_3$, except for the sample or standard dilution buffer which also contains 0.1% Tween-20. To each tube was added the following: (1) 100 μl of diluted sample, standard, or buffer; (2) 100 μl of ^{125}I-labeled LCAT (0.02 μg/ml); and (3) 100 μl of goat anti-human LCAT sera. The mixture was incubated overnight at 4°, then 100 μl of nonimmune goat serum diluted 1 : 100 as carrier and 300 μl of rabbit anti-goat IgG serum diluted 1 : 4 were added. After another overnight incubation, 1.0 ml of assay buffer was added. The tubes were spun at 3000 rpm for 20 min, then the supernatants were decanted. The precipitates were washed twice with 1.0 ml assay buffer and then counted in a gamma counter. The log/logic transformation was used to linearize the relationship between percentage antigen bound and log antigen dose.

Distribution and Levels of LCAT Mass. Most of the plasma LCAT appears to be associated with lipoprotein particles in the size range of HDL.[33] Approximately 50–75% of the LCAT mass is associated with HDL particles prepared by centrifugation.[29,30] The proportion of total plasma LCAT that is associated with HDL is highly dependent on centrifugation conditions. Approximately 1% of the plasma LCAT mass is associated with the LDL fraction.[33] No LCAT has been found to be associated wit the VLDL fraction. Plasma LCAT levels have been reported to be slightly higher in women than men (5.9 $\pm$ 1.0 μg/ml vs 5.5 $\pm$ 0.9 μg/ml).[34] Levels of LCAT mass have shown a slight positive association with age and relative body mass in women, and men who smoke cigarettes have significantly lower LCAT masses than do men who do not smoke cigarettes.[34,35] Subjects phenotypically LCAT deficient by clinical criteria and by the deficiency (10% or less of normal levels) or absence of LCAT

[33] C.-H. Chen and J. J. Albers, *Biochem. Biophys. Res. Commun.* **107,** 1091 (1982).

[34] J. J. Albers, R. O. Bergelin, J. L. Adolphson, and P. W. Wahl, *Atherosclerosis* **43,** 369 (1982).

[35] S. M. Haffner, D. Applebaum-Bowden, P. W. Wahl, J. Joanne Hoover, G. R. Warnick, J. J. Albers, and W. R. Hazzard, *Arteriosclerosis* **5,** 169 (1985).

activity have levels of LCAT mass well below the reference values: 0.73 ± 0.70 µg/ml, range 0 (undetectable levels, less than 0.1 µg/ml) to 2.65 µg/ml, $n = 20$. Obligate heterozygotes for familial LCAT deficiency, i.e., parents or children of LCAT-deficient subjects have reduced levels of plasma LCAT mass: 3.59 ± 0.69, range 2.59–4.61 µg/ml, $n = 19$. Thus far, familial LCAT deficiency can be categorized into three discrete classes: (1) Subjects with complete absence of enzyme mass and activity[36]; (2) subjects with low levels of LCAT mass and activity[37]; and (3) subjects with a reduced level of partly inactive or functionally defective enzymes.[38] Significant decreases in LCAT mass and activity can also occur in parenchymal liver disease and renal disease.

Measurement of LCAT Activity

LCAT activity refers to the amount of active enzyme in plasma or plasma fractions and therefore its measurement should be independent of plasma constituents. It is measured by the so-called common substrate methods. The activity is generally expressed in terms of molar activity of nanamoles of cholesterol converted to cholesteryl ester per hour per milliliter. LCAT activity can be expressed in terms of nanomoles of cholesterol esterified per hour per milligram of protein with purified enzyme preparations. LCAT activity can also be expressed in terms of the fractional activity on the percentage of cholesterol esterified per hour. All calculations of LCAT activity make the assumption that the enzyme source (usually plasma or serum) does not contribute to or affect the "common" substrate. The term LCAT activity is often confused with the measurement of the endogenous cholesterol esterification rate, which depends not only on the amount of active enzyme present but also on the LCAT substrates and cofactors and other plasma constituents in the plasma. The original method of measurement of LCAT activity was that of Glomset and Wright.[39] It involves using heat-inactivated plasma that has been subsequently labeled with radioactive unesterified cholesterol. In this procedure a small amount of test plasma is added as enzyme source to the common substrate of heated plasma. The assay values can vary from one sample to another, depending upon the method of preparation and the mode of introduction of the radioactive dispersion of the plasma. Usually an albumin-bound isotopic dispersion of unesterified

[36] J. J. Albers, C.-H. Chen, and J. L. Adolphson, *Hum. Genet.* **58,** 306 (1981).

[37] J. J. Albers and G. Utermann, *Am. J. Hum. Genet.* **33,** 702 (1981).

[38] J. J. Albers, E. Gjone, J. L. Adolphson, C.-H. Chen, P. Tenberg, and H. Torsvik, *Acta Med. Scand.* **210,** 455 (1981).

[39] J. A. Glomset and J. L. Wright, *Biochim. Biophys. Acta* **89,** 266 (1964).

cholesterol is prepared by mixing ^{14}C-labeled cholesterol dissolved in acetone or other organic solvent with an albumin solution. This original approach for the measurement of LCAT activity has a number of drawbacks: (1) Heated plasma is not comparable to the native lipoprotein substrate(s) in plasma; (2) it is difficult to maintain the uniformity of these heated plasma substrates; (3) it is difficult to standardized the heat plasma substrates.

Ideally, for the assay of LCAT activity, a stable, efficient, and uniform substrate that can be added to the enzyme source under conditions such that the concentration of substrate is not rate limiting is required. For this reason, the newer methods of measuring of LCAT activity in plasma have used artificial substrates containing lecithin–cholesterol liposomes into which the cofactor apoA-I has been incorporated by the cholate dialysis procedure.[12,40,41] This method permits the preparation of large amounts of stable, efficient homogeneous and well-defined substrate that is suitable for measuring the enzyme activity in plasma fractions.

Assay Method

Principle. LCAT activity is determined by measuring the conversion of radiolabeled cholesterol to cholesteryl ester after incubation of the enzyme and lecithin–cholesterol (including labeled cholesterol) liposome substrates containing apoA-I (a cofactor) prepared by the cholate dialysis procedure.

Reagents

Assay buffer: 10 mM Tris, 140 mM NaCl, 1 mM EDTA, pH 7.4

Egg yolk lecithin (Lipid Products, Surrey, England), 50 mg/ml in ethanol

Cholesterol, 1 mg/ml in ethanol

[4-^{14}C]Cholesterol, specific activity 54.5 mCi/mmol, 0.25 mCi/2.5 ml benzene

Sodium cholate, 314 mg/ml in assay buffer

Human serum albumin, essentially fatty acid-free, 2% solution in assay buffer

ApoA-I, isolated from fresh human plasma by a combination of ultracentrifugation and DEAE–Sepharose column chromatography.[5] ApoA-I is stored in the assay buffer at −70° at a concentration of 0.5 to 1 mg/ml

[40] H. J. Pownall, W. B. Van Vinkle, Q. Pao, M. Rohde, and A. M. Gotto, *Biochim. Biophys. Acta* **713**, 494 (1982).
[41] C. E. Matz and A. Jonas, *J. Biol. Chem.* **257**, 4535 (1982).

2-Mercaptoethanol, hexane, chloroform, petroleum ether, and ethyl
   acetate, all reagent grade

Procedure. For every 40 assays, the substrate is prepared by pipetting
0.154 ml of 50 mg egg lecithin/ml ethanol, 0.116 ml of 1 mg unlabeled
cholesterol/ml ethanol, and 0.108 ml of 1.8 mg [4-^{14}C]cholesterol (0.25
mCi)/2.5 ml benzene into a glass vial. The lipid mixture is evaporated to
dryness under a stream of nitrogen at room temperature. To these dried
lipids 2.5 ml of the assay buffer, 0.8 ml of 1.10 mg apoA-I/ml solution, and
0.3 ml of 7.25 mM sodium cholate solution are added, mixed for 1 min,
and then incubated at 24° in a shaking water bath for 20 min. The mixture
is dialyzed extensively against the assay buffer for 20 hr at 4°. The dialysate is then adjusted to 4 ml using the assay buffer.

Measurement of LCAT activity is carried out in screw-capped culture
tubes in which 0.235 ml of the assay buffer, 0.125 ml of 2% HSA solution,
and 0.1 ml of the liposome substrate containing apoA-I were added. The
substrate mixture was preincubated at 37° for 15 min, 0.025 ml of 100 mM
mercaptoethanol solution was added, and followed by an addition of 0.015
ml of plasma (or LCAT fractions) as LCAT source. This assay mixture
contained 250 nmol lecithin, 7.5 nmol unlabeled cholesterol, 5 nmol [4-
[14]C]cholesterol, 0.8 nmol (22 μg) apoA-I, 0.5% HSA, 5 mM mercaptoethanol, and 0.015 ml of enzyme in a final volume of 0.5 ml. The assay
mixture is mixed immediately on a vortex mixer and incubated in a 37°
water bath for 30 min. Control mixtures without enzyme were incubated
simultaneously. The enzymatic reaction is stopped by the addition of 2 ml
of ethanol, and the bottom phase extracted twice with 4 ml of hexane
containing 50 μg each of unlabeled cholesterol and cholesterol linoleate.
The hexane extract is then evaporated to dryness and dissolved in 0.5 ml
chloroform. Twenty-five microliters of this chloroform solution is spotted
onto a plate coated with silica gel H and cholesterol and cholesteryl ester
are separated by a solvent system of petroleum ether and ethyl acetate
(85 : 15, v/v). The radioactivity in each lipid region is determined in a
toluene/PPO/POPOP scintillation mixture, by a liquid scintillation
counter. LCAT activity is determined from the conversion of [4-^{14}C]cho-
lesterol to labeled cholesteryl ester.

The rate of the LCAT reaction can be expressed as both fractional
activity (percentage of cholesterol esterified/hr), and molar activity (nmol
cholesterol esterified/hour/ml plasma or per milligram purified enzyme).
The latter value is obtained by multiplying the fractional activity with the
concentration of cholesterol in the substrate mixtures.

Pownall *et al.*[40] reported preparation of complexes of phosphatidylcholine, cholesterol, and apoA-I by cholate treatment in the ratio of
100 : 2 : 1. This artificial substrate had been a more efficient substrate than

HDL, but had many of the physical properties of HDL, e.g., size and α-helical content. This substrate was also shown to be appropriate for the measurement of LCAT activity in plasma and plasma fractions.

If the LCAT activity, as determined by an artificial substrate, is an accurate reflection of the amount of active enzyme in plasma and all the enzyme in plasma has the potential to act on the substrate, then there should be a 1:1 relationship between LCAT mass and LCAT activity. Examination of 22 healthy adult subjects for LCAT activity, using the artificial substrate prepared by the cholate dialysis procedure, the LCAT mass, by radioimmunoassay, indicated that LCAT mass was very highly correlated with LCAT activity, $r = 0.98$, $n = 22$ (Fig. 1). Note that the line describing this relationship went through the origin. However, among some subjects with familial LCAT deficiency, the level of LCAT activity appears to be lower than that expected for the given level of LCAT mass in the plasma.[38,42] These data suggest that some LCAT-deficient subjects have functionally defective enzyme in their plasma. After examination of the parents and children of subjects with familial LCAT deficiency and functionally defective LCAT, it appears that these obligate heterozygotes also have slightly less enzyme activity than would be predicted by the amount of LCAT mass present. These facts suggest that the plasma of these subjects contains both normal and functionally defective enzymes.[42]

Measurement of Endogenous Cholesterol Esterification Rate

The endogenous cholesterol esterification rate refers to the rate of cholesteryl ester formation upon incubation of plasma or serum. It is determined by assaying the decrease in free cholesterol chemically or by measuring the formation of cholesteryl esters radiochemically. The endogenous cholesterol esterification rate reflects not only the amount of LCAT present but also the nature and amount of substrate and cofactors in the plasma. Thus, the endogenous cholesterol esterification rate is not necessarily proportional to the amount of enzyme LCAT mass. In healthy adult human subjects approximately 65% of the total variance in the endogenous cholesterol esterification rate could be accounted for by the amount of enzyme mass present in the plasma.[43] The apolipoprotein cofactors that determine the plasma cholesterol esterification rate are not clearly defined. Although apoA-I may not be an absolute requirement for plasma cholesteryl esterification mediated by LCAT (discussed earlier in this chapter), it is reasonable to conclude that under normal circum-

[42] J. J. Albers, C.-H. Chen, J. L. Adolphson, M. Sakuma, T. Kodama, and Y. Akanuma, Hum. Genet. 62, 82 (1982).
[43] J. J. Albers, C.-H. Chen, and J. L. Adolphson, J. Lipid Res. 22, 1206 (1981).

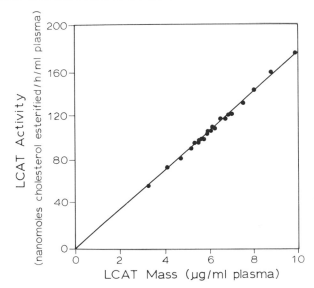

FIG. 1. Correlation of LCAT mass with LCAT activity. For details see text.

stances apoA-I plays a major role as a cofactor in plasma cholesterol esterification. A significant proportion of the unesterified cholesterol that is esterified is derived from lipoproteins other than HDL, particularly from the triglyceride-rich VLDL. Determination of the rate of serum or plasma cholesterol esterification *in vitro* can be of considerable clinical significance assuming that it reflects the turnover of cholesterol *in vivo*. Although the cholesterol esterification rates *in vivo* and *in vitro* have been reported to be comparable,[44] a rigorous comparison of the plasma cholesterol esterification rate and the cholesterol esterification rate *in vivo* has not been performed.

The method that is used extensively for estimation of the plasma cholesterol esterification rate is the radioassay procedure of Stokke and Norum[45] or its modifications. In this assay LCAT is first inhibited by means of DTNB, followed by equilibration of the plasma with radioactive cholesterol solubilized in an ethanol–albumin or in an acetone–albumin mixture. The inhibition is reversed by adding mercaptoethanol in excess to initiate esterification. Although this method or its various modifications have been used quite extensively, a number of objections have been raised regarding its potential pitfalls. During the preincubation period,

[44] B. J. Kudchodkar and H. S. Sodhi, *Clin. Chim. Acta* **68,** 187 (1976).
[45] K. T. Stokke and K. R. Norum, *Scan. J. Lab. Invest.* **27,** 21 (1971).

with DTNB some cholesterol esterification takes place. Second, neutral lipids are transferred between lipoproteins during the preincubation step, altering the LCAT substrates in plasma. This phenomenon can clearly take place in the absence of LCAT activity. Third, mercaptoethanol in excess could potentially alter the esterification rate.

To avoid some of these potential disadvantages we have recently developed a simple and reproducible radioassay, by labeling LCAT substrate at 4° without adding DTNB and mercaptoethanol to the plasma.[46]

Radioassay Method for Plasma Cholesterol Esterification[46]

Principle. The radioassay method for the determination of endogenous cholesterol esterification rate is assessed by measuring the rate of conversion of labeled cholesterol to cholesteryl ester after incubation of fresh plasma that is labeled with a trace amount of radioactive cholesterol by equilibration with [^{14}C]cholesterol–albumin mixture at 4°.

Reagents

[4-^{14}C]Cholesterol specific activity, 54.5 mCi/mmol, 0.25 mCi/2.5 ml benzene

Human serum albumin, essentially fatty acid free

Ethanol, hexane, chloroform, petroleum ether, and ethyl acetate, all reagent grade

Procedure. The [4-^{14}C]cholesterol–albumin mixture is prepared by injecting 0.5 ml of an acetone solution containing 5.4 μCi (100 nmol) of [4-^{14}C]cholesterol (specific activity 54.5 mCi/mmol) slowly into 1 ml of 2% HSA solution in 50 mM phosphate buffer, pH 7.4, stirring continuously. The acetone was carefully evaporated under a stream of nitrogen and the volume was brought up to 1 ml with 50 mM phosphate buffer, pH 7.4. Fresh plasma to be assayed is kept in ice. One hundred microliters of each plasma sample is pipetted into six screw-capped culture tubes that are precooled in ice. To each tube, 25 μl of the [4-^{14}C]cholesterol–albumin mixture containing 0.135 μCi (2.5 nmol) of [4-^{14}C]cholesterol is added, and the mixtures are incubated at 4° for 4 hr in a rotary shaker to allow equilibration to take place between labeled and unlabeled cholesterol. Then, three tubes are transferred into a 37° shaking water bath and incubated for 30 min, whereas the other three tubes are kept at 4° to serve as controls. The enzymatic reaction is stopped by the addition of 2 ml of ethanol and the cholesterol and cholesteryl ester are extracted, separated, and counted as described in the previous section. The rate of esterifica-

---

[46] C.-H. Chen and J. J. Albers, unpublished procedure (1984).

tion can be expressed as both fractional cholesterol esterification rate (percentage conversion of labeled cholesterol/hr) and molar cholesterol esterification rate (nmol cholesterol esterified/hr/ml of plasma). The latter value is obtained by multiplying the fractional cholesterol esterification rate by the number of nanomoles of unesterified cholesterol in 1 ml of plasma, which quantities are determined separately by an enzymatic colorimetric method as described in the following section.

In order to avoid the necessity of introducing a radioactive label to plasma, other investigators have developed methods for measuring the plasma cholesterol esterification rate by directly measuring the decrease in free cholesterol. Several methods have been used to measure the decrease of free cholesterol, including gas liquid chromatography, enzymatic colorimetric methods, and enzymatic fluorometric methods. Because the linearity of the plasma esterification rate is of short duration (usually from 40 min to 1 hr) the difference in the amount of free cholesterol is very small, ranging from 2 to 4%. Thus, a large number of very precise measurements of cholesterol are needed to obtain an accurate assessment of the plasma cholesterol esterification rate by direct chemical analysis of the decrease in free cholesterol. Procedures that do not use an extraction procedure prior to analysis such as the enzymatic colorimetric determination of free cholesterol are preferable because they permit a more precise determination of the small changes in free cholesterol. The following enzymatic procedure for determination of free cholesterol is currently used in the laboratory of Drs. Chen and Albers.[46]

Enzymatic Cholesterol Method for Plasma Cholesterol Esterification [10]

Principle. The plasma cholesterol esterification rate is assessed by directly measuring the decrease in the mass of endogenous unesterified cholesterol in the incubated plasma sample. The quantity of unesterified cholesterol in the control plasma sample (with the addition of iodoacetate to inhibit LCAT) and in the test plasma sample (without addition of iodoacetate) is determined by the enzymatic colorimetric method. To improve the sensitivity we have used 2-hydroxy-3,5-dichlorobenzene acid and sodium cholate, respectively, to substitute for phenol and Triton X-100 in our modified procedure.[46] This improved procedure is 4–5 times more sensitive than the previous method and the rate of the color reaction is twice as fast.

Reagents

Incubation buffer, 50 mM sodium phosphate, pH 7.4
Assay buffer, 50 mM sodium phosphate, pH 7.0

Iodoacetate, sodium salt, 150 mM in incubation buffer

Cholesterol standard solution (Boehringer-Mannheim, Indianapolis, IN), 100 mg/ml in isopropanol

4-Aminoantipyrine, 20 mg/ml in assay buffer

2-Hydroxy-3,5-dichlorobenzenesulfonic acid (Research Organies Inc., Cleveland, OH), 50 mg/ml in assay buffer

Sodium cholate, 2% solution in assay buffer

Cholesterol oxidase from *Streptomyces pseudomonas* (Calbiochem-Behring corp., La Jolla, CA), specific activity 18.5, 2.6 U/mg, 1 mg/ml in assay buffer

Peroxidase from horseradish, specific activity 300 U/mg, 2 mg/ml in assay mixture

Procedure. For every 25 assays, the enzymatic color reagent is made fresh daily by mixing 0.1 ml of 1 mg cholesterol oxidase/ml solution, 1 ml of 2 mg peroxidase/ml solution, 1 ml of 20 mg 4-aminoantipyrine/ml solution, 1 ml of 50 mg 2,4-dibromophenol/ml solution, 3.75 ml of 2% sodium cholate solution, and 30.65 ml of assay buffer. Forty microliters of fresh plasma is pipetted into glass tubes, in sextuplicate for both control and test samples. At zero time, 20 μl of 150 mM iodoacetate solution is added to each control sample to inhibit the LCAT reaction, whereas 20 μl of incubation buffer (50 mM phosphate buffer, pH 7.4) is added to each test sample. All samples are incubated at 37° for 40 min. At the end of this incubation, 20 μl of 150 mM iodoacetate solution is added to each test sample and 20 μl of incubation buffer is added to each control sample. Then, 1.5 ml each of color reagent is added to all samples and the mixtures are incubated for 10 min at 37°. The absorbance of the assay mixtures is measured spectrophotometrically at a wavelength of 500 nm. Unesterified cholesterol in each sample is determined by comparison against the color of cholesterol standard solution containing 1 to 100 μg of unesterified cholesterol in which color is developed in the same manner in the sample solution. The rate of cholesterol esterification is obtained by subtraction of the amount of cholesterol in the test samples from that in the control samples. The rate of cholesterol esterification is obtained by subtraction of the amount of cholesterol in the test samples from that in the control samples. The rate can be expressed as both fractional cholesterol esterification rate (percentage decrease in unesterified cholesterol/hr) and molar cholesterol esterification rate (nmol decrease in unesterified cholesterol/hr/ml plasma).

Over 20 years have passed since Glomset identified the LCAT enzyme mechanism as the vehicle for generating cholesteryl esters in human plasma. Since that time, hundreds of research reports and over a dozen review articles have appeared on this subject. Perhaps due to the complicated nature of the macromolecular substrate(s), the multitude of biochemical parameters that are involved in the regulation of enzyme/substrate interactions and the many unusual features of the enzyme molecule itself, progress in this area has been relatively slow. Consequently, our understanding of this enzyme system is still far from satisfactory. It is hoped that with a sharper focus on investigations that are likely to yield biochemically and physiologically relevant information, the elucidation of the mechanism of LCAT and its function *in vivo* will become a realistic goal of investigators in the not too distant future.

Acknowledgments

Research of J.J.A. and C.-H. Chen referred to in this chapter was supported by the Natural Heart, Lung, and Blood Institute (HL-30086), and that of A.G.L. by the National Institutes of Health (AG-03255), the American Heart Association (82-1043), and the Robert A. Welch Foundation (B-935).

[46] Mechanisms of Action of Lecithin–Cholesterol Acyltransferase

By CHRISTOPHER J. FIELDING

Introduction

Lecithin–cholesterol acyltransferase (LCAT) generates the major part of cholesteryl esters in human plasma from lipoprotein lecithin and unesterified cholesterol.[1] The enzyme protein has been isolated and many of its physical properties described[2] (see also Chapter [45], this volume). Less information is available on its mechanism of action. As with other enzymes active at lipid interfaces, the kinetics and substrate specificity of this enzyme probably depend as much upon the physical state of the substrate, and its overall lipid composition, as upon the active site sequence. As a practical matter, isolated LCAT is reactive to a significant extent only with the high-density lipoprotein (HDL) class of human plasma lipoproteins.[3] However, LCAT active in unfractionated native plasma mostly utilizes unesterified cholesterol derived from the low- and

[1] M. Dobiasova, *Adv. Lipid Res.* **20**, 107 (1983).

[2] K.-S. Chong, S. Hara, R. E. Thompson, and A. G. Lacko, *Arch. Biochem. Biophys.* **222**, 553 (1983).

[3] C. J. Fielding and P. E. Fielding, *FEBS Lett.* **15**, 355 (1971).

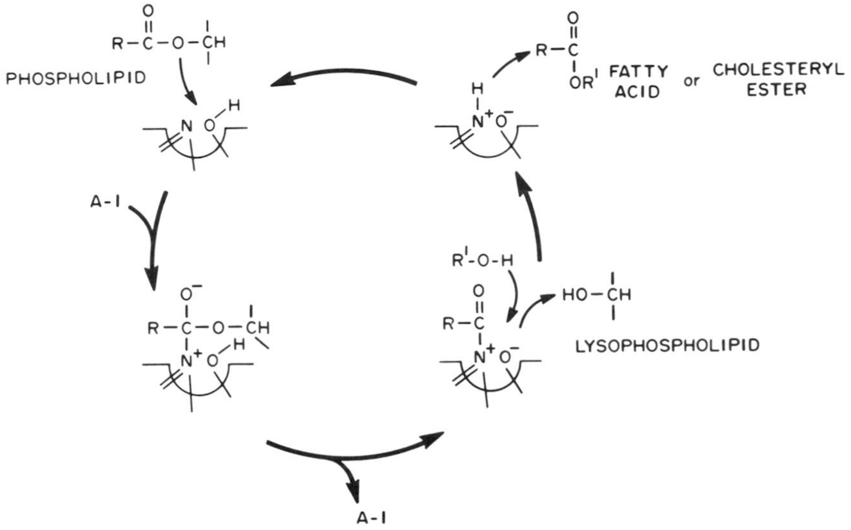

Fig. 1. A model for the role of apolipoprotein cofactor (apoA-I) as catalyst for LCAT activity.

very-low-density classes of lipoproteins (LDL and VLDL) and (in the presence of cells such as fibroblasts, endothelial cells, or smooth muscle cells) cholesterol from cell membranes.[4,5] Clearly, the mechanism of action of LCAT under physiological conditions is likely to be quite different from that determined with a fully defined system of synthetic lipids. Because information is insufficient at this point to fully integrate the data that have been obtained with synthetic and physiological substrates, these will first be considered separately. Finally, a comparison of the types of data will be made.

The Active Site of the Enzyme

Two kinds of data suggest that LCAT has the active site structure of a typical serine hydrolase (Fig. 1). First, the pH dependence of the reaction shows the classical bell-shaped curve with half-maximal reaction rates suggesting pK_a and pK_b values of 6.0 and 9.0, respectively.[6] Second, LCAT is inhibited by organophosphates, such as diethyl p-nitrophenyl phosphate,[6] as are other enzymes (for example, chymotrypsin) where the involvement of an active site serine has been well established. Other

[4] C. J. Fielding and P. E. Fielding, *J. Biol. Chem.* **256,** 2102 (1981).
[5] C. J. Fielding and P. E. Fielding, *Proc. Natl. Acad. Sci. U.S.A.* **78,** 3911 (1981).
[6] L. Aron, S. Jones, and C. J. Fielding, *J. Biol. Chem.* **253,** 7220 (1978).

inhibitors of the LCAT reaction are heavy metals and blocking agents, such as dithiobis(2-nitrobenzoic acid), likely to be active via covalent binding to enzyme sulfhydryl groups. This finding does not necessarily imply the formation of an enzyme-S-acyl intermediate at the active site and these reagents may exert their effects by steric hindrance.

Reaction of Purified LCAT with Synthetic Lipid Substrates

Substrate Specificity of LCAT. While LCAT is reactive in plasma with lecithin, other native and synthetic phospholipid acyl donors with the same net charge (e.g., phosphatidylethanolamine and dimethyl phosphatidylethanolamine) are also transesterified in synthetic lipid dispersions.[7,8] Significant rates are obtained with a wide range of lecithins containing different acyl chain lengths and degrees of saturation, although long chain saturated lecithins (e.g., di-$C_{18:0}$ lecithin) are relatively or else completely inactive,[9] probably because of the high transition temperatures of these lipids. While slightly different orders of relative reactivity are obtained when synthetic lipid substrates of LCAT are either dispersed individually or in an inert (diether analog) matrix,[8] such differences are probably not significant relative to the mixture of lecithins in native lipoproteins, since in plasma the mixture of cholesteryl esters generated appears to largely reflect the lecithin unsaturated fatty acids available for transesterification.[10]

LCAT shows a relative lack of specificity in terms of acyl acceptor. Water is an avid acceptor of lecithin-derived acyl groups.[6] The kinetics of LCAT-mediated phospholipase activity are closely similar to those for the transferase reaction in the presence of free cholesterol with the same lecithin. Accordingly, the rate-limiting step of the LCAT reaction overall must lie in the initial (acyl donor) phase of the reaction (Fig. 1). LCAT also reacts with lysolecithin as acyl acceptor (the lysolecithin acyltransferase or LAT reaction)[11] and with long chain alcohols and a variety of sterols.[12] While LCAT shows specificity among sterols (reaction as an effective LCAT substrate requires a 3β-hydroxyl group and planar A and

[7] C. J. Fielding, *Scand. J. Clin. Lab. Invest.* **33** (Suppl. 137), 15 (1974).

[8] H. J. Pownall, Q. Pao, and J. B. Massey, *Arteriosclerosis* **3**, 493a (1983).

[9] G. Assmann, G. Schmitz, N. Donath, and D. Lekim, *Scand. J. Clin. Lab. Invest.* **38** (Suppl. 137), 15 (1974).

[10] V. P. Skipski, *in* "Blood Lipids and Lipoproteins" (G. J. Nelson, ed.), p. 471. Wiley (Interscience), New York, 1972.

[11] P. V. Subbaiah, J. J. Albers, C.-H. Chen, and J. D. Bagdade, *J. Biol. Chem.* **255**, 9275 (1980).

[12] K. Kitabatake, U. Piran, Y. Kamio, Y. Doi, and T. Nishida, *Biochim. Biophys. Acta* **573**, 145 (1979).

B rings),[13,14] these requirements clearly relate to the conformation of the acyl acceptor at the lipid surface rather than to specificity at the enzyme active site, since they are identical to those of various nonenzymatic effects of sterols in membranes.[15]

Apolipoprotein Cofactors of LCAT. LCAT is among those lipoprotein metabolic factors with an evident requirement for a polypeptide activator. The most effective activator among lipoprotein apolipoproteins is apolipoprotein A-I (a major protein of HDL),[16] although several other apolipoproteins show reactivity with synthetic lipid dispersions.[17,18] Since apolipoproteins (and particularly apoA-I) are readily denatured in terms of their effectiveness in LCAT activation during isolation or storage, it is particularly important that their effects be considered relative to the catalytic constant of LCAT and apoA-I in native plasma, rather than relative to each other as isolated apolipoproteins. The catalytic constant based on the native enzyme in plasma[19] with native lipoproteins is about 0.3 sec^{-1}.

The mechanism of LCAT activator by polypeptides has been approached by three different kinds of studies: (1) comparisons of the activity of different natural polypeptides with identifiable sequence homologies; (2) by apoA-I modified chemically, or by apoA-I with a genetically modified amino acid sequence; and (3) by studies with model polypeptides with similar helical content and lipid binding properties to apoA-I, but with a sequence unrelated to that of the natural activator protein. Comparison of the sequences of apoA-I with apoA-IV and apoC-I (minor proteins of HDL with reported LCAT cofactor activity) shows that these polypeptides share some sequence blocks.[20] However, apoE, also reported to activate LCAT with similar synthetic lipid substrates, has not been reported to share these sequences.

Polypeptides with similar helical content and lipid binding capacity

---

[13] U. Piran and T. Nishida, *Lipids* **14**, 478 (1979).

[14] Unpublished experiments.

[15] K. A. Kemal, K. R. Bruckdorfer, and L. L. M. van Deenen, *Biochim. Biophys. Acta* **255**, 311; **255**, 321 (1972).

[16] C. J. Fielding, V. G. Shore, and P. E. Fielding, *Biochem. Biophys. Res. Commun.* **46**, 1493 (1972).

[17] G. F. Sigler, A. K. Soutar, L. C. Smith, A. M. Gotto, and J. T. Sparrow, *Proc. Natl. Acad. Sci. U.S.A.* **73**, 4122 (1976).

[18] C.-H. Chen and J. J. Albers, *Arteriosclerosis* **4**, 519a (1984).

[19] Normal plasma contains approximately 6 μg LCAT protein per ml; for a molecular weight of 61,000 (ref. 2), LCAT concentration is therefore 90 nM. The rate of activity of LCAT in native plasma is about 80 nmoles cholesterol esterified per ml per h at 37°C. The catalytic constant (k_{cat}) of LCAT under these conditions is therefore about 0.3 sec^{-1}.

[20] M. S. Boguski, N. Elsahourbagy, J. M. Taylor, and J. I. Gordon, *Proc. Natl. Acad. Sci. U.S.A.* **81**, 5021 (1984).

have quite different abilities to activate LCAT[21] and while a mutant apoA-I species containing deletion of a single amino acid residue had a significantly reduced activator capacity,[22] the general location of this deletion (position 107) in the known sequence of human apoA-I does not show particular homology with that of other activator proteins. At least two small polypeptides with sequences unrelated to that of apoA-I show detectable activator capacity.[23,24] Catalytic rates for LCAT activity relative to those for the same enzyme mass in plasma were not reported. These data, overall, suggest that LCAT activation is a specific function dependent upon the spatial organization of a particular and limited region of apoA-I, depending upon more than helical content and lipid binding, but the factors involved remain to be identified.

Reaction of LCAT in Native Plasma

Substrate Specificity of LCAT. LCAT in plasma is active with lipoprotein lecithin. Species such as $C_{16:0}$ lecithin are effective substrates when incorporated into native HDL, even when they are inactive in the form of dispersions of pure phospholipids.[13] In plasma containing lecithin of normal or abnormal fatty acid distribution, plasma cholesteryl ester fatty acid composition reflects that of lecithin unsaturated fatty acids.[10] Finally, phosphatidylethanolamine is only a minor constituent (2–3%) of plasma phospholipids, and its major unsaturated fatty acids are polyunsaturated species (20:6, 22:4, 22:5, and 22:6) that are poorly represented in plasma cholesteryl esters.[9] Phosphatidylethanolamine is therefore unlikely to be a quantitatively significant substrate for LCAT in native plasma. Dimethyl phosphatidylethanolamine (another effective substrate in highly defined synthetic assay systems) is not present to any significant extent in plasma. These data suggest that the acyl donor specificity of LCAT observed in synthetic systems is relatively unimportant in determining the spectrum of cholesteryl esters generated in native plasma.

While LCAT shows phospholipase activity with pure lecithin vesicles, free cholesterol is an effective competitive inhibitor such that little or no free fatty acid was generated when substrate free cholesterol/phospholipid molar ratio was >0.2 (equivalent to a mass ratio of 0.1).[6] All major

[21] A. Jonas, S. A. Sweeny, and P. N. Herbert, *J. Biol. Chem.* **259**, 6369 (1984).

[22] S. C. Rall, K. H. Weisgraber, R. W. Mahley, Y. Ogawa, C. J. Fielding *et al.*, *J. Biol. Chem.* **259**, 10063 (1984).

[23] S. Yokoyama, D. Fukushima, J. P. Kupferberg, F. J. Kezdy, and E. T. Kaiser, *J. Biol. Chem.* **255**, 7333 (1980).

[24] H. J. Pownall, A. Hu, A. M. Gotto, J. J. Albers, and J. T. Sparrow, *Proc. Natl. Acad. Sci. U.S.A.* **77**, 3154 (1980).

plasma lipoprotein classes contain significantly higher substrate ratios, and unfractionated native plasma containing LCAT generates little unesterified fatty acid when incubated *in vitro*. It is likely that LCAT does not function as a phospholipase *in vivo*. The significance of LAT (lysolecithin acyltransferase) activity is hard to assess, and activity in native plasma has not been conclusively demonstrated.

While cholesterol itself is the only significant sterol in normal plasma, in some patient groups esters of other sterols such as β-sitosterol or cholestanol can be identified.[25,26] It would be predicted that LCAT would esterify steroids such as estradiol in plasma, but such esters have apparently not been identified.

Significance of Apolipoprotein Cofactors of LCAT in Native Plasma. Even though several apolipoproteins have been identified as activators of LCAT in synthetic systems, several lines of evidence indicate that apoA-I is the only cofactor of physiological significance in plasma. First, immunoaffinity chromatography of native plasma using immobilized antibodies to apolipoprotein A-I quantitatively removes LCAT activity from plasma, while similar experiments with other potential activators such as apoE[27] were without effect. These data indicate that essentially the whole of LCAT in unfractionated native plasma is bound to apoA-I. Second, while other potential activators (apoC-I, apoE) are presently distributed between several plasma lipoprotein classes, LCAT protein is located almost exclusively in the HDL and very high-density lipoprotein classes[28] where apoA-I is the major apolipoprotein; the distribution of LCAT therefore reflects the distribution of apoA-I, but not that of the other apolipoproteins (the distribution of apoA-IV in unfractionated native plasma has apparently not been determined). Finally, among centrifugally derived isolated plasma lipoproteins, only HDL (containing apoA-I) show significant reactivity with LCAT, while other fractions containing apoA-IV, apoC-I, or apoE without apoA-I are without reactivity. These data suggest that apoA-I is the physiological activator of native plasma. In plasma from which apoA-I is genetically deficient, low but detectable levels of cholesterol esterification by LCAT are detectable. However, the presence of trace levels of apoA-I in substantially apoA-I-deficient plasma is indicated by the removal of such LCAT activity by anti-apoA-I affinity

[25] T. A. Miettinen, *Eur. J. Clin. Invest.* **10**, 27 (1980).

[26] V. Shore, G. Salen, F. W. Cheng, T. Forte, S. Shefer *et al., J. Clin. Invest.* **68**, 1295 (1981).

[27] P. E. Fielding and C. J. Fielding, *Proc. Natl. Acad. Sci. U.S.A.* **77**, 3327 (1980).

[28] J. J. Albers, J. L. Adolphson, and C.-H. Chen, *J. Clin. Invest.* **67**, 141 (1981).

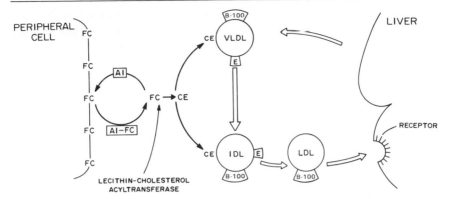

Fig. 2. Reaction steps in the transport of peripheral cell membrane and plasma lipoprotein cholesterol as cholesteryl ester to the liver. Closed arrows: molecular transfers of free and esterified cholesterol. Open arrows: metabolic conversions of lipoprotein particles.

chromatography.[29] These data suggest that even traces of apoA-I may be sufficient to catalyze detectable levels of LCAT activity *in vivo*.

A final consideration relates to the source of substrates of LCAT in native plasma. Since the majority of free cholesterol for LCAT activity is derived from surfaces other than that of HDL itself (mostly from surfaces not containing apoA-I), this free cholesterol must be transferred to surfaces containing both LCAT and apoA-I for esterification. Because of the significant rate of transfer of free cholesterol by diffusion through the aqueous phase,[30] it is likely that substrate is supplied to the enzyme down the chemical potential gradient generated by its own activity (Fig. 2). Other data[31] indicate that there is little or no such gradient between lipoprotein classes in normal plasma when LCAT has been inhibited. Product inhibition is prevented by the transfer away of cholesteryl ester product by one or more of the plasma transfer proteins. The most effective direct substrate for LCAT would therefore be one containing phospholipid, apoA-I only, and little cholesteryl ester or free cholesterol. Such a particle has been identified in normal plasma by immunoaffinity chromatography[5] and accumulates in congenital LCAT deficiency.[32] It is likely that

[29] Unpublished experiments.

[30] L. R. McLean and M. C. Phillips, *Biochemistry* **20**, 2893 (1981).

[31] C. J. Fielding, G. M. Reaven, G. Liu, and P. E. Fielding, *Proc. Natl. Acad. Sci. U.S.A.* **81**, 2512 (1984).

[32] C. Chen, K. Applegate, W. C. King, J. A. Glomset, K. R. Norum, and E. Gjone, *J. Lipid Res.* **25**, 269 (1984).

this is the primary substrate for LCAT *in vivo,* acting as the recipient of free cholesterol from cell membranes, VLDL, and LDL, and the source of the cholesteryl esters transferred away to the major lipoprotein classes of the plasma.

[47] Lysolecithin Acyltransferase of Human Plasma: Assay and Characterization of Enzyme Activity

By P. V. Subbaiah

Human plasma lecithin–cholesterol acyltransferase (LCAT, EC 2.3.1.43) can also esterify lysolecithin back to lecithin in the presence of low-density lipoproteins (LDL).[1,2] This latter enzyme activity was given the name lysolecithin acyltransferase (LAT).

Although human plasma produces more lysolecithin than cholesterol esters, the further metabolism of this cytolytic product in the plasma has not been determined. The demonstration of an enzyme activity in the plasma that acylates lysolecithin to lecithin proved that lysolecithin can undergo metabolic transformations in the plasma compartment itself.[3] This enzyme which was responsible for this activity was later found to be identical to LCAT.[2] However, unlike the esterification of cholesterol which requires apolipoprotein A-I for the activity, the acylation of lysolecithin requires LDL. The enzyme appears to transfer an acyl group from lecithin of LDL to lysolecithin, thus catalyzing the transfer of a fatty acid from one lecithin molecule to another on the lipoprotein surface. Although no net change in the mass of lecithin or lysolecithin results by this reaction, the molecular species composition of lecithin is altered.

Assay Method

Principle. 1-Acylglyceryl phosphorylcholine (lysolecithin), labeled either in the fatty acid or glycerol moiety, is dispersed in buffer and incubated with the enzyme preparation in presence of human plasma LDL. The total lipids are then extracted, the lysolecithin and lecithin are sepa-

---

[1] P. V. Subbaiah and J. D. Bagdade, *Biochim. Bipohys. Acta* **573,** 212 (1979).
[2] P. V. Subbaiah, J. J. Albers, C. H. Chen, and J. D. Bagdade, *J. Biol. Chem.* **225,** 9275 (1980).
[3] P. V. Subbaiah and J. D. Bagdade, *Life Sci.* **22,** 1971 (1978).

rated by thin-layer chromatography (TLC), and the amount of labeled lysolecithin converted to labeled lecithin is determined.

Reagents

10 mM Tris-HCl buffer, pH 7.4, containing 1 mM EDTA and 0.15 M NaCl (Tris/NaCl/EDTA)

[1-^{14}C]Palmitoyl glycerylphosphorylcholine (commercial) purified on TLC

Unlabeled 1-palmitoyl glycerylphosphorylcholine

Normal human LDL is prepared by centrifugation of whole plasma, first at 1.019 g/ml in KBr (100,000 g for 18 hr) followed by centrifugation of 1.019 infranatant at 1.060 g/ml in KBr for 18 hr. The 1.019–1.060 supernatant is recentrifuged at 1.060 g/ml to remove any higher density lipoprotein contamination. The final LDL preparation is dialyzed against 400 vol of 0.15 M NaCl, 1 mM EDTA, with at least three changes of dialysis solution, and concentrated by Amicon Centriflo filters to give a protein concentration of 7 mg/ml

2-Mercaptoethanol, reagent grade, in Tris/NaCl/EDTA

Chloroform, reagent grade

Methanol, reagent grade

TLC plates, silica gel G, 0.5 mm thickness (20 × 20 cm)

Liquid scintillation cocktail: Aquasol (New England Nuclear, Boston MA)

Preparation of Substrate

[^{3}H]Glycerol-labeled lysolecithin prepared from soybeans germinating in the presence of [^{3}H]glycerol,[2] or [1-^{14}C]palmitoyl glycerylphosphorylcholine obtained from commercial sources, can be used. The use of [1-^{14}C]palmitoyl lysolecithin is preferred because, with defined fatty acid composition, the molecular species of lecithin formed can be determined more easily. In our experience, if the two substrates are compared under identical conditions there is no significant difference in activity.

[1-^{14}C]Palmitoyl lysolecithin, obtained from New England Nuclear (Boston, MA) is purified on silica gel TLC plates using the solvent system of chloroform:methanol:water (65:25:4 by volume). The lysolecithin spot, identified with the help of a standard lysolecithin, is scraped from the plate and eluted immediately. The spots from one 20-cm plate are scraped into a 40-ml glass tube and 9.5 ml of the elution mixture (chloroform:methanol:water, 5:10:4 by volume) is added. The sample is vortexed thoroughly and kept mixing at room temperature for 15 min. The tube is centrifuged (2000 rpm, 5 min) and the supernatant is taken into a separate tube. The extraction of silica gel is repeated with another 9.5 ml of the same mixture. To the combined eluates 5.0 ml chloroform is added, followed by 5 ml of water. The mixture is vortexed and centrifuged to clear the chloroform layer. After the chloroform is transferred into another tube, the aqueous layer is extracted with another 5 ml of chloroform. The combined chloroform extracts are evaporated to dryness and the labeled lysolecithin is redissolved in chloroform : methanol (9 : 1) to give a final radioactivity of 800,000 dpm/ml. The sample is stored at $-20°$ and remains stable for at least 6 months under these conditions.

Unlabeled 1-palmitoyl lysolecithin, obtained from Sigma Chemical Co. (St. Louis, MO) is diluted in chloroform : methanol (9 : 1) to a concentration of 3.5 μmol/ml. One milliliter of labeled lysolecithin is mixed with 1 ml of unlabeled lysolecithin in a glass tube, and the solvent is evaporated under nitrogen at $25°$. After the lysolecithin is completely dry, 1.75 ml of Tris/NaCl/EDTA is added and the lipid is dispersed by vortexing for 1 min. The lysolecithin forms a clear dispersion under these conditions.

Enzyme Assay in Whole Plasma

For the assay of LAT, draw the blood in EDTA (7 mg/ml) or in acid-citrate dextrose. Heparinized plasma is not suitable for the assay, because LAT activity is inhibited by heparin. Separate the plasma from the cells by centrifugation at 2000 rpm for 20 min and keep the plasma frozen at $-20°$, unless assayed immediately.

For each plasma to be assayed, take four 13 × 100 mm disposable glass tubes and add to each, 100 μl of substrate, 20 μl of 0.1 M mercaptoethanol and 130 μl of Tris/EDTA/NaCl. After mixing the reagents by vortexing and incubating them at $37°$ for 10 min, add 1.0 ml methanol to two of the four tubes (0-min controls) and then 150 μl of test plasma to all the four tubes. Vortex the experimental tubes and incubate them at $37°$ for 60 min in a water bath with constant shaking. At the end of 60 min, stop the reaction by adding 1.0 ml methanol. To all tubes (experimental and 0 min controls) add 0.5 ml chloroform, vortex, and keep them for 1 min at room temperature. Add 0.5 ml more chloroform to split the phases, followed by 0.5 ml water to wash the water-soluble contaminants from the lipid extract. Mix thoroughly by vortexing after each addition. Centrifuge the tubes at 2000 rpm for 10 min to clear the phases, and take all the chloroform (bottom) layer with a Pasteur pipet into a centrifuge tube with conical bottom. Wash the aqueous layer with 0.5 ml of chloroform, and combine the chloroform extracts. Evaporate the combined chloroform extracts to dryness with nitrogen at 25° and redissolve them immediately in 100 μl of chloroform. Spot the entire extract onto a silica gel TLC plate that has been activated at 110° for 1 hr and scored into 2-cm lanes. Wash the tube with another 100 μl of chloroform and spot the wash on the same area of the plate. Run the plate in a TLC chamber containing the solvent mixture chloroform : methanol : water (65 : 25 : 4 by volume), until the solvent is about 1 in. from the top. After drying the plate in the fume hood, visualize the phospholipid spots by exposing them to iodine vapor. Identify the lysolecithin and lecithin spots with the help of standards run simultaneously on the same plate, and scrape the silica gel corresponding to each spot in a 20-ml scintillation vial. Add 1.0 ml water to the silica gel, followed by 10 ml Aquasol, and count for ^{14}C radioactivity in a liquid scintillation counter. It is also possible to use the minivials (6.5-ml capacity), instead of the 20-ml vials. Use 0.5 ml water and 5.0 ml Aquasol for each sample, if using the minivials. No significant loss of counting efficiency is noticed under these conditions.

For each plasma sample to be assayed, extract 0.1 ml plasma (in duplicate) by the same procedure as above and separate the phospholipids on a TLC plate. Estimate the lysolecithin spots from these extracts for lipid phosphorus,[4] and take into account this amount of lysolecithin while calculating the specific radioactivity of the substrate (see below).

Calculations

Calculate the specific radioactivity of substrate lysolecithin by dividing the radioactivity in the lysolecithin of the control sample by the total amount of lysolecithin added (0.2 μmol) and that present in 0.15 ml of plasma. Calculate the radioactivity of the lecithin formed by subtracting the average counts of lecithin in the control samples from the average counts of lecithin in the experimental samples. Then divide this value by the specific radioactivity of lysolecithin substrate to get the nanomoles of lysolecithin converted by lecithin by 0.15 ml of plasma in 60 min. The enzyme activity of the plasma is expressed as nanomoles of lysolecithin acylated/60 min/ml of plasma at 37°:

$$(L_e - L_c)/(S \times 0.15) = \text{nmol/hr/ml plasma}$$

where L_e = dpm in lecithin of experimental sample, L_c = dpm in lecithin of control sample, and S = specific radioactivity of lysolecithin in the reaction mixture.

[4] G. R. Bartlett, *J. Biol. Chem.* **234**, 466 (1959).

Properties of LAT in Whole Plasma

The enzyme activity is stable in whole plasma for at least 4 weeks if it is kept frozen at $-20°$. Repeated freezing and thawing, however, destroys the activity. The enzyme has a pH optimum of 7.4 and is completely inactivated when the plasma is heated at $58°$ for 20 min. Addition of more than 0.7 mM lysolecithin to the reaction mixture is inhibitory, possibly because of its surface active properties. Mercaptoethanol is not absolutely essential for the enzyme activity, but maximal activity is not obtained in its absence, especially if the plasma has been frozen.

Addition of LDL to normal human plasma can stimulate the activity by as much as 100%, indicating that the amount of LDL in normal plasma is rate limiting. Therefore, under pathological conditions like type II hyperlipidemia, where the LDL levels are higher, the LAT activity is also higher than in control populations, although there is no increase in the enzyme mass.[5]

Assay of Enzyme Activity in Purified Preparations

Because the LAT requires LDL for its activity, the assay mixture should contain LDL in addition to the enzyme, substrate, and mercaptoethanol. Add 100 μl of the substrate dispersion (prepared as described above) containing 0.2 μmol of labeled lysolecithin to the reaction tube, followed by 20 μl of 0.1 M mercaptoethanol and 100 μl of LDL solution containing 0.7 mg protein. Incubate this reaction mixture at $37°$ for 20 min and then add enzyme solution and Tris/NaCl/EDTA to make up the final volume to 0.4 ml. Continue incubation at $37°$ for 1 hr after addition of the enzyme and stop the reaction by adding 1 ml methanol. Extract the lipids, and do the TLC separation and the radioactivity determination exactly as described above for the enzyme from whole plasma. Run control reactions containing all the ingredients except the enzyme simultaneously with the experimental samples.

Calculations

Since the purified enzyme preparations and the added LDL contain negligible amounts of lysolecithin, the specific radioactivity of the added lysolecithin can be assumed not to change after the addition of an enzyme and LDL. However, each batch of LDL preparation should be checked for the amount of lysolecithin and any significant amount (more than 10 nmol lysolecithin/0.1 ml of preparation) should be included in the calcula-

[5] P. V. Subbaiah and J. T. Ogilvie, *Lipids* **19**, 80 (1984).

tion of specific radioactivity. The specific activity of the enzyme is expressed as nanomoles of lysolecithin converted to lecithin/mg protein/hr, which is calculated by the following formula:

$$\text{Specific activity of enzyme} = (L_e - L_c)/(SE)$$

where L_e = dpm in lecithin of experimental sample, L_c = dpm in lecithin of control sample, S = specific radioactivity of lysolecithin, and E = milligrams enzyme added per reaction.

Purification of the Enzyme

The purification of LAT activity from human plasma has been carried out by the same procedure used for the purification of LCAT, because both activities are carried out by a single protein. Details of the purification procedure are given in Chapter [45] by Albers *et al.*[6] Both activities are enriched to a similar extent after each step of purification.[2]

Properties of the Purified Enzyme and Mechanism of the LAT Reaction

The physical characterization of the enzyme is described by Albers *et al.* in Chapter [45].[6] Compounds that inhibit LCAT activity also inhibit LAT activity.[2] The only major difference between the two activities is the inhibition of LAT by heparin, which is obviously due to heparin binding of LDL. Serum albumin, which activates LCAT, inhibits LAT,[7] probably because it binds lysolecithin, the substrate. However, in whole plasma, addition of lysolecithin bound to albumin gives the same activity as does lysolecithin added to aqueous dispersion. This shows that the lysolecithin, which is carried by albumin in the plasma, is a good substrate for LAT.

The purified enzyme preparation uses both 1-acyl lysolecithin and 2-acyl isomers as substrates.[8] The unsaturation of the fatty acid in the 1-acyl lysolecithin does not have a significant effect on the enzyme activity. Lysophosphatidylethanolamine is acylated at a much lower rate (25%) than is lysolecithin of same fatty acid composition. Lyso platelet activating factor, which has an alkyl group instead of an acyl group at 1-position, is also esterified by the enzyme to a similar extent as lysolecithin. The

[6] J. J. Albers, C.-H. Chen, and A. G. Lacko, this volume [45].

[7] P. V. Subbaiah, C. H. Chen, J. J. Albers, and J. D. Bagdade, *Atherosclerosis* **45**, 181 (1982).

[8] P. V. Subbaiah, C. H. Chen, J. D. Bagdade, and J. J. Albers, *J. Biol. Chem.*, **260**, 5308 (1985).

major fatty acids used for the acylation of the lysolecithin are 18 : 2, 18 : 1, and 16 : 0, in decreasing order. However, the highest specific activity is obtained in the lecithin species acylated by 16 : 0 fatty acid. Thus when [1-^{14}C]palmitoyl lysolecithin was used as substrate, the highest specific activity was obtained in dipalmitoyl lecithin. With [1-^{14}C]linoleoyl lysolecithin, the highest specific activity was found in 1–18 : 2–2–16 : 0 lecithin species.[8] These results show that there is an enrichment of certain species of lecithin by the LAT reaction, depending on the species of lysolecithin used for acylation.

The acylation of lysolecithin by LAT is not stimulated by ATP and CoA,[3] neither does the transesterification of two lysolecithins to form one lecithin molecule occur.[2] On the basis of these results and the effects of inhibitors, we suggested that the mechanism of LAT reaction is similar to that of LCAT, probably involving an acyl–enzyme intermediate. Our surmise is that the 2-acyl group from the lecithin first forms an acyl–enzyme intermediate, and then the acyl group transfers to either cholesterol (LCAT), or lysolecithin (LAT), depending on the activator present and the concentration of the acceptor. If no acceptor is available, the acyl group is transferred to water and the activity manifests itself as phospholipase A_2 activity as reported in vitro.[9,10]

The above mechanism still does not explain why LAT activation requires the presence of LDL whereas the LCAT and the phospholipase A activity require apolipoprotein A-I. Our recent results[11] show that the apolipoprotein B as well as the phospholipids of LDL are essential for the stimulation of LAT. The arginine and lysine groups of the apolipoprotein B appear essential but the sulfhydryl groups are not. So far it has not been possible to substitute apoA-I for apoB in the activation of LAT. VLDL cannot substitute for LDL in the reaction using either purified enzyme or abetalipoproteinemic plasma.[12]

If the LAT transfers an acyl group from one lecithin molecule to another without affecting the net amounts of lysolecithin or lecithin, what is its physiological function? One possibility is that it changes the molecular species composition of lecithin by its preferential transfer of certain fatty acids. Since predominantly saturated lysolecithins are formed by the LCAT reaction and unsaturated lysolecithins are formed by the action of lipoprotein lipase, the acylation of these different lysolecithins even with the same fatty acid would give rise to different molecular species of leci-

[9] U. Piran and T. Nishida, J. Biochem. **80,** 887 (1976).
[10] L. Aron, S. Jones, and C. J. Fielding, J. Biol. Chem. **253,** 7220 (1978).
[11] P. V. Subbaiah, C. H. Chen, J. D. Bagdade, and J. J. Albers, Atherosclerosis **54,** 99 (1985).
[12] P. V. Subbaiah, Metabolism **31,** 294 (1982).

thin. Further variation in the molecular species composition can result from the enzyme's apparent specificity for 16:0, 18:1, and 18:2 fatty acids.

Recent studies indicate that disaturated lecithins may promote atherogenesis.[13] If LAT reaction results in an enrichment of disaturated lecithins, as our recent results indicate, it is possible that increased LAT activity, as observed in certain hyperlipemias,[5] can contribute to atherogenesis by increasing the concentration of disaturated lecithins.

[13] N. L. Gershfeld, *Science* **204**, 506 (1979).

[48] Isolation, Characterization, and Assay of Plasma Lipid Transfer Proteins

By JOHN H. TOLLEFSON and JOHN J. ALBERS

Introduction

The recent identification of specific plasma proteins which facilitate the exchange, and under the appropriate conditions the net mass transfer, of cholesteryl esters, triglycerides, and phospholipids between the various classes of the plasma lipoproteins has opened a new and important area of research for understanding the physiology and pathophysiology of lipid transport.

It has been postulated that a plasma neutral lipid transfer protein, together with high-density lipoprotein (HDL) (Oram, Chapter [38]),[1] and the enzyme lecithin–cholesterol acyltransferase (LCAT) (Albers *et al.*, Chapter [45])[2] plays a key role in removing unesterified cholesterol from extrahepatic tissues, including the arterial wall, and in transporting cholesterol to the liver. HDL accepts unesterified cholesterol from peripheral tissues. Through LCAT action, unesterified cholesterol of HDL is esterified and cholesterol ester is transferred to acceptor lipoproteins by a plasma lipid transfer protein for transport to the liver. This latter process permits additional transfer of unesterified cholesterol to HDL. Thus, LCAT and a lipid transfer protein create a gradient of unesterified cholesterol concentration between peripheral tissues and HDL that permits a net efflux of cholesterol from tissue or "reverse cholesterol transport." This process (Rothblat *et al.*, Chapter [37])[3] is believed to prevent lipid

[1] J. F. Oram, this volume [38].
[2] J. J. Albers, C.-H. Chen, and A. G. Lacko, this volume [45].
[3] G. H. Rothblat, M. Bamberger, and M. C. Phillips, this volume [37].

accumulation and retard atherogenesis. The plasma lipid transfer protein also plays an important role in removing surface components such as phospholipid, unesterified cholesterol, and protein from very low-density lipoproteins (VLDL) and chylomicron remnants during lipolysis. Furthermore, the plasma neutral lipid transfer protein may also redistribute neutral lipids not only between lipoproteins but also between lipoproteins and plasma membranes of cells.

This chapter describes the methodological approaches for the isolation and assay of the human plasma lipid transfer proteins.

Isolation of the Human Plasma Lipid Transfer Protein (LTP-I)

General Considerations

Most purification schemes[4-7] for the lipid transfer proteins exploit some physicochemical property of the lipid transfer protein such as hydrophobicity, isoelectric point, charge, size, or affinity for lipid. However, there is yet no consensus as to which biochemical techniques or what sequence of procedures are best for the purification of the plasma neutral lipid transfer protein. The first step in the isolation usually involves removal of most or all of the plasma lipoproteins by either ultracentrifugation at a nonprotein solvent density of 1.21 g/ml or 1.25 g/ml, or by various precipitation techniques. The "lipoprotein free" plasma is then adjusted to a very high ionic strength (2 to 4 M NaCl), then passed over a hydrophobic resin (Phenyl Sepharose, Pharmacia, Piscataway, NJ) which has a high affinity for both the LCAT enzyme (see Chapter [45])[2] and the plasma lipid transfer proteins. Both the LCAT enzyme and the lipid transfer protein are eluted in high yield (see discussion of inhibition of lipid transfer protein) with water.

Most commonly, one or more applications of ion-exchange chromatography are next employed, such as diethylaminoethyl (DEAE) Sepharose CL-6B (Pharmacia) at a physiological pH of 7.4. Both the lipid transfer proteins and the enzyme LCAT bind at low ionic strength, and are conveniently separated with a linear salt gradient (20 to 200 mM NaCl), with the lipid transfer protein eluting before the LCAT enzyme. (Fig. 1). Many investigators follow the anion exchange step with cation exchange chromatography on a resin such as Whatman CM-52 at an acid

[4] N. M. Pattnaik, A. Montes, L. B. Hughes, and D. B. Zilversmit, *Biochim. Biophys. Acta.* **530,** 428 (1978).

[5] R. E. Morton and D. B. Zilversmit, *J. Lipid. Res.* **23,** 1058 (1982).

[6] J. J. Albers, J. H. Tollefson, C.-H. Chen, and A. Steinmetz, *Arteriosclerosis* **4,** 49 (1984).

[7] A. R. Tall, E. Abreu and J. Shuman, *J. Biol. Chem.* **258,** 2174 (1983).

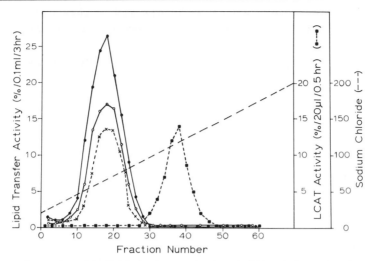

FIG. 1. Separation of the lipid transfer proteins from lecithin–cholesterol acyltransferase (LCAT). One hundred and fifty milliliters of a phenyl–Sepharose pool obtained from the d = 1.21 to d = 1.25 g/ml plasma fraction was dialyzed against 25 mM NaCl/10 mM Tris buffer, pH 7.4, then applied to a 1.6 × 30 cm column of DEAE–Sepharose equilibrated with the same buffer. After the sample was applied, a 1000-ml linear sodium chloride gradient (500 ml of 25 mM NaCl/10 mM Tris, pH 7.4, + 500 ml of 200 mM NaCl/10 mM Tris, pH 7.4) was run. Fractions were then assayed for lipid transfer activity. CE, Cholesteryl ester (●); TG, triglyceride (○); PC, phosphatidylcholine (×), LCAT (■).

pH (4.5) and low ionic strength. The lipid transfer protein can be conveniently eluted from the gel with a linear salt gradient (0 to 200 mM NaCl).

It is after the ion exchange steps where most laboratories diverge in their purification scheme for the lipid transfer protein. Some investigators proceed with chromatofocusing (Pharmacia), a separation procedure based on differences in the isoelectric point of proteins, followed by gel permeation chromatography on BioGel A0.5M (Bio-Rad, Richmond, California).[5] Others have observed the binding of the lipid transfer protein to Concanavalin A–Sepharose,[4,6,8] suggesting the attachment of carbohydrate moieties to the lipid transfer protein. However, one apparent drawback to Con-A affinity chromatography has been variable and low recovery of transfer activity.[4,6]

Another, possibly more specific exploitation of the hydrophobic–lipophilic nature of the plasma lipid transfer protein has been the incubation and binding of the lipid transfer protein with Intralipid[4,6] (Cutter Labora-

[8] J. Ihm, J. A. K. Harmony, J. Ellsworth, and R. L. Jackson, *Biochem. Biophys. Res. Commun.* **93**, 1114 (1980).

tories, Berkeley, California) or triglyceride-rich particles (TRP).[9] The TRP–LTP complex is easily separated from nonbound material by a brief ultracentrifugation in a swinging bucket rotor followed by dissociation of the TRP–LTP complex with sodium cholate. This procedure will selectively absorb a subpopulation of lipophilic proteins from a pool of predominantly hydrophobic proteins eluted from phenyl–Sepharose.

We have also observed the binding of the plasma lipid transfer protein to an organomercurial agarose gel (Affi-gel 501, Bio-Rad), which requires the formation of covalent mercaptide bonds. Transfer activity can then be eluted with dilute mercaptoethanol or dithiothreitol, indicating the lipid transfer protein contains —SH groups (J. J. Albers and J. H. Tollefson, unpublished observation).[10]

Preparation of Plasma Lipid Transfer Protein (LTP-I)[6]

One liter of fresh whole plasma, over-dried potassium bromide for adjustment of plasma nonprotein solvent density, and the following chromatographic gel media: phenyl–Sepharose CL-4B, DEAE–Sepharose CL-6B, heparin–Sepharose CL6B, Polybuffer 74, and Polybuffer exchanger PBE 94, Sephacryl S-200 (Pharmacia), carboxymethylcellulose CM-52 (Whatman, Clifton, NJ), and hydroxylapatite (Bio-Rad).

Procedure

The human plasma lipid transfer protein (LTP-I) is purified from whole plasma by a combination of two ultracentrifugation steps followed by hydrophobic interaction chromatography on phenyl–Sepharose, anion (DEAE–Sepharose) and cation (CM cellulose) exchange chromatography, affinity chromatography on heparin–Sepharose (which removed the plasma phospholipid transfer protein, LTP-II), and finally adsorption to hydroxylapatite. All procedures are performed at 4° unless otherwise specified. Lipid transfer activity is monitored as described below in the section, Assay of Plasma Lipid Transfer Activity.

Step 1. One liter of fresh whole plasma is adjusted to d 1.25 g/ml by the addition of dried solid KBr (0.399 g KBr/ml whole plasma), then centrifuged at 43K (50.2 Ti rotor) or 50K (60 Ti rotor) for 26 hr at 4° in an ultracentrifuge. The top one-half of each tube is harvested by tube slicing, then three parts of d = 1.25 g/ml "top" are mixed with two parts d = 1.19 g/ml KBr solution (v/v) and recentrifuged at d 1.21 g/ml for 48 hr as

[9] D. W. Erkelens, C.-H. Chen, C. D. Mitchell, and J. A. Glomset, *Biochim. Biophys. Acta* **665,** 221 (1981).
[10] J. J. Albers and J. H. Tollefson, unpublished observation.

described above. The top one-third fraction containing the plasma lipoproteins, and the bottom one-third fraction containing predominately albumin, are discarded, and the clear middle zone from each tube is collected.

Step 2. The pooled d = 1.21 to 1.25 g/ml "middle" fraction is applied directly (without prior dialysis) to a 1.5 × 30 cm phenyl–Sepharose column previously equilibrated with a buffer of high ionic strength (2.0 M NaCl, 10 mM Tris, pH 7.4), which promotes binding to the hydrophobic resin. Flow rates up to about 2.5 ml/min can be used with phenyl–Sepharose. The bulk of the protein applied (>90%) does not bind to the resin and is eluted with the 2.0 M NaCl/10 mM Tris, pH 7.4 equilibration buffer. After the A_{280} reads <0.3, the column is washed with 0.15 M NaCl/10 mM Tris, pH 7.4 to lower the ionic strength, which may elute additional protein without eluting detectable lipid transfer activity. Finally, the lipid transfer protein is eluted with distilled water in excellent yield. Some investigators[5,11] have reported >100% recovery for this procedure, suggesting the removal of some inhibitory factor of the lipid transfer protein (see discussion of lipid transfer protein inhibition).[12,13]

Step 3. Fractions containing lipid transfer activity are pooled and dialyzed against 25 mM NaCl/10 mM Tris, pH 7.4, then applied to a 1.5 × 30 cm column of DEAE–Sepharose CL-6B equilibrated with the same buffer, at a flow rate about 40 ml/hr. After the sample has been loaded, the column is developed with a 1000-ml linear sodium chloride gradient of 500 ml equilibration buffer + 500 ml of 250 mM NaCl/10 mM Tris, pH 7.4. This procedure effectively separates lecithin–cholesterol acyltransferase (LCAT) from the lipid transfer protein.[4,6]

Step 4. Fractions from the DEAE step containing lipid transfer activity (without significant A_{280} contamination) are pooled and dialyzed against several changes of 10 mM sodium acetate, pH 4.5, buffer.[12] A small (0.9 × 10 cm) column of carboxymethylcellulose is equilibrated against the same buffer. The sample is loaded onto the column at a flow rate of approximately 30 ml/hr, then eluted with a 200-ml linear sodium chloride gradient (0 to 200 mM) in 10 mM sodium acetate, pH 4.5. Each fraction is collected into 1.0 ml of 1.0 M Tris, pH 7.4, buffer to minimize exposure of partially purified lipid transfer protein to low pH.

Step 5. Fractions from the CM–cellulose step of highest specific activity are pooled and dialyzed against several changes of 50 mM NaCl/10 mM Tris, pH 7.4/1 mM EDTA, then applied to a (0.5 × 5 cm) heparin–

[11] Y. Ogawa and C. J. Fielding, this series, Vol. 111, p. 274.
[12] R. E. Morton and D. B. Zilversmit, *J. Biol. Chem.* **256,** 11992 (1981).
[13] Y. S. C. Son and D. B. Zilversmit, *Biochim. Biophys. Acta* **795,** 473 (1984).

Sepharose column equilibrated with the same buffer at a flow rate of 20 ml/hr. This step effectively binds a heat-sensitive phospholipid transfer protein (LTP-II) which does not transfer cholesteryl esters or triglycerides.[6] The heat-stable neutral lipid transfer protein (LTP-I, which also facilitates the transfer of phospholipid) does not bind to heparin–Sepharose, and is collected in the nonbinding fractions. A separate lipid transfer protein (LTP-II) which has some distinctly different physicochemical properties from LTP-I, is next eluted from the heparin–Sepharose column with 50 mM NaCl/10 mM Tris, pH 7.4/1 mM EDTA.

Step 6. The nonbinding material from the heparin–Sepharose column is pooled and dialyzed against 150 mM NaCl/4 mM sodium phosphate buffer, pH 6.9. A small (0.9 × 5 cm) hydroxylapatite column is prepared and equilibrated with the same buffer. The sample is loaded onto the column of a flow rate of about 10 to 15 ml/hr. Once the sample is loaded into the column, the sample is eluted with a 200-ml linear phosphate gradient (100 ml of equilibrated buffer plus 100 ml of 150 mM NaCl/60 mM sodium phosphate, pH 6.9). Gradient fractions (about 3 ml/fraction) must be dialyzed (against 0.15 M NaCl/10 mM Tris, pH 7.4) prior to assay for lipid transfer activity, as the concentration of phosphate will interfere with the lipoprotein precipitation procedure used to separate donor and acceptor lipoprotein particles.

The final lipid transfer protein product (LTP-I) is purified approximately 17,000-fold with a yield of about 5%. Because it is difficult to measure the total lipid transfer activity in the starting material due to the presence of plasma inhibitors of the lipid transfer protein, these figures are only approximations. As with the purification of many enzymes, the lipid transfer protein becomes increasingly unstable or labile as it is purified, thus making absolute recovery at each successive step increasingly difficult to calculate (see discussion of stability of plasma lipid transfer proteins.)

Characterization of the Human Plasma Lipid Transfer Protein (LTP-I)

Molecular Weight

The final product obtained from the above purification procedure reveals a single major staining band of M_r from 58,000 to 64,000 on sodium dodecyl sulfate–polyacrylamide gel electrophoresis (SDS–PAGE) under either reducing or nonreducing conditions, suggesting a monomeric nature of the purified plasma lipid transfer protein.[6] Under nonreducing conditions, some LTP preparations have a slightly lower apparent molecular weight than LTP preparations reduced prior to SDS–PAGE. By

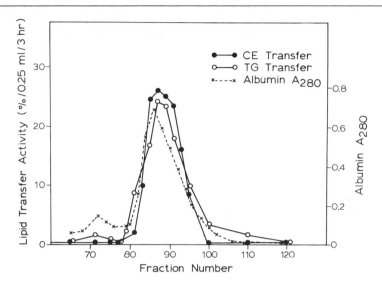

FIG. 2. Elution profile of LTP-I off Sephacryl S-200. Twenty-five milliliters of active LTP-I prepared by sequential chromatography of d = 1.21 to 1.25 g/ml middle plasma fraction on phenyl–Sepharose, DEAE, and heparin–Sepharose was concentrated to 1.0 ml, then applied in an ascending direction to a 1.5 × 90 cm column of Sephacryl S-200, at a flow rate of 8 ml/hr. Fractions (1 ml/fraction) with LTP activity were pooled, dialyzed, and concentrated to 1.0 ml. Human serum albumin (1 mg) was added to the active LTP-I preparation and then it was rechromatographed on the same column. Fractions were assayed for lipid transfer activity. CE, Cholesteryl ester (●); TG, triglyceride (○); HSA (A_{280}) (×).

breaking *intramolecular* —SH bonds, reduction could cause a slight unfolding of the lipid transfer protein. Complementary data as to the apparent molecular weight of the *purified* plasma lipid transfer has been obtained from gel filtration of LTP-I preparations on Sephacryl S-200. The elution profile reveals that the peak of lipid transfer activity elutes just slightly later than bovine serum albumin with an M_r of 67,200 (Fig. 2). Interestingly, chromatography of whole plasma on Sephacryl S-300 reveals that the lipid transfer protein elutes with particles of a much larger size (160,000 to 320,000 Da), suggesting that the lipid transfer protein may circulate as part of a lipoprotein complex in plasma.[14]

Isoelectric Point

The isoelectric point (pI) of the purified plasma lipid transfer protein has been shown by several investigators[5,6,11] to be in the pH 4.8–5.2

[14] J. J. Albers, M. C. Cheung, A. C. Wolf, and J. H. Tollefson, unpublished observation.

range. We have determined the p*I* by two different methods, chromatofocusing and analytical isoelectric focusing, and obtained very similar results (Fig. 3).

Thermal Stability

The human plasma lipid transfer protein (LTP-I) appears to be fairly stable to irreversable thermal denaturation, a unique feature which is not shown by either the LCAT enzyme or the heparin-bound phospholipid transfer protein (LTP-II).[6] We have shown LTP-I to lose little or no cholesteryl ester, triglyceride, or phospholipid transfer activity upon incubation at 56° for up to 2 hr (Fig. 4). However, higher temperatures (80°) will destroy all lipid transfer activity.

Substrate Specificity

LTP-I has been shown to be fairly nonspecific as to the lipid classes transferred. It will readily facilitate the transfer of cholesteryl esters, triglycerides, and phosphatidylcholine between all classes of the plasma lipoproteins in any lipoprotein combination, provided the optimal ratio of donor lipoprotein to acceptor lipoprotein has been established. Morton and Zilversmit[5] have shown LTP-I to have a preferential affinity for cholesteryl esters over triglycerides in either a lipoprotein donor (LDL) : lipoprotein acceptor (HDL) assay or an artificial liposome donor : lipoprotein acceptor assay (LDL).

Characterization of LTP-II

Three laboratories have partially purified a lipid transfer protein from human plasma which appears to be distinct from LTP-I, and which preferentially transfers phospholipids between plasma lipoproteins, or between phospholipid containing liposomes and HDL.[6,7,15]

Table I illustrates several of the divergent physical properties of LTP-II and LTP-I. It has recently been demonstrated that antibody to purified LTP-I precipitates and inhibits all detectable cholesteryl ester and triglyceride transfer activity in the $d > 1.21$ g/ml plasma fraction, but only about 30% of the observed phospholipid transfer activity, suggesting more than one protein in plasma is responsible for facilitated phospholipid transfer.[16] This is consistent with our estimation that approximately 50%

[15] J. Damen, J. Regts, and G. Scherphof, *Biochim. Biophys. Acta* **712**, 444 (1982).
[16] M. Abbey, S. Bastiras, and G. D. Calvert, *Biochim. Biophys. Acta* **833**, 25 (1985).

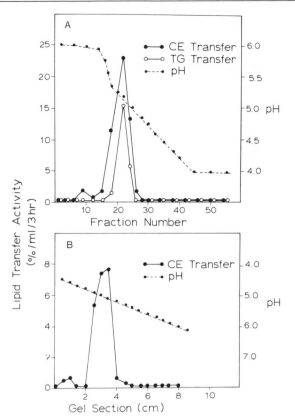

FIG. 3. (A) Chromatofocusing and (B) analytical isoelectric focusing (IEF) of LTP-I. (A) Thirty milliliters of postheparin–Sepharose material was dialyzed overnight against 25 mM imidazole, pH 7.4, and applied to a 0.9 × 10 cm column of PBE 94 polybuffer exchanger equilibrated with the same buffer, at a flow rate of 25 ml/hr. The pH gradient was generated with 300 ml of polybuffer PBE 7.4 (diluted 1:8 with deionized water). Fractions were assayed for cholesteryl ester and triglyceride transfer activity, and pH. (B) Portions (0.2 ml) of postheparin–Sepharose (LTP-I) material were applied to 16 gels and electrofocused for 18 hr. At the conclusion of the run, each gel was sliced into 0.5-cm segments, which were then sliced in half. One-half was incubated in 10 mM KCl for pH determination, the other half was incubated in 0.15 M NaCl/10 mM Tris, pH 7.4, buffer for assay of cholesteryl ester transfer activity. Segments from all gels were pooled according to pH.

of the total phospholipid transfer activity in plasma was heat sensitive, and not due to LTP-I.[6]

The functional properties of LTP-II are yet to be well defined, but it appears to promote the net mass transfer of phosphatidylcholine from either VLDL or phospholipid liposomes to HDL.[6,7,14] It may function to facilitate the removal of the "excess surface coat" of chylomicrons,

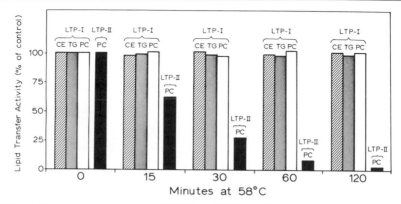

FIG. 4. Temperature sensitivity of LTP-I and LTP-II. Aliquots of active LTP-I (heparin–Sepharose nonbound) and active LTP-II (heparin–Sepharose retained) fractions were incubated from 0 to 120 min at 58° prior to assay. Results are plotted relative to control samples, which were not heated. Note the near total loss of LTP-II phospholipid transfer activity.

VLDL, and remnant lipoproteins during lypolysis. LTP-II may also facilitate the transport of phospholipids to and from cell membranes and thereby maintain cellular integrity.

Molecular Weight

The molecular weight of LTP-II is not yet well established. Tall reports an M_r of about 41,000 by SDS–PAGE, but about 58,000 by gel filtration, and suggests that the "PC transfer protein" may run anomalously on gel filtration, or perhaps may still be complexed with lipid and elute as a larger protein.[7] Preliminary results in our laboratory suggests that LTP-II has an M_r of approximately 65,000 by chromatography on Sephacryl S-200 and by SDS–PAGE. Damen *et al.* reported three molecular species in their most purified preparations: M_r 80,000, 75,000, and 40,000.[15]

TABLE I
COMPARISON OF LTP-I AND LTP-II

Property	LTP-I	LTP-II
Precipitated by anti-LTP-I	Yes	No
Transfers CE, TG	Yes	No
Heat stable	Yes	No
Affinity for heparin–Sepharose	No	Yes

Isoelectric Point

We have observed an isoelectric point (pI) of approximately 5.0 with chromatofousing (pH 7.4 to 4.0) for LTP-II, which is very similar to that observed for LTP-I.

Thermal Stability

Figure 4 compares that thermal stability of LTP-II or LTP-I. When incubated at 58° for up to 2 hr, LTP-I shows little or no loss of CE, TG, or PC transfer activity. However, after 15 min about one-third of the starting LTP-II activity is lost, increasing to about 75% loss after 30 min. Nearly 90% of LTP-II phospholipid transfer activity is destroyed after 60 min, while little or no loss of LTP-I phospholipid transfer activity is observed.

Substrate Specificity

We have observed that LTP-II has little or no ability to promote the transfer of cholesteryl esters or triglycerides, but instead facilitates the transfer of phosphatidylcholine between the plasma lipoproteins,[6] which is similar to what Tall *et al.* observed with their phospholipid transfer protein isolated from human plasma but using a phospholipid liposome donor · HDL acceptor assay system.[7] Damen *et al.*[15] observed a net mass transfer of labeled phosphatidylcholine from phosphatidylcholine liposomes to HDL. In their assay, transfer of radioactivity was paralleled by an equivalent increase in the acceptor (HDL) phospholipid mass, suggesting undirectional transfer of phosphatidylcholine.

Assay for Lipid Transfer Activity

General Considerations

Figure 5 illustrates the complexity of any assay of the lipid transfer protein in plasma (chylomicrons and remnant lipoproteins have been omitted for clarity). Not only are there multiple lipid donors and acceptors, but there are at least three lipid classes undergoing facilitated transfer: cholesteryl esters, triglycerides, and phospholipids. Furthermore, there are two types of transfer "reactions" occurring—an *exchange reaction* which does not alter lipoprotein composition, and a unidirectional transfer reaction of one lipid class from donor lipoprotein to acceptor lipoprotein which will alter the lipid composition of both the donor and acceptor lipoproteins. Finally, there are at least three other factors that exert either a direct or indirect effect on the "activity" of the lipid trans-

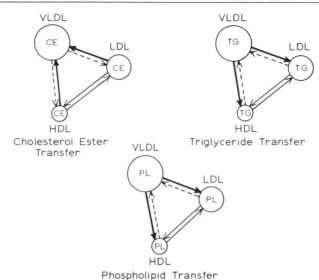

Fig. 5. Facilitated (protein-mediated) lipid transfers and exchanges between the major classes of the plasma lipoproteins. Net mass transfer of cholesteryl esters, triglycerides, and phospholipids primarily occurs between VLDL and HDL, or VLDL and LDL, as illustrated. Lipid transfer between LDL and HDL is primarily an exchange phenomenon. CE, Cholesteryl esters; TG, triglycerides; PL, phospholipids.

fer protein both *in vivo* and *in vitro*. First of all, it has been recently demonstrated by Tall *et al.*[17] that the enzyme lipoprotein lipase *enhances* LTP-I-mediated cholesteryl ester transfer from HDL to "VLDL." Second, LCAT activity in the *in vitro* assays for transfer activity will generate cholesteryl esters, which may dilute the radiolabeled cholesteryl esters with newly generated unlabeled cholesteryl esters. Third, it has been recently suggested[12,13] that plasma contains at least two inhibitory proteins of the lipid transfer process. These factors make it difficult to accurately determine lipid transfer activity in plasma.

Incubation of unfractionated whole plasma at 37°, relative to a control incubation at 0° (when all transfer process are arrested), with or without LCAT inhibition, followed by a convenient means of physically separating donor lipoproteins (e.g., HDL) from acceptor lipoproteins (e.g., VLDL and LDL), permits the chemical analysis of the lipid and apolipoprotein composition of both donor and acceptor lipoproteins, and the evaluation of the rate and direction of lipid transfer. One caveat for assay-

[17] A. R. Tall, D. Sammett, G. M. Vita, R. Deckelbaum, and T. Olivecrona, *J. Biol. Chem.* **259,** 9587 (1984).

ing unfractionated plasma is that the absolute lipid transfer activity (the facilitated lipid transfer independent of plasma components) cannot be *directly* determined, but rather, the observed plasma lipid transfer activity reflects the sum total of all factors present in that particular plasma which may exert an effect on the lipid transfer processes, such as donor and acceptor concentration, and the concentration of activators, inhibitors, and the lipid transfer protein. The chemical analyses must be sensitive enough to detect relatively small changes in lipid composition of the lipoprotein donor and lipoprotein acceptor. To obtain the necessary precision, many replicate analyses are required.[11,18]

Radiolabeled Lipid Transfer Assays

The assay of lipid transfer activity in whole or fractionated plasma fractions obtained during purification is most conveniently achieved by monitoring the *facilitated* transfer of the radiolabeled lipid of interest incorporated into a donor lipoprotein to an unlabeled acceptor lipoprotein. Following the incubation, the donor and acceptor lipoproteins are separated, and the percentage of the total starting counts in the donor incubated with acceptor in the *absence* of any added transfer activity is compared to the counts in the donor incubated with acceptor in the *presence* of added transfer activity. The percentage of radiolabeled lipid transferred reflects the activity of the added test sample. None of the radiolabeled assays can distinguish between net mass or an exchange process.

Preparation of [¹⁴C]Cholesterol Ester-Labeled HDL₃. Approximately 100 ml of fresh whole plasma is centrifuged at $d = 1.125$ g/ml (density is adjusted by the addition of solid KBr) for 40 hr at 45,000 rpm at 4°. The $d > 1.125$ g/ml fraction (HDL_3 + plasma proteins) is obtained by tube slicing, then dialyzed against 12 liters (4 liters × 3 changes) of 0.15 M NaCl/10 mM Tris, pH 7.4/1 mM EDTA. Approximately 20 μCi of [4-¹⁴C]cholesterol (52.5 mCi/mmol, New England Nuclear, Boston, MA) in benzene is dried under N_2, and dissolved in 50 μl of 95% ethanol. This ethanolic [¹⁴C]cholesterol solution is slowly added to the $d > 1.125$ g/ml plasma fraction with gentle stirring. The [¹⁴C]cholesterol–$d > 1.125$ g/ml plasma mixture is then incubated 18 hr at 37° to allow for cholesteryl ester formation via the LCAT reaction.[19] The [¹⁴C]cholesteryl ester–HDL_3 substrate is then isolated by ultracentrifugation at $d = 1.21$ g/ml (40,000 rpm/ 40 hr/4°). The top one-third of each tube is harvested by tube slicing and dialyzed extensively against 0.15 M NaCl/10 mM Tris/1 mM EDTA/ 0.05% (w/v) sodium azide, pH 7.4. For a typical preparation, >90% of all

[18] G. R. Warnick, this volume [6].
[19] J. A. Glomset, *J. Lipid. Res.* **9**, 155 (1968).

added cpm will float at $d = 1.21$, and >90% of the total counts represent [^{14}C]cholesteryl esters.

Preparation of [^{3}H]Triglyceride (TG)–HDL$_3$: HDL$_3$ is isolated from fresh plasma by sequential ultracentrifugation at d 1.125 g/ml and 1.21 g/ml (40,000 rpm/40 hr/4°), then dialyzed against 0.15 M NaCl/10 mM Tris/1 mM EDTA/0.05% sodium azide, pH 7.4 buffer. Approximately 25 μCi of 9,10-[^{3}H(N)]triolein (100–150 Ci/mmol, New England Nuclear, Boston, MA) is dried under N$_2$ and dissolved in 95% ethanol, the volume of which should be <0.10% of the HDL$_3$ volume. The ^{3}H-labeled TG ethanolic solution is slowly added beneath the surface of the HDL$_3$ with gentle stirring. The ^{3}H-labeled TG : HDL$_3$ mixture is then incubated 3 to 6 hr to allow for the equilibration of the exogenous labeled lipid with the lipoprotein. Removal of traces of ethanol are not mandatory, but can be achieved by overnight dialysis against the Tris/saline buffer described above. We have found that >95% of the added exogenous label is associated with the HDL$_3$ harvested upon reisolation by sequential ultracentrifugation at d = 1.125 to 1.21 g/ml.

Preparation of [^{3}H]Phosphatidylcholine (PC)–HDL$_3$. The [^{3}H]phosphatidylcholine (PC)–HDL$_3$ substrate is prepared exactly as described above for ^{3}H-labeled TG–HDL$_3$, except that 1 μCi of [^{3}H]PC(L-α-dipalmitoyl)choline–[*methyl*-^{3}H]phosphatidylcholine (27.0 Ci/mmol, New England Nuclear, Boston, MA) was added in an ethanol carrier (<0.1% of the HDL$_3$ volume) to the freshly isolated and dialyzed HDL$_3$.

Preparation of Acceptor Lipoproteins. Unlabeled acceptor lipoproteins of d < 1.060 g/ml are isolated from either fresh or pooled plasma by ultracentrafugation,[20] then dialyzed against 0.15 M NaCl/10 mM Tris/1 mM EDTA/0.05% sodium azide, pH 7.4.

General Assay Considerations. For each batch of substrate (CE, TG, or PC) and each batch of acceptor lipoproteins, it is necessary to determine the optimal ratio of donor/acceptor, and the total lipoprotein mass which will yield the maximal lipid transfer activity per incubation. Several groups have reported that under the appropriate conditions, HDL can inhibit lipid transfer.[4,13] Once the optimal donor/acceptor ratio and mass have been determined, one must next establish the range of linearity for a given preparation of donor and acceptor. Linearity is normally obtained up to approximately 30% transfer of the HDL-radiolabeled lipid to the unlabeled lipoprotein acceptor. The lipid transfer activity of a particular sample will most likely be underestimated if the percentage transfer is outside the linear range.

The incubation time for a lipid transfer assay is totally dependent on the activity of the sample being assayed. One must incubate long enough to

[20] R. J. Havel, H. A. Eder, and J. H. Bragdon, *J. Clin. Invest.* **34,** 1345 (1955).

express significant transfer above background, but not so long as to extend beyond the linear region of a given preparation of donor and acceptor. If enough sample to be assayed is available, it is advisable to assay the sample at several different concentrations, or alternatively, assay a single concentration at several different time intervals. Appropriate controls (both negative and positive) should be included in each assay. The negative control is required for the calculation of lipid transfer activity, and the positive control demonstrates that the assay is working.

It is possible to assay whole plasma directly, provided the mass of whole plasma lipid does not significantly alter the optimal lipoprotein donor: acceptor ratio previously established for a given assay. We have observed that if the *added* whole plasma lipid is <10% of the lipid in the donor and acceptor lipoproteins, assays of whole plasma are linear both with respect to volume of whole plasma assayed (normally 5–20 μl) and the incubation time, provided the linear range is not exceeded.

Cholesteryl Ester Transfer Assay

Principle. The transfer of radiolabeled HDL_3–cholesteryl esters to the $d < 1.060$ g/ml lipoproteins is monitored by incubation (1 to 18 hr) at 37° with and without added lipid transfer protein. The donor (HDL_3) and acceptor lipoproteins are separated by heparin : $MnCl_2$ precipitation[21] and the radioactivity in the supernatant (HDL_3) is determined.

Reagents

^{14}C-labeled HDL_3–cholesteryl ester (donor)
Unlabeled $d < 1.060$ g/ml lipoprotein (acceptor)
0.15 M NaCl/10 mM Tris buffer, pH 7.4
"Positive" control (a plasma fraction known to have neutral lipid transfer activity)
Test sample to be assayed
"Carrier" pooled whole plasma
Combined heparin : $MnCl_2$ reagent (1 : 1, v/v) (concentration : heparin, 5000 U/ml; 2 M $MnCl_2$)

Assay. Portions of donor (^{14}C-labeled CE–HDL_3) and acceptor lipoproteins ($d < 1.060$ g/ml plasma fraction) are transferred to 12 × 75 mm glass culture tubes at the optimal ratio for assay to transfer activity (approximately 1 : 5, HDL_3–CE mass: $d < 1.060$ g/ml fraction CE mass). Next, either buffer (negative control) or test sample (usually 50 to 400 μl) is added to the tubes. All tubes are then brought to a volume of 0.60 ml with 0.15 M NaCl/10 mM Tris buffer, pH 7.4, capped and incubated in a shaking 37° water bath for 1 to 18 hr (the appropriate time depends upon

[21] G. R. Warnick and J. J. Albers, *J. Lipid Res.* **19**, 65 (1978).

the level of transfer activity). After incubation, the tubes are chilled on ice, and the donor and acceptor lipoproteins are separated with the addition of 400 μl of chilled carrier whole plasma (which facilitates the heparin–MnCl$_2$ separation of VLDL and LDL from HDL). The tubes are vortexed briefly, followed by the addition of 0.125 ml of the combined heparin–MnCl$_2$ reagent. The tubes are vortexed once more, then incubated for 30 min on ice. The apolipoprotein B-containing lipoproteins : heparin : MnCl$_2$ aggregate is precipitated by centrifugation at 3000 rpm for 30 min at 4°. Portions (750 μl) of each supernatant (HDL) are transferred to 5-ml plastic counting vials, and 6 ml of scintillant (Aquasol II, New England Nuclear Boston, MA) is added. The samples are then counted in a scintillation counter to an accuracy of <2%.

Calculation of Lipid Transfer Activity: Transfer activity is expressed as percentage transfer/volume/incubation time and calculated as [(buffer control cpm − test sample cpm)/buffer control cpm]100. The 37° buffer control corrects for the nonfacilitated lipid transfer and incomplete separation of donor from acceptor lipoproteins.

Triglyceride and Phosphatidylcholine Transfer Assays

The principle for both of these assays is the same as that described in the section, Cholesteryl Ester Transfer Assay, except the appropriate radiolabeled donor is substituted for the ^{14}C-labeled CE–HDL$_3$. For the triglyceride transfer assay, we use a TG donor : acceptor mass ratio of about 1 : 50, and in the phospholipid transfer assay we use a phospholipid donor : acceptor mass ratio of about 1 : 5.

Lipid Transfer Protein in Other Animal Species

Many of the earlier studies on lipoprotein and extracellular lipid metabolism were carried out with the rat, although it has been clearly demonstrated that rat plasma contains a component which facilitates phospholipid transfer.[6,9,22] The rat was an unfortunate choice for evaluation of the neutral lipid transfer processes. Upon incubation of cholesteryl ester-labeled lipoproteins with rat plasma, it was observed that the cholesteryl esters did not transfer to other lipoproteins, and it was therefore concluded that the cholesteryl esters of plasma lipoproteins did not participate in exchange or transfer processes with other lipoproteins.[23] Although

[22] S. Eisenberg, *J. Lipid Res.* **19,** 229 (1978).
[23] P. S. Roheim, D. E. Haft, L. I. Gidez, A. White, and H. A. Eder, *J. Clin. Invest.* **42,** 1277 (1963).

several reports of neutral lipid transfer in other species appeared,[24,25] it was nearly a decade before it was generally accepted that neutral lipids may transfer and exchange among lipoproteins. At this time Zilversmit and colleagues clearly established the presence of a protein present in rabbit[26] and later human[4] plasma which facilitated the exchange of cholesteryl esters between all classes of the plasma lipoproteins.

Ha and Barter[27] later examined the cholesteryl ester transfer activity in 16 vertebrate species, using the animals' lipoprotein-free plasma as the source of transfer protein. Of the animals examined, they were able to group them into either a low-activity (rat, sheep, cow, pig, dog); intermediate-activity (guinea pig, chicken, turkey, lizard, toad, snake, man, wallaby); and high-activity group (possum, rabbit, trout). The low-activity group expressed a cholesteryl ester transfer activity about 20% of that seen in man. Other investigators have reported that the rat has negligible or no neutral lipid transfer activity.[8,28]

The interspecies differences may relate to the concentration and relative proportions of the lipid transfer protein plasma inhibitors of the lipid transfer protein.[12,13] Preliminary data from our laboratory has shown that several animals of the "low-activity" group (rat, dog, pig, sheep) have little or no apparent neutral lipid transfer activity when the $d > 1.21$ g/ml plasma fraction is assayed. However, after passage of the $d > 1.21$ g/ml fraction over a phenyl–Sepharose CL-4B gel matrix, significant neutral lipid transfer activity can be detected in the water eluted material, suggesting perhaps all mammals have the neutral lipid transfer protein (LTP-I), but *in vitro* (and perhaps *in vivo*) its activity is masked or made nonfunctional by the presence of the plasma inhibitors of the lipid transfer protein. Animals which have little apparent neutral lipid transfer activity in the plasma or the $d > 1.21$ g/ml fraction may not be an appropriate animal model for the study of experimental atherosclerosis.

Chemical Inhibition of the Plasma Lipid Transfer Proteins

Several reports[5,6,29] have shown the activity of the plasma lipid transfer proteins may be sensitive to various chemical agents, particularly

[24] A. V. Nichols and L. Smith, *J. Lipid Res.* **6,** 206 (1965).
[25] J. A. Glomset and K. R. Norum, "Advances in Lipid Research," Vol. II, p. 1. Academic Press, New York, 1973.
[26] D. B. Zilversmit, L. B. Hughes, and J. Balmer, *Biochim. Biophys. Acta* **409,** 393 (1975).
[27] Y. C. Ha and P. J. Barter, *Comp. Biochem. Physiol.* **71,** 265 (1982).
[28] Y. Oschry and S. Eisenberg, *J. Lipid Res.* **23,** 1099 (1982).
[29] G. J. Hopkins and P. J. Barter, *Metabolism* **29,** 546 (1980).

mercurial compounds. Although the transfer of triglyceride and cholesteryl ester is believed to be mediated by the same plasma protein,[5,6] several mercurial compounds (p-chloromercuriphenylsulfonate, 2 mM; p-hydroxymercuribenzoate, 1.0 mM; and ethyl mercurithiosalicylate, 0.2 mM)[5] inhibit triglyceride transfer, while those reagents may actually stimulate cholesteryl ester transfer. Some have argued that this dissociation of triglyceride and cholesteryl ester transfer suggests these processes are mediated by *two separate* proteins with remarkably similar physicochemical properties, or that there exists two or more nonidentical lipid binding domains on a single protein, one which binds triglycerides molecules and is blocked in the presence of the mercurial agents.[6] We have also demonstrated the parallel inhibition of triglyceride and cholesteryl ester transfer activities over a range of iodoacetate concentrations (5–40 mM).[10] Thus far, it has not been possible to chemically inhibit the neutral lipid transfer reactions without simultaneously inhibiting the LCAT enzyme.

Stability of Plasma Lipid Transfer Proteins

Partially purified preparations of LTP-I ($d > 1.21$ g/ml plasma, or fractions from phenyl–Sepharose derived from the $d > 1.21$ g/ml plasma) are remarkably stable over time. When stored at 4° (with 0.05% sodium azide) in 0.15 M NaCl/10 mM Tris, pH 7.4 buffer, partially purified preparations show little or no loss of lipid transfer activity for up to 2 months. More purified preparations (post CM-cellulose) have been reported to have lost >60% of their lipid transfer activity within 2 days, although these preparations could be partially stabilized upon the addition of 0.5% bovine serum albumin (>65% of the lipid transfer activity remained after 2 weeks).[5] More recently, it has been shown that storage of highly purified preparations of LTP-I were stable at 4° for up to 4 months when 4 M urea was included in the storage buffers.[16] This suggests that the loss of activity observed with highly purified preparations might be caused by self-association which is partially reversible in the presence of 4 M urea.

Most reports on the isolation or assay of the human plasma lipid transfer protein have not stressed the inclusion of protease inhibitors during the isolation or prolonged incubation of plasma lipoproteins with isolated plasma lipid transfer proteins. However, as a precaution, the use of protease inhibitors is recommended.

Cellular Origin of the Plasma Lipid Transfer Protein

To date relatively little is known as to the cellular origin of the plasma lipid transfer proteins, and even less is known about the metabolic regula-

TABLE II
COMPARISON OF MACROPHAGE AND PLASMA-DERIVED LIPID TRANSFER PROTEIN

Property	Macrophage-LTP	Plasma-LTP
Molecular weight	63,000	63,000
Isoelectric point	5.0	5.0
Hydrophobic (binds phenyl–Sepharose)	Yes	Yes
Affinity for heparin–Sepharose	No	No
Heat stable	Yes	Yes
Facilitates transfer of:	CE, TG, PC	CE, TG, PC

tion of its synthesis and secretion from the site(s) of origin into the plasma compartment. Two recent reports[30,31] have demonstrated that neutral lipid transfer activity was secreted from the perfused rabbit liver, but no distinction was made as to a parenchymal or nonparenchymal origin for the lipid transfer activity. It is possible that the lipid transfer protein may be secreted as part of high-density lipoprotein complex, as little or no "free" (M_r 63,000) LTP-I is present in normolipidemic plasma, but elutes with HDL-sized particles off Sephacryl S-300. However, in one subject with Tangiers disease, whoe congenitally has very little A-I and very low HDL levels, the bulk of the subject's lipid transfer activity eluted in a position where "free" transfer protein (M_r about 63,000) would elute.[10] These preliminary observations suggest that the lipid transfer proteins are normally circulated as part of an HDL-sized complex, but when the carrier particles are absent or significantly reduced in plasma (as in Tangiers disease), the lipid transfer proteins circulate uncomplexed to lipoproteins.

More recently, our laboratory has demonstrated that the human monocyte-derived macrophage synthesizes and secretes a lipid transfer protein with many physicochemical properties indistinguishable from the human plasma lipid transfer protein, LTP-I (Table II).[32] The human monocyte-derived macrophage in culture synthesizes and secretes a lipid transfer activity in a linear manner over 24 hr. This activity is stimulated by the macrophage activator phorbol myristate acetate (about 65% increase in total transfer activity over nonstimulated macrophages), and is blocked completely by cycloheximide, suggesting that protein synthesis is required for the secretion of the macrophage lipid transfer protein. If this observation from cultured cells applies to metabolic events *in vivo,* then it

[30] L. De Parscau and P. L. Fielding, *J. Lipid Res.* **25,** 721 (1984).
[31] M. Abbey, J. K. Savage, A. M. Mackinnon, P. J. Barter, and G. D. Calvert, *Biochim. Biophys. Acta* **793,** 481 (1984).
[32] J. H. Tollefson, R. Faust, J. J. Albers, and A. Chait, *J. Biol. Chem.* **260,** 5887 (1985).

is highly probable the macrophage-derived lipid transfer protein is homologous to the LTP-I found in plasma, as antibody to purified LTP-I inhibited all of the observed cholesteryl ester and triglyceride transfer activity in $d > 1.21$ g/ml plasma, suggesting that a single protein is responsible for neutral lipid transfer.[16]

Finally, it is also possible that other cells synthesize and secrete the lipid transfer protein into plasma. We have detected lipid transfer activity secretion into serum-free medium in two other macrophage-like established cell lines, human U937 and murine J774. Preliminary results suggest that the human hepatoma cell line Hep G2 also secretes neutral lipid transfer activity.

Conclusion

Many advances have been achieved in the past 10 years in the relatively new field of the plasma lipid transfer proteins, but many questions await further investigation before the physiological role of the lipid transfer proteins can be fully appreciated. As mentioned earlier in this chapter, little is known of the metabolic events which regulate the synthesis and secretion of the plasma lipid transfer proteins *in vivo*. Some additional questions are as follows: (1) What is the functional relationship of LTP-I and LTP-II? (2) Is there deranged lipid transfer in the hyperlipidemias? (3) How are the physiological functions of LCAT and the lipid transfer proteins regulated? (4) What is the physiological function of the lipoprotein particles on which the lipid transfer protein(s) reside? Answers to these and many other important questions concerning the role of the human plasma lipid transfer proteins in human lipoprotein pathophysiology await further research.

Acknowledgment

This research was supported by NIH Grant HL 30086.

[49] Synthesis of Ether Analogs of Lipoprotein Lipids and Their Biological Applications

By GIDEON HALPERIN, OLGA STEIN, and YECHEZKIEL STEIN

Introduction

In this chapter we shall review the synthesis of cholesteryl ethers, retinyl ethers, tri- and dialkyl glycerols, and dialkyl phospholipids, which are nonhydrolyzable analogs of the lipid moieties of lipoproteins. The biological applications of these lipid analogs will be limited also to studies on lipoproteins which have been carried out with the ultimate goal to further research on atherosclerosis. Additional information on ether lipids not included in this chapter can be found in the excellent monographs of Snyder,[1] Rosenthal,[2] and Mangold and Paltauf.[3]

Cholesteryl Alkyl Ethers

Synthesis

The structural similarities between cholesteryl alkyl ethers and cholesteryl esters have been demonstrated by studying their behavior with differential scanning calorimetry, polarized microscopy, and X-ray powder diffraction.[4] In biological studies the cholesteryl alkyl ethers displayed metabolic inertness, which made them suitable nonhydrolyzable probes for the study of cholesteryl ester metabolism.[5-8]

In order to follow the metabolic fate of a model molecule it is convenient to have a labeled compound of high specific activity and an unlabeled compound prepared on a larger scale. For the introduction of labeled model molecules into lipoproteins, a very pure high specific activity product is needed while the unlabeled compound is generally needed for chromatographic purposes and for experiments in which dilution of the labeled probes to a lower specific activity is desirable.

The methods for etherification of cholesterol may be divided into three general groups:

1. The Williamson type of etherifications, in which alcoholates react with alkyl halides or alkyl sulfates to form ethers.[9] This group of reactions includes the following methods: etherification of the potassium salt of cholesterol with alkyl halogenides,[10] or with alkyl *p*-toluene sulfonates[11];

[1] F. Snyder, "Ether Lipids, Chemistry and Biology." Academic Press, New York, 1972.
[2] A. F. Rosenthal, this series, Vol. 35, p. 429.
[3] H. K. Mangold and F. Paltauf, "Ether Lipids, Biochemical and Biomedical Aspects." Academic Press, New York, 1983.
[4] R. J. Deckelbaum, G. Halperin, and D. Atkinson, *J. Lipid Res.* **24**, 657 (1983).
[5] Y. Stein, G. Halperin, and O. Stein, *FEBS Lett.* **111**, 104 (1980).
[6] Y. Stein, G. Halperin, and O. Stein, *Biochim. Biophys. Acta* **663**, 569 (1981).
[7] T. Chajek-Shaul, G. Friedman, G. Halperin, O. Stein, and Y. Stein, *Biochim. Biophys. Acta* **666**, 147 (1981).
[8] G. L. Pool, M. E. French, R. A. Edwards, L. Huang, and R. H. Lumb, *Lipids* **17**, 448 (1982).
[9] A. W. Williamson, "Results of Research on Aetherification." B. A. Rep. II, 656 (1850); *Chem. Gaz.* **9**, 294 (1851); *Ann.* **1, 27**, 31 (1851); *J. Chem. Soc.* **4**, 292 (1852).
[10] W. Steikof and E. Blummer, *J. Prakt. Chem.* **84**, 460 (1911).
[11] H. Funasaki and J. R. Gilbertson, *J. Lipid Res.* **9**, 766 (1968).

condensation of cholesterol and alkyl methane sulfonate in the presence of potassium in boiling benzene.[8]

2. Methods based on the Stoll reaction,[12] in which cholesterol p-toluene sulfonate in solvolyzed with various alcohols under neutral conditions and the corresponding cholesteryl-O-alkyl ethers are obtained, retaining the 3β-configuration of the alkoxy side chain. This procedure was applied for cholesterol[12–15] and other 5-ene-steroids.[12,16] A large variety of unlabeled cholesteryl alkyl ethers was prepared in good yields using this method (see the table). However, since excess alcohol must be added in the solvolysis reaction, chromatographic purification of the product cannot be avoided. The method was also found unsuitable for the preparation of labeled short alkyl ethers, such as the methyl, ethyl, allyl, and butyl homologs. On the other hand, labeled long chain saturated or unsaturated alkyl ethers were conveniently prepared in good yields.[15]

3. Modification of the Bauman and Mangold procedure for glyceryl ethers,[17] in which cholesterol and powdered KOH in benzene[18] are stirred and refluxed. The water formed by the reaction is removed by an azeotropic distillation separator and the potassium salt of cholesterol formed is reacted with alkyl methane sulfonate. This reaction is formally similar to the Williamson type of etherification; however, since the anhydrous conditions are continuously controlled in the system, the application of this technique is more convenient, especially in the preparation of trace amounts of labeled ethers. The main by-product formed in this reaction is the alkyl-O-alkyl ether (R-O-R), which is chromatographically similar to the cholesteryl alkyl ethers. When 3-O-alkylcholesterol labeled in the cholesteryl moiety is prepared, milligram amounts of alkyl mesylate are reacted with tracer cholesterol. Under such conditions, substantial amounts of the R-O-R by-product are formed, making the efficiency of this method dependent on a reliable chromatographic step of purification. On the other hand, when unlabeled saturated O-alkyl ethers are prepared, equivalent amounts of cholesterol and alkyl methane sulfonate are condensed,[18] giving a product which can be purified merely by crystallization.

Miscellaneous Methods. Cholesteryl alkyl ethers were also prepared by condensation of cholesterol and alcohol in the presence of a dehydrat-

[12] W. Stoll, *Z. Phys. Chem.* **207**, 149 (1932).
[13] H. McKennis, Jr., *J. Am. Chem. Soc.* **70**, 675 (1948).
[14] E. Borgstrom, *Proc. Soc. Exp. Biol. Med.* **127**, 1120 (1968).
[15] G. Halperin and S. Gatt, *Steroids* **35**, 39 (1980).
[16] F. W. Heyl, M. E. Herr, and A. P. Centolella, *J. Am. Chem. Soc.* **71**, 247 (1949).
[17] W. J. Baumann and H. K. Mangold, *J. Org. Chem.* **29**, 3055 (1964).
[18] F. Paltauf, *Monatsh. Chem.* **99**, 1277 (1968).

YIELD AND PHYSICAL PROPERTIES OF CHOLESTERYL-3-O-ALKYL ETHERS OBTAINED BY MODIFIED STOLL PROCEDURE[15] COMPARED WITH THE mp AND $[\alpha]_D^{25}$ OF CORRESPONDING ESTERS

Number	Alkyl	Yield (%)[a]	Molecular ion (m/e)	Relative retention times[b]	mp (°C)	$[\alpha]_D^{25}$	Esters mp (°C)	Esters $[\alpha]_D^{25}$
I	2'-Propenyl	53.2	426	0.453[c]	78–79	−29.3		
II	Butyl	64.1	442	0.574[c]	81–82	−27.7	102–111[d]	−35.4[d]
III	Isoamyl	52.9	456	0.060	88–90	−27.9	109.8–110.5[d]	−34.6[d]
IV	Hexyl	57.5	470	0.080	69–71	−31.9	93.5–94.5[d]	−32.5[d]
V	Octyl	60.7	498	0.122	97–98	−28.1	104.6–106[d]	−31.6[d]
VI	Decyl	47.2	526	0.196	58–62	−25.0	91[d]	−28.7[d]
VII	Dodecyl	58.9	554	0.279	71–73 (73)[e]	−28.2 (−24.6)[e]	91–92[f]	−30.6[f]
VIII	Tetradecyl	46.4	582	0.405	47–49 (50)[e]	−25.3 (−22.8)[e]	70–70.5[f]	−26.9[f]
IX	Hexadecyl	46.5	610	0.636	57–59 (57)[e]	−22.9 (−22.0)[e]	77–78[f]	−24.8[f]
X	Octadecyl	66.8	638	0.850	65	−21.9	81.5–82.5[f]	−23.7[f]
XI	Eicosyl	45.8	666	1.225	61–64	−20.5	85–85.5[f]	−23.2[f]
XII	cis-9'-Hexadecenyl	54.1	608	0.598	18–24	−20.1		
XIII	trans-9'-Hexadecenyl	56.8	608	0.597	47–48	−22.9		
XIV	cis-9'-Octadecenyl	63.6	636	0.853	39–41 (35)[e]	−23.4 (−21.3)[e]	46.5–47[f]	−24.5[f]
XV	trans-9'-Octadecenyl	60.2	636	0.830	48–49	−22.3		
XVI	cis,cis-9',12'-Octadecadienyl	53.2	634	0.877	24–28	−20.6	42[f]	−24.4[f]
XVII	trans,trans-9',12'-Octadecadienyl	59.2	634	0.853	57–58	−21.9		

[a] Calculation based on crystallized ethers obtained from cholesterol p-toluene sulfonate.
[b] Relative to cholesteryl hexadecanoate on 3% QF-1 at 260° (absolute retention time 49.3 min).
[c] Relative to cholesteryl hexyl ether on 3% SE, 270° (absolute retention time 15.3 min).
[d] Swell and Treadwell.[23]
[e] Paltauf.[22]
[f] Mahadevan and Lundberg.[24]

ing agent.[19] Some sterols were alkylated with methyl iodide in the presence of silver oxide.[20] The latter was modified for the preparation of cholesteryl methyl and allyl ethers, as will be illustrated below.

Procedures for the Preparation of Cholesteryl Alkyl Ethers

In this section several convenient preparative procedures of groups 2 and 3 will be given. The silver oxide condensation method will also be illustrated as a route for the synthesis of cholesteryl methyl or allyl ethers.

Alcoholysis of Cholesterol p-Toluene Sulfonate,[15] According to Stoll[12]. Cholesterol p-toluene sulfonate (200 mg) and 500 mg of fatty alcohol (or 0.5 ml if liquid at room temperature) were placed in a 20-ml constricted glass tube. The tube was flushed with a stream of dry N_2, sealed, and kept in an oven at 110° for 150 min. The tube was opened, $NaHCO_3$ (0.5 g) and hexane (10 ml) were added, and the mixture was stirred and extracted three times with hexane which was applied to a column containing 15 g of silicic acid.[21] The silicic acid column was eluted with 50 ml each of the following solvents: hexane; 5%, 10%, and two 50-ml portions of 15% benzene in hexane. Aliquots of the fractions were analyzed by TLC on silica gel G plates (petroleum ether : ether, 98 : 2, v/v), using iodine vapor for visualization of the spots. The cholesteryl ethers were generally recovered from the first 15% benzene in hexane fraction; some alkyl ethers were also eluted with the 10% or the second 15% benzene in hexane fractions (R_f 0.31). The crude product was crystallized from acetone which was cooled to 0° before filtration. *cis*-9′-Octadecenyl and *cis, cis*-9′, 12′-octadecadienyl ethers (compounds XIV and XVI in the table[22-24]), were crystallized from acetone at −20° overnight in test tubes. The mixture was centrifuged for 5 min at 5° and the solvent was quickly decanted. The residue was transferred with petroleum ether into a tube, the solvent was removed with a stream of nitrogen at 50–60°, and the tube was tightly stoppered and kept at −20°. For extended storage, samples were kept in sealed ampoules in a nitrogen atmosphere at −20°. The 2′-propenyl ether was prepared by heating the p-toluene sulfonate in 1 ml of 2-propenyl alcohol and was crystallized from methanol without chromatographic purification.

[19] C. Bills and F. McDonald, *J. Biol. Chem.* **72**, 1 (1927).
[20] T. H. Campion and G. A. Morrison, *Tetrahedron* **29**, 239 (1973).
[21] J. Hirsch and E. R. Ahrens, *J. Biol. Chem.* **233**, 311 (1958).
[22] F. Paltauf, *in* "Ether Lipids" (H. K. Mangold and F. Paltauf, eds.), Ch. 9, p. 49, 177. Academic Press, New York, 1983.
[23] L. Swell and C. R. Treadwell, *J. Biol. Chem.* **212**, 141 (1955).
[24] V. Mahadevan and W. O. Lundberg, *J. Lipid Res.* **3**, 108 (1962).

Preparation of [7(n)-³H]Cholesteryl cis,cis-9',12'-Octadecadienyl Ether [Compound XVI (See the Table)]. A solution of [7(n)-³H]cholesterol (0.4 mCi, specific activity 9.5 Ci/mmol) in toluene (0.4 ml) was placed in a tube and the solvent was evaporated at 40° under a stream of N_2. Dry pyridine (1 ml) and *p*-toluene sulfonyl chloride (0.3 g) were added and the tube was kept at 38° for 20 hr. Ice (approximately 0.5 g) was added and after 30 min the mixture was extracted with hexane. The organic layer was successively washed with water, a saturated solution of $NaHCO_3$ and again with water, dried over Na_2SO_4, and evaporated. The residue was heated with *cis,cis*-9,12-octadecadienyl alcohol (0.4 ml) in a sealed tube, as outlined above, to give 0.13 mCi of labeled compound XVI (32.5% yield). When the product was analyzed by TLC using a solvent system of hexane : chloroform (7 : 3, v/v), >98% of the radioactivity was found in the cholesteryl ether zone (R_f 0.3, visualized with iodine vapor). The yield of cis-9'-octadecenyl ether compound (XIV), prepared similarly, was 29.6%.

Etherification of Cholesterol with Alkyl Methane Sulfonates in the Presence of Potassium Hydroxide

The procedure below illustrates the preparation of saturated long chain 3-*O*-alkyl cholesterol omitting the chromatographic purification step.

Preparation of Cholesteryl Hexadecyl Ether [Compound IX (See the Table)]. Benzene (250 ml), powdered KOH (5 g), and cholesterol (1.93 g) were added to a 500-ml three-necked flask fitted with a water-separating head and reflux condenser dropping funnel, nitrogen inlet, and outlet tubes.[17,18] The mixture was refluxed and magnetically stirred for 1 hr to remove water by azeotropic distillation. Hexadecyl methane sulfonate, 1.6 g in 200 ml of dry benzene, was added dropwise and refluxing continued for a further 10 hr; 400 ml of the solvent was removed by distillation and the residue was cooled and extracted with 500 ml petroleum ether, (30–60°). The solution was washed with water (four times), dried with sodium sulfate, filtered, and evaporated under reduced pressure. The crude residue (3.1 g) was crystallized overnight at room temperature from acetone (300 ml) to give 1.8 g (59% yield) of cholesteryl hexadecyl ether (mp 57–59°). The *cis*-9'-octadecenyl and *cis,cis*-9',12'-octadecadienyl ethers (compounds XIV and XVI, respectively; see the table) were prepared similarly and purified as follows: the crude extract was chromatographed on 50 g alumina (Riedel de Haen, aluminium oxid-S) with 500-ml fractions of petroleum ether (30–60°). The material, recovered from fractions 3–6, was crystallized from acetone at −20° as described above to give compounds XIV and XVI in 57% and 65% yields, respectively.

Preparation of Labeled Cholesteryl Ethers [Compounds XIV and XVI (See the Table)] of High Specific Activity. [7-³H]-, [1,2-³H₂]-labeled cholesterol (1–2 mCi) or [4-¹⁴C]cholesterol (100–200 μCi), powdered KOH (2–3 g), and alkyl methane sulfonate (0.04 ml or 40 mg of unsaturated or saturated alkyl methane sulfonates, respectively), were refluxed in benzene (100 ml) as described above. The product was chromatographed on 15 g silicic acid to give a product of above 98% radiochemical purity. However, the eluates usually contained 5–10 mg of unlabeled by-products, which were separated by chromatography on alumina columns (Riedel de Haen, aluminium oxid-S). As various alumina batches behaved differently, two chromatographic procedures were used for the purification of the product from the impurities. The pure labeled cholesteryl ether could be eluted with hexane (100-ml fractions) and recovered in fractions 4–6. When alumina columns were deactivated with 1% water and eluted with 10% benzene in hexane (100-ml fractions), the labeled cholesteryl ether was found in fractions 3–5. The yield ranged from 25–35%.

Preparation of Labeled Cholesteryl Methyl and Allyl Ethers. This procedure is a modification of a method for alkylation of sugars.[25] Into a flat-bottomed 50-ml flask equipped with a reflux condenser with a calcium chloride protecting tube and magnetic stirrer, were added the following: [7-³H]cholesterol (0.3 mCi), allyl bromide (5 ml), silver oxide prepared according to ref. 25 (300 mg), dried magnesium sulfate (800 mg), and peroxide-free dry ether (20 ml). The mixture was refluxed and stirred in the dark in a water bath at 50° for 20 hr. The product was filtered and washed with ether (100 ml) and the solvent was removed *in vacuo*. The residue was chromatographed on a silicic acid column (15 g) eluted with increasing concentrations of benzene in hexane such that the cholesteryl allyl ether was found in the 30% benzene in hexane fraction. The yield was 40% and radiochemical purity >98%; the yield and purity of [7-³H]cholesteryl methyl ether prepared similarly were 50 and >97%, respectively. If necessary, further purification can be obtained using TLC on silica gel G plates.

Physical Properties

Mass Spectra. The fragmentation of cholesteryl alkyl ethers is illustrated in Fig. 1. The molecular ion and the $M - CH_3$ fragment were found in all the cholesteryl ethers studied. The following fragmentations were elucidated by Funasaki and Gilbertson:[11] m/e 369 ($M - O$-alkyl side chain), 368 ($M - O$-alkyl—H), 370 ($M - O$-alkyl + H), $M - 113$, C_{17-20}

---

25 W. J. De Grip and P. H. M. Bovee-Geurts, *Chem. Phys. Lipids* 23, 321 (1979).

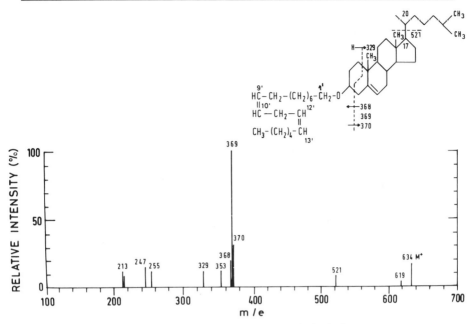

FIG. 1. Mass spectrum of cholesteryl linoleyl ether.

splitting of the side chain; 399 $C_{1'-2'}$ cleavage of the O-alkyl side chain. The last fragmentation was found only in the saturated cholesteryl ethers.[15] No significant differences were observed between the *cis–trans* isomers of the pairs XII–XIII, XIV–XV, or XVI–XVII (see the table).[15]

IR Spectra. The 965 cm^{-1} peak (*trans*-CH=CH—)[17] was present in compounds XIII, XV, and XVII (see the table). The 915 cm^{-15} peak (CH$_2$=CH—) was present in the 2'-propenyl ether I.

NMR Spectra in CDCl$_3$. The δ (ppm) 0.64, 0.70, 0.90, and 1.0 (18-CH$_3$, 19-CH$_3$, 21-CH$_3$, and 26,27-CH$_3$) were found in the spectra of all the cholesteryl ethers studied and of cholesterol; 1.2 (broad s, methylenic protons) was not shown by 2'-propenyl ether I; 3.4–3.6 (m, 3α-H and 1'-CH$_2$); 5.36 (m, 6-CH, and —CH=CH—). The following peaks were shown by compound I: δ = 3.96 (s, 3α-H), 4.06 (s, 2H, 1'-CH$_2$), 5.05, 5.24, and 5.36 (three centered m, 6-CH and CH$_2$=CH—).

The melting points and the negative optical rotations of the cholesteryl ethers given in the table are lower than the corresponding cholesteryl esters.

The properties of cholesteryl–alkyl ethers were also compared by differential scanning calorimetry, polarizing microscopy, and X-ray powdered diffraction.[4] The saturated compounds, excluding V and XI, melt into stable liquid crystal mesophases. Mesophase transition temperatures vary over a narrower range than in cholesteryl esters and are lower than those of the corresponding esters. Polarizing microscopy and X-ray diffraction showed structural similarities between the two groups of cholesterol derivatives.

Biological Applications

Introduction of Cholesteryl Linoleyl Ether (CLE) into Lipoproteins

Labeling of Donor HDL and Transfer Procedure.[26] *Relipidation:* 5 mg of lyophilized HDL protein per siliconized tube was delipidated three times with 5 ml heptane at $-10°$ for 30 min. The tubes were centrifuged, the heptane was evaporated, and the protein dried under N_2 and dissolved in 1 ml 0.01 M Tris-HCl buffer (pH 8.0). [^{3}H]Cholesteryl linoleyl ether (10 μCi) or [^{14}C]cholesteryl linoleate (1 μCi) was dissolved in heptane together with 300 μg of free cholesterol. The heptane was removed under N_2, the lipids were suspended in 1 ml Tris-HCl buffer, and the lipid mixture sonicated twice for 30 sec at $45°$ under N_2. THe HDL was then added to the sonicated lipid and cosonicated twice for 30 sec at $40°$, with the 4-mm tip immersed into the solution to prevent foaming. The sonicator used was a Braun-Sonic 3000 (Braun, Melsungen, FRG) and the output was 50% of the maximum. After the final sonication the suspension was clear to opalescent. Following addition of 40% sucrose to give a final concentration of 2%, 4.5 ml of the sonicated mixture were placed in SW-41 polyallomer tubes and overlayered with 6.0 ml saline. The tubes were centrifuged in a SW-41 rotor at 39,000 rpm for 60 min, cut in the middle to give the upper and lower phase, which was removed without disturbing the precipitate which was firmly attached to the bottom of the tube. From 57 to 68% was in the lower phase representing the partially relipidated HDL; recovery of protein was more than 90%. The free cholesterol content of this HDL was about 80% of the native when expressed per milligram protein. When relipidation was carried out in the absence of added free cholesterol, only traces of free cholesterol were detected by gas liquid chromatography.

Transfer to Chylomicrons.[7] The labeling mixture consisted of 2.0 mg protein of rat HDL labeled with cholesteryl linoleyl ether and $d > 1.25$ fraction of human serum containing the cholesteryl ester transfer protein to give 30 mg protein/ml and 10–20 mg of chylomicron triacylglycerol. Following incubation for 6–18 hr at $37°$ the mixture was placed in SW-41

[26] O. Stein, G. Halperin, and Y. Stein, *Biochim. Biophys. Acta* **620**, 247 (1980).

tubes overlayered with 0.9% NaCl and centrifuged for 60–90 min at 39,000 rpm. The recovery of the label in the top fraction containing the chylomicrons was 50–85%.

Transfer to VLDL.[26] To achieve the labeling of VLDL, HDL labeled with radioactive cholesteryl linoleyl ether and/or with cholesteryl linoleate was incubated with VLDL and $d > 1.25$ g/ml fraction of human serum. The concentration of partially delipidated HDL and $d > 1.25$ g/ml protein in the incubation mixtures was 1.5 and 30 mg/ml, respectively, while the concentration of VLDL triacylglycerol ranged between 0.2 and 1.8 mg/ml. The incubation was carried out at 37° for 18 hr and thereafter 40% sucrose was added to give a final concentration of 2%. Three milliliters of the mixture was placed in each centrifuge tube and overlayered with 7.5 ml 0.9% NaCl; the VLDL was reisolated by centrifugation for 18 hr at 39,000 rpm in an SW-41 rotor.

Transfer to LDL.[6] The delipidated HDL (5 mg protein/tube) was labeled with [3H]cholesteryl linoleyl ether (10 μCi) and [14C]cholesteryl linoleate (4 μCi) by cosonication. The opalescent suspension was centrifuged at 39,000 rpm for 60 min in an SW-41 rotor. The tube was cut to give 3 ml of upper and 8 ml of lower phase. The lower phase was then brought to $d = 1.063$ g/ml and centrifuged for 24 hr at 39,000 rpm in an SW-41 rotor. The tube was sliced to give 4 ml of top layer and 7 ml of lower layer (HDL), which was then dialyzed and used for the labeling of LDL. The labeling of LDL was carried out by incubation of 2–4 mg LDL protein with the labeled HDL and $d > 1.25$ fraction of human serum 1.5 and 30 mg protein/ml, respectively, for 18 hr at 37°, final volume 3–4 ml. The incubation mixture was brought to $d = 1.063$ g/ml and centrifuged in an SW-41 rotor at 39,000 rpm for 24 hr and the LDL was concentrated in the top 2 ml.

Labeling of Intralipid and Transfer Procedure. Labeling of Intralipid and Transfer to HDL[27]: [3H]Cholesteryl linoleyl ether in heptane (1.7 × 10^7 dpm) was placed in a siliconized glass tube, the solvent was dried under N_2, and 1 ml of 0.15 M NaCl was added. Intralipid (Vitrum Co., Stockholm, Sweden), 125 μg triacylglycerol, was labeled with [3H]cholesteryl linoleyl ether by cosonication under N_2 twice for 30 sec using a 4-mm tip and the Braun Sonic 3000 (Braun, Melsungen, FRG) at 20% output at room temperature. To label HDL with [3H]cholesteryl linoleyl ether, 10 mg of HDL protein was incubated with the labeled Intralipid, 125 μg triacylglycerol, and the $d > 1.25$ g/ml fraction of human serum, at final concentration of 30 mg protein/ml, for 18 hr at 37° under N_2. The incuba-

27 Y. Stein, Y. Dabach, G. Hollander, G. Halperin, and O. Stein, *Biochim. Biophys. Acta* **752**, 98 (1983).

tion mixture was brought to $d = 1.063$ g/ml with solid KBr and centrifuged in an SW-41 rotor at 190,000 g for 24 hr. The tube was sliced to remove about 1.5 ml of the top layer and the clear infranatant was brought to $d = 1.21$ g/ml and centrifuged at 190,000 g for 48 hr.

Labeling of Intralipid and Transfer to LDL and Acetylated LDL: Cholesteryl linoleyl ether was transferred from labeled Intralipid as described above, and the labeled LDL or acetylated LDL was reisolated at $d < 1.063$ g/ml.

Labeling Procedure of Rabbit Serum.[28] [³H]Cholesteryl linoleyl ether (10 μCi) dissolved in chloroform was transferred to a siliconized glass tube 13 mm in diameter. The solvent was evaporated under an N_2 stream and 1 ml of 0.15 M NaCl, containing 0.1 ml of the centrifuged rabbit serum, was added to the tube. The tube content was sonicated with a 4-mm tip for 1 min under N_2, without foaming, using the Braun Sonic 3000 on 10% output. This procedure resulted in a homogeneous dispersion of 90% or more of the label added to the tube. The sonicate was added to 5 ml of the serum and incubation was carried out under N_2 in sterile 50-ml plastic tubes covered with tin foil for 18 hr at 37° to achieve equilibration of the label between the different nonsonicated lipoproteins. Thereafter, the serum was brought to 6.5 ml with 0.15 M NaCl and centrifuged in a 40.3 rotor for 6.3 × 10⁶ g/min. The top 1 ml was cut and the infranatant was used for injection, after removal of aliquots for the determination of radioactivity.

Studies with CLE-Labeled Lipoproteins

Studies in Vivo

Chylomicrons.[29] Rat chylomicrons labeled with CLE and CE were used to study the role of lipoprotein lipase in the uptake of cholesteryl ester by lactating mammary gland *in vivo*. Rat mesenteric duct chylomicrons were labeled by transfer with [³H]cholesteryl linoleyl ether and [¹⁴C]cholesteryl linoleate. Following intravenous injection into rats, 50% of labeled chylomicrons were cleared from the circulation in 5 min and the ratio of both labels in the plasma and liver was similar to the injected material. The chylomicrons labeled with radioactive cholesteryl linoleyl ether were injected into rats on the fourth day postpartum and about 3% of the injected label was recovered in the lactating mammary gland. When

[28] Y. Stein, O. Stein, and G. Halperin, *Arteriosclerosis* **2**, 281 (1982).
[29] T. Chajek-Shaul, G. Friedman, G. Halperin, O. Stein, and Y. Stein, *Biochim. Biophys. Acta* **666**, 216 (1981).

the labeled chylomicrons were injected into postpartum rats that had been separated from their litters for 24 hr the uptake of labeled cholesteryl linoleyl ether by the mammary glands was very low and did not exceed 0.2% of injected dose and was not different from the uptake by virgin mammary glands. The uptake of cholesteryl linoleyl ether by lactating mammary glands was related to the activity of lipoprotein lipase of the mammary tissue, which was high in suckled and low in unsuckled rats. However, this correlation became apparent only above a certain threshold of lipoprotein lipase activity. These findings indicate that high local activity of lipoprotein lipase promotes cellular uptake of lipoprotein cholesteryl ester, as studied with the help of cholesteryl linoleyl ether.

VLDL. Rat VLDL labeled with CLE and CE were injected into rats to learn about the stability of the cholesteryl ether and hence its advantage to study extrahepatic sites of cholesteryl ester uptake.[26] The fate of VLDL labeled with [^{3}H]cholesteryl linoleyl ether and [^{14}C]cholesteryl linoleate was determined at 3- to 48-hr intervals after injection. The liver was the main organ in which radioactivity was recovered at various time intervals after injection. While the recovery of the labeled cholesteryl ether was about 90% at all time intervals studied, not more than 44–48% of the labeled cholesteryl ester was recovered at 3 and 6 hr after injection. A comparison between liver, spleen, lung, and heart showed an uptake of 12.2, 1.7, 0.5, and 0.15% of injected [^{3}H]cholesteryl linoleyl ether per gram of tissue. A relatively high uptake of the labeled cholesteryl ether was seen in the adrenals and amounted to 6% per gram of tissue; in the testes the uptake was less than 0.1% per gram.

The localization of the labeled [^{3}H]cholesteryl linoleyl ether in the liver was determined with the help of autoradiography. The autoradiographic reaction was seen over the entire liver lobule; it was associated with the cytoplasm of the hepatocytes and some label was seen also over nonparenchymal cells. There was little or no label over erythrocytes or the sinusoidal lumen.[26]

LDL. Low-density lipoprotein, the main carrier of plasma cholesterol ester in many species, is catabolized by two pathways, the receptor-mediated and scavenger pathways,[6] the anatomical localization of which is not completely defined. ^{125}I has been the conventional label used to follow the fate of LDL–protein, but as the $t_{1/2}$ of the lipoprotein is relatively long, recovery of nondegraded protein in the organs was difficult. The transfer of cholesteryl ester among lipoproteins is not catalyzed by rat plasma. Therefore LDL labeled with [^{14}C]cholesteryl ester, and its nondegradable analog [^{3}H]cholesteryl ether, has been injected into rats to learn about the fate of the main core lipid component of LDL in the intact animal. The $t_{1/2}$ of the LDL determined between 1 and 24 hr was 7 hr. Up to 4 hr, 88–95% of the ^{3}H or ^{14}C label in plasma was in cholesteryl ether or ester, respectively. Fifteen minutes after injection, 5% of the injected ^{3}H label and 3.8% of the ^{14}C label were recovered in the liver. The amount of ^{3}H label in the liver increased gradually up to 24 hr and was 28.8% of the injected dose, but only 8.9% of the injected ^{14}C label remained in the liver 24 hr after injection. One hour after injection, 92.8% of ^{3}H and 52% of ^{14}C radioactivity in the liver were recovered in cholesteryl ether and ester, respectively. The amount of label recovered in the carcass was 34.7 ± 1.3% of the injected dose, and thus the total recovery of the [^{3}H]cholesteryl ether 24 hr after injection was 89 ± 3.4%. The use of the cholesteryl linoleyl ether as a nondegradable marker of cholesteryl ester provided a unique opportunity to follow the fate of the core lipid of LDL.

Comparison of Fate of Cholesteryl Linoleyl Ether after Injection of Labeled Acetylated LDL or Chylomicrons. In a study in which chylomicrons labeled with [^{14}C]cholesteryl linoleyl ether and retinyl [^{3}H]hexadecyl ether were injected (see below), loss of both labeled compounds from the liver was encountered already 1 week after injection.[30] This finding prompted the investigation to determine whether the prolonged hepatic retention of labeled ether analogs depends on the type of cell which takes up the lipoprotein containing the labeled nondegradable lipid. The question was asked whether labeled cholesteryl ether disappears more quickly when injected as part of lipoprotein taken up preferentially by hepatocytes and persists much longer in the liver when injected as a part of acetylated LDL, which is taken up preferentially by nonparenchymal cells.

Rat mesenteric duct chylomicrons labeled with [^{3}H]cholesteryl linoleyl ether and human acetylated low-density lipoproteins labeled with [^{14}C]cholesteryl linoleyl ether were injected simultaneously into rats. Three hours after injection 80–90% of the injected radioactivity was recovered in the liver and the ratio of ^{3}H/^{14}C in the liver was the same as in the injected material.[31] The ^{3}H/^{14}C ratio declined gradually over a period of 18 days due to loss of [^{3}H]cholesteryl ether, which had been injected with the chylomicrons, and retention of the same compound injected bound to acetylated LDL. The loss from the liver of the chylomicron-bound cholesteryl linoleyl ether was shown to occur through the bile, and its elimination from the body was verified by monitoring fecal excretion. These results provide evidence that hepatic persistence of a nonhydrolyzable analog of cholesteryl ester is a function of the cell type which has

[30] DeW. S. Goodman, O. Stein, G. Halperin, and Y. Stein, *Biochim. Biophys. Acta* **750,** 223 (1983).
[31] Y. Stein, Y. Kleinman, G. Halperin, and O. Stein, *Biochim. Biophys. Acta* **750,** 300 (1983).

ingested the lipid. Thus, the uptake of labeled chylomicrons by hepatocytes results in a slow but progressive excretion of the nonhydrolyzable lipid through the bile, while the preferential uptake of acetylated LDL by nonparenchymal cells of liver and by the spleen leads to persistence of the lipid in the organ.

HDL. To compare clearance from the circulation and tissue uptake of HDL protein labeled with ^{125}I and of HDL labeled with [3H]cholesteryl linoleyl ether, labeled HDL was injected into donor rats and was screened for 4 hr.[27] ^{125}I-Labeled HDL was subjected to the same protocol as the [3H]CLE–HDL, including screening. The screened, labeled sera were injected into acceptor rats and the disappearance of radioactivity from the circulation was compared. The $t_{1/2}$ in the circulation of ^{125}I-labeled HDL was about 10.5 hr, while that of [3H]cholesteryl linoleyl ether–HDL was about 8 hr. The liver and carcass were the major sites of uptake of [3H]cholesteryl linoleyl ether–HDL and accounted for 29–41% (liver) and 30% (carcass) of the injected label. Maximal recovery of [3H]cholesteryl linoleyl ether in the liver was seen 48 hr after injection, and thereafter there was a progressive decline of radioactivity, which reached 7.8% after 28 days. The maximal recovery of ^{125}I-labeled HDL in the liver was about 9%. Pretreatment of the acceptor rats with estradiol for 5 days resulted in a 20% increase in the hepatic uptake of [3H]cholesteryl linoleyl ether–HDL and a 5-fold increase in adrenal uptake. These findings indicate that in the rat the liver is the major site of uptake of HDL cholesteryl ester. The loss of the [3H]cholesteryl linoleyl ether from the liver after 14–28 days was interpreted to indicate that the labeled [3H]cholesteryl linoleyl ether had been taken up by hepatocytes.[27]

In another study, cholesteryl linoleyl ether-labeled HDL, in which apoA-I was labeled with covalently linked [125I]tyramine cellobiose, was injected into rats. Selective delivery of cholesteryl ether to liver, adrenal, and gonads was found while other tissues took up apoA-I and cholesteryl ether at nearly equal rates.[32]

Studies with Rabbit Plasma Lipoproteins. Use was made of [3H]cholesteryl linoleyl ether (CLE), a nondegradable analog of cholesteryl ester (CE) to measure plasma lipoprotein CE influx into rabbit aorta.[28] Autologous serum labeled with [3H]CLE was injected into seven hypercholesterolemic rabbits, and more than 90% of the label was recovered in the plasma compartment 10 min after injection. Between 4 hr and 3 days the label was cleared from the circulation with a $t_{1/2}$ of about 24 hr. Between 4 and 24 hr the plasma lipoproteins isolated at $d < 1.006$, $d < 1.019$, and $d <$

[32] C. Glass, R. C. Pittman, D. B. Weinstein, and D. Steinberg, *Proc. Natl. Acad. Sci. U.S.A.* **80,** 5435 (1983).

1.063 approached similar specific activity, assuming that [³H]CLE had mixed with the lipoprotein CE pool. The rabbits were killed 7 to 14 days after injection when plasma radioactivity decreased to <0.03% of injected dose. Total recovery of the CLE ranged from 70 to 95% and 48 to 72% was found in the liver. The minimum influx of plasma CE into the aortic intima was determined by dividing the label found in the artery by the mean specific activity of the labeled compound in the plasma. The minimum influx into regions with atheromatous involvement ranged from 0.8 to 3.4 μg CE/cm²/hr. The rate of influx was highly correlated with the amount of CE mass in the intima and media, indicating that the bulk of aortic CE is derived from plasma lipoprotein CE.[28]

Studies in Vitro

Chylomicrons. In cell cultures derived from newborn rat hearts lipoprotein lipase can be easily regulated. Therefore, we have used the rat heart cell cultures and chylomicrons labeled with cholesteryl linoleyl ether to study the relation between lipoprotein lipase activity and the uptake of a chylomicron cholesteryl ester analog.[7,33] Rat heart cell cultures incubated with chylomicrons labeled with [³H]cholesteryl linoleyl ether and [¹⁴C]cholesteryl ester took up both lipids to the same extent, and the intracellular degradation of the labeled cholesteryl ester was inhibited by chloroquine. The uptake of the labeled cholesteryl ether from intact chylomicrons and from chylomicron remnants, prepared by a membrane-supported lipoprotein lipase, was the same. The uptake of the labeled cholesteryl ether from remnants prepared by milk lipoprotein lipase was significantly higher than that from native chylomicrons, and addition of milk lipoprotein lipase doubled the uptake of cholesteryl ether from remnants produced by membrane-supported lipoprotein lipase of heart cell cultures. However, remnants which had undergone completely lipolysis by bacterial lipase were not better donors of cholesteryl ether than native chylomicrons. These findings suggest that lipoprotein lipase may play a role in cellular uptake of chylomicron cholesteryl ester independently of hydrolysis of triacylglycerol.

LDL. Human skin fibroblasts from normal and apoB, E receptor-negative cells were used in order to determine the suitability of human LDL labeled with cholesteryl linoleyl ether to study receptor-mediated uptake of LDL. [³H]CLE–LDL was taken up via apoB, E receptor-mediated endocytosis in a manner very similar to ¹²⁵I-labeled LDL.[34] This was

[33] G. Friedman, T. Chajek-Shaul, O. Stein, T. Olivecrona, and Y. Stein, *Biochim. Biophys. Acta* **666,** 156 (1981).

[34] D. W. Higgs, D. R. van der Westhuyzen, W. Gevers, G. A. Coetzee, O. Stein, and Y. Stein, *Arteriosclerosis* **4,** 238 (1984).

demonstrated by similar rates of uptake of the two differently labeled LDL preparations, saturation kinetics of uptake with respect to [^{3}H]CLE–LDL concentration, regulation of [^{3}H]CLE–LDL uptake by procedures that up/down regulate the number of apoB, E receptors, and negligible uptake of [^{3}H]CLE–LDL by receptor-negative cell strains.[34]

HDL. Studies *in vivo* indicated that [^{3}H]CLE–HDL disappears from rat plasma at a more rapid rate than ^{125}I-labeled HDL. Such a finding suggested a preferential uptake by tissues of [^{3}H]CLE from HDL. In tissue culture it is possible to account for protein degradation products and thus compare in a closed system the uptake of HDL protein labeled with ^{125}I and of cholesteryl ester labeled with a nondegradable analog [^{3}H]CLE. Primary cultures of rat hepatocytes and adrenal cells were used to study this problem. Comparison of uptake of cholesteryl ester moiety represented by uptake of [^{3}H]CLE, and of protein moiety represented by metabolism of ^{125}I-labeled protein, was carried out using both native and apoE-free HDL. In all experiments, the ratio of ^{3}H/^{125}I representing cellular uptake exceeded 1.0. In cultured rat adrenal cells, the uptake of [^{3}H]CLE–HDL was stimulated 3- to 6-fold by 1×10^{-7} *M* ACTH, while the uptake of ^{125}I-labeled HDL increased about 2-fold. The ratio of ^{3}H/^{125}I representing cellular uptake was 2.0–3.0 and increased to 5.0 in ACTH-treated cells.[35]

Labeling of Liposomes with CLE

Small unilamellar liposomes were prepared by sonication. Phosphatidylcholine (2.66 μmol), 1.33 μmoles of free cholesterol (FC), and trace amounts of [^{3}H]- or [^{14}C]cholesteryl linoleyl ether were placed in a siliconized glass tube, the solvents were evaporated under N_2, and 4 ml of 0.15 *M* NaCl was added. The lipids were subjected to ultrasonic irradiation for 5 periods of 1 min, at room temperature under N_2, using the Braun Sonic 3000 instrument at 100 W output and a 9-mm probe. The clear-to-opalescent sonicate was diluted with 7 ml F_{10} culture medium containing 4% fatty acid poor bovine serum albumin, left overnight at 4°, and subjected to ultracentrifugation in an SW-41 rotor at 190,000 *g* for 3 hr. The tubes were sliced at 1.5 ml from the top and the clear infranatant was aspirated and sterile filtered through a 0.45-μm Millipore filter. The recovery of the phospholipids was 62–66%, of FC 32–35%, and labeled cholesteryl linoleyl ether 60%. The liposomes were studied by electron microscopy in preparations stained with 1% phosphotungstic acid and were found to

[35] E. Leitersdorf, O. Stein, S. Eisenberg, and Y. Stein, *Biochim. Biophys. Acta,* **796,** 72 (1984).

consist mainly of unilamellar particles with an average diameter of 250–350 Å.[36]

Studies with Liposomes Labeled with CLE

Studies in Vivo

Liposomes prepared from dioleyl ether phosphatidylcholine were labeled with [^{14}C]CLE and injected into rats; 1 hr thereafter 60% of injected radioactivity was still present in the circulation and declined gradually during the next 24 hr.[37] More than 70% of the injected material was recovered in the liver and was retained during the next 3 days. Thereafter, a slow decrease in liver [^{14}C]CLE occurred, so that after 10 days only 35% of the injected dose was still present in the liver.

Studies in Vitro

Unilamellar liposomes prepared from purified phospholipids and labeled cholesteryl linoleyl ether were used to study lipoprotein lipase-catalyzed transfer of cholesteryl ester into cells in culture.[36] In mesenchymal rat heart cell cultures, the transfer of cholesteryl linoleyl ether and cholesteryl linoleate was similar and related to the activity of endogenously produced lipoprotein lipase. In human skin fibroblasts transfer of labeled cholesteryl linoleyl ether was proportional to the concentration of milk lipoprotein lipase added to the incubation medium. Liposomes prepared from phosphatidylcholine or phosphatidylethanolamine were much better donors of cholesteryl linoleyl ether to normal and apolipoprotein B, E receptor-negative fibroblasts and to endothelial cells than those prepared from sphingomyelin. Attachment of lipoprotein lipase to the cell surface was mandatory for the transfer of cholesteryl ether and could be prevented by heparin.[36]

Retinyl Hexadecyl Ether

Synthesis

Retinyl hexadecyl ester, naturally present in the liver, has been used as marker for chylomicron metabolism.[38] As this compound is a hydrolyz-

---

[36] O. Stein, G. Friedman, T. Chajek-Shaul, G. Halperin, T. Olivecrona, and Y. Stein, *Biochim. Biophys. Acta* **750,** 306 (1983).

[37] Y. Stein, G. Halperin, E. Leitersdorf, Y. Dabach, G. Hollander, and O. Stein, *Biochim. Biophys. Acta* **793,** 354 (1984).

[38] A. C. Ross and D. B. Zilversmit, *J. Lipid Res.* **18,** 169 (1977).

able probe,[39] it seemed of interest to prepare retinyl hexadecyl ether, its nonhydrolyzable analog.[30]

Synthesis of Retinyl Hexadecyl Ester. Retinol has been previously etherified with butyl and methyl moieties by the following methods: (1) retinol was converted with phosphorus tribromide to the retinyl bromide, which was reacted with sodium butylate[40]; (2) the lithium salt of retinol, obtained from retinol and butyl lithium, was etherified with dimethyl sulfate.[41]

The alkyl methane sulfonate–KOH etherification method was found suitable for the preparation of both the labeled and unlabeled cholesteryl hexadecyl ethers[30]; however, since the retinol and its derivatives are extremely sensitive to light, air, or high temperatures, the solvent chosen was petroleum ether of low boiling point (30–60°) and all the procedures were carried out under nitrogen in a darkened room. Attempts to etherify tracer amounts of labeled retinol gave poor results, which was attributed to the lability of the retinol; yields of over 20% were achieved when the labeled alkyl methane sulfonate was reacted with milligram amounts of retinol.

All-*trans* retinol (100 mg), powdered KOH (1 g), and light petroleum ether (100 ml; bp 30–60°C) were stirred and refluxed in an oil bath at 40–45°. Water formed by the reaction of retinol with KOH was removed with an azeotropic separator. Hexadecyl methane sulfonate (112 mg) in 50 ml light petroleum ether was added dropwise, and the stirring and refluxing were continued for 10 hr. After cooling, the petroleum ether solution was washed three times with water, dried with anhydrous Na_2SO_4, and evaporated almost completely under nitrogen. The residue was dissolved in 4 ml hexane, and then chromatographed on a column of alumina-S deactivated with 1% water (20 g), using hexane as eluent. The eluted fractions (3 ml each) were analyzed by TLC, using plates coated with Merck basic alumina, and hexane : chloroform (9 : 1, v/v) as solvent. Sixty-seven milligrams of apparently pure reaction product (only one spot visible on TLC, with R_f 0.77) was recovered from fractions 7–15. Analysis by high-performance liquid chromatography (HPLC) (C_{18} μBondapak column, elution with methanol) showed that the product eluted as a single sharp peak, comprising more than 95% of eluted material (i.e., more than 95% pure). HPLC analysis of the retinyl ether, together with retinyl palmitoyl ester, showed that the ether was eluted separately from and after the retinyl

[39] DeW. S. Goodman, H. S. Huang, and T. Shiratori, *J. Lipid Res.* **6**, 390 (1965).

[40] O. Isler, A. Ronco, W. Guex, N. C. Hindley, W. Huber, K. Dialer, and M. Kofler, *Helv. Chim. Acta* **63**, 489 (1949).

[41] A. R. Hanze, T. W. Conger, E. C. Wise, and D. I. Weisblat, *J. Am. Chem. Soc.* **70**, 1253 (1948).

ester. The ultraviolet absorption spectrum of the retinyl hexadecyl ether was very similar to the retinyl ester, with a peak maximum at 325 nm.

Retinyl hexadecyl ether labeled with ^{3}H in the hexadecyl moiety was prepared as follows: [9,10-^{3}H]hexadecyl mesylate was synthesized from [9,10-^{3}H]palmitic acid as described by Paltauf and Spener.[42] The labeled mesylate was reacted with 33 mg retinol as described above, to yield 21% of the ^{3}H-labeled ether. The purity of the ether was determined by TLC on alumina plates and 92.3% of the ^{3}H label was recovered in the region of the retinyl ether. Higher purity was not achieved by two additional column chromatographic runs. It is likely that during TLC, changes occurred to a small extent due to the sensitivity of the retinol derivatives to oxygen and light (which were not totally excluded), and that the actual purity of the ^{3}H-labeled retinyl ether was greater than 92.3%. The retinyl [^{3}H]hexadecyl ether was stored in hexane in the dark under N_2 at $-20°$.

Biological Applications

Studies with Chylomicrons Labeled with Retinyl Hexadecyl Ether. On the morning of the experiment, the retinyl [^{3}H]hexadecyl ether was chromatographed on a small column of alumina (3 g) in order to provide freshly purified retinyl ether for each study. Chylomicrons were prepared labeled with both [^{14}C]cholesteryl linoleyl ether and with retinyl [^{3}H]hexadecyl ether, as described for cholesteryl linoleyl ether. After 18 hr of incubation of a mixture that contained delipidated HDL doubly labeled with both ethers, rat chylomicrons, and the $d > 1.25$ fraction of human serum, 75% of the labeled cholesteryl ether but only 36% of the labeled retinyl ether was transferred to the chylomicrons.[30]

Retinyl palmitate, which is incorporated into core lipids of chylomicrons during vitamin A and fat absorption, has been used as a marker for chylomicron metabolism.[38] After the entry of chylomicrons into the blood stream, both chylomicron cholesteryl and retinyl esters are mainly taken up by the liver.[39,43] Hydrolysis of the cholesteryl and retinyl esters then occurs within the liver,[39,44] followed by subsequent hepatic and extrahepatic metabolism of the compounds.

Rat lymph chylomicrons were labeled with ether analogs of [^{14}C]cholesteryl linoleate and retinyl [^{3}H]palmitate and injected into rats. Three hours after injection, 85–90% of the ^{14}C and 80–85% of the ^{3}H were recovered in the liver. Thereafter, the cholesteryl and retinyl ethers were lost progressively from the liver at slow but differing rates. After 4 days,

[42] F. Paltauf and F. Spener, *Chem. Phys. Lipids* **2,** 168 (1968).
[43] DeW. S. Goodman, *J. Clin. Invest.* **41,** 1886 (1962).
[44] O. Stein, Y. Stein, D. S. Goodman, and N. H. Fidge, *J. Cell Biol.* **43,** 410 (1969).

two-thirds of the [^{14}C]cholesteryl ether but only one-third of the ^{3}H-labeled retinyl ether were found in the liver. By 27 days, the hepatic recovery of ^{14}C had decreased to 20–25% and that of ^{3}H to about 3% of the injected dose. In contrast, the amount of [^{14}C]cholesteryl linoleyl ether recovered in the spleen did not change during 27 days, while the ^{3}H label in the spleen declined about 10-fold. There was no redistribution of either labeled compound from the liver to extrahepatic tissues, indicating that the labeled compounds were lost from the body, most probably through the bile.

Tri- and Dialkyl Glyceryl Ethers

Synthesis

Trialkyl Glyceryl Ethers. Trialkyl glyceryl ethers have been used as nonhydrolyzable probes in the study of intestinal absorption.[45,46] Procedures for preparation of radioactive triethers labeled either at the 3-O-alkyl side chains[47] or tritiated at C-2 of the glycerol moiety[48] will be described.

Preparation of labeled trialkyl glyceryl ether according to Paltauf and Spener.[47] 3-O-Tetrahydropyranylglycerol (0.88 mmol), synthesized according to Barry and Craig,[49] was alkylated with 3.8 g hexadecyl methane sulfonate and KOH as described in the section on preparation of cholesteryl alkyl ethers. The crude product was dissolved in 10 ml of ether and 10 ml of methanol, and 0.5 ml of concentrated HCl was added. The mixture was allowed to stand at room temperature for 1 hr. The crystalline precipitate formed was collected on a Buchner funnel and washed with ice-cold petroleum ether (40–60°). The yield was 2.5 g (92.6%). Recrystallization from petroleum ether yielded 2.1 g of pure 1,2-di-hexadecyloxy-3-propanol (78% yield), mp 59°.

When 1,2-di-*cis*-9'-octadecenyloxy-3-propanol was prepared, the oily layer formed by the hydrolysis was separated. The methanolic layer was mixed with 50 ml of water and extracted twice with 30-ml portions of peroxide-free ether. The oil and the extracts were combined, washed with water, and dried over anhydrous Na_2SO_4. The yield after removing the solvent *in vacuo* was 2.7 g (93%). The crude diether was purified by TLC, yielding 1.5 g (52%). Purification was also achieved by column chro-

[45] F. Spener, F. Paltauf, and A. Holasek, *Biochim. Biophys. Acta* **152,** 368 (1968).
[46] R. G. H. Morgan and A. F. Hofman, *J. Lipid Res.* **11,** 231 (1970).
[47] F. Paltauf and F. Spener, *Chem. Phys. Lipids* **2,** 168 (1968).
[48] G. Halperin, *J. Label. Compounds Radiopharmacol.* **20,** 269 (1982).
[49] P. J. Barry and B. M. Craig, *Can. J. Chem.* **33,** 716 (1955).

matography with 25 g of silicic acid (eluting mixture: 50% benzene in hexane).

[1-^{14}C]cis-9-Octadecenol was prepared from [1-^{14}C]oleic acid (200 × 10⁶ cpm, specific activity 36.9 mCi/mmol) which was diluted with 1.5 g (5.31 mmol) of unlabeled oleic acid and esterified with diazomethane. The methyl oleate was dissolved in 30 ml of diethyl ether, reduced to oleyl alcohol by 0.4 g of LiAlH₄, and suspended in 40 ml of ether. The yield and total activity were 1.35 g (93%) and 186 × 10⁶ cpm, respectively, 1.2 g of this product was mesylated with 1.0 g of methane sulfonyl chloride and 30 ml of dry pyridine to yield 1.4 g of crude product (138 × 10⁶ cpm) which was not purified further. Condensation of the mesylate with 1,2-di-cis-9'-octadecenyloxy-3-propanol (2.9 g) and powdered KOH (4 g) in dry benzene (130 ml) yielded 3.7 g of crude trialkyl ether. Purification by TLC yielded 2.0 g (59%) of pure product. Total activity 85 × 10⁶ cpm, specific activity 52.5 × 10³ cpm/mg. This procedure was also suitable for preparation of the high specific activity 1,2-di-cis-9'-octadecenyloxy-3-[9',10'-³H]-cis-9'-octadecenyloxypropane.

The labeled 9-cis-octadecenol was prepared as follows: [9',10'-³H]-oleic acid (1 mCi, 5.7 Ci/mmol) was esterified and reduced as described above. The product was purified on silicic acid columns (elution with 40% and 50% benzene in hexane) to give radiochemically pure [9,10-³H]cis-9'-octadecenol (52% yield) containing 2 mg of unlabeled oil. To 50 × 10⁶ dpm of this product in dry benzene (0.5 ml) under anhydrous conditions, in an ice bath, dry pyridine (0.04 ml) and methane sulfonyl chloride (0.02 ml) were added and the mixture was kept 60 min at 0°, and another 3 hr at room temperature. Ice (about 0.1 g) was added and the mixture was extracted with 100 ml of hexane. The solution was washed with water (30 ml), 0.1 M HCl (50 ml), water (30 ml), saturated solution of NaHCO₃ (50 ml), and again with water (30 ml). The solution was dried (Na₂SO₄), filtered, and evaporated in vacuo to a small volume (1–5 ml) and was kept under N₂ at −20° prior to use. The product (43 × 10⁶ dpm, >98% purity) contained <0.5 mg of unlabeled oil. This compound was condensed with 1,2-di-cis-9'-octadecenyloxy-3-propanol (12 mg), and the product was purified on 10 g silicic acid (eluting solvent 30 and 40% benzene in hexane), to give the labeled triglyceryl ether (yield 76%, radiochemical purity >98%, unlabeled oil <0.5 mg). Mesylation of the labeled oleyl alcohol in benzene solution avoided the presence of oils in the labeled products.

The synthesis of [C-2-³H]-tri-cis-9'-octadecenylglyceryl ether.[48] This synthesis is illustrated in Scheme 1. The starting material for this procedure, dipropionoxyacetone diethyl mercaptal,[49] was hydrolyzed under alkaline conditions and the dihydroxy compound was then etherified with alkyl mesylate–KOH to di-cis-9-octadecenyloxyacetone diethyl mercap-

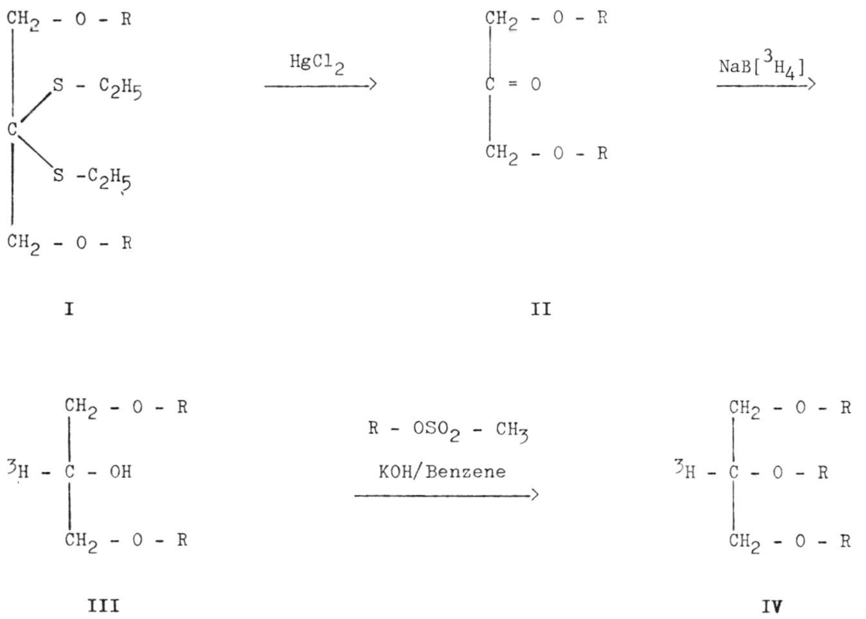

$$CH_2 - O - R$$

$$\overset{|}{\underset{C}{\diagup}} \overset{S - C_2H_5}{\underset{\diagdown S - C_2H_5}{}}$$

$$\overset{|}{CH_2} - O - R$$

$$\xrightarrow{\text{HgCl}_2}$$

$$CH_2 - O - R$$
$$\overset{|}{C} = O$$
$$\overset{|}{CH_2} - O - R$$

$$\xrightarrow{\text{NaB}[^3\text{H}_4]}$$

I   II

$$CH_2 - O - R$$
$$^3H - \overset{|}{C} - OH$$
$$\overset{|}{CH_2} - O - R$$

$$\xrightarrow[\text{KOH/Benzene}]{R - OSO_2 - CH_3}$$

$$CH_2 - O - R$$
$$^3H - \overset{|}{C} - O - R$$
$$\overset{|}{CH_2} - O - R$$

III   IV

SCHEME 1. Synthesis of [C-2-³H]trialkylglycerol ether.

tal (I), which was converted to the di-alkyloxyacetone (compound II), and then to the C-2 labeled di- and trialkyl glycerol (III and IV).

The advantages of this method are as follows: (1) Inexpensive NaB[³H]₄ is used for labeling; (2) only the last reaction is performed on a labeled intermediate; and (3) all the other steps are conventional chemical reactions, unhampered by the limitations involved in working with radioactive material.

Di-cis-9-octadecenyloxyacetone (compound II). Compound I (700 mg) was reacted with HgCl₂ (2.5 g) in acetone solution, as previously described,[48,49] to give 550 mg of crude product. Chromatography on silicic acid (16 g) and elution with 30% benzene in hexane yielded 300 mg of compound II, which crystallized in the refrigerator and melted at 23°. R_f in hexane : chloroform (1 : 1, v/v), 0.57; NMR (CDCl₃) δ (ppm) 0.867 (*m*, 6H, 17-CH₃), 1.277 (broad *s*, methylenic protons), 3.463 (*m*, 4H, O—CH₂—), 4.191 (*s*, 4H, CO—CH₂—O), 5.333 (*m*, 4H, —CH=CH—); IR 1750 (C=O), 1475, 1025 cm⁻¹; mass spectrum (*m/e*) 592 (M⁺). Analysis calculated for C₃₉H₇₄O₃: C, 79.32%; H, 12.54%; found: C, 79.61%; H, 12.26%.

1,3-Di-cis-9'-octadecenyloxy-2-propanol (compound III). NaBH₄ (40 mg) was added to a solution of compound II (145 mg) in isopropanol (5 ml). The solution was stirred overnight at room temperature, water (2 ml) was added, and the stirring was continued for 2 hr. The mixture was extracted with hexane (60 ml), and the organic phase was washed with water (50 ml) and evaporated under reduced pressure. The product was further purified on silicic acid column (14 g). Elution with 50% benzene in hexane gave 125 mg of compound III. A satisfactory chemical analysis was achieved after additional purification by preparative TLC: mp 17–18°; NMR (CDCl$_3$) δ (ppm) 0.863, 1.277, 1.942 (*m*), 3.386 (*m*, 8H, CH$_2$—O—), 4.370 (*m*, 1H, OH), 5.33 (*m*, 4H, —CH=CH—); IR 3520, 1765, 1425 cm^{-1}; mass spectrum (*m/e*) 592 (M$^+$). Analysis calculated for C$_{39}$H$_{76}$O$_3$: C, 79.05%; H, 12.84%; found: C, 79.30%; H, 12.98%.

[2-^{3}H]1,3-Di-cis-9'-octadecenyloxy-2-propanol: Crystalline NaB[^{3}H]$_4$ (~0.2 mg, specific activity 469 mCi/mmol) was added to a solution of di-*cis*-9'-octadecenyloxyacetone (31 mg) in isopropanol (6 ml), and stirred. Na$_2$CO$_3$ (5 mg) and water (0.5 ml) were added and the stirring was continued for another 2 hr. Extraction was carried out with toluene and the organic phase and washings contained 3.6 × 10^9 dpm and 4.1 × 10^9 dpm, respectively. The labeled residue from the organic phase was chromatographed on silicic acid column (14 g) with 50% benzene in hexane to give ~1.7 × 10^9 dpm of the diether (compound III) with radiochemical purity of 92.5%. Further purification of 160 × 10^6 dpm on preparative TLC (chloroform : ethyl acetate, 95 : 5) yielded 140 × 10^6 dpm of 98% pure compound, specific activity 118 mCi/mmol; however, the crude labeled 1,3-diether was used for further reactions.

[2-^{3}H]Tri-cis-9'-octadecenyloxy-propane: [2-^{3}H]1,3-Di-*cis*-9'-octadecenyloxy-2-propanol (13 × 10^8 dpm) and *cis*-9-octadecenylmethane sulfonate (20 mg) were refluxed in benzene with KOH under N$_2$. The labeled triether was purified on silicic acid column (as described above), to give 7.1 × 10^8 dpm of product being 98.4% pure.

Dialkylglyceryl Ethers. 1,2-O-Dialkylglyceryl ethers. Racemic 1,2-dialkylglyceryl ether can be conveniently prepared from 3-*O*-tetrahydro-pyranylglycerol.[47] For the synthesis of the optically active diether, 1,2-isopropylidene-*sn*-glycerol (compound I, Scheme 2) or 2,3-isopropylidene-*sn*-glycerol can be used as starting material. The 1,2-isomer has been prepared from D-mannitol[50] in three steps; this procedure was later modified.[51-53] 1,2-Isopropylidene-*sn*-glycerol, which had been considered configurationally unstable, is stable in the presence of potas-

[50] E. Baer, *Biochem. Prep.* **2**, 31 (1952).
[51] P. R. Bird and J. S. Chandha, *Tetrahedron Lett.* **4541** (1966).
[52] J. LeCocq and C. E. Ballou, *Biochemistry* **3**, 976 (1964).
[53] H. Eibl, *Chem. Phys. Lipids* **28,** 1 (1981).

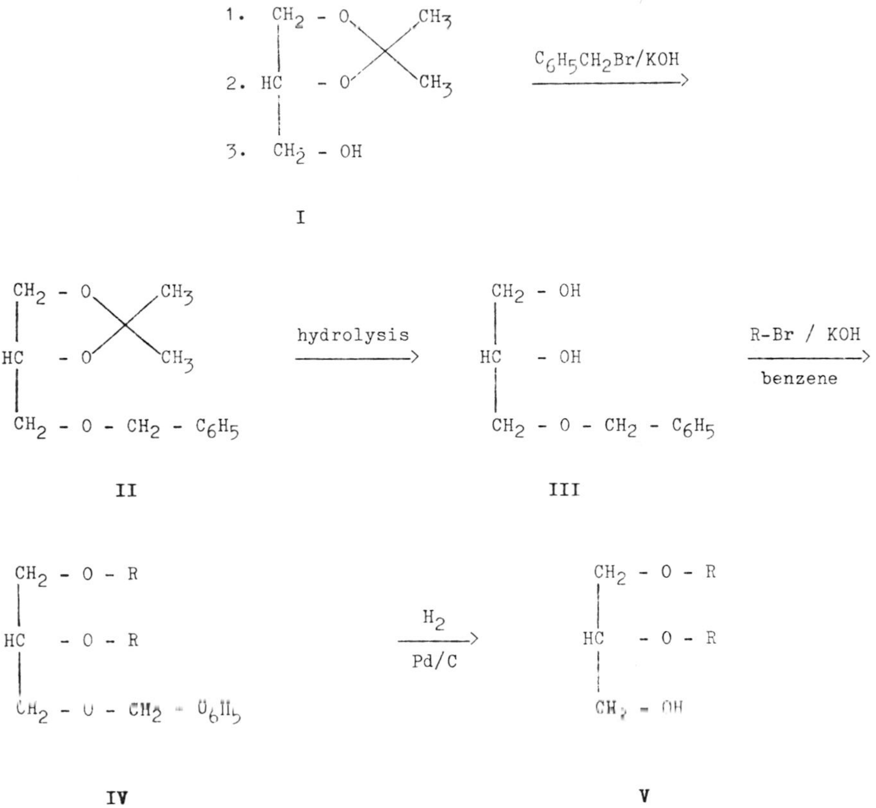

$$
\begin{array}{ll}
1. & CH_2 - O \\
& \quad\quad\quad \diagdown \quad CH_3 \\
& \quad\quad\quad \diagup\diagdown \\
2. & HC \quad - O \diagup \quad CH_3 \\
3. & CH_2 - OH
\end{array}
\qquad \xrightarrow{\; C_6H_5CH_2Br/KOH \;}
$$

I

$$
\begin{array}{l}
CH_2 - O \diagdown CH_3 \\
\quad\quad \diagtimes \\
HC \;\; - O \diagup CH_3 \\
CH_2 - O - CH_2 - C_6H_5
\end{array}
\qquad \xrightarrow{\; hydrolysis \;}
\begin{array}{l}
CH_2 - OH \\
\\
HC \;\; - OH \\
\\
CH_2 - O - CH_2 - C_6H_5
\end{array}
\qquad \xrightarrow[\; benzene \;]{\; R-Br \, / \, KOH \;}
$$

II   **III**

$$
\begin{array}{l}
CH_2 - O - R \\
\\
HC \;\; - O - R \\
\\
CH_2 - O - CH_2 - C_6H_5
\end{array}
\qquad \xrightarrow[\; Pd/C \;]{\; H_2 \;}
\begin{array}{l}
CH_2 - O - R \\
\\
HC \;\; - O - R \\
\\
CH_2 - OH
\end{array}
$$

IV   **V**

SCHEME 2. Synthesis of optically active 1,2-O-dialkylglycerol via 1- or 3-O-benzylglycerol.

sium hydroxide.[53] 2,3-Isopropylidene-sn-glycerol was obtained from L-arabinose[54,55] or from L-serine.[56]

In one of the early routes the optically active 1,2- or 1,3-dialkyl glycerols were prepared as described in Scheme 2.[57–60] The optically active isopropylidene glycerol (I) was etherified with benzyl chloride and KOH

[54] E. Baer and H. O. L. Fischer, *J. Biol. Chem.* **128**, 463 (1939).
[55] P. Kanda and M. A. Wells, *J. Lipid Res.* **21**, 257 (1980).
[56] C. M. Lok, J. P. Ward, and D. A. Dorp, *Chem. Phys. Lipids* **16**, 115 (1976).
[57] M. Kates, T. H. Chan, and N. Z. Stanacev, *Biochemistry* **2**, 394 (1963).
[58] M. Kates, B. Palameta, and L. S. Yengoyan, *Biochemistry* **4**, 1595 (1965).
[59] E. Baer and N. Z. Stanacev, *J. Biol. Chem.* **240**, 44 (1965).
[60] C. N. Joo, T. Shier, and M. Kates, *J. Lipid Res.* **9**, 782 (1968).

in benzene solution. Then compound II was hydrolyzed and etherified with alkyl bromide. At the final step, the benzyl group was removed by hydrogenolysis over palladium catalyst to give the dialkylglyceryl ether (V). This route is restricted to the dialkylglycerols containing two identical and saturated alkyl moieties.

A procedure for the preparation of saturated and unsaturated, as well as mixed 1,2-O-dialkyl-sn-glycerol, is given in Scheme 3. The starting compound for the synthesis is 3-O-alkyl-sn-glycerol (compound I) prepared by condensation of 1,2-isopropylidene-sn-glycerol with alkyl methane sulfonate and hydrolysis of the protection group.[17] Tosylation of the dialcohol and Walden inversion gave 1-O-alkyl-sn-glycerol (compound III),[61] which was protected at the 3-hydroxyl by tritylation.[62] Then compound IV was alkylated and the trityl protection group removed, to give the 1,2-dialkyl-sn-glycerol (compound VI).

It has been proposed[22] that 1-O-alkyl-sn-glycerol (compound III) prepared from 2,3-isopropylidene-sn-glycerol[55] should be used as the starting material for the synthesis, omitting the Walden inversion step, as the latter has been found to be incomplete.[53]

Preparation of 1-O-cis-9'-Octadecenyl-2-O-cis-9'-[9',10'(n)-³H]-octadecenyl-sn-glycerol. The starting compound 1-O-cis-9'-octadecenyl-3-O-trityl-sn-glycerol (compound IV, Scheme 4) was prepared according to Chacko and Hanahan.[61] However, the inversion of the ditosylate under reflux conditions was found to be slow and incomplete. Improvement was achieved when the reaction was carried out in an ampoule as follows: 1-O-cis-9'-octadecenyl-2,3-di-p-toluene-sulfonyl-sn-glycerol (compound II) (7.5 g), freshly fused potassium acetate (12.6 g), and ethanol (60 ml) were sealed in a 100-ml constricted tube and kept at 110° overnight. The product was extracted with peroxide-free ether and hydrolyzed with potassium hydroxide (5 g) in 90% ethanol (100 ml); 70 ml of the solvent was evaporated *in vacuo* and the residue was extracted with ether to give 3.8 g of crude material. Crystallization from hexane at −20°, centrifugation, and decantation (as described previously) yielded 3.6 g of 1-O-cis-9'-octadecenyl-sn-glycerol (compound III), which seemed pure by TLC analysis. This compound (38 mg), together with [9,10(n)-³H]cis-9-octadecenyl methane sulfonate (700 × 10⁶ dpm), and powdered KOH (1 g) in benzene (50 ml) under N_2 was refluxed for 10 hr and about 45 ml of the benzene was distilled off. The product was extracted with diethyl ether, the solution was washed three times with water, dried over Na_2SO_4, filtered, and evaporated. The residue was chromatographed on an alu-

[61] G. K. Chacko and D. J. Hanahan, *Biochim. Biophys. Acta* **164,** 252 (1968).
[62] W. J. Baumann and H. K. Mangold, *J. Org. Chem.* **31,** 498 (1966).

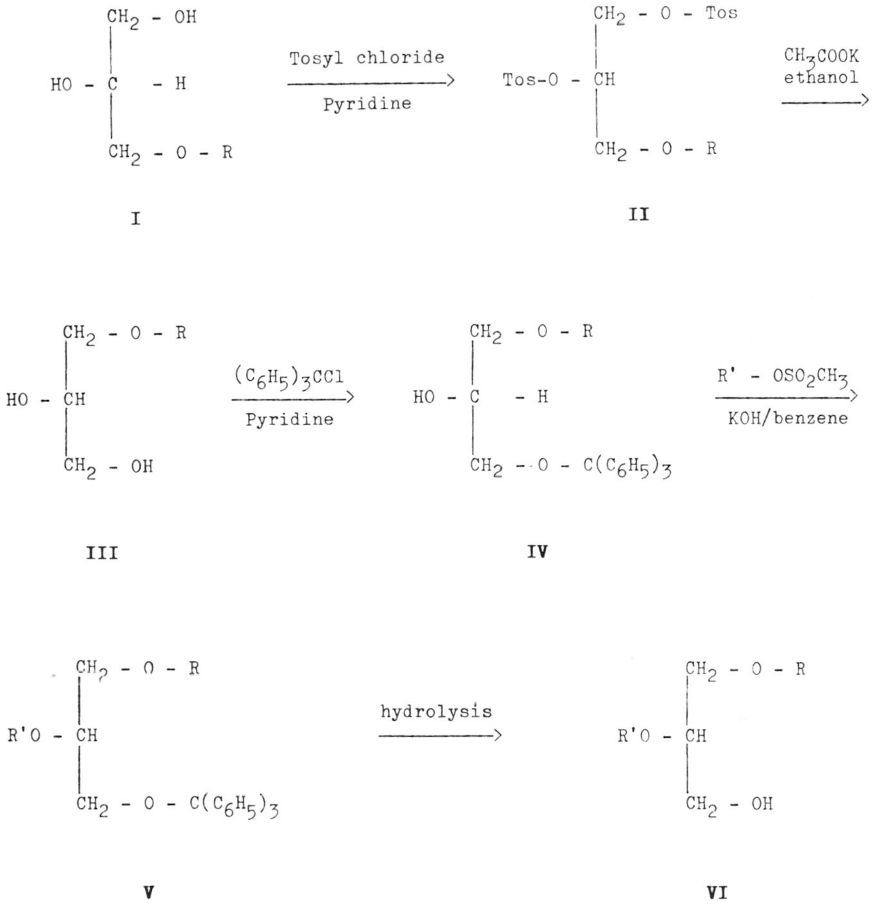

$$CH_2 - OH$$
$$HO - \underset{|}{\overset{|}{C}} - H \qquad \xrightarrow[\text{Pyridine}]{\text{Tosyl chloride}} \qquad Tos-O - \underset{|}{\overset{|}{CH}}$$
$$CH_2 - O - R$$

$$CH_2 - O - Tos$$
$$\xrightarrow[\text{ethanol}]{CH_3COOK}$$
$$CH_2 - O - R$$

I   II

$$CH_2 - O - R$$
$$HO - \underset{|}{\overset{|}{CH}} \qquad \xrightarrow[\text{Pyridine}]{(C_6H_5)_3CCl}$$
$$CH_2 - OH$$

$$CH_2 - O - R$$
$$HO - \underset{|}{\overset{|}{C}} - H \qquad \xrightarrow[\text{KOH/benzene}]{R' - OSO_2CH_3}$$
$$CH_2 - O - C(C_6H_5)_3$$

III   IV

$$CH_2 - O - R$$
$$R'O - \underset{|}{\overset{|}{CH}} \qquad \xrightarrow{\text{hydrolysis}}$$
$$CH_2 - O - C(C_6H_5)_3$$

$$CH_2 - O - R$$
$$R'O - \underset{|}{\overset{|}{CH}}$$
$$CH_2 - OH$$

V   VI

SCHEME 3. Synthesis of optically active mixed 1,2-O-dialkylglycerol.

mina column (17 g) with benzene in hexane. The product recovered from the fractions up to 7% benzene in hexane (220×10^6 dpm) was refluxed for 1 hr in 80% acetic acid (50 ml). The solvent was removed *in vacuo* and KOH (3 g) and 95% methanol (100 ml) were added to the residue. The solution was kept at room temperature overnight and approximately two-thirds of the solvent was removed *in vacuo*. The mixture was extracted with ether and the product was chromatographed on a silicic acid column (14 g) using chloroform in hexane for elution. The labeled compound III was found in the 50 and 60% fractions (140×10^6, radiochemical purity 95%). Purification to >98% was achieved by TLC in a solvent system of chloroform : ethyl acetate (95 : 5, v/v).

Biological Applications. In this paragraph, only those biological applications of alkyl glycerols will be reviewed which have a direct bearing on lipoprotein metabolism and atherosclerosis. The studies on fat absorption in which extensive use was made of these compounds have been thoroughly reviewed in a recent book.[3] Morgan and Hofman[63] studied the fate of labeled 1-hexadecyl-2,3-didodecylglycerol triether in young mice after intraperitoneal injection. The labeled compounds were found unchanged in adipose tissue, liver, and spleen up to 9 days.[63] We have used [³H]trioleylglycerol for long-lived labeling of phagocytic cells, which may become a part of an atheroma.[64] The labeled lipids were injected intraperitoneally into mice, either emulsified with Intralipid or after incorporation into acetylated LDL. The labeled lipid was taken up by peritoneal macrophages and was slowly transported into the liver. After intravenous injection of acetylated LDL labeled with [³H]trioleylglycerol, 90% of the radioactivity was recovered in the nonparenchymatous cells in the liver where it persisted for 10 weeks.[64] This was true also when the animals were subjected to hormonal perturbation with ethinyl estradiol of glucocorticoids.[65]

Ether Analogs as Enzyme Inhibitors

Retinyl ester hydrolase activity from rat liver was partially purified (approximately 200-fold) and was found to copurify with hydrolytic activities against cholesteryl oleate and triolein.[66] These three copurifying hydrolytic activities were found to have a number of similar properties (including pH optima, bile salt requirements, apparent molecular size, and hydrophobicity). However, differential solubility and inhibition studies clearly showed that the three lipid hydrolase activities are due to at least three different catalytically active centers, and at least two distinct and separable enzymes. The effects of a series of cholesteryl alkyl ethers and one of acylglyceride ethers, with different alkyl chain compositions, on partially purified rat hepatic retinyl palmitate hydrolase activity, were studied. Cholesteryl ether analogs were potent inhibitors of all three hydrolase activities with relative potencies for the series of ethers of linoleyl > oleyl = palmitoyl > *n*-butyl = *n*-propyl > ethyl = methyl. Retinyl palmitate hydrolase activity was strongly inactivated by this series of

[63] R. G. H. Morgan and A. F. Hofman, *J. Lipid Res.* **11**, 223 (1970).
[64] Y. Kleinmann, G. Halperin, O. Stein, and Y. Stein, *Atherosclerosis* **43**, 1 (1982).
[65] Y. Kleinman, O. Stein, and Y. Stein, *Isr. J. Med. Sci.* **18**, 819 (1982).
[66] W. S. Blaner, G. Halperin, O. Stein, Y. Stein, and DeW. S. Goodman, *Biochim. Biophys. Acta* **794**, 428 (1984).

analogs with 48–86% of the activity inhibited at cholesteryl ether levels of 1 μM. The acylglyceride ether analogs were much weaker inhibitors of the three hydrolase activities, with the triolein, diolein, and dipalmitin analogs showing similar inhibitory potency, greater than that of the mono-lein and monopalmitin analogs.

1,2-O-Dialkyl-sn-glycero-3-phosphocholine

Synthesis

An efficient method for the conversion of 1,2-dialkyl- or diacylglycerol to the corresponding phosphocholine has been introduced by Hirt and Berchtold[67] and Eibl et al.[68] The route (Scheme 4) includes phosphoesteri-fication with β-bromoethylphosphoryl dichloride and the bromoester (compound II) is reacted with trimethylamine to the corresponding phos-phocholine (compound V). The bromoester could also be directly ami-nated with excess of various amines to give the amino methylamino and dimethylamino analogs.[69] The β-bromoethylphosphoryl dichloride was prepared from phosphorus oxychloride and 2-bromoethanol.[67,70]

The procedure of Eibl et al.[68] will be given as an example for conver-sion of 1,2-O-dialkyl glycerol to the corresponding phosphocholine. Dry triethylamine (1.25 g, 12.5 mmol) and 1.25 g (5 mmol) of β-bromoethyl-phosphoryl dichloride were added to absolute chloroform (15 ml) at 0° While stirring and cooling in an ice water bath a solution of 1.9 g (3.5 mmol) of 1,2-O-dihexadecyl-rac-glycerol in chloroform (15 ml) was added dropwise. The mixture was allowed to stir for 6 hr at room temperature and then at 40° for 12 hr. The dark solution was cooled at 0° and the phosphomonochloride ester was hydrolyzed by addition of 15 ml of 0.1 M KCl with stirring. After 1 hr, 25 ml of methanol was added and the aque-ous phase was adjusted to pH 3 with concentrated HCl. The mixture was shaken and the residue in the organic phase was dried over P_2O_5 in a high vacuum.

For introduction of the trimethylammonium group the β-bromoethyl-phosphodiester in butanone (chloroform, acetone, or, in the case of ethers, methanol can also be used) was treated with trimethylamine; for example, when a solution of butanone (50 ml) and trimethylamine (10 ml)

[67] R. Hirt and R. Berchtold, *Pharmacol. Acta Helv.* **33**, 349 (1958).
[68] H. Eibl, D. Arnold, H. U. Welzien, and O. Westphal, *Justus Liebigs Ann. Chem.* **709**, 226 (1967).
[69] H. Eibl and A. Nicksch, *Chem. Phys. Lipids* **22**, 1 (1978).
[70] A. F. Rosenthal, this series, Vol. 35, p. 429.

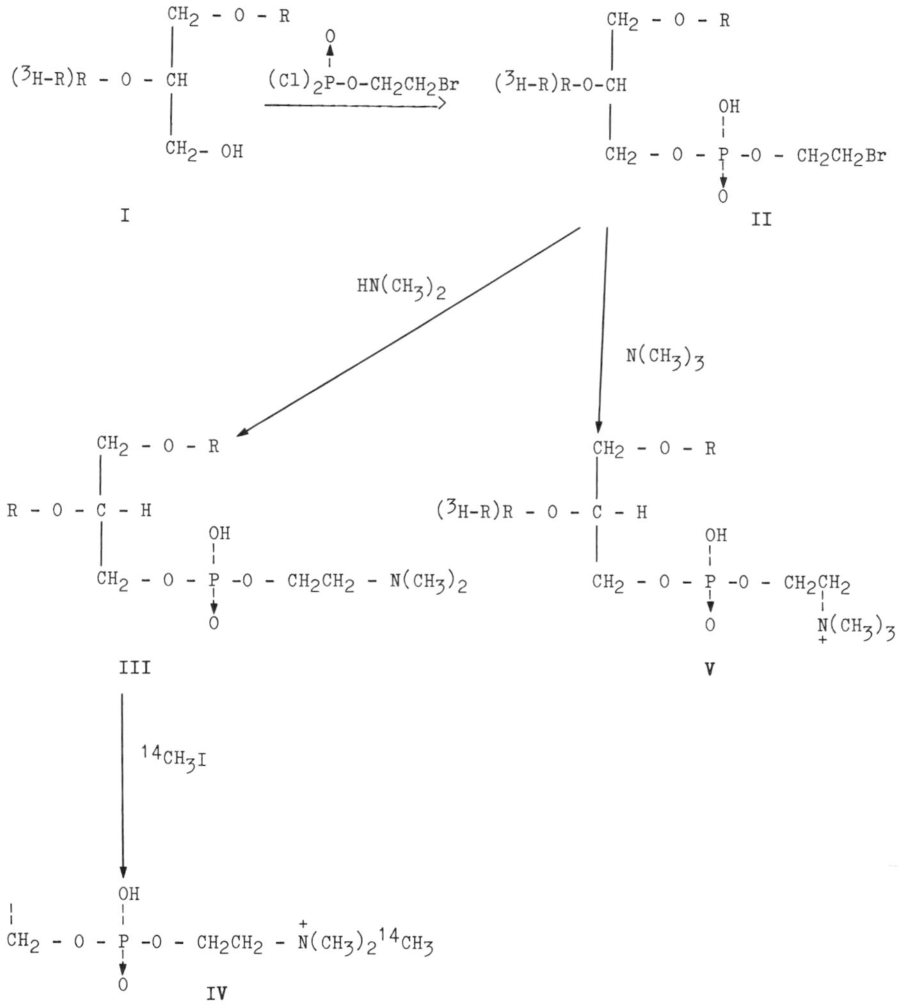

$$CH_2 - O - R$$
$$(^3H-R)R - O - CH$$
$$CH_2 - OH$$

I

$$\xrightarrow{(Cl)_2P-O-CH_2CH_2Br}$$
$$O$$

$$CH_2 - O - R$$
$$(^3H-R)R-O-CH$$
$$CH_2 - O - P -O - CH_2CH_2Br$$
$$OH$$
$$O$$

II

$$HN(CH_3)_2$$

$$N(CH_3)_3$$

$$CH_2 - O - R$$
$$R - O - C - H$$
$$CH_2 - O - P -O - CH_2CH_2 - N(CH_3)_2$$
$$OH$$
$$O$$

III

$$CH_2 - O - R$$
$$(^3H-R)R - O - C - H$$
$$CH_2 - O - P -O - CH_2CH_2$$
$$OH$$
$$O$$
$$\overset{+}{N}(CH_3)_3$$

V

$$^{14}CH_3I$$

$$CH_2 - O - P -O - CH_2CH_2 - \overset{+}{N}(CH_3)_2{}^{14}CH_3$$
$$OH$$
$$O$$

IV

Scheme 4. Procedures for preparation of 1,2-*O*-dialkyl-*sn*-glycero-3-phosphocholine labeled at the 2-alkoxy moiety or at the *N*-methyl.

was warmed for 12 hr at 55° and then cooled to 0°, the diether phosphocholine bromide crystallized and was filtered, washed with acetone, water, and again acetone. For complete purification of the product, it was stirred in 90% aqueous methanol with 1 g of silver acetate (or silver carbonate) for 30 min and chromatographed, if necessary, on silicic acid column. The yield of pure diether phosphocholine, recrystallized from butanone, was about 60%.

This procedure was modified using pyridine instead of triethylamine for β-bromoethylphosphorylation,[71,72] and the purification of the final product was carried out chromatographically when unsaturated alkyl analogs or/and tracer labeled compounds were prepared, as will be described below.

Preparation of 1-O-cis-9'-octadecenyl-2-O-[9',10'-³H₂]-cis-9'-octadecenyl-sn-glycero-3-phosphocholine³⁷ (Scheme 4). 1-O-cis-9'-Octadecenyl-2-O-[9',10'-H₂]-*cis*-9'-octadecenyl-*sn*-glycerol (compound I) (140×10^6 dpm) was converted to the β-bromoethylphosphoryl ester as described above, with β-bromoethylphosphoryl dichloride (0.2 ml) and triethylamine (0.2 ml redistilled from CaH_2) in chloroform (15 ml, freshly distilled from P_2O_5). The crude bromoester was transferred with chloroform into a constricted 20-ml tube. The solvent was evaporated under a stream of nitrogen, methanol (2 ml) was added, and the tube was cooled by liquid air. Trimethylamine (1.8 g), chilled to $-20°$ was added, and the tube was flushed shortly with N_2, sealed, and kept overnight at $60°$. The ampoule was opened in an ice–salt bath, the solvent was removed under reduced pressure, the residue was extracted with chloroform : methanol (2 : 1), and chromatographed on a silicic acid column (11 g) with increasing concentration of methanol in chloroform. The product (84×10^6 dpm) was eluted with 40% methanol and rechromatographed on an alumina column which was also eluted with increasing concentration of methanol in chloroform. The labeled compound V (60×10^6 dpm) was recovered from the 30% methanol fraction and kept in methanol solution under N_2 at $-20°$ prior to use (radiochemical purity, 99%).

Preparation of [N-¹⁴CH₃]-1,2-Di-O-cis-9'-octadecenyl-sn-glycero-3-phosphocholine.[73] 1,2-di-*O-cis*-9'-octadecenyl-*sn*-glycerol (compound I) (286 mg) was bromophosphorylated as described above, and the crude product was transferred into a constricted 20-ml test tube containing methanol (2 ml) and chilled dimethylamine (3.8 g). The tube was cooled with liquid air, sealed, and kept overnight at $60°$. The product was extracted as described above, and purified on silicic acid column (8 g) with increasing concentration of 100-ml fractions of methanol in chloroform. The product was recovered from the 5–15% fractions (227 mg) and was dissolved in acetone (2 ml), the solution was kept at $-20°$ for 4 hr and the solvent, containing impurities, was rapidly decanted. The residue exhib-

[71] F. Paltauf, *Biochim. Biophys. Acta* **176**, 818 (1969).

[72] W. J. Hansen, R. Murari, Y. Wedmid, and W. Baumann, *Lipids* **17**, 453 (1982).

[73] O. Stein, G. Halperin, E. Leitersdorf, T. Olivecrona, and Y. Stein, *Biochim. Biophys. Acta* **795**, 47 (1984).

ited a single spot on TLC in the system chloroform:methanol:water (150:50:8, v/v/v) with an R_f of 0.5. The spectrometric data of 1,2-di-O-cis-9'-octadecenyl-sn-glycero-3-phosphoethanol-N, N-dimethylamine (compound III) were similar to those reported by Stoffel et al.[74] NMR (CDCl$_3$) δ (ppm), 0.90 (m, 6H, 17-CH$_3$), 1.30 (broad s, methylenic protons), 2.02 (m, 8H, CH$_2$—CH=CH—), 2.88 (s, 6H, CH$_3$—N—CH$_3$), 3.30 (m, 2H, CH$_2$—N), 3.46 (m, CH$_2$—O), 4.40 (m, CH$_2$—OP), 5.36 (m, 4H, —CH=CH—); mass spectrum (m/e) 595 (M − C$_4$H$_{10}$O$_3$NP + H), 252 (C$_{18}$H$_{36}$), 112 (CH$_3$H$_2$PO$_4$), 98 (H$_3$PO$_4$), 94 (CH$_3$PO$_3$), 84, 72 [CH$_3$-CH=N(CH$_3$)$_2$], 70, 58 [CH$_2$=N(CH$_3$)$_2$]. The dimethylamine (compound III) (24 mg) was sealed with methanol (3 ml), potassium carbonate (25 mg), and ^{14}CH$_3$I (300 μCi) at liquid air temperature. The ampoule was kept at 70° for 6 hr and an additional 18 hr at room temperature. The product was extracted and chromatographed on a silicic acid column (9 g) as described for the ^{3}H-labeled compound V, Scheme 4. The 30% methanol fraction contained 276 × 10^6 dpm, which were rechromatographed on an alumina column (10 g) with the same system of solvents; the [N-^{14}CH$_3$]1,2-di-O-cis-9'-octadecenyl-sn-glycero-3-phosphocholine (compound IV) was recovered from the 35% methanol fraction (260 × 10^6 dpm). On TLC analysis of an aliquot, 99.1% of the radioactivity was found in the diether phosphocholine region (R_f 0.3).

Biological Applications

Studies with Dioleyl Ether Phosphocholine (DOEPC)

In Vivo. Liposomes prepared from DOEPC were labeled with [^{14}C]CLE and injected into rats; 1 hr thereafter 60% of injected radioactivity was still present in the circulation and declined gradually during the next 24 hr. More than 70% of the injected material was recovered in the liver and was retained during the next 3 days. Thereafter, a slow decrease in liver [^{14}C]CLE occurred, so that after 10 days only 35% of the injected dose was still present in the liver.[37]

In recent years, there has been considerable interest in the use of liposomes as carriers of drugs to be delivered to specific target sites. If liposomes are to be used to deliver therapeutic compounds to cells it may be important to use phospholipids which are not readily degraded by cellular or circulating enzymes. Another prospective use of liposomes could be to enhance reverse cholesterol transport from peripheral cells to

[74] W. Stoffel, D. LeKim, and T. S. Tchung, *Hoppe-Seyler's Z. Phys. Chem.* **352,** 1058 (1971).

the liver.[75] In such an approach it might also be advantageous to use phospholipids which are not a substrate for lecithin: cholesterol acyl transferase, in order to retain the cholesterol in its free form and thus facilitate its removal from the body via the liver and bile. Plasma half-lives of saturated diether phosphatidylcholines were shown to be longer than those of their corresponding diester phosphatidylcholines injected as model HDL into rats.[76] Labeled dioleyl ether phosphatidylcholine (DOEPC) was used to prepare 250- to 350-Å unilamellar liposomes, which also contained phosphatidylcholine (PC) and free cholesterol. Following intravenous injection into rats labeled PC was cleared from the plasma at a faster rate than DOEPC. The uptake of both labeled compounds by the liver increased up to 3 hr at which time there was about 40% of injected PC and 60% of DOEPC. The PC disappeared more rapidly than the DOEPC, so that 17 and 48% of injected label was present in the liver 24 hr after injection of PC and DOEPC, respectively. Ten days after injection of DOEPC, about 10% of the label was still present in the liver. During the first 5 days after injection of DOEPC, 10% of radioactivity was found in the gastrointestinal tract and about 20% in the carcass; no increase in carcass radioactivity occurred during the loss of label from the liver; 24 and 48 hr after injection of DOEPC, 40% of liver radioactivity was present in a neutral lipid, which on TLC comigrated with triacylglycerol. Since after alkaline hydrolysis this compound comigrated with diacyl glycerol it appears that the ether bond of DOEPC was not hydrolyzed, but after removal of phosphocholine, presumably by phospholipase C, the diether glycerol was reacylated.

In Vitro. [³H]Dioleyl ether phosphocholine (DOEPC) transfer between liposomes and red blood cells was studied in the presence of $d >$ 1.21 g/ml fraction of rat plasma which contains a phospholipid transfer protein. A slower exchange of DOEPC than of phosphatidylcholine was observed.[37] A 20–25% reduction in the rate of transfer of diether phosphatidylcholine, when compared to the diester form, was described when the phospholipid transfer protein was isolated from the rate liver.[77] The availability of DOEPC liposomes labeled in the fatty acid moiety helped to elucidate the mechanism of lipoprotein lipase (LPL)-mediated transfer of cholesteryl ester and CLE into cells.[73] It was found that LPL enhanced the uptake of CLE and of DOEPC. These findings provided a definitive proof that hydrolysis of liposomal PC is not needed for the LPL-

[75] O. Stein, J. Vanderhoek, and Y. Stein, *Biochim. Biophys. Acta* **431,** 347 (1976).
[76] D. L. Hickson, B. C. Sherrill, and H. J. Pownall, *Fed. Proc., Fed. Am. Soc. Exp. Biol.* **42,** 1818 (1983).
[77] P. Somerharju, H. J. Brockerhoff, and K. W. A. Wirtz, *Biochim. Biophys. Acta* **649,** 521 (1981).

catalyzed transfer of CLE and CE to cells. The lipids transferred by LPL to cells were localized in three compartments, trypsin releasable, resistant, and metabolic; the latter was a chloroquine-sensitive pool as evidenced by inhibition of cholesteryl ester hydrolysis. Labeled PC and, to a lesser extent, DOEPC, in the trypsin releasable pool was able to return to the medium, while CLE and CE required cholesteryl ester transfer protein for release. The transfer of CLE and CE into a trypsin-resistant compartment did not require metabolic energy and occurred also in formaldehyde-fixed cells. Metabolic energy was needed for the translocation of CLE and CE into the lysosomal compartment, presumably by a process of endocytosis.[73]

[50] Fluorescent Labeling of Lipoproteins

By DAVID P. VIA and LOUIS C. SMITH

Fluorescent compounds have been introduced into lipoproteins to study the dynamics of individual lipoprotein components and of the interaction of lipoproteins with cells *in vitro* and *in vivo*. This experimental approach is necessary because the naturally occurring lipids do not have chromophores that are suitable for quantification of the lipids, for studies of kinetic processes, or for cellular localization. In many cases the experimental design is simpler when fluorescence probes are used, provided there is some estimate of the extent to which the reporter group perturbs the properties of the native compound. Information obtained with antibodies is limited, since only the protein moiety of the lipoproteins is recognized. Mantulin and Pownall[1] have recently summarized studies of plasma lipoproteins that utilized fluorescence as a major experimental approach.

Three general categories of fluorescent compounds that have been utilized to label lipoproteins include lipid analogs, lipid-soluble dyes, and protein modifiers. These compounds are listed in the table. Incorporation of these fluorescent probes has been accomplished by (1) passive transfer from solid supports; (2) protein-mediated transfer from microemulsions and phospholipid dispersions; (3) dispersion from organic solvents; (4) delipidation and reconstitution with specific lipids; and (5) sonication with

[1] W. W. Mantulin and H. J. Pownall, *in* "Excited States of Biopolymers" (R. F. Steiner, ed.), p. 163. Plenum, New York (1983).

phospholipid dispersions and lipoproteins. In most cases, the appropriate studies have not been done to show that the procedures used to incorporate the fluorescent probe into the lipoprotein did not alter significantly the biological properties of the labeled lipoprotein. This consideration is particularly important when the experimental objective is to quantify differences in metabolism, i.e., cellular uptake of low-density lipoproteins (LDL), lipolysis of very-low-density lipoproteins (VLDL) from different subjects, or uptake of fluorescent lipoproteins *in vivo*.

Quantitative evaluation of fluorescent probes delivered by lipoproteins in both living and fixed cells is difficult, if not impossible, because of severe photobleaching. The spatial heterogeneity in photobleaching rates and the errors in apparent intensity values introduced by photographic recording of fluorescent images have been documented.[2] These problems are discussed briefly elsewhere in this volume.[3] The combination of reduced excitation light, computer-controlled shutters, a low light level camera, and millisecond digitization of the fluorescent image can circumvent these difficulties.

Sterols and Fatty Acids. Fluorescent analogs of cholesterol and fatty acids, like the parent compounds, desorb from the lipoprotein surface with relatively short half-times.[4–6] This rapid transfer limits their use to studies of transfer kinetics, as with 3'-pyrenylmethyl-23,24-*dinor*-5-cholen-22-oate-3β-ol (PMCA),[6] or to localization studies under conditions of equilibrium distributions, as with cholesta-5,7,9(11)triene-3β-ol and parinaric acids in lipoproteins[7–12] and N-(7-nitrobenz-2-oxa-1,3-diazole)-23,24-*dinor*-5-cholen-22-amine-3β-ol (NBD-cholesterol) in cultured cells.[13] These compounds show reasonable activity as cholesterol or fatty acid substitutes by several biochemical criteria with purified enzymes and in cell culture systems.

PMCA and other pyrene-containing lipids have spectral properties

[2] D. M. Benson, A. L. Plant, J. Bryan, A. M. Gotto, Jr., and L. C. Smith, *J. Cell Biol.* **100**, 1309 (1985).

[3] L. C. Smith, D. M. Benson, A. M. Gotto, and J. Bryan, this volume [51].

[4] S. Lund-Katz, B. Hammerschlag, and M. C. Phillips, *Biochemistry* **21**, 2964 (1982).

[5] M. C. Doody, H. J. Pownall, Y. J. Kao, and L. C. Smith, *Biochemistry* **19**, 108 (1979).

[6] Y. J. Kao, M. C. Doody, and L. C. Smith, *J. Lipid Res.*, in press (1986).

[7] J. M. Smith and C. Green, *Biochem. J.* **137**, 413 (1973).

[8] L. A. Sklar, M. C. Doody, A. M. Gotto, and H. J. Pownall, *Biochemistry* **19**, 1294 (1980).

[9] L. A. Sklar, I. F. Craig, and H. J. Pownall, *J. Biol. Chem.* **256**, 4286 (1981).

[10] F. Schroeder, E. H. Goh, and M. Heimberg, *J. Biol. Chem.* **254**, 2456 (1979).

[11] R. J. Bergeron and J. Scott, *J. Lipid Res.* **23**, 391 (1982).

[12] I. F. Craig, D. P. Via, B. C. Sherill, L. A. Sklar, W. W. Mantulin, A. M. Gotto, Jr., and L. C. Smith, *J. Biol. Chem.* **257**, 330 (1982).

[13] A. L. Plant, D. M. Benson, and L. C. Smith, *J. Cell Biol.* **100**, 1925 (1985).

FLUORESCENT COMPOUNDS USED IN STUDIES OF PLASMA LIPOPROTEINS

	Reference
A. Fluorescent lipid analogs	
1. Sterols	
Cholesta-5,7,9(11)triene-3β-ol	$b-e$
Ergosta-5,7,9(11)-22-tetraene-3β-ol	e, f
3'-Pyrenylmethyl-23,24-$dinor$-5-cholen-22-oate-3β-ol^a	g
N-(7-Nitrobenz-2-oxa-1,3-diazole)-23,24-$dinor$-5-cholen-22-amine-3β-ol^a	d, h, i
2. Steryl esters	
Cholesta-5,7,9(11)triene-3β-ol oleate	c, e, j, k
N-[7-Nitrobenz-2-oxa-1,3-diazole]-23,24-$dinor$-5-cholen-22-amine-3β-yl linoleate^b	d, h
1-Pyrenemethyl-23,24-$dinor$-5-cholen-22-oate-3β-yl oleate^a	$l-o$
Cholesteryl 12-O-[3'-methyl-5(6)carboxyfluorescein]ricinoleyl carbonate	p, q
3. Phospholipids	
1-Dodecanoyl-2[9-(3-pyrenyl)nanoanoyl]-phosphatidylcholine	r
N-Dansylphosphatidylethanolamine	s, t
N-Pyrenyl-glucocerebrosides	u
4. Other Lipids	
rac-1-Oleyl-2-[4-(3-pyrenyl)butanoyl]glycerol	v, w
cis and $trans$-Parinaric acid^a	j, k, x
16-(9-Anthroyloxy)palmitic acid^a	j
B. Lipophilic dyes	
1. Hydrocarbons	
1,6-Diphenyl-1,3,5-hexatriene	$c, e, j, x-z$
Pyrene	v, w, aa, bb
Benzo[a]pyrene	$i, bb-dd$
2. Dyes	
1,1'-Dioctadecyl-3,3,3',3'-tetramethylindocarbocyanine^a	$ee-ii$
5-(N-Hexadecanoylamino)fluorescein^a	d, j, k
C. Protein modifying agents	
Tetramethylrhodamine isothiocarbamate	jj

^{a} Molecular Probes, Inc., Junction City, OR; ^{b}J. M. Smith and C. Green, *Biochem. J.* **137,** 413 (1973); ^{c}F. Schroeder, E. H. Goh, and M. Heimberg, *J. Biol. Chem.* **254,** 2456 (1979); ^{d}I. F. Craig, D. P. Via, B. C. Sherrill, L. A. Sklar, W. W. Mantulin, A. M. Gotto, Jr., and L. C. Smith, *J. Biol. Chem.* **257,** 330 (1982); ^{e}R. J. Bergeron and J. Scott, *J. Lipid Res.* **23,** 391 (1982); ^{f}P. L. Yeagle, J. Bensen, M. Greco, and C. Arena, *Biochemistry* **21,** 249 (1982); ^{g}Y. J. Kao, M. C. Doody, and L. C. Smith, *J. Lipid Res.,* in press (1985); ^{h}I. F. Craig, D. P. Via, W. W. Mantulin, H. J. Pownall, A. M. Gotto, Jr., and L. C. Smith, *J. Lipid Res.* **22,** 687 (1981); ^{i}A. L. Plant, D. M. Benson, and L. C. Smith, *J. Cell Biol.* **100,** 1925 (1985); ^{j}L. A. Sklar, M. C. Doody, A. M. Gotto, and H. J. Pownall, *Biochemistry* **19,** 1294 (1980); ^{k}L. A. Sklar, I. F. Craig, and H. J. Pownall, *J. Biol. Chem.* **256,** 4286 (1981); ^{l}M. Krieger, L. C. Smith, R. G. W. Anderson, J. L. Goldstein, Y. J. Kao, H. J. Pownall, A. M. Gotto, Jr., and M. S. Brown, *J. Supramol.*

which are concentration dependent. The emission maximum for the excited state monomer, which is present in dilute solutions, occurs at 373 nm. At higher concentrations, the emission maximum for the excited state dimer (excimer) appears around 475 nm. When these lipids containing the pyrene moiety are incorporated into lipoproteins and mixed with unlabeled lipoproteins, the kinetics of transfer can be measured accurately without separation of the donor and acceptor lipoproteins. Changes in the relative intensities for pyrene monomer and excimer emission have provided a convenient fluorimetric method to characterize the mechanism of passive lipid transfer between lipoproteins[6,14,15] and to identify some of the physicochemical factors that influence the desorption rates.[16,17]

The location, within lipoproteins, of several fluorescent chromophores

[14] S. C. Charlton, J. S. Olson, K.-Y. Hong, H. J. Pownall, D. D. Louie, and L. C. Smith, *J. Biol. Chem.* **251**, 7952 (1976).
[15] S. S. Charlton, K. Y. Hong, and L. C. Smith, *Biochemistry* **17**, 3304 (1978).
[16] S. C. Charlton and L. C. Smith, *Biochemistry* **17**, 4023 (1982).
[17] J. B. Massey, D. Hickson, H. S. She, J. T. Sparrow, D. P. Via, A. M. Gotto, Jr., and H. J. Pownall, *Biochim. Biophys. Acta* **79**, 274 (1984).

Struct. **10**, 467 (1979); [m]R. G. W. Anderson, J. L. Goldstein, and M. S. Brown, *J. Recept. Res.* **1**, 17 (1980); [n]M. Krieger, M. S. Brown, and J. L. Goldstein, *J. Mol. Biol.* **150**, 167 (1981); [o]S. T. Mosley, J. L. Goldstein, M. S. Brown, J. R. Falck, and R. G. W. Anderson, *Proc. Natl. Acad. Sci. U.S.A.* **78**, 5717 (1981); [p]J. R. Falck, M. Krieger, J. L. Goldstein, and M. S. Brown, *J. Am. Chem. Soc.* **103**, 7396 (1981); [q]M. Kreiger, Y. K. Ho, and J. R. Falck, *J. Recept. Res.* **3**, 361 (1983); [r]J. B. Massey, D. Hickson, H. S. She, J. T. Sparrow, D. P. Via, A. M. Gotto, Jr., and H. J. Pownall, *Biochim. Biophys. Acta* **794**, 274 (1984); [s]J. D. Johnson, M. R. Taskinen, N. Matsuoka, and R. L. Jackson, *J. Biol. Chem.* **255**, 3461 (1980); [t]J. D. Johnson, M.-R. Taskinen, N. Matsuoka, and R. L. Jackson, *J. Biol. Chem.* **255**, 3466 (1980); [u]D. P. Via, J. B. Massey, S. P. Kundu, D. M. Marcus, H. J. Pownall, and A. M. Gotto, Jr., *Biochim. Biophys. Acta* **837**, 27 (1985); [v]S. C. Charlton, K. Y. Hong, and L. C. Smith, *Biochemistry* **17**, 3304 (1978); [w]S. C. Charlton and L. C. Smith, *Biochemistry* **17**, 4023 (1982); [x]A. Jonas, *Biochim. Biophys. Acta* **486**, 10 (1977); [y]E. Berlin and C. Y. Young, *Atherosclerosis* **35**, 229 (1980); [z]F. Schroeder and E. H. Goh, *J. Biol. Chem.* **254**, 2464 (1979); [aa]S. C. Charlton, J. S. Olson, K.-Y. Hong, H. J. Pownall, D. D. Louie, and L. C. Smith, *J. Biol. Chem.* **251**, 7952 (1976); [bb]L. C. Smith and M. C. Doody, *in* "Polynuclear Aromatic Hydrocarbons: Chemistry and Biological Effects" (M. Cooke and A. J. Dennis, eds.), p. 615. Battelle, Columbus, Ohio, 1981; [cc]H. P. Shu and A. V. Nichols, *Cancer Res.* **39**, 1224 (1979); [dd]J. F. Remsen and R. B. Shireman, *Cancer Res.* **41**, 3179 (1981); [ee]R. E. Pitas, T. L. Innerarity, J. N. Weinstein, and R. W. Mahley, *Arteriosclerosis* **1**, 177 (1981); [ff]R. E. Pitas, T. L. Innerarity, and R. W. Mahley, *Arteriosclerosis* **3**, 2 (1983); [gg]A. J. Hass, H. R. Davis, V. M. Elner, and S. Glagov, *J. Histochem. Soc.* **31**, 1136 (1983); [hh]J. C. Voyta, D. P. Via, C. E. Butterfield, and B. R. Zetter, *J. Cell Biol.*, in press (1984); [ii]H. Funke, J. Boyles, K. H. Weisgraber, E. H. Ludwig, D. Y. Hui, and R. W. Mahley, *Arteriosclerosis* **4**, 452 (1984); [jj]A. V. Mazurov, S. N. Preobrazhensky, V. L. Leytin, V. S. Repin, and V. N. Smirnov, *FEBS Lett.* **137**, 319 (1982).

listed in the table has been determined with resonance energy transfer.[7–12,18,19] As expected, the fluorescent cholesta-5,7,9(11)triene-3β-ol and parinaric acids are found in lipoprotein surface monolayers. The method of choice for incorporation of fluorescent steryl esters into lipoproteins utilizes the cholesteryl ester transfer protein present in human plasma. Resonance energy transfer has shown that some, and possibly all, other methods of incorporation produce inappropriate localization of the fluorescent steryl esters in lipoproteins.[12]

Steryl Esters. Two fluorescent analogs of cholesteryl esters, PMCA oleate and NBD-cholesteryl linoleate, were synthesized to allow visualization of the interaction of lipoproteins with cell surface receptors directly.[20–22] LDL reconstituted with PMCA oleate have been used to select cell mutants defective in expression of lipoprotein receptors[23] and to achieve targeted killing of cultured cells by receptor-dependent photosensitization.[24]

For visualization of PMCA oleate by fluorescence microscopy, good results are obtained using a 365-nm excitation filter, 395-nm dichroic beam splitter, and a 420-nm barrier filter. Under these conditions, a blue fluorescence image will be obtained. After binding of the labeled lipoproteins to the cells at 4 or 37°, cells may be fixed with 3% formaldehyde. Organic fixatives such as acetone are to be avoided since the compound will be extracted from the cell. If living cells are to be studied, the exposure time with UV excitation light must be minimized, since cells are readily damaged by light in this region of the spectrum. The products generated from the interaction of the pyrene-containing probe with the UV light can damage intracellular organelles such as lysozomes, leading to cell autolysis.

NBD-cholesteryl linoleate was synthesized[21] because the NBD-chromophore has many of the spectral properties of fluorescein, with an absorbance maximum at 465 nm and emission maximum at 545 nm. The yellow-green emission of the probe is easily visualized with the standard fluorescein filter package supplied with fluorescence microscopes. There

[18] A. Jonas, *Biochim. Biophys. Acta* **486,** 10 (1977).
[19] F. Schroeder and E. H. Goh, *J. Biol. Chem.* **254,** 2464 (1979).
[20] M. Krieger, L. C. Smith, R. G. W. Anderson, J. L. Goldstein, Y. J. Kao, H. J. Pownall, A. M. Gotto, and M. S. Brown, *J. Supramol. Struct.* **10,** 467 (1979).
[21] I. F. Craig, D. P. Via, W. W. Mantulin, H. J. Pownall, A. M. Gotto, Jr., and L. C. Smith, *J. Lipid Res.* **22,** 687 (1981).
[22] R. G. W. Anderson, J. L. Goldstein, and M. S. Brown, *J. Recept. Res.* **1,** 17 (1980).
[23] M. Krieger, M. S. Brown, and J. L. Goldstein, *J. Mol. Biol.* **150,** 167 (1981).
[24] S. T. Mosley, J. L. Goldstein, M. S. Brown, J. R. Falck, and R. G. W. Anderson, *Proc. Natl. Acad. Sci. U.S.A.* **78,** 5717 (1981).

is some interference from cellular autofluorescence in this spectral region and a slight concentration dependence for both excitation and emission wavelength maxima. When incorporated in lipoproteins, receptor-lipoprotein interactions *in vitro* and *in vivo* can be observed.[12,21] Additional applications of this probe include the analysis of the cell population distribution of lipoprotein uptake and the isolation of cells interacting with lipoproteins from mixed populations using the fluorescence-activated cell sorter.[25]

Cholesteryl 12-O-(3'-O-methylfluorescein)ricinoleate (FME), cholesteryl 12-O-[5'(6')-carboxyfluorescein]ricinoleyl carbonate, (CFC), and cholesteryl 12-O-[3'-methyl-5'(6')carboxyfluorescein]ricinoleyl carbonate (MMC) have been synthesized by Falck *et al.*[26] These compounds have the general fluorescent properties of fluorescein with excitation maxima between 455 and 500 nm and emission maxima at approximately 530 nm. MMC and FME are relatively stable to the acidic environment of the endocytic compartments and are suitable for studies of lipoprotein endocytosis. These probes reconstituted into LDL have been used to distinguish lymphocytes of normal and familial hypercholestermic patients by fluorescence-activated cell sorting.[27]

Phospholipids. 1-Dodecanoyl-2-[9-(3-pyrenyl)nonanoyl]phosphatidylcholine has been incorporated by protein-mediated transfer from cholesteryl ester-rich microemulsions[28] into all lipoprotein density classes.[17] The kinetics of transfer between isolated lipoproteins exhibits a linear correlation between the transfer half-time and the size of the donor lipoprotein, such that transfer from VLDL is 10 times lower than from high-density lipoproteins (HDL). *N*-Dansyl phosphatidylethanolamine, introduced into VLDL by sonication, is hydrolyzed by lipoprotein lipase.[29,30] The changes in fluorescence intensities are the basis of an assay for the enzyme.

Hydrocarbons. The rapid transfer of the fluorescent hydrocarbon probes and 5-(*N*-hexadecanoyl)aminofluorescein, as well as the sensitivity of their fluorescence properties to the immediate microenviron-

[25] D. P. Via, unpublished experiments.

[26] J. R. Falck, M. Kreiger, J. L. Goldstein, and M. S. Brown, *J. Am. Chem. Soc.* **103**, 7396 (1981).

[27] M. Kreiger, Y. K. Ho, and J. R. Falck, *J. Recept. Res.* **3**, 361 (1983).

[28] D. P. Via, I. F. Craig, G. W. Jacobs, B. Van Winkle, S. C. Charlton, A. M. Gotto, Jr., and L. C. Smith, *J. Lipid Res.* **23**, 570 (1982).

[29] J. D. Johnson, M. R. Taskinen, N. Matsuoka, and R. L. Jackson, *J. Biol. Chem.* **255**, 3461 (1980).

[30] J. D. Johnson, M.-R. Taskinen, N. Matsuoka, and R. L. Jackson, *J. Biol. Chem.* **255**, 3466 (1980).

ment, have restricted the use of these probes to kinetic studies. By contrast, 1,1'-dioctadecyl-3,3,3',3'-tetramethylindocarbocyanine perchlorate (DiI)[31] does not transfer because the two octadecyl moieties make the compound extremely hydrophobic. The spectral properties of DiI are very similar to those of tetramethylrhodamine, with an excitation maximum at about 540 nm and an emission maximum at 556 nm. The fluorescent emission can be visualized with filter sets optimized for rhodamine fluorescence. When incorporated into lipoproteins, it has been used to visualize the interaction of lipoproteins with cells both *in vitro*[32–34] and *in vivo.*[35] In addition, the isolation of specific cell types such as endothelial cells from mixed cultures of primary cells by cell sorting is readily accomplished.[36] The probe is very stable and does not photobleach rapidly.[2] In addition, it is ideally suited for visualizing the uptake of lipoproteins in living cells since the excitation wavelength is well tolerated by cells for extended periods of time.

Protein Modifiers. Labeling of the protein moiety of lipoproteins has involved conjugation to rhodamine.[37] The ε-amino groups of lysine, which may not be labeled randomly by these reagents, are essential in the ligand–receptor interaction of LDL. If such modifications are used, very careful documentation is necessary to show that receptor–ligand interactions have not been altered.

Methods of Incorporation

Method 1: Cholesteryl Ester-Rich Microemulsions and Plasma Lipid Transfer Protein[12]

Solutions
50 mg ml^{-1} cholesterol
50 mg ml^{-1} trioleylglycerol

[31] P. J. Sims, A. S. Waggoner, C. H. Wang, and J. F. Hoffman, *Biochemistry* **13,** 3314 (1974).
[32] R. E. Pitas, T. L. Innerarity, J. N. Weinstein, and R. W. Mahley, *Arteriosclerosis* **1,** 177 (1981).
[33] R. E. Pitas, T. L. Innerarity, and R. W. Mahley, *Arteriosclerosis* **3,** 2 (1983).
[34] A. J. Hass, H. R. Davis, V. M. Elner, and S. Glagov, *J. Histochem. Soc.* **31,** 1136 (1983).
[35] H. Funke, J. Boyles, K. H. Weisgraber, E. H. Ludwig, D. Y. Hui, and R. W. Mahley, *Arteriosclerosis* **4,** 452 (1984).
[36] J. C. Voyta, D. P. Via, C. E. Butterfield, and B. R. Zetter, *J. Cell Biol.* **99,** 2034 (1984).
[37] A. V. Mazurov, S. N. Preobrazhensky, V. L. Leytin, V. S. Repin, and V. N. Smirnov, *FEBS Lett.* **137,** 319 (1982).

50 mg ml^{-1} cholesteryl oleate or linoleate
50 mg ml^{-1} 1-palmitoyl-2-oleylphosphatidylcholine
1–2 mg fluorescent probe
$\rho > 1.21$ human plasma prepared from fresh plasma (<48 hr old)
25 mM Tris-HCl, pH 7.4, containing 0.15 M NaCl and 0.3 mM EDTA
2-Propanol dried over Linde type 2A molecular sieves for at least
    48 hr

Equipment

Hamilton syringe, 100 μl, blunt tip
Hamilton syringe water jacket (Model 875-00)
Water-jacketed receiving vessel
Stirring motor
Thermostatted circulating water bath

Procedure

Preparation of Microemulsions. Lipid components in suitable organic solvents (1.41 mg phospholipid, 0.7 mg triglyceride, 2.37 mg cholesteryl ester and up to 50% as fluorescent probe) are combined in a small vial and the solvents removed with a stream of dry N_2. The dried lipids are desiccated under mechanical vacuum for at least 2 hr. The lipid is dissolved in 200 μl of dry 2-propanol and heated to 60° in a water bath. Aliquots of 66 μl are pulled into the Hamilton syringe, which is maintained at 55° by the water jacket. The tip of the syringe is placed below the surface of 1.1 ml of rapidly stirring Tris buffer. The lipids are injected smoothly into the solution by simultaneously releasing the syringe plunger and allowing the 325-g brass weight to fall on the plunger. The 2-propanol is removed by centrifugation of the solution through a buffer-depleted column of Sephadex G-25 in a 5-ml syringe.[38] When preparing the microemulsion, it is crucial that the temperature of the jacketed syringe be maintained precisely at 55° and that the 2-propanol be dry. In addition, solutions of phospholipid and triglyceride should not be kept longer than 2 weeks.

Labeling of Lipoproteins. Incorporation of cholesteryl ester and phospholipid probes into lipoproteins utilizes plasma cholesteryl ester and phospholipid transfer activities. Freshly prepared microemulsions, 1.0 ml containing 1.5 mg total lipid, is mixed with 1 mg of LDL in 2.0 ml of freshly prepared $\rho > 1.21$ plasma. The mixture is incubated for 24 hr at 37° before the microemulsion removed by ultracentrifugation at ρ 1.006–

---

[38] D. W. Fry, J. C. White, and I. D. Goldman, *Anal. Biochem.* **90,** 809 (1978).

1.019. The LDL is isolated by centrifugation at ρ 1.019–1.063. HDL is also labeled and can be isolated by centrifugation at ρ 1.063–1.21.

It is also possible to label lipoproteins in plasma directly without prior isolation of the lipoproteins. One to 5 ml of microemulsion is mixed with 20 ml of fresh plasma and incubated for 24 hr at 37°. Microemulsions, LDL (ρ 1.109–1.063) and HDL (ρ 1.063–1.21) are separated by centrifugation as above.

Incorporation of fluorescent labels into VLDL, chylomicrons, and chylomicron remnants is not possible by this protocol since these lipoproteins and the microemulsions have similar flotation rates. To label these lipoproteins, 1 mg of the triglyceride-rich lipoproteins is mixed with 1 mg of previously labeled LDL in 5 ml of fresh $\rho > 1.21$ plasma. After 24 hr at 37°, ultracentrifugal separation of the donor and acceptor particles is easily achieved.

General Comments. As much as 40% of the cholesteryl ester content of the LDL can be replaced in a single incubation. The method is gentle and yields a particle that is biologically indistinguishable from normal LDL by *in vitro* binding studies and *in vivo* turnover studies.[12] In previous studies with lipoproteins enriched with specific lipid components, it has been assumed that the exogenously supplied lipid occupied the same region as the lipid incorporated biologically. In most cases, however, no structural studies were performed to establish the exact location of the exogenous component in the lipoprotein. Initial spectroscopic experiments in this laboratory with LDL labeled with a fluorescent cholesteryl ester by the solvent injection method indicated that exogenous and endogenous cholesteryl esters were not in the same locations.[12]

Method 2. Delipidation and Reconstitution

The details of this method are given elsewhere in this volume.[39] This procedure works well with PMCA oleate and NBD-cholesteryl linoleate. In this method, the components are dissolved in benzene rather than hexane and are reconstituted into the delipidated lipoprotein by incubation at 4° instead of −20°. This method allows 100% replacement of the cholesteryl esters in LDL. In our experience, when LDL reconstituted by this method are injected intravenously into the rat, a large portion of the r-LDL is cleared from circulation at an abnormally rapid rate (2–5 min). This finding suggests that the reconstituted LDL are heterogeneous and should be used with caution, for any *in vivo* studies.

---

[39] R. G. W. Anderson, this volume [12].

Method 3. Solvent Delivery

Solutions

3 mg ml^{-1} DiI-C18(3) in DMSO

$\rho > 1.21$ plasma, sterile

5–10 mg ml^{-1} lipoprotein, sterile

0.05 M Tris-HCl, pH 7.4, containing 0.15 M NaCl and 0.3 mM EDTA

Procedure. Two milligrams of lipoprotein is added to 4 ml of $\rho > 1.21$ plasma. The mixture is gently vortexed while 100 μl of DiI in DMSO is added. After incubation for 8–16 hr at 37° in the dark, the solution is adjusted to $\rho = 1.063$ and centrifuged for 24 hr at 45,000 in an SW-50.1 rotor. The top fluorescent band is removed, dialyzed against four changes of Tris buffer, and sterilized by filtration. If more dilute solutions can be used, separation of the labeled lipoproteins can be achieved on a 1.5 by 90 cm column of Sephacryl S-200 in approximately 4 hr. The lipoprotein elutes ahead of all other fluorescent material.

General Comments. This procedure yields an extremely fluorescent lipoprotein. The method is suitable for labeling all lipoprotein classes. More dilute solutions of lipoprotein can be mixed with the dye, but the amount of dye incorporated per particle declines unless the dye concentration is increased. It is also necessary that the lipoprotein be mixed with the lipoprotein-deficient plasma before addition of the dye. Addition of the dye directly to lipoproteins can produce aggregation and precipitation. It should be noted that lipoproteins labeled by this method have not been examined for heterogeneity by *in vivo* turnover studies in rats. Based on cell culture studies, this method of labeling seems to be satisfactory. A recent study suggests that the receptor specificity of DiI-labeled lipoproteins is maintained *in vivo*. [R. E. Pitas, J. Boyles, R. W. Mahley, and D. M. Bissell, *J. Cell Biol.* **100,** 103 (1985).]

[51] Digital Imaging Fluorescence Microscopy

By Louis C. Smith, Douglas M. Benson, Antonio M. Gotto, Jr., and Joseph Bryan

In this article, we provide an overview of the unique features of digital imaging microscopy and suggest several research areas in lipoprotein metabolism where the technology can be applied to great advantage.

The relationship between the plasma lipoproteins and atherosclerosis has remained elusive because of the complexity of interactions between lipoproteins and the different cell types in the normal blood vessel wall.

The situation is even more complex in and around vascular lesions. In the last decade, significant progress has been made toward understanding the molecular and physicochemical basis of lipoprotein structure and function, as documented in Volume 128.[1,2] Studies of the interaction of lipoproteins with cultured cells have demonstrated great diversity in the cellular uptake of lipoproteins[3] and considerable effort is now directed toward determining the amino acid sequence and conformation of both the lipoprotein receptors and the apolipoprotein ligands. The receptor for low-density lipoproteins (LDL) that interacts with both apolipoprotein E-3,4 and apolipoprotein B-100 is the one which has been best characterized.[4]

The molecular regulation of receptor expression and function is poorly understood.[5] Most of the information about the role of lipoproteins in regulation has come from studies of ^{125}I-labeled protein components of lipoproteins, even though cholesterol, or a cholesterol derivative, is thought to be the important regulatory component of lipoproteins.[6] Receptor recognition of protein ligands is the key event that leads to internalization of lipoproteins and the accumulation of intracellular lipid. Action of the lysosomal enzymes on LDL is necessary for subsequent regulatory processes. Since the protein moiety must be degraded, regulation involves the lipid components. Although lipid–protein interactions can be studied directly in the test tube using spectroscopic techniques, similar studies of the regulatory role of lipids *in situ* are technically impossible using conventional biochemical methods. One strategy is to combine the use of fluorescent lipid analogs with fluorescence microscopy. The best approach is to apply existing hardware and software to digitize images of living cells exposed to a particular fluorescent lipid.

Through digitization, an image of a cell is mapped to a matrix of numbers with the rows and columns corresponding to discrete horizontal and vertical picture elements, termed pixels. The value of each pixel is proportional to the fluorescence intensity or brightness at that location. The ability to convert a fluorescence image to a digital image permits an objective, accurate measurement of the intensity value for discrete cellular locations with the resolution of the light microscope. Modern digital imaging and microscopic equipment allows the excitation and detection of

[1] A. M. Gotto, H. J. Pownall, and R. J. Havel, this series, Vol. 128 [1].

[2] D. M. Driscoll and G. S. Getz, this series, Vol. 128 [2].

[3] R. W. Mahley and T. L. Innerarity, *Biochim. Biophys. Acta* **737**, 19 (1983).

[4] T. Yamamoto, C. G. Davis, M. S. Brown, W. J. Schneider, M. L. Casey, J. L. Goldstein, and D. W. Russell, *Cell* **39**, 27 (1984).

[5] G. J. Schroepfer, Jr., *Annu. Rev. Biochem.* **50**, 585 (1981).

[6] L. C. Smith and A. M. Gotto, Jr., in "Regulation of HMG-CoA Reductase" (B. Preiss, ed.), p. 221. Academic Press, New York, 1985.

multiple fluorophores and the determination of kinetic parameters on several cells simultaneously. Furthermore, there is a rigorous analytic approach, developed as part of the United States space program, for acquiring, processing, and analyzing images stored in a numerical format.[7-16] The importance and significance of the analytic approach applied to biological problems has been reviewed.[17,18]

Our approach has been to develop a digital image processing system that combines a low light level camera to detect weak fluorescent signals, an image processor to digitize images at video frame rates, and the necessary computer hardware and software. Objective criteria have been developed to document, on a daily basis, the performance of the microscope and the camera with respect to stability, reproducibility, resolution, and sensitivity. This strategy provides an objective means to control day-to-day variables such as lens alignment, lamp stability, and focus. The specific configuration for our initial system is described by Benson et al.[19] Similar systems for digital fluorescence microscopy have been described by others.[20-29] These systems furnish a means to initiate the development

[7] K. R. Castleman, "Digital Image Processing." Prentice-Hall, New York, 1979.

[8] W. K. Pratt, "Digital Image Processing." Wiley, New York, 1978.

[9] W. B. Green, "Digital Image Processing, A Systems Approach." Van Nostrand-Reinhold, New York, 1983.

[10] A. Rosenfeld and A. C. Kak, "Digital Picture Processing," 2nd Ed., Vol. 1. Academic Press, New York, 1982.

[11] A. Rosenfeld and A. C. Kak, "Digital Picture Processing," 2nd Ed., Vol. 2. Academic Press, New York, 1982.

[12] D. H. Ballard and C. M. Brown, "Computer Vision." Prentice-Hall, New York, 1982.

[13] Y. Talmi, "Multichannel Image Detectors," Vol. 2. American Chemical Society, Washington, D.C., 1983.

[14] A. Rosenfeld, "Image Modeling." Academic Press, New York, 1981.

[15] J. Serra, "Image Analysis and Mathematical Morphology." Academic Press, New York, 1982.

[16] T. Pavlidis, "Algorithms for Graphics Image Processing." Computer Science Press, Rockville, MD, 1982.

[17] P. H. Bartels and G. B. Olson, in "Methods of Cell Separation" (N. Catsimpoolas, ed.), p. 1. Plenum, New York, 1980.

[18] R. Walter and M. Berns, "Digital Image Processing and Analysis." Plenum, New York, in press, 1985.

[19] D. M. Benson, J. Bryan, A. L. Plant, A. M. Gotto, and L. C. Smith, J. Cell Biol. **100**, 1309 (1985).

[20] E. Kohen, B. Thorell, and P. Bartick, Exp. Cell Res. **119**, 23 (1979).

[21] L. Tanasugarn, P. McNeil, G. T. Reynolds, and D. L. Taylor, J. Cell Biol. **98**, 717 (1984).

[22] D. Axelrod, D. E. Koppel, J. Schlessinger, E. Elson, and W. W. Webb, Biophys. J. **16**, 1055 (1976).

[23] R. J. Walter and M. W. Berns, Proc. Natl. Acad. Sci. U.S.A. **78**, 6927 (1981).

[24] P. M. Martin, H. P. Magdelenat, B. Benyahia, O. Rigaud, and J. A. Katzenellenbogen, Cancer Res. **43**, 4956 (1976).

of radiometric and geometric fluorescence standards for quantitative fluorescence microscopy.

Problems Unique to Fluorescence Microscopy

The ability to make quantitative fluorescence measurements depends on the resolution of three major experimental problems: (1) elimination of photobleaching, (2) calibration with fluorescent standards, and (3) elimination of out-of-plane contributions to the fluorescence intensities. These are discussed in the following paragraphs.

Photobleaching

One major complication in quantitative fluorescence microscopy is photobleaching of the fluorophores. The photochemical process produces a loss of fluorescence intensity during intermittent or constant illumination. High numerical aperture objectives maximize spatial resolution and improve the limits of detection, but also accelerate photobleaching. A practical solution to this problem is to establish an analytic approach to measure photobleaching reactions at each pixel.[19] An algorithm has been developed to allow determination of the rate constant, initial and final fluorescence intensities, and the corresponding standard deviations and correlation coefficients on a pixel-by-pixel basis for an image composed of 256 columns by 240 rows, or 61,440 pixels. The input data consist of 20 fluorescence images acquired at variable time intervals from samples under continuous illumination.

The equation for calculating the photobleaching rate constants is

$$I_{(x,y,t)} = B_{(x,y)} \exp(-k_{(x,y)}t) + A_{(x,y)}$$

where $I_{(x,y,t)}$ is the fluorescence at pixel (x,y) at time t
$A_{(x,y)}$ is the fluorescence at $t = \infty$
$A_{(x,y)} + B_{(x,y)}$ is the fluorescence at $t = 0$; and
$k_{(x,y)}$ is the rate constant for the change in fluorescence at pixel (x, y)

[25] G. L. Wied, P. H. Bartels, H. E. Dytch, F. T. Pishotta, and M. Bibbo, *Anal. Quant. Cytol.* **4,** 257 (1982).

[26] T. W. R. Macfarlane, S. Petrowski, L. Rigutto, and M. R. Roach, *Blood Vessels* **20,** 161 (1983).

[27] S. Bradbury, *J. Microsc.* **131,** 203 (1983).

[28] K. P. Roos and A. J. Brady, *Biophys. J.* **40,** 233 (1982).

[29] H. J. Tanke, M. J. Van Driel-Kulker, C. J. Cornelisse, and J. S. Ploem, *J. Microsc.* **130,** 11 (1983).

PHOTOBLEACHING RATE CONSTANTS FOR FLUOROPHORES[a]

	Rate constants within the cell perimeter	
Fluorophore	Range of values (sec^{-1})	Whole cell average (sec^{-1})
Benzo[a]pyrene		
Cells	0.0–0.125	0.044 ± 0.006
Microemulsion	0.018–0.030	0.025 ± 0.003
NBD-cholesterol		
Cells/POPC	0.0–0.196	0.116 ± 0.019
Acridine orange		
Cells	0.03–0.194	0.134 ± 0.026
FITC		
Cells	0.009–0.021	0.012 ± 0.001

[a] D. M. Benson, J. Bryan, A. L. Plant, A. M. Gotto, and L. C. Smith, *J. Cell Biol.* **100**, 1309 (1985).

Using this algorithm, we found that the localization and microenvironment of a probe within cells had a marked effect on photochemical processes. The photobleaching rate constants for several fluorophores, studied in microemulsions and in both fixed and living cells, are given in the table. The cellular distribution of photobleaching and the spatial heterogeneity of rate constants can only be appreciated by determining cellular maps of the rate constants. Presentation of the rate constant data as an average of the entire cell obscures the differences in photobleaching that actually exist, as shown by the range of values for the rate constants. Data acquired by using photomultipliers as sensing devices in microscopy systems underestimate the magnitude of photobleaching even more, since the values for the extracellular areas are included. Thus, in a hypothetical example, a probe localized in lysosomes may appear to be present at a higher concentration than in an endocytic vesicle due to differences in the photobleach rate which are a function of the pH.

The pattern of photobleaching of fluorescently labeled cells was determined for 0.065 μm^2 areas of the cells. A pixel by pixel comparison of photobleaching rates for NBD-cholesterol, a fluorescent cholesterol analog, reveals extensive spatial heterogeneity. The complexity of the photobleaching is shown by the rate constant map (Fig. 1), by the isometric projections of the rate constant map, and by the two-dimensional histogram of initial fluorescence intensity values vs rate constants (Fig. 2). The cells were labeled by spontaneous transfer of NBD-cholesterol from 1-palmitoyl-2-oleylphosphatidylcholine vesicles.[30] The effect of cellular

[30] A. L. Plant, D. M. Benson, and L. C. Smith, *J. Cell Biol.* **100**, 1295 (1985).

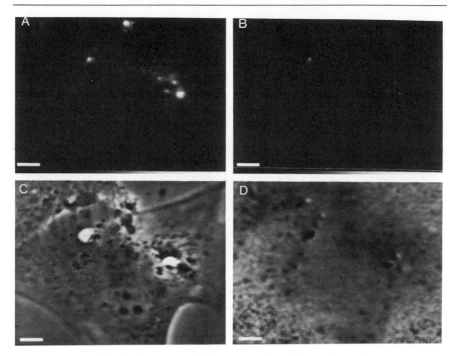

Fig. 1. Photobleaching of NBD-cholesterol in fibroblasts. NBD-cholesterol was transferred spontaneously from 1-palmitoyl-2-oleoylphosphatidylcholine vesicles [D. M. Benson, J. Bryan, A. L. Plant, A. M. Gotto, and L. C. Smith, *J. Cell. Biol.* **100,** 1309 (1985)]. (A) Calculated initial fluorescence image. (B) Calculated final fluorescence image. (C) Rate constant map for photobleaching of NBD-cholesterol. The range of rate constants was 0–0.199 sec^{-1}, with a mean value of 0.122 ± 0.021 within the cell perimeter. The bar = 10 μm. (D) The phase image of the fibroblast containing NBD-cholesterol.

compartmentalization is illustrated by the plot of the rate constants vs the initial fluorescence intensity values (Fig. 2). The height of the plot represents the number of pixels that have that particular combination of rate constants and initial intensity values. Software routines can display either the population of photobleach rate constants for designated cellular compartments in the phase image or the areas in the phase image of the cell that correspond to a designated population of rate constants in the histogram (Fig. 3). Inspection of the overlays of the rate constants on the phase images shows that photobleaching rates are significantly different in cellular compartments which separated by no more than a few microns.

Photobleaching has been shown experimentally to be proportional to the excitation energy and involves the formation of radicals. The rate of depletion of the ground state fluorophore depends on the microenviron-

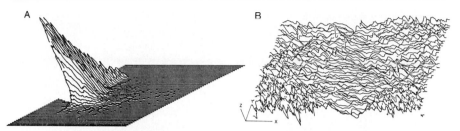

FIG. 2. Heterogeneity of rate constants for NBD-cholesterol photobleaching. (A) Isometric projection of the rate constant map [W. M. Newman and R. F. Sproull, "Principles of Interactive Computer Graphics," 2nd Ed. McGraw-Hill, New York, 1979]. (B) Isometric projection of the plot of rate constants as a function of initial fluorescence intensity values. Calculated initial fluorescence intensity values are plotted on the x axis; the values for the rate constants on the y axis; and the frequency of pixels with given values for intensities (x) and rate constants (y) on the z axis. The scale for the rate constants on the y axis is 0 to 0.255 sec^{-1}. For the initial intensity values on the x axis, the scale is 0 to 255 gray levels. For the number of pixels at each location, the scale is 0 to 255 on the z axis.

ment and varies by up to 65-fold within a single cell.[19] With such differences in photobleaching rates, long-term integration techniques such as photographic recording will systematically underestimate the fluorescence intensity in rapidly bleaching areas. Radical scavengers, such as propyl gallate, reduce photobleaching, as measured by entire field integration.[31] Whether or not these radical scavengers eliminate the spatial heterogeneity is not known.

The important finding for quantitative fluorescence measurements that emerges from the fitting algorithm is the necessity to recognize and make the appropriate correlations for differences in photobleaching rates. The exposure time, as well as the excitation intensity, can be decreased so there is no significant loss in fluorescence. The absence of photobleaching can be verified objectively and unambiguously using this protocol. The heterogeneity of these photochemical reactions suggests that attempts to quantify fluorescence intensities by extrapolation, assuming there is only a single rate constant, back to initial intensity distributions may also be hazardous, since differences in fluorescence quantum yields and lifetimes for a given fluorophore exist in different subcellular compartments.[32]

An important factor under direct experimental control is the selection of fluorophore. As has been shown in the table, different fluorophores show a wide range of photostability. Tetramethylrhodamine and 1,1'-dioctadecyl-3,3,3',3'-tetramethylindocarbocyanine, for example, have 10- and 50-fold longer photobleaching half-times than fluorescein.[19]

[31] H. Giloh and J. W. Sedat, *Science* **217,** 1252 (1982).
[32] B. S. Packard, K. K. Karukstis, and M. P. Klein, *Biochim. Biophys. Acta* **769,** 201 (1984).

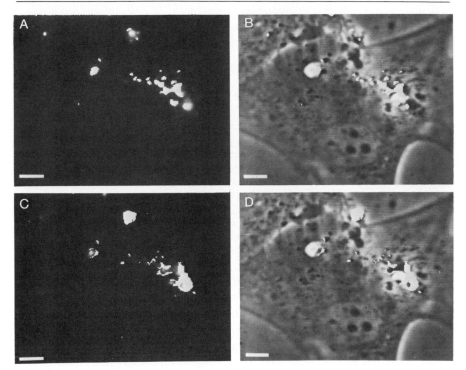

FIG. 3. Heterogeneity of rate constants for NBD-cholesterol photobleaching. (A) The white area is a mask which identifies rate constant values between 0 and 0.100 sec⁻¹ and fluorescence intensity values greater than 100 gray levels. (B) Overlay of the upper left mask on the phase image. (C) The white area is a mask which identifies rate constant values between 0.101 and 0.255 sec⁻¹ and fluorescence intensity values greater than 100 gray levels. (D) Overlay of the lower left mask on the phase image. The bar = 10 μm.

Calibration

Fluorescence measurements and absorbance measurements in microscopy systems differ significantly. With fluorimetry, there is no simple relation between the intensity of the excitation light (I_{exc}) or the fluorescent light (I_{fl}) and the fluorophore concentration.[33] Usually only I_{fl} is measured and therefore calculation of concentration is not possible. More rigorous attempts at quantification, i.e., knowledge of absolute values of concentrations, require, as a minimum, comparison with known standards of microscopic dimensions and a range of fluorophore concentra-

[33] F. Ruch, *in* "Introduction to Quantitative Cytochemistry" (G. L. Weid and G. F. Bahr, eds.), Vol. 2, p. 431. Academic Press, New York, 1970.

tions. Within a cell, the observed two-dimensional distribution of fluorescence depends on the plane of focus, the accessible volume within the cell, local solvent conditions, local fluorophore concentration, and fluorescence quenching. With adequate standards, the fluorescence images of individual cells in different samples and at different times can be compared quantitatively. At present, it is possible to measure the kinetics of changes in relative fluorescence intensities within cells,[30] provided there is no photobleaching.[19] These studies are based on the assumption that the quantum yield of a fluorophore in a cellular compartment does not change with changes in the concentration. Thus, the rate of change in fluorescence intensity values can be informative, even though the precise concentrations of the fluorophore are not known.

In fluorescence, there is no theoretical lower limit to the size of isolated particles that can be detected. For example, LDL have diameters of 20 nm. Individual LDL containing 35 molecules of 1,1'-dioctadecyl-3,3,3'3'-tetramethylindocarbocyanine (diI) have been analyzed,[34,35] even though the diameter of LDL is at least 10-fold below the limits of resolution for light microscopy. The lack of a lower limit of detection must not be confused with the resolving power of the microscope, which is similar in absorption and fluorescence. Fluorescent light from self-luminous objects, independent of particle size, enters the microscope objective. The size of this fraction, given by the numerical aperture of the objective and its location relative to the focal plane, has no influence on the relative fluorescence measurements. In practice, the limit of detection is given by the quantum yield of the substance, the sensitivity of the detector, the intensity of the excitation light, stray light and fluorescence in the optic components, the extent of spectral isolation of excitation and emitted light, and photodecomposition of the fluorophore.[33]

Fluorescence measurements in the microscopy system also differ from conventional fluorimetry.[36] The lower limits of detectability of macrofluorimetry and microfluorimetry can differ by a factor of about 10^9. The difference in sample volume is about 10^9, i.e., a cuvette is 10^{-3} liter and a cell is 10^{-12} to 10^{-14} liter. Typical concentration differences are 10^{-8} to 10^{-12} M, which corresponds to 10^{-11} to 10^{-15} mol in conventional fluorimetry. By contrast, with microfluorimetry, the concentrations are 10^{-5} to 10^{-6} M, which correspond to 10^{-17} to 10^{-18} mol. The number of molecules observed in each system at the limit of detection differ by about 9 orders

[34] L. S. Barak and W. W. Webb, *J. Cell Biol.* **90,** 595 (1981).

[35] L. S. Barak and W. W. Webb, *J. Cell Biol.* **95,** 846 (1982).

[36] G. von Sengbusch and A. A. Thaer, *in* "Quantitative Fluorescence Techniques as Applied to Cell Biology" (A. A. Thaer and M. Sternetz, eds.), p. 31. Springer-Verlag, Berlin, 1973.

of magnitude, 10^{12} vs 10^3, respectively. With more intense light sources[37,38] and more sensitive photon counting cameras,[39,40] single molecule detection is possible.[41,42]

The most promising reference standards are rare earth ions in glass microspheres or fibers, which are stable under the irradiation conditions produced by microscope optics.[43] They have emission characteristics in the green and red spectral regions, which overlap FITC and TRITC emission regions. These materials are mixtures of ions [intraconfigurational $f^* \rightarrow f$ electronic transitions of Tb(III)/Eu(III)]. When adequately characterized, they will be suitable for daily calibration procedures and to certify system performance.

Out-of-Plane Fluorescence

Agard[44] has reviewed in detail the rationale for and current problems associated with three-dimensional optical imaging of biological specimens. He notes that for very thin specimens (less than or equal to the depth of field) it is possible to reconstruct the object by recording images at many different specimen tilts, but also that this method cannot be applied to thicker samples. The recorded images no longer represent true projections of the specimen as is required by the tilted view reconstruction method. The alternative is optical sectioning. With an accurate z-axis drive, the focal plane is moved incrementally to obtain a series of images of the object, which can be as thick as 1 mm for fluorescence. Each recorded image is the sum of in-focus information from the focal plane and out-of-focus information from the remainder of the specimen, much of the out-of-focus information can be removed computationally. Several approaches to removal of out-of-plane fluorescence are available. Castleman[7] used an approximation which involves numerical blurring of adjacent optical sections and subtracting those from the plane of focus con-

[37] F. Docchio, R. Ramponi, C. A. Sacchi, G. Bottiroli and I. Freitas, *J. Microscopy* **134**, 151 (1984).
[38] F. van Geel, B. W. Smith, B. Nicolaissen, and J. D. Winefordner, *J. Microsc.* **133**, 141 (1984).
[39] E. F. Zalewski, J. Geist, and R. A. Velapoldi, in "New Directions in Molecular Luminescence," p. 103 ASTM Publ. Code No. 04-822000-39, 1983.
[40] E. F. Zalewski and C. R. Duda, *Appl. Opt.* **22**, 2867 (1983).
[41] R. A. Velapoldi, J. C. Travis, W. A. Cassatt, and W. T. Yap, *J. Microsc.* **103**, 293 (1975).
[42] T. Hirschfeld, *Appl. Opt.* **15**, 3135 (1976).
[43] T. Hirschfeld, *Appl. Opt.* **15**, 2965 (1976).
[44] D. A. Agard, *Annu. Rev. Biophys. Bioeng.* **13**, 191 (1984).

taining the object. Agard and Sedat[45] improved this approach, using a calculated defocus function for the optical system and transforming the images to frequency space to perform the deconvolution. They also implemented an iterative approach for a solution with deblurred images which, when summed, gave a total intensity equal to that of the sum of the observed images. Fay *et al.*[46] used a similar approach, but modeled the defocus function with fluorescent spheres of diameters smaller than the diffraction limits of the light microscope. Scattering and adsorption of the excitation and emission light of endogenous chromophores must also be taken into account.[47] Algorithms for computer reconstruction of three-dimensional biological structures from serial sections has been described.[48-55]

Fluorescent Lipoproteins

The usefulness of the information obtained with fluorescent probes depends on the extent to which they have been characterized biologically. It is mandatory to show that incorporation of the fluorescent reporter group into the parent molecule does not change significantly the usual biochemical properties of the compound. For example, N-(2-naphthyl)-23,24-*dinor*-5-cholen-22-amin-3β-ol was 50–70% as active as cholesterol in the enzymatic reaction catalyzed by lecithin cholesterol acyltransferase.[56] By contrast, N-[4',4'-(1'-oxyl-2',2',6',6'-tetramethylpiperidinyl)]-23,24-*dinor*-5-cholen-22-amid-3β-ol, a cholesterol analog with a spin label in the 17β side chain, was inactive as a substrate for the acyltransferase. The hyperfine splitting of the ESR spectrum showed that the nitroxide group was in the aqueous phase and that the analog was improp-

[45] D. A. Agard and J. W. Sedat, *Nature (London)* **302**, 676 (1983).
[46] F. S. Fay, K. E. Fogarty, and J. M. Coggins, in "Optical Methods in Cell Physiology" (P. DeWeer and B. Salzburg, eds.). Wiley, New York, in press, 1985.
[47] D. M. Benson and J. A. Knopp, *Photochem. Photobiol.* **39**, 495 (1984).
[48] R. W. Ware and V. LoPresti, *Int. Rev. Cytol.* **40**, 325 (1975).
[49] A. Veen and L. D. Peachy, *Comput. Graph.* **2**, 135 (1977).
[50] E. R. Macagano, C. Levinthal, and I. Sobel, *Annu. Rev. Biophys. Bioeng.* **8**, 323 (1979).
[51] I. Sobel, C. Levinthal, and E. R. Macagno, *Annu. Rev. Biophys. Bioeng.* **9**, 347 (1980).
[52] P. B. Moens and T. Moens, *J. Ultrastruct. Res.* **73**, 131 (1981).
[53] E. M. Johnson and J. J. Capowski, *Comput. Biomed. Res.* **16**, 79 (1983).
[54] Y.-M. M. Wong, R. P. Thompson, L. Cobb, and T. P. Fitzharris, *Comput. Biomed. Res.* **16**, 580 (1983).
[55] L. G. Briarty and P. H. Jenkins, *J. Microsc.* **134**, 121 (1984).
[56] Y. J. Kao, A. K. Soutar, K.-Y. Hong, H. J. Pownall, and L. C. Smith, *Biochemistry* **17**, 2689 (1978).

erly oriented in the phospholipid bilayer. NBD-cholesterol is incorporated as the ester into intracellular lipid droplets in both fibroblasts and macrophages,[57] whereas 3'-pyrenylmethyl-23,24-*dinor*-5-cholen-22-oate-3β-yl oleate (PMCA oleate) incorporated in LDL is hydrolyzed in fibroblast and macrophage lysosomes to PMCA, but the free sterol is not reesterified in either cell line.[58] Incorporation of fluorescent lipids into lipoproteins, particularly LDL, must be done with care to avoid denaturation, loss of receptor recognition, and decreased clearance times *in vivo*.[59]

The use of a single fluorescent lipid, like acquiring data at one concentration at a single time point, greatly limits the interpretation of results. A better approach is to utilize a series of homologous fluorescent compounds that differ systematically in structure. Comparison of kinetic parameters as a function of CH_2 content, for example, provides information about the properties of the transition state for the desorption process from phospholipid surfaces, independent of the fluorescent moiety.[60-63] These studies have shown that the hydrophobicity of a phospholipid measured by reversed-phase HPLC is directly correlated with the rate of spontaneous transfer of several series of fluorescent or radiolabeled phospholipids.[63]

The retention time of HPLC can be used to predict the rate of transfer of any homologous lipid from any lipoprotein or phospholipid vesicle. Thus, an experimental strategy to study the kinetics of passive transfer of cholesterol between intracellular membranes in a fixed cell may be the following. Both PMCA and NBD-cholesterol are reasonable cholesterol analogs by physical and biological criteria. These compounds and cholesterol have different desorption rates from HDL, LDL, and single-walled phosphatidylcholine vesicles. In these model systems in the test tube, both the rates of transfer and the amount of each sterol transferred can be measured directly. From information about the relative rates of transfer, the initial concentrations, and the increases in fluorescence intensities of the two fluorescent sterols measured in the cell by digital fluorescence microscopy, the rates of cholesterol transfer can be calculated. Separate biochemical measurements with 10^4 to 10^6 cells would provide the experi-

---

[57] I. F. Craig, D. P. Via, W. W. Mantulin, H. J. Pownall, A. M. Gotto, Jr., and L. C. Smith, *J. Lipid Res.* **22,** 687 (1981).

[58] D. P. Via, unpublished observations.

[59] D. P. Via and L. C. Smith, this volume [50].

[60] H. J. Pownall, D. Hickson, and L. C. Smith, *J. Am. Chem. Soc.* **105,** 2440 (1983).

[61] J. B. Massey, A. M. Gotto, Jr., and H. J. Pownall, *J. Biol. Chem.* **257,** 5444 (1982).

[62] J. B. Massey, A. M. Gotto, Jr., and H. J. Pownall, *Biochemistry* **21,** 3630 (1982).

[63] J. B. Massey, D. Hickson, H. S. She, J. T. Sparrow, D. P. Via, A. M. Gotto, Jr., and H. J. Pownall, *Biochim. Biophys. Acta* **794,** 274 (1984).

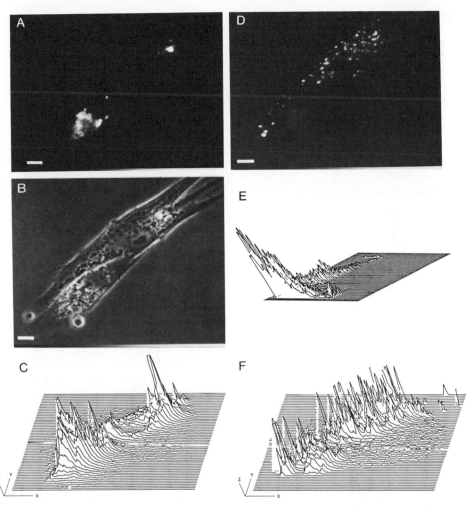

FIG. 4. Differences in cellular uptake of LDL containing diI and of NBD-cholesterol from phosphatidylcholine single-walled vesicles. Fibroblasts previously induced for LDL receptors were incubated for 6 hr at 37° with both 50 μg ml^{-1} LDL containing diI and 30 μm 1-palmitoyl-1-oleylphosphatidylcholine containing 30 mol% NBD-cholesterol [A. L. Plant, D. M. Benson, and L. C. Smith, *J. Cell Biol.* **100**, 1295 (1985)]. Cells were washed two times with 0.15 *M* NaCl and fixed with 2% HCHO before viewing. To compare the distribution of two fluorescent probes, the phase images for the respective fluorescent images were superimposed by computer alignment of two memory planes. This procedure was necessary to correct misalignment associated with the nonparfocal nature of the filter cubes. (A) NBD-cholesterol fluorescence. The gray level values ranged between 80 and 255. (B) Phase image of the cell. (C) Isometric projection of NBD-cholesterol fluoresence. (D) diI fluorescence. The gray level values ranged from 150 to 255. The bar = 10 μm. (E) Isometric projection of the plot of NBD-cholesterol fluorescence intensity values (*y* axis) as a function of diI fluorescence intensity values (*x* axis). The scale for the *y* axis is 0–255 and 0–255 for the *x* axis. The *z* axis is the number of pixels from each *x*,*y* location and ranges from 0 to 255. (F) Isometric projection of diI fluorescence.

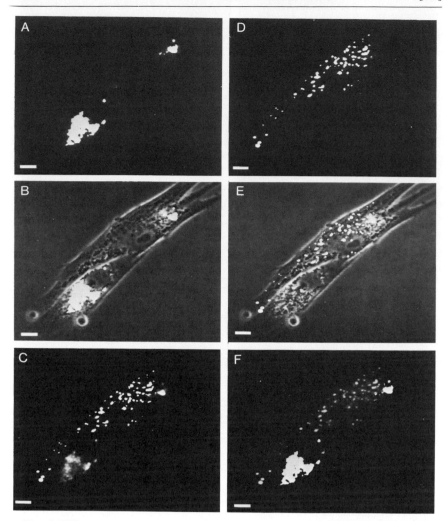

Fig. 5. Differences in cellular uptake of LDL containing diI and of NBD-cholesterol from phosphatidylcholine single-walled vesicles. Analysis of the two-dimensional histogram (Fig. 4, lower right) of the fluorescence intensities gives the regions of the image where the intensity values of the two fluorescent probes are the largest. There is no cross-over of fluorescence of the two probes with these filter sets. (A) NBD-cholesterol fluorescence. The mask identifies the areas where the relative fluorescence intensity of NBD-cholesterol fluorescence is greater than 100 and that of diI is less than 100. (B) NBD-cholesterol fluorescence. Overlay of the upper left mask on the phase image. Most of the NBD-cholesterol fluorescence is associated with the lower cell. (C) NBD-cholesterol fluorescence. The mask identifies the areas where the relative fluorescence intensity of NBD-cholesterol fluorescence is less than 100 and that of diI is greater than 100. (D) diI fluorescence. The mask identifies the areas where the relative fluorescence intensity of diI fluorescence is greater

mental data for comparison with that obtained by fluorescence microscopy with a small number of cells.

Uptake of benzo[a]pyrene by living cultured cells has been visualized
in real time using digital fluorescence imaging microscopy.[30] Benzo[a]py-
rene was noncovalently associated with lipoproteins, as a physiological
mode of presentation of the carcinogen to cells. When incubated with
either human fibroblasts or murine P388D$_1$ macrophages, benzo[a]pyrene
uptake occurred in the absence of endocytosis. The half-time for this
process was about 2 min, regardless of the identity of the delivery vehicles, which were HDL, LDL, VLDL, and 1-palmitoyl-2-oleoylphosphati-
dylcholine single-walled vesicles. Thus, cellular uptake of benzo[a]py-
rene from these hydrophobic donors occurs by spontaneous transfer
through the aqueous phase. Moreover, the rate constant for uptake, the
extent of uptake, and the intracellular localization of benzo[a]pyrene
were identical for both living and fixed cells. In these studies, the intracellular location of benzo[a]pyrene at equilibrium was coincident with a
fluorescent cholesterol analog, NBD-cholesterol. Similar rate constants
for benzo[a]pyrene uptake and efflux from cells to extracellular lipoproteins suggests that the plasma membrane is involved in the rate-limiting
step. Thus, benzo[a]pyrene uptake by cells is a simple partitioning phenomenon, controlled by the relative lipid volumes of extracellular donor
lipoproteins and of cells, and does not involve lipoprotein endocytosis as
an obligatory step.

An example of experimental results that would have been lost using
conventional biochemical techniques is shown in Fig. 4. The cells were
incubated simultaneously with LDL containing diI and phosphatidylcholine vesicles containing NBD-cholesterol. Different cells accumulated
LDL and the fluorescent lipid, information that would have been averaged by other techniques. The two-dimensional plot of the NBD-cholesterol fluorescence vs diI fluorescence has several components (Fig. 4,
lower right). Analyses of these components are shown in Fig. 5. The areas
with the highest NBD-cholesterol fluorescence were identified with a
mask (Fig. 5, upper left), which was superimposed on the phase image of
the cells. Most of the NBD-cholesterol was associated with only one cell
(Fig. 5, center left). By contrast, the areas with the highest diI fluorescence, identified by the mask in the upper right panel of Fig. 5, were found

---

than 100 and that of NBD-cholesterol is less than 100. (E) diI fluorescence. Overlay of the
upper right mask on the phase image. Most of the diI fluorescence is associated with the
upper cell. (F) diI fluorescence: The mask identifies the areas where the relative fluorescence intensity of diI fluorescence is less than 100 and that of NBD-cholesterol is greater
than 100. The bar = 10 μm.

in the upper cell in this field. The differences in the uptake of LDL and NBD-cholesterol by neighboring cells can be explained by the suppression of synthesis of the LDL receptor in the fibroblast-containing lipid deposits.

The information about subcellular compartmentalization of lipophilic compounds may be lost by homogenization of the cells for biochemical analysis, depending on the rate of spontaneous transfer. For example, the time for disruption of cell membranes and for separation of subcellular fractions is sufficiently long that new equilibrium distributions of benzo[a]pyrene between lipid domains are reached at each step.[30] The data finally collected on the separated fractions would be meaningless.

Future Directions

Monoclonal antibodies have the specificity to dissect the complex architecture of cells and tissue. Since digital fluorescence imaging can provide a multispectral analysis, several different fluorescent antibodies and lipids can be used simultaneously. For living cells, microinjection of antibodies and fluorescent proteins into individual cells[64] or populations of cells[65] is a strategy to establish that specific components of the cytoskeleton are required for intracellular transfer of endocytic vesicles. For example, the itinerary of the vesicles can be determined by changes in the kinetics of movement and by spatial coincidence of the fluorescence probes at several time points. It should be possible to circumvent the limits of light microscopy through resonance energy transfer between the appropriate donor–acceptor pairs to identify molecular interactions *in situ* that range between distances of 2–7 nm.[66–71] For absorption measurements, imaging microscopy systems with a monochromator to select the desired wavelengths at 2-nm intervals have been described.[72–75] The abil-

[64] T. E. Kreis and W. Birchmeier, *Int. Rev. Cytol.* **75,** 209 (1982).
[65] P. L. McNeil, R. F. Murphy, F. Lanni, and D. L. Taylor, *J. Cell Biol.* **98,** 1556 (1984).
[66] R. H. Fairclough and C. R. Cantor, this series, Vol. 48, p. 347.
[67] L. Stryer, *Annu. Rev. Biochem.* **47,** 819 (1978).
[68] E. F. Ullman, M. Schwarzberg, and K. E. Rubenstein, *J. Biol. Chem.* **251,** 4172 (1976).
[69] S. Damjanovich, L. Trön, J. Szöllösi, R. Zidovetzki, W. L. C. Vaz, F. Regateiro, and D. J. Arndt-Jovin, *Proc. Natl. Acad. Sci. U.S.A.* **80,** 5985 (1983).
[70] L. Trön, J. Szöllösi, and S. Damjanovich, *Biophys. J.* **45,** 939 (1984).
[71] J. Szöllösi, L. Trön, S. Damjanovich, S. H. Helliwell, D. Arndt-Jovin, and M. Jovin, *Cytometry* **5,** 210 (1984).
[72] D. W. Johnson, J. A. Gladden, J. B. Callis, and G. D. Christian, *Rev. Sci. Instrum.* **50,** 118 (1979).
[73] G. D. Christian, J. B. Callis, and E. R. Davidson, in "Modern Fluorescence Spectroscopy" (E. L. Wehry, ed.), Vol. 4, p. 111. Plenum, New York, 1981.

ity to acquire images at optimum excitation and emission wavelengths is an obvious development to increase significantly the analytic power of digital fluorescence imaging.

An understanding of regulatory mechanisms at the cellular level requires measurements of both transcription and translation of specific genes in individual cells. For *in situ* hybridization, biotin-containing nucleotides can be incorporated into cDNA by nick translation and into synthetic oligonucleotides.[76–81] Modified nucleotides containing antigenic determinants other than biotin and with the appropriate combination of fluorescent antibodies and lipophilic probes should provide the information to reach some understanding about the relationships between gene expression, lipoprotein endocytosis, and the amounts of cholesterol in specific cellular locations.

At a tissue level, one of the most impressive characteristics is the heterogeneity of cell types. Within a lesion, individual cell morphology is highly variable. Various lipid inclusions, including cholesterol crystals and cholesteryl ester droplets, are a distinguishing feature. The origin of this compartmentalization and its involvement in lesion progression and regression is unknown. The extent to which tissue organization and dysfunction is based on the properties and interactions of the plasma membrane of individual cell types remains to be explained. The interactions between different cell types within both normal and dysfunctional tissues are still unknown, and the role of lipoproteins as essential participants in the progressing heterogeneity of the lesion development is speculative. The multispectra analysis of a series of optical sections, suitably labeled for lipid deposits and individual cell types, ultimately should provide information about the three-dimensional organization of cells and lipids involved in this complex disease.[26,82]

[74] I. M. Warner, M. P. Fogarty, and D. C. Shelly, *Anal. Chim. Acta* **109,** 361 (1979).
[75] M. L. Gianelli, D. H. Burns, J. B. Callis, G. D. Christian, and N. H. Andersen, *Anal. Chem.* **55,** 1858 (1983).
[76] R. H. Singer and D. C. Ward, *Proc. Natl. Acad. Sci. U.S.A.* **79,** 7331 (1982).
[77] L. Manuelidis, P. R. Langer-Safer, and D. C. Ward, *J. Cell Biol.* **95,** 619 (1982).
[78] C. E. Gee and J. L. Roberts, *DNA* **2,** 157 (1983).
[79] J. J. Leary, D. J. Brigati, and D. C. Ward, *Proc. Natl. Acad. Sci. U.S.A.* **80,** 4045 (1983).
[80] A. B. Studencki and R. B. Wallace, *DNA* **3,** 7 (1984).
[81] A. Murasugi and R. B. Wallace, *DNA* **3,** 269 (1984).
[82] L. C. Smith, D. M. Benson, A. L. Plant, and A. M. Gotto, Jr., *in* "New Aspects in Lipoprotein Metabolism" (G. Schettler and J. Augustin, eds.). Thieme, Heidelberg, in press, 1986.

Author Index

Numbers in parentheses are footnote reference numbers and indicate that an author's work is referred to although the name is not cited in the text.

A

Aamodt, R. L., 352, 379, 380(20), 426, 427(9), 428(9), 429, 431(9), 433(9), 434(9), 435(9), 436(9), 437(9), 442(9)
Abano, D. A., 765, 769(7), 770(7), 771(7), 772(7)
Abbcy, M., 804, 814(16), 815, 816(16)
Abell, L. L., 96, 100(46), 105
Abeln, G. J. A., 681, 682
Abelson, J., 252
Abreu, E., 477, 798, 804(7), 805(7), 806(7), 807(7)
Aburatani, H., 73, 78(23)
Adams, G. H., 3, 43
Adams, P. W., 391, 402, 408
Adamson, G. L., 33(7), 34, 36, 39(7)
Adelberg, E. A., 249, 252
Adelman, M. R., 286
Adler, G., 769
Adolphson, J. L., 80, 178, 773, 774, 775, 778, 788
Agard, D. A., 866, 867
Aggerbeck, L. P., 536
Agostini, B., 23
Ahrens, E. R., 820
Aisaka, K., 63
Akamatsu, Y., 249
Akanuma, Y., 73, 78(23), 730, 778
Åkesson, B., 729
Akita, H., 63
Alaupovic, P., 23, 43, 127, 131, 412, 430, 435(17), 436(17), 437(17), 438(17), 440, 441, 442(17), 483, 569, 740
Albers, J. J., 80, 85, 86, 87, 88, 91, 92(36), 93(38), 96, 102, 106, 108, 111(15), 113(15), 131, 133, 140, 142, 143, 170, 178, 181(45), 186, 187, 399, 402, 409, 415, 629, 634(6), 716, 764, 765, 766, 768, 769(14), 772, 773, 774, 775, 776(12), 778,
780, 781(10), 786, 787, 788, 790, 795, 796(2), 797, 799(6), 800, 801(6), 802(6), 803, 804(6), 805(6, 14), 807(6), 811, 812(6), 813(6), 814(6), 815
Albers-Schonberg, G., 232
Alberts, A. W., 232, 603, 605(64)
Alexander, C. A., 294
Alfarsi, S., 391, 402, 408
Allain, C. C., 104, 713
Allen, R. C., 378, 403, 405, 500
Allgyer, T. T., 753
Alper, C., 81
Altschuh, D., 174
Andersen, J. M., 684, 685
Anderson, A. P., 627
Anderson, D. W., 26, 36(5), 37(5), 70, 72(20), 95, 164, 430, 440(14), 442(14)
Anderson, N. G., 3, 7(7), 270
Anderson, N. H., 873
Anderson, N. L., 270
Anderson, R. G. W., 201, 203, 205, 206, 207(17, 18), 209, 211, 212(10), 213, 215(10), 216, 224, 240, 253, 550, 674, 852, 856
Anderson, R. M., 593
Anfinsen, C. B., 187, 190(6), 716
Angel, A., 441, 443(32), 662
Angelin, B., 146
Appelmans, F., 288
Applebaum-Bowden, D., 774
Applegate, K., 125, 126(3), 360, 536, 789
Arbeeny, C. M., 205, 209(14), 609
Arbogast, L. Y., 244, 245(32), 629, 632, 642, 643(2)
Archambault-Schexnayder, J., 273
Arias, J. M., 501
Arky, R. A., 471, 472(1), 480(1)
Armstrong, W. B., 627
Arndt-Jovin, D. J., 872

Subject Index

A